FOREWORD

The 27th National Heat Transfer Conference was held in Minneapolis, Minnesota, July 28–31, 1991. A total of 10 sessions were sponsored by the American Institute of Chemical Engineers:

Transport Phonomena In Nuclear Containments - I
Transport Phonomena In Nuclear Containments - II
Heat Transfer and Fuel Behavior in Nuclear Reactor Accidents I
Heat Transfer and Fuel Behavior in Nuclear Reactor Accidents II
Heat Pipes for Space Applications
Heat Transfer and Fluid Flow In Microgravity
Engineering Models and Instrumentation for Multiphase Flow
Heat Transfer in Advanced Nuclear Systems
Nonlinear Dynamics and Chaos in Multiphase Flow and Heat Transfer
Phase Change and Heat Transfer in Porous Media and Heat Exchangers

A total of 47 papers were presented. This volume contains 45 papers and two abstracts. The papers are grouped according to sessions and session sequence are consistent with the 27th National Heat Transfer Conference Program.

We gratefully acknowledge the dedicated work of AIChE Symposium Series Production Manager Maura Mullen in the preparation of this volume. We also thank the session chairmen for their timely contributions. The reviewers of the AIChE papers for the 27th National Heat Transfer Conference are listed on the following page.

Salim B. Yilmaz, *editor*
Rockwell International Corporation
Rocketdyne Division
Canoga Park, CA 91303

Brian G. Volintine, *editor*
U.S. Department of Energy
Washington, D.C. 20585

Reviewers

D.R. Adkins
C.M. Allison
Z.I. Antoniak
H. Arastoopour
K. Arnamalai
G. Berna
F. Best
B. Buescher
I. Catton
R. Chiang
C.F. Chen
J.L. Chen
L. Cheng
D.H. Cho
R. Cole
E. Coomes
E.W. Coryell
S. Dingman
P.E. Domagala
R.S. Dougall
A. Faghri
D. Gidaspow
R.G. Gido
F.S. Gunnerson
D.L. Hagrman
J.K. Hohorst
C. Hwang
M. Ishii
A.E. Jones
D.C. Jones
M. Karasulu
M. Kazimi
J. Kelley
E. Keshock
M. Khonsari
A.C. Klein
P. Kuan
R.T. Lahey
S. Leipziger
A.E. Levin
R.A. Lorenz
P. Lottes
R.W. Lyczkowski
P. Maciejewski
J. March-Levhe
M.J. Martinez
H. Mecte
M.A. Merrigan
P.E. Minton
A.C. Mueller
D.J. Osetek
K. Pasamehmetoglu
S.-J. Peng
A.C. Peterson
M.Z. Podowski
T. Rabas
K. Reil
J.L. Rempe
J. Rest
R.D. Rovang
R.C. Schmidt
L.J. Siefken
S.L. Soo
S. Sunder
A. Szeri
R.P. Taleyarkhan
Y.S. Tang
T.G. Theofanous
R.W. Tilbrook
N.E. Todreas
D. Warinner
S.W. Webb
T. Wei
C. Weirman
W. Wulff
S.C. Yao
G.L. Yoder

CONTENTS

NONLINEAR DYNAMICS AND CHAOS IN MULTIPHASE FLOW AND HEAT TRANSFER

Michael Z. Podowski and Robert W. Lyczkowski, *co-chairs*

PHASE CHANGE/HEAT TRANSFER IN POROUS MEDIA AND HEAT EXCHANGERS

V.K. Dhir and Allan E. Jones, *co-chairs*

STEADY STATE THERMAL ANALYSIS OF EXTERNAL COOLING OF A PWR VESSEL LOWER HEAD

Hyunjae Park and Vijay K. Dhir ■ Mechanical, Aerospace and Nuclear Engineering Department, School of Engineering and Applied Science, University of California, Los Angeles, Los Angeles, CA 90024

Flooding of reactor vessel cavity is one of the many accident management strategies being proposed to manage severe accidents in light water reactors. The steady state analysis of a PWR vessel lower head containing molten core material is numerically investigated in this paper using a two dimensional finite difference scheme. Results are obtained for the vessel shell temperatures, pool temperatures for steady state conditions. Calculations are made for various values of the emissivity of the molten pool surface, the vessel wall and the upper structure, and the temperature of the upper structure. For a certain set of parameters, nucleate boiling occurs on the outside surface of the vessel wall, which is found to be effective in lowering the temperature of the inner wall of the vessel below the melting temperature of the steel.

One of the management strategies currently being considered for severe accidents is to flood the cavity so as to cool the outside of the reactor vessel lower head. The idea here is that cooling of the reactor vessel wall will prevent the molten core material contained in the vessel lower head from melting through the vessel. The earliest study on the effectiveness of cooling the outer wall of the vessel in containing the relocated core in the vessel is that of Condon et al [1]. Their study was focused on a Boiling Water Reactor (BWR) and the pool containing molten core material was assumed to be in the form of stratified layers of UO_2 and steel. One dimensional steady state analysis with a specified temperature for the vessel outer wall was carried out. It was shown that for realistic decay heat generation rates, the flooding of the reactor cavity will prevent the vessel melt through. It was also concluded that for Browns Ferry plant, RHRSW pumping capacity will have to be upgraded to flood the drywell in 30 minutes (the time elapsed between beginning of melting and major core slumping calculated with MARCH code). Also, an issue was raised with respect to overpressurization of the drywell which could prevent flooding unless the containment was vented.

The purpose of the present work is to investigate the effectiveness of flooding the cavity of a Pressurized Water Reactor (PWR) in preventing the vessel melt through in case melting and relocation of the core in the vessel lower head occur. Two dimensional steady state analysis is carried out including the heat loss by radiation to the upper regions of the reactor vessel and the unwetted portion of the vessel lower head. Heat loss from the molten pool surface to its surroundings occurs mainly through radiation heat transfer. Due to the high temperature of the pool surface, heat transfer mechanisms such as natural convection or conduction are negligible compared to radiation heat transfer. Radiative heat transfer was not included in the analysis of Condon et al. Also in the present work various constituents of the core material are assumed to be mixed homogeneously.

ANALYSIS

In carrying out the analysis, it is assumed that the reactor cavity is flooded prior to slumping of the core. The insulation around the vessel is in a degraded state such that water can seep through it and can directly contact the vessel surface. The temperature of water is taken to be 323 K. The slumping of the core material into the vessel lower head occurs instantaneously. Outside the vessel, water is still present, which will cool the vessel lower head.

Since the contact temperature between the relocated core material and the vessel inner wall is below the freezing temperature of corium, a crust will form on the inner wall of the vessel immediately after core slumping. In the present analysis, equilibrium pool temperature was determined by equating the heat generation rate in the pool with the heat loss rate. The decay heat in the corium was taken to be about 0.9 % of full power. The thermophysical properties of corium at elevated tempratures have considerable uncertainty. The extreme conditions under which core meltdown occurs make the measurements and calculations of properties extremely difficult. For this reason, an estimate of these properties can only be made when the structure and the constitution of the molten core materials are known [2]. Earlier studies [2,3,4] in this area were reviewed. Knowing the possible range of temperature of interest, mean values were determined. Figure 1 shows a schematic diagram of the core material relocated in the vessel lower head. Various assumptions made in carrying out the analysis are:

(1). The average heat transfer coefficients due to free convection in the molten pool can be used to estimate the heat transfer into upper and lower parts of the molten pool.

(2). The temperature of the upper surface of the pool is uniform.

(3). The radiative transport of heat is independent of wavelength and direction.

(4). The radiation shape factors are calculated by assuming that structural elements between the pool free surface and the grid plate do not participate.

Several equilibrium states are characterized by the steady state pool temperature and the thickness of the crust separating the molten pool from the vessel.

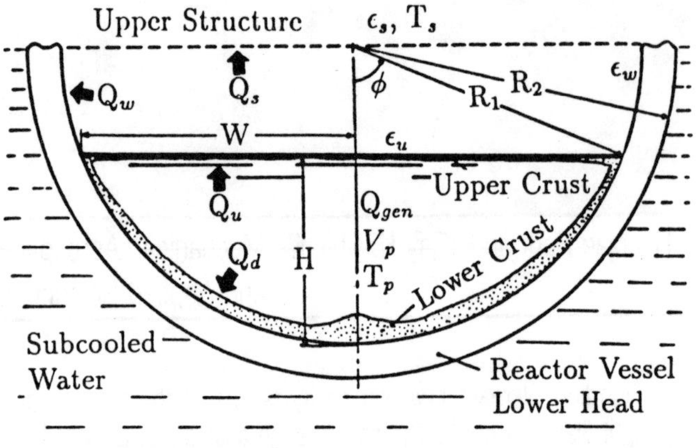

W : Radius of Free Surface of the Pool
R_1 : Inner Radius of Vessel Lower Head
R_2 : Outer Radius of Vessel Lower Head
H : Pool Depth
ϕ : Pool Angle
V_p : Pool Volume
T_p : Pool Temperature
Q_{gen} : Heat Generation in the Pool
Q_u : Heat Transfer to the Top of the Pool
Q_s : Heat Transfer to the Upper Structure
Q_d : Heat Transfer to the Lower Part of the Pool
Q_w : Heat Transfer to the Dry Wall of the Vessel Lower head

Figure 1. Schematic diagram of the core material relocated in the vessel lower head.

Governing Equation of the Vessel Shell

The differential equation for the two-dimensional steady-state temperature in the spherical shell is given by

$$\frac{1}{r^2}\frac{\partial}{\partial r}\left(r^2\frac{\partial T_w}{\partial r}\right)+\frac{1}{r^2 sin\theta}\frac{\partial}{\partial\theta}\left(sin\theta\frac{\partial T_w}{\partial\theta}\right)=0 \quad (1)$$

The corresponding boundary conditions are

$$-k_w\frac{\partial T_w}{\partial r}=k_c\frac{\partial T_c}{\partial z}, \quad 0\le\theta\le\phi, \quad r=R_1, z=0$$

$$-k_w\frac{\partial T_w}{\partial r}=q_{w,rad}, \quad \phi<\theta\le\pi/2, \quad r=R_1$$

$$-k_w\frac{\partial T_w}{\partial r}=h_L(T_w-T_L), \quad r=R_2$$

$$\frac{\partial T_w}{\partial\theta}=0, \quad \theta=0,\pi/2$$

Gauss-Seidel iterative method in conjunction with central-difference approximations for the derivatives in Equation (1) is used to solve the above partial differential equation. The boundary conditions along the vessel wall are determined as follows:

The outside vessel wall temperature, $T_{wo}(\theta)$, varies along the hemispherical surface of the vessel lower head. Heat transfer coefficients for natural convection, nucleate, transition and film boiling at different outer wall temperatures are needed to calculate the heat loss into the subcooled water. In this study water was assumed to have a subcooling of 50 K at one atmosphere pressure.

Several correlations for heat transfer coefficients during natural convection around a sphere are available in the literature [5,6]. The heat transfer coefficient for a submerged hemisphere was obtained from the general correlation suggested by Churchill [6]. The heat transfer coefficients in the nucleate and transition boiling are obtained by extrapolating the experimental results of Dhir [7]. When vessel outside wall temperature exceeds the minimum film boiling temperature, completely failure of the vessel wall is expected to occur.

Heat transfer from the pool to the crust/pool interface is assumed to take place by convection. For free convection with internal heat sources, the correlations developed by Gabor et al [12], Kulacki et al [13] and Jahn et al [14] are employed. The local heat transfer heat transfer coefficient along the hemispherical wall is assumed to vary sinusoidally and expressed as

$$h_d(\theta) = h_{d,av}\left(B_1 sin^2\theta + B_2\right) \qquad (2)$$

where

$$B_1 = \frac{12[1-0.24(1-cos\phi)]}{8-9cos\phi+cos3\phi}, \qquad \phi \neq 0$$

and

$$B_2 = 0.24$$

Assuming that the surfaces are gray, the relations for radiative heat exchange using the enclosure theory can be written as

$$Q_i - \sum_{i=1}^{n} F_{ij}(1-\epsilon_j)(\frac{\epsilon_i A_i}{\epsilon_j A_j})Q_j = \epsilon_i A_i \sigma \left(T_i^4 - \sum_{j=1}^{n} F_{ij} T_{ij}^4\right) \qquad (3)$$

for i, j = u, w, s

Using T_w as the average temperature along the unwetted vessel wall, the desired radiation heat transfer rates and the shape factor are obtained. The obtained value of Q_w can be used as the boundary condition along the dry vessel wall and Q_u can be used to determine the pool temperature for a certain set of parameters of ϵ_u, ϵ_w, ϵ_s, and T_s.

For a uniform temperature of the upper surface of the pool, the temperature of the pool without the upper crust is obtained as follows:

$$Q_{gen} = A_d h_{d,av}(T_p - T_m) + A_u h_{u,av}(T_p - T_u) \qquad (4)$$

or

$$T_p = \frac{Q_{gen,p} + A_d h_{d,av} T_m + A_u h_{u,av} T_u}{A_d h_{d,av} + A_u h_{u,av}} \qquad (5)$$

Equilibrium Crust Thickness

At any location along the wetted vessel wall, one-dimensional steady state heat conduction equation with a distributed heat source in the crust is given as

$$\frac{d^2 T_c}{dz^2} + \frac{q_{gen,c}}{k_c} = 0 \qquad (6)$$

The corresponding boundary conditions are

$$T_c = T_w, \qquad z = 0, r = R_1$$
$$T_c = T_m, \qquad z = \delta_c$$

The equation governing the equilibrium crust thickness is

$$k_c \frac{dT_c}{dz}\,|_{z=\delta_c} = h_d(T_p - T_m) \qquad (7)$$

The heat flux into the wetted vessel wall is obtained as

$$q_{wi} = k_c \frac{dT_c}{dz}\,|_{z=0} = \frac{q_{gen,c}}{2}\delta_c + (T_m - T_{wi})\frac{k_c}{\delta_c} \qquad (8)$$

In the above equations T_{wi} varies along the vessel inner wall.

RESULTS AND DISCUSSION

Steady state temperatures in the reactor vessel shell were obtained when full core was assumed to drop into the dry vessel lower head. Calculations were performed by varying between 0.1 and 0.8 the emissivity, ϵ_w, of the vessel, the emissivity, ϵ_s, of the upper structure and the emissivity, ϵ_u, of the upper surface. The upper structure temperature was assumed to be either 800 or 1600 K.

Steady state pool temperatures when vessel lower head contains the full core are plotted in Figure 2 as a function of emissivity of the pool free surface. The predictions are made for several assumed emissivities of the structure. It is seen that the pool temperature decreases as the pool free surface and structure emissivities increase. This is due to the fact that as emissivities increase the heat loss from the pool also increases.

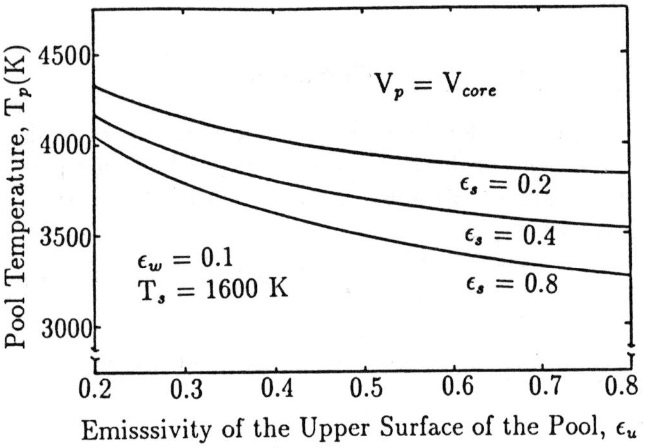

Figure 2. Variation of steady state pool temperature with pool and structure emissivities.

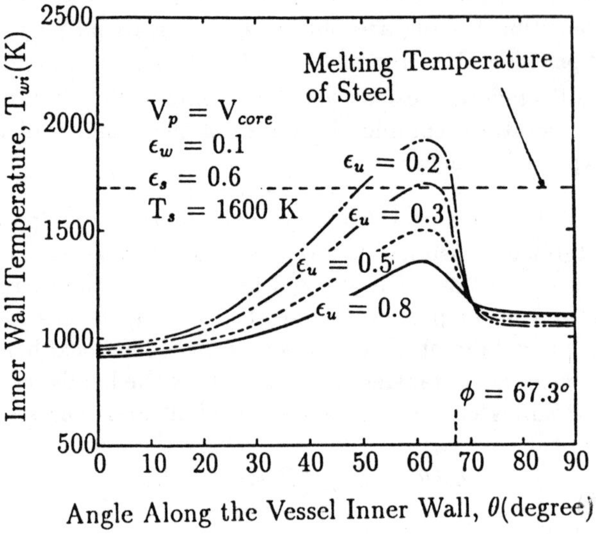

Figure 3. Effect of the emissivity of the upper surface of the pool (ϵ_u) on the inner wall temperature.

The effect of emissivity of pool free surface, of structure and of the unwetted region of vessel on the vessel inner wall temperature is shown in Figures 3, 4 and 5 respectively. From Figure 3, it is seen that the vessel wall temperature over the region wetted by the pool increases as the emissivity of the free surface of the pool decreases. However, the effect on the unwetted portion of the wall is opposite in that the vessel wall temperature decreases with a reduction in the emissivity of the pool free surface. This is due to the fact that as the pool free surface emissivity decreases, the reduced heat loss to unwetted vessel wall and to the upper structure causes the pool temperature to increase. For $\epsilon_u = 0.2$, some melting of the vessel wall at the pool, the wetted surface and the unwetted surface-triple interface is predicted to occur. The results plotted in Figure 4 show that structure emissivity plays a role similar to that played by emissivity of the pool free surface. However, for structure emissivity of 0.2, the vessel inner wall temperature increases monotomically with the angular position. This demonstrates that as pool temperature increases because of reduced heat loss to the structure, the radiative heat transfer from the pool to unwetted vessel wall becomes comparable to the heat transferred by natural convection in the wetted region. It is also noted that for $\epsilon_s = 0.2$, substantial melting of the unwetted portion of the vessel can occur.

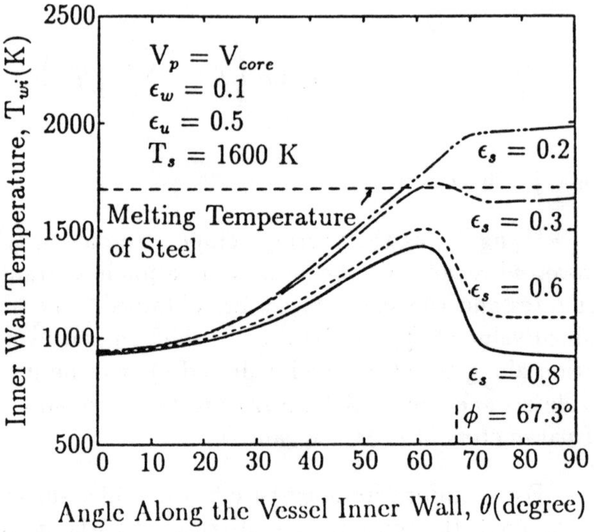

Figure 4. Effect of the emissivity of the upper structure (ϵ_s) on the inner wall temperature.

The effect of emissivity of the vessel on the temperature of the unwetted portion of the inner wall is opposite to the effect of emissivity of the structure. With increase in emissivity, the temperature of the unwetted wall increases. However, there is a very small decrease in the temperature of the wetted wall. As the emissivity increases, more energy is intercepted by the unwetted wall by radiation. Again for $\epsilon_w = 0.2$, substantial portion of the unwetted wall can melt.

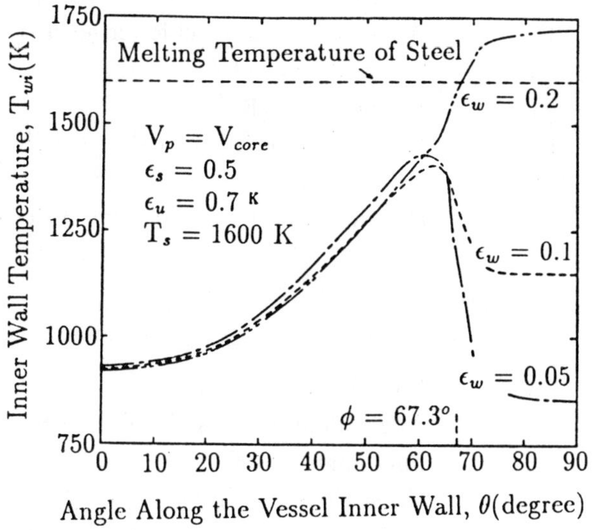

Figure 5. Effect of the emissivity of the unwetted wall (ϵ_w) on the inner wall temperature

The effect of structure temperature on the vessel inner wall temperature is shown in Figure 6. It is found that increased structure temperature causes an increase in the temperature of both the wetted as well as the unwetted portion of the vessel wall. The increase can be attributed to the higher steady state pool temperature with reduced radiative heat loss to the structure.

It appears that for the range of emissivities and structure temperatures studied in this work, molten core material can be contained in the reactor vessel indefinitely (from thermal considerations alone) if the reactor vessel is cooled from outside. In the absence of cooling with water, the calculations show that reactor vessel will fail even when $V_p = 1/16 V_{core}$.

Under steady state conditions, the fraction of

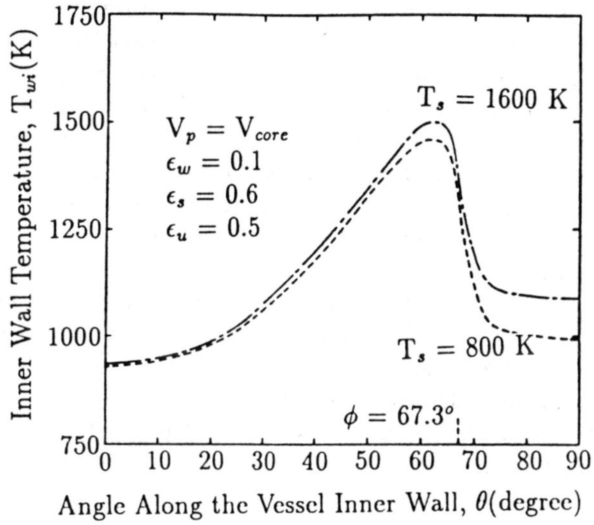

Figure 6. Effect of the upper structure temperature (T_s) on the inner wall temperature.

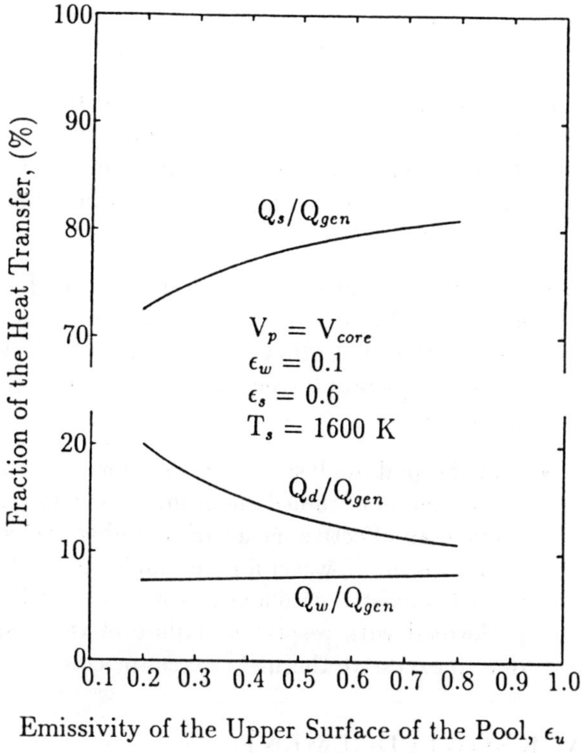

Figure 7. Partition of the heat loss from the molten pool.

generated energy that goes into structure, the unwetted and the wetted portion of the vessel wall are plotted in Figure 7. It is noted that between 70 to 80 % of the decay heat is absorbed in the structure that does not relocate with the core. This energy must get redistributed in the remainder of the reactor coolant system (RCS) by convection. As such this raises the possibility of the failure of the hot leg nozzle or the steam generator tubes. A break in the hot leg or the steam generator tubes will open a new path for the escape of radioactivity into the containment. The time for a break to develop in the RCS and further evolution of this mode of the severe accident have not yet been analyzed. The failure of the reactor vessel due to excessive thermal stresses or due to creep have also not been looked into.

SUMMARY AND CONCLUSIONS

- Two dimensional steady state thermal analysis of the reactor vessel shell has been performed after core material relocates into the vessel bottom head with cavity flooded with water. In the analysis the emissivities of the pool free surface, the structure and the vessel wall were varied parametrically. Calculations were performed for structure temperature of either 800 K or 1600 K.
- The emissivities of the pool free surface, the structure and the vessel play an important role. However the structure temperature has a small effect on the temperature distribution around the vessel inner wall.
- Melting of the unwetted portion of the vesel wall is predicted to occur for vessel wall emissivities greater than 0.2. However melting is expected not to propagate further than half of the vessel thickness.
- The thermal analysis shows that for the range of parameters studied, flooding of cavity may provide an effective means of retaining the core in the vessel. However a potential for a break in the RCS exists. Also analysis has not yet been performed with respect to failure of the vessel due to excessive thermal stresses or creep.

ACKNOWLEDGEMENT

This work received support from NRC under contract no. NCR-04-89-349

NOTATION

A	Area (m^2)
F	Radiation shape factor
h	Heat transfer coefficient (W/m^2 K)
k	Thermal conductivity (W/m K)
Q	Rate of Heat transfer (W)
Q_{gen}	Heat generation (W)
q	Heat flux (W/m^2)
q_{gen}	Volumetric heat generation (W/m^3)
R_1	Inner radius of vessel lower head (m)
R_2	Outer radius of vessel lower head (m)
r	Radial distance (m)
T	Temperature (K)
T_m	Melting temperature of the pool (K)
V	Volume (m^3)
z	Distance from the vessel inner wall (m)

Greek Letters

ϵ	Emissivity
θ	Angle along the vessel wall (deg)
ϕ	Pool angle (deg)
δ	Crust thickness (m)
σ	Stefan-Boltzmann constant (W/m^2 K^4)

Subscripts

av	Average
c	Crust
d	Lower
L	Subcooled water
p	Pool
s	Upper structure
u	Upper
w	Vessel shell
wi	Vessel inner wall
wo	Vessel outer wall

LITERATURE CITED

1. W. A. Condon, S. R. Greene, R. M. Harrington and S. A. Hodge, "SBLOCA Outside Containment at Browns Ferry Unit One - Accident

Sequence Analysis," NUREG/CR-2672, Vol. 1, ORNL/TM-8119/V1, p. 197 (1982).

2. S. Nazare, G. Ondracck and B. Schulz, "Properties of Light Water Reactor Core Melts," Nuclear Society, Vol. 32, p. 239 (1977).

3. Modular Accident Analysis Program (MAAP) User's Manual Version 3.0, Fauske and Associates, Inc. (1989).

4. S. S. Dosanjh, "Melt Propagation in Dry Core Debris Beds," Nuclear Technology, Vol. 88, p. 30 (1989).

5. J. H. Lienhard, "On the Commonality of Equations for Natural Convection from Immersed Bodies," Int. J. Heat Mass Transfer, Vol. 16, p. 2121 (1973).

6. S. W. Churchill and R. U. Churchill, "A Comprehensive Correlating Equation for Heat and Component Transfer by Free Convection," AIChE Journal, Vol. 21, No.3, p. 604 (1975).

7. V. K. Dhir, "Study of Transient Transitional Boiling Heat Fluxes from Spheres Subjected to Forced Vertical Flow," International Heat Transfer Conference, Toronto, Canada (1978).

8. V. K. Dhir and G. P. Purohit, "Subcooled Film-Boiling Heat Transfer from Spheres," Nuclear Engineering and Design, Vol. 47, p. 49 (1978).

9. F. S. Gunnerson and A. W. Cronenberg, "On the Minimum Film Boiling Conditions for Spherical Geometries," Journal of Heat Transfer, Vol. 102, p. 335 (1980).

10. E. Marschall and L. C. Farrar, "Film Boiling from a Partly Submerged Sphere," Int. J. Heat Mass Transfer, Vol. 18, p. 875 (1975).

11. R. Siegel and J. R. Howell, Thermal Radiation Heat Transfer, McGraw-Hill Book Company, NY, (1981).

12. J. D. Gabor, P. G. Ellison and J. C. Cassulo, "Heat Transfer from Internally Heated Hemispherical Pools," Presented at the 19th National Heat Transfer Conference, Orlando, FL (1980).

13. F. A. Kulacki and A. A. Emara, "Heat Transfer Correlations for Use in PAHR Analysis and Design," Trans. American Nuclear Society, 22, p. 447 (1975).

14. M. Jahn and H. H. Reineke, "Free Convection Heat transfer with Internal Heat Sources," Proceedings of the Fifth International Heat Transfer Conference, Vol. 3, p.74 (1974).

ESTIMATES OF LIMITS TO FUEL/COOLANT MIXING

G.C. Park and M.L. Corradini ■ University of Wisconsin, Department of Nuclear Engineering and Engineering Physics, Madison, WI 53706

The vapor explosion process can involve the mixing of fuel with coolant prior to the explosion. A number of analysts have suggested limits to the amount of fuel/coolant mixing that could occur within the reactor vessel following a core melt accident. Past models are reviewed and a simplified approach is suggested to estimate a limit on the amount of fuel/coolant mixing possible. The approach uses concepts first advanced by Fauske in a different way. The results indicate that fuel temperature, ambient pressure and in particular the mixing length scale D_{mix} are important parameters. For large values of D_{mix} the fuel mass mixed in-vessel is limited to the range of 1–12 metric tons (1–10% of the core mass.)

INTRODUCTION

Prolonged failure of normal and emergency coolant flow in a light water reactor (LWR) would cause melting of the reactor fuel, fuel slumping and eventual contact with residual water in- or ex-vessel. This fuel/coolant contact would allow the fuel to mix with the coolant producing steam and hydrogen as the fuel slowly fragments while in film boiling. This "premixing" phase of the fuel/coolant interaction (FCI) also sets up the initial conditions for a possible vapor explosion that may follow. It is our intent to estimate possible limits to this fuel/coolant mixing. Identification of mixing limits is of interest because it would suggest a restriction on the size of large-scale vapor explosions; the implication of this result would be to limit the steam explosion damage potential in a reactor-containment system.

PAST WORK IN FUEL-COOLANT MIXING

Past research (1-15) into fuel/coolant mixing (sometimes called premixing) has been mainly directed at predicting the physical limits for which mixing could or could not occur. Fauske (1) and Henry and Fauske (2) originally proposed that the fuel/coolant interface temperature upon liquid-liquid contact must exceed the spontaneous nucleation temperature to allow premixing. The physical picture was that stable film boiling is established above this limit for a liquid-liquid system, and this allows the fuel time to penetrate and mix within the coolant. Henry and Fauske (4,5) subsequently proposed the concept that for the fuel to exist in a premixed configuration within the coolant, a conceptual picture (as Figure 1) must be achieved and sustained. Note that this picture has one liquid (the 'fuel') as discrete particles intermixed in a continuous phase of the other liquid coolant with vapor present. If this configuration breaks down by driving the water coolant away from the molten fuel, discrete fuel particles may coalesce into larger particles as the coolant liquid is depleted from the mixing region.

Theofanous et al., (7) also addressed the question of a limit of fuel/coolant mixing and took a different approach. He concluded that the mixing process would be driven by the hydrodynamics of transient jet breakup as the fuel pours into a water coolant pool (Figure 2). Order-of-magnitude calculations indicated that only a few percent of the available fuel mass could mix with the water coolant for in-vessel core melt situations. Theofanous et al., (15) also developed a two-dimensional model for mixing employing the K-FIX computer model with substantial modifications of the constitutive relations. The basic simplifying assumptions of the computer analysis were exactly the same as those of Bankoff [13]. Recently Theofanous et al. (17) have developed a three-fluid model (PM-ALPHA) to remove the restriction of the coolant being treated as a homogeneous fluid. The major limitation of the analysis is that the fuel is modelled as a collection of droplets entering at a prefragmented user-specified size. The predicted fuel mass mixed for similar conditions was reduced compared to the past study where now the mixed mass was 4000 kg for a PWR lower plenum compared to 5000 kg previously. The reason for the difference seemed to hinge on the allowed slip between the vapor and liquid coolant where a larger liquid coolant depleted zone forms in the central region of the mixture (void > 75%).

In an attempt to remove the restriction of a constant fuel diameter during mixing Chu and Corradini

G. C. Park is a Visiting Professor from Seoul National University, Seoul, Korea.

(16) developed a technique to predict the dynamic fragmentation of the fuel due to relative velocities. This was incorporated into a transient, multifluid, one-dimensional model developed to allow one to consider mixing between fuel and coolant with dynamic fuel fragmentation. Chu and Corradini (18) based this model on the TEXAS code (Young (19)), a fuel-coolant interaction model for LMFBR safety. The key constitutive relation is a fuel fragmentation model based on Rayleigh-Taylor instabilities. This constitutive model considers the fuel to be dynamically fragmented from its initial entry diameter to smaller sizes. In the model parallel velocity shear forces (e.g., Kelvin-Helmoltz instabilities) are neglected because of their limited effect with a vapor film present (Fauske (20)). For LWR safety applications, Chu and Corradini (18) considered an in-vessel situation of a corium fuel jet similar to those of Bankoff (13) and Theofanous (15). The results indicate that for the base case, less than 20% of the available submerged mass mixes with coolant where the vapor void is less than 50% (~350 kg, where one-third of the pool has a void fraction >.5). Thus according to the original concept of Fig. 1 less than half of the fuel mass is "mixed" with a continuum of liquid coolant. Quite recently, Young (21) has also considered the concept of dynamic mixing and, with separate dynamic mixing models by Pilch, developed a two-dimensional model for fuel-coolant mixing, IFCI.

MIXING AND CRITERIA FOR A WELL-MIXED SYSTEM

In our discussion of the past work on fuel-coolant mixing it becomes obvious that "fuel mixed with coolant" is a statement that does not have a precise meaning. In fact although many investigators have similar conceptual pictures their interpretation of test data and subsequent computational analysis may differ based on their meaning of "fuel mixed with coolant." Since our purpose in this work is to present an analysis which estimates the possible bounds on fuel-coolant mixing it is important that we first define what we mean by "fuel mixed with coolant."

The first difficulty with any such definition is that one lacks precise quantitative values to establish mixing criteria. Although many investigators have proposed mixing models and analyses the available data (10-12; 22-24) is quite sparse. Two possible criteria have already been mentioned: 1) stable film boiling should exist between the two liquids (1,2) and 2) mixing energy requirements are small compared to the fuel thermal energy (3). In the event of a vapor explosion we would propose two other necessary criteria for "fuel mixed with coolant": 3) the local length scale for the discrete liquid phase,l, should be much smaller than the system length scale, L, (l<< L) and 4) the continuous phase in this local region should be one of the two liquids. This latter condition is based on a continuous liquid volume fraction ($\alpha_c > 50\%$ or $\alpha_f > 50\%$) as Figure 1 illustrates and may be necessary for a vapor explosion to occur.

These criteria are based on the premise that for the vapor explosion to be initiated, a trigger is available and the vapor explosion escalation/propagation process can proceed. The spontaneous triggering of the explosion is empirically (10) known to be retarded if the ambient pressure increases or the void fraction increases (due to saturated conditions or noncondensible gas). The detailed mechanisms for the explosion escalation and propagation are not well understood. However, it seems to be accepted that for the explosion to be triggered the fuel and coolant liquids must be in close proximity (small length scales) and with a minimum of vapor present (i.e., vapor film collapse and liquid-liquid contact, Ref. 1-5).

Consider the criterion that the discrete liquid phase should have a length scale, l, much less than the overall system length scale, L. If our estimate for a fuel mixing limit is to make sense, the fuel 'mixed' must participate in the vapor explosion. This implies two points. First all of the fuel should be in close proximity to the coolant liquid, and should breakup to a size where the mixture of the fuel, coolant and vapor appear homogeneous at the system length scale (Fig. 1). Without this occurring not all of the fuel would be dispersed in the coolant and we would have a heterogeneous mixture situation approaching stratified liquids. Second, this local length scale for the discrete mass should be small enough that the fuel can be completely fragmented and quenched during the explosion escalation and propagation/expansion time. This characteristic time would be that interval in which the explosion shock wave grows in strength and traverses the system length scale as the fuel coolant mixture expands (e.g., a factor of two expansion). If this local fuel length scale is too large some of the fuel will remain unquenched during the explosion expansion and would not contribute to the explosion energy release. By analogy to fuel-oxidant mixing in diesel engine combustion (25), it is not correct to assume that all the diesel fuel injected into the engine chamber is mixed with the oxidant; rather it must be only that fuel which has atomized to small enough length scales to undergo vaporization and eventual combustion during the burning and expansion phase (large fuel droplets are not mixed and actually hamper the combustion process efficiency).

The final criterion for fuel-coolant mixing also has an impact on the vapor explosion. Consider first the effect of the vapor void fraction on the explosion from strictly a thermodynamic perspective. As the void fraction increases the ideal work potential can decrease (e.g., a 50% decrease in efficiency is possible for a void fraction increase to 90% at equal fuel and coolant liquid volumes). This is because the mixture has already expanded to allow for the vapor void. In addition, a void fraction increase may suppress the kinetics of the explosion, and this effect may be most significant when the threshold from a continuous liquid to continuous vapor field is crossed; $\alpha_v > 0.5$. One reason is that the explosion should be much more difficult to trigger. This was noted in the parametric calculations by Fletcher et al (26), where initial trigger pressures in excess of about 50-

100 MPa were needed for voids of 70-90%. Our past work on explosion triggering (27-29) also indicates that to initiate liquid-liquid contact and to begin rapid fuel fragmentation in the explosion, the vapor volume fraction should be small (α_v < 50%). These observations do not imply that one could not trigger at higher vapor volume fractions (smaller liquid volume fractions), but that the necessary trigger energy (or impulse) would increase substantially (27-29). Another reason is that the physical concept of 'liquid-liquid' contact which drives the vapor explosion loses meaning when the vapor becomes the continuous phase. One now has to propose that the thermal interaction of dispersed droplets of fuel and coolant would escalate and propagate an explosive heat transfer process. It would be our judgement that explosion propagation with liquid-liquid contact and rapid heat transfer would be significantly reduced or suppressed at higher vapor void fractions. Another effect of having a continuous liquid phase during mixing would be to allow for an inertial constraint from the fuel-coolant mixture itself. If the void fraction is too large it is not clear that the explosion will collapse these large vapor voids and propagate without simply venting through this dilute mixture (30, 31). In the analysis by Fletcher et al (26) the explosion propagated through the high vapor fraction mixture (70-90% void) by the assumption of a large trigger (100MPa) and a large parametric liquid-liquid heat transfer coefficient ($10^7 W/m^2K$).

In our past work (e.g., Chu et al (18) and Corradini et al (32)) we have been consistent with this definition of a fuel coolant mixture. Theofanous (15, 17) and Bankoff (13) have also considered similar definitions when analyzing their multidimensional calculations. In his recent work (15, 17) Theofanous and coworkers have considered fuel-coolant mixtures with vapor void fractions as high as 75-90% to estimate the possible fuel mass mixed, and have considered this to be a conservative limit. However, one should realize these are bounding values chosen to assure the real fuel mixture mass is well below such limits. Also, the values for fuel mixed mass, when α_v < 50%, were not significantly smaller.

One might argue that the explosion could begin in a low void region of the mixture and propagate into a highly voided region. This has never been conclusively shown to occur experimentally. For example, in the FITS experiments (10,11) it was usually observed that the explosion was triggered near the base and propagated upward. Probably, the void increased in the central region of the mixture with height in the pool, but it is not clear what this void was and if the explosion propagated through this vapor rich region. Also, based on the parametric results of Fletcher et al (26) it appears that the explosion pressure needed for such a sustained propagation through a highly voided region is much larger that any measured explosion pressure data. These proposed criteria are based on our analysis of a sparse set of data. Clearly more data is needed to assess these criteria and the derived model on which it is based.

FUEL-COOLANT MIXING MODEL

Our objective is to provide an estimate on the limits to fuel-coolant mixing given the previous definition. In this analysis we neglect the effect of solid structure within the coolant pool on mixing even though it may be realistically present. This is the case for the in-vessel lower plenum core/diffuser plates and support structure (Figure 3). We discuss the possible effects of the structure after the analysis is presented along with other realistic considerations. Comparing the work of the previous investigators, the mixing energy criterion does not limit the amount of fuel that can mix with the coolant; rather, it is hydrodynamic considerations either due to boiling processes (4, 5) or transient jet breakup (6, 7), or some combination of these two effects (8, 9) that limit mixing. A second observation is that for a given set of conditions (e.g., 1 atm saturated water, PWR geometry neglecting structure), these past studies on mixing limits yield a range of results.

The limit to mixing would correspond to the point when the coolant liquid is fluidized by the vapor upflow. One could recast the model originally proposed by Fauske in these terms to be

$$\frac{6m_f}{\rho_f D_{mix}} q''_{drop} = \rho_v i_{fg} A_{ch} V_{SC} \qquad (1)$$

where

$$q''_{drop} = h_{film} (T_f - T_{sat}) + \sigma_r (T_f^4 - T_{sat}^4) \qquad (2)$$

and h_{film} is the film boiling heat transfer coefficient from the drop. Note the energy removed per unit volume is $\rho_v i_{fg}$ and V_{SC} is the superficial critical (SC) steam velocity that will fluidize the water coolant. Now, using Kutataladze's or Zuber's model (33) for CHF in pool boiling, Fauske estimated V_{SC} to be

$$V_{SC_{CHF}} = 0.14 \left[\frac{\sigma g (\rho_c - \rho_v)}{\rho_v^2} \right]^{1/4} \qquad (3)$$

where the constant 0.14 was determined empirically from pool boiling data. Borishanskii (34) indicated that this empirical constant can be explained in terms of the hydrodynamic size of the liquid and vapor as CHF is reached and their relative velocities. Let us also use this picture here in our analysis where the fuel mixes within the coolant pool with an ever increasing mass, until fluidization of the coolant liquid is reached. This fluidization would first occur at the top of the pool where all the vapor flow exits and where the vapor void fraction would first become large enough (α_v > .5) that coolant droplets are formed and could be swept away by the vapor flow. At this point if more fuel is added to the coolant pool the coolant vapor void fraction would rise further and coolant liquid could be carried out of the mixture. Based on our proposed criteria the actual mass of fuel mixed would decrease because more and more mass resides in a vapor rich region where the explosion can be difficult to trigger and propagate and where its own inertial constraint is significantly reduced. It is the fuel mass mixed at this limit that we seek.

Consider a coolant droplet existing at the top of the pool with a diameter, D_c; if this droplet were to become fluidized by the surrounding vapor flow, a simple force balance gives us this fluidization velocity V_{fl}:

$$V_{fl} = \left[\frac{4}{3}\, g\, \frac{D_c}{C_D}\, \frac{(\rho_c - \rho_v)}{\rho_v}\right]^{1/2} \tag{4}$$

where all the terms have been previously defined. Now, for a collection of coolant droplets, the local vapor velocity would increase because the volume occupied by the coolant liquid decreases the available flow area and the local drag coefficient would also increase due to this same effect. Using the empirical results from Wallis (35) on the drag coefficient in an array of particles, the SC fluidization velocity becomes

$$V_{SC} = \alpha_v \left[\frac{4}{3}\, g\, \frac{D_c}{C_D(\alpha_v)}\, \frac{(\rho_c - \rho_v)}{\rho_v}\right]^{1/2} \tag{5}$$

where

$$C_D(\alpha_v) = \frac{C_{Ds}}{(\alpha_v)} 2.65 \tag{6}$$

and where C_{Ds} is the drag coefficient for a single sphere.

The next question to consider is what is the proper value for the coolant characteristic length scale, D_c. The coolant is the continuous phase just before fluidization occurs at the pool surface. Based on geometry considerations as the liquid coolant volume fraction decreases below about 50% due to an increase in the amount of coolant vapor, the coolant liquid will become discontinous between the intermixed fuel droplets and the vapor film surrounding them. The characteristic length scale would be to consider the coolant surrounding any individual fuel droplet as a single coolant droplet at the point of fluidization.

$$D_{c_I} = D_f \left(\frac{\alpha_c}{\alpha_f}\right)^{1/3} \tag{7}$$

Substituting this value into Eq. (5), one obtains

$$V_{SC_I} = \alpha_v \left[\frac{4}{3}\, g\, \frac{D_f}{C_D(\alpha_v)} \left(\frac{\alpha_c}{\alpha_f}\right)^{1/3} \left(\frac{\Delta\rho}{\rho_v}\right)\right]^{1/2} \tag{8}$$

Now, this coolant length scale is an upper bound on the size of a coolant droplet. This droplet is the largest diameter that could be formed based on the geometry, and it neglects the fact that as the coolant liquid begins to fluidize, it would fragment in the vapor flow. If one considers droplet fragmentation then the coolant liquid will break up in the vapor flow down to sizes given by the critical Weber number

$$D_{cII} = \frac{We_{crit}\sigma_c}{\rho_v V_{SC}^2} \tag{9}$$

where σ_c is the coolant surface tension. Substituting this length scale into Eq. (5), yields

$$V_{SC_{II}} = \alpha_v^{1/2} \left[\left(\frac{4}{3}\right) \frac{We_{crit}}{C_{D(\alpha_v)}} \left[\frac{g\, \sigma \Delta\rho}{\rho_v^2}\right]\right]^{1/4} \text{(14)} \tag{10}$$

Comparing this to the CHF velocity (Eq. 3) one notes the similar functional form where 0.14 is replaced by the vapor volume fraction, α_v, the critical Weber number, We_{crit}, and the local drag coefficient (given by Eq. 6).

Note that the actual coolant length scale would be between these two bounds because the actual fuel/coolant mixing process is probably a transient phenomenon. For the case of light water reactor fuel in saturated water at one atmosphere the ratio of D_{cI}/D_{cII} is about 5-10. This transient aspect can be demonstrated by considering the time it takes the coolant with a characteristic length given by D_{cI} to break apart due to coolant vapor flow down to the characteristic length given by D_{cII}. This is a classic case of hydrodynamic breakup where the time for breakup is approximately given by

$$T^+ = 3 \text{ to } 6, \tag{11}$$

where, as before, T^+ is defined as

$$T^+ = \frac{V_{SC_I} t_B}{D_{cI}} \left[\frac{\rho_v}{\rho_c}\right]^{1/2} \tag{12}$$

For our example case assuming $T^+ = 4$, with a pour velocity of 5 m/s and D_{cI} of 30 mm, one finds the time for breakup t_B is less than a second. This is about the time it takes the fuel to pour by gravity into the coolant pool and reach the bottom of the chamber, both for in-and ex-vessel situations.

Now the final point to make is that to use this model one must know the vapor void fraction, α_v, at the point of fluidization. Since all of the vapor produced must eventually exit the pool, the location of greatest vapor flow occurs at the pool surface. Fluidization would also first occur at this location. To estimate this void fraction and the associated liquid fractions (α_c, α_f), this fluidization point is calculated based on the drift-flux model and the flooding criterion by Wallis (35). In the drift flux model, an average velocity of vapor ,V_v, is given as,

$$V_v = C_o j + V_{gj} \tag{13}$$

where j is the volumetric flux and V_{gj} is the average drift velocity.

For churn-turbulent flow, the average drift velocity is approximately equal to the terminal rise velocity ,U_t, given by

$$V_{gj} = U_t = 1.53 \left[\frac{\sigma_c g (\rho_c - \rho_v)}{\rho_c^2} \right]^{1/4} \tag{14}$$

where σ_c is the coolant surface tension. Since $j_c = 0$ at the flooding point and $V_v = j_v/\alpha_v$, Eqs. (13) and (14) can be combined to yield.

$$\left(\frac{1}{\alpha} - C_0\right) j_v = 1.53 \left[\frac{\sigma_c g(\rho_c - \rho_v)}{\rho_c^2} \right]^{1/4} \tag{15}$$

The general flooding criterion given by Wallis is

$$j_v^* = j_v \sqrt{\frac{\rho_v}{g D_H (\rho_c - \rho_v)}} = 0.9 \tag{16}$$

Now from Eqs. (15) and (16), we can obtain the fluidization void fraction at the top of the mixture to be equal to

$$\alpha_v = \frac{1}{C_0 + a} \tag{17}$$

where

$$a = 1.7 \left[\frac{\sigma_c \rho_v^{\,2}}{\rho_c^2 g D_H^2 (\rho_c - \rho_v)} \right]^{1/4} \tag{18}$$

Within this model is the hydraulic diameter for the coolant flow, D_H. This length scale can also be related to the fuel diameter. We consider the fuel particles to be stationary relative to the coolant vapor flow out of the pool, since this velocity is much larger in magnitude. The coolant hydraulic diameter, D_H is given by the standard relation of the ratio of the flow area to the wetted perimeter as

$$D_H = \frac{4\left[A_{ch} - m_f / \rho_f H_{mix}\right]}{P_w + 4 m_f / (\rho_f H_{mix} D_f)} \tag{19}$$

where P_w is the chamber perimeter, H_{mix} is the depth of the pool and m_f is the fuel mass mixed in the pool.

RESULTS AND DISCUSSION

To illustrate the use of the proposed model we have estimated the fuel-coolant mixing limit that would occur in the lower plenum of a pressurized water reactor. The German PWR, BIBLIS-B, was chosen as the specific design for the geometric details in the analysis (Figure 3). As the figure indicates there are three solid structures in the lower plenum; the lower core plate, the core mixing plate and lower plenum support structure. The first two structures are immediately below the reactor core and represent relatively thin-walled structures, which have relatively narrow passages into the open plenum region below. The lower plenum support structure is quite massive, although it does not extend throughout the plenum. Considering this geometry and in the event of a severe accident, we would expect that the core melt would pour from the core region through these upper two plates without being impeded. If an explosion is not triggered by melt contact with the solid structure the greatest effect of these mixing plates would be to subdivide a large pour into smaller melt streams (if an explosion is triggered in this region the fuel melt mixed would be quite limited and we have conservatively neglected this possibility). Once the fuel pours through these structures it will see a relatively open lower plenum pool of water coolant. Because of the shape and position of the lower support structure it does not seem it will have a major impact in mixing, particularly in the overall limit as considered here.

The remaining initial conditions needed for the analysis depend to a large extent on the severe accident sequences considered, especially on the dominant accident sequences. For illustrative purposes of the model the following ranges for initial conditions were chosen;

Pool Depth:	$1.0 < H_{mix} < 2.5$ m
Ambient Pressure:	$0.1 < P_\infty < 1$ MPa
Fuel Temperature:	$2000 < T_f < 3000$ K
Water Temperature:	Saturated at P_∞

The pool depth values represent a range of depths from the bottom of the vessel to the middle of the massive lower plenum support structure, or to above the core plate, where for most severe accident conditions the coolant is saturated at the vessel ambient pressure. The nominal value chosen for the analyses is 1.5m and this represents a water coolant depth from the bottom of the vessel to just below the core mixing and diffuser plates. The cross-sectional chamber area chosen ($9.5 m^2$) represents the complete core region. The ambient pressure is also quite sequence specific so a range of 0.1 to 1.0 MPa was chosen to represent likely values for low pressure sequences. Because of the increased difficulty to trigger explosions at higher ambient pressures (27-29) and the current views that suggest depressurizing the RPV under severe accident conditions is a reasonable strategy, $P_\infty = 1.0$ MPa seemed a reasonable upper limit with $P_\infty = 0.3$ MPa as a nominal case, for illustrative purposes. The nominal value of 3 bars represents two effects. First, during severe accident sequences the containment will begin to pressurize due to primary system blowdown and further steam generation. It is not unusual for the containment pressure to rise to over 2 bars due to this effect. Secondly, there will be some pressurization in the lower plenum when the fuel enters the coolant pool due to steam generation during mixing. This was noted by past investigators (13, 17, 18) and can be on the order of a bar. Finally, the range of fuel temperatures was chosen to bound actual values during the accident, with 2700 K taken as the nominal value from past severe accident analyses (6).

The final initial condition that one must specify for the model is the fuel mixing diameter, $D_f = D_{mix}$. In the previous section it was proposed that if this model is to be used to estimate the upper bound on the fuel mass mixed this diameter should be of an appropriate size that all of the droplet mass would be rapidly fragmented during the explosion propagation/expansion time. If this fuel local length scale is too large, some of the fuel would still remain unquenched during the explosion expansion, and not participate in the energy transfer or contribute to the explosion mechanical energy release. Thus, although it is present in the coolant pool it is not involved in the explosion, and thus is not really an active part of the mixture. Therefore, the final parameter needed to estimate a mixing limit for the vapor explosion is a range of values of D_{mix} (sometimes called the coupling length scale) that will completely fragment during the explosion. In the past analyses (e.g., Ref. 15, 17) the value for D_{mix} has been chosen to be between 10-100 mm, where some investigators feel the real value is near 10 mm and 100 mm is considered an extreme upper bound. The dynamic fragmentation models employed in the TEXAS model (Chu and Corradini, 16, 18) also indicate that in the region where the fuel is "mixed" with a continuum of liquid water (Table I) the fuel fragments to sizes near 10 mm. As Table I indicates for a variety of initial conditions (water depth, pour rates and jet number) the fuel particles "mixed" within the coolant pool have diameters between 5-15 mm. As the coolant begins to fluidize in vapor rich regions the fuel liquid (and coolant) fragments further to smaller sizes (fuel mass "dispersed," 1-4 mm). This model developed by Chu was also used to analyze the FITS experiments conducted by Sandia (e.g., Ref. 10-12). For these experiments the TEXAS model also predicted mixing diameters on the order of 10 mm. This was verified to some extent for the few non-explosive FITS tests where the recovered fuel debris average diameters were on the same order (<10 mm). This estimate should not be taken as a completely validated prediction since no mixing diameters have been precisely measured during any experiments. Recent experimental work has been completed by Bang et al (36-38) in which this length scale was measured for horizontally stratified layer explosions. The measured and predicted length scale was much less than 10 mm for small scale and large scale tests (2-4 mm for 20-50 cm apparatus sizes). For the LWR initial conditions specified here, this model suggests 10 mm to be the appropriate estimate; this assumes this length scale is similar for a mixed system as for a stratified system. However, to be conservative we also chose a value of 30 mm for the D_{mix} to bound this best estimate. We should clearly point out that in the absence of more data this value for the coupling length scale is not a certainty. This suggested range represents a good illustrative example based on a first set of data and mechanistic models that directly attempts to determine this parameter.

The results of the analysis are presented in Figures 4-11. One should note that curve marked "alpha-dept diameter" refers to the use of D_{CI} for the coolant length scale (Eq. 11) while "We crit" refers to the use of D_{CII} for the coolant length scale (Eq. 9). Since the coolant liquid fluidization involves droplets formed from the continuous coolant liquid phase one can only bound this length scale. Based on the models discussed we believe that coolant droplet fragmentation would rapidly occur as fluidization proceeds. This would favor the lower limit line in the model, D_{CII}. However, because this effect is uncertain one can only bound its effect here. Thus the values for fuel mass mixed would lie within this region. As the figures indicate the estimated fuel mass mixed is about 1-12 metric tons of fuel over the range of initial conditions considered. For our nominal values of $T_f = 2700$ K and $P_\infty = 0.3$ MPA the range is a few metric tons. One should note that the vapor void fraction at the top of the pool at the point of coolant liquid fluidization is predicted to be greater than 50% (55-65%) and thus, these mixing values may be slight overestimates. Also note that the amount of fuel which can mix with the coolant under these conditions increases as the water coolant depth and ambient pressure increases or as the fuel temperature decreases. The increase in the water depth allows a larger coolant volume within which the fuel can mix. However, because the vapor produced throughout the volume must exit the top of the pool, only a small depth of water (~ 0.5 m) is needed before fluidization occurs and the vapor void fraction becomes large ($\alpha_v > 50\%$). The net result is that water depth has a relatively small effect above this minimum depth. Although not shown, an increase in the pool cross-sectional area results in a linear increase in mass mixed since more coolant volume is linearly added. The increase in ambient pressure increases the vapor density and thus decreases the local void fraction proportionately; therefore, more fuel can mix before coolant liquid is fluidized by its vapor. As the fuel temperature decreases more fuel can mix because less coolant vapor is produced per unit surface area of fuel, since the incident heat flux from the fuel decreases. Thus, more fuel mass (fuel surface area) can exist within the pool before the vapor produced fluidizes the liquid in the pool. Counterbalancing this effect is the fact that as the temperature decreases the fuel thermal energy decreases proportionately as would the explosion work output per unit of fuel mass. The final effect to consider is the initial mixing diameter. Note that as D_{mix} increases the amount of fuel mixed also increases. The reason for this directly relates to the reduction in vapor production due to a reduction in the fuel surface area. Note though that if

D_{mix} is too large not all the fuel mass will participate in the explosion, thus the fuel mass would not be efficiently mixed.

When this model was first presented in the previous section it was pointed out that this integral approach to the problem of a limit to fuel mixing with coolant was appealing because it bypassed the complexities of a multidimensional analysis. Both the original model by Fauske (4,5) and this proposed model include the whole coolant pool volume as available for mixing with the fuel and the whole pool surface area available for coolant vapor outflow. It is our belief that the multidimensional aspects of fuel-coolant mixing may be bounded by this approach. To make such a claim, the available mixture volume must be larger and the surface area for vapor outflow must be also larger. This always implies the steam flow velocity is minimized and the volume of fuel mixed is maximized. Now the only way to conclusively prove this hypothesis is to perform well-instrumented mixing experiments and supporting analyses that demonstrate this to be the case. In the absence of such empirical evidence let us consider some of the possible multidimensional configurations that the model neglects and examine if they are bounded by the model.

One possible effect is that in the actual accident case for the reactor the fuel may pour into the central region of the pool and as the fuel-coolant mixture dynamically grows (axially and radially) the vapor flows up axially through the fuel eventually exiting at the pool surface. Qualitative observations from tests (10-12) and some simulations of this test data during mixing (13, 15, 17) suggest mixing occurs at the edge of the mixture and a vapor rich region exists internal to the mixture. In reactor scale simulations (e.g., Ref 17) similar behavior is demonstrated for finite fuel masses. In both cases the dynamic lateral movement of the mixture volume progressively allows more mixing. However, this effect would not allow more mixing compared to the whole pool surface area and pool depth because this volume is always greater than the growing multi-dimensional mixture volume. In addition, the surface area for vapor out flow is always larger, minimizing the vapor velocity. Thus, by taking the complete pool surface area we have bounded this geometric effect. A second configuration would be the fuel mass enters as a series of jets spread across the pool surface instead of one large jet. In the limit this situation actually approaches our one dimensional picture, and the model approximations become unimportant.

Another effect to consider is fuel-coolant remixing during the explosion expansion. Past CSQ calculations (30, 31) indicated that a steam channel (void fraction, 35%) could be collapsed by the explosion propagation and expansion (note, that this case falls within our definition of mixed fuel). This concept may allow for more mixing between the liquids as the large voids collapse. In this case one must compare two fuel masses; the one that initially mixes in our control volume approach and the other being that mass mixed in the more realistic case where part is in a vapor rich region. If the initial available pool volume for fuel mixing is conserved between these two cases, then our conceptual picture always bounds this situation since the integral analysis considers coolant liquid is the continuous phase in the pool volume and all the fuel in the pool would be initially mixed. No additional coolant and fuel liquids would be available to remix during the explosion expansion.

A final consideration is the possibility of multiple explosions, where the first explosion is too weak to disperse the fuel but of sufficient strength to aid in fragmentation. These multiple explosions were observed in some Sandia experiments (10-12). The proposed model in a certain sense could include this possibility since no specific path is considered by which fuel and coolant liquids become initially intermixed. In the last few paragraphs we considered three different paths by which the mixture configuration could develop. In this situation the mixture may not significantly expand and disperse from an energetic FCI and yet the fuel fragments to smaller sizes. Now there are two possibilities. First, the expansion is appreciable enough that the mixture void rises significantly. In our analysis the average void is slightly greater than 50%, in the range of 55-65%. A factor of two expansion would increase the average void to about 65-75%. (A factor of ten expansion corresponds to a void of about 90%.) We would contend this mixture of fuel and coolant could not efficiently undergo a vapor explosion for the reasons we mentioned previously. Even if this mixture could undergo a vapor explosion in which all the fuel participates the additional fuel mass that could be mixed at a given D_{mix} is not much greater. Figure 11 shows a sensitivity study where we plot the mixed fuel mass for an arbitrary pool void fraction of 65% and 90% to show the effect of this void limit. As one can see this only increases the fuel mass mixed by 25%. A second possibility is that the FCI expansion is minimal. In this case the pool volume remains about constant. If the fuel mixing diameter decreases then the vapor velocity would increase and fluidization would drive away more coolant liquid and the mixed fuel mass decreases as the mixture expands. The model considers the amount of fuel that can mix only to the point where fluidization begins within the pool. One can investigate the effect of different mixing diameters or fuel masses by parametrically varying these values in the analysis, and determining the maximum amount of fuel that can enter before liquid is expelled by vapor flow.

Now one should note that the difference between these two scenarios is the time it takes for the fuel-coolant mixture to expand after the fuel has been pre-fragmented by a mild FCI. However, this expansion time would be short in all cases (<< 1 sec) because the heat flux between fuel and coolant is quite large for LWR conditions (3-4 MW/m^2 based on eqn. 6 and representative properties and temperatures). Under the conditions of approximately constant pressure in the mixture the expansion time, Δt, is given by

$$\Delta t = H_{mix} A_{ch} \alpha_v \left[\frac{6q''_{drop} m_f}{\rho_f i_{fg} D_{mix} \rho_v} \right]^{-1} \quad (20)$$

where $H_{mix}A_{ch}\alpha_v$ is the vapor volume and all other quantities are defined. For nominal conditions ($m_f \sim 1000$ kg, $D_{mix} \sim 10$ mm, P_∞ - 3 bar) and a volume expansion of a factor of two, the expansion time, Δt, is less than 100 msec. Thus, the transient time between these bounding cases of a multiple explosion is short and the simple model is applicable at the bounds.

One effect that has not been directly examined in this study is the effect of coolant liquid subcooling. In some accident sequences the coolant may be subcooled and this can affect the fuel mass mixed by reducing the amount of coolant vapor produced by the mixing process. Instead a portion of the fuel thermal energy may go into sensible heat, by raising the temperature of the liquid coolant in the mixing region. One can bound this effect by finding the amount of fuel energy required to raise the coolant liquid within the pool to saturation, thus reducing the vapor produced from fuel energy transfer. For the nominal case of D_{mix} = 10 mm as shown in Figure 7 one finds the fuel mass mixed increasing by about 40 % for a subcooling of 25K (Figure 14). Thus, this is a second order effect if the degree subcooling is small.

CONCLUSION

The vapor explosion process involves the mixing of fuel with coolant prior to the explosion. In this work we have reviewed past models to describe the evolution of the mixing concept and the various analytical approaches. We have also tried to compare the results of these models. With this as a basis a simplified approach is proposed here to estimate the limits to fuel-coolant mixing. The approach uses the concept first advanced by Fauske but directly derives the fluidization point. The results indicate that decreasing fuel temperature and increasing ambient pressure are important parameters in increasing the fuel mass mixed with the coolant; the water coolant depth has a minor effect above some minimum depth (~0.5 m). As expected fuel mixing diameter is a key factor in determining this limit to mixing and our analysis here considers a range of fuel diameters up to 30 mm. Based on the nominal values for these initial conditions we find that the range of fuel masses mixed can range from 1-12 metric tons with increasing pressure and subcooling causing this to increase by about a factor of two. It is important to realize that experiments and supporting analyses are eventually needed to verify this integral approach, and until that time one must consider a range of initial conditions to get a proper assessment of mixing limits.

ACKNOWLEDGEMENTS

We would like to thank Dr. H. Jacobs and Dr. G. Kessler for their suggestions and detailed discussions about this work. Also the criticisms and suggestions of Dr. M. Berman are gratefully acknowledged. This work was supported by a research grant from KfK Nuclear Research Center.

NOMENCLATURE

a - drift flux parameter
A - area
c_p - specific heat
C_o - drift flux void distribution parameter
C_D - drag coefficient
D - diameter
D_H - hydraulic diameter
g - gravitational acceleration constant
h - heat transfer coefficient
H - depth
i_{fg} - latent heat of vaporization
j - volumetric flux
k - thermal conductivity
m - mass
P_w - wall perimeter
P_∞ - ambient pressure
q" - heat flux
t - time
t_B - breakup time
T - temperature
T^+ - dimensionless time
U_t - terminal velocity
V - velocity
V_{gj} - average drift velocity
Vol - volume
We - Weber number
α - volume fraction
μ - viscosity
ρ - density
σ - surface tension
σ_r - Stefan-Boltzman constant

SUBSCRIPTS

c - coolant
mix - mixture
ch - chamber
o - initial value
crit - critical
sat - saturated
d - droplet
s - single sphere
drop - droplet
v - vapor
f - fuel
film - film value

SUPERSCRIPTS

fl - fluidization
j - jet
*dimensionless quantity

REFERENCES

1. Fauske, H.K., "Some Aspects of Liquid-Liquid Heat Transfer and Explosive Boiling," Proc Conf Fast Reactor Safety, Beverly Hills, CA, CONF-740401 (1974).
2. Henry, R.E. and H.K. Fauske, "Nucleation Characteristics in Physical Explosions," Proc 3rd Specialists' Mtg Sodium-Fuel

Interactions in Fast Reactors, Tokyo, Japan, CONF-760328 (March 1976).
3. Cho, D.H., H.K. Fauske, and M.A. Grolmes, "Some Aspects of Mixing in Large-Mass, Energetic Fuel-Coolant Interactions," Proc Int Mtg Fast Reactor Safety and Related Physics, Chicago, IL, Vol 4, p 1852 (October 1976).
4. Henry, R.E. and H.K. Fauske, "Core Melt Progression and the Attainment of a Permanently Coolable State," Proc Reactor Safety Aspects of Fuel Behavior, Sun Valley, Idaho, Vol 2, p 481, American Nuclear Society (August 1981).
5. Henry, R.E. and H.K. Fauske, "Required Initial Conditions for Energetic Steam Explosions," Fuel-Coolant Interactions, HTD-V19, American Society of Mechanical Engineers (1981).
6. Meyer, J.F., "Preliminary Assessment of Core Melt Accidents at the Zion and Indian Point Nuclear Power Plants and Strategies for Mitigating Their Effects, Appendix B," NUREG-0850), Vol 1, U.S. Nuclear Regulatory Commission (1981).
7. Theofanous,T.G. and M. Saito, "An Assessment of Class-9 (Core-Melt) Accidents for PWR Dry Containment Systems," Nuclear Engr Des, V66, pp 301-332 (1981).
8. Corradini, M.L., "Proposed Model for Fuel-Coolant Mixing During a Core Melt Accident," Proc Int Mtg Thermal Nuclear Reactor Safety, Chicago, IL, NUREG/CP-0027, U.S. Nuclear Regulatory Commission (September 1982).
9. Corradini, M.L. and G.A. Moses, "A Dynamic Model for Fuel-Coolant Mixing," Proc Int Mtg Light Water Reactor Severe Accident Evaluation, Cambridge, MA, Vol 1 p 6.3, American Nuclear Society (August 1983).
10. Berman, M., LWR Safety Quarterly and Semi Annual Progress Reports," Sandia National Laboratories, SAND-80-1304, (Jan-Mar, Apr-June, July-Sept, Oct-Dec, 1980); SAND81-1216 (Jan-Mar, 1981); SAND82-006 (Apr-Sept, 1981); SAND82-1572 (Oct, 1981-Mar, 1982); SAND 83-1576 (Apr-Sept, 1982)' SAND85-2500 (Oct 1983-Mar 1984); SAND84-1072(2) (Apr-June 1984); SAND84-1072(3) (July-Sept 1984); SAND84-1072(4) (Oct-Dec 1984); SAND85-1606(1) (Jan-June 1985); SAND85-1606(2) (July-Dec 1985); SAND 86-2752(1) (Jan-June 1986); SAND86-2752(2); (July-Dec 1986); SAND87-2411(1) (Jan-June 1987); SAND87-2411(2) (July-Dec 1987).
11. Mitchell, D.E., N.A. Evans and M.L. Corradini, Intermediate Scale Steam Explosion Phenomena: Experiments and Analysis, SAND81-0124, NUREG/CR-2145, Sandia National Laboratories (Sept 1981).
12. Mitchell, D.E. and N.A. Evans, "The Effect of Water to Fuel Mass Ratio and Geometry on the Behavior of Molten-Core-Coolant Interaction at Intermediate Scale," Proc Int Mtg Thermal Nuclear Reactor Safety, Chicago, IL, NUREG/CP-0027, U.S. Nuclear Regulatory Commission (August 1982).
13. Bankoff, S.G. and A. Habid, "Application of a User-Friendly Code to Nuclear Thermal-Hydraulic Reactor Safety Problems," Proc of Int'l Mtg NUPTHO-1 Taipei, Taiwan (Oct 1984).
14. Fletcher,D.F. and A. Thyagaraja, "A Mathematical Model of Premixing," ANS Proc of 1988 Nat'l Heat Transf Conf, Houston, TX p 184 (July 1988).
15. Abolfadl, M.A. and T.G. Theofanous, ":An Assessment of Steam-Explosion Induced Containment Failure: Part II Premixing Limits," Nuclear Sci Engr, V 97, pp 282-295 (Dec 1987).
16. Chu, C.C. and M.L. Corradini, "One-Dimensional Transient Model for Fuel-Coolant Fragmentation and Mixing Analysis," ANS/ENS Int'l Topical Mtg on Thermal Reac Safety, San Diego, CA (Feb 1986).
17. Amarasooriya, W.H. and T.G. Theofanous, "Premixing of Steam Explosions: A Three Fluid Model," ANS Proc of 1988 Nat'l Heat Tran Conf, Houston, TX p 191 (July 1988).
18. Chu, C.C. and M.L. Corradini, "One-Dimensional Transient Fluid Model for Fuel-Coolant Interaction Analysis," Nucl Sci Engr, V101, pp 48-71 (Jan 1989).
19. Young, M.F., "The TEXAS Code for Fuel-Coolant Interaction Analysis," Proc ANS/ENS Fast Reactor Safety Conf, Lyon, FR (July 1982).
20. Epstein, M. and H.K. Fauske, "Steam Film Instability and Mixing of Core Melt Jets and Water," ANS Proc of 1985 Nat'l Heat Trans Conf, Denver, CO (August 1985).
21. Young, M.F., IFCI: An Integrated Computer Model for All Phases of Fuel Coolant Interactions, Sandia National Laboratories, NUREG/CR-5084, SAND87-1048 (1987).
22. Mitchell, D.E. and N.A. Evans, Steam Explosion Experiments at Intermediate Scale: FITSB Series, Sandia National Laboratories, NUREG/CR-3983, SAND83-1057, (February 1986).
23. Marshall, B.W. Jr., "Recent Fuel-Coolant Interaction Experiments Conducted in FITS Vessel," 1988 ASME-AIChE National Heat Transfer Conference, Houston, TX, HTC-Vol 3, pp 265-278, (July 1988).
24. Marshall, B.W. and M. Berman, "Mixing of Isothermal and Boiling Molten-Core Jets with Water: The Initial Conditions for Energetic FCIs," Proceedings of the International ENS/ANS Conference on Thermal Reactor Safety, Vol 1, pp 117-127, Avignon, FR, (October 1988).
25. Bower, G., S.K. Chang, M. El-Beshbeeshy, J. Koo, J. Martin, and M.L. Corradini, "Physical Mechanisms for Atomization of Jet Spray: A Comparison of Models and Experiments," SAE Paper No. 881318 (Sept 1988).
26. Fletcher, D.F. and A. Thyagaraja, "A Mathematical Model of Melt/Water Detonations," Appl Math Modelling, Vol 13, pp 339-347 (June 1989).
27. Corradini, M.L., "Phenomenological Modelling of the Triggering Phase of Small-Scale Steam Explosion Experiments," Nucl Sci Engr, V78, pp 154-170 (1981).
28. Kim, B.J., Heat Transfer and Fluid Flow Aspects of Small-Scale Single Droplet FCIs, PhD Thesis, University of Wisconsin (August 1985).
29. Kim, B.J. and M.L. Corradini, "Recent Film Boiling Calculations: Implication on Fuel-Coolant Interactions," Fifth Int'l Mtg on Thermal Nuclear Reactor Safety, V2, Karlsruhe, FRG (October 1985).
30. Corradini, M.L. and D.V. Swenson, Probability of Containment Failure Due to Steam Explosions Following a Postulated Core Meltdown in an LWR, Sandia National Laboratories, NUREG/CR-2214, SAND80-2132 (June 1981).
31. Corradini, M.L., "Analysis and Modelling of Large-Scale Steam Explosion Experiments," Nucl Sci Engr, V 82, pp 429-447 (1982).
32. Corradini, M.L. and G.A. Moses, "Limits to Fuel/ Coolant Mixing," Nucl Sci Engr, V90, pp 19-27 (1985).
33. Zuber, N., "Bubble Growth in Variable Pressure Fields," Journal of Heat Transfer, V 453, ASME Transactions (April 1978).
34. Borishanskii, Z.M. and Shurn Tekh Fiz, V26, 452 (1956); see also Sov Phys-Technical Phys V1, 438 (1957) (English Translation).
35. Wallis, G.B., One-Dimensional Two-Phase Flow, McGraw-Hill Book Company, Inc., New York (1981).
36. Bang, K.H. and M.L. Corradini, "Stratified Vapor Explosion Experiments," ANS Proc of Nat'l Heat Transfer Conf, Houston, TX (July 1988).
37. Bang, K.H. and M.L. Corradini, "An Experimental Study of Vapor Explosions in a Stratified Contact Mode," Proc ANS/ENS Thermal Reactor Safety Mtg, Avignon, FR (Oct 1988).
38. Bang, K.H., Stratified Vapor Explosions, PhD Thesis, University of Wisconsin (Dec 1988).

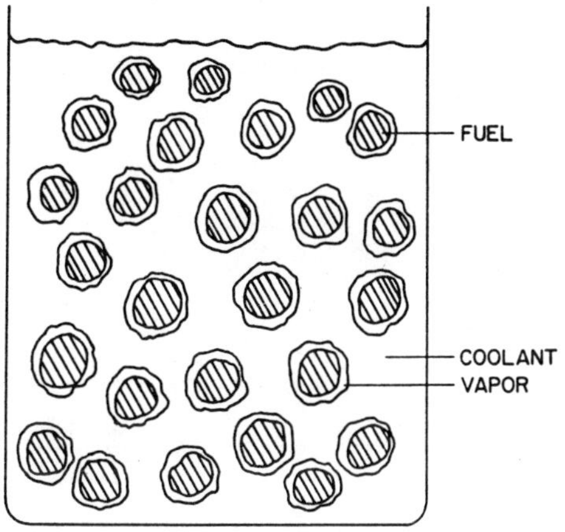

Figure 1. Conceptual picture of fuel/coolant mixing.

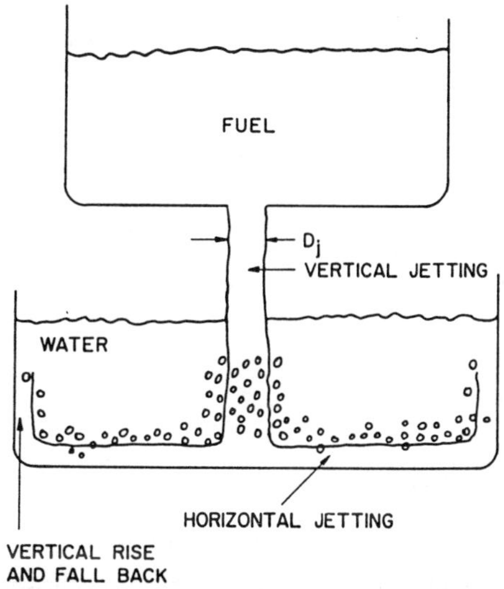

Figure 2. Geometry of fuel pouring into the lower plenum of the vessel.

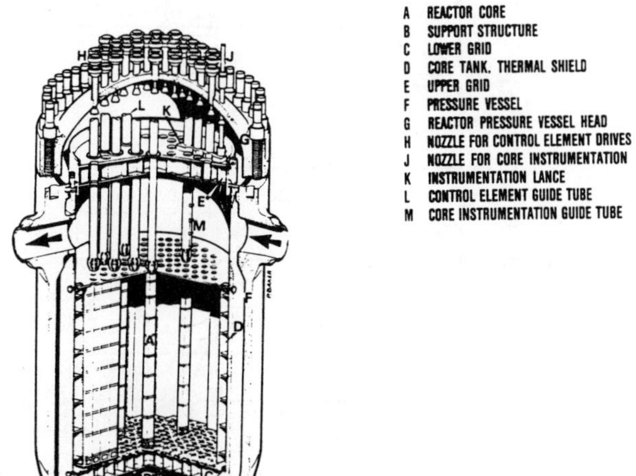

Figure 3. Geometric details of BIBLIS-B reactor.

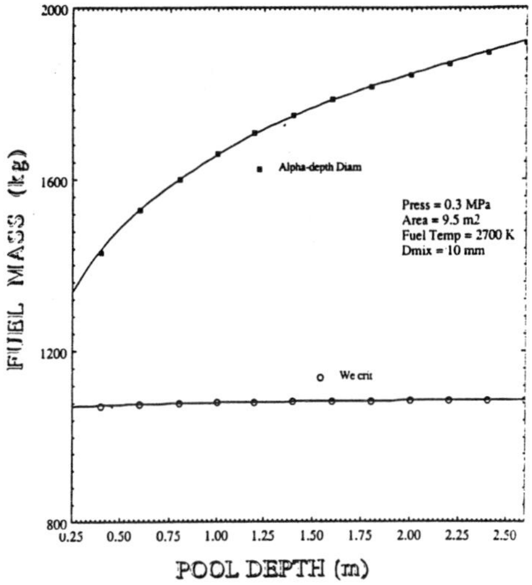

Figure 4. Amount of fuel mixed vs. pool depth for d_{mix} = 10mm.

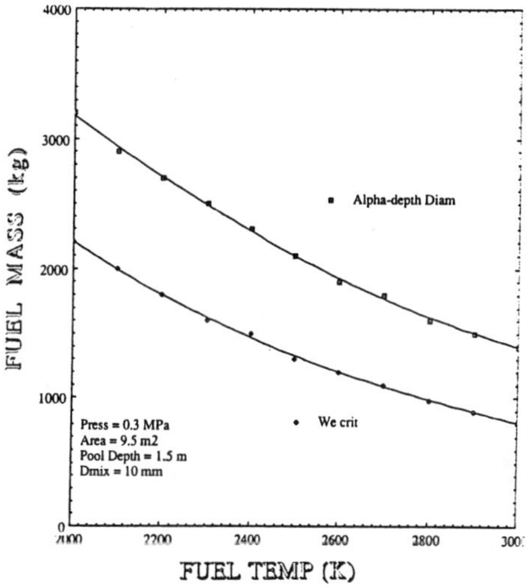

Figure 5. Amount of fuel mixed vs. fuel temperature for d_{mix} = 10mm.

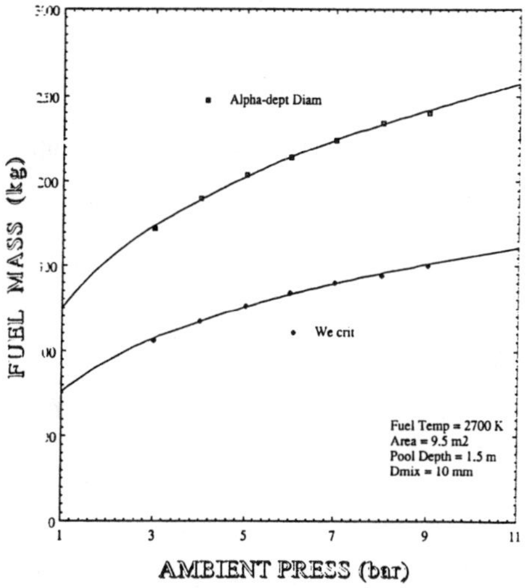

Figure 6. Amount of fuel mixed vs. coolant pressure for d_{mix} = 10mm.

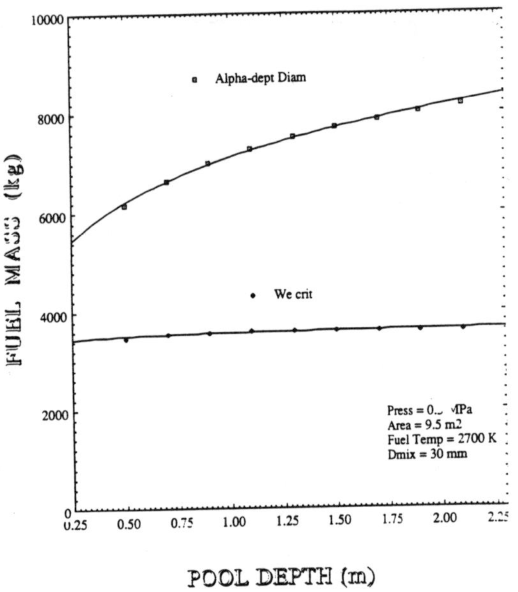

Figure 7. Amount of mixed fuel vs. pool depth for d_{mix} = 30mm.

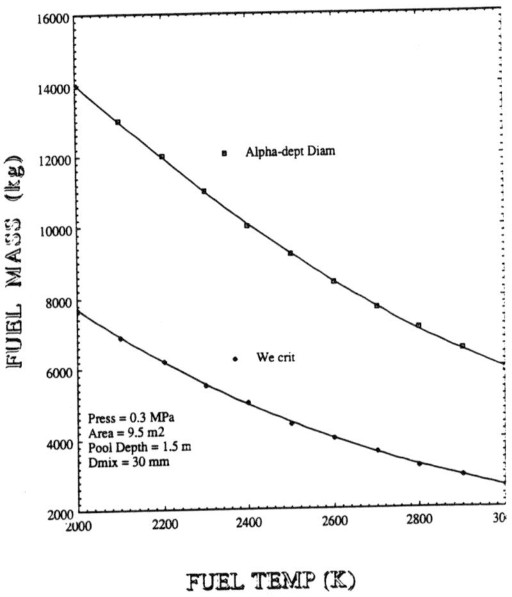

Figure 8. Amount of mixed fuel vs. fuel temperature for d_{mix} = 30mm.

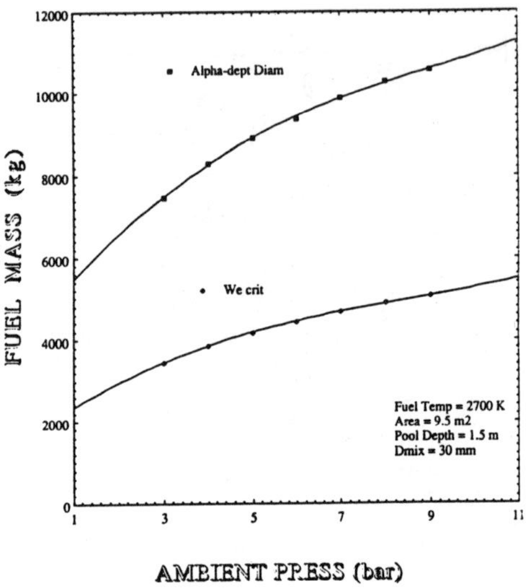

Figure 9. Amount of mixed fuel vs. coolant pressure for d_{mix} = 30mm.

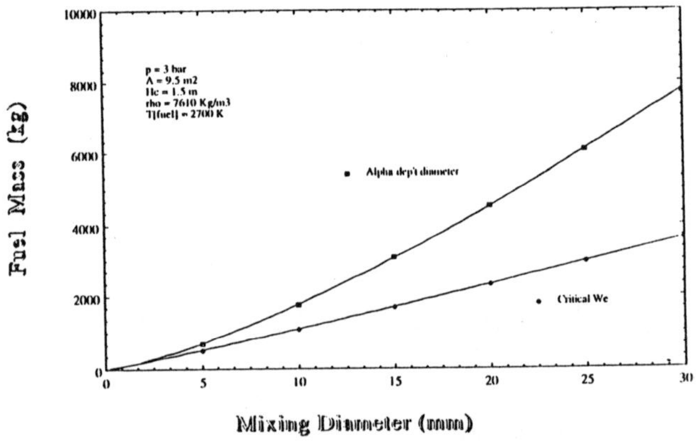

Figure 10. Effect of varying the fuel mixing diameter.

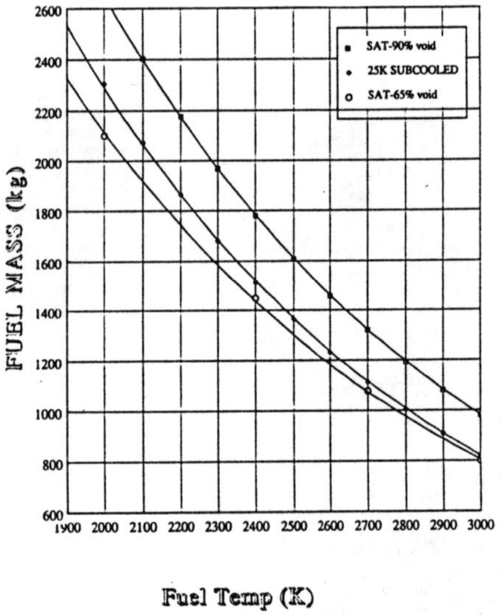

Figure 11. Sensitivity study of void and subcooling effect for d_{mix} = 10mm and D_{cII}.

CONDENSATION EXPERIMENTS IN STEAM-AIR AND STEAM-AIR-HELIUM MIXTURES UNDER TURBULENT NATURAL CONVECTION

A.A. Dehbi, M.W. Golay and M.S. Kazimi ■ Massachusetts Institute of Technology, Cambridge, MA 02139

An experimental investigation has been conducted to determine the condensation heat transfer coefficient of steam in the presence of air or air and helium. The experimental test section consisted of a 3.5 m long water cooled copper tube enclosed in a large pressure vessel. Known amounts of noncondensable gases were admitted inside the vessel while steam was produced at the bottom of the vessel by a set of heaters with a total capacity of 36 kW. The pressure vessel was fully insulated so that the heat produced by steam condensation was entirely removed by the coolant. The gas-steam mixture was in a steady state in which natural circulation currents were the only driving mechanism. In the steam-air investigation, three sets of isobaric experiments were performed at pressures of 1.5, 3.0, and 4.5 atmospheres, respectively. The air mass fraction varied between 25% to 90% while the wall temperature subcooling varied between 15 to 50°C. The heat transfer coefficient decreased slowly with subcooling, increased mildly with height, and significantly with pressure. At low pressures, the developed correlation predicts heat transfer rates similar to those reported by Tagami and Uchida. In the steam-air-helium investigation, three sets of experiments were conducted at pressures around 3.0 atmospheres. In each set, the helium mass fraction was held constant at 1.7%, 4.7% and 8.3%, respectively. The heat transfer coefficient (normalized to the steam-air experimental value) deteriorated with increasing helium fraction, especially at low total noncondensable mass fractions. Data was taken only when all the gases were relatively well mixed. At high enough helium-to-air mass ratios, sharp stratification patterns could be observed, with the helium migrating to the top of an air-steam mixture settling in the lower region of the vessel.

The inhibitive effect of noncondensable gases on steam condensation is a well known phenomenon. The quantification of such an effect is of prime interest for the prediction of the response of a nuclear reactor containment in accidental conditions. A typical loss of coolant accident (LOCA) would produce a rapid injection of hot steam in a quiescent containment air atmosphere. Following the initial blowdown period, a natural circulation pattern of the steam-air mixture will prevail, and steam would condense on cold containment walls. The extent of this condensation determines the pressure rise in the containment, and therefore has a direct impact on containment integrity. Many experimental investigations have been performed to correlate the overall condensation heat transfer coefficient of steam in the presence of noncondensable gases. Most of these experiments [1,2] dealt with low noncondensable mass fractions (<25%), short condensing lengths (<20 cm), and near atmospheric pressures. Although useful from a qualitative point of view, the latter experiments do not directly represent typical containments conditions where lengths of several meters and air fractions greater then 50% are usual. For long term post-LOCA analysis of containments, the most widely used correlations are due to Uchida [3] and Tagami [4]. These correlations relate the condensation heat transfer coefficient simply to the noncondensable fraction in the mixture. Although attractive for their simplicity, these correlations do not include important physical parameters. The objective of the present experimental investigation is to determine the dependence of the condensation heat transfer on pressure, wall temperature subcooling, length, and the noncondensable content of the mixture. The first part of this study deals with steam-air mixtures while the second part investigate the behavior of steam-air-helium mixtures.

STEAM-AIR EXPERIMENTS

Experimental Apparatus

Referring to Figure 1, the condensing section consists of a 3.5 m long, 3.8 cm in diameter, copper cylinder immersed in a 5 m long, 0.45 m in diameter stainless steel vessel able to withstand pressures of up to 10 atmospheres. Steam was supplied at the bottom of the vessel by a set of immersion heaters with a total capacity of 36 kW. Air was injected in the vessel from an air supply source. The vessel was fully insulated so that condensation takes place only on the cold copper wall. In steady state, the condensation latent heat is absorbed by the water coolant flowing inside the tube. The horizontal pipe sections supporting the copper cylinder were also insulated so that

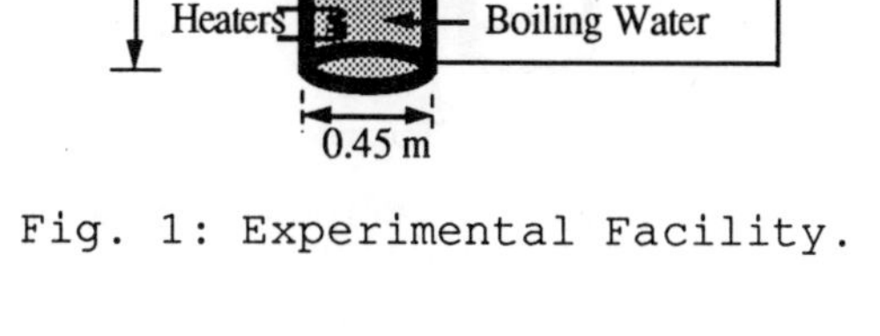

Fig. 1: Experimental Facility.

the condensation occurred exclusively along the vertical direction.

The 3.5 m copper section was subdivided into six equal 0.58 m sections so that axial variation of the heat flux could be determined. Wall, coolant, and gas space thermocouples were placed as shown in Figure 2. Coolant thermocouples were horizontally inserted through the copper wall via compression fittings. The air space thermocouples were similarly placed through the stainless steel vessel wall. The thermocouple tips were located far enough from the boundary layer in order for the readings to indicate the mixture bulk temperature at each station. The copper wall thermocouples tips were sandwiched between the copper wall and small square pieces of 0.15 mm thick copper, as shown in Figure 3. The copper wall was so thin that the metal temperatures on the cold and hot side were essentially equal for the heat fluxes encountered in this experiment. Two high precision pressure gauges were installed on the pressure vessel in order to measure the total system pressure.

Cooling water was allowed to enter from the bottom of the copper section in order to prevent flow instabilities that are likely to occur with a downward flowing coolant. The coolant flowrate and pressure were controlled by valves located at the inlet and the outlet of the condensing section. The coolant flowrate was readily measured by two carefully calibrated flowmeters with different ranges.

Water used for boiling was introduced into the bottom of the vessel. The water level was displayed by a U-tube indicator. The chosen level of 0.4 m was high enough to safely cover the heaters and shallow enough to allow a 0.3 m space between the boiling water surface and the start of the condensing section.

The pressure vessel had three view ports in order to see the nature of the condensation taking place at the copper wall (Figure 4). In all experimental runs, a downward flowing thin film was observed on the condensing wall. No evidence of film turbulence could be detected. The mist in the boundary layer could clearly be seen drifting towards the copper surface.

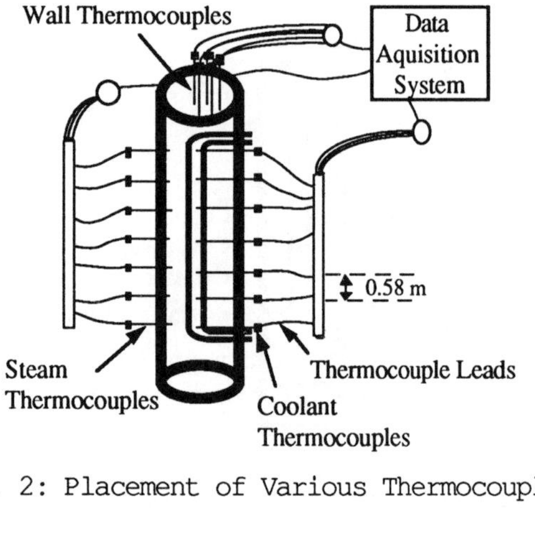

Fig. 2: Placement of Various Thermocouples.

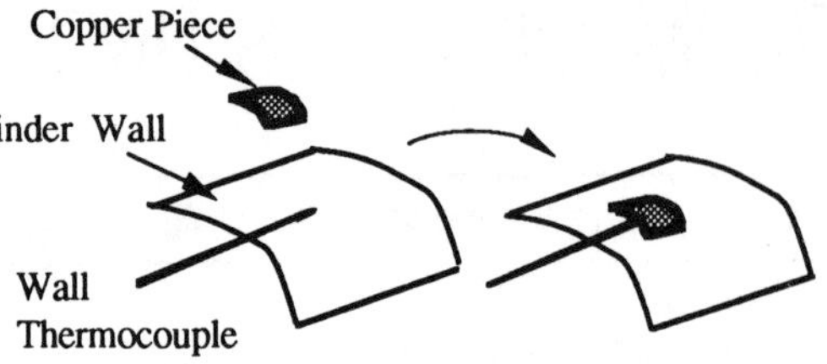

Fig. 3: Placement of the Wall Thermocouples.

A PID controller monitored the temperature of the boiling water. Controlling the boiling water temperature indirectly sets the total pressure of the system. Indeed, in steady state, the boiling water is saturated at a preset temperature corresponding to the desired total pressure. This leads to a steam-air temperature that is less than that of the boiling water, because the steam is saturated at its partial pressure.

At any vertical position, the air partial pressure was found using the total pressure readings from the gauges, then subtracting the partial pressure due to steam. The latter was obtained using the local steam temperature and steam tables, assuming saturation conditions. The ideal gas law was then applied to calculate the mean air mass fraction.

Temperature measurements were made with J-type thermocouples. The data acquisition system consisted of a scanner with a 40 channel per second sweeping time, and a processor linked to a personal computer. After steady state was achieved, the processor recorded data for a two minute period, then performed averages and standard deviation calculations on the raw experimental data.

Proedure

The steam-air experiments involved three isobaric sets of runs, at pressures of 1.5, 3.0, and 4.5 atmospheres, respectively. In each of these sets, about 40 runs were performed. The air mass fraction varied from 25% to 90%.

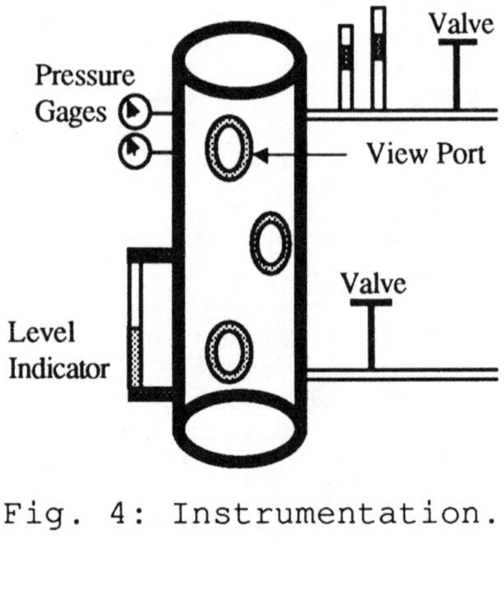

Fig. 4: Instrumentation.

At the start of each run, the pressure vessel was filled with room temperature air until the pressure reading indicated about 90% of the desired equilibrium pressure. This step insured that the mixture would be in the air rich region at the start of the run.

After selecting the desired boiling water temperature (and indirectly the total pressure), the heaters were turned on at full power. The water heated up while steam gradually mixed with the air atmosphere. After about 15 minutes, the water reached its preset temperature. At that point, the coolant valve was opened enough to allow the desired flow rate. After a 30-minute period, the system reached steady state as judged by stable temperature and pressure readings. Temperature data was then collected for a 2-minute period. The flowrate and pressure values were visually recorded.

The selection of the coolant flow rate was critical for the achievement of low experimental error results. Increasing the flow rate produced a high wall subcooling, which is warranted for an accurate determination of the heat transfer coefficient. On the other hand, the flow rate had to remain low enough if reasonable coolant temperature rises were to be obtained.

A significant limitation in the selection of a suitable water circulation rate was related to the flow regime inside the copper section. Early on, it was thought that a coolant flow rate leading to a 5°C coolant temperature rise between adjacent stations would be adequate, and would yield reasonably constant wall temperatures. However, the early shakedown runs indicated that a constant wall temperature could be achieved only at the expense of large fluctuations in the coolant temperature. These fluctuations were caused by the fact that the heated coolant was in the transition mixed convection flow regime in which the temperature profile might have a dip near the centreline of the tube. It was decided, therefore, to use a very small flowrate (Re < 1500) corresponding to a turbulent natural convection flow regime [5]. In these conditions, it was possible to obtain very stable temperature readings at the expense of a non-uniform wall temperature profile. The coolant temperature rise along the copper section was typically between 40 and 90°C, and correspondingly, the wall temperature rise was 10 to 30°C. The wall subcooling will, therefore, be defined on the section averaged basis.

After the run was completed, the next air fraction setting was obtained by opening up a valve and evacuating some of the steam-air mixture. This relieved mass was directed toward a cold tank where the steam condensed. In the process, the air-steam temperature increased as less air was present. This procedure did not permit precise control over the air content, but when the evacuation time was made short enough, target air to steam ratios were reasonably achieved.

Results and Discussion

Varying mixture temperature in the vessel. After steady state was reached, it was observed that the axial mixture temperature decreased systematically as one moved away from the steam supply point. The temperature profile was nearly linear as shown in Figure 5. The bulk temperature difference across the entire section varied from 2 to 10°C, with the smallest value corresponding to the lowest air fractions in this experiment. This temperature variation, along with isobaric conditions in the vessel, indicated that the steam concentration decreased with height while the air concentration increased as one moved away from the steam source.

If no heat removal were present, and the air-steam mixture allowed to come to equilibrium, two plausible outcomes can be expected. In one situation, there would be no concentration gradient in the gas and temperature would be uniform across the vessel. Another alternative would be a stratified configuration in which the lighter gas (steam) concentration would increase with height due to buoyancy effects, and conversely, the heavier gas (air) would be more concentrated as the bottom of the vessel is approached. In this situation the mixture temperature would increase with height. These two alternatives pertain to a static equilibrium conditions. In the present experiment, however, the equilibrium is dynamic rather than static. Steam enters the gas space from bottom of the vessel with a velocity less than 0.05 m/sec. As it moves upward, some of it condenses on the copper wall. The net effect is a gradual decrease in steam concentration with height. The concentration gradient driving mechanism, which tends to produce a flat concentration profile, never gets a chance to enter into

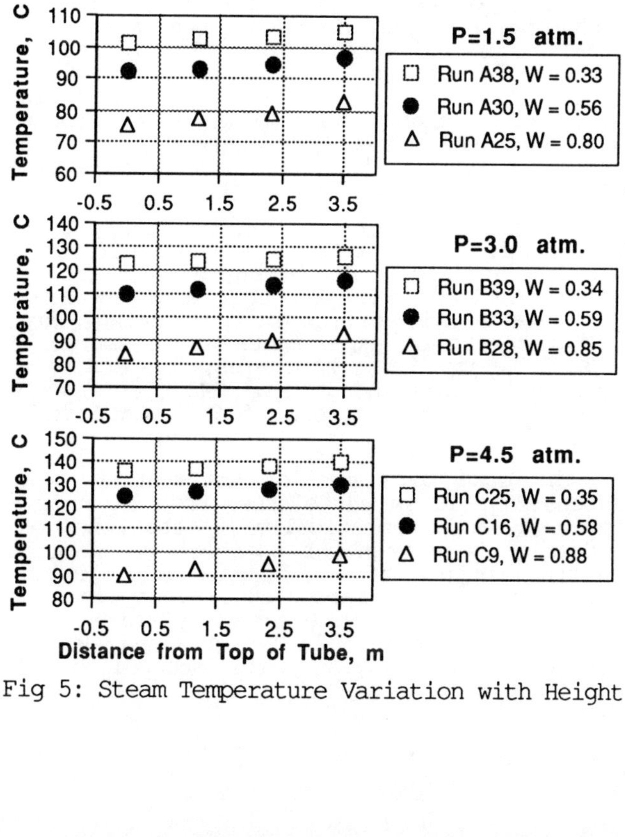

Fig 5: Steam Temperature Variation with Height

play because steam is continuously being supplied from the bottom and condensed along the cold walls.

The axial mixture temperature profile can be used to figure out the concentration variation of air. Assuming steam was at or near saturation at any given temperature, the air partial pressure at that location is given by:

$$P_A = P_T - P_S(T_z)\,. \tag{1}$$

Using the ideal gas laws for steam and air, the ratio of the mass of air to that of steam is:

$$\frac{M_A}{M_S} = \left(\frac{P_A}{P_S\,(T_z)}\right)\left(\frac{MW_A}{MW_S}\right) = \left(\frac{P_A}{P_S\,(T_z)}\right)\left(\frac{29}{18}\right)$$
$$= 1.61\left(\frac{P_T - P_S(T_z)}{P_S\,(T_z)}\right)\,. \tag{2}$$

The mass fraction of air is therefore:

$$W(z) \equiv \frac{M_A}{M_A + M_S} = \frac{\left(\frac{M_A}{M_S}\right)}{1 + \left(\frac{M_A}{M_S}\right)}\,.$$

Therefore:

$$W(z) = \frac{1.61\left(\frac{P_T - P_S(T_z)}{P_S\,(T_z)}\right)}{1 + 1.61\left(\frac{P_T - P_S(T_z)}{P_S\,(T_z)}\right)}\,. \tag{3}$$

Given the pressure and steam axial temperatures, the latter formula was used to compute the air fraction profile for typical runs. For each pressure, three runs were selected, each one representative of the air rich, steam rich, and intermediate region, respectively. The results are shown in Figure 6. It is seen that the air fraction increases linearly with height. This is a direct consequence of the linear temperature profile of the mixture. The stratification is more pronounced in the steam rich region.

The experimental correlation. Early trial runs indicated that the pressure effect on the condensation heat transfer coefficient was significant. It was therefore necessary to isolate this effect by performing three isobaric sets of experiments at pressures of 1.5, 3.0, and 4.5 atmospheres, respectively. These values are thought to span the range of interest to post accident containment conditions.

The next parameter of interest is wall subcooling. It is established both experimentally [1] and theoretically [6] that the condensation heat transfer coefficient of steam in the presence of a noncondensable gas decreases with wall subcooling. Given this, it was necessary to devise a way by which local information could still be obtained despite longitudinally varying wall subcooling and air fraction.

For a length L of the condensing cylinder, the average condensation heat transfer coefficient of steam is defined as follows:

$$(\bar{h}_L)_{cyl} \equiv \frac{q''}{(\bar{T}_\infty - \bar{T}_w)}$$

Because the radius of curvature of the condensing section is small, the heat transfer is enhanced compared to the flat plat geometry case. To make the correlation valid for a flat surface, a correction must be applied to the experimental heat transfer in order to remove the enhancement due to curvature. Let β be defined as:

$$\beta \equiv \frac{\bar{h}_L}{(\bar{h}_L)_{cyl}}$$

where h_L is the flat plate heat transfer coefficient. Therefore,

$$\bar{h}_L \equiv \beta\,\frac{q''}{(\bar{T}_\infty - \bar{T}_w)}$$

Both Sparrow [7] and Cebeci [8] have numerically calculated this factor for a variety of conditions in natural convection. They showed that it is an increasing function of the Prandlt number and a decreasing function of ζ defined as:

$$\zeta = \frac{2\sqrt{2}}{(Gr_z)^{0.25}}\left(\frac{z}{r_o}\right)$$

For the conditions of this experiment, ζ was in the range 0.82 to 0.86. The selected value was 0.8 for all runs, so that systematically

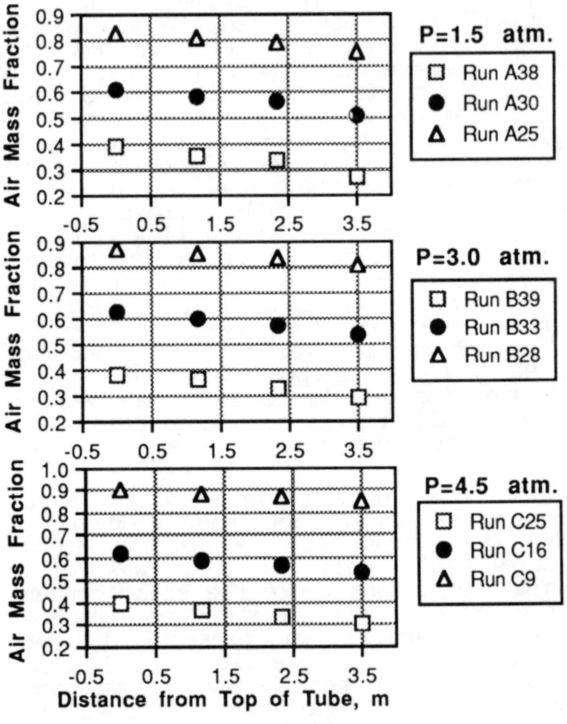

Fig. 6: Air Fraction Variation with Height.

conservative estimates of the heat transfer coefficient are reported. The validity of this correction factor for film condensation in the presence of an inert gas was theoretically confirmed by Dehbi [9]. In this study, the gas boundary layer equations in cylindrical form were solved using the turbulent kinetic energy (TKE) model for turbulence, while the condensate film was treated according to the simple Nusselt analysis. The model, which agrees very well with the data reported here, confirms that 0.8 is good correction factor for the conditions of these experiments.

At steady state, the condensation heat is absorbed by the water coolant. Therefore:

$$\overline{h}_L = \beta \frac{\dot{m}c_p(\Delta T)_{coolant}}{A(\overline{T}_\infty - \overline{T}_w)} \tag{4}$$

For any length of the section L, the average wall and mixture temperatures is obtained by simply averaging local experimental values over the length of interest. The averaged air fraction is calculated according to the following expression:

$$\overline{W(L)} = \frac{1.61\left(\dfrac{P_T - P_S(\overline{T}_\infty)}{P_S(\overline{T}_\infty)}\right)}{1 + 1.61\left(\dfrac{P_T - P_S(\overline{T}_\infty)}{P_S(\overline{T}_\infty)}\right)}$$

For the purpose of obtaining average heat transfer coefficients, the condensing copper tube can be regarded as six separate tubes with lengths varying from 0.58 m to 3.5 m. For each run, therefore, it was possible to determine six average heat transfer coefficients. To each of these coefficients corresponds a length, an average wall subcooling, and an average air fraction.

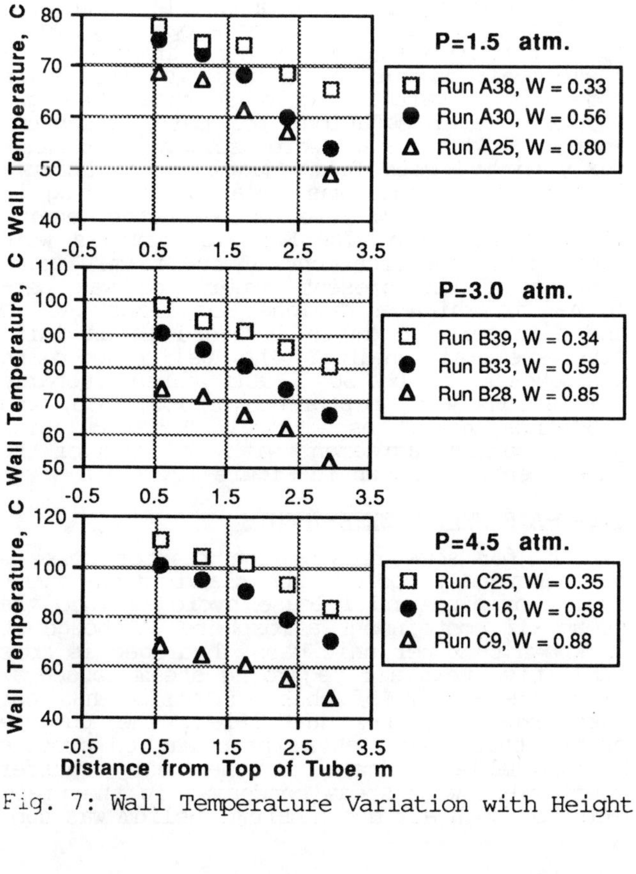

Fig. 7: Wall Temperature Variation with Height.

Range of conditions in these experiments. Each experimental run is associated with a specific average air fraction. For a pressure of 1.5 atmosphere, 41 runs were performed. For 3.0 and 4.5 atmospheres, 40 and 27 runs were achieved, respectively. The section averaged wall subcooling ranged from 10 to 50°C, while the local subcooling varied between 10 and 65°C. The wall temperature increased almost linearly with height as shown in Figure 7.

Data reduction. In analogy with the Nusselt formula for pure condensation, the heat transfer coefficient is assumed to depend on length and subcooling. Moreover, initial tests revealed that the system pressure and air content have strong impact on the heat transfer coefficient. So:

$$\overline{h}_L = L^\alpha(\overline{T}_\infty - \overline{T}_w)^\delta F(W,P) \tag{5}$$

where α and δ and F are to be determined experimentally.

To calculate the average heat transfer coefficient, two lengths were selected, corresponding to L = 1.16 m, and L = 3.5 m. These were thought to be sufficient to obtain the length effect. The steps for obtaining each isobaric experimental correlation were as follows:

- For each run, calculate the two average heat transfer coefficients with their corresponding average subcooling and air mass fraction.
- For each length, plot the grouping:

$$F(W,P) = \frac{\overline{h}_L}{L^\alpha(\overline{T}_\infty - \overline{T}_w)^\delta}$$

as a function of air fraction.

The experimental data points collapsed around a smooth curve with $\alpha = 0.05$ and $\delta = -0.25$. Therefore,

$$F(W,P) = \frac{\overline{h}_L}{L^{0.05}(\overline{T}_\infty - \overline{T}_w)^{-0.25}} \tag{6}$$

The reduced experimental data is shown in Figure 8. As expected, the shape of the curve is logarithmic, with a higher slope as the steam rich region is approached. As noted from the graph, the dependence of the heat transfer on pressure is significant. Next, F(W,P) was least square fitted to a logarithmic function for each of the three pressures. The correlations obtained are:

$F(W,P) = 48.0 - 3081.6\ \text{Log}(W)$ for P = 1.5 atm.

$F(W,P) = 89.1 - 3898.8\ \text{Log}(W)$ for P = 3.0 atm.

$F(W,P) = 133.4 - 4456.8\ \text{Log}(W)$ for P = 4.5 atm.

where Log is the logarithm in base 10.

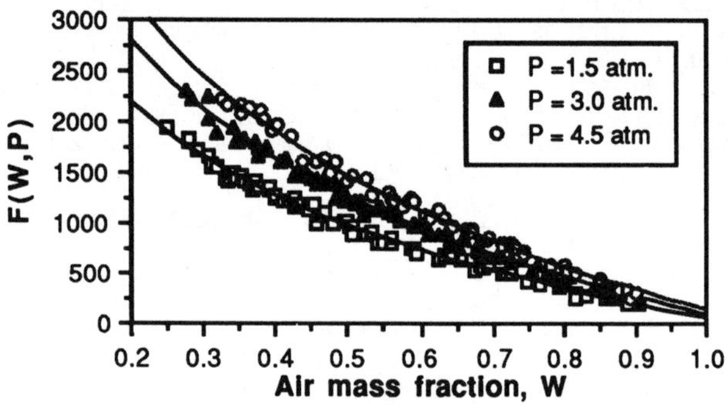

Fig. 8: Plot of the Experimental Correlation Function, F.

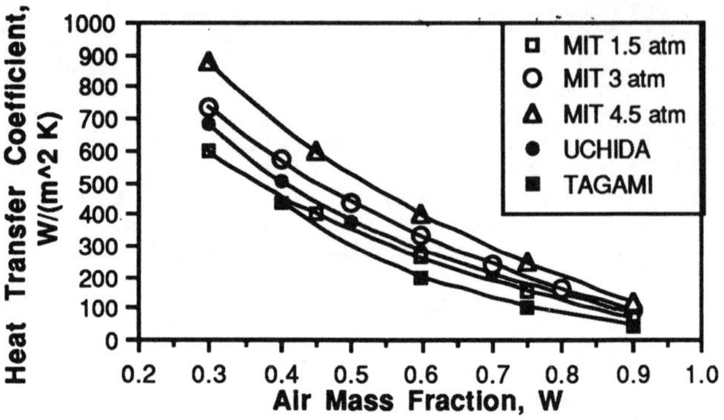

Fig.9: Comparison with Other Correlations.

The regression R^2 is greater than 0.99 for all three correlations. The latter equations can be written in the more compact form as:

$$F(W,P) = (3.7+28.7P) - (2438+458.3P)\mathrm{Log}(W)$$

Finally, the heat average heat transfer coefficient can be expressed as:

$$\bar{h}_L = \frac{L^{0.05}[(3.7+28.7P) - (2438+458.3P)\mathrm{Log}(W)]}{(\bar{T}_\infty - \bar{T}_w)^{0.25}} \quad (7)$$

The fact that the heat transfer coefficient decreases with wall subcooling confirms the experimental findings of Al-Diwani and Rose [1]. This effect has also been determined theoretically by Corradini [10] who based his model on the heat and mass transfer analogy. The small dependence on length is typical of natural convection heat transfer off vertical surfaces. The theoretical model by Dehbi [9] predicts boundary layer thicknesses on the order of 3 cms, which is much less than the radius of the vessel (22.5 cms), so that the applicability of Eq. 7 is not limited to the geometry of the present experiment. The heat transfer improves significantly with pressure as already predicted by Corradini [10].

Comparison with the Tagami and Uchida correlations. A comparison between the correlation developed in this work and those obtained by Tagami and Uchida is warranted, since the latter have been widely used in containment safety analysis. Both experiments involved condensation on the surface of a vertically positioned cylinder with a length 0.3 m and a diameter of 0.2 m. Tagami's data was obtained in natural convections conditions. The exact conditions of the Uchida experiment were not reported, but it is known that the air-steam mixture velocities were small, and therefore a comparison with natural convection data is meaningful. In metric units, the Uchida and Tagami results were fitted to the following curves:

$$\bar{h}_{Uchida} = 380\left(\frac{W}{1 - W}\right)^{-0.7} \quad (8)$$

$$\bar{h}_{Tagami} = 11.4 + 284\left(\frac{1 - W}{W}\right) \quad (9)$$

No effort was made in either experiments to retain isobaric conditions, and pressure was allowed to vary between 1 and 3 atmospheres. It was therefore decided to compare Eq. 8 and Eq. 9 to the developed correlation at pressures of 1.5 and 3.0 atmospheres, and with a typical subcooling of 50°C. As seen in Figure 9, the Uchida correlation lies between the 1.5 and 3.0 atmospheres curves for all the range of air fractions considered. This indicates that for typical post LOCA pressures (P > 2 atmospheres) the Uchida correlation is indeed conservative, especially in view of the fact that low temperature subcoolings are expected after the initial blowdown period. The Tagami correlation is below the present data for all above atmospheric pressures, which confirms the view that the Tagami correlation gives lower bound estimates of the heat transfer for all imaginable containment conditions.

Comparison with the boundary layer theory. A theoretical model was developed by Dehbi [9] for the prediction of heat transfer from a steam-air mixture in a natural circulation mode off a vertical cold surface. The gas boundary layer equations for mass, momentum, energy, and species were solved simultaneously with liquid film equations. Turbulence in the gas mixture was treated according to the one equation turbulent kinetic energy (TKE) formulation. The Nusselt formulas were adopted for the treatment of the film. A simulation of the present experiment was performed in which a cylinder of length 1.0 m and diameter 3.8 cm is immersed in an infinite steam-air medium. The wall subcooling was chosen to be 30°C, and three isobaric simulations were performed at 1.5, 3.0, and 4.5 atmospheres, respectively. The simulation gave excellent agreement with the data of the experiment, as shown in Figure 10.

STEAM-AIR-HELIUM EXPERIMENTS

If, for some reason, the reactor core is not adequately cooled, the cladding may eventually oxidize and release hydrogen into the steam-air containment atmosphere. Although it is known that per unit mass, hydrogen is more inhibitive than air regarding steam condensation, the extent of this inhibition has not been experimentally quantified. The purpose of the this experimental part was to provide a reasonable estimate of the heat transfer coefficient when steam condenses in the presence of both air and helium. Helium was sub-

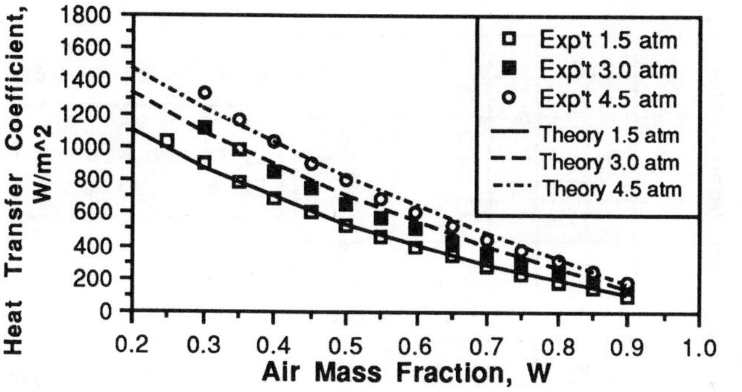

Fig 10: Data vs Boundary Layer Theory Predictions

stituted for hydrogen because of the many similarities between the two gases. Moreover, it is experimentally very demanding to handle hydrogen because of its potential for combustion.

Procedure

The experimental setup was the same as the one used for the air experiments, except for the addition of a high pressure helium supply source. Early on, it was found that the injection of air and helium from the top of the vessel produced sharp helium stratification pattern in the upper region as evident by an almost constant coolant temperature in that region. The bottom region consisted of an air and steam mixture with heat transfer rates similar to those found earlier in experiments with air alone. The goal of the present investigation was to arrive at conditions where the gases are well mixed. It was therefore decided to inject air at high velocity from the top of the vessel while helium was introduced from the lower end below the boiling water surface. These steps were effective in eliminating stratification and providing reasonable mixing.

At the start of each run, the boiling water temperature was selected at the value corresponding to the desired total pressure. As vapor was produced, the relief valve was opened in order to purge the system from any noncondensable gas. After 15 minutes of continuous venting, pure steam filled the vessel.

Helium was then slowly introduced from beneath the water level. The pressure increase due to the added helium mass was obtained from the change in the total system pressure. Few moments later, air was injected from the top of the vessel, and the change in total pressure was again noted. The coolant water line was then activated, and the system allowed to reach steady state. When the various temperatures settled around a mean value, data was recorded using the data acquisition system. A 0.5°C standard deviation in the temperature reading was judged acceptable.

Experimental Results

The range of physical parameters was chosen to correspond to typical values expected in post LOCA conditions. In order to separate the effects of air and helium, it was decided to keep the helium mass fraction constant and conduct a series of about 25 runs with air mass fractions ranging from 0.35 to 0.85. Three sets of experiments were performed with average helium mass fractions of 1.7%, 4.7%, and 8.3%.

Since the determination of the various gas mass fractions relied on the change in the system pressure upon the introduction of the gases, it was very important to perform these runs at high enough pressures to minimize the impact of the error associated with gauge reading. Therefore, the total pressure in the helium experiments varied between 2.7 and 3.5 atmospheres. The average temperature subcooling was kept below 40°C for all the helium tests.

Mixture Temperature Profile. As with the steam-air experiments, the bulk gas temperature decreases axially away from the boiling water surface, as shown in Figure 11. In the top half of the condensing section, the steam temperature is essentially constant, while in the bottom section the profile is sharper. Unlike the experiments with air only, the local concentrations of the gases could not be deduced from the local mixture temperature

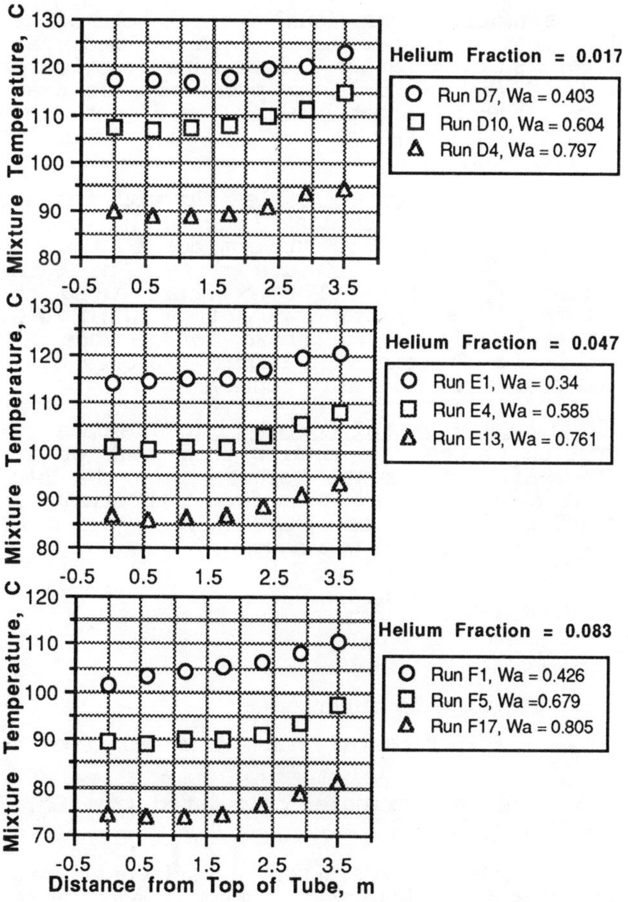

Fig. 11: Longitudinal Variation of the Steam Temperature

and the gas partial pressures. An attempt was made to use a gas analyzer to find the concentration profiles, but steam condensed in the capillary tubes and reliable data could not be obtained. The average gas fractions could however be calculated from the average partial pressure and bulk temperature using the ideal gas law:

$$W_{He} = \frac{4P_{He}}{4P_{He} + 18P_S + 29P_A} \quad (10)$$

$$W_S = \frac{18P_S}{4P_{He} + 18P_S + 29P_A} \quad (11)$$

$$W_A = \frac{29P_A}{4P_{He} + 18P_S + 29P_A} \quad (12)$$

The average steam partial pressure is:

$$P_S = P_T - P_A - P_{He}$$

Another way of checking this result was to use the average steam temperature, and assume that steam is in the saturation state:

$$P_S = P_{sat}(\overline{T_\infty})$$

The two methods yielded close values for the steam partial pressure.

As mentioned previously, three sets of helium tests were performed. Each set was characterized by a target helium mass fraction of 1.7%, 4.7%, and 8.3%, respectively. In practice, these target values were met only within ± 10% due to the inherent precision limitations.

Wall Temperature Profiles. As with the air experiments, the coolant tube wall temperature was not axially uniform due to the restriction in the flowrate range. The total variation in the wall temperature along the length of the section varied from 10 to 20°C. Typical profiles are shown for various air and helium fractions in Figure 12. Once again the curves are slightly concave.

The Experimental Correlation. The data reduction process is simpler than that of the steam-air experiments because only the section averaged heat transfer coefficient is calculated for every run. A correlation of the following form was sought:

$$G(W_{He}, W_A) \equiv \frac{\overline{h}_{exp}}{\overline{h}_A}$$

where:

G = correlation function

h_{exp} = experimental heat transfer coefficient

h_A = the heat transfer calculated from Eq. 7 with $W = W_A + W_{He}$.

In other words, the experimental heat transfer coefficient is normalized to the value that would be obtained if the entire noncondensable mass fraction consisted of air under the same total pressure. It is assumed

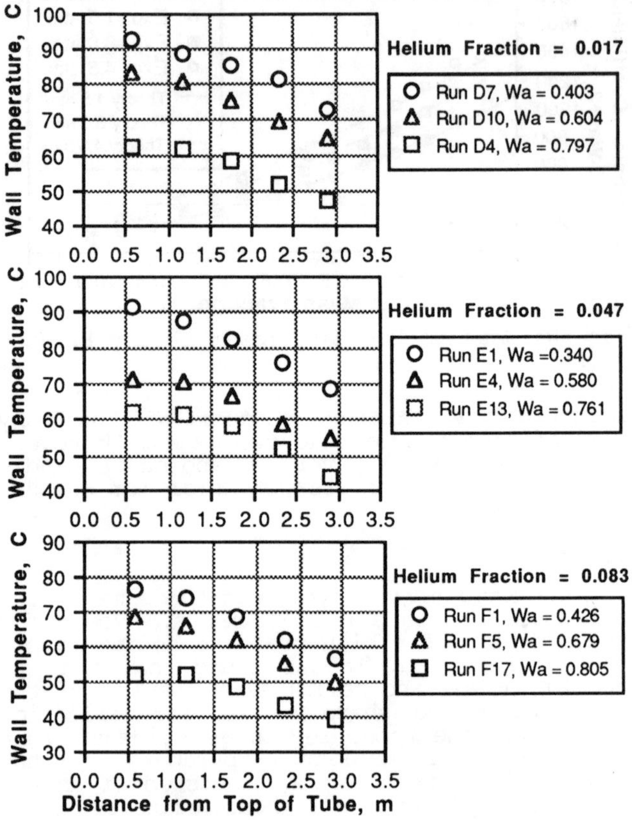

Fig. 12: Longitudinal Variation of the Wall Temperature

that this normalized value would be a function of the air and helium mass fractions.

The experimental data is shown in Figure 13. The total noncondensable mass fraction is the dependent variable while the y-axis represents the correlation function G defined earlier. As expected, the effect of helium is to further deteriorate the heat transfer coefficient. For the same noncondensable mass fraction, the heat transfer is improved when the helium fraction is decreased. This effect is especially pronounced in the steam rich region. A least squares fit of the data gave the following relations:

$$G(W_{He}, W_A) = 0.792 + 0.137(W_{He} + W_A) \text{ for } W_{He} = 1.7\%$$

$$G(W_{He}, W_A) = 0.464 + 0.418(W_{He} + W_A) \text{ for } W_{He} = 4.7\%$$

$$G(W_{He}, W_A) = 0.216 + 0.626(W_{He} + W_A) \text{ for } W_{He} = 8.3\%$$

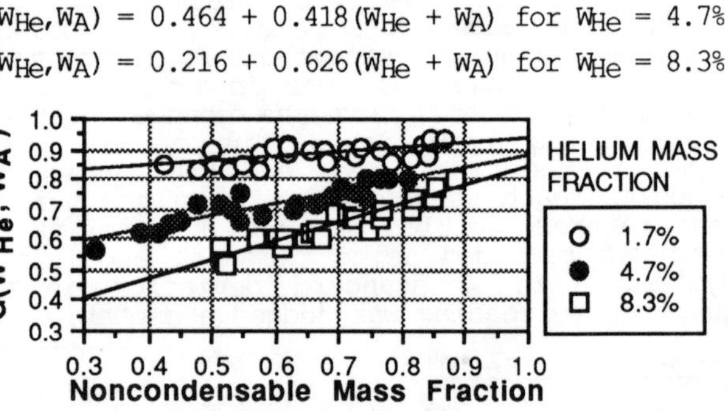

Fig. 13: Air-Helium Experiments, Results Based on Mass Fraction.

The preceding can be written in a more compact form as:

$$G(W_{He},W_A) = 0.948 - 8.67W_{He} + 7.36W_{He}(W_{He}+W_A)$$

Finally, the condensation heat transfer coefficient in the presence of air and helium is given by the following correlation:

$$\bar{h}_L = \frac{L^{0.05}[(3.7+28.7P) - (2438+458.3P)\,Log(W_{He}+W_A)]}{(\bar{T}_\infty - \bar{T}_w)^{0.25}}$$

$$* \; [0.948 - 8.67W_{He} + 7.36W_{He}(W_{He}+W_A)] \qquad (13)$$

where all the variables are in S.I. units, except for the pressure which is in atmospheres.

The data was taken for total noncondensable mass fractions greater than 40%. Below this level, stratification could not be avoided. For the experimental set corresponding to the highest concentration of helium (8.3%), the air-helium content had to be at least 50% to prevent stratification. It is also instructive to look at the data on a molal basis. Using the ideal gas law, the mole fraction N_j of any gas in the air-helium-steam mixture can be written as:

$$N_j = \frac{n_j}{n_A + n_{He} + n_S} = \frac{\frac{W_j}{MW_j}}{\frac{W_A}{MW_A} + \frac{W_{He}}{MW_{He}} + \frac{W_S}{MW_S}}$$

where n is the number of moles, MW stands for molecular weights, and j = {Air, Helium, Steam}. Knowing the gas mass fraction is equivalent to knowing the mole fraction for every gas in the mixture. It turns out that the three sets of equal helium mass fractions can be represented by average helium mole fractions of 9.23%, 22.8%, and 35.2%, respectively.

Let us define a non-dimentional function H as:

$$H(N_{He},N_A) \equiv \frac{\bar{h}_{exp}}{\bar{h}_A}$$

where h_A is the value that would be obtained if all the helium moles in the mixture were replaced by air moles.

Using the ideal gas law, it is straightforward to show that the mole fraction and the mass fraction of any species j are related by the following formula:

$$W_j = \frac{MW_j N_j}{MW_{He}N_{He} + MW_A N_A + MW_S N_S}$$

When all helium moles are replaced by air moles, the new air mass fraction in the mixture:

$$W_A = \frac{29(N_{He} + N_A)}{29(N_{He} + N_A) + 18N_S}$$

The experimental data plotted on a molal basis is shown on Figure 14. It appears that

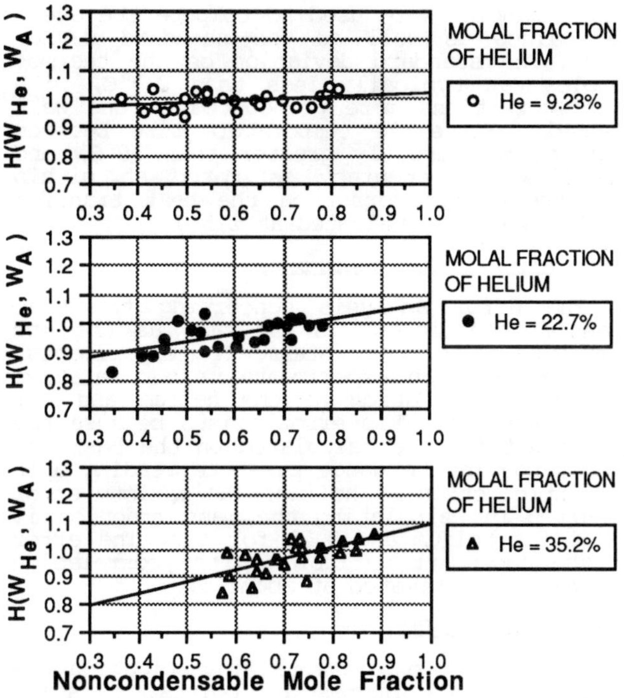

Fig 14: Air Helium Experiments, Results Displayed on Molal Basis.

helium is a little more inhibitive than air, especially at low air fractions. This seems to indicate that the helium buoyancy effect, which tends to oppose turbulence, is stronger than the helium diffusive effect, which tends to enhance steam mass transfer. The data sets can be well represented by straight lines whose equations are given be:

$$H(N_{He},N_A) = 0.951 + 0.066(N_{He}+N_A) \quad \text{for } N_{He} = 9.23\%$$

$$H(N_{He},N_A) = 0.799 + 0.266(N_{He}+N_A) \quad \text{for } N_{He} = 22.7\%$$

$$H(N_{He},N_A) = 0.667 + 0.422(N_{He}+N_A) \quad \text{for } N_{He} = 35.2\%$$

These relations can be summarized in a more compact form as follows:

$$H(N_{He},N_A) = 1.051 - 1.149\,N_{He} - 0.0553\,N_A + 1.371\,(N_{He})^2 + 1.371\,N_{He}N_A \qquad (14)$$

The heat transfer coefficient of an air-helium mixture in which the mole fractions are known is hence given by:

$$\bar{h}_L = L^{0.05}[(3.7+28.7P) - (2438+348.3P)\,Log\left(\frac{29(N_{He}+N_A)}{29(N_{He}+N_A)+18N_S}\right)] / (\bar{T}_\infty - \bar{T}_w)^{0.25}$$

$$* \; [1.051 - 1.149N_{He} - 0.0553N_A + 1.371(N_{He})^2 + 1.371N_{He}N_A] \qquad (15)$$

EXPERIMENTAL ERROR

Steam-Air Experiments

As previously mentioned, the temperature readings were taken for a period of 2 minutes at steady state. For each channel, more than thirty samples were used to compute the average values as well as the standard deviations. The standard deviation on the thermocouples was typically less than 0.7°C, with the fluctuations more pronounced in the wall channels. Since the temperature rise between stations was usually greater than 10°C, and the flowmeter accuracy estimated at within ±3%, the maximum error on the heat transfer coefficient was evaluated at ±15%.

Steam-Air-Helium Experiments

The error analysis concerning the heat transfer coefficient is the same as the one performed for the steam-air experiments. There is also an experimental error relative to the determination of the helium and air mass fractions. This error arises because the mass fractions are calculated on the basis of the change in the gauge pressure reading upon the introduction of helium. The maximum error incurred in calculating the mass fraction is 18.6% for helium and 5.2% for air. The error in the dependent variable of the correlation (W_A+W_{He}) is estimated at about 5%.

CONCLUDING REMARKS

The present experimental investigation exhibits several interesting points. Firstly, it demonstrates that if condensation is taking place in a steam-gas mixture, the emigration of the lighter gas to the upper regions is not always true, but depends on the boundary conditions in the system. Secondly, the improvement of the heat transfer with pressure is significant. This is especially relevant for long term post LOCA analysis in which natural circulation currents dominate and wall subcoolings are not too great. The developed correlations imply that a negative feedback mechanism exist whereby as pressure increases, steam condensation is accelerated, which then causes the pressure to decrease. These experiments also show the extent to which a unit mass of a light gas such as helium obstructs condensation more severely than than a unit mass of air.

NOMENCLATURE

A = Area
C_p = water heat capacity
F = correlation function
G = correlation function
Gr = Grashoff number
h = condensation heat transfer coefficient
L = length in m
m = water mass flowrate
M = mass
MW = molecular weight
P = pressure in atmospheres
ρ_o = radius of curvature
S = perimeter
T = temperature in °C
W = air mass fraction
z = longitudinal position
β = correction for curvature effects

Subscripts

A = air
He = helium
S = steam
T = total
w = wall
∞ = bulk of steam-air mixture
α,δ = constants

REFERENCES

1. Al-Diwani, H.K. & J.W. Rose, "Free Convection Film Condensation of Steam in the Presence of Non-Condensing Gases," *Intl J. Heat Mass Transfer* **16**, 1359 (1973).

2. Kroger, D.G. & W.M. Rohsenow, "Condensation Heat Transfer in the Presence of a Noncondensable Gas," *Intl J. Heat Mass Transfer* **11**, 15 (1968).

3. Uchida, H., "Evaluation of Post Accident Cooling Systems of LWR's," *Proc. Intl. Conf. Peacful Uses of Atomic Energy,* **13** , 93 (1965).

4. Tagami, T., "Interim Report on Safety Assessments and Facilities Establishment Project for June 1965," No 1 Japanese Atomic Energy Research Agency, unpublished work (1965).

5. Maitais, B. & E.R.G. Eckert, "Forced, Mixed and Free Convection Regimes," *J. Heat Transfer,* ser. C, **86** , 295 (1964).

6. Sparrow, E.M., W.J. Minkowsky & M. Saddy, "Forced Convection Condensation in the Presence of Noncondensables and Interfacial Resistance," *Intl J. Heat Mass Transfer* **10** , 1829 (1967).

7. Minkowycz, W.J. & E.M. Sparrow, "Local Non-Similar Solutions for Natural Convection from a Vertical Cylinder," *J. Heat Transfer,* 178, (May 1974).

8. Cebeci, T., "Laminar Free Convection Heat Transfer from the Outer Surface of a Vertical Slender Circular Cylinder," *5th Intl Heat Transfer Conf.* (1974).

9 Dehbi, A.A., "Analytical and Experimental Investigation of the Effects of Noncondensable Gases on Steam Condensation under Turbulent Natural Convection Conditions," Ph.D Thesis, Dept. of Nuclear Engineering, MIT, Jan. 1991. To be published.

10. Corradini, M.L., "Turbulent Condensation on a Cold Wall in the Presence of a Noncondensable Gas," *Nuclear Technology,* Vol. **64** , Feb 84.

A THEORETICAL MODELLING OF THE EFFECTS OF NONCONDENSABLE GASES ON STEAM CONDENSATION UNDER TURBULENT NATURAL CONVECTION

A.A. Dehbi, M.W. Golay and M.S. Kazimi ■ Massachusetts Institute of Technology, Cambridge, MA 02139

This paper presents a condensation model for the prediction of natural convection heat and mass transfer in the presence of a noncondensable gas. The conservation boundary layer equations for mass, momentum, energy, and species are written for the steam-gas mixture boundary layer. Turbulence in the gas layer is treated according to the Turbulent Kinetic Energy (TKE) one equation formulation. The model equations are reduced to nonlinear ODE's through similarity transformations. Condensate film modelling follows the simple Nusselt analysis. The condensation rate is obtained iteratively by matching the interface gas and film heat fluxes. Previous turbulent natural convection models have systematically underpredicted the condensation rate of steam in the presence of noncondensables. To remedy this bias, the present model postulates that the effect of condensation induced suction is to eliminate the sublaminar layer, and hence permit turbulent heat and mass transfer at the gas-film interface. The model predictions compared well with the data obtained in laminar conditions by Al Diwani and Rose. The model also simulated very well the turbulent natural convection experiments performed by the authors, and showed close agreement with the Tagami correlation.

Several investigators have developed models for predicting the steam condensation rate in the presence of a noncondensable gas. Sparrow and co-workers [1] used the boundary layer approach in which film and gas conservation equations are solved separately, then an iterative process on the interface conditions is performed in order to extract the heat transfer rate. The analysis yielded reasonable agreement with the scarce experimental data available then, but was restrictive in the sense that it considered laminar natural convection in the gas. Similarly, Denny et al [2]. presented a numerical solution for the case of a laminar steam-air flow forced down a vertical plate. Later, Corradini [3] used the heat and mass transfer analogy to develop a simple algebraic model for the gas boundary layer under turbulent natural convection conditions. Since the enhancement effect due to suction is not addressed by the analogy formulation, Corradini used correction factors according to the methodology outlined by Bird [4]. Despite this correction, the predicted heat transfer coefficients were well below experimental values such as those reported by Uchida [5] and Tagami [6]. This led Corradini to infer that the experiments were probably done in the forced flow regime. There might well have been some forced flow in the Uchida and Tagami experiments, but the correlations obtained are regarded as conservative and natural convection models should predict values comparable to what was reported in these investigations. More recently, Vernier and Solignac [7] tested the heat and mass transfer analogy model against data obtained in laminar and turbulent natural convection. Although the data displayed a large scatter, the general trend indicated that the analogy model predictions were too conservative. Lately, Kim and Corradini [8] proposed a set of condensation models in which the liquid film waviness is used to explain the enhancement of heat transfer. In this study, a model based on the boundary layer approach is proposed. It is intended to make reasonably accurate predictions of steam-gas boundary layer behavior in laminar and turbulent natural convection conditions.

THEORETICAL MODEL

In pure condensation, only the liquid film boundary layer needs to be analyzed. The presence of a noncondensable gives rise to an additional gas boundary layer characterized by an accumulation of the noncondensable gas near the interface, and a corresponding reduction in the condensation rate. Since the gas-liquid interface conditions are not known a priori, both film and gas boundary layer equations have to be solved simultaneously in order to determine the heat transfer rate.

Liquid Boundary Layer

The liquid boundary layer is treated according to the simple Nusselt analysis for a laminar film. This is appropriate given that film turbulence is not expected to occur for practical applications. The film parameters needed for the subsequent analysis are expressed as follows:

a) The film thickness:

$$\delta(x) = \left[\frac{4k_\ell\mu_\ell(T_i - T_w)x}{g\rho_\ell(\rho_\ell - \rho_v)h_{fg}}\right]^{1/4} \tag{1}$$

b) The condensation heat transfer coefficient:

$$h_c(x) = \frac{k_\ell}{\delta} = \left[\frac{g\rho_\ell(\rho_\ell - \rho_v)k_\ell^3 h_{fg}}{4\mu_\ell(T_i - T_w)x}\right]^{1/4} \quad (2)$$

c) The longitudinal film velocity at the liquid-gas interface:

$$u_\ell(y) = \left[\frac{g(\rho_\ell - \rho_v)k_\ell(T_i - T_w)x}{\rho_\ell\mu_\ell h_{fg}}\right]^{1/2} \quad (3)$$

As seen from the latter expressions, the condensation rate can be found if the interface temperature T_i is known. The whole object of solving the gas boundary layer equations is to determine T_i by an iterative process explained later on.

Gas Boundary Layer

Steam and the noncondensable gas are treated as a single mixture, with the same temperature and velocity at any location. The constant density assumption is made, except for the buoyancy term (Boussineq assumption). The local concentrations of the gas and the vapor are defined as:

$$W_g \equiv \frac{\rho_g}{\rho} \quad \text{and} \quad W_v \equiv \frac{\rho_v}{\rho} \quad (4)$$

where ρ is the total density of the gaseous mixture. From these definitions, the concentration of the two components are related by:

$$W_v + W_g = 1 \quad (5)$$

Therefore, only one concentration equation needs to be solved. In the remainder of this analysis, the gas concentration is solved, and the subscript g is dropped.

To make this model as general as possible, the conservation equations are written in cylindrical coordinates. The gas boundary layer conservation equations of a mixture moving along a vertical cold surface can be written as follows:

$$\frac{\partial(ru)}{\partial x} + \frac{\partial(rv)}{\partial r} = 0 \quad (6)$$

$$u\frac{\partial u}{\partial x} + v\frac{\partial u}{\partial r} = (1 - \frac{\rho_\infty}{\rho})g + \frac{1}{r}\frac{\partial}{\partial r}\left[(\nu + \varepsilon)r\frac{\partial u}{\partial r}\right] \quad (7)$$

$$u\frac{\partial T}{\partial x} + v\frac{\partial T}{\partial r} = \frac{1}{r}\frac{\partial}{\partial r}\left[(\alpha + \alpha_t)r\frac{\partial T}{\partial r}\right] \quad (8)$$

$$u\frac{\partial W}{\partial x} + v\frac{\partial W}{\partial r} = \frac{1}{r}\frac{\partial}{\partial r}\left[(D + D_t)r\frac{\partial W}{\partial r}\right] \quad (9)$$

The fluid properties are assumed constant and evaluated at the average temperature and gas fraction. The treatment of the buoyancy term is made using the ideal gas law, that is:

$$\frac{\rho_\infty}{\rho} = \frac{M_\infty}{M}\frac{T}{T_\infty}$$

The molecular weight of the gas-vapor mixture is found using mixture rules:

$$M = \frac{M_v M_g}{M_g - (M_g - M_v)W}$$

Therefore, the buoyancy term in the momentum equation is:

$$1 - \frac{\rho_\infty}{\rho} = 1 - \frac{M_g - (M_g - M_v)W}{M_g - (M_g - M_v)W_\infty}\frac{T}{T_\infty} \quad (10)$$

Hence, the momentum equation can be expressed as:

$$u\frac{\partial u}{\partial x} + v\frac{\partial u}{\partial r} = \left[1 - \frac{M_g - (M_g - M_v)W}{M_g - (M_g - M_v)W_\infty}\frac{T}{T_\infty}\right]g + \frac{1}{r}\frac{\partial}{\partial r}\left[(\nu + \varepsilon)r\frac{\partial u}{\partial r}\right] \quad (11)$$

Turbulence Modelling

There are a few models available for the analysis of turbulent free convection boundary layers over vertical surfaces. The simplest one is the Prandlt mixing length (PML) formulation which is successful in predicting forced convection boundary layer behavior. The model relates the turbulent viscosity to the longitudinal velocity gradient which implies a null turbulence level at the maximum velocity location. This is not physically intuitive in free convection because the maximum velocity occurs inside the boundary layer. At the other end of the spectrum, the K-ε model uses two rate equations to determine the turbulent viscosity. A large number of empirical constants - six - must be supplied. To verify the applicability of the constants obtained from forced convection studies requires a lot of computational experimentation and validation. As a tradeoff between accuracy and simplicity, it was decided to use the turbulent kinetic energy (TKE) model which uses a single equation to determine the turbulent viscosity. Mason and Seban [9] showed that the TKE formulation yields good agreement with pure natural convection data.

The TKE equation is arrived at by a simplification of the Navier-Stokes equation relative to the turbulent kinetic energy E defined as:

$$E \equiv \frac{1}{2}\left(\overline{u'^2} + \overline{v'^2}\right)^{1/2}$$

The TKE equations is [9]:

$$u\frac{\partial E}{\partial x} + v\frac{\partial E}{\partial r} = \frac{1}{r}\frac{\partial}{\partial r}\left[(\nu + \frac{\varepsilon}{\sigma})r\frac{\partial E}{\partial r}\right] + \varepsilon\left(\frac{\partial u}{\partial r}\right)^2 - \frac{b}{\ell}E^{3/2} \quad (12)$$

where b and σ are constants to be supplied and ℓ is the Prandtl mixing length.

The usual convective and diffusive terms appear in the TKE expression. The second and third terms on the right denote the production and dissipation of the turbulent kinetic energy, respectively. The TKE model expresses the turbulent kinematic viscosity in terms of the turbulent kinetic energy:

$$\frac{\varepsilon}{\nu} = \frac{c\ell\sqrt{E}}{\nu} \qquad (13)$$

where c is a constant, and ℓ is the mixing length specified according to van Dreist formulation. For the laminar sub-layer region:

$$\ell = \kappa y\left[1 - \exp\left(-\frac{\sqrt{\tau_w\rho}\,y}{26\mu}\right)\right] \qquad (14)$$

For the turbulent core region:

$$\ell = \lambda y_{out} \qquad (15)$$

where y_{out} is the thickness of the boundary layer, $\ell = 0.09$ and $\kappa = 0.435$.

The TKE model requires the specification of three constants: b, c, and σ. Mason and Seban performed an extensive numerical investigation to determine the set of constants which produced the best agreement with air data in pure natural convection up to a Grashoff number of $5\text{x}10^{11}$. They recommended the following values:

$$c = 0.55\ ,\ b = 0.164\ ,\ \sigma = 1.0$$

These values are subsequently used for all simulations.

The Numerical Scheme

The set of partial differential equations was first reduced into ordinary differential equations, using the new independent variables suggested by Minkowycs and Sparrow [10]:

$$\xi \equiv \frac{2x^{1/4}}{r_o\left[\dfrac{g\beta(T_\infty - T_w)}{4\nu^2}\right]^{1/4}} \equiv \frac{2x^{1/4}}{ar_o} \qquad (16)$$

$$\eta \equiv \left[\frac{g\beta(T_\infty - T_w)}{4\nu^2}\right]^{1/4}\frac{r^2 - r_o^2}{2r_o^2x^{1/4}} \equiv a\,\frac{r^2 - r_o^2}{2r_o^2x^{1/4}} \qquad (17)$$

In the case of large curvatures, ξ tends towards zero, and η reduces to the flat plate similarity variable encountered in natural convection solutions. The stream function is defined as:

$$\Psi \equiv 4\nu r_o a x^{3/4}F(\xi,\eta) \qquad (18)$$

From this definition, one can readily obtain an expression for the velocity components u and v in such a way that mass conservation is automatically satisfied:

$$u \equiv \frac{1}{r}\frac{\partial\Psi}{\partial r} = 4\nu a^2x^{1/2}F' \qquad (19)$$

$$v \equiv -\frac{1}{r}\frac{\partial\Psi}{\partial x} = -3\nu a x^{-1/4}(1 + \xi\eta)^{1/2}F \qquad (20)$$

where

$$F' \equiv \frac{\partial F}{\partial\eta}$$

In their study of the effect of curvature on pure natural convection boundary layers, Minkowycs and Sparrow [10] showed numerically that the inclusion of the $\partial/\partial\xi$ terms had very little effect on the overall solution accuracy, although it made the equations more complicated to solve. Accordingly, the $\partial/\partial\xi$ terms are neglected and $\partial/\partial\eta$ is denoted by a prime. For convenience, a dimensionless temperature is defined as:

$$\theta(\xi,\eta) \equiv \frac{T_\infty - T}{T_\infty - T_w}$$

When the change of variables is performed, the working set of equations becomes:

$$\begin{aligned}&\left(1 + \frac{\varepsilon}{\nu}\right)(1 + \xi\eta)F''' \\ &+ \left[\left(1 + \frac{\varepsilon}{\nu}\right)\xi + (1 + \xi\eta)\left(\frac{\varepsilon}{\nu}\right)' + 3F\right]F'' - 2F'^2 \\ &+ \left\{1 - \frac{M_g - (M_g - M_v)W}{M_g - (M_g - M_v)W_\infty}\left[1 - \left(1 - \frac{T_w}{T_\infty}\right)\theta\right]\right\}g \\ &/\,(4\nu a^2) = 0\end{aligned} \qquad (21)$$

$$\begin{aligned}&\left(Sc^{-1} + Sc_t^{-1}\frac{\varepsilon}{\nu}\right)(1 + \xi\eta)W'' \\ &+ \left[\left(Sc^{-1} + Sc_t^{-1}\frac{\varepsilon}{\nu}\right)\xi + Sc_t^{-1}\left(\frac{\varepsilon}{\nu}\right)'(1 + \xi\eta) + 3F\right]W' \\ &= 0\end{aligned} \qquad (22)$$

$$\begin{aligned}&\left(Pr^{-1} + Pr_t^{-1}\frac{\varepsilon}{\nu}\right)(1 + \xi\eta)\theta'' \\ &+ \left[\left(Pr^{-1} + Pr_t^{-1}\frac{\varepsilon}{\nu}\right)\xi + Pr_t^{-1}\left(\frac{\varepsilon}{\nu}\right)'(1 + \xi\eta) + 3F\right]\theta' \\ &= 0\end{aligned} \qquad (23)$$

$$\begin{aligned}&\left(1 + \frac{1}{\sigma}\frac{\varepsilon}{\nu}\right)(1 + \xi\eta)E'' \\ &+ \left[\left(1 + \frac{1}{\sigma}\frac{\varepsilon}{\nu}\right)\xi + \frac{1}{\sigma}\left(\frac{\varepsilon}{\nu}\right)'(1 + \xi\eta) + 3F\right]E' \\ &+ \nu a^8r_o^4\,\xi^4(1 + \xi\eta)\left(\frac{\varepsilon}{\nu}\right)F''^2 \\ &- \left(\frac{br_o^2\,\xi^2}{4\ell\nu}\right)E^{3/2} = 0\end{aligned} \qquad (24)$$

The closure is provided by the relation between the turbulent viscosity and the turbulent kinetic energy:

$$\frac{\varepsilon}{\nu} = \frac{c\ell\sqrt{E}}{\nu}$$

These latter equations were further reduced to a system of nine first order ordinary differential equations with boundary conditions at both ends of the boundary layer. As can be noted, the change of variables has decoupled the spatial coordinates. All the derivatives are functions of the longitudinal variable η only, allowing for a marching solution in the x direction. The relaxation method was selected to solve the model equations. These were first discretized using the mid-point scheme which assures second order accuracy. The difference equations were then linearized according to Newton's method. The matrix inversion required at each iteration is accomplished very efficiently by a block elimination scheme [11]. The grid used in the η direction needed to be dense in the vicinity of both ends of boundary layer. This requirement could be satisfied by the following spacing [9]:

$$\eta(i) = \eta_{out} \sin^3\left[\frac{\pi}{2}\left(\frac{i}{I}\right)\right]$$

where I is the number of intervals, and i an index ranging from 1 to I. The grid spacing is displayed in Figure 1 for I = 100 and η_{out} = 10.

An earlier attempt was made to solve the model differential equations using the shooting method. However, the high degree of non linearity caused divergence in many instances. Therefore, the shooting method was discarded in favor of the more robust relaxation scheme.

Nature of the mass and heat at the interface: In turbulent convection, the boundary layer is usually modelled as consisting of a thin laminar sub-layer near the wall and a turbulent core away from the wall. When mass transfer occurs at the wall, the thickness of the laminar sub-layer decreases, and heat transfer is enhanced. When the Reynolds analogy approach is used, this enhancement is captured through the use of ad hoc corrections factors. It was stated earlier that this procedure predicted heat transfer coefficients which were systematically lower than experimental data for turbulent natural convection. In order to remedy this bias, we make here the rather critical assumption that the thin laminar sublayer disappears altogether due to condensation induced suction, and that the entire boundary layer consists of a turbulent core, as shown in Figure 2. This assumption eliminates the need for specifying the length of the laminar sub-layer, and allows turbulent heat and mass transfer at the interface. This model hypothesis cannot be fully justified on theoretical grounds, and only a thorough agreement with experimental data can show its validity.

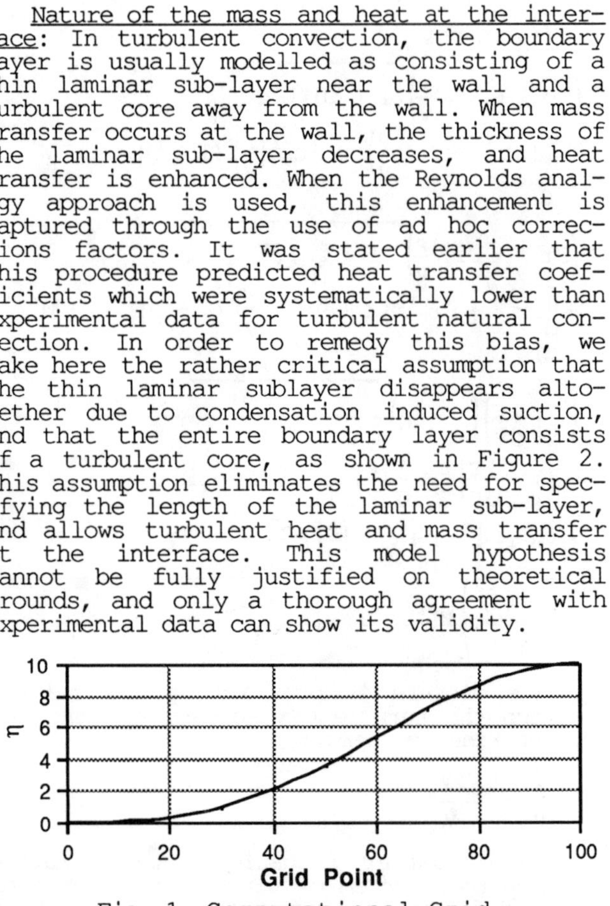

Fig. 1: Computational Grid.

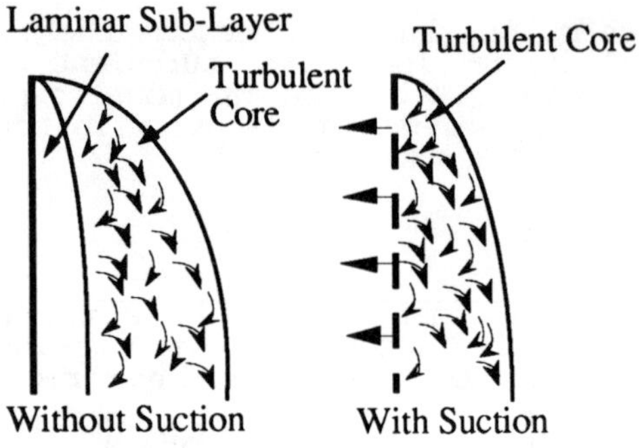

Fig 2: Schematic of Assumed Boundary Layer With and Without Suction.

Boundary conditions. Nine boundary conditions need to be satisfied, four at the liquid-gas interface and five at the gas-bulk interface. These are as follows:

- At the liquid interface:

1) equal liquid and gas mixture temperatures,

2) the vapor concentration, $W_{v,i}$ corresponds to the saturation temperature, T_i,

3) equal liquid and gas longitudinal velocity,

4) the liquid is impermeable to the noncondensable gas,

5) the turbulent kinetic energy is obtained by equating the production and dissipation terms in Eq. 24.

- At the outer edge of the boundary layer ($y = y_{out}$):

6) mixture temperature = bulk temperature,

7) noncondensable gas concentration = bulk temperature,

8) longitudinal velocity = 0,

9) turbulent kinetic energy = 0.

The reasoning behind boundary condition (5) is that the velocities at the interface are small enough to make the convection terms in the TKE equation negligible. Moreover, the distribution of the turbulent kinetic energy is postulated to reach a minimum at the interface, so that the diffusion term is also negligible. With these simplifications, the TKE equation states that the production and dissipation of turbulent terms in the TKE equation are equal at the interface.

Solution Procedure

The solution procedure steps are as follows:

a) guess and interface temperature T_i,

b) calculate the interface gas concentration. This follows from boundary condition (2):

$$W(0,\xi) = \frac{\frac{M_a}{M_v}\left(\frac{P}{P_{v,i}} - 1\right)}{1 + \frac{M_a}{M_v}\left(\frac{P}{P_{v,i}} - 1\right)} \tag{25}$$

c) calculate $F'(\xi,0)$, which is implied by the velocity continuity:

$$F'(0,\xi) = \frac{\left[\dfrac{g(\rho_\ell - \rho_v)k_\ell(T_i - T_w)}{\rho_\ell\mu_\ell h_{fg}}\right]^{1/2}}{4\nu a^2} \quad (26)$$

d) calculate the turbulent kinetic energy at the interface when turbulence is present. According to boundary condition (5):

$$E(0,\xi) = \frac{4c\nu a^8 r_o^2 \xi^2 (1+\xi\eta)\ell^2 F''^2(0,\xi)}{b} \quad (27)$$

e) calculate $F(\xi,0)$ to satisfy impermeability condition:

The impermeability condition is arrived at by equating the diffusive and convective mass flux of the noncondensable gas.

$$m''_{g,i} = \rho(D + D_t)\left(\frac{\partial W}{\partial r}\right)_{y=0} - \rho_{g,i} v = 0$$

therefore,

$$F(0,\xi) = -\frac{\left(Sc^{-1} + Sc_t^{-1}\dfrac{\varepsilon}{\nu}(0,\xi)\right)W'(0,\xi)}{3W(0,\xi)} \quad (28)$$

f) satisfy conditions (6) through (9) by letting:

$$\theta(\xi,\eta_{out}) = 0 \;;\; W(\xi,\eta_{out}) = W_\infty \;;$$

$$F''(\xi,\eta_{out}) = 0 \;;\; E(\xi,\eta_{out}) = 0 \;.$$

g) solve the gas boundary layer equations.

h) calculate the heat flux at the gas-liquid interface:

The heat flux at the interface is due to both condensation and conduction. Therefore:

$$q''_i = m''_{v,i}h_{fg,i} - (k + k_t)\left(\frac{\partial T}{\partial r}\right)_{y=0}$$

$$= m''_{v,i}h_{fg,i} - ax^{-1/4}k\left(1 + \frac{Pr}{Pr_t}\frac{\varepsilon}{\nu}\right)(T_\infty - T_w)\theta'(0,\xi) \quad (29)$$

At the interface, the vapor flux has both a diffusive and a convective component. Neglecting film curvature, the vapor flux is:

$$m''_{v,i} = -\rho(D + D_t)\left(\frac{\partial W}{\partial r}\right)_{y=0} - \rho_{v,i} v$$

$$= -\mu a x^{-1/4}\left(Sc^{-1} + Sc_t^{-1}\frac{\varepsilon}{\nu}\right)W'(0,\xi) + 3\mu a x^{-1/4}\left(1 - W(0,\xi)\right)F(0,\xi) \quad (30)$$

i) calculate the improved interface temperature, T_i:

$$T_i = T_w + \frac{q''_i\delta}{k_\ell} \quad (31)$$

j) if T_i has converged, go to step (k). Otherwise, go back to step (b).

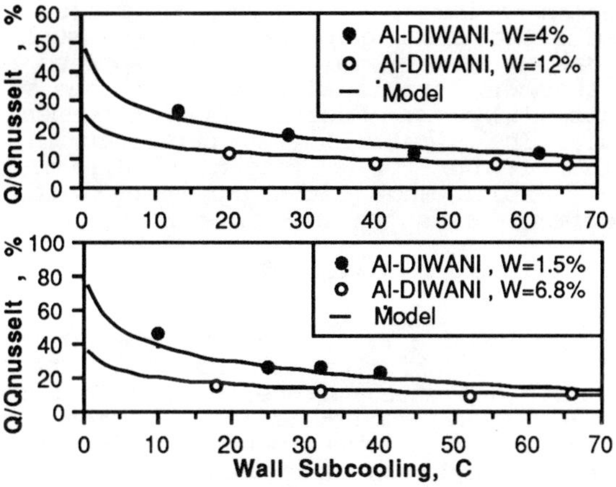

Fig. 3: Model Prediciton versus Laminar Gas Flow Data by Al-Diwani at Atmospheric Pressure.

k) calculate the local heat transfer coefficient:

$$h(\xi) \equiv \frac{q''_i}{T_\infty - T_w}$$

and move on to the next axial location, i.e.,

$$\xi_{new} = \xi_{old} + \Delta\xi$$

l) stop if the end of the section is reached. Otherwise, go to step (a).

The various gas mixture properties were evaluated at the average temperature and mass fractions. The mixture density was obtained using the ideal gas law . The mixture viscosity and thermal conductivity were calculated according to the Wilke [12] and Lindsay-Bromley methods [13], respectively. The diffusion coefficient was calculated following the procedure outlined by Bird [14]. A value of 0.80 was retained for the turbulent Prandlt and Schmit numbers, in agreement with what is generally recommended in the literature. It should be noted that one must specify the initial distribution of the turbulent kinetic energy at the laminar to turbulent transition point. One way to accomplish this is to assume that the rates of production and dissipation of the turbulent kinetic energy are equal at the onset of turbulence.

SIMULATIONS

The model was used to make heat transfer predictions as well as simulate several experiments performed in laminar and turbulent conditions. A summary of these simulations is given in the next sections.

Laminar Gas Flow Simulations

As mentioned earlier, several investigators have studied the condensation of steam in the presence of noncondensable gases in laminar conditions. Al-Diwani and Rose [15] performed several free convection experiments with small air mass fractions at near atmospheric conditions. The condensing surface consisted of a square condensing plate of 0.1 m side length immersed in a chamber filled with a quiescent steam-gas mixture. Wall subcooling varied from 10 to 70°C. The reported heat transfer was normalized in terms of the value predicted by the Nusselt film condensation theory. The simulation was performed in the laminar regime for a length of 0.1 m. The model results show close agreement with the experiments as shown in Figure 3.

Turbulent Flow Simulations

Experiments by Dehbi, et al. Dehbi, et al. [16] developed correlations for the condensation heat transfer coefficient of steam in steam-air and steam-air-helium mixtures under turbulent gas flow conditions. The condensing section in these experiments consisted of a 3.5 m long cylinder with a radius of 1.9 cm. The following section presents a wide range of theoretical simulations of these experiments. To provide a more general insight into the effects of noncondensables on steam condensation, the predictions were carried out beyond the range of conditions reported in the experiments.

Air-steam experiments simulations. For the air experiments, the reported wall subcooling temperatures ranged from 10 to 50°C, while the air mass fractions ranged from 25% to 90%. For a flat plate geometry, the experimental correlation is given by:

$$\overline{h}_L = \frac{L^{0.05}\left[(3.7 + 28.7P) - (2438 + 458.3P)\,\mathrm{Log}(W)\right]}{\left(\overline{T}_\infty - \overline{T}_w\right)^{0.25}} \qquad (32)$$

where L is in meters, P is in atmospheres, and the subcooling in °C. The logarithm is in base 10.

Because the radius of the condensing cylinder is small, the heat transfer is enhanced compared to the flat plate case by a factor which the authors estimated at 1.25. Therefore, Eq. 32 must be multiplied by 1.25 in order to obtain the original correlation for the cylinder of the experiment. Hence, the condensation coefficient off the 1.9 cm radius cylinder is:

$$\overline{h}_L = \frac{L^{0.05}\left[(4.62 + 35.6P) - (3047.5 + 572.9P)\,\mathrm{Log}(W)\right]}{\left(\overline{T}_\infty - \overline{T}_w\right)^{0.25}}$$

For the purpose of this simulation, the chosen geometry is a cylinder with a length of 1.0 m and a radius of 1.9 cm. The wall subcooling is kept constant at 30°C. Three isobaric sets of simulations were performed with the pressures reported in the experiments, i.e., 1.5, 3.0, and 4.5 atmospheres, respectively. For all external boundary layers, there is always an initial short region where the flow is laminar. The laminar to turbulent transition point is usually specified in terms of a critical Grashoff number. In the present investigation, this matter is complicated by the presence of more than one fluid as well as by the suction effect due to condensation. To avoid any arbitrariness associated with the selection of the transition point, the working assumption was that turbulence starts after a vertical length of 1 cm. The short laminar region is merely used to provide the necessary boundary conditions for the turbulent flow calculation. The simulation of the air experiments for a one meter long cylinder is shown in Figure 4. The agreement between theory and experiment is very good.

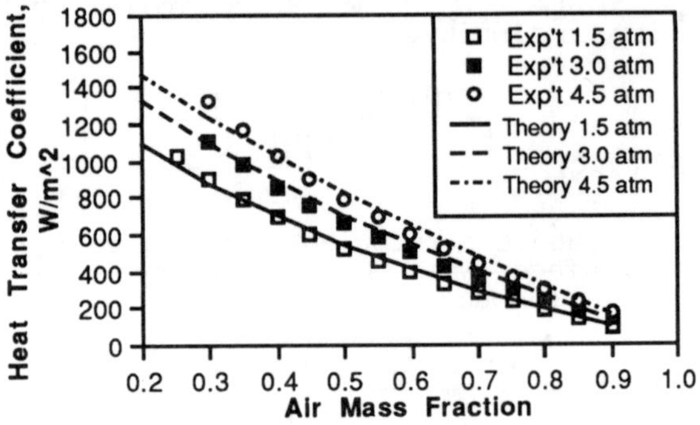

Fig. 4: Comparison of the Model with the Turbulent Gas Flow Correlation by Dehbi, et al.

Parametric study of the dependence of the heat transfer coefficient. The dependence of the heat transfer coefficient on various parameters in the correlation by Dehbi, et al., is addressed in this section. Each physical parameter will be varied while holding the others fixed. The fixed values for the various parameters are as follows:

Flat plat length = 1.0 m

Temperature subcooling = 30°C

Pressure = 2 atmospheres.

Length effect. The local heat transfer coefficient is shown in Figures 5 and 6 for several air fractions ranging from 0.05 to 0.90. The general trend is for the heat transfer coefficient to drop initially then start to increase mildly with height. This trend can be explained in terms of the film resistance to the heat transfer. At the top of the condensing section, there is very little turbulence in the gas boundary layer. Therefore, the local dependence of the heat transfer coefficient on height is similar to what is predicted by the Nusselt theory, i.e., $h \sim x^{-0.25}$. As turbulence in the gas boundary layer increases, the former effect is damped until it is overwhelmed, at which point the heat transfer coefficient starts to increase slowly with length.

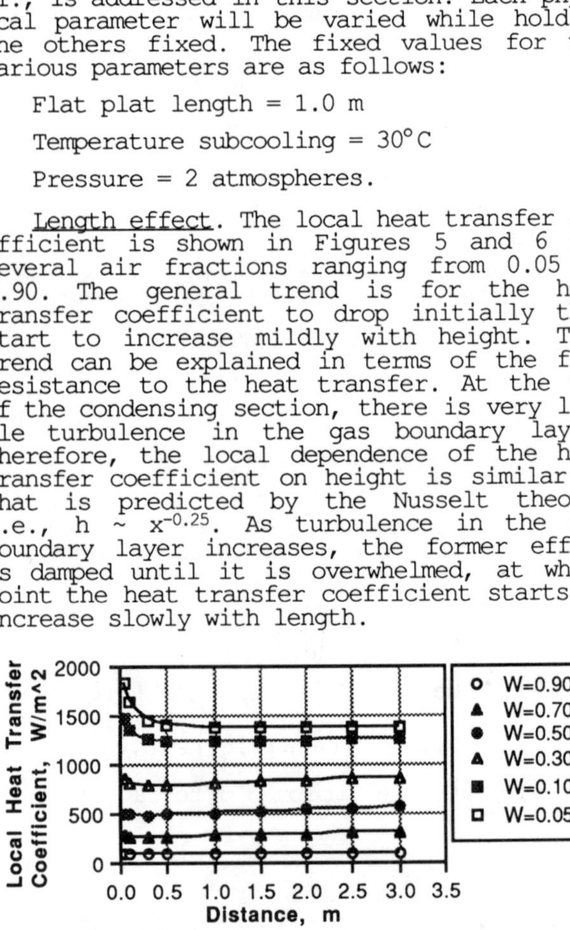

Fig. 5: Predicted Effect of Length on the Heat Transfer Coefficient.

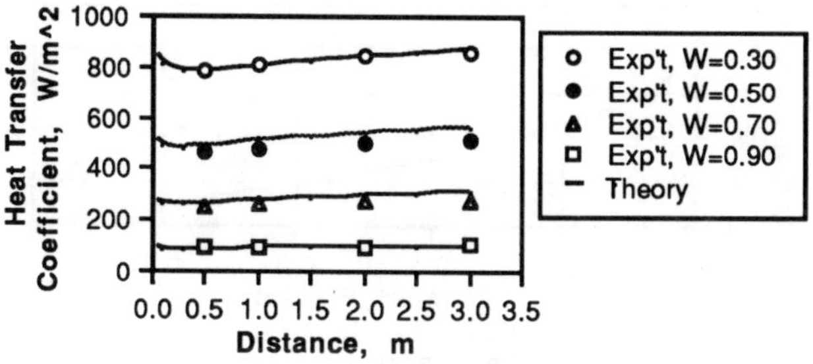

Fig. 6: Comparison of the Model with the Turbulent Gas Flow Correlation by Dehbi, et al., Length Effect.

It can be noted that the region over which h decreases is longer as one approaches the pure steam case. This is caused by the film thickness resistance which becomes larger as the air fraction decreases. Therefore, it takes more gas turbulence (i.e., a greater length) before the initial trend can be reversed.

Figure 6 compares the theoretical and experimental length effect predictions for the air mass ratios of the experiment (0.25 to 0.90). The theoretical prediction is well within the scatter of the experimental correlation (reported at ±15%).

Wall subcooling effect. As seen in Figure 7, the average heat transfer coefficient decreases with wall subcooling in analogy with pure condensation. At a given mixture bulk temperature, decreasing the wall temperature causes a drop in the steam partial pressure at the interface and a corresponding drop in the steam density. From Eq. 30, this leads to a lower steam mass flux, and hence a lower heat transfer coefficient. The slope of the curves is steeper in the steam rich region, in confirmation with the experiment by Al-Diwani and Rose [15]. A comparison with the experimental correlation shows close agreement as exhibited in Figure 8.

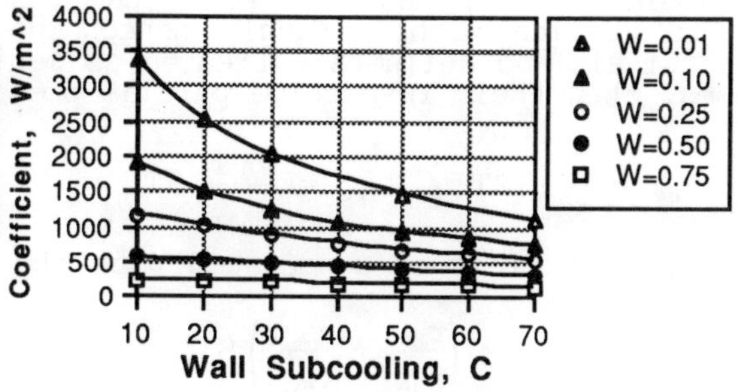

Fig. 7: Predicted Effect of Wall Subcooling on the Heat Transfer Coefficient.

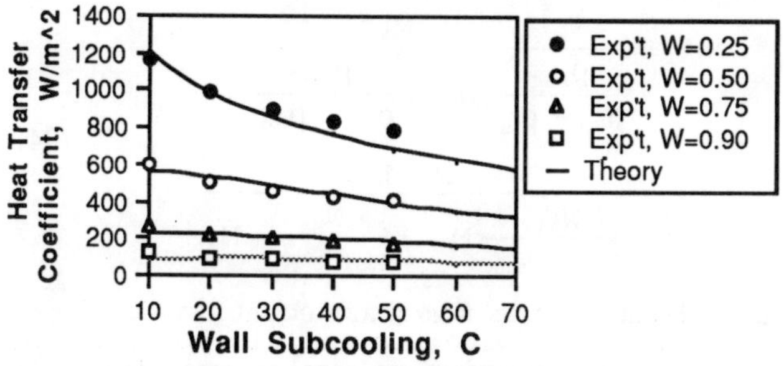

Fig. 8: Comparison of the Model with the Turbulent Gas Flow Correlation by Dehbi, et al., Subcooling Effect.

Pressure effect. If the wall subcooling and W_∞ are fixed, increasing the total pressure of the system amounts to raising T_∞, and hence raising T_w. Referring to Eq. 30 again, the latter step causes an increase in the interface vapor mass flux, with a corresponding improvement in the heat transfer coefficient.

The pressure effect on the heat transfer coefficient versus air fraction curves is shown on Figure 9. The curves start up concave, then approach straight lines as pressure increases. Figure 10 shows the theoretical prediction versus the experimental correlation. Despite the fact that the data was taken between 1.5 and 4.5 atmospheres, Figure 10 demonstrates that the correlation agrees very well with theory for pressures between 1 and 7 atmospheres. This indicates the possible use of the correlation beyond the pressures where data was collected.

Curvature effect. In many applications, the chosen condensing surface may be a tube rather than a flat plate. For that reason, the model was written in cylindrical coordinates. In Figure 11, the y axis refers to the ratio of the averaged heat transfer coefficient over a 1.0 meter tube normalized to the value obtained for the flat plate geometry. The abscissa represents the radius of curvature in cm. The effect of curvature becomes less important as the air content decreases. This is explained by the fact that the film heat transfer resistance becomes a sizable

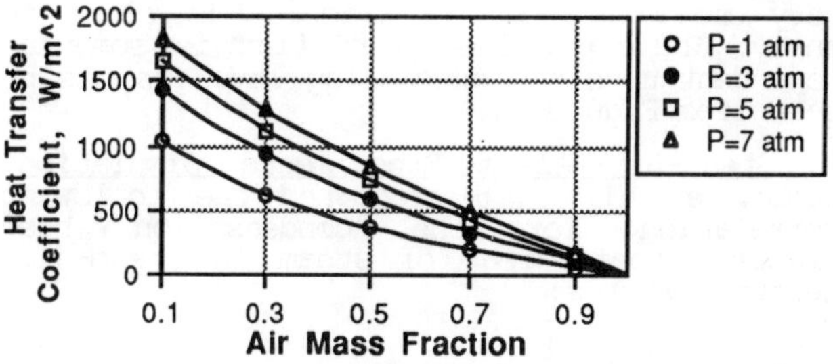

Fig. 9: Predicted Effect of Pressure on the Heat Transfer Coefficient.

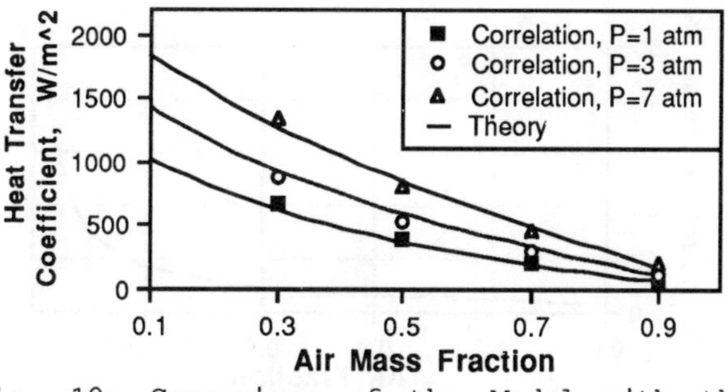

Fig. 10: Comparison of the Model with the Turbulent Gas Flow Correlation by Dehbi, et al., Pressure Effect.

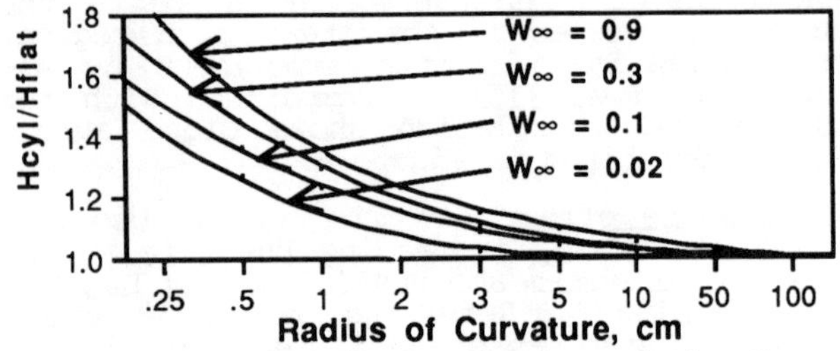

Fig. 11: Theoretical Prediction of the Curvature Effect on the Heat Transfer Coefficient.

fraction of the total resistance as air is depleted. In the limit where pure steam condensation takes place, the film thickness becomes the only resistance, and curvature will have no effect on the heat transfer coefficient for radii of curvature greater than the film thickness.

Contribution of conduction. Conduction heat transfer was included in order to make the model as general as possible. The contribution of conduction to the overall heat transfer is shown in Figure 12. Up to an air fraction of 80%, conduction can be neglected. Above that value, the curve begins to increase sharply, and the conduction fraction tends asymptotically towards 1.0 as the pure air region is approached.

Comparison of the heat transfer with the pure steam values. To evaluate the heat transfer degradation in the presence of noncondensable gases, a plot of the heat flux normalized to the pure steam Nusselt value is shown in Figure 13. As expected, the normalized heat flux tends asymptotically towards 1.0 as the air fraction becomes very small. At the other end, the heat transfer becomes very small as the air fraction is close to unity. The normalized heat flux is somewhat dependent upon the subcooling level as can be seen from Figure 14.

Steam-Air-Helium Experiments Simulation. Dehbi, et al., [16] reported the following correlation for the condensation heat transfer coefficient of steam in steam-air-helium mixtures:

$$G(W_{He}, W_A) \equiv \frac{\bar{h}}{\bar{h}_A} \tag{33}$$

$$= \left[0.948 - 8.67W_{He} + 7.36W_{He}(W_{He}+W_A)\right]$$

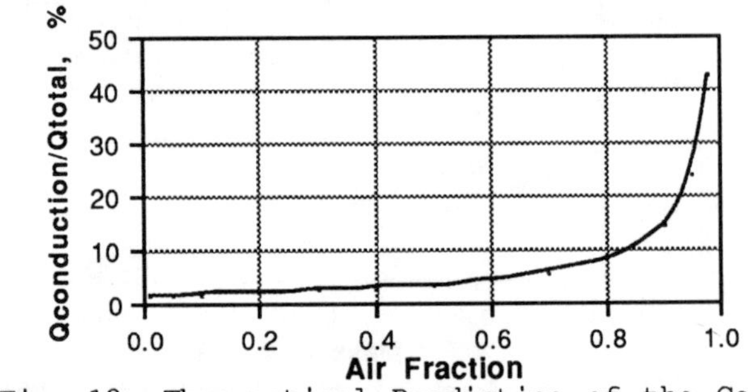

Fig. 12: Theorectical Prediction of the Contribution of Conduction to the Total Heat Transfer.

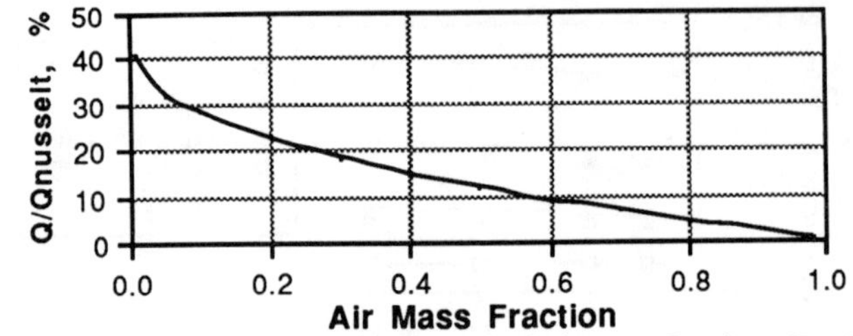

Fig 13: Theoretical Predicition of the Heat Transfer Rate Normalized to the Nusselt Pure Steam Value.

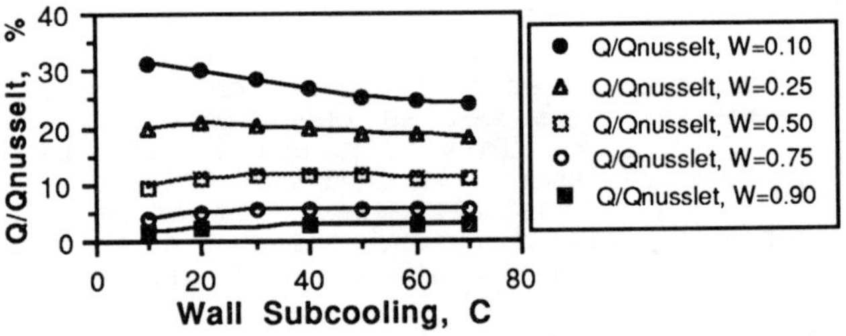

Fig. 14: Theorectical Prediction of the Normalized Heat Transfer Rate as a Function of Wall Subcooling.

where h is the experimental heat transfer coefficient and h_A is obtained from Eq. 32 with W replaced by the total noncondensable fraction. This correlation is valid for helium fractions between 1.7% and 8.3%, and air fractions between 40% and 85%.

The model does not have the provision to address a situation in which more than one noncondensable gas is hindering steam condensation. However, for the case of two noncondensable gases, one can derive an effective molecular weight, and thus treat the problem as if dealing with a single noncondensable gas. If air and helium are treated as perfect gases, then their respective partial pressure is:

$$P_{He} = \frac{\rho_{He}}{M_{He}} RT \;;\; P_{Air} = \frac{\rho_{Air}}{M_{Air}} RT$$

The partial pressure associated with the two gases is:

$$P_{Air} + P_{He} = \left(\frac{r_{He}}{M_{He}} + \frac{r_{Air}}{M_{Air}}\right) RT$$

One can therefore define an effective molecular weight as:

$$M_{Eff} \equiv \frac{M_{He}M_{Air}}{\dfrac{\rho_{Air}}{\rho_{He}+\rho_{He}} M_{He} + \dfrac{\rho_{He}}{\rho_{He}+\rho_{He}} M_{Air}} \tag{34}$$

$$= \frac{M_{He}M_{Air}}{\dfrac{W_{Air}}{W_{He}+W_{He}} M_{He} + \dfrac{W_{He}}{W_{He}+W_{He}} M_{Air}}$$

Going back to the momentum equation developed earlier, the driving buoyancy is specified by Eq. 10. The effective molecular weight of the air-helium mixture can replace the M_g term with the additional condition that the gas concentration at infinity is equal to the sum of the helium and air concentrations at infinity. It is clear that the boundary layer solution is meaningful as long as the value of Eq. 10 is positive. In other words, the effective molecular weight of the air-helium mixture must be greater than that of steam. If this is not the case, the boundary layer will simply separate and another approach will have to be developed. One must also find a way to specify an effective diffusion coefficient for the air-helium mixture in steam. No known method is available to provide such information. Therefore, an approximation is proposed in which the effective diffusion coefficient is obtained by weighing by molal fraction the individual helium and air diffusion coefficients in steam. Hence,

$$D_{Eff} \equiv \frac{N_{He}}{N_{He} + N_{Air}} D_{Helium\text{-}Steam} + \frac{N_{Air}}{N_{He} + N_{Air}} D_{Air\text{-}Steam} \quad (35)$$

where N represents the mole fraction and j = {helium, air}

The other air-helium mixture properties were computed following the same logic. The simulation was performed for the cases where the noncondensable gas was heavier than steam. The results of the simulation are shown in Figure 15. The normalized heat transfer coefficient is always less than unity because when some air is replaced by an equal mass of helium, the net result is that more noncondensable molecules are present. Therefore the inhibition of steam condensation is more severe. The agreement with data is surprisingly good given the simplification associated with lumping the noncondensable gases is a single entity.

Comparison with the Tagami Correlation. The Tagami correlation is used as a lower bound estimate of the heat transfer coefficient in turbulent natural convection. In the Tagami experiments [6], the wall temperature was kept constant at 49°C, while the air fraction varied from 0.4 to 0.9. A proposed fit for the Tagami data is:

$$\bar{h}_{Tagami} = 11.4 + 284 \left(\frac{1 - W}{W}\right) \quad (36)$$

A simulation of the experiment at atmospheric pressure is shown in Figure 16. The experimental correlation falls under the theoretical curve for most of the air fractions, but the values are in reasonably good agreement.

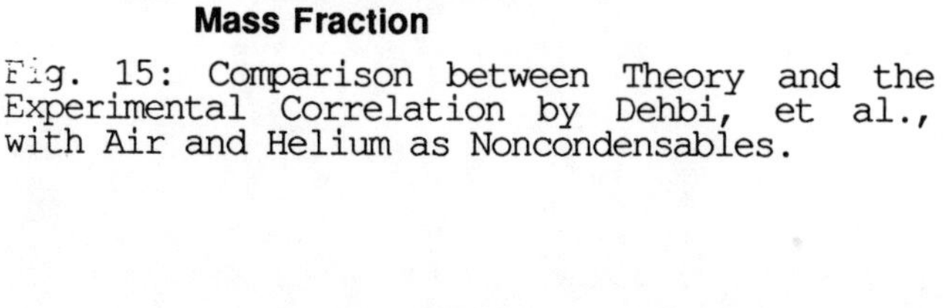

Fig. 15: Comparison between Theory and the Experimental Correlation by Dehbi, et al., with Air and Helium as Noncondensables.

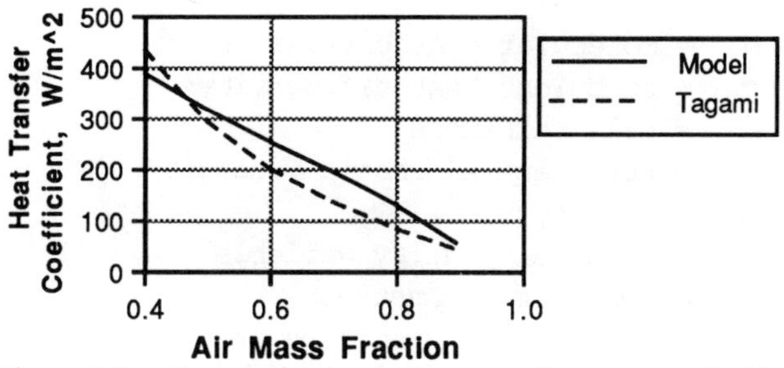

Fig. 16: Comparison between Theory and the Tagami Correlation.

CONCLUSION

In the present paper, a predictive model for the condensation heat transfer of steam in the presence of a noncondensable gas has been put forth. The agreement with various data taken in both laminar and turbulent natural convection is very good. This lends strong support to the central assumption of the model, which states that the condensation induced suction eliminates the laminar sublayer near the wall and causes turbulent heat and mass transfer at the interface between the film and the gas boundary layers.

NOMENCLATURE

- a = similarity transformation constant
- b = TKE constant
- c = TKE constant
- D = molecular mass diffusion coefficient
- D_t = turbulent mass diffusion coefficient
- E = turbulent kinetic energy
- F = transformation function
- g = gravity acceleration
- h = heat transfer coefficient
- h_{fg} = condensation enthalpy
- k = thermal conductivity
- ℓ = Prandlt mixing length
- M = molecular weight
- N = molar fraction
- P = pressure
- Pr = Prandlt number
- Pr_t = turbulent Prandlt number
- q'' = heat flux
- r = radial length = wall radius of curvature + y = r_o + y
- Sc = Schmit number
- Sc_t = turbulent Schmit number
- T = Temperature
- u' = turbulent longitudinal velocity
- u = longitudinal velocity component
- v = axial velocity component
- v' = turbulent axial velocity
- W = mass fraction
- x = longitudinal length
- y = axial length

α = molecular heat diffusivity
α_t = turbulent heat diffusivity
δ = film thickness
ε = turbulent kinematic viscosity
ρ = density
η = axial similarity variable
κ = von Karman constant
λ = ratio of outer mixing length to boundary layer thickness
μ = molecular viscosity
ν = kinematic viscosity
θ = dimentionless temperature
ρ = density
σ = TKE constant
τ = shear stress
ξ = longitudinal similarity variable
ψ = steam function

Subscripts

A = air
g = gas
He = helium
i = interface
ℓ = liquid
v = vapor
w = wall
∞ = bulk

REFERENCES

1. Sparrow, E.M. & S.H. Lin, "Condensation Heat Transfer in the Presence of a Noncondensable Gas," *J. Heat Transfer*, 431, 1964.

2. Denny, V.E., A.F. Mills & V.J. Jusionis, "Laminar Film Condensation from a Steam-Air Mixture Undergoing Forced Flow Down a Vertical Surface," *J. Heat Transfer*, p. 297-304, Aug. 1971.

3. Corradini, M.L., "Turbulent Condensation on a Cold Wall in the Presence of a Noncondensable Gas," *Nuclear Technology*, Vol. **64** , Feb 1984.

4. Bird, R.B., et al., *Transport Phenomena*, Wiley and Sons, NY, 1960.

5. Uchida, H., "Evaluation of Post Accident Cooling Systems of LWR's," *Proc. Intl Conf. Peacful Uses of Atomic Energy*, **13** 93, 1965.

6. Tagami, T., "Interim Report on Safety Assesments and Facilities Establishment Project for June 1965," No. 1 Japenese Atomic Energy Research Agency, unpublished work, 1965.

7. Vernier, P. & P. Solignac, "A Test of some Condensation Models in the Presence of a Noncondensable Gas against the Ecotra Experiment," *Nuclear Technology*, Vol. **77**, p. 82, 1987.

8. Kim, M.H. & M.L. Corradini, "Modeling of Condensation Heat Transfer in a Reactor Containment," *Nuclear Engineering and Design*, 118 p. 193-212, 1990.

9. Mason, H.B. & R.A. Seban, "Numerical Predictions for Turbulent Free Convection from Vertical Surfaces," *Intl J. Heat and Mass Transfer*, Vol. **17**, p 1329, 1974.

10. Minkowycz, W.J. & E.M. Sparrow, "Local Non-Similar Solutions for Natural Convection from a Vertical Cylinder," *J. Heat Transfer*, 178, May, 1974.

11. Ferzire, J.H., *Numerical Methods for Engineering Applications*, Wiley Interscience, 1981.

12. Wilke, C.R., *J. Chem. Physics*, Vol. **18**, p. 157, 1950

13. Lindsay, A.L. & L.A. Bromley, *Indus. Eng. Chem.*, Vol **42**, p. 508, 1950.

14. Bird, R.B. & J.O. Hirschfelder, *Molecular Theory of Gases and Liquids*, Wiley, NY, 1954.

15. Al-Diwani, H.K. & J.W. Rose, "Free Convection Film Condensation of Steam in the Presence of Non-Condensing Gases, *Intl J. Heat Mass Transfer* **16**, 1359, 1973.

16. Dehbi, A.A., M.W. Golay & M.S. Kazimi, "Condensation Experiments in Steam-Air and Steam-Air-Helium Mixtures under Turbulent Natural Convection," Natl Heat Transfer Conf., *AICHE Proceedings*, Minneapolis, July 1991.

CONDENSATION IN THE PRESENCE OF A NONCONDENSABLE GAS: THE EFFECT OF SURFACE ORIENTATION

I. Huhtiniemi, A. Pernsteiner, and M.L. Corradini ■ University of Wisconsin, 1500 Johnson Drive, Madison, WI 53706

An experimental investigation to examine the effect of surface orientation on the condensation of steam in the presence of a noncondensable gas is reported. An air-steam mixture is directed into a rectangular flow-channel over a condensing aluminum surface which has a painted wetting surface finish. In order to simulate a flat plate geometry as closely as possible, measurements were done in the region where flow was still developing. In this series of experiments the orientation of the condensing surface was varied from 0–90° (plate surface facing downwards to 0°), with a variable air-steam mass ratio of 0.24–0.8, and a mixture velocity of 1–3 m/s. The heat transfer coefficient was measured as well as making visual observations of the condensation process. The heat transfer coefficient varied from 100 to 600 W/m^2K. In comparison to the previous set of experiments with a bare polished aluminum surface the results showed an increase of 50–200% from upward facing bare surface, but lower than the downward facing bare surface more than 30%. The observed differences are attributed to the mode of condensation. A filmwise condensation mode prevailed throughout the current tests. The heat transfer coefficient decreased with the angle and increased with the flow velocity. The results are compared with some previously published data and a condensation model which show reasonable agreement with the data trends.

1. Introduction

In the event of a reactor accident involving a pipe break, a large amount of steam may be released into the containment building. Unless measures to reduce the pressure are taken, it may exceed the structural strength of the containment and lead to a release of the containment atmosphere which may contain radioactive material. However, condensation of steam on cold containment surfaces will act as a passive heat sink and mitigate this pressurization.

If one is to estimate the condensation heat transfer rates in such a scenario, the presence of a large amount of noncondensable gas and its suppression effect on heat transfer rates must be taken into account. This effect is due to the formation of a layer of noncondensable gas adjacent to the wall through which the steam must diffuse to condense. Therefore, the heat transfer rates are controlled by a diffusion phenomenon as well as heat conduction into the wall. Condensation may also affect the deposition rates of aerosols that can be present during accident. In order to review briefly the factors that affect the condensation rates, some previous separate effects experiments are discussed.

In most of the experiments with mixtures of noncondensable gases and vapor, *the mass ratio* of gas to the vapor has been low (m_{gas}/m_{tot} = 0.0001 - 0.1). Generally, the heat transfer coefficients are observed to decrease dramatically with increased noncondensable gas mass fraction [1,2]. Cho and Stein [3] and Kroger and Rohsenow [4] studied condensation on a small horizontal plate facing down with stagnant conditions. Both research groups reported that a one-dimensional molecular diffusion theory gives results that agree well with the experimental ones. Both workers noted the effect of *buoyancy forces* caused by different molecular weights of participating gases.

Turner and Mills [5] studied the condensation of mercury vapor in the presence of air. *A forced convection flow* resulted in higher heat transfer coefficients, because of the sweeping effect of the flow. Higher heat transfer coefficients due to forced convection were also reported by Kutsuna, Inoue and Nakanishi [6], who also studied the deceleration effect in a channel flow caused by condensation. Forced flow also induces interfacial instabilities that increase the heat transfer rates. Barry [7] studied the effects of interfacial structure caused by shear. He reported that enhanced mixing, which is caused by the interfacial film structure, mitigated the effect of the noncondensable gas.

The surface finish has a major effect on the mode of the condensation [8]. The wetting characteristics of the surface ultimately determine the mode of condensation. Dropwise condensation is likely to exist on non-wetting surfaces and filmwise is likely on wetting surfaces. In dropwise condensation mode with polished metal surfaces the heat transfer characteristics are likely to change due to oxidation of the surface or tarnishing [9]. Thus one cannot precisely know the wetting characteristics as surface aging occurs.

Several workers have investigated the effect of *the system pressure* with stagnant flow (no forced convection). The heat transfer rates are reported to increase by system pressure [1,13]. Cho and Stein [3] reported that an increase in the system pressure (0.31 MPa to 1.24 MPa) had a complicated effect on the mode of condensation in the presence of helium.

Containment safety concerns have prompted integral type of studies of the effect of pressure with a high noncondensable gas content. The effect of pressure in the range of 2.72-6.17 bars for air/steam mixture was studied by Robinson and Windebank [10]. The noncondensable gas mass fraction varied from 24% to 88%. Recently, similar experiments were also conducted by Dehbi, Golay and Kazimi [11]. The pressure was varied in the range of 1.5-4.5 bars and the noncondensable mass fraction ranged from 25% to 90%. Both investigations showed that the pressure had a significant effect on condensation. Robinson and Windebank also noted that the velocity field had an effect on their results, although no detailed measurements were done.

The geometry and, in particular, *the orientation of the surface* affects the dynamics of the condensation phenomena. The two important factors that have an effect on the dynamics are: 1) the stability of condensate interface which is dependent on the interfacial shear and the orientation of the surface, 2) buoyancy forces which are dependent on the gas mixture density differences and orientation of surface. Only a few studies have been performed with a different condensing surface orientation. Malkov [12] studied the film condensation of pure steam on surfaces with orientation from 5.74 degrees (referenced to a horizontal position and face up) to vertical. Gerstmann and Griffith [13] studied the condensation of the pure, stagnant Freon-113 and water vapor at atmospheric pressure. Heat transfer measurements and visual observations of the interfacial behavior were made. Gerstmann and Griffith observed several distinct flow regimes in the condensate film depending on the angle of inclination. The behavior in these regimes changed with inclination angles between 7.5-20 . No changes were observed with the angles greater than 20 .

All the factors discussed above have to be considered if an accurate analysis of the condensation rates is required. Most of the experiments that have been conducted to address the condensation phenomena inside the containment building under accident conditions have been integral in nature [10,11]. Several important factors like velocity field and orientation that affect the scaling of the results have not been included, thus making predictions of suggested correlations questionable outside the range of experimental conditions. On the other hand, the detailed separate effects tests have not been performed under the range of conditions that are likely to exist in a reactor accident.

This paper describes the second part of a continuing experimental research effort to address the effect of condensing surface orientation. The first part of experimental series [14] was performed with a bare polished aluminum plate. The orientation of the plate was varied from a horizontal position (face down) to 72 degrees (nearly vertical). The experimental conditions were chosen to be representative of accident conditions.

2. Experimental Facility

The experimental facility that was used in the test series consists of a test section and necessary auxiliary equipment; a boiler, air blower and secondary condenser. The process diagram is given in Fig 1.

The steam flow is provided by Sussman ES-72 boiler. The power of the boiler is 72 kW and it is rated to produce 0.027 kg/s steam at 100 C (with 10 C feedwater). A centrifugal blower was used for air flow. Chilled water for cooling of the test plate was provided by Neslab HX-150 refrigeration unit which is rated to supply a cooling power of 4500 W at 20 C. This cooling enables one to maintain a 27.68 kW/m^2 average heat flux through the test wall. A building water supply was used with the experiments of mass ratio of 0.4 or below.

The test section consists of a 109.50 cm long sandwiched clear polycarbonate-plates. These plates form a square duct of 15.24*15.24 cm aside. Dimensions of the major components are given in Figs. 2 and 3. The test section was designed to withstand test temperatures of 140 C and test pressures up to 2 bars.

The first 84 cm of the test section serves as an entrance length. This length is calculated to be long enough so that the turbulence profiles are well developed (the free stream turbulence intensities were spatially uniform and considerably smaller than in the boundary layer close to wall). Some measurements of turbulence intensity with air were performed to confirm that the entrance length was sufficient. The total length of the test section is scaled so that the boundary layers of the opposite walls do not interact. Both design considerations are important in order to satisfy the objective of simulating a planar developing flow geometry. The condensation takes place in the section following the entrance length.

The condensing section consists of a cooled 3.81 cm thick aluminum plate mounted as the ceiling of the duct. The arrangement with instrumentation locations is illustrated in Figs.4 and 5. This section is simulating a portion of the containment wall. The aluminum plate is 106.68 cm long and the effective width of the heat transfer area (the area exposed to air/steam flow) is 15.24 cm. This results in an effective heat transfer area of 0.1626 m^2. The surface finish on the plate was applied by Westinghouse Electric. The surface was sandblasted and then painted with the same coating that Westinghouse Electric plans to use in their AP-600 design for the interior of the containment steel shell.

The cooling of the test wall is provided by seven independent coolant plates mounted on the backside of the aluminum plates. The coolant plates were arranged symmetrically on top of heat flux meter stations. In order to ensure an even heat flux profile through the test plate, a spiral coolant channel geometry was adopted.

The experimental facility has a variety of instruments in order to set the test conditions and to acquire data and make observations. The overall goal in performing the measurements was that the measurement accuracy of heat fluxes and heat transfer coefficients would be within ten percent. The local overall condensation heat transfer coefficients are calculated from the heat flux measurements

$$h_{con} \equiv \frac{q''}{T_{mix}-T_{wall}} \qquad (1)$$

A simple analysis using the error propagation formula showed that flow rate had to be set so that the temperature rise of the coolant was high enough for the error of the temperature measurement to be acceptable. The heat flux measurements are accomplished using two different methods; 1) with Heat Flux Meters (**HFM**) and 2) calculating the Coolant Energy Balance (**CEB**). The principle behind both methods is straightforward.

The HFM's consist of an array of four type-E thermocouples (wire 0.254 mm dia). The thermocouples are placed in the grooves machined on 1.59 mm thick teflon plates. The teflon strip provides support for the assembly. The locations of the thermocouple junctions on the teflon strips were recorded carefully (0.05 mm). The strips were placed into slits on the backside of the test wall. The arrangement is illustrated in Fig. 5. The temperature gradient in the wall can be obtained from the thermocouple readings. Therefore, if the thermal conductivity of the aluminum is known, and if one-dimensional heat flow and steady state conditions are assumed, the heat fluxes are given by

$$q'' = k\frac{\Delta T}{\Delta x} \qquad (2)$$

A least-squares-fit was performed with the measured temperatures. The deviations from the perfect linear fit were caused either by longitudinal heat flow inside the plate or the instrument accuracy. The observed deviations through the tests were less than 3 percent which was well within the measurement accuracy objective.

The CEB-method involves measuring the volumetric flow rates (V) through the individual coolant loops. In addition to flow rates, the ΔT_{cool} between the inlet and outlet of each coolant loop is required. Now writing the steady state energy balance equation, an expression for local overall heat flux can be obtained

$$q'' = \frac{Q}{A} = \frac{V\rho c_p \Delta T_{cool}}{A} \qquad (3)$$

where Q is the total heat removed and A is the cooled area in the test section. Thus q'' is the area-averaged heat flux over one of these 15 by 15 centimeter square coolant plates and h_{con} can be obtained as before.

3. Results and Observations

The heat transfer tests performed so far are given in Table 1. These tests were essentially a repeat of test matrix performed earlier with the different condensing surface [14]. By comparing these two sets of experiments the effects of the surface finish on the condensate film behavior (condensation mode) through the heat transfer results, can be determined. The runs with low mass ratio and high velocities were not performed, because the mixing of the air and steam flow would have resulted a superheated steam, unless an intermediate cooling of the air or steam flow was provided. The current facility does not have this capability yet.

During the test runs it became evident that the boiler heater cycle created an oscillation of the steam mass flow rate which, consequently, resulted in an oscillation of the mass ratio (m_{air}/m_{tot}) and saturation temperature. However, the experimental setting was chosen so that this effect was minimized. Using the data available, it was estimated that the deviation of the mass flow of steam was always less than 14 percent. Time histories of heat flux meter readings did indicate that the effect of the variation of the mass ratio on heat transfer rates was averaged out.

During the test runs periodic checks of traverse and longitudinal temperature profiles and overall energy balances were made. The traverse temperature profiles at the HFM-1 station were measured to confirm that no temperature stratification existed due to a possibility of inadequate mixing of air and steam flows. The mixing of the air/steam flows was less efficient with low flow velocities as expected. A special flow control system was installed into the steam diffusers to keep the temperature deviations less than 5 percent with low flow rates. The temperature deviation was less than 2 percent for most data.

A Vaisala thin-film humidity probe was used to measure the relative humidity and bulk temperature in three locations. These measurements indicated that the mass ratio in the bulk was not affected by condensation in the test section.

Table 1. Test Matrix of Heat Transfer Measurements

		VELOCITY		
T_{mix}	m_{air}/m_{tot}	1 m/s	2 m/s	3 m/s
70	0.78	0 ,6 ,12 ,45 ,90	0 ,6 ,12 ,45 ,90	0 ,6 ,12 ,45 ,90
80	0.65	0 ,6 ,12 ,45 ,90	0 ,6 ,12 ,45 ,90	0 ,6 ,12 ,45 ,90
90	0.42	0 ,6 ,12 ,45 ,90		
95	0.24	0 ,6 ,12 ,45 ,90		

Some tests were repeated to check that the data was consistent (within error margins). The result of such a repeated test is given in Fig. 6 where both the heat flux meter (HFM) and coolant energy balance (CEB) data sets are compared. The repeatability is well within the error margins. The difference between the heat flux meters and the coolant energy balance method is consistent and also within the goal error margin (10%).

A summary of the tests is given in Figs. 7, 8 and 9. The results confirm the previously published observations that the heat transfer coefficient increases with increasing velocity and decreasing noncondensable gas mass fraction.

The effect of angle is relatively small and it is close to the magnitude of measurement accuracy. However, the results indicate that the heat transfer coefficient decreases as the inclination angle increases from a horizontal orientation. The results and observations of the condensate behavior strongly suggest that the higher heat transfer coefficient at the horizontal position is caused by enhancement of mixing in boundary layer by a) pendant droplets, b) droplet rain-out and c) buoyancy forces.

The results were compared (Fig. 10) with a simple condensation model for a flat plate developed by Kim and Corradini [15]. This one-dimensional model uses the Reynold's analogy of heat and mass transfer to model the condensation heat flux. Currently, this model does not have a provision to take into account any of these factors that are believed to cause the enhanced heat transfer rates in the downward facing orientation. However, the results seem to compare reasonably well with the vertical results where the above mentioned effects are not important and the boundary layer configuration is stable.

The model calculations suggested that the effect of a condensate film on the heat transfer rates is negligible in the range of test parameters. A straightforward calculation, using measured heat transfer rates, showed that film thicknesses in these runs were in the submillimeter range. A simple hand calculation can be done to obtain a stable film thickness when the condensing surface is facing down. The upper bound for the thickness was estimated to be on the order of a millimeter, which was confirmed by visual observation. Unless a significant amount of water is being dragged downstream by the flow, this steady state film thickness would be significantly higher than that for the inclined plate. This might be the reason why with low flow velocity and high mass fluxes of the inclined plate positions causes slightly higher heat transfer coefficients (Fig. 7). In this case the effect of enhanced mixing by droplets would be compensated by a thicker film.

The condensate droplets were observed to be drifting downstream, because of the drag of the imposed gas flow. The size of the pendant droplets also corresponded to the calculated value of about one centimeter. The size of droplets can be estimated using the Taylor wavelength [13].

The effect of surface finish is shown in Fig. 10. The current results are compared with previously published data of Barry [7] and Huhtiniemi [15]. Barry's data is for a bare polished aluminum plate facing up. The dashed lines indicate the range of heat transfer coefficients obtained with the same apparatus as Barry's [7] and orientation ranging from the horizontal (face down) to 72 degrees. The significant effect of angle on heat transfer coefficients was attributed to a change in the mode of condensation, from dropwise to filmwise (rivulets) as the inclination angle was increased [14]. The current results seem to confirm this observation. The current results are close to Barry's data, because the well wetting surface allows a thin film to exist even at the horizontal position.

4. Conclusions

The results confirmed the previously published findings that the heat transfer coefficient increases with velocity and decreasing air mass fraction. The data and observations suggest that the variation of heat transfer coefficient with angle is caused by enhancement of mixing in boundary layer by a) pendant droplets, b) droplet rain-out, c) buoyancy forces. A condensation model developed by Kim and Corradini [15] agree reasonably well with the data obtained at the vertical position where the boundary layer configuration is stable. The model does not include buoyancy force terms and the effect of the inclination angle. Simple modifications can be included to account the effect of inclination on the condensate film. However, in order to model the effect of buoyancy forces in gas phase a two-dimensional model is required. A set of experiments with local flow measurements is going to be performed in the future to quantify the effect of mixing and aid the development of a two-dimensional model.

LITERATURE CITED

1. L. Slegers, R.A. Seban, *Laminar Film Condensation of Steam Containing Small Concentrations of Air*, Int. J. Heat Mass Transfer, vol. 13, pp. 1941-1947, 1970
2. C.L. Henderson, J.M. Marchello, *Film Condensation in the Presence of a Noncondensable Gas*, Trans. ASME J. Heat Transfer, vol. 91(3), 447-450
3. D. H. Cho, R. P. Stein, *Steam Condensation on the Underside of a Horizontal Surface*, Proc. of Third Int. Topical Meeting on Nuclear Power Plant Thermal Hydr. and Operations, Nov. 1988
4. D. G. Kroger, W. M. Rohsenow, *Condensation Heat Transfer in the Presence of a Non-condensable Gas*, Int. J. Heat Mass Transfer, vol. 11, pp. 15-26, 1968
5. R. H. Turner, A. F. Mills, V. E. Denny, *The Effect of Noncondensable Gas on Laminar Film Condensation of*

Liquid Metals, Trans. ASME J. Heat Transfer, vol. 95(1), pp. 6-11, 1973

6. H. Kutsuna, K. Inoue, S. Nakanishi, *Filmwise Condensation of Vapor Containing Noncondensable Gas in a Horizontal Duct*, Int. Symp. on Heat Transfer, Beijing 1987
7. J. J. Barry, *Effects of Interfacial Structure on Film Condensation*, PhD Thesis, University of Wisconsin, 1987
8. A. Yamauchi, S. Kumagai, T. Takeyama, *Condensation Heat Transfer on Various Dropwise-Filmwise Coexisting Surfaces*, Trans. of JSME, vol. 51(B), pp. 2606-2613, 1985
9. T. G. Sundararaman, T. Venkatram, *Heat Transfer during Dropwise Condensation of Steam in the Presence of Non-condensable Gases-Effects of Geometrical Shape of the Surface Reversal of Cooling Water Flow & Orientation*, Indian Journal of Technology, vol. 14, pp. 313-321, 1976
10. J. A. Robinson, S.R. Windebank, *Measurement of Condensation Heat Transfer Coefficients in a Steam Chamber Using a Variable Conductance Heat Pipe*, Proc. 2nd. UK Natl. Conf. on Heat Transfer, vol.1, pp. 617-637, Sep. 1988
11. M.S. Kazimi, Private Communications, 1990
12. N. V. Suryanarayna, G. L. Malchow, *Film Condensation on Inclined Plane Surfaces*, Trans. ASME J. Heat Transfer, vol. 97, pp. 79-82, 1975
13. J. Gerstmann, P. Griffith, *Laminar Film Condensation on the Underside of Horizontal and Inclined Surfaces*, Int. J. Heat Mass Transfer, vol. 10, pp. 567-580, 1967
14. I. K. Huhtiniemi, J. J. Barry, M.L. Corradini, *Condensation in the Presence of a Noncondensable Gas: The Effect of Surface Orientation*, Proc. 26th National Heat Transfer Conference, vol. 85(269), pp. 205-210, 1989
15. M. H. Kim, M. L. Corradini, *Modeling of Condensation Heat Transfer in a Reactor Containment*, Nuclear Engineering and Design, vol. 118, 1990.

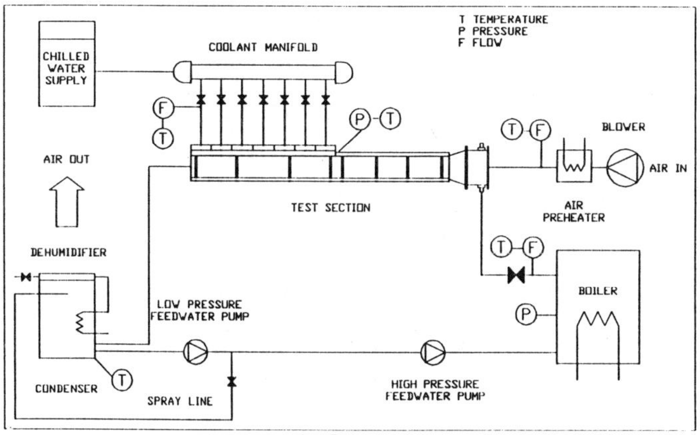

Figure 1. Process diagram.

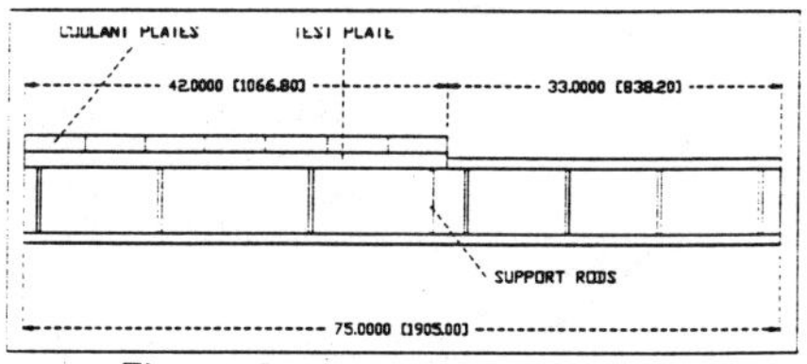

Figure 2. Testsection, sideview.

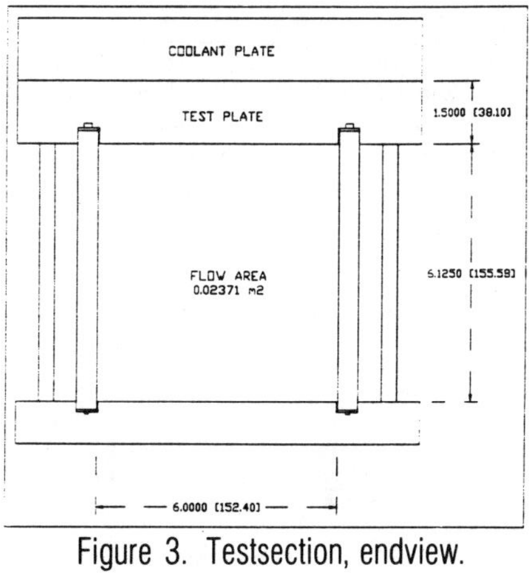

Figure 3. Testsection, endview.

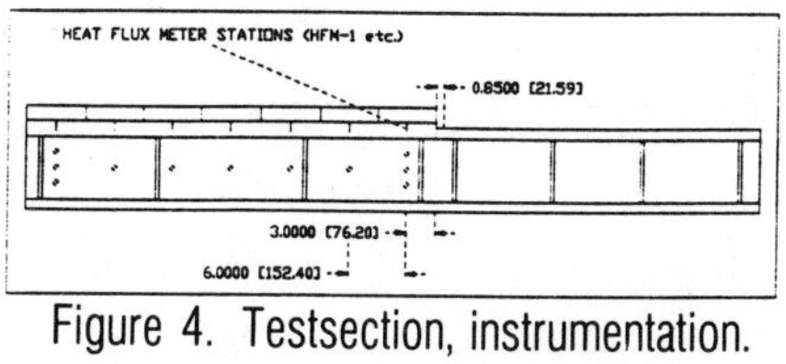

Figure 4. Testsection, instrumentation.

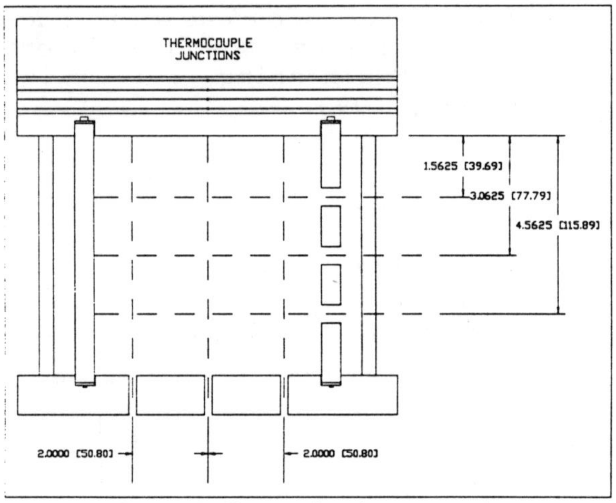

Figure 5. Testsection, instrumentation.

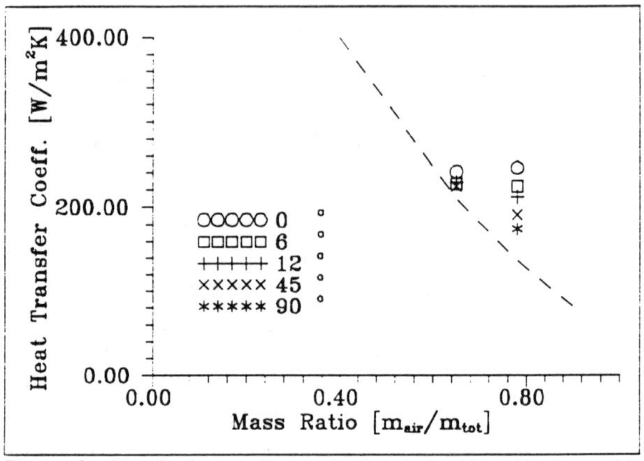

Figure 8. Results for medium velocity, dashed line Kim's model [15].

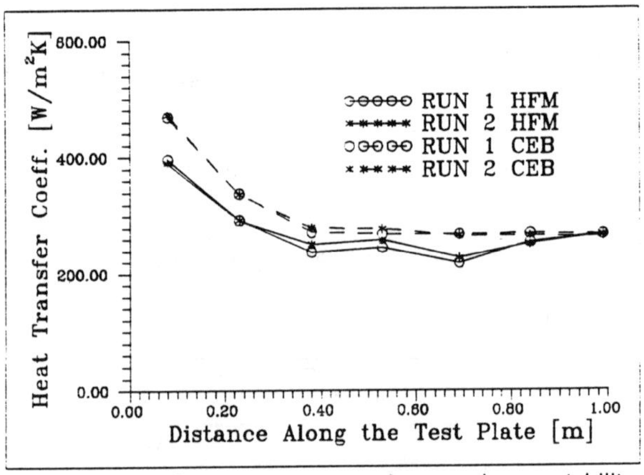

Figure 6. Heat flux meter comparison and repeatability of experiments.

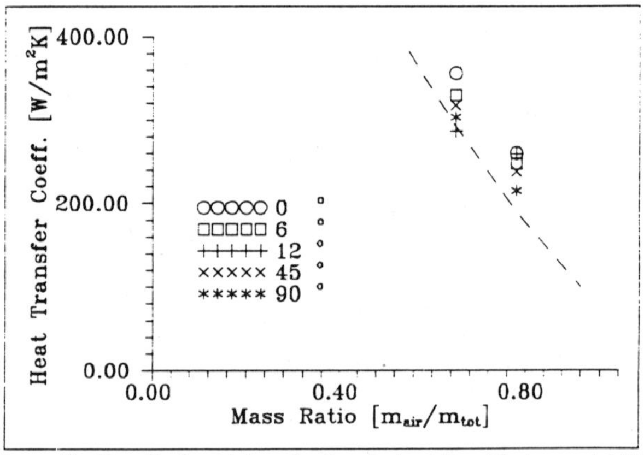

Figure 9. Results for high velocity, dashed line Kim's model [15].

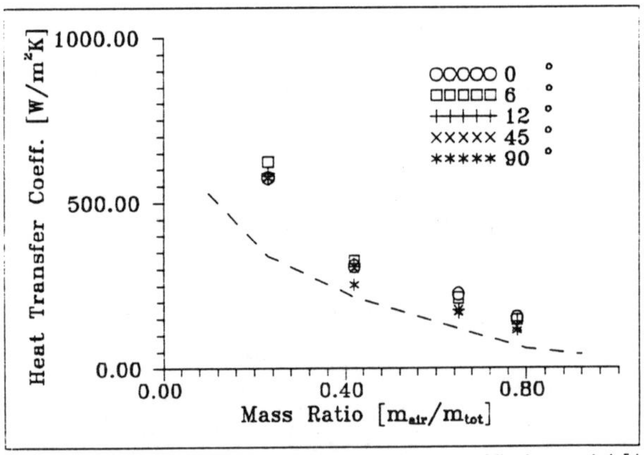

Figure 7. Results for low velocity, dashed line Kim's model [15].

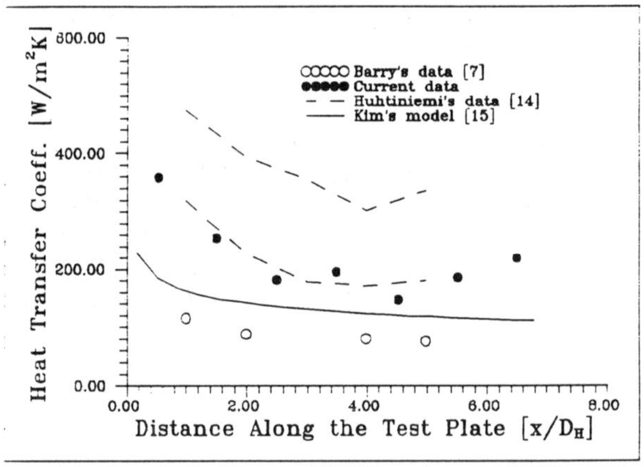

Figure 10. Results for T_{mix} = 70 C (m_{air}/m_{tot} = 0.79), medium velocity.

CONTAINMENT ANALYSIS WITH GOTHIC

Thomas L. George, Marvin J. Thurgood,
Lawrence E. Wiles, and Carl L. Wheeler ■ Numerical Applications, Inc., 825 Goethals Dr., Richland, WA 99352
Mati Merilo ■ Electric Power Research Institute, 3412 Hillview Avenue, Palo Alta, CA 94303

The separated field equations for three phase flow with multiple gas components as solved by the GOTHIC computer program are presented. Representative comparisons with experimental data for high energy line breaks and hydrogen dispersion in containment geometries are included.

GOTHIC is a general purpose thermal-hydraulics computer program for design, licensing, safety and operating analysis of nuclear containments and other confinement buildings. The code is the result of a long term effort that began with the development of COBRA-NC [1], continued with the addition of modeling features and capabilities in FATHOMS [2], and is now under EPRI sponsorship for further development and assessment. GOTHIC is currently being used by many utilities and research facilities in a wide variety of applications. The code package includes a graphical pre- and post-processor that makes GOTHIC especially useful on PCs and engineering workstations.

GOTHIC solves mass, momentum and energy balances for three separate phases, vapor, liquid and drops. The vapor phase can be a mixture of steam and noncondensable gases and a separate mass balance is solved for each component of the vapor mixture. An ice phase is also included for ice condenser containment analysis. The phase balance equations are coupled by mechanistic models for interface mass, energy and momentum transfer that cover the entire flow regime from bubbly flow to film/drop flow as well as single phase flows. The interface models allow for possibility of thermal nonequilibrium between the phases and unequal phase velocities.

The code can be used for three-dimensional analysis of the thermal-hydraulic behavior of containment atmospheres and structures. Containment compartments can be modeled using a 1-, 2- or 3-dimensional rectangular grid. Many containment problems cover long periods of real time so that a finely noded multidimensional model may be impractical. For these problems a simpler lumped parameter analysis may be used. Lumped parameter volumes are connected by junctions that employ a one-dimensional model for flow between containment compartments. A combination of lumped parameter and three-dimensional analysis is also possible, with regions of special interest modeled in more detail while lumped parameter volumes are used to model the remaining regions.

GOTHIC includes full treatment of the momentum transport terms in multidimensional models with an optional one parameter turbulence model for turbulent shear and mass and energy diffusion. Thermal conductors model heat transfer surfaces in the containment. Wall heat transfer correlations are incorporated for a wide range of heat transfer situations, including condensation heat transfer with the effects of noncondensable gases. Special models for engineered safety equipment including pumps, fans, valves, heat exchanges, coolers and vacuum breakers are included. Trip logic is used to control the action of the safety equipment in response to changes in the containment atmosphere.

The code assessment program includes a wide range of analytic tests, small and large scale experimental tests that include hydrogen dispersion, suppression pool performance, single and multicompartment water and steam blowdowns, and superheated steam blowdown. Representative results from some of these simulations are presented in this paper.

EQUATION SET

Conservation Equations

The conservation equations are presented in integral form because this form is closely related to the finite volume numerical method used to solve the equations. The equations are written for a fixed volume V, bounded by area A, as shown in Figure 1. The volume may be the entire region of interest or a portion of the total volume, although, in practice, V corresponds to one of the finite volumes of the computational grid. The equations below are for multidimensional analysis but reduce to the lumped parameter equations if the volume is not subdivided.

Mass Conservation

Mass conservation equations are solved for four phases; drops, liquid, ice and steam, and for each noncondensable gas component. The steam/gas mixture, herein referred to as the vapor phase, can be in the form of bubbles or a continuous vapor region. The liquid phase can be in the form of pools, films or large chunks found in slug flow. The drop phase results from two phase break discharges, entrainment off films and spray nozzles. The general form of the mass balance is given by

$$\frac{\partial}{\partial t}\int_V \alpha_\phi \rho_{\phi\zeta} dV = -\int_A \alpha_\phi \rho_{\phi\zeta} \vec{U}_\phi \cdot \vec{n} dA + \int_{A_f} \alpha_\phi \varepsilon^c_{\phi\zeta} \rho_\phi \nabla\left(\frac{\rho_{\phi\zeta}}{\rho_\phi}\right) \cdot \vec{n} dA \qquad (1)$$

$$+\int_{A_w} s^c_{\phi\zeta} dA + \int_{A_{I_\phi}} S^c_{\phi\zeta} dA_{I_\phi} + E^c_\phi$$

The subscript ϕ refers to the phase and takes on the values v (vapor), l (liquid), d (drops) and i (ice). The subscript ζ refers to a component of the steam/gas mixture. Only the vapor phase is made up of more than one component. For the other phases, the component subscript can be ignored. α is the volume fraction, ρ is the density, $\vec{U}$ is the velocity, $\vec{n}$ is outward normal to the surface dA, A_f is that portion of the total surface area in contact with adjacent fluid volumes, ε^c is the mass diffusion coefficient, including turbulence effects, s^c is the mass source per unit area generated at or passing through bounding wall A_w, S^c is the mass source per unit area coming from the interfacial area, A_I, and E^c is the source from engineered safety equipment.

For the liquid, drop and ice phases, there is no contribution to the mass balance due to mass diffusion. Further, there are no convection terms included in the mass balance for the ice.

Energy Conservation

Energy conservation equations are solved for three fluid phases; drops, liquid and vapor, and for conductors. The fluid energy equation is

$$\frac{\partial}{\partial t}\int_V \alpha_\phi \rho_\phi e_\phi dV = -\int_A \alpha_\phi (\rho_\phi e_\phi + P)\vec{U}_\phi \cdot \vec{n} dA + \int_{A_f} \alpha_\phi \varepsilon^e_\phi \nabla T_\phi \cdot \vec{n} dA \qquad (2)$$

$$+\sum_\zeta \int_{A_f} \alpha_\phi \varepsilon^c_{\phi\zeta} \rho_\phi \nabla\left(\frac{\rho_{\phi\zeta} h_{\phi\zeta}}{\rho_\phi}\right) \cdot \vec{n} dA$$

$$+\int_{A_w} s^e_\phi dA + \int_{A_{I_\phi}} S^e_\phi dA_{I_\phi} + E^e_\phi$$

where e is the internal energy, P is the static pressure, ε^e is the thermal diffusion coefficient, s^e is the energy source per unit wall area, S^e is the energy source per unit interfacial area and E^e_ϕ is the equipment energy source. Kinetic energy and other energy forms not explicitly represented above are neglected. Viscous dissipation is also neglected. All components of the vapor are assumed to be at the same temperature. The energy transported with the mass through mass diffusion is included only for the vapor energy balance. The ice energy equation is eliminated by assuming that the ice remains at its initial temperature.

The energy equation for conductors is

$$\int_{V_c} \rho_c c_c \frac{\partial T_c}{\partial t} dV_c = \int_{A_i} \varepsilon^e_c \nabla T_c \cdot \vec{n} dA + \int_{A_b} s^e_c dA \qquad (3)$$

where the subscript c refers to a particular conductor, V_c is the conductor volume or portion of a conductor, A_i is the surface area of V_c internal to the conductor, ε^e_c is thermal conductivity and A_b the external bounding surface surface area of the conductor which may be in contact with one or more of the fluid phases.

Momentum Conservation

Momentum conservation equations are solved for three phases; drops, liquid and vapor. The general form of the equation is

$$\frac{\partial}{\partial t}\int_V \alpha_\phi \rho_\phi \vec{U}_\phi dV = -\int_A \alpha_\phi \rho_\phi \vec{U}_\phi (\vec{U}_\phi \cdot \vec{n}) dA \qquad (4)$$

$$+\int_{A_f} \alpha_\phi \overleftrightarrow{\sigma}_\phi \cdot n dA + \int_{A_w} s^m_\phi dA + \int_{A_{I_\phi}} S^m_\phi dA_{I_\phi} + E^m_\phi$$

$\overleftrightarrow{\sigma}$ includes the static pressure and viscous and Reynolds stress terms, s^m is the momentum source per unit wall area, S^m is the momentum source per unit interfacial area and E^m is the momentum source from equipment. All components of the vapor are assumed to move at the same velocity.

Equations Of State

The densities of the vapor, liquid and drop phases are all assumed to be functions of temperature and pressure. The vapor components are all assumed to exist at the same temperature. Tabular data and curve fits are used for the water and steam properties. The perfect gas model is used for the noncondensable gases and for the conductivity and viscosity of the steam/gas mixture.

Boundary Source Terms

External mass, energy and momentum sources, such as a specified source from a line break, are included in the convective terms of Equations 1 through 4. The boundary source terms in the conservation equations include heat and momentum sources and sinks due to bounding walls, e.g., convection heat transfer from conductor surfaces and wall drag on the fluid.

Surface mass source terms include mass sources and sinks due to phase change at a non-fluid surface. For example, condensation heat transfer on conductors results in a mass sink for the steam and a mass source for the liquid.

The boundary source terms for the fluid energy equations include convection and radiation heat transfer from walls and the energy associated with any surface mass source terms. It is assumed that all wall heat transfer is between the walls and the liquid and vapor phases. There is no heat transfer directly between the walls and the drops.

For the conductors, the surface heat flux is either specified or calculated from known conductor and fluid temperatures and heat transfer coefficients. The boundary source term for the conductor energy equation includes sensible heat flux and latent heat associated with phase change on the conductor surface.

The boundary momentum source includes friction and form drag due to walls, orifices and obstructions.

Interface Source Terms

The interface source terms include the mass, energy and momentum transfer from one phase to another due to vaporization, condensation, drop entrainment and deposition and ice melting. They can be expressed as

$$\int_{A_I} S_\phi^c dA_I = \Gamma_\phi^+ - \Gamma_\phi^- + \gamma_\phi^+ - \gamma_\phi^- \quad (5)$$

$$\int_{A_I} S_\phi^e dA_I = \Gamma_\phi^+ h_{\phi_{sat}} - \Gamma_\phi^- h_\phi + \gamma_\phi^+ h_\phi^* - \gamma_\phi^- h_\phi + Q_\phi^L \quad (6)$$

$$\int_{A_I} S_\phi^m dA_I = \Gamma_\phi^+ \vec{U}_\phi^* - \Gamma_\phi^- \vec{U}_\phi + \gamma_\phi^+ \vec{U}_\phi^* - \gamma_\phi^- \vec{U}_\phi \quad (7)$$

where Γ_ϕ^+ and Γ_ϕ^- are the rate of mass addition to and removal from phase ϕ due to vaporization and condensation at the phase interfaces. γ_ϕ^+ and γ_ϕ^- are the rate of mass addition to and removal from phase ϕ due drop entrainment and deposition. The superscript * indicates that the value is determined by the phase contributing the added mass. Q_ϕ^L is the latent heat released or consumed in the phase change and is added or subtracted from the phase causing the phase change. For example, when superheated vapor causes vaporization at the liquid interface, the required latent heat is removed from the vapor phase.

The interface heat and mass transfer rates depend on the interfacial area and the interface heat and mass transfer coefficients. These, in turn, depend on the flow regime, fluid conditions and relative motions. GOTHIC contains a set of correlations for interface heat and mass transfer for the flow regimes relevant to containment analysis. The flow regime is selected based on the local phase fractions and the relative velocities. Details of the flow regime map and the interface correlations are provided in the GOTHIC Technical Manual [3].

Junction Equations

Junctions are used to hydraulically connect any two computational cells. The junctions may represent doorways, pipe or instrument penetrations, pipes, ductwork, and so forth. These hydraulic connections are in addition to the axial and lateral flow connections for 1-, 2- and 3-dimensional meshes that give the velocities at subvolume faces. They allow lumped parameter volumes to be interconnected or connected to subdivided volumes. For each junction, one-dimensional momentum equations for the vapor, droplets and liquid are solved. The junction momentum equations are consistent with those used for the subvolume face velocities except that the effects of viscous and turbulent shear are not included.

Figure 2 shows a junction between two volumes, labeled 1 and 2, with overflow liquid from the pool in volume 2 displacing the vapor in volume 1. Although the figure shows vapor and liquid residing in the junction, it is important to note that the mass of fluid in the junction is not accounted for in the overall mass balance for the system. For purposes of mass balancing, there is no mass stored in the junctions and mass taken out of one volume via a junction is immediately placed in the connecting volume. However, for the purpose of calculating the inertia of the junction flow, an approximate junction fluid mass is calculated.

The momentum equation solved for each phase is

$$AL\frac{d(\hat{\alpha}_\phi\hat{\rho}_\phi W_\phi)}{dt}=\hat{\alpha}_\phi(P_1+PL_{1\phi}-P_2-PL_{2\phi})-D_{w_\phi}-D_{i_\phi} \qquad (8)$$

$$+\hat{\alpha}_\phi\hat{\rho}_\phi g(el_1-el_2)+E^m_\phi+(MF_{1\phi}-MF_{2\phi})$$

where W_ϕ is the phase velocity and $\hat{\alpha}_\phi$ and $\hat{\rho}_\phi$ are the effective junction volume fraction and density. P_1 and P_2 are the static pressures at the vertical centers of volumes 1 and 2. PL_1 and PL_2 are the static heads between the cell center and the junction end elevations. MF_1 and MF_2 are the dynamic forces at the ends of the junctions. These momentum forces are included only if the junction connects two subdivided volumes. E^m is the momentum source due to equipment (pumps), and D_w and D_i are the wall and interface drag, respectively.

Junction Parameters and Terms

For the purpose of determining the donor phase fractions for flow out of a cell through a junction, the fluid in the computational cell is assumed to be in the form of a pool with drops and vapor above and, possibly, bubbles in the pool. This representation is used to calculate an effective pool height and pool vapor fraction from the cell geometric parameters and the current cell phase volume fractions. The pool vapor fraction is estimated using the Yeh correlation [4].

The parameter $\hat{\alpha}_\phi$ represents the volume fraction of phase ϕ in the junction for purposes of calculating the inertia, gravitational head and pressure forces. $\hat{\alpha}_\phi$ is intended to represent the volume fraction of phase ϕ in the junction if a mass balance was maintained for the junction. For the vapor phase, $\hat{\alpha}_v$ is controlled by

$$\frac{d\hat{\alpha}_v}{dt}=\frac{|W_v|}{L}(\alpha_v^*-\hat{\alpha}_v)-\frac{|W_l|}{L}(\alpha_l^*-\hat{\alpha}_l)-\frac{|W_e|}{L}(\alpha_d^*-\hat{\alpha}_d) \qquad (9)$$

where α_ϕ^* is the donor cell phase fraction and L is the length of the junction. The junction volume fractions for the liquid and drop phases are then calculated from

$$\hat{\alpha}_l=(1-\hat{\alpha}_v)(1-\eta) \quad and \quad \hat{\alpha}_e=(1-\hat{\alpha}_v)\eta \qquad (10,11)$$

where $\eta = \alpha_d/(\alpha_d+\alpha_l)$, averaged over the twc cells connected by the junction. The junction phase density, $\hat{\rho}_\phi$ is simply the average of the densities in the adjoined volumes. Equations 9 through 11 are formulated so that the junction phase fractions tend toward the upstream cell phase fractions.

The mass flow through the junction is

$$F_\phi = A\,W_\phi\bar{\rho}_\phi\alpha_\phi^* \qquad (12)$$

α_ϕ^* is determined by the direction of the phase velocity and the location of the pool surface relative to the junction end elevation and spanning height. If a junction end is completely submerged, flow into the junction through that end will be all liquid. If a junction end is completely above the pool surface, flow into the junction through that end will be a mixture of vapor and drops as determined by the drop fraction in the vapor region of the volume providing the flow. If the pool surface level falls within the region defined by the junction elevation and the junction end height, then vapor, drops and liquid will all be allowed to flow through the junction and the magnitude of each will be determined by the phase momentum balances.

The momentum flux terms in the junction momentum equation are included so that junctions can be used to connect compartments with 1-, 2- or 3-dimensional meshes while maintaining consistency with the momentum equations for the subdivided mesh. In addition, momentum transverse to the direction of flow through the junction can be transported from one subdivided volume to another.

Critical Flow Model

The flow through a junction may be limited by that calculated using a critical flow model. The model used in GOTHIC is based on analytic and semi-analytic models for water and steam mixtures, modified to account for the presence of noncondensable gases. Three different models are used, depending on the upstream fluid conditions: the Henry [5] model for the subcooled region, Moody [6] in the two-phase region and the perfect gas model for single-phase vapor. The Henry and Moody models are applicable to water and steam mixtures, but their use has been extended to include flows with noncondensable gases. To account for the noncondensable gases, an equivalent mixture enthalpy is defined as

$$h_{eq}=\min[h_f,h_m]+x_f h_{fg} \qquad (13)$$

where h_m is the thermodynamic mixture enthalpy in the upstream volume and x_f is the flowing quality. A two phase mixture in thermal equilibrium with an enthalpy of h_{eq} would have approximately the same vapor volume fraction as the actual steam/gas/water mixture and it is assumed that the two systems would have similar critical flow behavior.

ASSESSMENT

GOTHIC has been assessed against a wide variety of analytic solutions and experimental data. Simple single volume thermodynamic and heat transfer problems provide a firm basis for the assessment of the conservation and state equations. Experimental tests assess the constitutive models and the overall operation of the code. The assessment program includes water and steam blowdown tests in HDR [7], Battelle Model Containment [8], CVTR [9], LACE [10] and the Marviken [11] test facility. Tests from HDR, Battelle Model Containment and the HEDL [12] test facility have been simulated to assess capabilities for modeling noncondensable gas dispersion.

Three tests that are representative of the assessment effort were selected for presentation in this paper. These simulations were all completed with the benefit of having the test results in hand. The initial comparisons led to improvements in the code modeling capabilities and some minor changes in the input models.

Thermal-Hydraulic Response in LACE

The Light Water Reactor Aerosol Containment Experiments (LACE) investigated aerosol behavior for certain postulated accident situations. LACE test LA5 simulated containment failure and blowdown following a 500 minute heatup and pressurization period. Although the focus of the LACE test series was to study aerosol behavior under various containment conditions, thermal-hydraulic data were also recorded during the tests that could be used for benchmarking thermal-hydraulic codes.

Test Description. Figure 3 shows the important features of the facility. The vessel is an insulated, carbon steel cylinder with a total volume of 852 m^3. The containment was initially full of air at 1 atm and 25°C. An internal tank, half full of water, was included in the vessel to simulate a flashing pool. There was also about 1 m^3 of water in the bottom head.

There were three primary heat sources to the vessel; direct steam injected through a pipe standing vertically alongside the internal tank, submerged steam-fed heating coils at the bottom of the internal tank, and steam and nitrogen fed through the aerosol delivery line. In addition to these sources, the combined effect of lights and cooling for instrumentation resulted in a small amount of net cooling that was significant relative to the difference between the mass/energy sources and the heat loss from the outer surface of the vessel.

The containment atmosphere was heated to a temperature of 124°C and a pressure of 4.4 bar prior to blowdown. The water temperature in the internal tank was maintained near the boiling point during the heating period to maximize flashing during blowdown. Immediately prior to blowdown the temperature of the liquid in the internal tank was 145°C. After these desired conditions were established, the discharge pipe was opened and rapid venting of the vessel occurred. Concurrently, water in the internal tank flashed to steam or was simply ejected from the tank as liquid.

GOTHIC Model. LA5 was simulated using a lumped parameter model. This was consistent with the level of detail in the test reports and proved to be adequate for predicting the thermal-hydraulic response. The GOTHIC model consists of three lumped parameter volumes; one to model the internal tank, one for the bottom head of the vessel and one for the upper portion of the vessel. A separate computational volume was used for the bottom head to accommodate the large differences in the water and atmosphere conditions in the sump and upper containment regions and for better modeling of the pool surface area and pool depth in the bottom head.

Conductors were used to model the internal tank and the vessel walls and heads. The measured temperature of the outer surface of the insulation provided a thermal boundary condition for the vessel conductors.

A cooler was placed in the upper vessel volume to represent net heat removal by the instrumentation coolers and the lights. A heater was placed in the internal tank to model the steam coil.

GOTHIC Data Comparisons. Figure 4 shows the vessel temperature and pressure response during the heatup period. The sump temperature and volume are shown in Figure 5. The pressure and temperature response during the blowdown phase are shown in Figure 6. GOTHIC did an excellent job predicting the transient with this very simple model. The close match with the pressure and temperature during the heatup phase requires an accurate prediction of the heat leaving the vessel through the walls and heads. The condensation heat transfer, predicted by the Gido-Koestel correlation [13], had to be correct to accurately predicted the collection of water in the sump. The sump temperature is somewhat overpredicted early in the heatup. A more detailed subdivided model would be necessary to correctly predict the cooling of the condensate as it runs down the wall into the sump. The temperature and pressure during the blowdown phase agree very well with the data and provides a good verification of the critical flow model in the perfect gas regime.

Hydrogen Distribution in BFMC

During and after a loss-of-coolant accident in a light water reactor, hydrogen gas may form from chemical reactions and radiolytic decomposition of water. Flammable concentrations of hydrogen are of particular concern. The Battelle-Frankfurt model containment (BFMC) test program was designed to study transport of hydrogen by convection and diffusion through an air-filled containment. BFMC Test 20 [14] is among several hydrogen distribution tests selected to assess the ability of GOTHIC to accurately predict transport of noncondensable gases.

Test Description. The test geometry for BFMC Test 20 is shown in Figure 7. There was a total of $234\,m^3$ in six compartments.

The initial containment atmosphere was air at a pressure of 1 atm. Rooms on the upper level, rooms R2, R5, and R7, were preheated to 37.0, 39.5 and 38.0°C, respectively. Initial temperatures for the lower level, rooms R1, R6, and R8, were 18.0, 19.0 and 17.0°C, respectively. A gas mixture of 67% hydrogen and 33% nitrogen by volume was injected at -3°C in room R6 for five hours. The injection velocity was very low to avoid forced mixing.

GOTHIC Model. The lower half of Figure 7 shows a schematic of the noding used for this test. In this figure, the annular rooms are unwrapped and the connections to the inner cylinder, modeled with junctions, are indicated by the solid lines. Rooms 5, 6, 7 and 8 were each subdivided into two-dimensional 2h by 3w grids and R1 was subdivided into a 5h by 3w grid. A lumped parameter volume was used to model room 2. The elevations of horizontal grid lines within individual compartments were specifically set so that they fell on planes that cut through the entire containment. This is important for modeling stratification in buoyancy dominated flows because the grid lines must coincide with the constant density lines in stratified atmospheres. Other noding arrangements can cause numerically induced flows that dominate the solution. The subdivided model was necessary for this test because the initial temperature distribution results in significant stratification and circumferential variation in the hydrogen concentration.

Conductors are used to model concrete and steel surfaces in each volume. The conductors were assigned initial temperatures corresponding to the preheat conditions. A turbulent natural convection heat transfer coefficient was used.

GOTHIC Data Comparisons. Comparison with test data for the six rooms are shown in Figures 8, 9 and 10. Except for the concentration in room 6, GOTHIC compares very well with the data. Room 6 is where the hydrogen was injected and it is likely that the data for that room reflects some locally high hydrogen concentration in the vicinity of the source since the hydrogen concentration is not correspondingly overpredicted elsewhere in the containment. Vertical stratification of hydrogen and the delay for hydrogen penetration into the upper rooms is well predicted.

Variations on the base model for this test showed that the distribution of hydrogen was extremely sensitive to changes as small as 2°C in the initial temperature of the walls and structures.

Steam/Water Blowdown in Marviken

The Marviken power plant was built as a boiling water reactor and was completed up to the commissioning tests, but was never loaded with fuel. The reactor pressure vessel and containment system were subsequently used in an experimental role [11][15][16]. GOTHIC has been used to simulate a number of tests performed at the site. These simulations demonstrate the ability of GOTHIC to predict a full scale steam/water blowdown from a pressure vessel into an actual containment geometry. Specifically, the simulations test the overall capability and adequacy of GOTHIC models to predict condensation heat transfer on walls, and mass and energy transport within and between multi-compartment geometries and the special models used to predict bubble condensation and heat transfer in suppression pools.

Test Description. The wetwell is Marviken room 105. All rooms other than the pressure vessel and the wetwell are collectively referred to as the drywell.

The reactor vessel, initially at 295°C,

48 atm, provided the blowdown source to room 122. Thermal-hydraulic communication throughout the drywell was maximized by securing valves and doors in their fully open positions whenever possible.

The drywell includes the upper dome, identified as room 124 in Figure 11. Several drywell rooms are defined between elevations 102.5 and 115 m. During a blowdown, a great deal of water collects on the floor at 102.5 m. Steam that does not condense in the drywell flows to the wetwell through pipes connecting rooms 104 and 106. This connection consists of 2 parallel pairs of 1.2-m-diameter pipes referred to as 'blowdown channels'. The vent pipe header is connected to the wetwell through 58 nearly identical vertical vent pipes, 5.3 meters long, 0.3 meters in diameter and are submerged 2.8 meters in the wetwell.

The initial containment pressure for each test is about 1 atm. Initial temperatures range from about 15°C in the wetwell and lower rooms of the drywell, to above 60°C in the annular region around the pressure vessel. The test was terminated within 3 minutes following the primary break.

GOTHIC Model. The lumped parameter noding for the containment is shown in Figure 11. This is a typical schematic noding diagram constructed with the GOTHIC interactive graphical preprocessor. Volumes are indicated by the dashed rectangles, junctions marked by the solid lines and conductors by the small rectangles. Hydraulic and thermal connection data are automatically defined as the noding diagram is constructed.

The GOTHIC model consists of 21 lumped parameter control volumes. A flow boundary condition, marked 1F, to model the break is shown connected to volume 8, the break room. Volume 1 is the upper dome. Volume 16 models the blowdown channels, volume 18 models the vent pipe header, and volume 17 models the vent pipes. Volume 19 represents the wetwell. Heat transfer surfaces within the containment include concrete, steel lined concrete and steel components of many different thicknesses.

GOTHIC Data Comparisons. Results for test 17 are presented in Figures 12 through 14. Figure 12 shows the break room and wetwell pressure response and the wetwell to drywell differential pressure. The initial spike observed in the differential pressure data and predicted by GOTHIC is due to the clearing of water from the vent pipes.

Figure 13 shows temperatures in the break room and upper dome. The break room temperatures are slightly over predicted and the dome temperature is closely matched. Figure 14 shows the temperature response of the vapor and liquid and the wetwell. As indicated by the data, there is significant thermal stratification in both the vapor and liquid regions. The simple lumped volume model used for the wetwell allows only one temperature for the vapor and one for the liquid. The predicted vapor temperature represents very well the measured average vapor temperature. GOTHIC overpredicts the average liquid temperature in the wetwell. Apparently, GOTHIC predicts less condensation and water retention in the drywell than occurred in the experiment. The superheating of the liquid relative to the wetwell vapor is well predicted by GOTHIC.

CONCLUSIONS

The separated three phase model, together with the wall and interface constitutive relations, make GOTHIC applicable to a wide range of containment analysis problems. Combined lumped parameter and multi-dimensional noding allow flexible and effective input models to be developed. With the interactive pre- and post-processing capabilities available with the GOTHIC code package, creation, execution and review of the models is done quickly and accurately. The selected set of comparisons with experimental data are a small part of the qualification effort put into GOTHIC, but represent the accuracy that can be expected when the required input is well defined.

LITERATURE CITED

1. Wheeler, C.L., et al., "COBRA-NC:A Thermal Hydraulic Code for Transient Analysis of Nuclear Reactor Components", Volumes 1,2,4 and 7, NUREG/CR-3262, PNL-5515, March 1986.

2. George, T.L. and M.J. Thurgood, "CAP, Containment Analysis Package User's Manual", Numerical Applications, Inc., 1987.

3. George, T.L., et al., "GOTHIC Containment Analysis Package: Technical Manual", To be published, Numerical Applications, Inc.

4. Cunningham, J.P. and H.C. Yeh, ANS Trans., Vol. 17, 369-370, 1973.

5. Henry, R.E., Nuc. Sci. and Eng., 41, 1970 p. 336.

6. Moody, F.J, J. of Heat Transfer, Trans ASME, 87, No. 1, February 1965.

7. "Design Report for the HDR Containment Experiments V21.1 to V21.3 and V42 to V44 with Specifications for the Pre-Test Computations." Report No. 3.280/82. January, 1982.

8. "Investigation of the Phenomena Occurring within a Multi-Compartment Containment after Rupture of the Primary Cooling Circuit in Water-Cooled Reactors". Technischer Bericht BF RS 50-30-D1. Battelle-Institut e.V., 6000 Frankfurt am Main 90, Am Römerhof 35, Postfach 900160. March, 1977.

9. Schmitt, R.C., et al., 1970, "Simulated Design Basis Accident Tests of the Carolinas Virginia Tube Reactor Containment Final Report," IN-1403, Idaho Nuclear Corporation, Idaho Falls, Idaho.

10. Dickinson, D.R., et al., 1988, "Final Report of Experimental Results of LACE Tests LA5 and LA6 - Rapid Containment Depressurization," LACE TR-026, Westinghouse Hanford Company, Richland, Washington.

11. "The Marviken Full Scale Containment Response Tests, Second Series, Interim Report of Summary Report". AB Atomenergi Sweden. MXB-301. March, 1977.

12. Bloom, G.R., et al., 1983, "Hydrogen Mixing and Distribution in Containment Atmospheres," EPRI NP-2669, Electric Power Research Institute, Palo Alto, California.

13. Gido, R.G. and A. Koestel, "Containment Condensing Heat Transfer", Second International Topical Meeting on Nuclear Reactor Thermal Hydraulics, Santa Barbara, California, January 1983.

14. Langer, G., et al., "Experimental Investigation of the Hydrogen Distribution in a Model Containment (Preliminary Experiments II)". BF-R-64.036-3. Battelle-Institut e.V., 6000 Frankfurt am Main 90, Am Römerhof 35, Postfach 900160. March, 1977. (German with English summary)

15. "The Marviken Full Scale Containment Experiment, Second Series, Description of the Test Facility". AB Atomenergi Sweden. MXB-101. March, 1977.

16. "The Marviken Full Scale Containment Response Tests, Second Series, Interim Report of Blowdown 17 Test Results". AB Atomenergi Sweden. MXB-217. April, 1976.

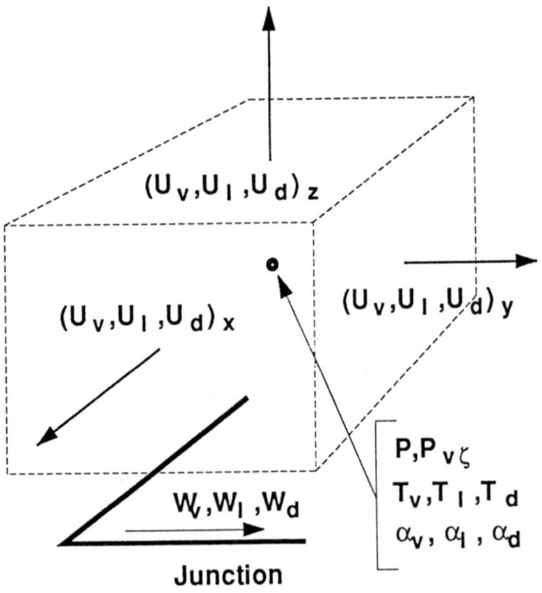

Figure 1. System control volume and variables.

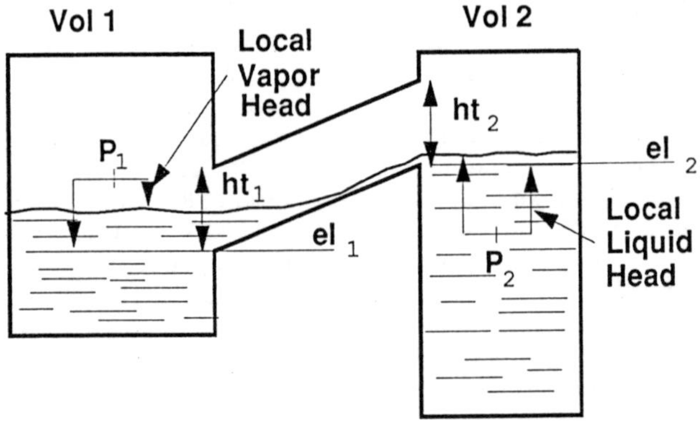

Figure 2. Junction parameters.

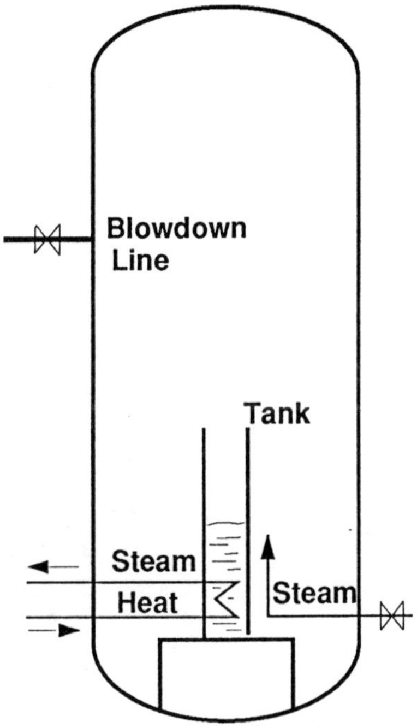

Figure 3. Containment configuration for LACE test LA5.

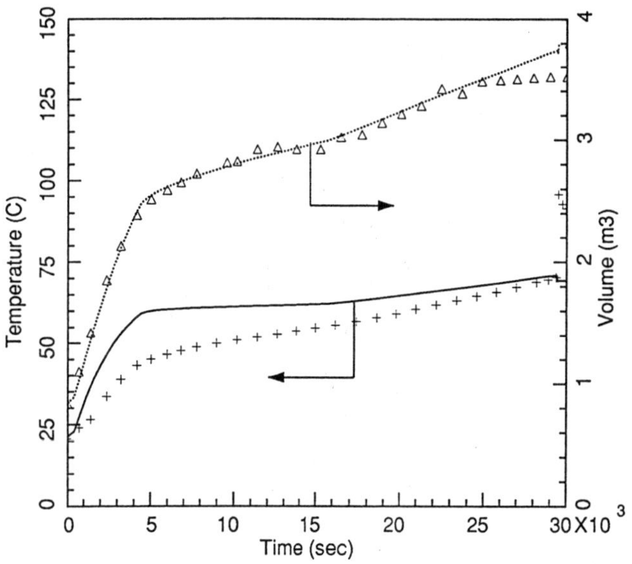

Figure 5. Sump water temperature and volume; LACE LA5.

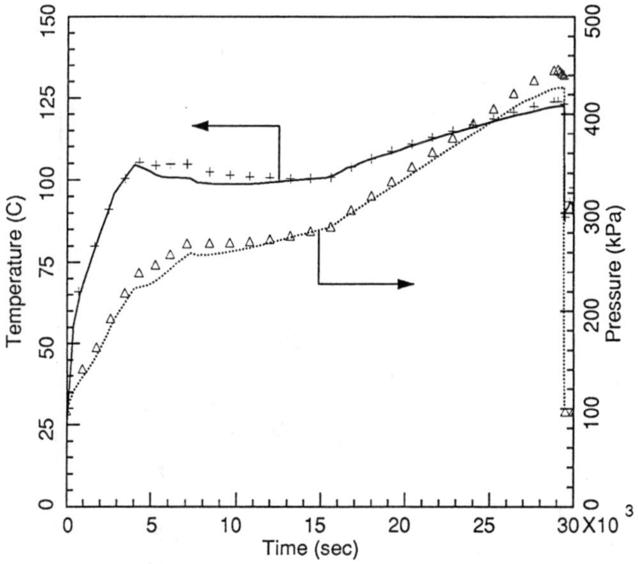

Figure 4. Vessel heatup temperature and pressure; LACE LA5.

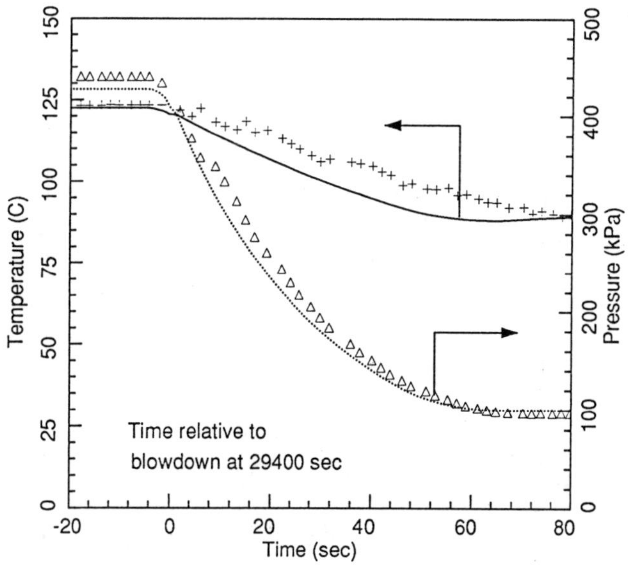

Figure 6. Vessel blowdown temperature and pressure; LACE LA5.

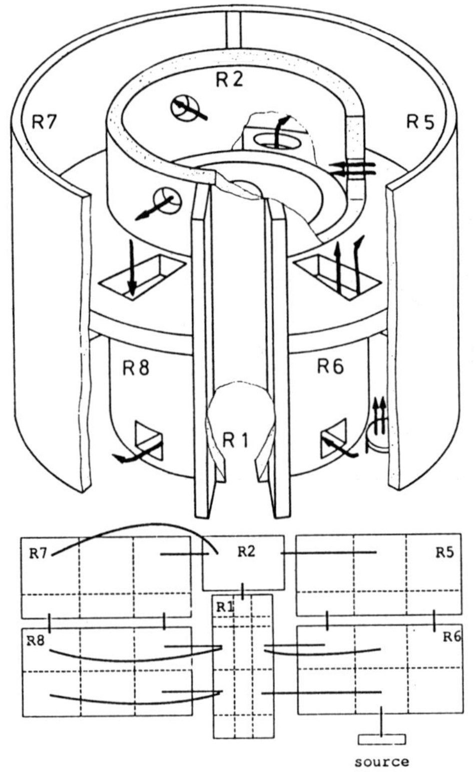

Figure 7. Battelle Frankfurt model containment and GOTHIC model.

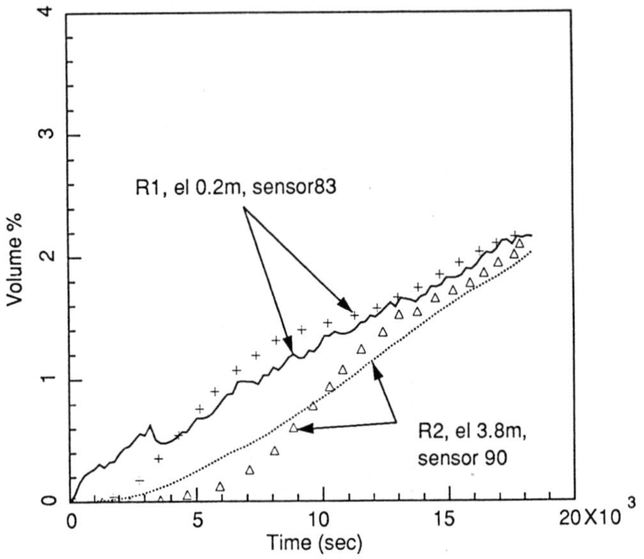

Figure 9. Hydrogen concentration in rooms R1 and R2; BFMC Test 20.

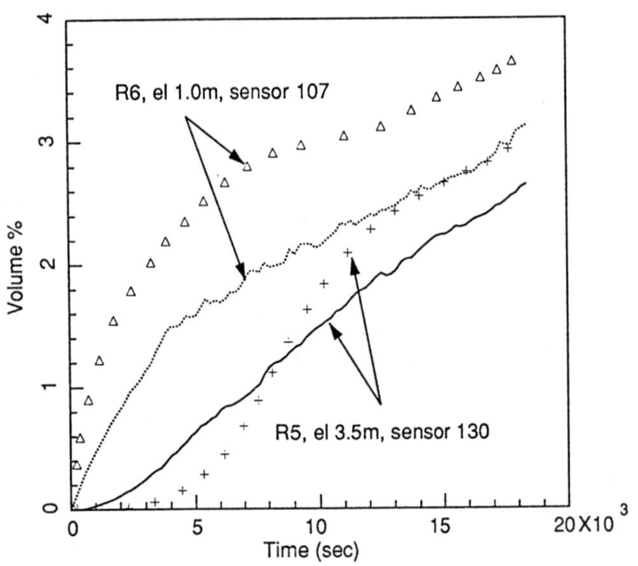

Figure 8. Hydrogen concentration in rooms R5 and R6; BFMC Test 20.

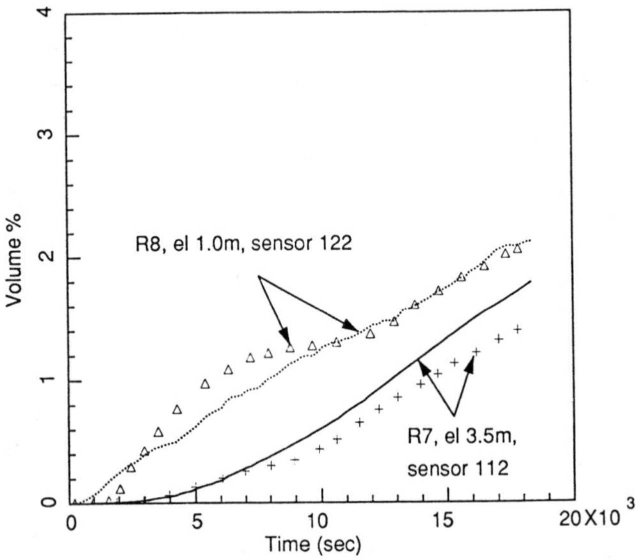

Figure 10. Hydrogen concentration in rooms R7 and R8; BFMC Test 20.

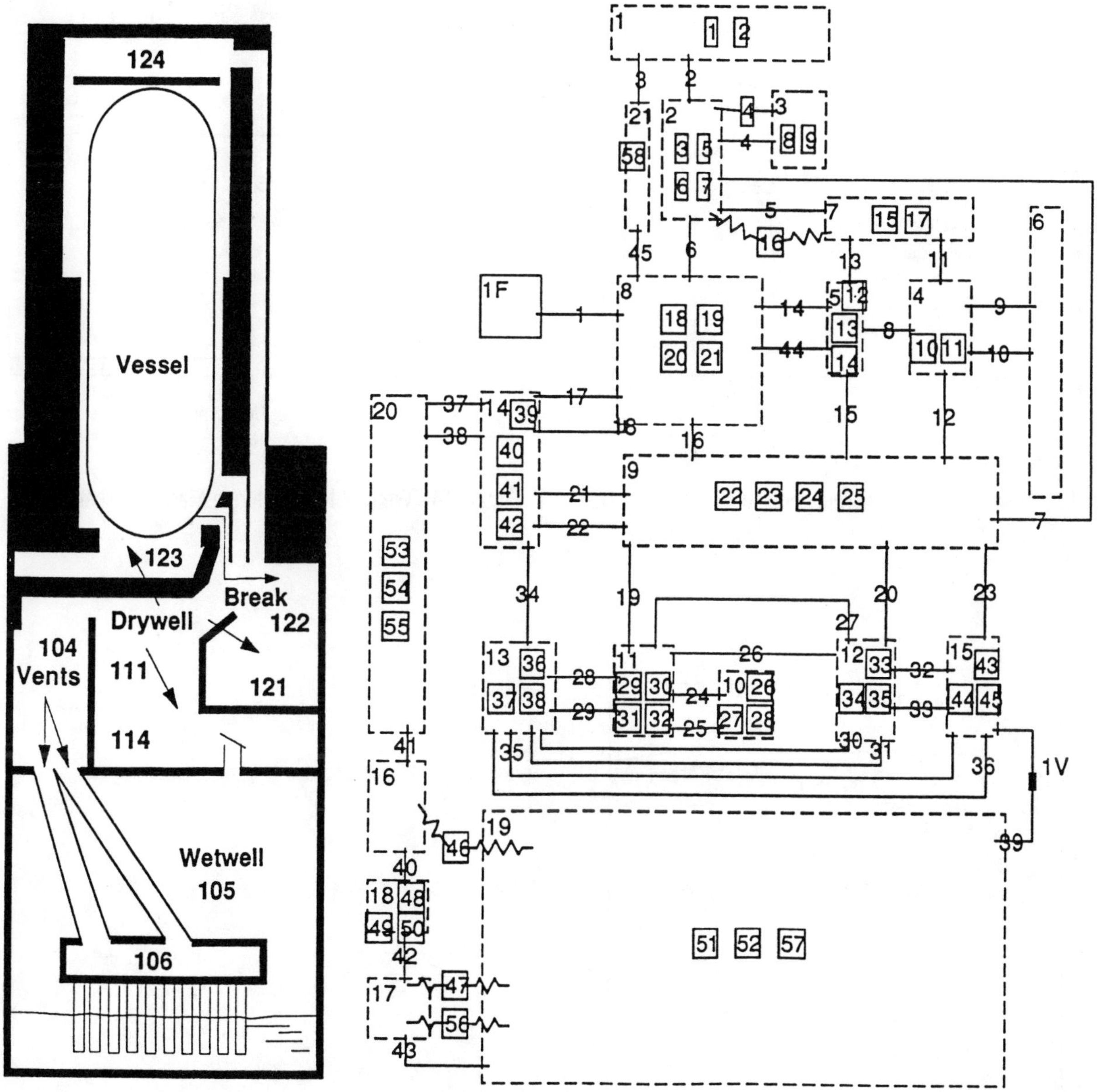

Figure 11. Marviken containment and GOTHIC model, Test 17.

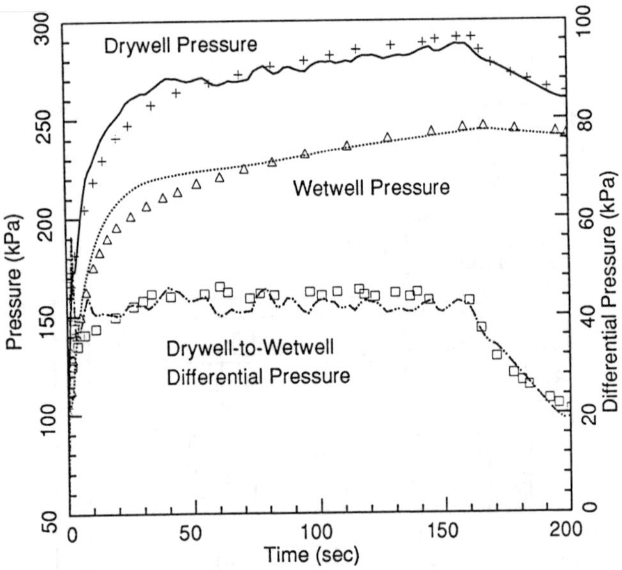

Figure 12. Break room and wetwell pressure; Marviken Test 17.

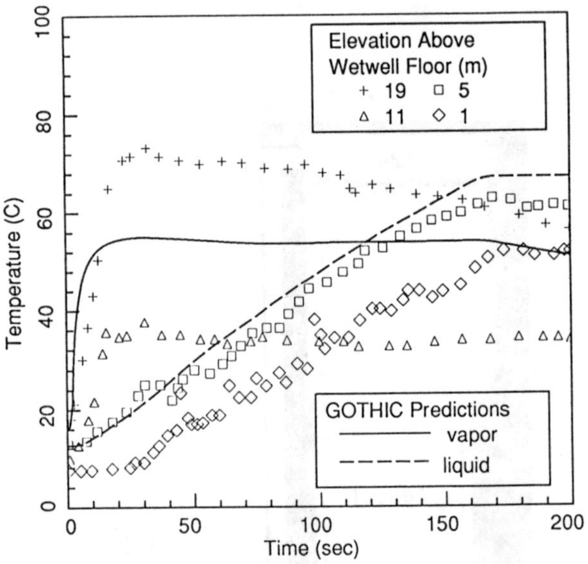

Figure 14. Wetwell temperatures; Marviken Test 17.

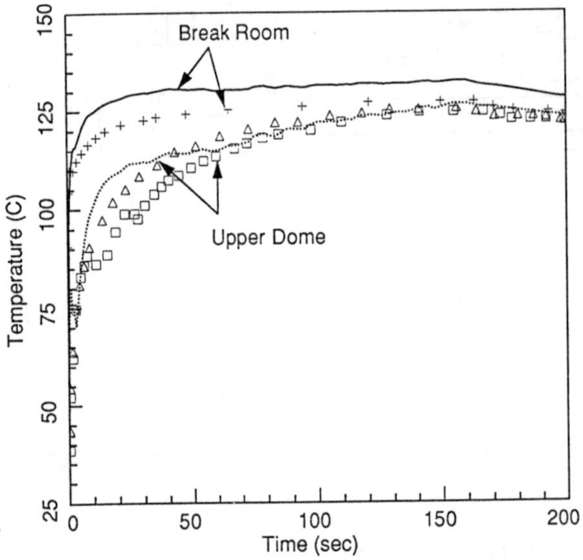

Figure 13. Break room and upper dome temperature; Marviken Test 17.

MAAP PREDICTION OF THE HDR T31.5 CONTAINMENT EXPERIMENT

Chan Y. Paik, Sung Jin Lee, and Martin G. Plys ■ Fauske & Associates, Inc., 16W070 West 83rd Street, Burr Ridge, IL 60521

The Modular Accident Analysis Program (MAAP) was used to predict the containment response to a simulated large LOCA and hydrogen distribution test (Test T31.5) in the decommissioned German Heiss Dampf Reaktor (HDR) facility. In MAAP, 22 nodes and 66 flow paths were used to model the highly compartmentalized containment. Each node and two types of heat sinks. Wall, ceilings, and floors were modeled as on-dimensional slabs with temperature distribution along their thickness. Metal equipment was modeled as separate lumped parameter heat sinks with single temperatures. Both convective and condensation heat transfer between the gas and the heat sinks were considered. To calculate gas flow rates between compartments, the model considers pressure driven unidirectional flows and buoyancy-driven counter-current flows. The MAAP predictions of the containment pressure, temperatures, and hydrogen/helium concentration distribution are compared here with experimental results.

INTRODUCTION

During a severe accident in a nuclear reactor, hydrogen gas may be generated by oxidation of zirconium after prolonged fuel uncovery. Steam and hydrogen are released to the containment at various times with rates that are dependent on the scenario in question. The subsequent distribution of these gases in the containment determines the potential for flammability, and is therefore an important safety consideration for nuclear reactors.

Hydrogen mixing experiments conducted in the decommissioned HDR nuclear reactor in Germany provide large scale experimental data against which we can test computer models. One such experiment, designated as T31.5, simulated a large LOCA steam blowdown into the containment followed by the injection of a hydrogen/helium gas mixture. In this paper, MAAP (Modular Accident Analysis Program) was used to simulate the T31.5 experiment and its results are compared with experimental data. This is a post-test comparison.

HDR FACILITY

As shown in Figure 1, HDR is a 50 m high and 20 m diameter cylindrical containment with a hemispherical dome. The containment is characterized by relatively large dome space, which constitutes 5,000 m³ of 11,000 m³ total free space inside the containment, and highly compartmentalized rooms. There are about 70 rooms interconnected by numerous flow junctions. Nevertheless, a dominant flow path (not visible in Figure 1) can be identified by a pair of staircases and equipment shafts running from low elevations up to the dome at diametrically opposite positions.

TEST T31.5

In Test T31.5, a large LOCA was simulated by blowdown of the HDR pressure vessel (30,000 kg of water and steam mixture in 50 seconds) into Room 1704, which is situated at an elevation of 14.25 m with respect to the ground level, followed by a superheated steam release (2.2 kg/s between roughly 21 min and 36 min after the beginning of the blowdown). A hydrogen/helium gas mixture was released between 36 and 48 minutes after the beginning of the blowdown at a rate of 0.24 kg/s. Figure 2 shows the blowdown mass flow rate. The flow rate quickly rose to 2,300 kg/s in less than 3 seconds and then gradually decreased to about 400 kg/s in 15 seconds. The late steam and hydrogen/helium gas mixture injection flow rate is shown in Figure 3.

MAAP MODELS

In MAAP, mass and energy are conserved in each node. In particular, MAAP tracks the individual mass of water, suspended water, steam, H_2, O_2, N_2, CO, and CO_2 gas. There is one energy equation for water and one for gas (which includes the suspended water). In each time step, temperature and pressure in each node are determined from mass and energy. Then, mass and heat transfer rates within each node and between nodes through junctions are calculated. These rates are integrated over the time step to update mass and energy.

Heat Transfer

Heat sinks in each node are placed in two categories. Concrete walls with relatively low thermal conductivity are modeled as distributed-parameter heat sinks using a finite difference solution method for the temperature profile. Metal internals with high thermal conductivity, on the other hand, are modeled as lumped masses with a single temperature for each mass.

Radiation heat transfer, natural convection heat transfer, and condensation heat transfer between heat sinks and gas are considered. Newton's method is used to find the condensate-film/gas interface temperature at which the gas-to-condensate and the condensate-to-wall heat transfer rates become equal. That is, at the condensate film outside temperature T_{FO},

$$Q_{AIR} = Q_{FILM} \tag{1}$$

where the heat transfer rates are given by:

$$\frac{Q_{AIR}}{A} = \left(H_{CONV} + H_{RAD}\right)\left(T_G - T_{FO}\right) + H_m M_w \left[PP_{STG} - P_{SAT}\left(T_{FO}\right)\right]\left(h_{STG} - h_{WCD}\right) \tag{2}$$

and

$$\frac{Q_{FILM}}{A} = H_{CD}\left(T_{FO} - T_{SW}\right) \tag{3}$$

H_{CONV}, H_{RAD}, and H_{CD} are convective, radiative and condensate-to-wall heat transfer coefficients. For the convective heat transfer, only natural convection heat transfer is considered. H_m is the mass transfer coefficient for diffusion of steam through non-condensible gas and M_w is the molecular weight of steam. h_{STG} and h_{WCD} are the enthalpies of steam and of condensate. PP_{SAT} (T_{FO}) is the saturation pressure of steam at the interface temperature. T_G, T_{FO}, and T_{SW} are the temperatures of the bulk gas, the interface, and the wall, respectively. The mass transfer coefficient is evaluated using an analogy between heat and mass transfer. That is,

$$H_m = \frac{H_{CONV}}{c_{pg} P_{am} M_w} \left(\frac{Pr}{Sc}\right)^{0.66} \tag{4}$$

The convective heat transfer coefficient H_{CONV} is given by,

$$\frac{H_{CONV} L}{k_g} = a\ (Gr\ Pr)^n \tag{5}$$

where the Grashof, Prandtl, and Schmidt numbers are respectively defined as

$$Gr = \frac{\rho_g^2\ g\ \left|T_G - T_{FO}\right|\ L^3}{T_G\ \mu_g^2} \tag{5a}$$

$$Pr = \frac{\mu_g\ c_{pg}}{k_g} \tag{5b}$$

$$Sc = \frac{\mu_g}{\rho_g D} \tag{5c}$$

with L as the wall height and P_{am} the log mean pressure difference between the partial pressure of steam in the bulk and at the interface:

$$P_{am} = \frac{PP_{STG} - P_{SAT}\ (T_{FO})}{\ln \dfrac{P - P_{sat}\ (T_{FO})}{P - PP_{STG}}} \tag{5d}$$

The constants "a" and "n" in Equation (5) are given values ranging 0.021 to 0.555 for "a" and 0.25 to 0.4 for "n" depending on the orientation of the heat transfer area [2].

Given the wall temperature T_{SW} and the gas bulk temperature T_G, the condensate interface temperature T_{FO} is computed iteratively to satisfy Equation (1). From the condensation and convective heat transfer rate, the energy of the condensate is subtracted and the remaining energy is assumed to be transferred to the wall. The average temperature of the film is given by

$$T_{film} = T_{SW} + 0.25\ \left(T_{FO} - T_{SW}\right) \tag{5e}$$

MAAP does not keep track of mass and energy of condensate formed on the surface of cold walls; all condensate formed is moved immediately to water pools in the node and a quasi-steady Nusselt film thickness is assumed.

Fluid Flow

Two types of flow through junctions are considered in MAAP. First, Bernoulli flows are calculated at each junction based on the pressure difference across the junction. The pressure in each node is defined as the pressure at the bottom of the floor. The differential pressure which drives the flow through each junction is then evaluated by considering the node pressure, hydrostatic pressure due to elevation of the junction with respect to the floor, and hydrostatic pressure inside the vertical junction as shown in Figure 4. Therefore, the Bernoulli flow through a junction j is determined by solving:

$$\left(P_2 - \rho_{g_2} g H_{j_2}\right) - \left(P_1 - \rho_{g_1} g H_{j_1}\right) + \rho_{g_2} g L_j = K \frac{W_u^2}{2 \rho_{g_2} A^2} \quad (6)$$

where:

P_1, P_2 = nodal pressure for receiver and donor nodes, respectively,

ρ_{g_1}, ρ_{g_2} = gas densities in the receiver and donor nodes,

H_{j_1}, H_{j_2} = the height of the junction with respect to the bottom of the floor,

L_j = the height of the vertical junction (should be zero for horizontal junctions),

K = the local loss coefficient = 1.78 (equivalent to discharge coefficient C_D =0.75),

W_u = the unidirectional Bernoulli flow rate,

A = the flow area of the junction.

Moreover, because flow affects pressure substantially, an implicit scheme is used where end of time step values are used for the nodal pressures in Equation (6). Bernoulli equations are solved simultaneously for all junctions.

Another type of flow considered in MAAP is a buoyancy driven counter-current flow through a single junction. In this case, there is no net volumetric flow through the junction, but two nodes connected by a single junction exchange gas if heavier gas is located above lighter gas. This exchange flow is responsible for distributing gas into dead-end nodes. MAAP uses Epstein's correlation [3] which is in the form of the Froude number versus the length-to-diameter ratio of the opening:

$$\frac{Q'_{cc}}{(D^5 g \, \Delta\rho/\bar{\rho})^{1/2}} = \frac{0.055 \, (B_1)^{1/6}}{\left[1 + 0.00527 \, (B_1)^{1/2} \, (B_2)^{3/4}\right]^{1/3}} \quad (7)$$

where:

Q'_{cc} = volumetric counter-current flow rate,

D = diameter of the junction,

L = length of the junction,

$\Delta\rho$ = gas density difference between the two nodes,

$$B_1 = [1 + 400 \, (L/D)^3] \quad (7a)$$

$$B_2 = [(L/D)^6 + 117 \, (L/D)^2] \quad (7b)$$

However, one-way forced flow can reduce or prevent the counter-current flow. Hence, the flooding volumetric flow rate Q_P is first evaluated:

$$Q_p = 0.75 \, C_D \sqrt{D^5 \, g \, \Delta\rho/\bar{\rho}} \quad (8)$$

If the unidirectional flow rate is greater than the critical value above, then the exchange flow is suppressed completely. Otherwise, the effective exchange flow is reduced via the following correlation [3]:

$$Q_{cc} = Q'_{cc} \left(1 - \left|\frac{Q_u}{Q_p}\right|\right)^{1.6} \quad (9)$$

where $Q_u \triangleq W_u/\rho_g$ is the volumetric unidirectional flow rate. Note that separate correlations are used for horizontal junctions (flow in a vertical direction) and vertical junctions (flow in a horizontal direction) and (7) through (9) are for horizontal junctions.

Nodalization

To simulate test T31.5 , the HDR facility was modeled using 22 nodes as shown in Figure 5. Rooms near the break location are modeled as separate nodes, and regions further away from the break are lumped together. The entire dome region was modeled as one node and the lower region (1400 level and below) is also modeled as one node. To connect 22 nodes, 66 flow paths (junctions) were defined. The air gap between the steel shell and outside concrete wall was not modeled, and the outside of the steel shell is modeled as an environment.

Input

Since the initial blowdown is two-phase critical flow mixture, MAAP assumed the mixture to be in thermal equilibrium at the blowdown pipe exit and after the expansion to the containment. Assuming adiabatic expansion, new quality of the two-phase mixture was calculated based on the containment pressure. Then, the steam and water blowdown rates were added to mass and energy rate equations for gas and water assuming instantaneous fall out respectively in the break node. Water carry-over by steam is not modeled. For the simulation of T31.5, it was assumed that every node except the bottom node (which has a sump) has an ideal drain system in the floor such that as soon as liquid appears in any node, it is drained to the sump node instantaneously.

RESULTS AND DISCUSSION

MAAP predictions of the containment pressure, temperatures, and hydrogen/helium concentration distribution were compared with experimental results. Figures 6 and 7 show short term and long term pressure comparisons. MAAP over-predicted the short term and peak containment pressures by 10% (about 0.25 bar). The predicted peak pressure occurred at about 23 seconds, from the beginning of the blowdown, which is similar to the data. During the initial 23 seconds, MAAP predicted a higher pressurization rate than the data. Pressure overprediction may be attributed to MAAP not considering forced convection heat transfer. However, the depressurization rate after the peak pressure was similar to the data. The over-prediction was reduced to 2% for long transient. In general, the predicted trend was similar to that of the data.

The predicted gas temperatures in different elevations are compared to data as shown in Figures 8 and 9. In the dome area, the MAAP prediction falls between the measured temperatures. Also, the peak temperature in the dome area was predicted quite well. However, for regions below the dome, the results showed that MAAP, in general, over-predicted the temperature. The source room gas temperature was noticeably over-predicted. Figure 9 shows a comparison of predicted gas temperatures and the data below the break/injection location along the staircase. It can be seen that there is a temperature gradient along the staircase. MAAP predicted a steeper temperature gradient.

During the experiment, a strong natural convection flow was reported inside the annular gap between the steel shell and the outer concrete wall. MAAP potentially could have under-predicted the heat transfer to the steel shell because this was neglected. Below the dome, the steel shell is normally covered by insulation and an inner concrete wall. However, during our inspection of the HDR facility we observed that insulation had been degraded so that there was an interior air gap of about 1/4" between the steel shell and the concrete through which the steel shell was exposed to condensing steam. The interior air gap was not modelled and thus MAAP might under-predict the heat transfer to the steel shell.

The hydrogen/helium injection started at about 36 min. and lasted for 12 min. Figure 10 shows the comparison of predicted hydrogen concentrations with the data at two different axial levels, dome area and room 1604 which is below the source room. The hydrogen concentration in the dome was well predicted. However, for the hydrogen concentration in the staircase below the injection, MAAP overestimated the build-up of hydrogen in the early part of transient, although the steady-state concentration was closely predicted. In general, MAAP predicted slower buildup of hydrogen for the rooms above the injection and faster buildup for the regions below the injection compared to data. This indicates that MAAP predicted too much mixing. MAAP predicted a complete mixing at about five hours from the beginning of the blowdown, but data showed some stratification even at 20 hours.

In this simulation, MAAP predicted relatively small counter-current flows - less than 1 kg/sec - and the effect of the counter-current flows to the overall prediction seems to be small. For this large break blowdown test, most counter-current flows start long after all injections are over. As shown in Figure 11, in the earlier stage large unidirectional purge flows effectively suppressed counter-current flows. Nevertheless, counter-current flows should play a crucial role in distributing gases into dead-end nodes. Also, they will be more important in longer and slower transients such as in small LOCA simulations.

CONCLUSIONS

1. MAAP predicted the pressure accurately.

2. MAAP substantially overpredicted the temperature in the source room, indicating a non-homogeneous condition in the source room. MAAP uses lumped parameter models in each node.

3. In general, MAAP did well in predicting temperatures above the break location. However, for lower nodes where natural circulation plays an important role in distributing hot gas into the lower part of the containment, MAAP predicted much steeper temperature gradients.

4. MAAP predicted better mixing than indicated by data. As a consequence, the hydrogen concentration in the lower node rose more rapidly than indicated by data. Also, MAAP predicted complete homogenization after about 18000 sec. Data indicates stratification of hydrogen persisting up to the end of the transient.

ACKNOWLEDGEMENT

This study was supported by the Electric Power Research Institute with Robert Ritzman as a project manager. We would like to express our thanks to Dr. Ritzman for his continuing interest and support for this project.

NOTATION

Roman Letter

A - Heat/mass transfer area.

B_1 - Parameter defined in Equation (7a).

B_2 - Parameter defined in Equation (7b).

c_p - Specific heat.

C_D - Discharge coefficient.

D - Diameter of the junction.

D - Diffusion coefficient of steam in the gas.

g - Gravitation acceleration.

Gr - Grashof number.

H - Height of the junction relative to the floor of the node.

H - Heat/mass transfer coefficient.

K - Loss coefficient.

k - Thermal conductivity.

L - Vertical length of the junction.

L - Characteristic height of heat sink.

M - Molecular weight.

p - Pressure.

p_{am} - Log mean pressure difference.

PP - Partial pressure.

Pr - Prandtl number.

Q - Heat transfer rate.

Q - Volumetric flow rate.

Sc - Schmidt number.

T - Temperature.

W - Mass flow rate.

Greek Letter

ρ - Density.

μ - Viscosity.

Subscript

1 - Receiver node of a flow through junction.

2 - Donor node of a flow through junction.

AIR - Air.

cc - Counter-current exchange flow.

CD - Condensate.

CONV - Convective heat transfer.

FILM - Condensate film.

FO - Outer surface of the condensate film.

g - Bulk gas.

G - Bulk gas.

j - Junction-j.

m - Mass transfer.

p - Purge flow.

RAD - Radiative heat transfer.

SAT - Saturation.

STG - Steam.

SW - Surface of heat sink.

u - Unidirectional flow (Bernoulli flow).

w - Water.

WCD - Condensate.

LITERATURE CITED

1. L. Wolf and L. Valencia, "Hydrogen Mixing Experiments in the HDR - Facility," 17th Water Reactor Safety Meeting, Rockwell, MD, Oct. 1989.

2. MAAP User's Manual, Fauske & Associates, Burr Ridge, Illinois, 1990.

3. M. Epstein and M. A. Kenton, "Combined Natural Convection and Forced Flow Through Small Openings in a Horizontal Partition, with Special Reference to Flows in

Multicompartment Enclosures," Transactions of the ASME, Vol. iii, Nov. 1989.

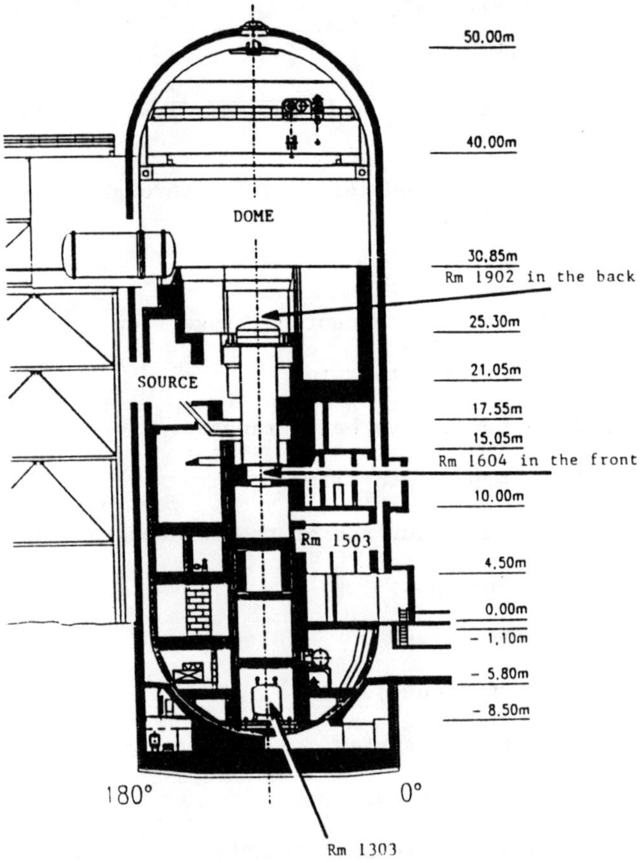

Figure 1 HDR Facility

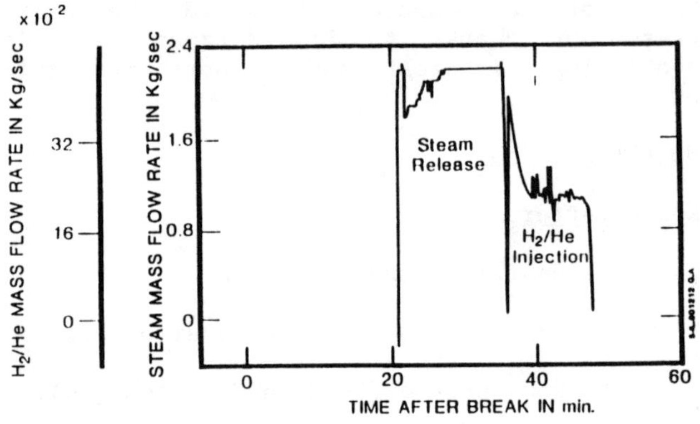

Figure 3 Late Steam and H_2/He Mass Flow Rate

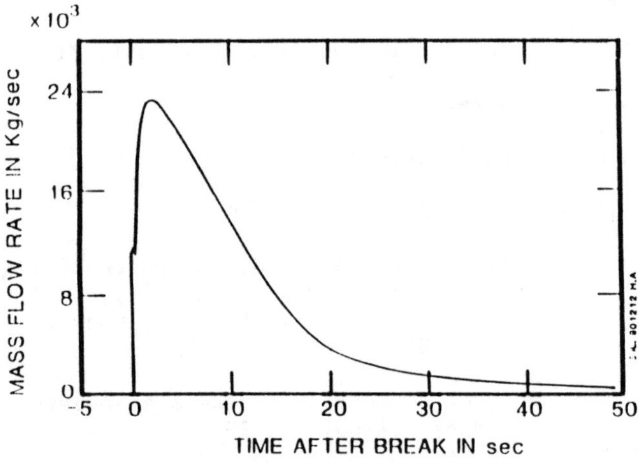

Figure 2 Blowdown Mass Flow Rate Into the HDR-Containment

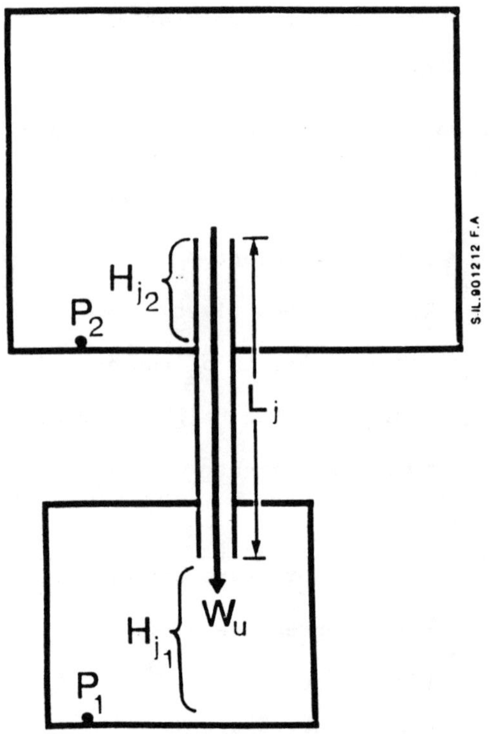

Figure 4 Bernoulli Flow Through a Vertical Junction

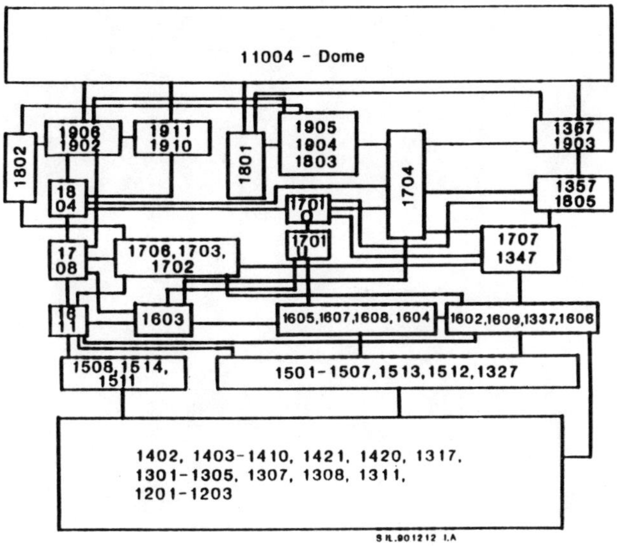

Figure 5 22 Node Model

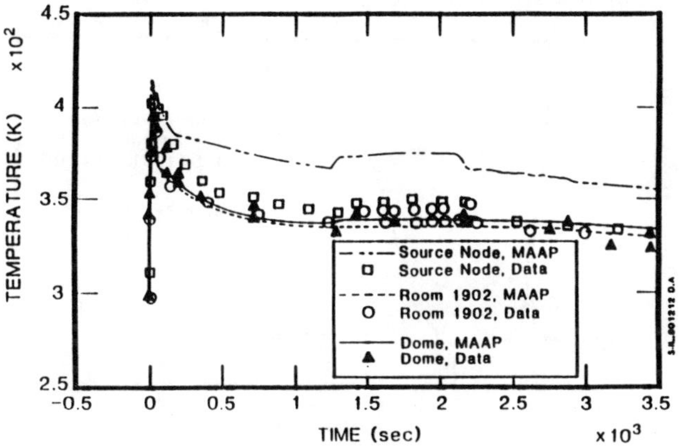

Figure 8 Temperature Transient Above the Break Location

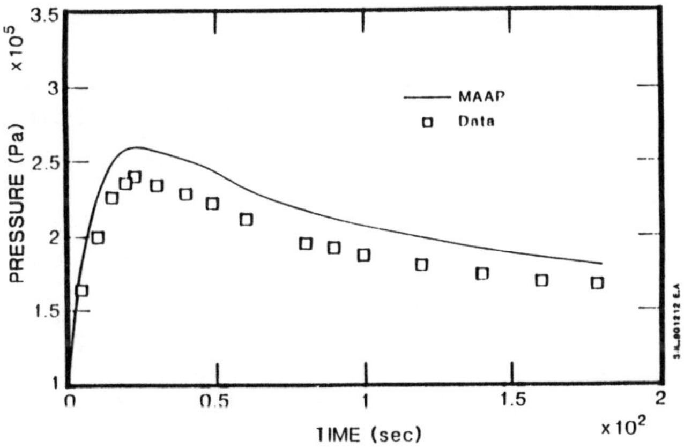

Figure 6 Short Term Pressure Transient

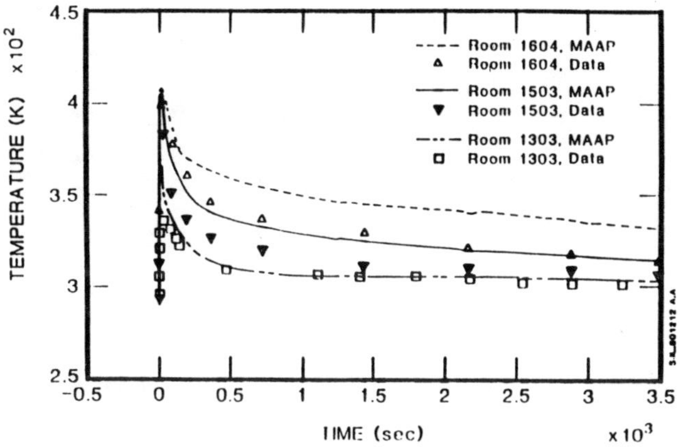

Figure 9 Temperature Transient Below the Break Location

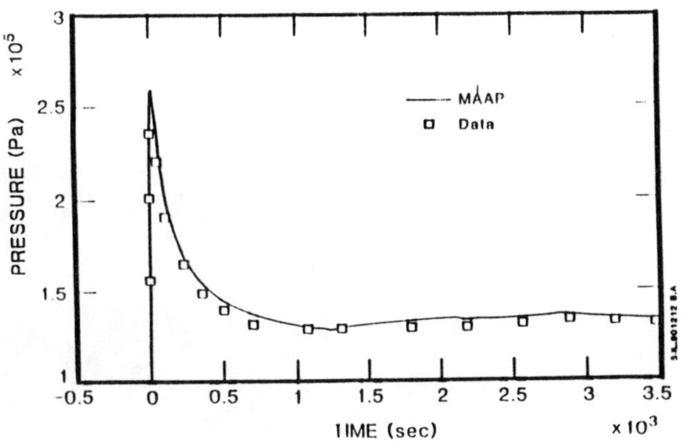

Figure 7 Long Term Pressure Transient

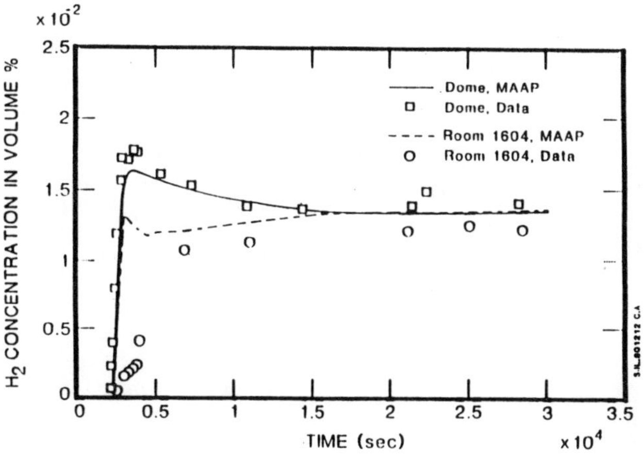

Figure 10 Hydrogen Concentration

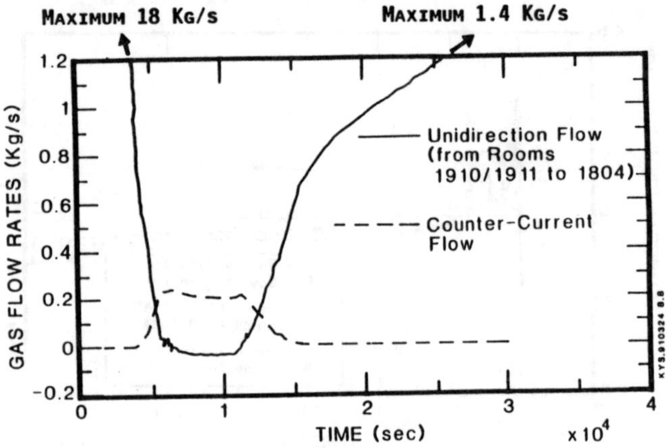

Figure 11 Gas Flow Rates Through the Connecting Room 1910/1911 and 1804

LIQUID-METAL HEAT TRANSFER FOR THE FIRST WALL AND LIMITER/DIVERTOR PLATES OF FUSION REACTORS

Mohammad Z. Hasan and Tomoaki Kunugi ■ Institute of Plasma and Fusion Research, 44-139 Engineering IV, University of California, Los Angeles, CA 90024-1597

Liquid-metal heat transfer in a coolant channel in the first wall and limiter/divertor plates of a fusion reactor is significantly affected by the circumferentially varying heat flux and transverse magnetic field. These effects are estimated by numerically solving the heat transfer problem in an electrically insulated circular tube in the thermally developing and fully-developed flow regions. The transverse magnetic field is assumed to be at an arbitrary angle with the direction of plasma radiation incident in the coolant channel. At the point where the transverse magnetic field is normal to the tube wall, the steady-state Nusselt number (*Nu*) can be increased as much as by a factor of 2 compared with 4.36 for non-MHD flow (parabolic velocity profile) and uniform surface heat flux. The nonuniformity of surface heat flux, on the other hand, can reduce the *Nu* at the same location (also the point of maximum heat flux) to about 3.0. Nusselt number at the point of peak heat flux is further reduced when the transverse magnetic field is not parallel to the plasma radiation. This reduction is about 40% when the heat flux is almost uniform and only 10% when the heat flux is highly peaked, *i.e.*, when the incident radiation heat flux is very large. When the transverse magnetic field is parallel to the plasma radiation, the nonuniformity of surface heat flux and transverse magnetic field combined can increase the thermal entry length by a factor of 4.5 compared with the non-MHD, uniform heat flux case. However, when they are not normal, the extent of increase of the entry length is lowered, by as much as 30% for highly peaked surface heat flux. Neglect of this decrease in *Nu* can result in an underestimation of the film temperature drop by 46% to 64%. The increase in entry length is the smallest when the transverse magnetic field is normal to the plasma radiation.

The first wall and limiter/divertor plates of a fusion reactor face the plasma. The plasma facing side of the coolant channels in these components receives radiation heat flux , thus giving rise to a circumferential variation of surface heat flux. For a circular tube, the distribution of radiation heat flux can be assumed cosine on the plasma-facing surface. The presence of a transverse magnetic field affects the velocity profile of an electrically conducting coolant such as liquid metals. The velocity profile is, in general, flattened under the effect of a transverse magnetic field. The extent of flattening depends upon the strength of the transverse field and varies circumferentially, being largest at the location where the field is normal to the surface of the tube and smallest where it is parallel.

Hasan [1] investigated the effects of the nonuniformity of surface heat flux and flattening of velocity profile on heat transfer for fully developed laminar flow in a circular tube. He assumed a power velocity profile to cover from parabolic velocity profile to slug flow. Gardner [2] investigated the effect of fully developed laminar pipe flow in a transverse magnetic field on steady state heat transfer with uniform heat flux. Reynolds [3] analyzed the effect of the nonuniformity of surface heat flux on heat transfer for fully developed laminar, non-MHD flow in a circular tube.

*Japan Atomic Energy Research Institute, Tokai-mura, Naka-gun, Ibaraki-ken, 319-11 Japan.

Recently, Hasan and Kunugi [4] have reported numerical results of heat transfer in fully-developed and thermally developing laminar flows of liquid metals in a circular coolant tube in a plasma-facing component of a fusion reactor. The plasma radiation and transverse magnetic field were assumed parallel. Similar results for non-MHD laminar and turbulent flows have also been reported by Kunugi and Hasan [5]. All these studies show a very significant effect of the nonuniformity of surface heat flux and/or the presence of a transverse magnetic field on convective heat transfer.

Both the nonuniformity of surface heat flux and the presence of a transverse magnetic field affect convective heat transfer. The effect on Nusselt number (Nu) results from the asymmetry of the temperature profile caused by the circumferential variation of the surface heat flux and the asymmetry and flattening of the velocity profile due to the presence of transverse magnetic field. The asymmetry of the temperature profile is expected to delay the attainment of steady-state condition and hence increase the entry length. The reduction of Nu has adverse effects on the thermal-hydraulic design of the plasma facing components. An increase in thermal entry length is expected to facilitate thermal-hydraulic design if the increase is substantial. It is, therefore, important to quantify the reduction of Nu and the increase in entry length due to the nonuniformity of surface heat flux and the presence of transverse magnetic field.

When the transverse magnetic field and plasma radiation are parallel, the peak heat flux and the most flat part of the velocity profile occur at the same circumferential location on the coolant tube. They are at different circumferential locations when the plasma radiation and the transverse magnetic field are not parallel. Detail results in the former case have been reported in reference [4]. For a typical layout of coolant channels in the first wall of a tokamak reactor, the plasma radiation and the transverse magnetic field are generally orthogonal. For the divertor plate, however, they may be inclined at an arbitrary angle.

The objective of this paper is to investigate the effects of the intensity of plasma radiation and a transverse magnetic field inclined at an arbitrary angle to the plasma radiation on the liquid-metal heat transfer in a circular coolant tube in the plasma-facing components of a fusion reactor. The wall of the tube is assumed either electrically non-conducting or insulated from the coolant. The flow is assumed fully developed and laminar. The results are obtained numerically in the thermally developing as well as in the steady-state regions.

FORMULATION OF THE PROBLEM AND NUMERICAL SOLUTION

A schematic of the problem is shown in Figure 1. In this figure, q''_R is the uniform plasma radiation heat flux incident on the side of the coolant tube facing the plasma, d is the tube diameter, L is the tube length, and (r, θ, z) are the spatial coordinates. The incident heat flux is assumed axially uniform. The magnetic field B is inclined at an angle θ_m to the plasma radiation, q''_R. The case $\theta_m = 0$ has been treated in reference [4]. The heat transfer solutions are obtained numerically using a transient, three-dimensional computer code.

Governing Equations

The transient , three-dimensional energy equation in tensor form can be written as

$$\frac{\partial(\rho c_p T)}{\partial t} + \frac{\partial(\rho c_p u_j T)}{\partial x_j} = \frac{\partial^2(kT)}{\partial x_j x_j} - \frac{\partial\langle \rho c_p u_j T'\rangle}{\partial x_j} + Q, \qquad (1)$$

where Q is volumetric heat source, t is time, T and T' are temperature and temperature fluctuation, respectively, u_j are the velocity components (u, v, w), x_j are

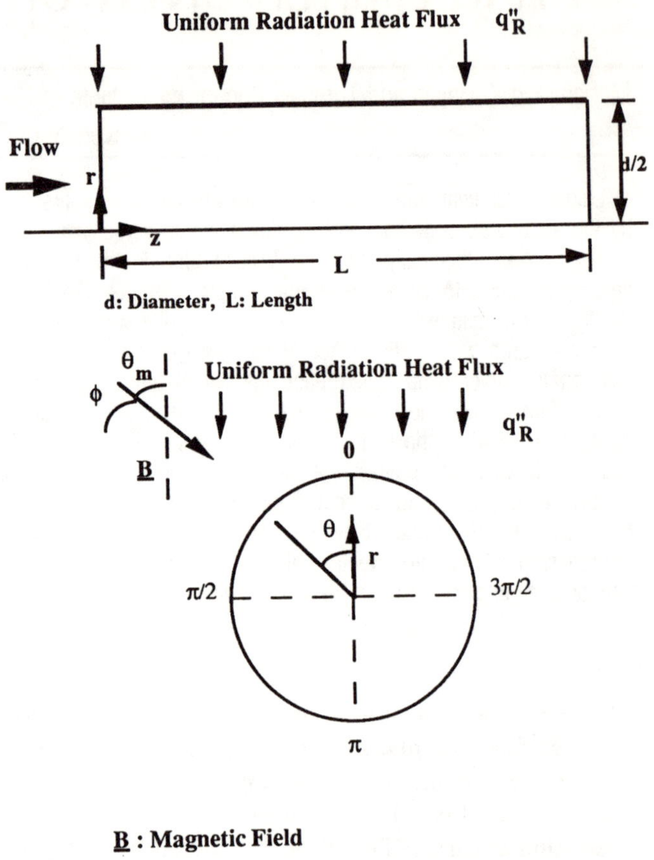

Figure 1: A schematic of the problem showing the geometry and the directions of the plasma radiation and the transverse magnetic field.

the spatial coordinates (r, θ, z), k, ρ, and c_p are the thermal conductivity, density, and specific heat capacity of the coolant, respectively, and $\langle - \rangle$ denotes ensemble average of the bracketed quantity. The second term in Equation (1) is the turbulent heat flux term. For laminar flow, this term is dropped. Volumetric heat generation is assumed to be zero in this problem.

The initial condition is $T(r, \theta, z) = T_\circ$ at $t = 0$, and the temperature at the entrance of the tube is always kept at $T_\circ$. For $\theta_m = 0$, there are symmetry conditions at $\theta = 0$ and $\theta = \pi$. For θ_m greater than 0, there are no symmetry conditions on θ, and the problem is solved for $\theta = 0$ to 2π. The boundary conditions are

$$q''(\theta) = \begin{cases} A + B\cos\theta & ;-\pi/2 \le \theta \le \pi/2 \\ A & ;\pi/2 \le \theta \le 3\pi/2, \end{cases} \qquad (2)$$

where $B = q''_R$ and A is the uniform component of the heat flux such as could arise from neutron heating in the wall of the tube. For $A = 0$, the backside of the tube is adiabatic. The radiation heat flux on the tube wall is assumed to follow a cosine distribution. A measure of nonuniformity or peakedness of the surface heat flux (and also of the intensity of the incident plasma radiation) is given by the ratio B/A. The surface heat flux is uniform for $B/A = 0$, and highly peaked for B/A large.

The conventional definition of Nu is used in the present analysis. Since the surface heat flux and wall/coolant interface temperature vary circumferentially, Nu also varies with θ.

$$Nu(\theta, z) = \frac{q''(\theta)d}{k[T_w(\theta, z) - T_b(z)]}, \quad (3)$$

where $T_w(\theta, z)$ is the wall-coolant interface temperature and $T_b(z)$ is the coolant bulk temperature.

The velocity profile is assumed to be fully developed so that $u(r, \phi) > 0$, and $v = w = 0$. Gold [6] has provided an analytical solution for the velocity profile for fully developed laminar flow in a circular tube with insulated wall and in the presence of a transverse magnetic field.

$$u(r, \phi) = -\frac{K}{4\alpha}\sum_{n=0}^{\infty} \varepsilon_n \frac{I'_n(\alpha)}{I_n(\alpha)} I_n(\alpha r)\cos(n\phi) \left[e^{-\alpha r \cos\phi} + (-1)^n e^{\alpha r \cos\phi}\right]. \quad (4)$$

Here $\phi = (\theta - \theta_m)$, $\alpha = Ha/4$, $K = (d^2/4U\eta)(\partial p/\partial z)$, U is a characteristic velocity, Ha is Hartmann number defined as $Ha = \mu_o B d(\sigma_e/\eta)^{1/2}$, μ_o, σ_e, and η are the magnetic permeability, electrical conductivity, and viscosity of the coolant, respectively. I_n is the modified Bessel function of order n, $\varepsilon_n = 1$ for $n = 0$ and $\varepsilon_n = 2$ for $n > 0$. Here r is normalized by the tube radius.

Numerical Solution

Equation (1) with the initial and symmetry conditions, if any, the boundary conditions in Equation (2), and the velocity profile given by Equation (4) is solved by the computer code CONDIF which has been developed as a transient, three-dimensional fluid flow and heat transfer equations solver [7,8]. Equation (1) is discretized by a finite/control volume method in this code. The code uses Euler implicit method for time advancing and central differencing scheme for the spatial derivatives. The convective terms are approximated by the CONDIF method [9]. All the variables are defined on the well-known staggered grid system. The grid size used in the calculations presented in this paper is $10 \times 30 \times 50$ in the r, θ, and z dimensions, respectively.

RESULTS AND DISCUSSIONS

All the results obtained by this code are for liquid-lithium flowing through a tube of 8 mm in diameter. The same coolant and tube size were used in the thermal-hydraulic design of the first wall and divertor plate of the TITAN-I reversed-field pinch conceptual fusion reactor [10]. The results have been obtained for steady-state and thermally developing conditions. The parameter ranges covered are: $Ha = 0$ to 400, $B/A = 0$ to 50, and $\theta_m = 0$ to $\pi/2$. The velocity profile is symmetric in each quadrant.

Steady-State Nusselt Number

The general effects of the transverse magnetic field and circumferential variation of surface heat flux are shown in Figure 2. These results are for $\theta_m = 0$. At small Hartmann number, Nu is almost uniform around the tube and its value is close to that for non-MHD Poiseuille flow case, 4.36. As the Ha increases, the velocity profile becomes more and more flat, and Nu increases. Since the velocity profile becomes more flat at $\theta = 0°$ than at $\theta = 90°$, the increase in Nu is much higher at the former location than at the latter. The symmetry of Nu about 90°

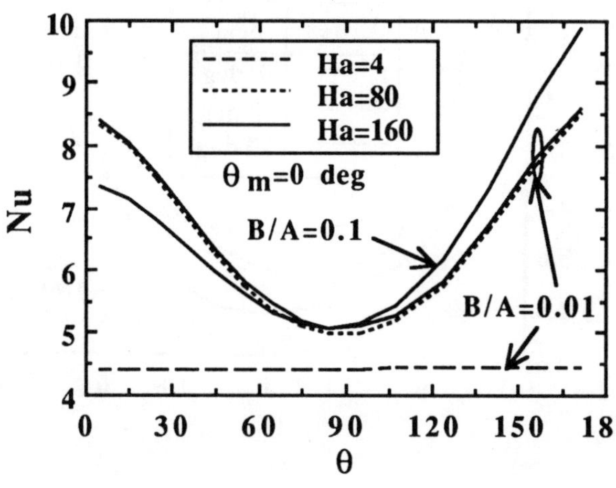

Figure 2: Circumferential variation of Nu for different values of Ha. The transverse magnetic field is parallel to the plasma radiation ($\theta_m = 0°$).

is due to the symmetry of the velocity profile in each quadrant. The curves for Ha=80 and 160 show that the effect of the transverse magnetic field becomes saturated at $Ha \sim 100$. In a fusion reactor environment, the transverse Ha is generally much larger than 100. Therefore, to reflect the maximum effect of the transverse magnetic field, all the remaining results are obtained for $Ha = 160$ (>100). The effect of nonuniformity of surface heat flux is shown by the curves for $Ha = 160$ and B/A=0.01 and 0.1. The nonuniformity of heat flux reduces Nu by a larger factor at $\theta = 0°$ than at $\theta = 90°$. Note that Nu becomes asymmetric. This is due to the circumferential variation of heat flux and the resulting asymmetry of the temperature profile in the coolant.

In Figure 3, circumferential variation of Nu near $\theta = 0°$ is shown for the three cases, $\theta_m = 0°$, 40°, and 90°. The Hartmann number is 160 and B/A is 50 which signifies a strong plasma radiation and highly peaked heat flux on the surface of the coolant tube. For both $\theta_m = 0°$ and $\theta_m = 90°$, the Nu is symmetric about $\theta = 0°$. The Nu at $\theta = 0°$ is also smaller for $\theta_m = 90°$ than for $\theta_m = 0°$. This is because, for $\theta_m = 90°$, the velocity profile is less flat at $\theta = 0°$ than for $\theta_m = 0°$. The symmetry of Nu for the two cases $\theta_m = 0°$ and $\theta_m = 90°$ is due to the symmetry of the velocity profile in each quadrant. The Nu for θ_m = 40° is asymmetric about $\theta = 0°$ and has intermediate magnitudes with respect to those for $\theta_m = 0°$ and

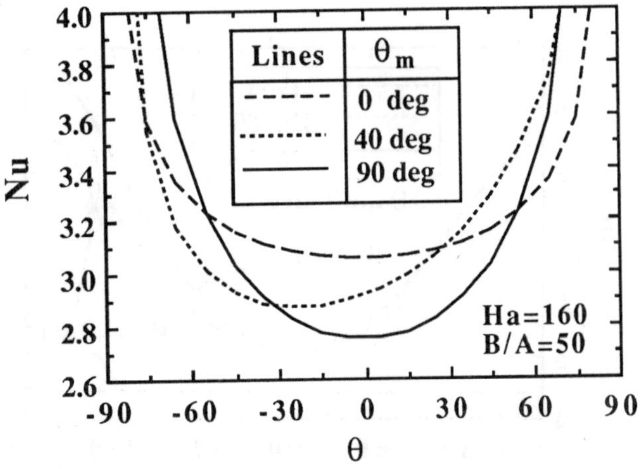

Figure 3: Circumferential variation of Nu for different inclinations (θ_m) of the transverse magnetic field to the plasma radiation (Ha=160 and B/A=50).

$\theta_m = 90°$. Figure 4 shows that the Nu at $\theta = 0°$ is nearly independent of θ_m for highly nonuniform surface heat flux (large B/A) and large transverse Ha. The Nu at $\theta = 0°$ changes only slightly, from 3.1 for $\theta_m = 0°$ to 2.7 for $\theta_m = 90°$. However, the variation of Nu at $\theta = 0°$ with θ_m is large for nearly uniform heat flux (small B/A). In this case, Nu at $\theta = 0°$ changes from 8.4 for $\theta_m = 0°$ to 5.1 for θ_m = 90°.

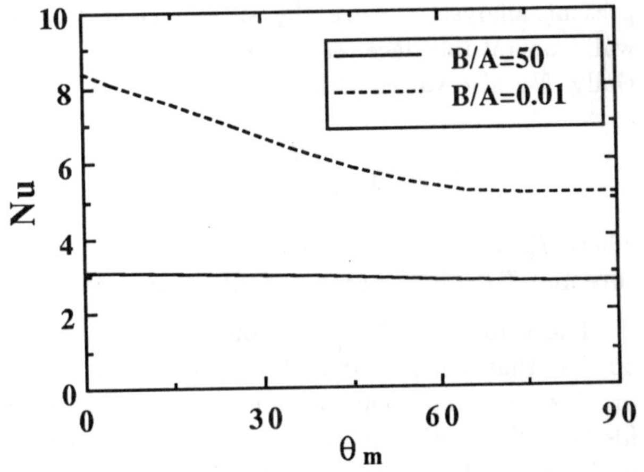

Figure 4: Variation of Nu at $\theta = 0°$ with θ_m for B/A= 0.01 and 50, and Ha=160.

The variation of Nu with B/A is shown in Figure 5 for $Ha = 160$ and $\theta_m = 0°$, 40°, and 90°. Since the heat flux is the highest at $\theta = 0°$, Nu at/near this location is important for the thermal-hydraulic design of the plasma-facing components. The Nu near $\theta = 0°$ drops rapidly with B/A and becomes almost constant beyond $B/A = 10$. For $\theta = 0°$, this decrease in Nu is about 64%. The corresponding reduction is 46% for $\theta_m = 90°$. The effect of rotating the transverse magnetic field with respect to the plasma radiation on Nu at $\theta = 0°$ is large for nearly uniform surface heat flux ($B/A \sim 0$) and small for highly peaked heat flux (B/A large). As θ_m is changed from 0° to 90°, Nu is reduced by $\sim$ 40% at $B/A = 0.01$ and by $\sim$ 10% at $B/A = 50$. It should also be noted that the theoretical result with power velocity profile (nearly slug flow) in reference [1] predicts Nu with good accuracy at this location for $\theta_m = 0°$, especially for large values of B/A.

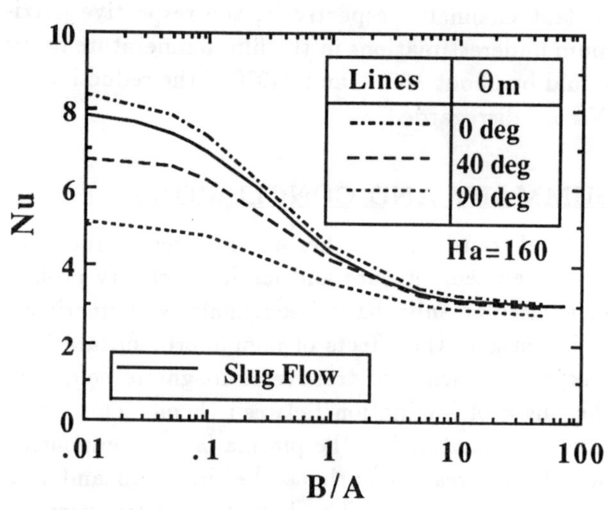

Figure 5: Variation of Nu with B/A and θ_m for Ha = 160.

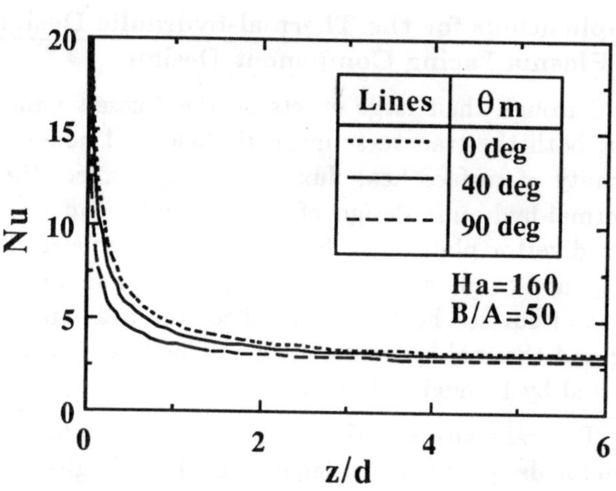

Figure 7: Dependence of Nu in the entry region upon θ_m ($B/A = 50$ and $Ha = 160$).

Heat Transfer in Entry Region

Nusselt numbers in the thermally developing region are shown in Figures 6 and 7 for $Ha = 160$. In both of these figures, Nu is at $\theta = 0°$. In Figure 6, Nu is plotted against z/d for $B/A = 50$ and $\theta_m = 0°$, 40°, and 90°, and in Figure 7 the parameters are $\theta_m = 90°$ and $B/A = 0$, 0.05, and 50. These figures show that the entry length decreases with θ_m but increases with B/A. It has been shown in reference [4] that the entry length is increased by both the transverse magnetic field and circumferential variation of surface heat flux, more significantly by the latter than the former. A measure of the change of entry length with θ_m and B/A is shown in Figure 8 where the entry length, L_e (z/d), is plotted against B/A for $\theta_m = 0°$, 40°, and 90°, and $Ha = 160$. The entry length is defined as the axial distance in terms of tube diameter at which the Nu is 99% of its steady-state value. The entry length increases with B/A but decreases with θ_m. It increases by a factor of about 4.5 when the transverse magnetic field is parallel to the plasma radiation ($\theta_m = 0°$) and by a factor of 3.5 when they are normal.

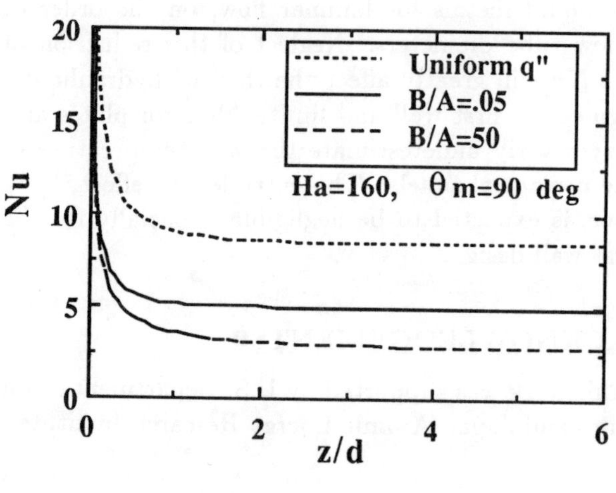

Figure 6: Dependence of Nu in the entry region upon B/A ($Ha = 160$ and $\theta_m = 90°$).

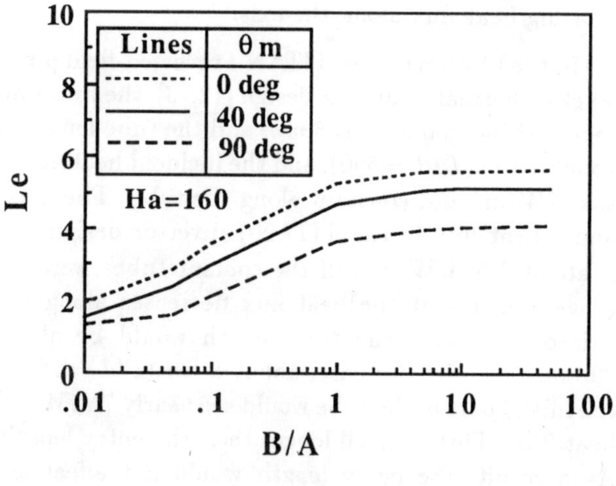

Figure 8: Variation of entry length with B/A for θ_m = 0°, 40°, and 90°. The Ha is 160.

Implications for the Thermal-hydraulic Design of Plasma Facing Component Design

Through their large effects on the Nusselt number, both the transverse magnetic field and nonuniformity of surface heat flux will strongly affect the thermal-hydraulic design of the first wall and limiter/divertor plates of a fusion reactor. For electrically non-conducting coolants, only the nonuniformity of surface heat flux will affect the heat transfer and thermal-hydraulic design. This has been reported by Kunugi and Hasan [5,11] elsewhere.

Typical transverse Hartmann numbers in a fusion reactor design are much larger than 100. Neglect of the nonuniformity of surface heat flux in thermal-hydraulic design of the first wall and limiter/divertor plates using liquid metals as coolants will result in a significant underestimation of the maximum structural temperature. For a circular coolant channel, the film temperature drop can be underestimated by about 46% to 64% depending upon whether the transverse magnetic field is normal or parallel to the plasma radiation, respectively. The thermal entry length, where the Nu can be significantly higher, is very short for liquid-metal coolants, on the order of one tube diameter. Therefore, entry length will not affect the design of the first wall and limiter/divertor plates if the heat flux is constant along the axis. In a conventional divertor, however, the heat flux decreases rapidly from its peak value along the field line. Some advantage of higher Nu in the short entry length can be realized if the coolant channels are arranged to have the peak heat flux at the entrance and then decreasing heat flux along the axis.

In the lithium cooled TITAN-I reversed-field pinch reactor thermal-hydraulic design [12,13], the first wall coolant tube diameter is 8 mm and the tube length is about 4.4 m ($L/d = 550$), and the incident heat flux is 4.6 MW/m^2 and constant along the tube. The maximum heat flux in the TITAN-I divertor design [14] is about 7.5 MW/m^2. If the coolant tubes were arranged such that the heat flux decreases along the axis of the tube, the tube length would be about 20 cm. Even in this case, about 8 cm ($z/d = 16$) of the initial part of the tube would see nearly 7 MW/m^2 heat flux. This is much longer than the entry length. As a result, the entry length would not affect the thermal-hydraulic designs of the first wall and divertor of TITAN-I. However, at 4.6 MW/m^2 and 7.0 MW/m^2 heat fluxes on the first wall and divertor coolant channels, respectively, the respective maximum underestimations in the film temperature drops would be about 160°C and 240°C if the reductions in Nu are disregarded.

SUMMARY AND CONCLUSIONS

Heat transfer in the thermal entry region and fully-developed temperature and laminar velocity profiles in a circular tube have been analyzed numerically to investigate the effects of nonuniform surface heat flux, the presence of transverse magnetic field, and the angle of inclination between them. This situation is encountered in the plasma facing components of a fusion reactor such as the first wall and limiter/divertor plates. The heat flux distribution on the tube is assumed cosine. The coolant is liquid lithium and the tube wall is assumed electrically insulated. With uniform surface heat flux, a transverse magnetic field can increase the Nusselt number by a factor of 2. However, the circumferential variation of heat flux adversely affect the heat transfer, decreasing Nu by more than a factor of 2. The net effect is to reduce the Nu at the point of maximum heat flux ($\theta = 0°$) by 46% to 64% depending upon whether the transverse magnetic field is perpendicular or parallel to the plasma radiation, respectively. The entry length has been shown to be increased by both the transverse magnetic field and circumferentially varying heat flux [4]. This increase can be up to a factor of 4.5 when the transverse magnetic field is parallel to the plasma radiation and by a factor of 3.5 when it is normal. However, the entry length is still short for liquid metals for laminar flow, on the order of a few tube diameters. Neglect of this reduction of the Nu will greatly affect the thermal-hydraulic design of the first wall and limiter/divertor plates and may severly underestimate the wall temperature of the coolant channels. The entry length effect, however, is expected to be negligible, especially for the first wall design.

ACKNOWLEDGEMENTS

This work was supported by U.S. Department of Energy and Japan Atomic Energy Research Institute.

LITERATURE CITED

[1] Hasan, M. Z., *Fusion Technology*, **16**, 44 (1989).

[2] Gardner, R. A., *Int. J. Heat Mass Transfer*, **11**, 1076 (1968).

[3] Reynolds, W. C., *Trans. ASME*, 108 (1960) .

[4] Hasan, M. Z. and Kunugi, T., "Entry-Length Effect on the Thermal-Hydraulic Design of Plasma Facing Components of Fusion Reactor, Part-II: MHD Flow," to be published in *Fusion Technology* (1991).

[5] Kunugi, T. and Hasan, M. Z., "Entry-Length Effect on the Thermal-Hydraulic Design of Plasma Facing Components of Fusion Reactor, Part-I: Non-MHD Flow," to be published in *Fusion Technology* (1991).

[6] Gold, R. R., *J. Fluid Mech.*, **13**, 505 (1962).

[7] Kunugi, T., Yokokawa, M. and Chiba, T., "Vectorization of a Three Dimensional Thermal-Hydraulic Analysis Code 'STREAM', " *JAERI-M 88-119* (1988). [in Japanese]

[8] Kunugi,T. and Yokokawa, M., "Numerical Analysis for Lid-Driven Cavity Flow Using a 'CONDIF' Approach," *Trans. JSME*, **B**, **55** (1989) 1823. [in Japanese]

[9] Runchal, A. K., "CONDIF: a Modified Central-Differencing Scheme with Unconditional Stability and Very Low Numerical Diffusion," *Int. Heat Transfer Conference*, 403 (1986).

[10] Najmabadi, J., Conn, R. W., *et. al.*, "The TITAN Reversed Field-Pinch Fusion Reactor Study: Final Report," Joint Report of UCLA, GAT, LANL, and RPI, *UCLA-PPG-1200* (1989).

[11] Kunugi, T. and Hasan, M. Z., "Heat Transfer with Nonuniform Surface Heat Flux and Thermal-Hydraulic Design of the Plasma Facing Components of Fusion Reactor," *16th Symp. on Fusion Technology*, London, September (1990).

[12] Hasan, M. Z., Blanchard, J. P. and Ghoneim, N. M., *Fusion Eng. Design*, **9**, 431 (1989).

[13] Hasan, M. Z., *et. al.*, "Thermal-Hydraulic Design for the Lithium-Cooled TITAN Reversed-Field Pinch Reactor," *Trans. IEEE 12th Symp. on Fusion Engineering*, **2**, 1054 (1987).

[14] Cooke, P. I. H., *et al.*, "Divertor Design for the TITAN Reversed-Field-Pinch Reactor," *Trans. IEEE 12th Symp. on Fusion Engineering*, **2**, 1050 (1987).

SCALING STUDIES ON SPREADING OF MOLTEN CORIUM IN MARK-1 BOILING WATER REACTORS

Hee Cheon NO ■ Department of Nuclear Engineering, Korea Advanced Institute of Science and Technology, P.O. Box 150, Cheongryang, Seoul, Korea
Mamoru Ishii ■ School of Nuclear Engineering, Purdue University, West Lafayette, IN 47907

A new scaling method called step-by-step integral scaling approach is proposed and applied to scaling analysis of BWR MARK-1 corium-shell interactions. Two group mechanisms are identified : the spreading group mechanism and water-corium-concrete interactions. In the first step the former is treated considering the mass and momentum transfer of the molten corium. Two spreading mechanisms are identified : gravitational force-dominant spreading and surface tension-dominant spreading is derived using the concept of open channel flow. The second step is divided into two substeps : interactions before and after solidification starts. In the first substep thermal convection is considered and an analytical solution for a variation of the corium temperature is obtained using the characteristic method. In the second substep governing equations for crust formation at the top and the bottom of the corium are analytically solved. In the real conditions of the prototype the time constant and relative importance of each phenomenon are examined. It is found out that time constants related to water-corium-concrete interactions are much shorter than those related to spreading. Initial superheat, thermal convection rates, water and concrete cooling, and latent heat of corium solidification turn out to have high relative importance. The scaling ratios of geometical and operating conditions are suggested for initial input of one-tenth scaling ratio of the radius of the inpedestal floor with lead as a corium simulant.

1. INTRODUCTION

Previous containment analysis of BWR MARK-1 has considered the γ -mode of containment failure, that is, overpressure failure of the drywell liner as the dominant mode. However, a possible mode of early containment failure has been identified.[1] In this mode the possible rapid spreading of the melt onto the drywell floor following the ejection of core melt materials causes thermal attack by the high-temperature corium through the direct contact of the molten corium with the steel liner.

Several experimental and analytical investigations have been performed in order to deal with the safety concern of the mode. The mechanism of melt spreading was experimentally studied by BNL.[2] Their studies revealed five melt spreading flow regimes characterized by hydrodynamically-limited and heat transfer-limited phenomena. Based on their observation, a single parameter dimensionless correlation was suggested for identifying the spreading regimes. Issues on the dynamic interactions between the corium and water were studied by Fauske Associates.[3] Their experiments designed with a one-tenth liner scale for geometry used iron thermite as a corium simulant. They reported that the most important part of the experiments is energy removal rates provided by water and quenching rates far exceeded critical heat flux values in early transients. The thermal response of the shell was studied by ORNL using the three-dimensional transient heat transport computer code, HEATING-6.[4] They found out that the most important parameters are melt superheat, melt depth, and melt-to-crust heat transfer coefficients. A detailed computer code for analysis of spreading of the molten corium is under development at ANL.[5] Their scoping calculation showed that the prediction of corium penetration is strongly sensitive to the superheat of corium but weakly sensitive to the volumetric heat source in the corium. Kazimi evaluated the cooling of the spreading corium and the heat transfer to the liner using his simple model.[6] He devised the map showing the regions of unlikely liner failure in terms of pour rate, pour superheat, pour material, and the presence of water. Moody developed a simplified theoretical method for estimating the maximum temperature of the shell considering the relative importance of each phenomenon through nondimensional analysis.[7] They found that the maximum shell temperature is directly proportional to the volumetric heat generating rate and water cooling is effective in preventing temperatures from reaching the shell melting value.

In view of complexities of phenomena involved in the mode we need properly-scaled experiments for studying interactions between spreading and crust formation by cooling. In order to generate scaling criteria, several studies have been performed on the base of Buckingham's π theorem[8–10], the method of nondimensional differential equations[11], and the integral method[12,13].

The most extensively used scaling approach is Buckingham's π theorem. Its application to the present problem is not very useful because of its main characteristics requiring to guess governing parameters and generating too many similarity groups without clear physical meanings. The method of nondimensional differential equations based on physical conservation laws is very useful to provide similarity groups with physical significance. For multiphase flow this method is applicable only for local scaling due to a large change of scales over wide ranges. The integral method has been successfully used for scaling analysis of LOCA. In this method differential equations are integrated over system implementing initial and boundary conditions in order to obtain integral response function. Therefore, scaling criteria generated from nondimensionalized integral response function are applicable for system.

However, it is very difficult to apply this method to a problem with poorly understood and very complicated phenomena because of difficulties in obtaining its integral response function. As a result, a new scaling method called step-by-step integral scaling approach, which is basically the integral method, is suggested here and applied to generating scaling criteria for the present problem.

2. DESCRIPTION OF STEP-BY-STEP INTEGRAL SCALING METHOD

The basic strategy of the step-by-step integral scaling method is first to identify and group all the potentially important phenomena involved in a problem, and then to deal with each group mechanism separately and sequentially starting from the most important group mechanism. As a result, we can generate scaling parameters governing each group mechanism.

In the proposed method, first of all, two identifications are needed ; system identification and phenomena identification. Whole system is divided into subsystems with several components. The initial and boundary conditions of the system are identified in order to isolate the present system and problem from other system and accident scenario. Also, through the identification of interfacial conditions between subsystems, each subsystem can be separately dealt with.

For each subsystem potentially important phenomena are carefully identified and ranked in terms of their relative importance and order of events. Possible bifurcation phenomena and feedback mechanisms are identified. Bifurcation phenomena have considerable effects on physical processes because sudden transitions can occur at bifurcation points. Through the identification of feedback mechanisms it can be determined whether phenomena related to them should be treated as the same group mechanism. Several phenomena can be put in a group. Use of group mechanisms enables us not only to deal with feedback mechanisms but also to reduce the number of steps taken by the present method.

After identification scaling analysis is done step by step. In the first step the most important group mechanism is chosen and its governing equations are written down with its initial and boundary conditions, and transition criteria. These balance equations are solved analytically or expressed as nonlinear integral response function. Their results are nondimensionalized to generate integral scaling parameters governing the group mechanism. In the second step the second most important group mechanism is chosen. The same procedure as that in the first step is taken independently of the first step but results from the first step are used in this step. In the third step interactions between the first and the second step are considered ; checking the existence of bifurcation phenomena and determining how strong feedback exists. If strong feedback between two steps exists, we may need to combine analysis or feedback results from the second step into the first step. Also, the relative importance of each phenomenon related to each group mechanism is considered choosing reference conditions or comparing scaling parameters.

After all the important group mechanisms are considered, the whole set of scaling parameters and transition criteria are generated as a final product. They can be used to check whether the same relative importance of each phenomenon is preserved for a prototype and its scaled-down model. If all the requirements of similarity can not be satisfied, possible scale distortion is carefully investigated based on the relative importance of each scaling parameter. Also, code-scaling capabilities can be evaluated checking whether codes include important phenomena and show equivalent relative importance for each phenomenon at several different scales.

3. APPLICATION OF THE PROPOSED METHOD

3.1 Identification of system

Whole system is divided into three subsystems as shown in Table 1 : vessel subsystem, floor subsystem, and shell subsystem. Each subsystem has its own components. For example, the floor subsystem consists of 4 components : sump, inpedestal floor, doorway, and expedestal floor. Identifying initial and boundary conditions, and interfacial conditions between subsystems as shown in Table 2, the problem can be separated from in-vessel accident scenario and each subsystem can be separately dealt with. Five possible bifurcation phenomena are identified : critical flow (hydraulic jump phenomena), film boiling and quenching, solidification, crust stability, and minimum corium thickness for spreading. Concrete ablation and stable crust formation are identified as important feedback mechanisms. Steam and gas produced by concrete ablation chemically react with metallic components in the corium and produce chemical energy. More ablation is possible due to an increase in the corium temperature by the produced

Table 1 Identification of System

Subsystem	Component	Geometry		Materials
vessel subsystem	reactor vessel	location of breach		
		height from floor	H_{rv}	
	below-vessel structure	distributed area of corium	A_{dc}	below-vessel materials
		height from floor	H_{bv}	
floor subsystem	sump	cross-sectional area	A_{su}	floor concrete materials
		height	H_{su}	
	inpedestal floor	area	A_{in}	
		outer radius	R_{in}	
	doorway	width	W_{dw}	
		length	L_{dw}	
	expedestal floor	outer radius	R_{ex}	
shell subsystem	shell	thickness	t_{sh}	shell materials
		angle	θ_{sh}	
	outer-shell air layer	thickness	t_{ar}	
	outer-shell concrete structure	thickness	t_{osc}	outer-shell concrete materials

Table 2 Initial and Boundary Conditions and Interfacial Conditions between Subsystems

Subsystem	Initial and boundary conditions		Interfacial conditions
vessel subsystem	pouring-out rate	Q_{co}	Q_{co}, T_{co}, X_{co}
	corium temperature	T_{co}	
	corium composition	X_{co}	
floor subsystem	containment pressure	P_{ct}	corium thickness (t_{co})
	water content	M_w	
	water temperature	T_w	water thickness (t_w)
	component temperature	T_{su}	heat flux from corium to shell ($q''_{co \to sh}$)
		T_{in}	
		T_{dw}	
		T_{ex}	
shell subsystem	component temperature	T_{sh}	shell temperature (T_{sh})
		T_{air}	shell stress (σ_{sh})
		T_{osc}	

Table 3 Potentially Important Mechanisms and Rank

Interacting Materials	Sub-Systems	Phenomena		Rank
Water	S, I, D, E	(1) boiling	T	2
Water-Corium Interface	S, I, D, E	(2) entrainment	H	2
		(3) film boiling	T	1
		(4) bubbling	H	2
		(5) quenching	T	1
		(6) radiation	T	1
Corium	S	(7) filling	H	1
	I	(8) spreading	H	1
	D	(9) spreading	H	1
	E	(10) spreading	H	1
	S, I, D, E	(11) crust formation & stability	T, H	1
		(12) bubbling	H	2
		(13) chemical reaction	C, T	2
		(14) decay heat	T	2
		(15) conduction	T	3
		(16) convection	T, H	2
Corium-Concrete Interface	S, I, D, E	(17) film heat transfer	T	1
		(18) radiation	T	3
Concrete	S, I, D, E	(19) ablation	C, T	1
		(20) conduction	T	2
		(21) melting	T	3
Shell	E	(22) conduction	T	1
		(23) melting	T	3
Shell-Corium Interface	E	(24) solidification	T	1
		(25) convection	T, H	1
		(26) conduction	T	2
Shell-Air-Concrete Interface	E	(27) radiation	T	1
		(28) air convection	T, H	2
		(29) concrete conduction	T	3

S : Sump, I : Inpedestal floor
D : Doorway, E : Expedestal floor
T : Thermal-dominant phenomena
H : Hydraulic-dominant phenomena
C : Chemical-dominant phenomena

chemical energy. However, more cooling on the top of the corium is expected by bubble agitation and corium entrainment. Stable top crust formation considerably retards spreading and, as a result, thickens crust. Stable bottom crust formation can generate mixed results : no chemical energy and no enhancement of water cooling.

Twenty-nine potentially important phenomena are identified and ranked as shown in Table 3. Spreading and crust formation are considered as phenomena with high relative importance.

These phenomena are grouped and sequence of analysis is decided according to their importance and order of events. Three group mechanisms are identified : spreading group mechanism , water-corium-concrete interactions and corium-shell interactions. Here, the former two group mechanisms are studied.

3.2 Spreading Group Mechanism

The spreading group mechanism is chosen as the most important group mechanism because it consists of main driving mechanisms to transfer the molten corium to the shell. In the first step spreading and filling phenomena in the sump, inpedestal floor, doorway, and expedestal floor are analyzed to generate scaling parameters governing the spreading group mechanism. As only mass and momentum transfer are considered, scaling parameters involved in the hydrodynamics of the molten corium can be generated.

In the sump

It can be thought that the molten corium leaving the reactor vessel first fills the sump below it. The filling mechanism in the sump is simply modeled using mass balance :

$$\frac{dh_{su}}{dt} = \frac{Q_p}{\pi R_{su}^2(1-\alpha_{su})}\ , \tag{1}$$

where h_{su} and α_{su} : corium thickness and averaged void fraction in the sump, respectively,

R_{su} : radius of the sump,

Q_p : volumetric pouring-out rate of the corium from the reactor vessel.

Equation (1) is nondimensionalized using the following nondimensional variables and parameters :

$$t^* = \frac{t}{\tau_p},\quad h^*_{su} = \frac{h_{su}}{H_{su}}\ ,\quad N_{su} = Q_p\frac{\tau_p}{V_{su}}\ ,\quad \frac{dh^*_{su}}{dt^*} = \frac{N_{su}}{(1-\alpha_{su})}\ , \tag{2}$$

where τ_p : time interval in which pouring-out of the corium from the reactor vessel continues,

V_{su} : volume of the sump,

N_{su} : dimensionless sump number.

The dimensionless sump number N_{su} represents the ratio of the total volume of the pouring-out corium to the total sump volume, noting that

$$(M/\rho) = Q_p\tau_p\ , \tag{3}$$

where M and ρ : total mass and density of the pouring-out corium, respectively.

If the sump is covered by steel plates during operation, the corium will initially flow over the cover plates. If the time constant related to their complete melting is large enough compared to the time constant of corium spreading, the scaling analysis of the sump can be simply neglected. However, for the short melting time constant the effect of the cover plates on scaling analysis can be neglected. Let us define the subscript R as the ratio between the model and the prototype in the following generalized form :

$$\Psi_R = \frac{\Psi_m}{\Psi_p} = \frac{\Psi \text{ for model}}{\Psi \text{ for prototype}}. \quad (4)$$

Similarity between the model and the prototype requires that Ψ_R should be equal to 1. For similarity in the sump,

$$(N_{su})_R = (\frac{M}{\rho})_R/(V_{su})_R = 1. \quad (5)$$

In the Inpedestal Region

After the sump is filled with the molten corium, spreading over the inpedestal floor is mainly governed by two competing forces : gravitational force and surface tension. Gravitational force on the spreading corium is produced by the difference of corium thickness between upstream and downstream. If the gravitational force becomes low enough to balance surface tension on the leading edge of the spreading corium, the corium will uniformly spread maintaining its minimum thickness. Therefore, spreading over the floor is divided into two regimes : gravitational force-dominated spreading and surface tension-dominated spreading.

After the front of the corium reaches the wall of the floor, filling is a more dominant phenomenon than spreading in the same way as water pours into the large empty-flat bowl. It can be observed that first water spread out and then its height in the bowl almost instantaneously increases.

Gravitational force-dominant spreading

Boundary conditions at the radius of the sump R_{su} such as corium velocity and thickness are needed for gravitational force-dominant spreading over the inpedestal floor. After the sump is filled with the pouring-out molten corium, its height h_p continues to increase without any spilling over the sump until gravitational force resulting from the potential difference between h_p and h_s overcomes the surface tension on the leading edge of the corium. After then, the volumetric flow rate leaving the sump region Q_s continues to increase until it becomes equal to Q_p. The corium velocity at R_s, v_s, can be approximately obtained using Bernoulli's equation:

$$v_s = \sqrt{2\, g\, (\, h_p - h_s\,)}. \quad (6)$$

The volumetric flow rate at R_s, Q_s , is expressed as

$$Q_s = 2\, \pi\, R_{su}\, h_s \sqrt{2\, g\, (\, h_p - h_s\,)}. \quad (7)$$

At the maximum draining rate from the sump region, which is equal to Q_p, the following relationship between h_s and h_p can be obtained differentiating Eq.(7) with h_s :

$$h_s = \frac{2}{3}\, h_p. \quad (8)$$

Putting Eq.(8) into Eq.(7) yields

$$h_s = a_1\, (\, Q_p \,/\, R_{su}\,)^{2/3}, \quad (9)$$

$$v_s = a_2\, (\, Q_p \,/\, R_{su}\,)^{1/3}, \quad (10)$$

where $a_1 = (\, 1 \,/ 4\pi^2 g\,)^{1/3}$, $a_2 = (\, g \,/\, 2\pi\,)^{1/3}$.

As noticed later, Eqs. (9) and (10) are equivalent to the critical conditions for open channel flow. Corium spreading is expected to be analyzed by equations governing the dynamics of open channel flow. Since spreading is at quasi-steady at the maximum draining rate from the sump, the following steady-state macroscopic mass and momentum balance equations for a fixed control volume are used:

$$\text{mass} : \int_A \rho\vec{v}\cdot\vec{n}\, dA = 0, \quad (11)$$

$$\text{momentum} : \int_A \rho\vec{v}\vec{v}\cdot\vec{n}\, dA = \int_V \rho\vec{g}\, dV + \int_A \vec{t}_{(n)} dA, \quad (12)$$

where A and V : control surface and volume, respectively,

$\vec{t}_{(n)}$: stress vector acting on the control surface.

Then, Eq.(11) becomes

$$h\, \frac{d\langle v\rangle}{dr} + \langle v\rangle\, \frac{dh}{dr} + \frac{\langle v\rangle\, h}{r} = 0, \quad (13)$$

where $\langle v\rangle = \int_0^h v dz \,/ \int_0^h dz$.

Equation (13) is the other expression of steady-state mass balance

$$Q_p = 2\, \pi\, r\, h\, \langle v\rangle. \quad (14)$$

Integrating Eq.(12) over the control surface and control volume yields

$$[\, 1 - \frac{C}{\cos\delta}\, N_{Fr}\,]\, \frac{dh}{dr} + \frac{h}{2\cos\delta}\, \frac{d}{dr}\, \cos\delta$$
$$= \frac{C}{\cos\delta}\, \frac{h}{r}\, N_{Fr} + \tan\delta - \frac{1}{2\cos\delta}\, f\, N_{Fr}, \quad (15)$$

where N_{Fr} : Froude number defined as $N_{Fr} = \langle v\rangle^2 \,/\, gh$,

δ : floor angle,

f : wall friction coefficient.

According to the Manning correlation[15] for friction coefficients in open channel flow, f can be expressed as

$$f = a_f\, \frac{n^2}{h^{1/3}}, \quad (16)$$

where a_f : constant

n^2 : coefficient depending on the material of the substrate and its roughness.

Let us nondimensionalize Eq.(15) for turbulent flow and the flat surface of the floor using the following nondimensional variables and parameters:

$$r^* = \frac{r}{r_r},\ h^* = \frac{h}{h_r},\ \langle v\rangle^* = \frac{\langle v\rangle}{\langle v\rangle_r},$$

$$(N_{Fr})^{in} = \frac{\langle v\rangle_r^2}{gh_r},\ (N_G)^{in} = \frac{h_r}{Rin},\ (N_f)^{in} = \frac{n_{in}^2 r_r}{h_r^{4/3}}$$

$$[\, 1 - (N_{Fr})^{in}\, \frac{1}{r^{*2}h^{*3}}\,]\, \frac{dh^*}{dr^*} =$$
$$(N_{Fr})^{in}\, \frac{1}{r^{*3}h^{*2}} - a_f\, (N_f)^{in}\, (N_{Fr})^{in}\, \frac{1}{h^{*3\frac{1}{3}} r^{*2}}, \quad (17)$$

where R_{in} and n_{in} : radius and coefficient of the inpedestal floor.

The subscript r denotes the reference value.

As Eq.(17) is a nonlinear differential equation, its integral response function is expressed in the following nonlinear form:

$$h^* = h_s^* + \int_{r_s^*}^{r^*} f_2\,[h^*(r^*),\, r^*,\, (N_{Fr})^{in},\, (N_f)^{in}]dr^*, \tag{18}$$

where the subscript s represents properties at R_s, and h_s^* and r_s^* are defined as $h_s^* = \frac{h_s}{h_r}$ and $r_s^* = \frac{R_{su}}{R_{in}}$, respectively.

Based on Eq.(18), similarity between the prototype and the model for corium spreading over the inpedestal floor can be maintained if the following boundary conditions and scaling parameters are satisfied:

boundary conditions:

$$h_s^* = \frac{h_s}{h_r},\ r_s^* = \frac{R_{su}}{r_r} = (N_G)^{in},\ r_{in}^* = \frac{R_{in}}{r_r}$$

scaling-parameters:

$(N_{Fr})^{in}$, $(N_f)^{in}$

If R_{in}, h_s, and v_s are used for reference values of R_{in}, h_r, and v_r, respectively, h_s^*, r_{in}^*, and $(N_{Fr})^{in}$ become 1. Then, the following two scaling parameters are required to be satisfied for similarity:

$$(N_G)_R^{in} = \left[\frac{R_{su}}{R_{in}}\right]_R = 1, \tag{19}$$

$$(N_f)_R^{in} = \left[\frac{n_{in}^2\, R_{in}}{h_s^{4/3}}\right]_R = 1. \tag{20}$$

The scaling parameter related to the time constant of spreading can be obtained using the following nondimensional characteristic equation which represents the propagation of mass and momentum of the spreading corium:

$$\frac{dr^*}{dt^*} = (N_d)^{in} \langle v\rangle^*, \tag{21}$$

where $(N_d)^{in}$: nondimensional dynamic time constant in the inpedestal floor defined as

$$(N_d)^{in} = \frac{\tau_r}{\tau_d^{in}}$$

$$\tau_d^{in} = \frac{R_{in}}{v_r}$$

$$t^* = \frac{t}{\tau_r}$$

If τ_d^{in} is used for τ_r, $(N_d)_{in}$ automatically satisfies the similarity requirement. However, since each process has its own time constant, the total pouring-out time τ_p is used as the reference time constant for all the processes. Then, $(N_d)_R^{in}$ becomes

$$(N_d)_R^{in} = \frac{1}{(R_{in})_R}\left[\frac{(M/\rho)\tau_p^2}{R_{su}}\right]_R^{\frac{1}{3}} = 1 \tag{22}$$

Surface tension-dominant spreading

If the gravitational force becomes low enough to balance surface tension on the leading edge of the spreading corium, the corium will uniformly spread maintaining its minimum thickness. The minimum thickness, h_{min}, is determined considering the static equilibrium condition between gravitational force and surface tension.

With the assumption that the corium uniformly spreads over the floor keeping its minimum, surface tension-dominant spreading can be simply modelled as follows:

$$Q_p\,(t - t_o) = \pi\, r^2\, h_{min}, \tag{23}$$

where t_o : initial spreading time.

In the nondimensional form Eq.(23) becomes

$$(t^* - t_o^*) = (N_{dm})^{in}\, r^{*2}, \tag{24}$$

where

$$(N_{dm})^{in} = \frac{\tau_{dm}^{in}}{\tau_p},\ \tau_{dm}^{in} = \frac{\pi\, R_{in}^2\, h_{min}}{Q_p}.$$

For similarity

$$(N_{dm})_R^{in} = \left[\frac{R_{in}^2\, h_{min}}{(M/\rho)}\right]_R = 1. \tag{25}$$

The following transition criterion between gravitational force-and surface tension-dominant spreading can be used assuming that the spreading regime is decided at the inlet of the inpedestal floor:

$$N_s^m = \frac{h_{min}}{h_s}.$$

If N_s^m is larger than 1, that is, h_{min} is larger than h_s, surface tension-dominant spreading takes place because the corium thickness can not be smaller than h_{min}.

For similarity

$$(N_s^m)_R = \left[\frac{h_{min}}{h_s}\right]_R = 1. \tag{26}$$

Filling in the inpedestal region

After the front of the corium reaches the wall of the inpedestal floor, its spreading will stop at the doorway. While part of the corium passes through the doorway, the remaining corium continues to accumulate in the inpedestal floor and the sump. The rate of an increase in the corium thickness in both regions is governed by the following mass balance equation:

$$\frac{dh_{in}}{dt} = \frac{Q_p - Q_d}{\pi R_{in}^2}, \tag{27}$$

where h_{in} : uniform thickness of the corium in the inpedestal region which includes both of the inpedestal floor and the sump,

Q_d : volumetric draining rate of the corium passing the doorway.

In the same way as the estimation of the corium thickness at R_s, Q_d can be estimated:

$$Q_d = bwh_d^{3/2}, \tag{28}$$

where h_d : corium thickness in the doorway,

w : width of the doorway,

b : constant defined as

$$b = \left(\frac{2}{3}\right)^{\frac{3}{2}}\sqrt{g}.$$

Maximum h_d and h_{in}, h_{dm} and h_{inm}, are obtained from Eq.(27) with Eq.(28) :

$$h_{dm} = (Q_p/bw)^{\frac{2}{3}},$$

$$h_{inm} = \frac{3}{2}(Q_p/bw)^{\frac{2}{3}}.$$

At maximum drainage the corium velocity v_{dm} becomes

$$v_{dm} = Q_p/(w\, h_{dm}).$$

Integrating Eq.(27) over a time interval from t to t_o and nondimensionalizing it yields

$$t^* - t_o^* = 2(\frac{2}{3})^{\frac{3}{2}} N_f^{in} [f(x) - f(x_o)], \tag{29}$$

where $x = \sqrt{h_{in}^*}$, $x_o = \sqrt{h_{in0}^*}$, $h^* = h_{in}/h_{inm}$, $h_{in0}^* = h_{in0}/h_{inm}$,

h_{in0} : initial corium thickness in the inpedestal region at t_0,

t_0 : time at which filling starts in the inpedestal region,

$$(N_f)^{in} = \frac{\tau_f^{in}}{\tau_p} = \frac{\pi R_{in}^2 h_{min}}{(M/\rho)},$$

$$f(x) = \frac{1}{6}\ln\left[\frac{1 - x^3}{(1 - x)^3}\right] - \frac{1}{\sqrt{3}}\tan^{-1}(\frac{2x + 1}{\sqrt{3}}).$$

The filling time constant of the inpedestal region represents how long it takes to fill the inpedestal region up to the maximum corium thickness without drainage. In the practical case, as t_o and $f(x_o)$ can be safely neglected, Eq.(43) can be expressed in the nonlinear form :

$$h_{in}^* = h_{in}^*(t^*, (N_f)^{in}) \tag{30}$$

In view of Eq.(30) the filling process in the inpedestal region is determined by N_f^{in}.

Therefore, for similarity

$$(N_f)_R^{in} = (R_{in}^2)_R \left[\frac{1}{w^2 (M/\rho)\, \tau_p^2}\right]_R^{\frac{1}{3}}. \tag{31}$$

In the expedestal region

Basically, the same phenomena as those in the inpedestal region are expected to take place in the expedestal region. The main difference between them is the existence of two-dimensional flow like a jet flow coming out of the small slit. Also, the traveling distance of the corium up to the shell is much shorter than its circumferential traveling distance. Therefore, even though the front of the corium reaches the shell, spreading in the circumferential direction will occur until all the corium is solidified on its way in the circumferential direction.

Gravitational force-dominant spreading

As it is not easy to model two-dimensional spreading in the expedestal region, analysis is done through simple two-dimensional treatment considering leakages in the y direction out of the control volume. The following three governing equations are used :

mass :

$$\int_A [v_x\vec{i} + v_y\vec{j}]\cdot\vec{n}\, dA = 0 \tag{32}$$

momentum :

$$\text{x – direction} : \int_A \rho v_x [v_x\vec{i} + v_y\vec{j}]\cdot\vec{n}\, dA = \int_V \rho g_x\, dV - \int_A \vec{i}\cdot\vec{n} p\, dA + \int_A \vec{i}\cdot\vec{n}\cdot\overline{\overline{\tau}}\, dA \tag{33}$$

$$\text{y – direction} : \int_A \rho v_y (v_x\vec{i} + v_y\vec{j})\cdot\vec{n}\, dA = \int_V \rho g_y\, dV - \int_A \vec{j}\cdot\vec{n} p\, dA + \int_A \vec{j}\cdot\vec{n}\cdot\overline{\overline{\tau}}\, dA \tag{34}$$

Let us express nondimensionalized integral mass and momentum equations obtained in the same way as that in the inpedestal region in the nondimensional form :

$x^* = x/L_{ex}$, $y^* = y/w$, $v_x^* = \langle v_x\rangle/v_r$, $h_x^* = \langle h_x\rangle/h_r$, $h_o^* = h_o/h_r$,

$h_Y^* = h_Y/h_r$, $(N_G)^{ex} = w/L_{ex}$, $(N_G)^{exl} = h_r/L_{ex}$, $(N_{Fr})^{ex} = v_r^2/(gh_r)$

$$\frac{dv_x^* h_x^*}{dx^*} + \frac{2}{(N_G)^{ex}} h_Y^* v_y^* = 0, \tag{35}$$

$$\left[1 - (N_{Fr})^{ex}\frac{v_x^{*2}}{h_x^*}\right]\frac{dh_x^*}{dx^*} = 2\frac{(N_{Fr})^{ex}}{(N_G)^{ex}}\frac{v_x^* v_y^* h_Y^*}{h_x^*} - a_{fx}\frac{(N_f)^{exl}(N_{Fr})^{ex}}{(N_G)^{exl}}\frac{v_x^{*2}}{2h_x^{*4/3}}, \tag{36}$$

$$(N_{Fr})^{ex}\frac{dv_y^*}{dx^*} = -\, a_{fy}\frac{(N_f)^{exl}(N_{Fr})^{ex}}{(N_G)^{exl}}\frac{v_y^{*2}}{2h_x^{*4/3}} + \frac{1}{(N_G)^{ex}}\frac{(h_o^{*2} - h_Y^{*2})}{h_x^*} \tag{37}$$

where a_{fx} and a_{fy} : constants related to the Manning correlation for x- and y-directional friction,

$$(N_f)^{exl} = (n_{ex}^2/h_r^{\frac{1}{3}}).$$

The integral function of Eqs.(35), (36) and (37) can be expressed in the following nonlinear form :

$$h_x^* = h_d^* + \int_0^{x^*} g\, [h_x^*, h_o^*, h_Y^*, x^*, (N_G)^{ex}, (N_{Fr})^{ex}, (N_f)^{ex}]\, dx^*, \tag{38}$$

where the subscript d represents properties at the exit of the doorway, h_d^* is defined as $h_d^* = h_d/h_r$, and $(N_f)^{ex} = L_{ex}(n_{ex}^2/h_r^{4/3})$.

Assuming that h_o^* and h_Y^* are functions of not only variables and parameters in Eq.(38) but also h_{min}^*, we obtain the following integral function from Eq.(38) :

$$h_x^* = h_d^* + \int_o^{x^*} g\, [h_x^*, h_{min}^*, x^*, (N_G)^{ex}, (N_{Fr})^{ex}, (N_f)^{ex}]\, dx^*, \tag{39}$$

where $h_{min}^* = h_{min}/h_r$.

The above assumption is equivalent to saying that the front of the spreading corium in the circumferential direction in the expedestal region maintain its minimum thickness. Based on Eq.(39), similarity between the prototype and the model for corium spreading over the expedestal floor is maintained if the following boundary condition and scaling parameters are satisfied :

boundary conditions :

$$h_d^* = \frac{h_d}{h_r},\ x_{ex}^* = \frac{L_{ex}}{L_{ex}} = 1,\ h_{min}^* = \frac{h_{min}}{h_r} = (N_m)^{ex},$$

scaling parameters :

$(N_G)^{ex}$, $(N_{Fr})^{ex}$, $(N_f)^{ex}$.

If h_{dm} and v_{dm} are used as reference values for h_r and v_r, respectively, and the same material as that of the inpedestal floor is used for the expedestal floor, h_d^*, $(N_{Fr})^{ex}$, and $(N_f)^{ex}$ are automatically satisfied. Then, for similarity

$$(N_G)_R^{ex} = \left[\frac{w}{L_{ex}}\right]_R = 1, \tag{40}$$

$$(N_{min})_R^{ex} = (h_{min})_R \left[\frac{w\tau_p}{(M/\rho)}\right]_R^{2/3}. \tag{41}$$

A scaling parameter related to a time constant of corium spreading can be obtained in the same way as that in the inpedestal region :

$$\frac{dx^*}{dt^*} = \frac{1}{(N_d)^{ex}} v_x^*, \tag{42}$$

where $(N_d)^{ex} = \frac{\tau_d^{ex}}{\tau_p}$, $\tau_d^{ex} = L_{ex}/v_r$.

Therefore, for similarity

$$(N_d)_R^{ex} = \frac{1}{(L_{ex})_R}\left[\frac{(M/\rho)\tau_p^2}{w}\right]_R^{\frac{1}{3}} \tag{43}$$

In the doorway the same analysis as that in the expedestal region can be done except the consideration of leakages in the y-direction ;

$$(N_G)_R^{dw} = (\frac{w}{L_{dw}})_R = 1, \tag{44}$$

$$(N_d)_R^{dw} = \frac{(\tau_p)_R^{4/3}}{(L_{dw})_R}\left[\frac{w}{(M/\rho)}\right]_R^{1/3} \tag{45}$$

Surface tension-dominant spreading

In the same way as that in the inpedestal floor, surface tension-dominant spreading in the expedestal region is simply modeled as follows :

$$Q_p(t - t_o) = h_{min}\frac{\Theta}{2}[R_{ex}^2 - (R_{in} + w)^2], \tag{46}$$

where t_o : time at which the front of the corium just passes through the doorway,

Θ : spreading angle in the circumferential direction,

R_{ex} and R_{in} : distances from the center of the floor to the walls of the expedestal floor and of the inpedestal floor, respectively.

In the nondimensional form Eq.(46) becomes

$$(t^* - t_o^*) = (N_{dm})^{ex}\Theta^*, \tag{47}$$

where

$\Theta^* = \Theta/2\pi$,

$(N_{dm})^{ex} = \frac{\tau_{dm}^{ex}}{\tau_p} = \frac{A_{ex}h_{min}}{(M/\rho)}$,

$\tau_{dm}^{ex} = (A_{ex}h_{min})/Q_p$,

A_{ex} : total area of the expedestal floor

For similarity

$$(N_{dm})_R^{ex} = \left[\frac{A_{ex}h_{min}}{(M/\rho)}\right]_R = 1. \tag{48}$$

For similarity for the doorway,

$$(N_{dm})_R^{dw} = \left[\frac{wL_{dw}h_{min}}{(M/\rho)}\right]_R = 1. \tag{49}$$

3.3 Water-corium-concrete interactions

In the first step the spreading group mechanism is treated considering the mass and momentum transfer of the corium. In the second step both of water-corium interactions and concrete-corium interactions as cooling mechanisms are considered for analysis of the energy transfer of the corium. The second step is divided into two substeps : interactions before solidification starts and interactions after solidification starts. The main difference between two substeps is thermal convection. While thermal convection is very important in the first substep, it is negligible in the second substep because the corium has reached solidification temperature in the latter. Although crust formation and mushy zone formation should be distinguished when solidification is considered, the rapid solidification process of the corium due to high energy loss to the coolant and/or the concrete enables us to neglect the difference between them.

Interactions before solidification starts

A simplified internal energy equation can be derived using the following assumptions:

- no void fraction in the corium,
- uniform volume source and material constants,
- z-direction dominant conduction,
- uniform velocity distribution in the z-direction.

$$(\rho c_p)_{co}\left[\frac{\partial \overline{T}_{co}}{\partial t} + v_{co}(r)\frac{\partial \overline{T}_{co}}{\partial r}\right] = q_{co}''' - \frac{1}{h}(q_c'' + q_w''), \tag{50}$$

where $\overline{T}_{co}$: axial-averaged corium temperature defined as

$\overline{T}_{co} = \int_0^h T_{co}dz/h$,

q_c'' and q_w'' : heat fluxes from the corium into the concrete and the overlying water, respectively, defined as

$q_c'' = h_c(\overline{T}_{co} - T_a)$,

$q_w'' = h_w(\overline{T}_{co} - T_w)$,

T_a and T_w : ablation temperature and water temperature, respectively.

By the characteristic method the following two nondimensional equations are obtained:

$$dt^* = N_d dr^* \tag{51}$$

$$dT_{co}^* = \frac{hc}{h^*v^*}\left\{N_{vh}h^{*N} - T_{co}^*(1 + N_{cw}) - N_{cc} - 1\right\}dr^*, \tag{52}$$

where

$t^* = \frac{t}{\tau_p}$, $r^* = \frac{r}{R}$, $v^* = \frac{v}{v_r}$, $h^* = \frac{h}{h_r}$, $T_{co}^* = \frac{\overline{T}_{co} - T_s}{\Delta T_w}$,

$\Delta T_w = T_s - T_w$, $\Delta T_a = T_s - T_a$, $\tau_d = R/v_r$,

$N_d = \frac{\tau_d}{\tau_p}$, $N_{hc} = \frac{h_w\tau_d}{(\rho c_p)_{co}h_r}$, $N_{vh} = \frac{q_{co}''' h}{h_w\Delta T_w}$,

$N_{cw} = \frac{h_c}{h_w}$, $N_{cc} = \frac{h_c\Delta T_c}{h_w\Delta T_w}$.

The integral function form of Eq.(52) is expressed as

$$T^*_{co} = T^*_{coo} + \int_{r_o}^{r^*} f\,[T^*_{co},h^*,v^*,N_{hc},N_{vh},N_{cw},N_{cc}]\,dr^*, \qquad (53)$$

where T^*_{coo} :nondimensional corium temperature at $r^* = r_0^*$, defined as

$$T^*_{coo} = \frac{\Delta T_{sup}}{\Delta T_w} = N_{sup}$$

$$\Delta T_{sup} = \overline{T}_{coo} - T_s.$$

Assuming that v and h are nearly constant, we can get the following analytical solution from Eq.(52):

$$T^*_{co} = -p^* + (N_{sup} + p^*)\exp[-N_{hc}(1 + N_{cw})(r^* - r^*_o)], \qquad (54)$$

where $p^* = (1 + N_{cc} - N_{vh})/(1 + N_{cw})$.

If water cooling is dominant over concrete cooling and internal volumetric heat generation, Eq.(54) is simplified as

$$T^*_{co} = -1 + [N_{sup} - 1]\exp[-N_{hc}(r^* - r^*_o)]. \qquad (55)$$

In this case the thermal process before solidification starts is mainly governed by two scaling parameters, N_{sup} and N_{hc}, which are related to initial superheat and thermal convection, respectively. Note that, if T^*_{co} becomes zero, solidification starts and the second substep should be taken.

If the corium thickness is greater than the water depth, no convective water cooling is available, radiation from the top of the corium to structural materials in the containment is a main cooling mechanism at the top of the corium. The following transition criterion between water cooling and radiation cooling only is used:

$$N_{wd} = \frac{h_{wd}}{h}, \qquad (56)$$

where h_{wd} : overlying water depth.

As a bifurcation phenomenon quenching should be considered. If the temperature difference between the corium temperature and the water temperature drops up to the sputtering temperature difference which depends on the containment pressure, quenching starts. In the similar way to the definition of DNBR, the following sputtering number N_{sp} is defined:

$$N_{sp} = \frac{\Delta T_w}{\Delta T_{sp}}, \qquad (57)$$

where $\Delta T_{sp} = T_{sp} - T_w$.

Interactions after solidification starts

After the corium temperature reaches its solidification temperature, rapid solidification occurs at both of its top and bottom. Using the energy balance equation and boundary conditions for the top crust,

$$(\rho C_p)_t \frac{\partial T_t}{\partial t} = k_t \frac{\partial^2 T_t}{\partial x^2} + q_t''', \qquad (58)$$

at z = 0,

$$-k_t \frac{\partial T_t}{\partial x}\Big|_{z=0} = -q_w'' = -h_w(T_{to} - T_w),$$

at $z = \delta_t$,

$$T_t(z = \delta_t) = T_s,$$

$$-k_t \frac{\partial T_t}{\partial z}\Big|_{z=\delta_t} = -q_b'' - (h_{fs}\rho)_t \frac{d\delta_t}{dt},$$

the following nondimensional equation can be derived:

$$T_t^* = \frac{\overline{T}_t - T_s}{\Delta T_w},\ \delta_t^* = \frac{\delta_t}{h_r},\ t^* = \frac{t}{\tau_p},\ N^t_{Bi} = \frac{h_w h_r}{k_t},\ N^t_{vh} = \frac{q_t''' h_r}{h_w \Delta T_w},$$

$$N^t_{sol} = \frac{(h_{fs}\rho)_t h_r}{\tau_p h_w \Delta T_w},\ N^t_{co} = \frac{q_t''}{h_w \Delta T_w},\ N^t_{pe} = \frac{(\rho C_p)_t h_r}{\tau_p h_w},$$

$$\frac{d\delta_t^*}{dt^*}\left\{-N^t_{sol} + N^t_{pe}\left[T_t^* + \delta_t^* \frac{N^t_{Bi}}{(1 + N^t_{Bi}\delta_t^*)} \times \left(-T_t^* - \frac{1}{2} + \frac{2}{3}N^t_{vh}\delta_t^* + \frac{1}{4}N^t_{Bi}N^t_{vh}\delta_t^{*2}\right)\right]\right\}$$

$$= -N^t_{co} - N^t_{vh}\delta_t^* + (1 + \frac{1}{2}N^t_{Bi}N^t_{vh}\delta_t^{*2})/(1 + N^t_{Bi}\delta_t^*). \qquad (59)$$

If τ^t_{cr} is defined as a time constant of top crust formation in which the corium with h_r is completely solidified by water cooling only, N^t_{sol} is expressed in terms of τ^t_{cr}:

$$N^t_{sol} = \frac{\tau^t_{cr}}{\tau_p}, \qquad (60)$$

where $\tau^t_{cr} = (h_{fs}\rho)_t h_r/(h_w \Delta T_w)$. The integral response form of Eq.(59) is expressed in the following nonlinear form:

$$\delta_t^* = \int_{t_o^*}^{t^*} g_t\,[\delta_t^*, N^t_{Bi}, N^t_{vh}, N^t_{co}, N^t_{sol}, N^t_{pe}]\,dt^*, \qquad (61)$$

Note that, if all the requirement of similarity in the first substep are satisfied, the requirement of similarity on t_o^* is automatically satisfied. Then, for similarity

$$(N^t_{Bi})_R = \left[\frac{h_w h_r}{k_t}\right]_R = 1,\ (N^t_{vh})_R = \left[\frac{q_t''' h_r}{h_w \Delta T_w}\right]_R = 1,$$

$$(N^t_{sol})_R = \left[\frac{(h_{fs}\rho)_t h_r}{\tau_p h_w \Delta T_w}\right]_R = 1,\ (N^t_{co})_R = \left[\frac{q_t''}{h_w \Delta T_w}\right]_R = 1,$$

$$(N^t_{pe})_R = \left[\frac{(\rho c_p)_t h_r}{\tau_p h_w}\right]_R = 1.$$

Assuming that N^t_{Bi} and N^t_{vh} are much less than 1, we have the following analytical solution:

$$\frac{N^t_{pe}N^t_{Bi}}{N^t}\delta_t^* + \frac{1}{N^t}\left[N^t_{sol} - \frac{N^t_{pe}N^t_{Bi}(N^t_{co} - 1)}{N^t}\right]\ln\left[1 + \frac{N^t}{N^t_{co} - 1}\delta_t^*\right]$$

$$= -(t^* - t_o^*), \qquad (62)$$

where $N^t = N^t_{vh} + N^t_{Bi}$.

If $\left[\frac{N^t}{N^t_{co} - 1}\right] \ll 1$, Eq.(62) is simplified as follows:

$$\delta_t^* = \left[\frac{1 - N^t_{co}}{N^t_{sol}}\right](t^* - t_o^*). \qquad (63)$$

In the similar way scaling parameters for bottom crust formation can be obtained.

If properties of top and bottom crust are the same, we have

$$N^b_{Bi} = N^t_{Bi},\ N^b_{vh} = N^t_{vh},\ N^b_{sol} = N^t_{sol},\ N^b_{pe} = N^t_{pe}.$$

Also, if $N^b_{co} = N^t_{co}$, there are no additional scaling parameters to govern the process of bottom crust formation.

Let us estimate heat fluxes, q_t'' and q_b''. Using the energy conservation equation and boundary conditions in the molten corium expressed as

$$(\rho c_p)_{co}\left(\frac{\partial T_{co}}{\partial t} + v_{co}\frac{\partial T_{co}}{\partial r}\right) = k_{co}\frac{\partial^2 T_{co}}{\partial z^2} + q_{co}''', \tag{64}$$

$$\text{at } z = 0,\ T_{co}=T_s \text{ and } -k_{co}\frac{\partial T_{co}}{\partial z}=-q_b'',$$

$$\text{at } z = l,\ T_{co}=T_s \text{ and } -k_{co}\frac{\partial T_{co}}{\partial z}=-q_t'',$$

the following dimensional form is obtained:

$$N_{co}^t = N_{co}^b = \frac{1}{2}N_{vh}\frac{\left[1-\frac{(1+N_{cc})}{N_{sol}h^*}(t^*-t_0^*)\right]}{\left[1-\frac{N_{vh}}{N_{sol}}(t^*-t_0^*)\right]} \tag{65}$$

Note that no additional scaling parameters are added to express N_{co}^t and N_{co}^b.

4. APPLICATIONS

Let us examine time constants and relative importance of scaling parameters to govern each group mechanism using information on typical severe accident parameters given in Table 4 used in Ref. 7. In Table 4 heat transfer coefficients for water cooling include effects by radiation and convection. For estimation of convective heat transfer coefficients before bottom crust formation the film boiling coefficient of Berenson[16] is multiplied by a factor of three in order to account for the effect of gas bubbles on convective heat transfer coefficients including interfacial area enhancement and corium entrainments. However, after bottom crust formation the Berenson correlation without any multiplication factor is used. Two simple models are used for estimating heat transfer coefficients of concrete cooling. While the Berenson correlation for slag film boiling is used before bottom crust generation, radiation is considered after bottom crust generation. Limestone is selected for a prototype concrete material and its thermophysical properties given in Ref.(17) are used. While the volumetric internal heat generation rates under gas injection include chemical and decay heat, decay heat only is considered after bottom crust formation. Note the relatively high superheat of the corium which will result in its more spreading and, as a result, produce more conservative results.

For the sump without gas injection due to stable bottom crust formation, the following filling time and scaling parameter N_{su} are calculated:

$\tau_{su}=696.3$ sec and $N_{su}=0.29$.

The critical thickness h_c at R_{su} and the minimum thickness h_{min} are $h_c=3.57\times10^{-3}$ m and $h_{min}=5.8\times10^{-3}$ m. As h_{min} is greater than h_s, spreading is governed by surface tension-dominant spreading. In this regime the corium velocity at R_s is calculated using the mass balance equation:

$v_s=0.114$ m/s.

Now, we have time constants and scaling parameters related to spreading:

$\tau_{dm}^{in}=35$ sec, $\tau_{dm}^{dw}=0.72$ sec, $\tau_{dm}^{ex}=54$ sec,
$(N_{dm})^{in}=1.5\times10^{-2}$, $(N_{dm})^{dw}=3\times10^{-4}$, $(N_{dm})^{ex}=2.27\times10^{-2}$.

As the time constant in the doorway τ_{dm}^{dw} is very short, it can be said that the doorway has a negligible effect on spreading. For filling we have

$h_{inm}=4.25$ cm, $\tau_f^{in}=256.5$ sec, $(N_f)^{in}=0.11$,
$h_{dm}=2.83$ cm, $\tau_f^{ex}=359.8$ sec, $(N_f)^{ex}=0.15$.

The time constants related to filling is much larger than those related to spreading due to the small pouring-out rate, Q_p. The time constants of spreading are needed to be compared with those related to energy transfer and crust formation. If the latter are much shorter than the former, crust formation considerably retards spreading and filling, and spreading may be a dominant phenomenon over filling.

For interaction before solidification starts we have

$N_{Bi}=0.15$, $N_{vh}=0.017$, $N_{cc}=0.62$, $N_{sup}=0.29$, $N_{cw}=2.5$, $N_{hc}=1.09$.

According to Fig.1, it can be said that thermal convection and concrete cooling have considerable effects on the variations of the corium temperature while the effect of internal heat generation rates is very small. We can see the importance of the superheat number, N_{sup}, clearly if its sensitivity is obtained differentiating Eq.(54) with N_{sup}:

$$\frac{\partial T_{co}^*}{\partial N_{sup}} = \exp\left[-N_{hc}(1+N_{cw})(r^*-r_0^*)\right].$$

The sensitivity of the superheat number is 1 at $r^*=r_0^*$ and then exponentially decreases with corium movement. Initially, N_{sup} has high relatively importance.

For interaction after solidification starts we have

$N_{vh}=5.8\times10^{-3}$, $N_{Bi}=0.11$, $N_{Pe}=0.018$, $N_{sol}=5.5\times10^{-3}$,
$N_{co}=2.9\times10^{-3}$, $N_{cc}=0.47$, $N_{cw}=1.87$.

As N_{Bi} and N_{vh} are much less than 1, the analytical solution, Eq.(62), can be used for determining relative importance of each scaling number. Figures 2 and 3 tell us that N_{sol}, N_{Bi}, and N_{vh} have the highest, intermediate, and the lowest relative importance for crust formation, respectively. Based on Figs.(1) through (3), the time constants related to thermal convection and crust formation, τ_{tc} and τ_{cr}, respectively, are estimated:

$\tau_{tc}=1$ sec and $\tau_{sol}=4.6$ sec.

Since the time constants related to the thermal precess are much shorter than those related to the hydraulic process, scaling parameters related to the former have considerable effects on the process of corium transfer and spreading is more important than filling.

Let us design a scale model based on information given in Table 4, the one-tenth scaling ratio of R_{in}, and use of lead as a corium simulant. The solidification temperature of lead is 600 K. From Eq.(19)

$$(N_G)_R^{in} = \left[\frac{R_{su}}{R_{in}}\right]_R = 1\ ;\ (N_{su})_R = 1.$$

From Eq.(25)

$$(N_{dm})_R^{in} = \left[\frac{R_{in}^2 h_{min}}{(M/\rho)}\right]_R = 1\ ;\ (M/\rho)_R = 5.4\times10^{-3},$$

where $(h_{min})_R=0.54$.

From Eq.(22)

$$(N_d)_R^{in} = \frac{1}{(R_{in})_R}\left[\frac{(M/\rho)\tau_p^2}{R_{su}}\right]_R^{\frac{1}{3}} = 1\ ;\ (\tau_p)_R = 0.136.$$

From Eq.(31)

$$(N_f)_R^{in} = (R_{in}^2)_R\left[\frac{1}{w^2(M/\rho)\tau_p^2}\right]_R^{\frac{1}{3}} = 1\ ;\ (w)_R = 0.1.$$

From Eq.(5)

$$(N_{su})_R = \left[\frac{(M/\rho)}{V_{su}}\right]_R = 1\ ;\ (V_{su})_R = 5.4\times10^{-3}.$$

From Eq.(20)

$$(N_f)_R^{in} = \left[\frac{n^2 R_{in}}{h_s^{4/3}}\right] = 1\ ;\ (n)_R = 2.1,$$

where $(h_s)_R = \left[\frac{Q_p}{R_{su}}\right]_R^{2/3} = 0.54.$

From Eqs.(40) and (44)

$$\left[N_G\right]_R^{ex} = \left[\frac{w}{L_{ex}}\right]_R = 1\ ;\ (L_{ex})_R = 0.1,$$

$$\left[N_G\right]_R^{dw} = \left[\frac{w}{L_{dw}}\right]_R = 1\ ;\ (L_{dw})_R = 0.1.$$

The remaining scaling parameters, $(N_d)_R^{ex}$, $(N_{min})_R^{ex}$, and $(N_{dm})_R^{ex}$ are automatically satisfied:

$$(N_d)_R^{ex} = (N_{min})_R^{ex} = (N_{dm})_R^{ex} = 1.$$

While all the geometrical scaling parameters have the same scaling ratio, 0.1, all the scaling ratios related to the corium thickness have the same but relatively high scaling ratio, 0.54, resulting from the scaling ratio of the minimum thickness. This relatively high scaling ratio of the corium thickness causes difficulties in thermal scaling because more cooling in the model is needed than that in the prototype due to relatively high thermal capacity and latent heat in the model. For thermal scaling

$$\left[N_{sup}\right]_R = \left[\frac{\Delta T_{sup}}{\Delta T_w}\right] = 1\ ;\ \left[\Delta T_{sup}\right]_m = 65.7\text{ K},$$

$$[N_{hc}(1+N_{cw})]_R = \left[\frac{(h_c+h_w)\tau_d}{(\rho c_p)h_r}\right]_R = 1\ ;\ (h_c+h_w)_R = 1.53,$$

$$\left[\frac{(1+N_{cc})}{(1+N_{cw})}\right] = \left[\frac{h_w + h_c\frac{\Delta T_c}{\Delta T_w}}{h_c+h_w}\right]_R = 1\ ;\ \left[h_w + h_c\frac{\Delta T_c}{\Delta T_w}\right] = 1.53,$$

$$(N_{sol})_R = \left[\frac{h_{fc}\rho h_r}{(h_w\Delta T_w + h_c\Delta T_c)\tau_p}\right] = 1\ ;\ \frac{(h_c\Delta T_c)_m}{(h_w\Delta T_w)_p} = 6.17.$$

It is impossible to satisfy the above three cooling requirements simultaneously. As the time constant related to interactions before solidification starts is much shorter than that related to interactions after solidification, it is better to distort the scaling ratios related to thermal convection.

The scaling ratio of the transition criterion of spreading is automatically satisfied:

$$(N_s^m)_R = 1.$$

From the transition criterion of water cooling and radiation cooling only,

$$(N_{wd})_R = \left[\frac{h_{wd}}{h_r}\right]_R = 1\ ;\ (h_{wd})_R = 0.54.$$

The sputtering scaling ratio is very low because of the low solidification temperature of lead:

$$(N_{sp})_R = 0.11.$$

However, the important thing is whether quenching can take place during the solidification process. By rough estimation, before the top crust thickness reaches 40 cm, quenching will not occur in the model. Table 5 shows the final output of the model. It is noted that the considerable amount of lead and enormous concrete cooling are needed in the model.

Table 4 Typical Severe Accident Parameters

Boundary parameters:
τ_p = 40 min $Q_p = 8.33\times10^{-3}$ m^3/s $\dot{m}$ = 75 Kg/s $\left[M/\rho\right] = 20$ m^3
$M = 1.8\times10^5$ Kg
Geometrical parameters:
R_{su} = 2 m R_{in} = 4 m L_{dw} = 0.91 m w = 1.03 m L_{ex} = 2.09 m
H_{su} = 43 cm V_{su} = 5.8 m^3
Corium parameters:
k = 30 W/mK ρ = 9000 Kg/m^3 C_p = 480 J/KgK $h_{fs} = 2.5\times10^5$ J/Kg
T_s = 2100 K T_{sup} = 500 K σ = 1.48 Kg/s^2
q''' = 4 Mw/m^3 before initial generation of bottom crust
q''' = 1 Mw/m^3 after initial generation of bottom crust
Concrete cooling reference parameters:
T_a = 1670 K
h_c = 1940 W/m^{2}K before initial generation of bottom crust
h_c = 1077 W/m^{3}K after initial generation of bottom crust
Water cooling reference parameters:
T_w = 373 K, h_{wd} = 30 cm
h_w = 777.1 W/m^{2}K before initial generation of bottom crust
h_w = 574.5 W/m^{3}K after initial generation of bottom crust

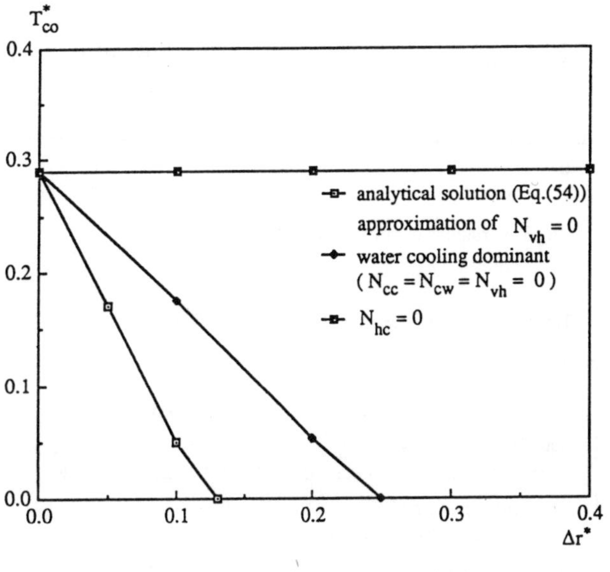

Fig. 1 Variations of corium temperature

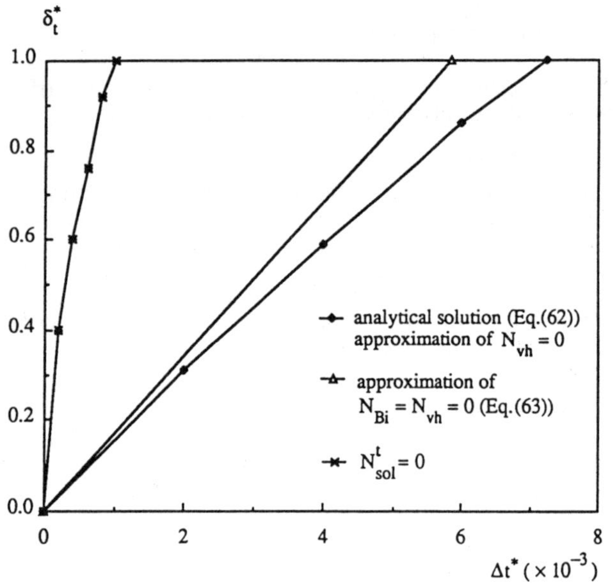

Fig. 2 Variations of top corium thickness

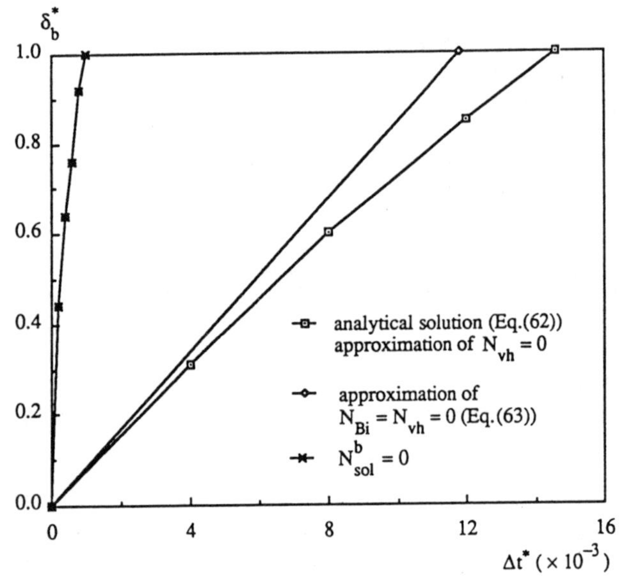

Fig.3 Variations of bottom corium thickness

Table 5 Output of the Model

Geometry :
R_{su}=20 cm, R_{in}=40 cm, w=10.3 cm, L_{ex}=20 cm, H_{su}=25 cm
Boundary conditions :
M = 1135 Kg, τ_p=5.4 min, Q_p=3.33×10^{-4} m^3/s, $\dot{m}_p$=3.5 Kg/s
Initial superheat : 65.7 K
Water depth : 16.2 cm
Thermal boundary condition : q_c'' = 10^6 w/m^2

5. SUMMARY AND CONCLUSIONS

Based on the review of the previous scaling methods a new scaling method called step-by-step integral scaling approach is proposed and applied to the scaling analysis of BWR MARK-1 shell corium interaction. The following items are its summary and conclusions:

1. Two group mechanisms are identified and considered in the first and second steps : the spreading group mechanism and water-corium-concrete interactions.
2. Two spreading mechanisms are identified and analyzed : gravitational force-dominant spreading and surface tension-dominant spreading.
3. The second step for scaling analysis of water-corium-concrete interactions is divided into two substeps : interactions before and after solidification starts.
4. Time constants related to water-corium-concrete interactions are much shorter than those related to spreading.
5. Initial superheat, thermal convection rates, concrete cooling, and latent heat of corium solidification have high relative importance.
6. Thermal scaling is much more difficult than spreading scaling because of the relatively high scaling ratio of the corium thickness.
7. Much more cooling in the model is required because of the relatively high scaling ratio of the thickness.
8. The proposed method is a useful tool for scaling analysis of a not well-understood and complex problem.

ACKNOWLEDGEMENT

This work was performed while one of the authors(H.C. NO) was on sabbatical leave at Purde University. The authors would like to express their sincere appreciation to Dr. Novak Zuber of NRC for valuable discussions and support on the subject. This work was performed under a research grant from the U.S. Nuclear Regulatory Commission to the Purdue Research Foundations.

REFERENCE

1. G. A. Greene, K. R. Perkins, and S. A. Hodge, "MARK-1 Containment Drywell : Impact of Core / Concrete Interaction on Containment Integrity and Failure of the Drywell Linear," Proc. Int. Symp. Source Term Evaluation for Accident Conditions, Columbus, Ohio, Oct. 28 - Nov. 1, 1986, IAEA-SM-281/36, International Atomic Energy Agency (1986)
2. G. A. Greene, C. Finfrock, J. Klages, C. E. Schwarz, and S. B. Burson, "Experimental Studies on Melt Spreading, Bubbling Heat Transfer, and Coolant Layer Boiling," Proc. Sixteenth Water

Reactor Safety Information Meeting, Gaitherburg, Maryland, Oct. 24 - 27, 1988, Vol. 5, pp. 341 - 358, NUREG/CP - 0097, U.S. Nuclear Regulatory Commission (1988)

3. R. E. Henry, R. J. Hammersley, B. Malinovic, R. Sehgal, and E. L. Fuller, "Experiments Relating to Drywell Shell-Core Debris Interactions," ibid, Vol. 5, pp. 317 - 340.
4. T. S. Kress and J. C. Cleveland, "Analysis of the Thermal Response of a BWR MARK-1 Containment Shell to Direct Contact by Molten Core Materials," ibid, Vol. 5, pp. 429 - 451.
5. J. J. Sienicki, M. T. Farmer, and B. W. Spencer, " Spreading of Molten Corium in MK I Geometry Following Vessel Meltthrough," Fourth Proc. of Nuclear Thermal Hydraulics, 1988 ANS Winter Meeting, Oct. 30 - Nov. 4, 1988, Washington, D.C., pp. 128 - 138 (1988)
6. M. S. Kazimi, "On the Linear Failure Potential in MARK-1 Boiling Water Reactors," Nucl. Sci. & Eng., V. 103, pp. 59 - 69 (1989)
7. P. J. Moody, R. Muralidharam, and S. S Dua, "MARK-1 Drywell Shell Temperature Response of the MARK-1 Drywell Shell," Proc. Sixteenth Water Reactor Safety Information Meeting, Gaithersburg, Maryland, Oct. 24 - 27, 1988, Vol. 5, pp. 525 - 550, NUREG/CP-0097, U.S. Nuclear Regulatory Commission (1988)
8. J. A. Boure, "A Method to Develop Similarity Laws for Two Phase Flow," ASME paper No. 70-HT-25 (1970)
9. R. W. Lockhart and R. C. Martinelli, "Proposed Correlation of Data for Isothermal Two-Phase, Two-Component Flow in Pipes," Chem. Eng. Progress, V. 45 (1949)
10. S. Y. Ahmad, "Fluid to Fluid Modeling of Critical Heat Flux : A Compensated Distortion Model," Int. J. Heat Mass Transfer, V. 16 (1973)
11. B. Spindler, "Rational Simplification of the Equations Governing the Liquid Film Flow over a Plane with Heat Flux at the Wall and Interfacial Phase Change," in Scaling in Two-Phase Flows Presented at the Winter Annual Meeting of ASME, Chicago (1980)
12. M. Ishii and N. Zuber, "Thermally Induced Flow Instabilities in Two-Phase Mixtures," 4th Int. Heat transfer Conf., Paris, Paper B. 5. 11 (1970)
13. M. Ishii and I. Kataoka, "Scaling Laws for Thermal-Hydraulic System under Single Phase and Two-Phase Natural Circulation," Nucl. Eng. and Des. 81 (1984)
14. G. A. Greene, C. Finfrock, and S. B. Burson, "Phenomenological Studies on Molten Core-Concrete Interactions," Nucl. Eng. Design, 108, pp. 167 - 177 (1988)
15. R. Manning, "On the Flow of Water in Open Channels and Pipes," Trans. Inst. Civ. Engrs, Ireland, 20:161 (1891)
16. P.J. Berenson, "Film Boiling Heat Transfer from a Horizontal Surface," Trans. ASME, Ser. C. V. 83, pp. 351 - 358 (1961)
17. M.L. Corradini, "A Transient Model for the Ablation and Decomposition of Concrete," Nucl. Tech., V. 62, pp. 263 - 273 (1983)

DIRECT CONTAINMENT HEATING EXPERIMENTS IN A ZION-LIKE GEOMETRY

Robert E. Henry and Robert J. Hammersley ■ Fauske & Associates, Inc., 16W070 West 83rd St., Burr Ridge, IL 60521
George T. Klopp ■ Commonwealth Edison Company, 1400 Opus Place, Suite 400 Downers Grove, IL 60515

ABSTRACT

The ongoing IPE studies for the Commonwealth Edison PWR designs (Zion, Byron, and Braidwood) require evaluations of accident phenomena that have been perceived to potentially challenge containment integrity including direct containment heating (DCH). Analyses and scaled experiments performed to date, suggest that the lower containment structures would play a substantial role in mitigating the extent of DCH. Since the geometry is judged to be of major importance, linearly scaled experiments were conceived and carried out to evaluate the role of such structures in a Zion-like system.

The experimental apparatus was a 5% linear scale of the reactor cavity, instrument tunnel, and seal table as well as the lower (inside the missile shield and below the operating floor) and upper containment compartments. The volume of the reactor coolant system (RCS) was scaled on a similar basis and iron thermite (temperature ~ 2400K) was used to represent the high temperature core debris. Since scaled experiments have a greater propensity for freezing of debris on structure walls, the tests were performed with 20 kg of thermite to assure the availability of sufficient melt.

Four experiments comprised the test matrix, the first three were performed with nitrogen and the fourth with steam as the pressurizing medium. The experiments included tests with a dry reactor cavity, a thin layer of water (~ 1 cm) in the cavity and one in which the cavity was half filled with water (~ 7.5 cm). In addition, the simulated lower compartment floor always had water as would be the case for virtually all accident sequences.

The results from the four tests are very consistent and demonstrate virtually no direct heating of the atmosphere. The most important result of these experiments is the lack of atmospheric heating due to debris dispersal. If any such heating is observed, it is less than 3% efficient for these experiments. The major pressurization mechanism observed is via rapid steam generation from dynamic interactions between the molten debris and water within the reactor cavity and on the lower compartment floor.

INTRODUCTION

This condition of debris dispersal and DCH was identified in the Zion Probabilistic Safety Study (CECo 1981) and the Indian Point Probabilistic Safety Study (PASNY/Con.Ed. 1982) and has been the subject of numerous experiments and analyses for example: (Bergeron 1986, Marx 1989, Spencer 1987 and Tarbell 1987). Bounding type analyses have shown that if this high temperature debris could be intimately dispersed into the containment atmosphere, substantial pressurization could occur (Ginsberg and Tutu, 1988). This would be the result of debris-gas heat transfer, oxidation of unreacted metals in the melt and combustion of hydrogen in the containment atmosphere. The combination of these physical processes has been called direct containment heating (DCH).

The potential for finely particulating and dispersing the debris into the gaseous atmosphere is directly related to the potential for debris entrainment in the reactor cavity. However, the debris could be removed (dispersed) from the reactor cavity/instrument tunnel region through a combination of droplet entrainment and the displacement of a relatively contiguous liquid mass on comparative time scales (Henry, 1989). These analyses suggest that in the reactor system there is a limited entrainment capability when a high velocity gas stream interacts with a dense liquid melt. In addition, both the entrained and non-entrained melt would encounter additional structures, such as the seal table, as the material left the instrument tunnel region. Structures which intercept a major part of the debris flow path would act as efficient separators of the debris from the high velocity gas stream. The influence of structures was reported by Ginsberg and Tutu (1988) in that models of the Zion containment "suggest that the structures exert a strong influence on the flow which issues from the cavity, redirecting the flow and, very likely, changing its droplet size characteristics." Two other studies (Spencer, et al., 1983a; Spencer, et al., 1987) observed that the structures in the containment lower compartment had a major effect on the ability of the debris to remain airborne outside the lower containment compartment.

SCALING RATIONALE

Extent of Entrainment

For scaled experiments to provide a meaningful representation of the reactor condition, the experiment should exhibit the same relative rates of debris entrainment and displacement, i.e. the ratio of the rates should be the same. As discussed in Henry (1989), one can use the Ricou-Spalding (1961) correlation to represent the entrainment rate.

$$\dot{m}_e = C_1 A_s \sqrt{\rho_D \rho_g} U_c \quad (1)$$

The displacement rate can be viewed as the mass of core debris over the displacement time.

$$\dot{m}_D \approx m_c / t_D \quad (2)$$

Assuming a constant acceleration (constant pressure differential), the displacement time can be represented as,

$$t_D = \sqrt{\frac{2L_p}{a}} = \sqrt{\frac{2\rho_D L_p L}{\Delta P}} \quad (3)$$

where a is the acceleration ($a = \Delta P/(\rho_D L)$) and the terms L and L_p are defined in Figure 1. The pressure differential can be related to the gas dynamic head, $\Delta P \approx \rho_g U_c^2/2$, such that

$$\dot{m}_D \approx 0.5 A_c \sqrt{\frac{L_p}{L}} \sqrt{\rho_D \rho_g} U_c \quad (4)$$

Formulating the ratio between the two rate dependent processes yields,

$$\frac{\dot{m}_e}{\dot{m}_D} = \frac{C_1}{0.5} \sqrt{\frac{L_p}{L}} \left(\frac{A_s}{A_c}\right) \quad (5)$$

If the ratios of the cavity lengths and areas are maintained, the relative importance of entrainment verse displacement will be retained. These ratios are maintained by a linearly scaled experiment.

De-Entrainment Scaling (Due to the Change in Flow Direction)

Sandia National Laboratory (Walker, 1987) developed a model for estimating the likelihood of debris particles not deflecting with the flow due to a change in flow direction, thus impacting structural boundaries of the flow path. The model was developed for the Zion-type seal table structure overhanging the instrument chase entrance into the lower containment space. The fraction of fine particulate trapped by the directional change is:

$$\alpha = \frac{1}{\lambda W} \ln \left[1 + \lambda W \left[\frac{W}{L_h} - \left(\frac{W}{L_h} - 1\right) \alpha\right]\right] \quad (6)$$

where

$$\lambda = \frac{3}{4} \left[\frac{1}{d}\right] \frac{\rho_g}{\rho_D} C_D \quad (7)$$

and the dimensions W and L_h are defined in Figure 2.

The ratio W/L_h in Equation (6) does not change with the linear scale. However, because λ does not necessarily vary with the scale, the term λW will, therefore, change with the scale.

Time of Flight Scaling

One of the issues to be addressed for scaled experiments is the influence of the "time of flight" for airborne debris which could be exchanging heat with the local atmosphere or oxidizing. Fractional scaled experiments have shorter lengths, which means the airborne debris is in flight for a shorter time in small scale tests compared to the reactor system. This is an inherent shortcoming of small scale experiments and must be addressed when evaluating the likely plant response.

To address this for the reactor system, we will evaluate the energy transfer from the finely particulated mass of airborne debris, which is not trapped by the structures and flows into the atmosphere, by assuming that the debris equilibrates with the containment atmosphere. This eliminates the need for any consideration with respect to the time of flight.

DESCRIPTION OF EXPERIMENTS

Based on the above scaling rationale these experiments, which are documented in (FAI, 1990), were conceived and designed to have a 5% linear scale of the Zion containment reactor cavity, instrument tunnel, lower compartment and upper compartment. Due to physical limitations of the facility, the upper and lower containment compartments were represent by two separate vessels connected by a 15 cm (6") pipe as illustrated in Figure 3. The smaller vessel (lower compartment) has a volume of 1.45 m^3 and the larger vessel, which simulates the upper compartment, has a volume of 12.2 m^3. Structures simulated at the exit of the instrument tunnel included the seal table and its structural support. This entire assembly was mounted in the smaller vessel as illustrated in Figure 4. Two steam generators and two reactor coolant pumps were also represented, at their scaled distant from the exit to the instrument tunnel, and the refueling canal wall (Figure 4). With the reactor cavity and instrument tunnel assembly in place, the free board volume of the lower compartment was about 1.33 m^3.

The reactor coolant system (RCS) was simulated by a 0.3 m (12 in.) diameter by 0.6 m (24 in.) long steel vessel which contained the iron-aluminum thermite mixture in a commercially available, high density, ceramic cone and carbon steel upper extension. To satisfy the 5% (1/20th) linear scaling, the 300 m^3 RCS volume is approximately 0.04 m^3 when divided by the 8000 linear scaling factor. Following a similar logic for the core debris, there are approximately 100,000 kg of fuel and 18,000 kg of Zircaloy in a 1000 MWe PWR system, and scaled on a volume basis this corresponds to an iron-thermite mass of 8.5 kg. Since there is a greater propensity for freezing molten material on the reactor cavity walls in small scale experiments, this mass was increased substantially (to 20 kg) to insure that the mass released to the lower compartment was a value equal to or exceeding that required for the scaled experiment.

The scaled debris mass can also be compared in terms of the energy inventory. In the absence of additional oxidation and with 100% core materials molten, the energy loading to the containment for the reactor system would be approximately 2 x 10^{11} J if the debris were quenched. With the 8000 volumetric scaling factor this corresponds to a energy inventory in the experiment of 2.5 x 10^7 J (25 MJ). As experienced in the Mark I tests (Malinovic, 1989), 20 kg of molten thermite contains at least 44 MJ above water saturation. Hence, the mass of thermite used in these experiments represents about twice the scaled debris mass and energy content. For reference, if 100% of the debris were equilibrated with the containment gas atmosphere with no oxidation of the iron, the gas temperature and pressure would be 1850K and 0.62 MPa respectively.

Once ignited, the self-sustaining reaction propagated through the 20 kg mixture in about 30 seconds. At the bottom, it melted a 2.5 cm (1 in.) lead plug in the seal between the pressurized RCS and the reactor cavity. Molten debris was then discharged into the reactor cavity under pressure, followed by the blowdown of the simulated RCS.

The experiments were performed with an initial RCS pressure of ~ 2.9 MPa in the crucible (the maximum safe operating pressure for the melt generator) and a containment pressure of 0.1 MPa. Debris dispersal is determined by the gas velocity through the reactor cavity, during RCS blowdown. This is a function of the gas mass flow rate, energy transfer to the gas and the reactor cavity dimensions. If we assume the gas behaves isothermally, the gas velocity is given by

$$U_c = \eta_T \left(\frac{P_v}{P_c}\right) \left(\frac{A_v}{A_c}\right) U_v \tag{8}$$

Since η_T and U_v are dictated by the RCS gas blowdown, these are similar between the reactor system and the experiment. The containment pressure is also similar. Hence, the gas velocity in the experiment is representative of a reactor accident condition which may have a different RCS pressure. For the same gas velocity through the reactor cavity, the experiment and the reactor system are approximately related as

$$P_{o,exp} \left(\frac{A_v}{A_c}\right)_{exp} = P_{o,R} \left(\frac{A_v}{A_c}\right)_R \tag{9}$$

The failure size for a localized penetration failure would be about 0.3 m in diameter (IDCOR, 1983). As will be discussed below, the failure size controlling blowdown in these tests had a diameter of about 0.03 m. Hence, the failure areas differ by a factor of 100 and the reactor cavity cross-sectional areas differ by 400. This means that the pressure in the experiment is representative of a value four times as large for the reactor system, i.e. about 12 MPa. This is in the range of values anticipated for station blackout-like conditions and small break loss-of-coolant accidents.

The lower containment compartment had water on the floor for all tests to protect the vessel and to represent the debris quenching which would be available under severe accident conditions. For Tests FAI/DCH-1, FAI/DCH-2 and FAI/DCH-4 water was added to the reactor cavity to simulate that which could exist in the Zion cavity. When water was added to the large vessel for protection, the upper compartment free board volume was reduced to 10.9 m^3. With the combined volume of the upper and lower compartments (12.23 m^3), this represents a scaled volume of 97,800 m^3 (3.4 x 10^6 ft^3) which is only about 28% larger than the Zion containment building.

Table 1 summarizes the various conditions including the RCS pressurization medium (driving gas). As noted, Test FAI/DCH-1, 2 and 3 used nitrogen and FAI/DCH-4 used steam. Figure 5 shows the apparatus for steam pressurization. In Tests FAI/DCH-2, 3 and 4, fire brick was added to the floor of the simulated lower compartment such that the water depth over most of the floor area was about 1 cm, (47 kg of water), as compared to the 91 kg used in FAI/DCH-1. This depth would scale to a depth of 20 cm (8 inches) on the Zion containment floor.

Instrumentation for these experiments includes:

- melt generator pressure,
- two pressure and temperature measurements (instrumented through-bolts) in the reactor cavity/instrument tunnel,
- two pressure measurements in the lower compartment,
- one pressure measurement in the upper compartment,
- temperature of the thermite igniter coil,
- temperature of the lead plug in the bottom of the melt generator,
- two measurements of the gas temperature in the lower compartment,
- one temperature measurement of the accumulated water on the lower compartment floor (located on the opposite side from the seal table), and
- one gas temperature measurement in the upper compartment at the location where the 15 cm pipe enters the large vessel (upper containment compartment).

Gas samples were taken in the upper containment compartment after the experiment to determine if additional gases (hydrogen, carbon monoxide and carbon dioxide) had been generated due to oxidation of the molten iron during the quenching process.

TEST RESULTS

The results for all the tests, which are documented in FAI (1990), are very similar. Therefore, in the interest of space, we will present FAI/DCH-1 in detail and show comparisons to the other two where appropriate.

Figure 6 illustrates the simulated RCS pressure transients for all four tests and shows that the gas blowdown of the simulated primary system was virtually the same for all tests and was completed within ~ 1.5 seconds. All signals are referenced to the time at which debris discharge was first observed are reported for 1 second prior to (-1 sec) this time. RCS blowdown is determined by the resulting gas discharge area at the bottom of the crucible, which was measured after the test to have a diameter somewhat larger than 3 cm for all tests. (Because of ablative melt attack, this parameter must be measured after the test.) Figure 7 illustrates the measured pressures in the instrument tunnel and shows a pressure increase above 0.3 MPa within about 0.5 seconds and then a decrease to about 0.16 MPa, again within 0.5 seconds. The lower compartment (lower vessel) pressures, Figure 8, were observed to be very similar to that in the reactor cavity.

It is interesting to investigate the detailed structures of the pressure response in the lower compartment for all tests, which is shown in Figure 9 with an expanded time scale. The lower compartment pressure for FAI/DCH-1 is observed to increase about 0.25 MPa within 0.5 seconds yet the temperature increase over this interval is only about 50°C, which would correspond to a 17% pressure increase if this temperature increase were uniform over the lower compartment. Primary system depressurization also occurs over this interval and would account for some of the pressurization. However, even if the blowdown is assumed to be restricted to the lower compartment, the pressure increase, including the temperature increase, would be about 0.1 MPa. Therefore an additional mechanism must exist for pressure increase.

One such mechanism is rapid steam generation, which may be either explosive or non-explosive, but in either case it is very rapid and has a first order effect on the containment response. Water initially in the cavity could be co-dispersed with the debris. For FAI/DCH-1 is about 0.7 kg or 0.04 kg moles, which if vaporized, would cause a pressure increase of 90 kPa (13 psig) in the lower compartment. Hence, more steam must be generated to explain the lower compartment pressurization. Additional steam could be generated when the molten debris contacted the water on the containment floor.

Close examination of the lower compartment pressure shows a rapid rise of 70 kPa (10 psi) in the initial ~ 100 msec and lasting 200 msec. The pressurization substantially slows for 200 msecs and then increases 166 kPa over the next 300 msecs, which likely is indicative of the gaseous blowdown from the RCS and debris quenching in the water on the lower compartment floor. It should be noted that substantial steam generation would occur as the debris contacts and interacts with the water (Benz, 1979).

In this regard, it should be noted that the initial rise of 70 kPa in the lower compartment pressure occurs over an interval in which the simulated RCS pressure has only decreased about 400 kPa (60 psi). Since the molten iron thermite is in the bottom of the crucible, this small pressure decrease corresponds to the melt discharge. Hence, the measured pressurization is not due to blowdown of the primary system. Dynamic interaction between the molten debris and water on the reactor cavity floor caused this pressurization; the magnitude of which requires that all of the 0.7 kg of water initially in the reactor cavity is vaporized by the thermal interaction. To generate this much steam requires approximately 4% of the energy in the entire melt. Early and simultaneous steam generation is likely the result of co-dispersal of debris and water caused by debris-water thermal interactions in the reactor cavity some of which may be localized steam explosions. Hence, relatively small amounts of water can have a significant influence on the debris configuration and dispersal as well as on the overall containment response, i.e., the steam partial pressure.

To further pursue the role of water, FAI/DCH-2 was performed with more water in the reactor cavity and FAI/DCH-3 was carried out with a dry reactor cavity. The FAI/DCH-2 results are almost identical with FAI/DCH-1 except the peak pressures are slightly greater. As illustrated in Figure 9, Test FAI/DCH-3 did not exhibit the early (~ 100 msec) pressurization. Also, the peak pressures in FAI/DCH-3 were slightly less than those of FAI/DCH-1 and FAI/DCH-2. Comparing the dry cavity and wet cavity cases clearly illustrates the importance of water in terms of rapid steam generation and as a mechanism to eject debris from the reactor cavity.

The gas temperature measurements in the lower compartment for all tests are shown in Figure 10. This sensor, is located behind the refueling canal wall for protection, began to increase within 200 msecs of the time when debris was released into the reactor cavity. During the 1 second pressurization-depressurization transient, the gas temperature increases to about 130°C and then slowly increases to the maximum value over the next 7 seconds. This long term temperature increase is due to heating of the lower compartment atmosphere by debris frozen to the compartment structures and not by airborne debris. Note that the test without water in the cavity (Test 3) has less heating.

Figure 11 shows the measured water temperature on the floor of the lower compartment. As illustrated, this shows a rapid increase from 16°C to about 50°C in 1.5 seconds. This increase in sensible heat corresponds to a very rapid energy addition to this 45 kg water pool; representing a transfer of about 1/7 of the melt energy into this water pool. The measurement was taken on half of the lower compartment water pool, and assuming that the other half behaves in a similar manner, (equal debris masses were found on each side) quenching in the water in this 1-1/2 seconds period accounted for about 30% of the energy removal from the distributed debris. (The steel structures submerged in the water were not included in this calculation.) If we translate the debris-to-water energy transfer rate into a heat flux averaged over the floor area, a value of 8.5 MW/m^2 is calculated. Figures 12 and 13 illustrate the measured pressures and temperatures in the upper compartment. As shown, these are similar for all the tests. Typically, the peak pressure was about 0.7 MPa (9.5 psig). The temperature increases demonstrated a rapid rise from about 20 to 50°C in about 0.5 seconds, with a slower increase to about 60°C over the next 4.5 seconds. Over the long term, both the pressures and temperatures decreased as steam condensed on the cold walls of the upper compartment vessel.

DEBRIS CHARACTERIZATION AND GAS SAMPLES

Virtually all of the debris was expelled from the reactor cavity in all four tests. This is consistent with other tests which represented a Zion like reactor cavity (Spencer 1983a, Spencer 1983b, Spencer 1987, Tarbell 1984, Tarbell 1986, Tarbell 1987 and Tutu 1988). The debris distribution after the dispersal was very similar for each test, hence only the FAI/DCH-1 results are presented in Table 2.

The debris deposited on the simulated containment floor was sieved to determine a particle size distribution for the material which fell to the floor. (The remainder of the debris was generally frozen as a thin layer on the respective surfaces listed in Table 2.) Typically, over 70% of the debris was very coarse and would not pass through a 6.3 mm square grid and more than 90% had a dimension larger than 2 mm. Hence, the debris was very coarse, a characterization which agrees with the very limited entrainment capability as discussed in the Scaling Rationale. Limited entrainment translates directly to a limited energetics (pressurization) associated with the debris dispersal.

Table 3 lists the results of the post-test samples for FAI/DCH-1. As shown, there is very little carbon monoxide and carbon dioxide detected, i.e., there is very little concrete erosion by the discharge and dispersal process. The dominant gas produced is hydrogen. This concentration in the experiment is relatively small compared to that which would create global combustion. Specifically, this is a small increase compared to the steam produced during the quenching process that could rapidly inert the containment atmosphere.

As listed in the table, the dry cavity Test FAI/DCH-3 has the smallest amount of hydrogen produced followed by FAI/DCH-1 and FAI/DCH-2 (increasing water mass in the reactor cavity), with the FAI/DCH-4 steam driven test having the greatest amount of hydrogen production. For this last test, about 15% of the metal was oxidized, which is more than the fraction estimated to be finely particulated. Additional oxidation likely occurred during the quenching process on the containment floor.

INTERPRETATION OF EXPERIMENTAL MEASUREMENTS

Pressurization of the simulated containment volumes could be due to the sum of several contributions, including blowdown, steam generation, heating of the non-condensable gases to the saturation temperature, and direct heating. Each of these is quantified for test FAI/DCH-1 in terms of the experimental system in Table 4.

If the high temperature thermite debris is assumed to thermally equilibrated with the containment atmosphere, considering no oxidation of the iron, the pressure would be 0.62 MPa. Hence, the 13 kPa maximum pressure increase which could be due to direct heating represents about a 2.5% efficient process. This is consistent with the observations of (Spencer, 1987) in a 3% linear scale geometry. These are the only two studies which used high temperature debris and modeled the containment structures outside of the reactor cavity, instrument tunnel and seal table.

This result is substantially different from the results reported in Tarbell, et al., (1987) and Tarbell, et al., (1988). Besides the experimental scale there is one major difference between these tests, i.e. the SNL DCH-1 test did not represent the containment geometry downstream of the reactor cavity. Unquestionably, this provides a longer "time of flight" for debris to transfer energy to the gas. However, since all the influential containment structures were not represented in a linear manner, the scaled relationship of entrained (finely particulated) debris to that which was displaced out of the cavity is greatly distorted. Moreover, without the influential structures, even the displaced debris, which would have a considerable velocity when exiting the instrument tunnel, would continue to breakup and transfer heat in free flight. Hence, the SNL DCH-1 and DCH-2 tests have a much greater potential for debris-gas heat transfer than either the FAI DCH tests or the Zion reactor system.

Application of the entrainment rate model to the 5% linear scaled experiments results in a mass of 5.5 kg entrained. Assuming that this is comprised of the frozen films on the steam generators, the refueling canal wall, the reactor coolant pumps and the underside of the operating deck, Table 2 shows this to be about 2.1 kg. Thus, the entrainment rate approach would appear to overestimate the extent of entrainment. Application of the de-entrainment model results in 0.17 kg remaining airborne for the experiment. This is one percent of the debris dispersed and, if this assumed to be equilibrated with the atmosphere (infinite flight time), the contribution to direct heating would be slightly more than 1%, which is consistent with the measured pressurization. Therefore, the structures appear to have a first order mitigative influence on the atmospheric heating.

Application of the entrainment/de-entrainment modeling to the reactor system with full primary system pressure at the time of blowdown results in an entrained mass of 26,700 kg, with a particle size of about 110μm, and a de-entrainment fraction of 0.3. Hence, the calculated mass entering the atmosphere as finely particulated debris would be about 18,700 kg. Following the thermal equilibrium calculations of Ginsberg and Tutu (1988) this would not threaten containment integrity even if hydrogen combustion is assumed. Here again, the structures have a first order mitigative influence on the atmospheric heating.

As part of analyzing any experiment, one must consider uncertainties in the interpretation. If the final temperature (60°C) (upper compartment) is used as being representative, the saturated steam partial pressure would be 20 kPa. With this contribution, the total pressure increase would be 165 kPa, which is the measured value. Hence, at the higher value of the measured temperature, the containment pressurization can be accounted for without any contribution to direct heating of the atmosphere. This provides some quantification on the uncertainty in the evaluation.

CONCLUSIONS

These measurements of the influence of debris dispersal in a scaled Zion reactor cavity and instrument tunnel as well as the major structures above the instrument tunnel and in the lower compartment show very little, if any, contribution to direct heating as a result of the material dispersal. Pressurization in the reactor cavity and instrument tunnel, as well as the lower compartment are essentially due to debris-water thermal interactions in the reactor cavity, the blowdown of the melt generator and steam generated in the debris quenching process on the lower compartment floor. Consequently, there was essentially no direct heating of the atmosphere; an observation consistent with the measured gas temperatures in the lower compartment for each of the tests. This demonstration in a scaled geometry shows (1) only a second order effect of direct heating, (2) a first order effect of rapid steam generation and (3) a first order effect of containment structures.

NOMENCLATURE

A_c Cavity cross-sectional flow area.

A_s Cavity surface area available for entrainment.

A_V Breach area for reactor vessel blowdown.

a Acceleration.

C_1 Constant.

C_D Drag coefficient.

d Debris particle diameter.

L Characteristic length of debris.

L_h Cavity height.

L_P Path length for debris displacement.

m_c Mass of core debris.

$\dot{m}_D$ Rate of debris displacement.

$\dot{m}_e$ Rate of debris entrainment.

P_c Pressure in the reactor cavity.

P_v Pressure in the reactor vessel.

ΔP Pressure differential.

t_D Time required for displacement.

U_c Gas velocity in the cavity.

U_v Isothermal sonic velocity.

W Aperture through which the particles travel as they turn to leave the reactor cavity.

α Fraction of particles failing to make the turn.

η_T Isothermal critical pressure ratio.

ρ_D Debris density.

ρ_g Gas density.

LITERATURE CITED

Benz, R., et al., 1979, "Melt-Water Interactions in Tank Geometry: Experimental and Theoretical Results", Proc. of 4th CSNI Specialists Mtg. on Fuel-Coolant Interactions in Nuclear Reactor Safety, CSNI Report No. 37, Vol. 2, Bournemouth, UK, pp. 363-386.

Bergeron, K., et al., 1986, "Development and Applications of the Interim Direct Containment Heating Model for the CONTAIN Computer Code", Transactions of the Fourteenth Water Reactor Safety Information Meeting, NUREG/CP-0081, Gaithersburg, MD.

Commonwealth Edison Company (CECo), 1981, Zion Probabilistic Safety Study, Chicago, Illinois.

Fauske & Associates, Inc. (FAI), 1990, "FAI/CECo Direct Containment Heating Experiments for a Zion-like Geometry", FAI/90-60, report submitted to Commonwealth Edison Company.

Ginsberg, T. and Tutu, N. K., 1988, "Progress in Understanding of Direct Containment Heating Phenomena in Pressurized Light Water Reactors", Invited paper presented at the Third International Topical Mtg. on Nuclear Power Plant Thermal Hydraulics and Operations, Seoul, South Korea.

Henry, R. E., 1989, "An Evaluation of Fission Product Release Rates During Debris Dispersal", PSA '89, Proc. of the ANS/ENS Intl. Topical Mtg.: Probability, Reliability and Safety Assessment, pp. 375-383.

IDCOR Program Report, 1983, "Debris Coolability, Vessel Penetration, and Debris Dispersal", IDCOR Technical Report 15.2B.

Malinovic, B., et al., 1989, "Experiments Relating to Drywell Shell-Core Debris Interaction", National Heat Transfer Conf., AIChE Symposium Series, Heat Transfer, Vol. 85, Philadelphia, pp. 217-222.

Marx, K. D., 1989, "A Computer Model for the Transport and Chemical Reaction of Debris in Direct Containment Heating Experiments", Nuclear Sci. & Eng., 102, 391-407.

Power Authority of the State of New York (PASNY) and the Consolidated Edison Company (Con.Ed.) of New York, Inc., 1982, "Indian Point Probabilistic Safety Study".

Ricou, F. G. and Spalding, D. B., 1961, "Measurements of Entrainment by Axisymmetrical Turbulent Jets", Journal of Fluid Mechanics, 11, pp. 21-32.

Spencer, B. W., et al., 1983a, "Corium/Water Dispersal Phenomena in Ex-Vessel Cavity Interactions", Paper TS-15.5, Proc. Intl. Mtg. on LWR Severe Accident Evaluation, Vol. 2, Cambridge, MA.

Spencer, B. W., et al., 1983b, "Overview and Recent Results of ANL/EPRI Corium-Water Thermal Interaction Investigations", Paper TS-15.2, Proc. Intl. Mtg. on LWR Severe Accident Evaluation, Vol. 2, Cambridge, MA.

Spencer, B. W., et al., 1987, "Hydrodynamics and Heat Transfer Aspects of Corium-Water Interactions", EPRI NP-5127, Argonne National Laboratory.

Tarbell, W. W., et al., 1984, "High-Pressure Melt Streaming (HIPS) Program Plan", SAND82-2477, NUREG/CR-3025, Sandia National Laboratories.

Tarbell, W. W., et al., 1986, "Pressurized Melt Ejection into Scaled Reactor Cavities", SAND86-0153, NUREG/CR-4512, Sandia National Laboratories.

Tarbell, W. W., et al., 1987, "Results from the DCH-1 Experiment", SAND86-2483, NUREG/CR-4817, Sandia National Laboratory.

Tarbell, W. W., et al., 1988, "DCH-2: Results from the Second Experiment Performed in the Surtsey Direct Heating Tests Facility", SAND87-0976, NUREG/CR-4917, Sandia National Laboratory.

Tutu, N. K., et al., 1988, "Debris Dispersed from Reactor Cavities During High-Pressure Melt Ejection Accident Scenarios", NUREG/CR-5146, BNL-NUREG-52147.

Table 1
TEST MATRIX

Test	RCS Gaseous Medium	Thermite Mass	Water Distribution Cavity	Water Distribution Lower Cont. Floor
FAI/DCH-1	N_2	20 kg	1.6 cm	10 cm
FAI/DCH-2	N_2	20 kg	7.5 cm	1 cm
FAI/DCH-3	N_2	20 kg	Dry	1 cm
FAI/DCH-4	Steam	20 kg	1.6 cm	1 cm

Table 2
FAI/DCH-1 FINAL DEBRIS DISTRIBUTION

Location	Mass (kg)
1. Compartment Floor/South	6.77
/North	7.64
2. Frozen on Stm. Gen., Etc.	1.26
3. Frozen in the Cavity	2.02
4. Laying Near Keyway Exit	1.25
5. On the Lid/Lower Compt.	0.74
6. Frozen in 6" Pipe	0.1
TOTAL RECOVERED	**19.78**

Table 3
GAS MOLAR FRACTIONS (PERCENT)

Gas	FAI/ DCH-1	FAI/ DCH-2	FAI/ DCH-3	FAI/ DCH-4
H_2	2.4%	3.4%	0.8%	5.4%
CO	0.2%	0.2%	0.3%	1.1%
CO_2	0.3%	0.2%	0.1%	0.2%
N_2	Balance*	Balance*	Balance*	Balance*
O_2	8.4%	9.2%	11.4%	3.9%

*Steam has been condensed, such that these represent percentages in a dry atmosphere.

Table 4
COMPONENTS OF THE FAI/DCH-1 PRESSURE INCREASE

Total Pressure Change = 65 kPa	
1. N_2 from the crucible.	9,800 Pa
2. Steam forcing N_2 from the lower compartment.	13,500 Pa
3. Hydrogen generation.	2,700 Pa
4. Heating the non-condensables to 323K by steam generation.	14,000 Pa
5. Steam partial pressure (50°C).	12,000 Pa
6. Max. DCH contribution.	13,000 Pa
	52,000 Pa

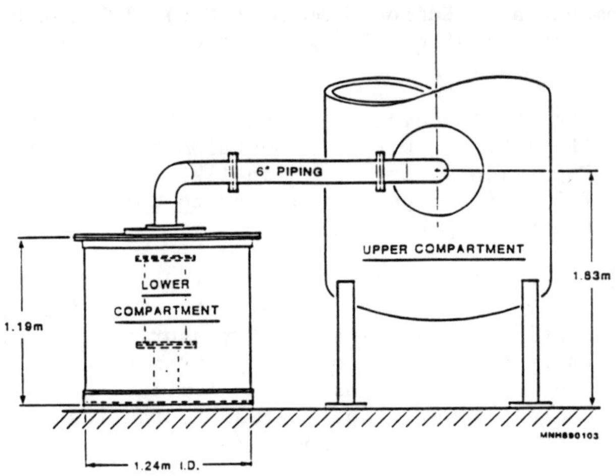

Figure 2. Definition of the structure dimensions.

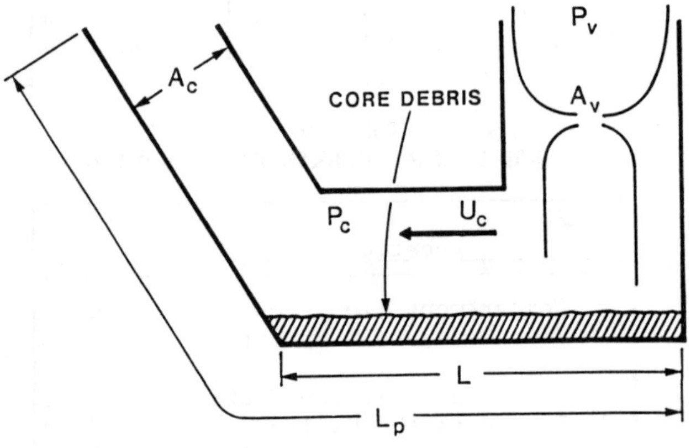

Figure 1. Characteristic debris dimensions.

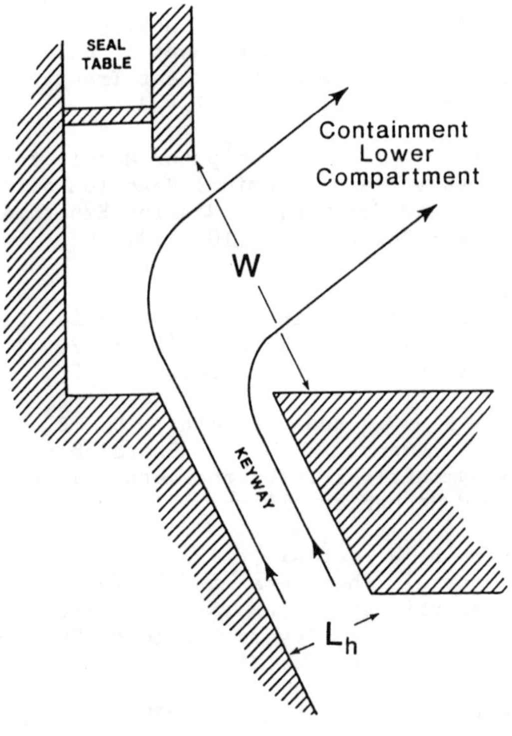

Figure 3. Elevation view of the vessel interconnections.

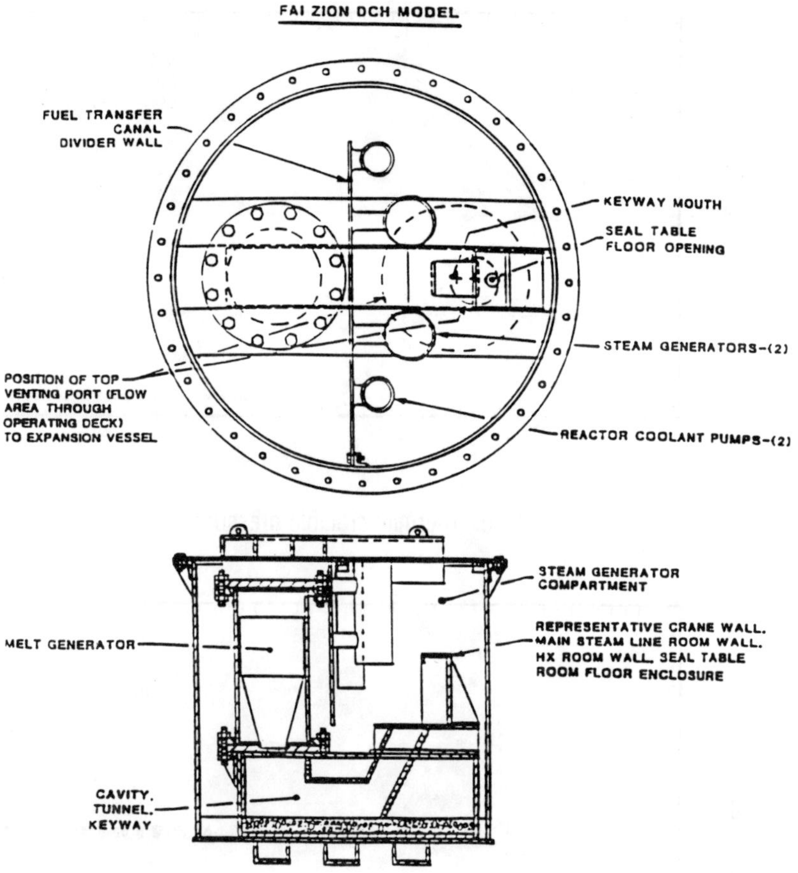

Figure 4. Cross-section of the melt generator, lower compartment structures.

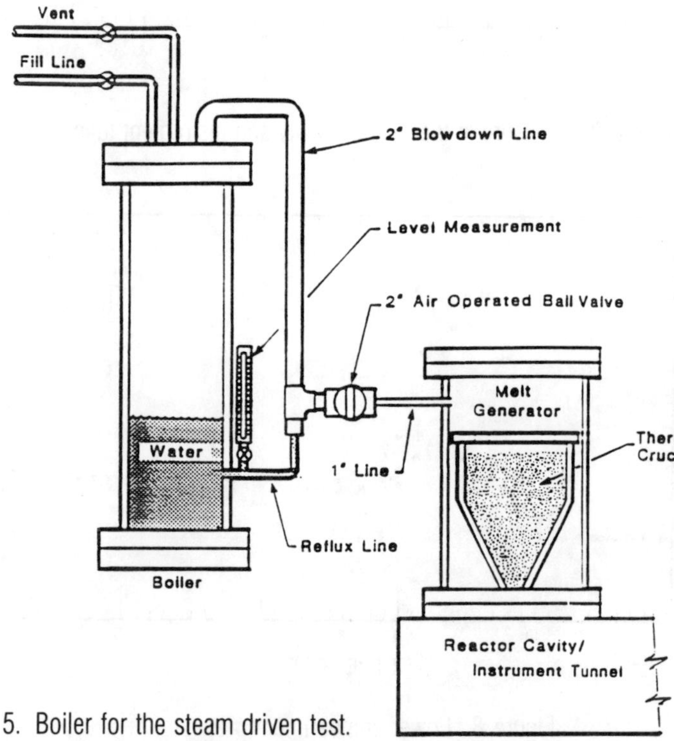

Figure 5. Boiler for the steam driven test.

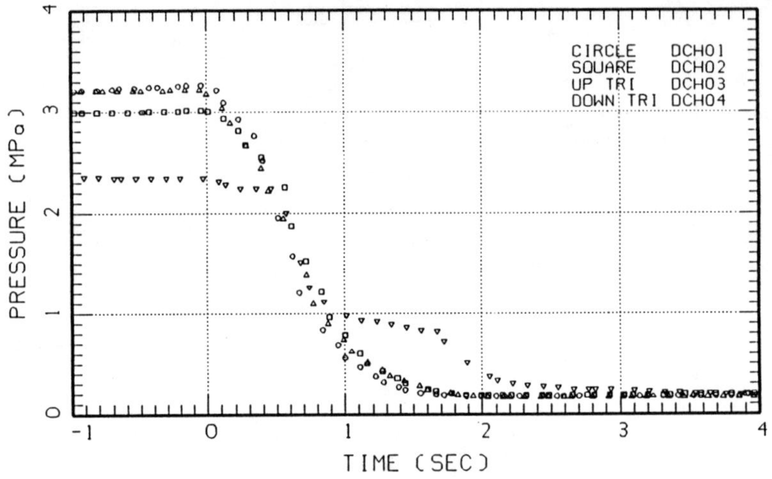

Figure 6. Thermite crucible pressure.

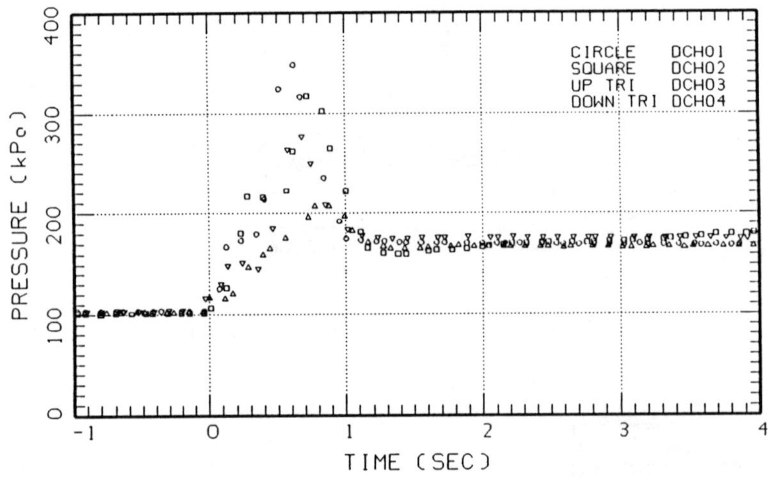

Figure 7. Pressure in the reactor cavity and instrument tunnel.

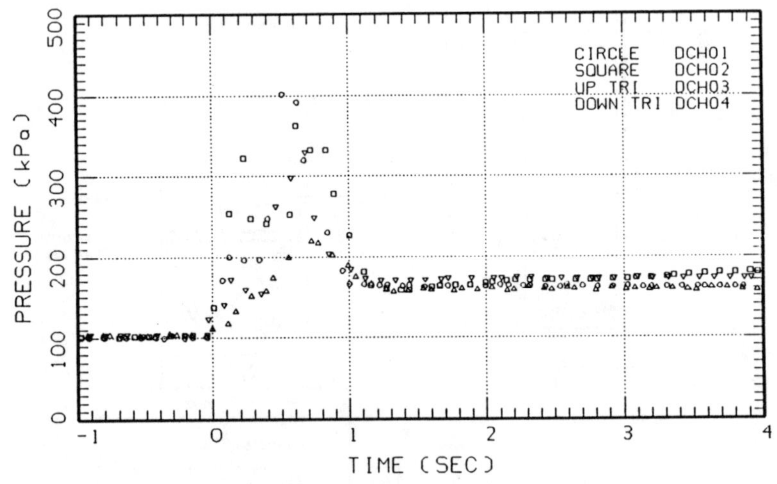

Figure 8. Lower compartment pressure.

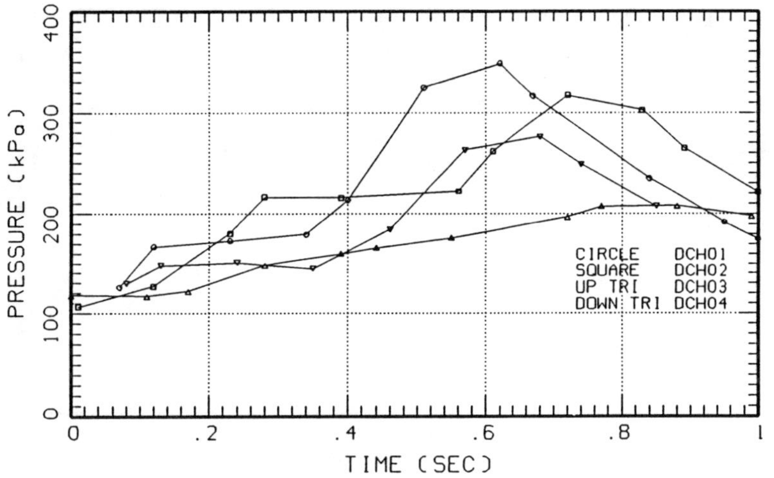

Figure 9. Lower compartment pressure.

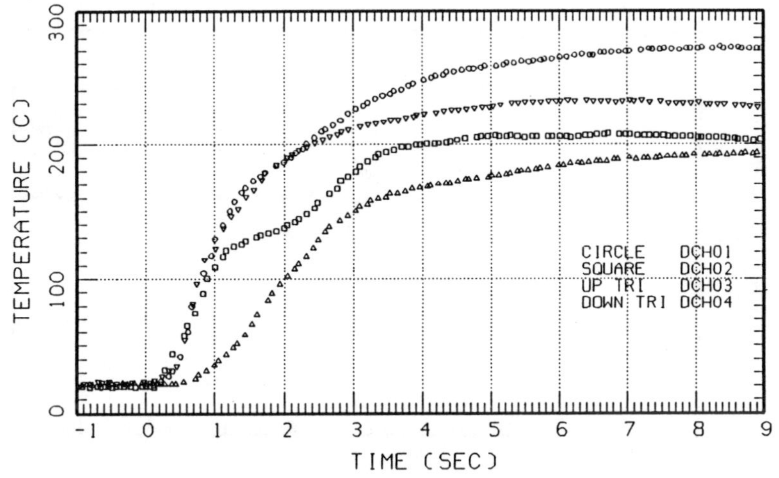

Figure 10. Lower compartment temperature.

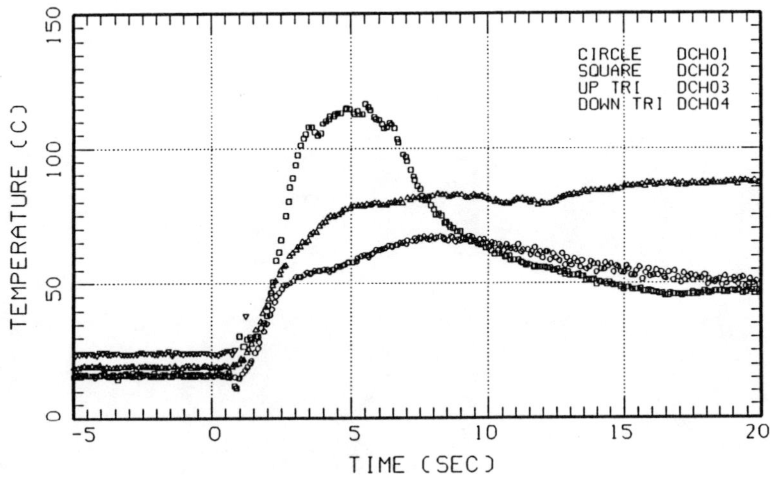

Figure 11. Water temperature on the lower compartment floor.

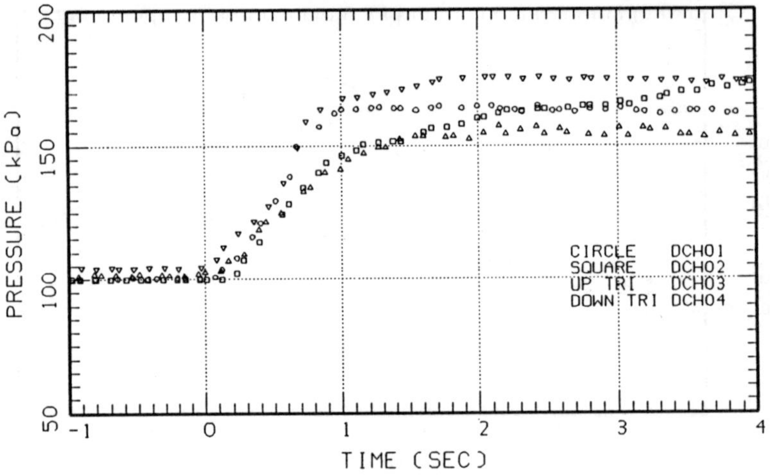

Figure 12. Pressure measurement in the upper compartment.

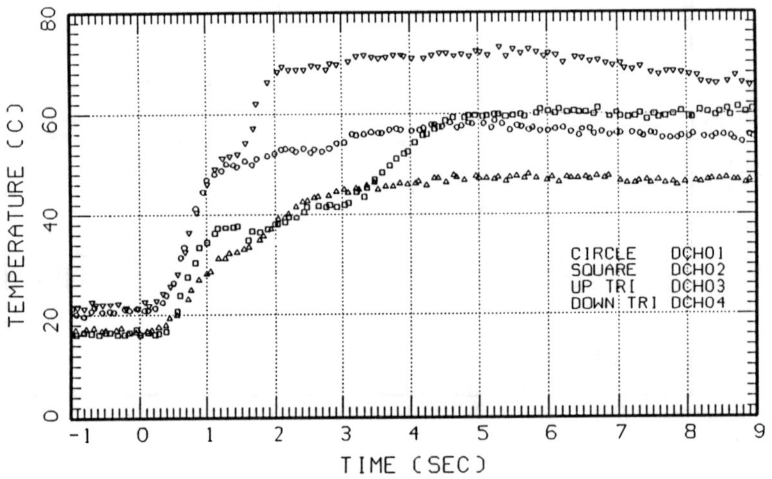

Figure 13. Temperature measurement in the upper compartment.

MODELING DIRECT CONTAINMENT HEATING PHENOMENA WITH CONTAIN 1.12

R.O. Griffith, N.A. Russell, and K.E. Washington ■ Sandia National Laboratories, Albuquerque, NM 87185

CONTAIN is a detailed mechanistic computer code developed at Sandia National Laboratories for the integrated analysis of light water reactor severe accident containment phenomena. The most recent version of the code, CONTAIN 1.12, incorporates models for the phenomena of high pressure melt ejection (HPME) and the subsequent processes collectively known as Direct Containment Heating (DCH). CONTAIN 1.12 was used to model the Limited Flight Path 8A (LFP8A) experiment conducted at the Surtsey test facility at Sandia National Laboratories. In the experiment, 50 kg of molten thermite was injected into a scale model of the Surry cavity and then blown into the Surtsey vessel by high pressure steam. A seven-cell best-estimate CONTAIN model, using only a minimum of measured data, was used to simulate the LFP8A experiment. A comparison of the experimental and calculated results indicated that CONTAIN 1.12 was accurately modeling the physical processes involved in DCH phenomena, but that the method of injecting the molten debris into the cavity in the CONTAIN model was causing the code to over-predict the chemical reaction and heat transfer rates between the molten debris and the system atmosphere. CONTAIN 1.12 predicted the peak vessel pressure to within less than 2% of the experimental value, but missed the timing on the pressure peak by approximately 1.75 s over the course of a 10 s calculation.

CONTAIN is a detailed mechanistic computer code developed at Sandia National Laboratories for the integrated analysis of light water reactor severe accident containment phenomena [1]. The code was developed under the sponsorship of the USNRC, and includes state-of-the-art mechanistic models for a wide variety of physical and chemical processes that occur within the nuclear reactor containment during a severe accident. The most recent version of the code, CONTAIN 1.12, incorporates models for the phenomena of high pressure melt ejection (HPME) and the subsequent processes that contribute to containment loading [2]. The HPME and the subsequent physical and chemical processes within the containment produce an efficient transfer of energy from the dispersed core debris to the containment atmosphere, resulting in a potential threat to containment integrity [3]. Collectively, these phenomena are referred to as direct containment heating (DCH). This paper reports on the modeling of DCH phenomena with CONTAIN 1.12, and discusses the results of a comparison between best-estimate code calculations and a limited flight path test conducted at the Surtsey facility at Sandia National Laboratories.

Sandia National Laboratories, Albuquerque, New Mexico. This work is supported by the United States Nuclear Regulatory Commission and was performed under FIN A1198 at Sandia National Laboratories, which is operated for the U.S. Department of Energy under contract number DE-AC04-76DP00789.

One of the first major tests of the DCH modeling capabilities in CONTAIN 1.12 is the comparison to experimental results from the limited flight path (LFP) separate effects tests currently being conducted at the Surtsey facility at Sandia National Laboratories. In these experiments, molten corium is ejected into a simulated reactor containment in various configurations. CONTAIN 1.12 has been used to model the Surtsey LFP test configuration to simulate the test results and identify weaknesses in our current understanding of DCH phenomena. Of particular interest is the ability of the new models in CONTAIN to simulate the transport of gases and debris within the Surtsey vessel. Several of the new DCH models present in CONTAIN 1.12 incorporate parametric elements, and the experimental results have been used to give valuable insight into the performance of these correlations.

CONTAIN 1.12 DCH MODELS AND CAPABILITIES

The new DCH capabilities in CONTAIN 1.12 include models for the flow of debris and gases, debris de-entrainment on structures, debris chemistry involving Zr, Fe, O_2, and H_2O, and convection and radiation heat transfer between debris and the environment within containment [2]. Also included are mechanistic models for the initial melt ejection, ablation of the vessel hole by escaping core material, gas blowthrough and mixed gas/core debris ejection, entrainment of debris from the cavity floor, de-entrainment of debris on cavity structures and surfaces, and dispersal of debris into the main containment atmosphere. These mechanistic models were taken from the codes CORDE, a HPME and debris dispersal modeling code developed at Winfrith in the UK, and from GASBLOW2, a HPME modeling code developed at Sandia National Laboratories [2]. For some of these processes, such as debris entrainment from the cavity floor, a number of correlations developed through experimental and analytical investigations have been incorporated into CONTAIN 1.12.

The CORDE and GASBLOW2 models that were incorporated into CONTAIN 1.12 were tested on a suite of standard test problems. However, all of these test problems were full-scale simulations of nuclear reactor systems. When the CORDE/GASBLOW2 modules were applied to the scaled geometry in the LFP8A experiment, a number of unexpected difficulties were encountered with the models. Preliminary investigations indicate that these difficulties are scale-dependent, in the sense that they were not encountered in large or full-scale simulations. Debugging efforts are currently underway to permit the use of the CORDE/GASBLOW2 models in future simulations of the LFP experiments.

A complete description of the DCH models in CONTAIN 1.12 is well beyond the scope of this paper. However, debris chemical reactions and debris trapping are two DCH phenomena that are of particular importance in the LFP8A experiment and are described in more detail. Chemical reactions can occur in the system atmosphere during a DCH event if Zr and/or Fe metal is present and the atmosphere contains oxygen and/or steam. In CONTAIN 1.12, it is assumed that all debris particles in a cell have the same size and the same temperature. Therefore, debris chemical reactions are only computed once for each cell and the effect is multiplied for the number of droplets in the bulk field for that cell. Another important aspect of the chemistry model is that the constituents in a cell are assumed to be evenly divided among all droplets in the bulk field of that cell.

Models are included for the following chemical reactions:

1) $Zr + O_2 \longrightarrow ZrO_2$

2) $2\,Fe + O_2 \longrightarrow 2\,FeO$

3) $Zr + 2\,H_2O \longrightarrow ZrO_2 + 2\,H_2$

4) $Fe + H_2O \longleftrightarrow FeO + H_2$

These reactions are treated in a hierarchical fashion, where Zr is favored over Fe in any reaction, and where the oxygen reactions are favored over the steam reactions. As a result, the model treats reactions in the order they are listed above. All reactions are assumed to go to completion, except for iron/steam.

The chemical reaction of Zr and Fe metal in the debris field is modeled in CONTAIN using a rate-limiting mechanistic approach. In this approach, the reaction rate is governed by the rate at which oxygen and steam can be transported to the drop (gas-side) and then transported within the drop as it chemically reacts (drop-side). The details of the calculation of the drop and gas-side reaction rate constants, and the relationship between these calculations and the hierarchial chemical reaction scheme, are presented in the CONTAIN 1.12 documentation [2].

Trapping is the process of debris removal resulting from interaction with containment structures and/or gravitational fallout. The trapping process is not well-understood, and experiments have shown that particle trapping is subject to scale distortions [4]. A variety of methods for modeling trapping are available in CONTAIN 1.12. These models are mostly parametric, although there is some degree of mechanism associated with some of the approaches. All models for trapping are based on the assumption that debris is removed according to

$$\frac{dM}{dt} = -F_{trap}\,M \qquad (1)$$

Equation (1) is integrated over time to model the removal of debris from the atmosphere.

Trapped debris is held in a repository and allowed to cool at a user-specified rate. The energy lost during this cooling is, by default, assumed to be deposited in the floor if one is present. The user may distribute this energy to structure surfaces or may specify that it be lost from the problem. CONTAIN input parameters allow the user to control the trapped debris heat loss rate as well as the repository for the energy loss.

The difference between the various trapping models present in CONTAIN 1.12 is the way in which F_{trap} is determined. The following four models are available:

1	USER	User-specified F_{trap} values
2	GFT	Gravitational fall time
3	TFI	Time of first impact
4	TOF/KU	Time of flight/Kutateladze

USER Trapping Model

The simplest and most parametric trapping model is the USER model. In this model, the user simply specifies the fractional removal rate for debris trapping for each cell. A constant can be specified or a parameter table can be used to vary the trapping rate as a function of time or particle temperature.

GFT Trapping Model

In this model, the trapping rate is taken to be the gravitational fall rate of a sphere in the cell atmosphere.

$$F_{trap} = \frac{V_t}{L_{gft}} \tag{2}$$

The terminal fall velocity, V_t, is computed by CONTAIN using the properties of the gases in the cell evaluated at T_g. The fall height, L_{gft}, is by default the cell height, although other values may be specified. The GFT trapping rate will also be computed and used in the TOF/KU model if velocities are sufficiently high that the Kutateladze criterion for particle sticking is not met by the second impact.

TFI Trapping Model

In the TFI trapping model, debris is trapped at a rate equal to the transport time required for debris to reach the first structure in the cell. The debris jet enters the cell through an orifice of diameter d_0, and travels a distance L_{tfi} before striking a structure in the cell. The velocity upon impact is calculated from the jet expansion law

$$W_x = W_o \cdot \frac{\alpha L_{tfi}}{d_o} \left(\frac{\rho}{\rho_o}\right)^{1/2} \tag{3}$$

The initial velocity is assumed to be the gas flow velocity from a specific flow path into the cell. The flow path is specified in the CONTAIN input. The trapping fraction rate is calculated by

$$F_{trap} = \frac{V_{jet}}{L_{tfi}} \tag{4}$$

If a flow path is not specified or if the gas flow through the specified path is not into the cell, the GFT model will be used. In addition, the TFI trapping rate is limited so that it will not be slower than the GFT rate. The distance to first impact, L_{tfi}, is by default the cell height. However, the value may also be specified by the user.

TOF/KU Trapping Model

The TOF/KU model is a combination of the GFT model, TFI model, and a Kutateladze criterion based model; hence the name "time of flight/Kutateladze" model. In this model, debris is conditionally removed from the cell atmosphere depending on the Kutateladze number of the impact of debris upon the first structure in the cell. The velocity and distance to first impact are determined the same way as for the TFI model. If the first impact does not cause de-entrainment, then a second impact on another structure is considered. The velocity of the second impact is assumed to be an average cell gas/debris cloud velocity. If both conditions fail, GFT trapping is used in the cell.

The Kutateladze number for debris entrainment represents the ratio of kinetic energy to surface tension forces and is defined as

$$Ku \approx \frac{\rho_g V_g^2}{(\rho_d g\sigma)^{1/2}}, \quad \rho_d >> \rho_g \tag{5}$$

Using the gas entrainment law, the gas density at the point of impact with the first structure may be written as

$$\rho_g\big|_{\text{at structural impact}} = \rho_1 = \frac{\rho_o \rho_a (w/w_o)}{\rho_a + \left(\frac{w}{w_o} - 1\right)\rho_o} \tag{6}$$

where w/w_o is determined from the TFI model.

The Kutateladze criterion at the point of impact with the first structure in the cell can be written as

$$Ku_1 = \frac{\rho_1 V_1^2}{(\rho_d g\sigma)^{1/2}} \tag{7}$$

where V_1 is the gas velocity upon impact as calculated by the TFI model.

If $Ku_1 \le Ku_{T1}$, where Ku_{T1} = threshold value for first impact, then t_{tr} from the TFI model is used. The kinetic energy of the incoming jet was not able to overcome the surface tension forces of the debris, and the debris remained de-entrained upon the first structure. However, if $Ku_1 > Ku_{T1}$, then TFI de-entrainment will not occur and an impact with the second structure in the cell is considered.

For impact with the second structure, the Kutateladze criterion is written as

$$Ku_2 = \frac{\rho_a V_2^2}{(\rho_d g\sigma)^{1/2}} \tag{8}$$

where V_2 = average cell gas/debris cloud velocity.

If $Ku_2 < Ku_{T2}$, where Ku_{T2} = threshold value for second impact, and $Ku_1 > Ku_{T1}$, then the trapping rate is governed by the time it takes to make the second impact:

$$\lambda_{tr} = (t_1 + t_2)^{-1} \tag{9}$$

However, λ_{tr} is limited by the value calculated from the GFT model. The GFT model is also used if $Ku_2 > Ku_{T2}$. The TOF/KU model is the most mechanistic of the CONTAIN trapping models.

DESCRIPTION OF EXPERIMENTAL SETUP AND PHENOMENA

Physical Description

The LFP experimental facility consisted of a 1:10 scale model of the Surry reactor cavity connected to the bottom of the Surtsey pressure vessel. Simulated core debris was liquified in a melt generator attached to the cavity and then forced into the cavity by a high-pressure steam source. After the molten debris was ejected into the cavity, it was swept up into the Surtsey vessel by high pressure steam passing through the cavity and out into the main containment atmosphere. The experimental setup is shown in Figure 1. A complete report on the LFP test series is currently being prepared [5].

In the LFP8A experiment, a generic structure was placed within the Surtsey vessel at a nominal height of 8 m above the floor of the vessel. The structure was a 5.16 cm thick concrete slab that measured 2.44 m by 2.44 m and was placed perpendicular to the vertical axis of the vessel. In addition, a 30 cm steel overhang was attached to each side of the concrete slab. The overhang was used to limit the amount of debris that was swept up above the slab in the vessel. The slab was intended to simulate a subcompartment structure in the Surry Plant, where debris exiting the cavity would strike the Residual Heat Removal Platform and either be trapped on the concrete ceiling, deflected back to the floor, or be re-entrained in the gas flow and travel above the structure.

The initial conditions for the LFP8A experiment were measured just prior to the initiation of the test. The initial temperature and pressure in the cavity and vessel were 293 K and 0.16 MPa, respectively. The atmospheric composition was 99.5 mol.% Ar, 0.38 mol.% N_2 , and 0.08 mol.% O_2. The molten thermite was at a temperature of approximately 2500 K. It consisted of 20.0 kg of Al_2O_3 , 24.6 kg of Fe, and 5.4 kg of Cr, for a total thermite mass of 50 kg. The steam accumulator contained approximately 4.17 kg of superheated steam at a temperature of 570 K and a pressure of 2.8 MPa. The hole connecting the melt generator to the cavity body had a diameter of 3.5 cm, which remained constant for the duration of the experiment [5]

Observed Phenomena

As the debris in the melt generator reached the molten state, a brass plug separating the thermite from the cavity vaporized almost instantaneously. This event marked time t=0 for the recorded phenomena. The molten debris was ejected into the cavity by pressurized steam contained in the accumulator. All of the thermite in the melt generator was ejected into the cavity within approximately 0.4 s after the vaporization of the brass plug. Very little mixing of the thermite and the driving gas was observed during the initial ejection, although the molten material was mixed violently within the cavity after the initial ejection was complete. The driving gas entered the cavity at approximately 0.4 s, following the ejected thermite [6]. The molten debris was entrained by the steam passing through the cavity and swept up into the Surtsey vessel. The debris jet then struck the concrete platform suspended within the Surtsey vessel, where debris particles either stuck to the surface or rebounded and fell to the bottom of the vessel. A small amount of the debris was re-entrained and swept above the concrete structure towards the upper head of the containment volume. In the LFP8A experiment, approximately 60% of the initial debris remained trapped in the cavity region, and was not ejected into the Surtsey vessel.

CONTAIN MODEL

Discretization

The CONTAIN 1.12 model of the LFP8A experiment consisted of seven computational cells. A diagram of the model is presented in Figure 2. The cavity and steam supply system were subdivided into four CONTAIN cells: the melt generator/steam accumulator, the horizontal body of the cavity, the cofferdam connecting the cavity body to the chute, and the chute connecting the cavity to the Surtsey vessel. The Surtsey vessel was divided into three CONTAIN cells: one for the region below the concrete slab, and two cells for the region above the slab. The slab was placed 7.7 m above the floor of the Surtsey vessel, with equal flow areas around each side.

The cell descriptions are as follows:

Cell 1:	horizontal body of the cavity
Cell 2:	cofferdam connecting cavity body and cavity chute
Cell 3:	chute connecting cavity body to the Surtsey vessel
Cell 4:	Surtsey vessel beneath the suspended concrete slab
Cell 5:	Surtsey vessel above the suspended concrete slab
Cell 6:	Surtsey vessel above the suspended concrete slab
Cell 7:	melt generator/steam accumulator

Modeling Assumptions and Approximations

A number of critical assumptions and approximations were made in modeling the LFP8A experiment with CONTAIN 1.12. However, it should be noted that only two pieces of experimental data were used in constructing the CONTAIN model of the LFP8A experiment; the timing of the ejection of thermite into the cavity, and the fraction of debris that was dispersed from the cavity. Experimental observations indicated that all of the thermite was injected into the cavity during the first 0.4 s of the test. This fact was used while constructing the debris source mechanism described below. The experiment also showed that only 39.2% of the debris actually entered the Surtsey vessel, so the mass of dispersed debris input to the CONTAIN simulation was adjusted to reflect this result. With these two exceptions, the CONTAIN 1.12 model of the LFP8A experiment was essentially a best-estimate blind post-test analysis.

The first modeling assumption in the CONTAIN simulation was that the molten thermite was injected into the cavity at a constant rate over the first 0.4 s of the calculation. The top of the melt generator was connected to a steam accumulator. When the brass plug separating the molten debris and the cavity was vaporized, the connection between the steam accumulator and melt generator opened, and the pressurized gas ejected the thermite into the cavity. The volume of the steam accumulator was much greater than the volume of the melt generator, leading to an almost constant pressure on the thermite as it was forced into the cavity. In addition, little or no mixing of the driving steam and thermite was observed in the LFP8A experiment; the material was essentially ejected as a homogeneous "plug" of molten debris. The assumption neglects friction and acceleration effects, but is consistent with experimentally observed behavior.

As a second modeling assumption, the molten debris ejected from the melt generator was treated in the CONTAIN simulation as a debris source term in the cavity. The CONTAIN model of the LFP8A experiment initially used the debris entrainment and dispersal models from CORDE/GASBLOW2 to simulate the ejection of the molten debris from the cavity. However, unexpected difficulties with these code modules prevented their use. Debugging efforts are currently underway to allow the use of CORDE/GASBLOW2 in future simulations of the LFP experiments. However, based on experimental observations, the violence of the ejection process quickly dispersed the material into the cavity. This dispersal created a situation similar to that which was assumed by the use of the debris source model in CONTAIN. A debris particle size of 0.8 mm was chosen for the injected debris. This particle size was based on the measured particle sizes from previous Surtsey HPME experiments [6].

Next, the melt generator and steam accumulator volumes were added together to form a single CONTAIN cell. The volume was then treated as a single pressurized cell containing the initial mass of steam at the experimentally measured initial temperature and pressure. A flow path between this cell and the cavity was opened at t=0.4 s, after all of the molten debris had been injected into the cavity cell. Experimental observations show that virtually all of the thermite is ejected from the melt generator ahead of the driving gas, so permitting the driving steam to be injected into the cavity simultaneously with the molten debris would not accurately reflect the experimentally observed behavior. In a HPME accident in a nuclear plant, it is hypothesized that the corium would first be deposited on the cavity floor and then swept out of the cavity by the blowdown gas.

A final modeling approximation involved the substitution of Zr for the Cr present in the melt. CONTAIN 1.12 does not have Cr chemistry models, but the thermite used in the LFP experiments contained a significant amount of Cr. Because the anticipated Cr reactions are highly exothermic, neglecting these chemical reactions could introduce significant and non-conservative errors into the calculation. Since CONTAIN 1.12 has the ability to model Zr chemistry, the Cr present in the melt was replaced by Zr. The mass of Zr substituted for the Cr was selected to produce roughly the same amount of energy that would be released by the Cr in the chemical reactions. The approximation is not ideal, but the Zr chemistry model at least calculates the correct gross exothermic energy release expected from the Cr. The approximation was certainly preferable to neglecting the presence of Cr entirely.

The behavior of debris ejected into the Surtsey vessel atmosphere represented an extremely important aspect of the simulation. Of particular importance was the length of time the debris particles remained airborne in the Surtsey vessel, since this time period directly influenced the amount of heat transfer and chemical reactions that took place between the debris and the containment atmosphere. As described in Section 2, CONTAIN 1.12 has a number of models available for simulating the trapping process, with the most mechanistic being the TOF/KU (time-of-flight/Kutateladze) model. The TOF/KU model was selected for use in the LFP8A model because it was the most mechanistic trapping model available in CONTAIN 1.12 and was the best suited for modeling the expected trapping behavior. The debris was assumed to enter cell 4 in the Surtsey vessel and impact the horizontal portion of the suspended concrete slab. The steel overhang attached to the edges of the slab was then specified as the second structure in the cell for the TOF/KU trapping model.

DISCUSSION OF RESULTS

Figure 3 presents the calculated pressure in the Surtsey vessel vs. the results of the LFP8A experiment. Only a single curve is presented for the CONTAIN results, since the pressures in cells 4, 5, and 6 were essentially identical. The CONTAIN model predicted a small and almost linear pressure increase over the first 0.4 s of the calculation, while the debris was being injected into the cavity cell. When the high pressure steam began entering the cavity at 0.4 s, the calculated pressure increased sharply and reached a peak value of approximately 0.336 MPa at about 2 s into the test. The calculated pressure in the vessel then gradually decreased to approximately 0.292 MPa after 10 s. The experimental results behaved in a similar manner, with the pressure quickly rising to a peak value of 0.330 MPa, and then gradually decreasing to about 0.292 MPa at a time of 10 s. However, the peak pressure value measured in the LFP8A experiment appeared at about 3.75 s into the test, which was approximately 1.75 s after CONTAIN predicted a peak pressure.

The calculated and experimental pressure curves in the vessel were quite similar in magnitude and shape, but appeared to be somewhat out of phase. CONTAIN predicted the correct magnitude and general behavior of the DCH pressurization, but did a poor job of modeling the timing of the phenomena. It was theorized that the method of injecting the molten debris into the cavity was responsible for the discrepancy between the calculated and experimental results. In the calculation, the steam injected into the cavity at 0.4 s encountered almost the full inventory of molten debris in dispersed form. As a result, chemical reactions between the debris and steam occurred quite rapidly, and the debris was quickly ejected into the Surtsey vessel atmosphere.

In the experiment, a sizable portion of the molten debris was deposited on the floor, walls, and ceiling of the cavity region prior to the entrance of the driving steam. Although the molten debris that escaped being trapped in the cavity was quickly entrained and ejected into the vessel atmosphere, heat transfer and chemical reactions between the steam and debris were certainly less efficient and rapid than the CONTAIN model predictions. CONTAIN 1.12 postulated a perfectly dispersed debris field in the cavity as the steam entered. As a result, the debris ejected from the cavity in the experiment took longer to interact with the injected steam and reach the Surtsey vessel atmosphere than was predicted by CONTAIN. Although no time-resolved debris ejection information was gathered from the experiment, this theory conforms to physical expectations and explains the timing discrepancy between the calculated and experimental results.

The calculated and experimental values for the cavity pressure further supported the theory that the method of injecting the debris into the cavity in the CONTAIN model was responsible for the timing discrepancy in the results. The pressures in the cavity region vs. time are presented in Figure 4. As the debris began entering the cavity, both the calculated and experimental results exhibited a short spike in the cavity pressure. CONTAIN then calculated a large pressure spike immediately after the high pressure steam began entering the cavity cell. The pressure in the cell reached approximately 0.81 MPa and then quickly dropped back as the gas and debris exited the cavity. However, the experimental results did not indicate any sudden spike as the steam began entering the cavity. These results support the theory that the dispersed debris in the CONTAIN model reacted rapidly and efficiently with the incoming steam, while in the experiment the steam/debris interaction was much slower.

Figure 5 presents the temperatures in the vessel vs. time. The calculated temperatures increased only slightly over the first 0.4 s, then increased rapidly after steam began to enter the cavity and peaked at about 2.0 s. The temperature below the concrete slab peaked at about 676 K, while the temperature above the slab reached around 400 K. The experimentally recorded temperatures matched the calculated results quite poorly; the peak experimental temperature below the slab was about 513 K, and occurred at about 5 s into the test. No temperature readings were taken above the slab. The difference in the calculated and experimental temperature magnitudes probably resulted from the fact that CONTAIN calculated a cell average temperature, while the experimental values were taken from discrete points along the wall of the Surtsey vessel. The temperatures along the center of the vessel were probably much higher than the experimental results indicated, since in the experiment the vessel atmosphere was not perfectly mixed [6]. The discrepancy in the timing of the temperature peak was attributed to the difference in the debris interaction rates described above.

The mass of airborne debris vs. time in the model is presented in Figure 6. As can be seen in Figure 6a, the total mass of airborne debris increased approximately linearly during the first 0.4 s of the calculation, peaking at approximately 18 kg. The linear behavior reflected the CONTAIN debris source model in cell 1 in the cavity. During the initial source period, a small amount of airborne debris appeared in cell 4 in the Surtsey vessel. As the debris was injected into the cavity, heat transfer from the hot particles caused the cavity gas to heat and expand into the vessel. Since CONTAIN uses a no-slip assumption for the gas/particle mixture, a portion of the debris in the cavity moved out into the vessel with the cavity gas prior to the injection of the driving steam. At 0.4 s, the driving steam entered the cavity and quickly forced the debris out into the Surtsey vessel. As expected, most of the debris remained in cell 4 below the concrete slab. However, as shown in Figure 6b, a small amount of debris appeared to have been airborne in cells 5 and 6, above the slab, between 0.5 s and 1.7 s.

The airborne debris was removed from the vessel atmosphere by the trapping process. In the first 0.4 s of the calculation, a small amount of debris was trapped in cell 4. This represented the gravitational fallout of the debris that was moved into the vessel as the cavity gas expanded. After the high pressure steam entered the cavity at 0.4 s, debris that entered the Surtsey vessel was rapidly trapped in the cell below the concrete slab. Virtually all of the debris entering the vessel rebounded from the slab and drifted to the lower floor of cell 4. A small amount of debris was transported above the slab, and then trapped on the dome and upper slab surface. Post-experiment analysis of the test apparatus revealed that approximately 9% of the debris was trapped on the underside of the slab, 2% of the debris was trapped above the slab, and the remaining 89% of the debris fell to the lower head of the Surtsey vessel [5]. The CONTAIN results using TOF/KU trapping model default values predicted that no debris was trapped on the underside of the slab, 4% of the debris was trapped above the slab, and the remaining 96% of the debris fell to the lower head of the Surtsey vessel.

A significant amount of hydrogen was generated in the cavity during the early stages of the test. For the first 0.4 s of the calculation, no hydrogen was present in the system. Immediately after the high pressure steam began entering the cavity, the hydrogen concentration in the three cavity cells increased to approximately 75.0 mol.%. However, as the entering steam swept the debris out through the cavity and into the vessel, the hydrogen concentration rapidly dropped, virtually disappearing by 0.75 seconds into the calculation. An analysis of gas grab bottles opened in the cavity between 0.5 s and 2.5 s revealed a peak hydrogen concentration of 54.1 mol.%. However, this value may have actually been higher, since the grab sample apparently leaked [5].

CONCLUSIONS

The severe reactor accident containment modeling code CONTAIN 1.12 was used to model a HPME experiment performed at the Surtsey test facility at Sandia National Laboratories. The experiment, LFP8A, consisted of the injection of 50 kg of molten thermite into a 1:10 scale model of the Surry cavity connected to the Surtsey containment vessel. A seven-cell best-estimate CONTAIN model, developed using only a few pieces of experimental data, was used to simulate the experiment. In addition to the new debris chemistry models, the CONTAIN simulation employed the TOF/KU trapping model to simulate the splashing of debris off of a concrete slab suspended in the Surtsey vessel to simulate a generic containment structure.

The magnitude of the pressure response calculated by CONTAIN for the Surtsey vessel agreed extremely well with the experimentally determined values, differing by less than 2%. In addition, the TOF/KU trapping model correctly predicted that the majority of the debris injected into the Surtsey vessel would rebound from the concrete slab and drift to the bottom of the vessel. However, the calculated and experimental results for the transient vessel pressure behavior differed significantly. CONTAIN predicted a much more rapid pressurization of the Surtsey vessel than was indicated by the experimental measurements. This was probably due to the fact that CONTAIN predicted a much more rapid interaction of molten debris with the injected steam and the vessel atmosphere than occurred in the experiment.

Because of problems with the mechanistic debris entrainment and dispersal modules from CORDE/GASBLOW2, the CONTAIN model was forced to simulate the injected molten debris with a particle source in the cavity cell. Based on the experimentally observed behavior of the system, debris was injected in the form of dispersed particles into the cavity cell at a constant rate over the first 0.4 s of the test. In the physical situation, the debris is partly de-entrained on various cavity surfaces and must be re-entrained by the high velocity steam moving through the cavity before significant chemical and heat transfer processes can take place. By injecting the debris in the form of airborne particles, CONTAIN over-predicted the rate at which chemical interactions and heat transfer would take place between the debris and the surrounding gas during the first seconds of the test. The result was the observed phase shift between the pressurization curves for the experimental and calculated results.

The fact that CONTAIN correctly predicted the magnitude of the vessel pressurization indicated that the code was accurately modeling heat transfer and chemical reactions with the molten debris. In addition, because the CONTAIN calculation used a best-estimate model primarily with default parameters in the various chemistry, heat transfer, and trapping models, the close agreement between the experimental and calculated results gives some confidence in CONTAIN 1.12's ability to model a HPME event. The verification of the TOF/KU trapping model for the simple LFP8A geometry was an important result, since the model had not been previously tested against experimental results. Although the discrepancy in the timing of the pressurization curves indicated the importance of a more mechanistic approach in providing debris sources in CONTAIN, the overall agreement between the calculated and experimental results showed that CONTAIN 1.12 is a useful tool for the analysis of containment response during DCH scenarios.

ACKNOWLEDGEMENTS

The authors would like to thank the following individuals for their help and support in this work: Al Notafrancesco, our USNRC monitor for this project; Mike Allen and Marty Pilch, who conducted the LFP experiments; Dave Williams and Ken Murata, for technical expertise and general advice; and Dick Gido and Bob Gross, for their careful review of the manuscript.

NOTATION

d_o = orifice hole size, m
F_{trap} = trapping fraction rate, s^{-1}
g = gravitational acceleration, m/s^2
Ku = Kutateladze number
L = fall height, m
L_{tfi} = distance to first impact, m
M = mass of airborne debris, kg
T = temperature, K
T_g = bulk gas temperature in a cell, K
t = time, s
t_1 = time of flight to first structure, s
t_2 = time of flight from first to second structure, s
V_g = gas velocity, m/s
V_{jet} = velocity of debris jet, m/s
V_t = terminal fall velocity, m/s
w = mass flux in cell, kg/m^2-s
w_o = mass flux at the entering orifice, kg/m^2-s
w_x = mass flux at point of impact x, kg/m^2-s

Greek Symbols

α = user input parameter describing the rate of jet expansion
λ_{tr} = trapping fraction rate for TOF/KU model
ρ = bulk gas density, kg/m^3
ρ_a = gas density in cell for trapping, kg/m^3
ρ_d = debris droplet density, kg/m^3
ρ_g = gas density, kg/m^3
ρ_o = density of entering gas, kg/m^3
σ = liquid surface tension,

Subscripts

gft = gravitational fall time
tfi = time of first impact

LITERATURE CITED

1. Murata, K.K., et. al., "User's Manual for CONTAIN 1.1: A Computer Code for Severe Nuclear Reactor Accident Containment Analysis," NUREG/CR-5026 and SAND87-2309, Sandia National Laboratories, Albuquerque, New Mexico, 1989.

2. Washington, K. E., and Griffith, R. O., "CONTAIN Direct Containment Heating User's Manual and Reference Manual Change Document," Letter Report to the U.S. Nuclear Regulatory Commission, Sandia National Laboratories, Albuquerque, New Mexico, 1990.

3. "Estimates of Early Containment Loads from Core Melt Accidents," NUREG-1079, U.S. Nuclear Regulatory Commission, Containment Loads Working Group, 1985.

4. Tarbell, W.W., et. al., "Direct Containment Heating and Aerosol Generation During High Pressure Melt Ejection Experiments," SAND88-1504c, ANS/ENS International Meeting, Washington, DC, October 31-November 4, 1988.

5. Allen, M. D., and Pilch, M., "Letter Report on the Limited Flight Path Experiment," Letter Report to the U.S. Nuclear Regulatory Commission, Sandia National Laboratories, Albuquerque, New Mexico, 1990.

6. Allen, M. D., *Personal Communication*, Sandia National Laboratories, Albuquerque, New Mexico, 1990.

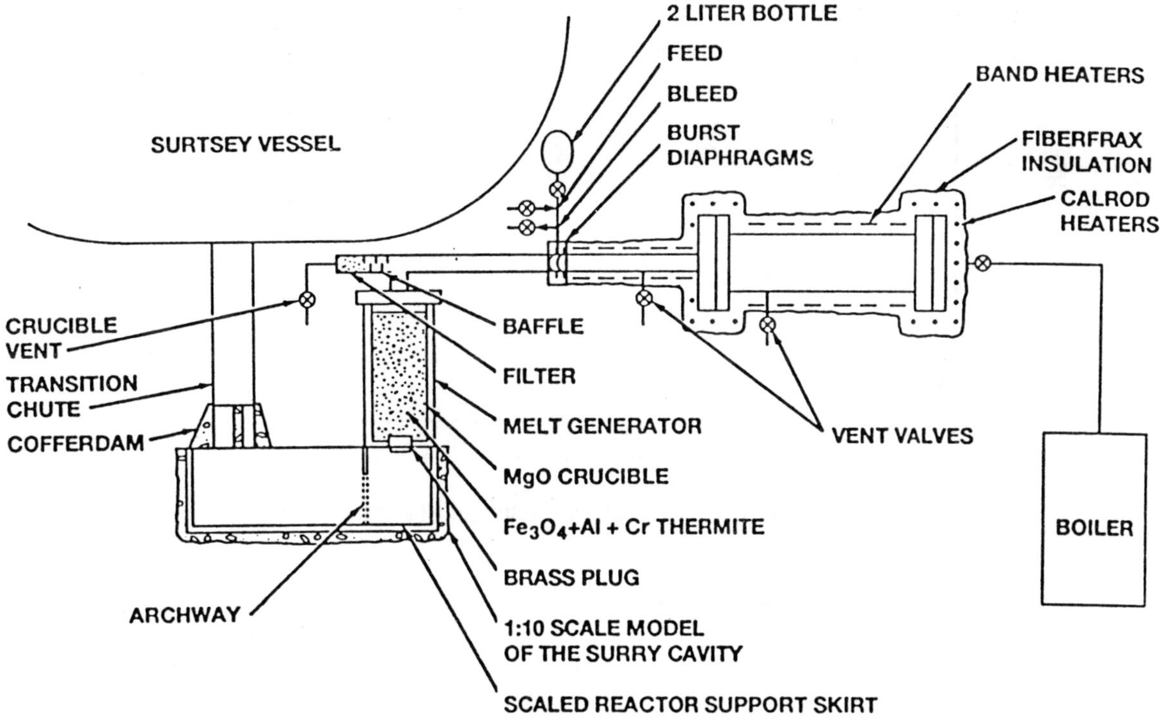

Figure 1. Experimental apparatus for the LFP experiments.

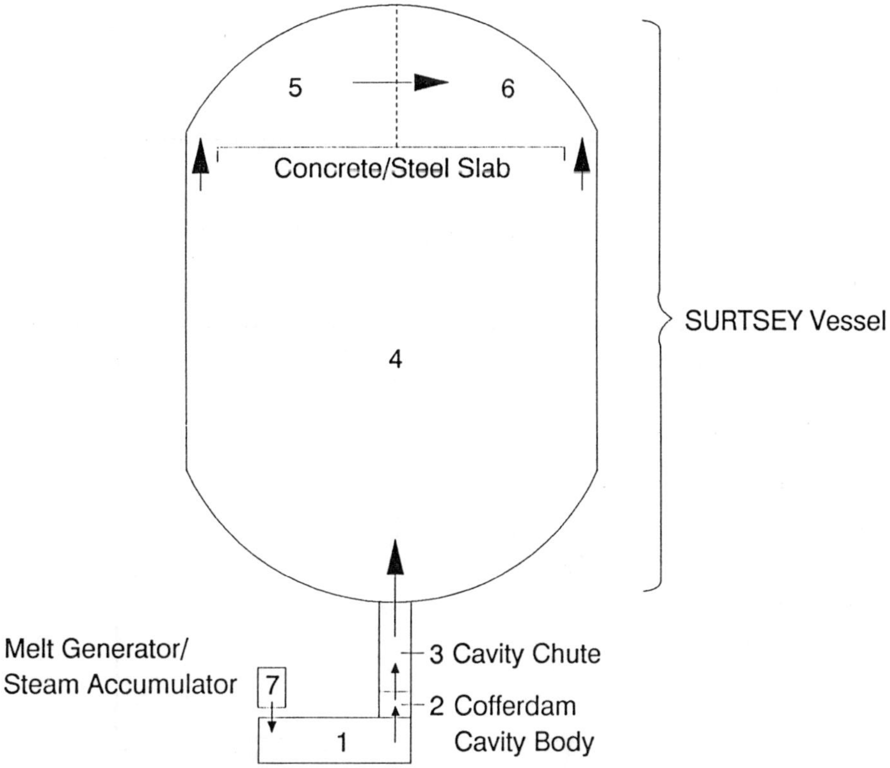

Figure 2. CONTAIN 1.12 model of the LFP8A experiment.

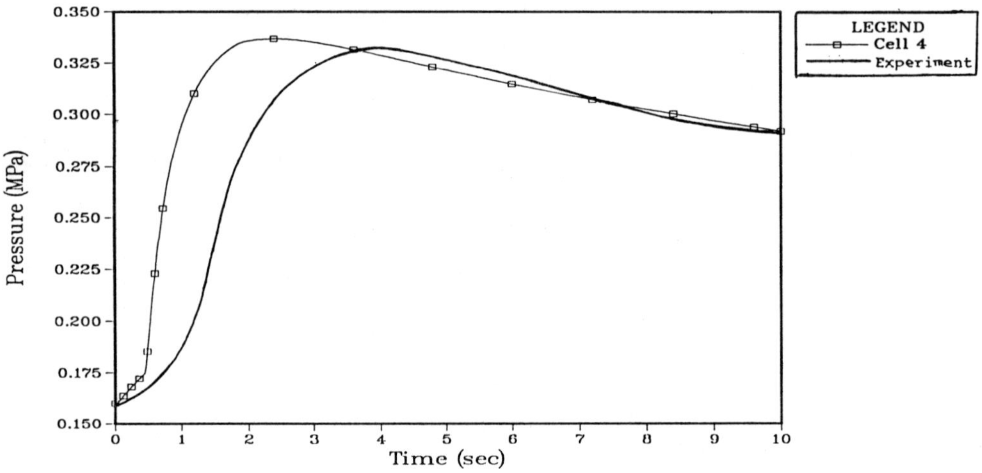

Figure 3. Pressure vs. time in the Surtsey vessel.

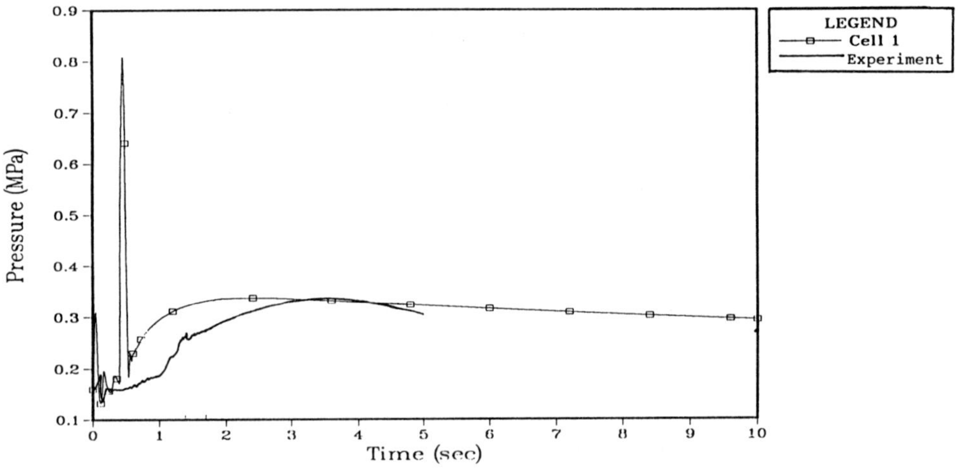

Figure 4. Pressure vs. time in the cavity.

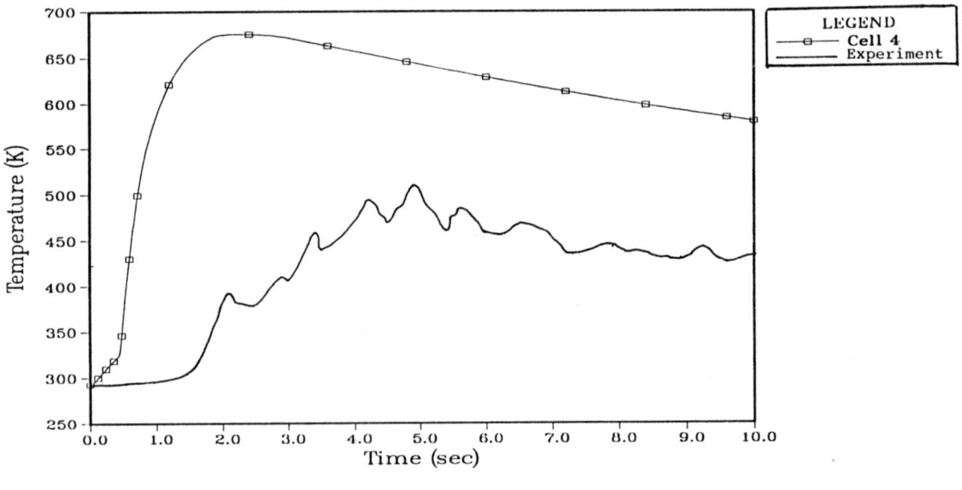

Figure 5. Gas temperature vs. time.

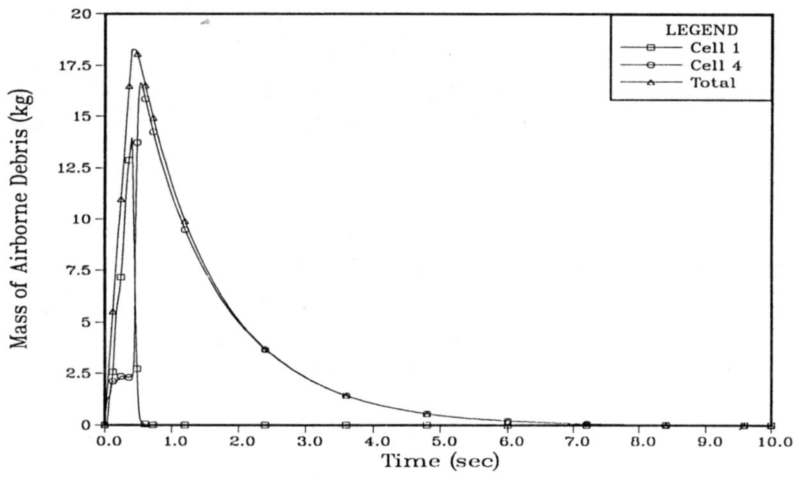

Figure 6a. Mass of airborne debris vs. time.

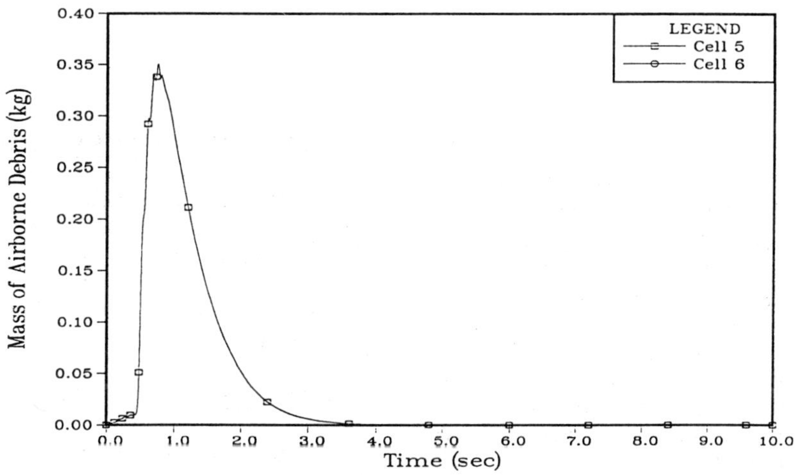

Figure 6b. Mass of airborne debris vs. time in cells 5 and 6.

ANALYSIS OF THE TMI-2 CORE DEGRADATION USING ICARE2 CODE

P. Chatelard, R. Gonzalez, and A. Porracchia ■ Commissariat à l'Energie atomique (CEA), Institut de Protection et de Sureté Nucléaire (IPSN), Centre d'Etudes nucléaires de Cadarache, 13108 St Paul-lez-durance - FRANCE

Since the accident at the TMI-2 reactor in 1979, a world-wide effort has been undertaken for the study of the severe core damage phenomena in LWRs. The development and the validation of ICARE2 code, devoted to core degradation analysis, are a part of the French analytical program. Since the TMI-2 accident has not been duplicated in an experimental facility, TMI-2 provides a unique opportunity to assess the capability of ICARE2 to simulate an accident in a full scale Nuclear Power Plant. This paper deals with a brief description of the ICARE2 code used for this calculation, its current status of development and validation, and the results obtained for the core heat-up and degradation phase of the accident.

Since the accident at the TMI-2 reactor in 1979, a world-wide effort has been undertaken for the study of the severe core damage phenomena in LWRs. In France, a large Severe Fuel Damage program both analytical (ICARE [1]) and experimental (in-pile PHEBUS SFD and FP [2]) is presently conducted at the Institute for Protection and Nuclear Safety of CEA, in order to clearly understand the main physical mechanisms occuring during the core degradation of a PWR. The PHEBUS SFD experimental program is now completed, while the PHEBUS FP program developed in collaboration with the European Community and many other countries is just starting now. Concerning the analytical program, a detailed mechanistic approach is adopted by coupling the thermal-hydraulic RCS code CATHARE2 [3], the fuel degradation code ICARE2 and the aerosols and fission products behavior code TRAPF [4]. The results obtained for the core heat-up and degradation phase of the TMI-2 accident, with the stand-alone version 2 of ICARE2, will be analyzed, after a brief description of the code version used for this calculation and its current status of development and validation.

ICARE2 CODE DESCRIPTION

ICARE2 models the progression of reactor core damage until crucible formation and may be applied to different core configurations such as PHEBUS SFD and FP, PBF [5], CORA [6] and PWR core.

Programming Conception

ICARE2 uses a special data organization technique called SIGAL [1], and is composed of about 30,000 statements written in Fortran 77. ICARE2 consists of a small driver program that calls in series the main modules, the data exchanges between them being performed through SIGAL data base ; this modular structure makes it possible new modules to be developed independently and, after self governing tests, to be easily introduced in ICARE2. The coupling between the modules is generally explicit in time. However, an implicit coupling between fluid-dynamics, thermics and oxidation reactions has been choosen to allow the computation with large time-steps.

Physical Models

The domain of interest is meshed with a one dimensional axial grid. In the radial direction, the meshing is obtained by the definition of different cylindrical structures and independent fluid channels. The intersection between the space meshing and a structure or a channel generates metallic and fluid meshes. A metallic mesh can be composed of several materials.

Fluid-dynamics

The thermalhydraulic model is derived from the 1D two-phase flow model of the CATHARE1 code ; the set of balance equations is discretized with finite difference scheme using axial staggered meshes and a fully implicit technique. The non-linear equations system is solved by a Newton-Raphson iterative method.

However, regarding to CATHARE1, the liquid phase is not available but some improvements allow the study of severe fuel damage transients :

- The gas phase flowing into a channel is a mixture of steam and a non-condensible gas which is generally hydrogen (may be also either helium or argon).
- In presence of oxidation reactions, a starvation model may limit the available oxygen flux at the wall.
- The geometry changes due to flow down and relocation of molten materials are considered.
- In a blocked mesh by refrozen materials, volume and cross section used to solve fluid equations never fall below 10% of initial values (minimum value linked with missing of cross-flow) and oxidation processes are interrupted.
- The radial temperature profile in the core can be approached with a multi-channels modeling. This treatment is done without any cross-flow between parallel channels but with inlet mass flow rates redistribution according to independent flow blockages occurence.

Thermics in all structures

The conservation of energy for the metallic components is written in a lumped form. The solution is obtained assuming that nuclear, chemical and candling powers are explicit while exchanged powers due to conduction, convection and radiation are semi-implicit in time (implicit temperature, other values explicit). These non linear equations, are solved by an iterative Newton method. The phase changes and the corresponding latent heat of the materials are taken into account.

Heat generation

According to given axial and radial power profiles as well as to power history, the code computes, at the beginning of a calculation, the power generated per mass unit at an axial level. Then, during the transient, this value is used to generate the nuclear power in a metallic mesh as a function of its mass of UO_2.

Electrical power may be generated for facilities such as CORA. Electrical current is supplied in parallel heated rods, each of them composed of conductive structures in series.

Heat transfers

A thermal resistance is computed for each face of a metallic mesh. It is a function of the geometry as well as the average thermal conductivity. This latter value depends on the thermal conductivity of the materials and on their geometrical arrangement (stratified or mixed).

The convective wall to fluid heat transfer is expressed in the usual form with the Nusselt number chosen as the maximum value of three flow regimes : laminar or turbulent forced convection and free convection.

A two dimensional radiation heat transfer model, with a participative gas, is available for bundles and large reactor cores ; it is based on the Hottel's method. The transmissivity of the gas between walls is expressed in each spectral absorbing band (5 for water vapor) using the statistical Goody-Mayer model. Geometric mean beam lengths are estimated with an analytical formula, while view factors are computed using the Hottel's crossed-string method. View factors for intact experimental bundles such as PHEBUS, PBF and CORA are predefined in the code. In reactor case, a specific formulation for view factors between concentric rod rings is used.

Mechanics

Only thermal expansions are taken into account. However, a simple loss of integrity model leads to define three possible physical states for a material : compact, cracked or dislocated. The change from one state to another is performed through input data criteria ; these criteria concern physical parameters of the materials such as temperature, layer thickness or temperature evolution with time, and may be combined together.

The concept of cracked state acts essentially for internal cladding oxidation by steam and for fission products release into the channel.

Materials movement and relocation

A mixture (molten or dislocated material) can only flow down if it is no longer enclosed by compact materials. The movement can take place either in an axial level (decanting model) or downwards from an axial level to another (candling model).

The first model allows to put, with an infinite velocity, dislocated fuel pellets on zircaloy cladding or Ag-In-Cd alloy on absorber guide tube in order to compute the downward relocation of a mixture containing solid debris by the candling model.

This latter one is applied after the decanting model during a time step and assumes a constant candling velocity. The candling module solves the mass transport of mixtures from the top to the bottom, with possible mass freezing in axial levels deduced from the solution of an energy balance equation taking into account phase changes.

Decanting and candling models modify the mass and chemical composition of the structures. During this process, the associated heat transfers due to conduction and enthalpic exchanges linked to mass transport are taken into account during the solution of energetic conservation by the thermic module which computes structure temperatures.

Chemical reactions

A specific treatment is done for zircaloy and stainless steel oxidation where steam plays a role. The amount of oxygen mass available for these two reactions is computed by the fluid-dynamics module, taking into account the maximum value requested by each reaction (in unlimited steam conditions). Therefore, oxidation modules are called two times during a time step : the first time to inform the fluid-dynamics module on the maximum oxygen mass needed, and the second time to deal with oxygen diffusion within metallurgical layers according to the oxygen mass provided by the thermal-hydraulic module.

Zircaloy oxidation : in case of unlimited steam conditions, the oxidation is governed by diffusion and is represented by parabolic correlations giving zirconia layer growth and total oxygen mass gain (in alpha-Zr(O) and ZrO_2). Cathcart, Urbanic and Prater correlations are available. The oxygen consumption in the zirconia phase is related to layer thickness variation. The layer thickness of the alpha-Zr(O) phase is then obtained assuming the oxygen atomic fraction to be known in this phase. Under limited steam conditions, it is assumed that available oxygen supplies preferentially the alpha-Zr(O) layer for its growth.

Stainless steel oxidation : as for zircaloy, the process is governed by oxygen diffusion. During the process, iron, chromic and nickel oxides are produced and assumed to be perfectly mixed. Only oxygen mass uptake is given by parabolic correlations.

Interaction between UO_2 and solid zircaloy : contact between cladding and fuel, leads to fuel liquefaction if the temperature is greater than 1473 K. The UO_2 reacts with beta-Zr to form the eutectic U-Zr located between two alpha-Zr(O) layers. The growth rates of these three layers are estimated according to Hofmann parabolic correlations.

UO_2 and ZrO_2 dissolution by molten zircaloy : these two chemical interactions are treated simultaneously. Zirconia dissolution is considered to be diffusion-controlled (parabolic correlation), owing to experimental works of Hofmann. For UO_2 dissolution, three models (introducing two different physical approaches) are available : either the process is governed by diffusion, (Hofmann and Kim Olander kinetics are available), or the dissolution phenomenon is assumed to be convection-controlled (as shown by Kim-Olander) and the model consists in the solution of a transport equation giving the UO_2 concentration in the UO_2-Zr mixture. The alloy U-Zr-O, resulting from dissolution processes, is assumed to be a mixture of solid UO_2 particles, solid ZrO_2 particles and liquid alpha-Zr(O) and is thermally treated as an ideal mixture. However, the U-Zr-O ternary phase diagram is used to provide the solubility limit of the created alloy.

Fission products and aerosols

The CORSOR-M model is used to simulate the release of fourty species of fission products and structural materials from fuel and control rods.

Verification Performed by Development Team

During the code development, verification process is performed. It aims at checking the models of the code in different situations in order to verify if no major points have been omitted. The tests choice philosophy leads to select : tests with different bundle geometries, Separate Effects Tests, tests with control rods and tests at different scales. Following these principles, the code is being assessed on : PHEBUS SFD C3, PBF Scoping Test, CORA5, PBF SFD 1.4 and TMI-2 accident.

Moreover, independent verification is being performed on all PHEBUS SFD tests, including a french participation to PHEBUS ISP 28.

Development Status

The version 2 has been released to Spain and CCR Ispra (in the framework of PHEBUS project).

Up to now, to perform calculations of severe accident in a reactor, ICARE2 runs in parallel with the RCS CATHARE code, i.e., without feedback between the two codes. For this year, the major development consists of the implementation of ICARE2 in the CATHARE code ; ICARE2 will be an optional core module of CATHARE to compute simultaneously the core degradation and the rest of the circuit. Moreover, the resulting integrated code CATHARE/ICARE/TRAPF will provide boundary conditions for containment calculations.

Late phase degradation (debris bed, fuel coolant interaction, vessel rupture) and works for BWR applications are analyzed too.

TMI-2 SEVERE ACCIDENT PHASE ANALYSIS

A calculation of the TMI-2 severe phase accident is performed as a validation test of ICARE2, in order to verify the ability of this tool in large core degradation transients. This work has been done also in the framework of the CSNI TMI-2 standard problem [7], allowing the comparison of the previous version with other codes dealing with severe core degradation. Indeed, progressively with code developments, the computation of the TMI-2 severe accident sequence is performed in order to check the modifications of the new code version. This powerful exercise highlights the code deficiencies, and directs the future code developments in parallel with the interpretation work done in PHEBUS tests analyses.

TMI-2 Accident Summary

The TMI-2 accident has been divided in several phases according to important events. Only the two first phases will be presented here.

Phase 1

The first phase can be characterized as a small break LOCA without core damage. It is initiated by the turbine trip, until the turn-off of the A-loop reactor coolant pumps at 100 mn. This phase begins by the loss of steam generator feed water ; this event leads to reactor scram and an increase of the primary pressure until the pressure set point of the PORV (Pilot Operated Relief Valves) in the top of the pressurizer is reached. Then, these valves open (small break in the top of the pressurizer). The other main event of this phase is the turn-off of primary pumps in loop B at 73 mn.

Phase 2

The second phase which is characterized by core heat-up and degradation, begins with the final pump shutdown at 100 mn until the start of the primary pump B at 174 mn. The consequence of pump trips is the phase separation in the RCS and a decrease of the liquid level in the core ; this decrease leads, at about 107 mn, to the beginning of core uncovery and to a subsequent core heat-up. The primary pressure starts to increase at about 125 mn, due mainly to a strong zircaloy oxidation which involves an increase of the heat source and a reduction of natural circulation flow and steam generator heat transfers. Despite the shutting of the PORV at 144 mn, core heat-up progresses to cladding melt, fuel dissolution and relocation. A crust composed of previously molten materials appears in the lower core region. The cooling required to form and maintain the crust is provided by the remaining water in the core. This crust forms a crucible that collects additional molten materials, flowing downwards in the core. This phase concludes with a significant blockage in the central core region.

ICARE2 calculation conditions

Before severe fuel damage, the transient is a pure thermalhydraulic problem which can be handled by CATHARE. This code, enhanced with some models of ICARE (water vapor properties at high temperatures, simplified radiation heat transfer model in the core ...), has computed the TMI-2 transient until 154 mn [7].

The ICARE2 computation starts at 107 mn (beginning of core uncovery) according to the results of the above mentionned CATHARE/ICARE calculation.

The ICARE2 calculation is stopped at the end of the phase 2, because the core reflooding which occurs at the beginning of the phase 3 cannot be computed with the fluid-dynamics module of the present code version.

Reactor Modeling

The vessel is modeled between the top and the bottom of the core. The axial grid is shown in one of the front views (fig.5). The core domain is divided in 23 meshes with one coarse mesh under the final location of the water level and several fine meshes in the zone where the crucible composed of refrozen materials will be evolved. In radial direction, the core is divided in 6 concentric rings of equal thickness, each of them containing a fluid channel, a representative fuel rod and a representative control rod. From the outside part of the core to the external side of the vessel, 5 other structures are represented, corresponding to the baffle plate, the core barrel, the thermal shield, the reactor vessel and the vessel insulator. Two other fluid channels are also modeled : one corresponds to the core bypass and the other to the downcomer [1].

Boundary Conditions and Physical Assumptions

Thermalhydraulics

The core thermalhydraulic boundary conditions (pressure, mass flow rate, water level), are taken from the above mentionned CATHARE/ICARE calculation until 154 mn and extrapolated after this time [1].

The froth water level decrease is modeled owing to a boundary condition applied to a time dependent axial zone of the metallic structures. In this zone, the structure temperatures are imposed to be equal to the water saturation temperature corresponding to the pressure in the core. The total mass flow rate versus time through the core is supplied in input data. Its distribution in each channel is done automatically, assuming that the vapor velocity at the entrance of the core is constant radially. Therefore, the velocity is the same at the inlet of the 6 channels until a flow blockage occurence in one or several of them. In this case, mass flow rates through the blocked channels are drastically reduced, and the part not injected in these channels automatically distributed in the remaining non blocked channels.

Thermal losses

The temperature outside the vessel insulator is the ambient temperature and the heat transfer coefficient is deduced assuming natural convection with the air of the containment.

Radiation heat transfer

The gas is supposed to be transparent because, at present time, there is no satisfactory model programmed in the code, taking into account the radiation exchange in an axial domain with a non isothermal, non homogeneous gas in the radial direction (in this calculation, they are 6 fluid meshes in each axial enclosure). Nevertheless, heat radiation exchanges between the structures of the 6 rings are considered.

Chemical reactions

The correlations have been choosen as follows :

- Zircaloy oxidation by steam : Urbanic-Heidrick correlation.
- Stainless steel oxidation by steam : White correlation.
- UO_2 dissolution by molten zircaloy : Kim-Olander correlation.
- ZrO_2 dissolution by molten zircaloy : Hofmann correlation.

Break criteria

Concerning the fuel rod claddings, it is assumed that a zirconia layer becomes dislocated if one of the two following criteria are verified :

- temperature greater than 2250 K and thickness less than 300 microns,
- temperature exceeding 2500 K.

The former criterion means that a zirconia layer can maintain in place molten beta-Zr and alpha-Zr(O) (T > 2250 K) if it is sufficiently thick (> 300 μm). The latter criterion assumes that zirconia layer breaks, whatever its thickness, when temperature exceeds 2500 K (embrittlement).

The fuel pellets are assumed to be fragmented when the fuel rod cladding melts and eventually flow down. In such a case, the UO_2 fragments fall downwards if the fuel rod cladding disappears in an axial location. However, in the present code version, it is impossible to allow the downward relocation of fuel pellets which are no longer supported by fuel pellets located below them and which have disappeared. It means that the fuel movement can take place if and only if the neighbouring cladding is absent in the axial mesh.

Concerning the absorber rod guide tubes in zircaloy, a chemical attack by the Ag-In-Cd alloy occurs when these materials are in contact. This phenomenon appears when the stainless steel cladding of the absorber rod melts (temperature greater than 1730 K). Therefore, when the guide tube temperature exceeds 1730 K, it is assumed that the guide tube is instantly dissolved by the absorber alloy.

Flow down velocity

There is no specific module computing the velocity of materials which flow down, because some sensitivity studies have shown that this parameter seems to be not important for the calculation of relocated materials. Indeed, the flow down problem is essentially thermally controlled if the amount of flowing materials is not too high. As a consequence, a value of 0.6 m/s, whose magnitude seems to be realistic, is used for any composition of flowing materials.

In addition, it is assumed that an axial fluid mesh is never completely blocked by refrozen materials. The porosity assumed for the crucible and the rubble bed above is 10 %.

Comments on the Results

General overview

The calculation starts at the beginning of core uncovery (107 mn in the accident time). As soon as an axial mesh becomes uncovered, the temperatures increase rapidly (fig.1 and fig.2). Two heat transfer modes contribute to extract nuclear power :

- the enthalpic transport by the gas in axial direction, only because there are no cross-flows radially,
- The radiation heat transfer between hot core to cold structures surrounding it in radial direction.

But the low values of the view factors between rod rings imply to have high temperature differences between the rings in order to remove a significant fraction of the energy. Consequently, the radial temperature profile becomes very steep, which explains why the temperature in the central part of the core is much higher than in the external part.

In addition, the low vapor mass flow rate through the core is not sufficient to remove the nuclear power. The temperature increases with elevation, and the resulting axial profile is a function of the axial power distribution and the gas mass flow rate. The cladding and the gas temperatures have practically the same value in each location.

Oxidation phenomena

When temperature exceeds 1000 K, zircaloy oxidation becomes significant with an escalation after 1850 K, due to the allotropic transformation of tetragonal ZrO_2 to cubic ZrO_2 predicted by the Urbanic-Heidrick correlation. The energy supplied by this chemical reaction is very important, and of course accelerates the core heat-up while the oxidation of the absorber rod stainless steel claddings does not contribute significantly to core heat-up (fig.3).

These two oxidation reactions result in a large hydrogen release in the channels. The total amount of hydrogen generated by oxidation is about 250 kg, essentially produced by zircaloy oxidation, while stainless steel oxidation contributes only to about 15 kg (fig.4).

Degradation aspects

In order to illustrate the degradation phase, different front views of one half of the 6 rods are shown.

The fuel rods degradation begins about 35 mn after the beginning of the core uncovery in the hot point of the core located in its centre near the top. The fig.5 which shows ,at the instant 151 mn, material compositions in the 6 representative fuel rods, illustrates four steps in the degradation :

- in rod 6, the zirconia layer is growthing with the alpha-Zr(O) layer but the oxidized zone is not greatly extended due to the pronounced axial temperature profile.
- in rod 5, no rupture has occured, but the melting temperature of alpha-Zr(O) and beta-Zr phases is reached at elevation 3.6 m and the liquid mixture is maintained in place by the zirconia whose thickness is greater than 300 μ m. Therefore, the liquid zircaloy dissolves the fuel pellets with the kinetics given by Kim-Olander correlation. The oxidation of liquid zircaloy continues during this period until the zirconia layer dislocation.
- in rod 4, the cladding rupture has occured at elevation 3.6 m, due to high temperature criterion, and flowdown of dislocated materials has just begun. The relocation of molten materials leads, at elevation 2.6 m, to an increase of the temperature, and consequently to an increase of UO_2 dissolution by molten zircaloy.
- in rods 1,2 and 3, the degradation is more advanced ; the zirconia layer has broken in several hot meshes when temperature exceeds 2500 K. The mixture of liquid zircaloy, fragments of zirconia and fuel pellets has flown down and refrozen below in a colder zone.

When an axial mesh is blocked by the refreezing of molten materials, then the mass flow rate through the corresponding channel is decreased. Consequently, the contribution of the gas to remove energy in the zone is reduced in the same time and the rod damage becomes more pronounced. Due to the present parallel channels modeling without cross-flows, the redistribution of mass flow rate in non blocked channels contributes to mitigate the heat-up of external channels. Nevertheless, only the degradation of the more external rod ring can be avoided : after a heat-up phase, its temperature is reduced, owing to the injection in the corresponding channel of the overall core mass flow rate when the other five channels are blocked.

The final front view at 174 mn shows, through the temperature field, the shape of the debris bed in the core supported by a crucible composed of refrozen materials (fig.6). This figure indicates that the external ring is intact and that the temperature level reached in the upper part of the debris bed is very high. In consequence, it is necessary to take into account radiative heat transfers from the top of the rubble bed to the surrounding structures. This modeling is not done at present time, but will be implemented in parallel with a specific module dealing with the evolution in a debris bed.

The composition of the rubble bed and the crucible at this time is very instructive. The rubble bed above the crucible is mainly composed of UO_2 fragments with an average mass fraction about 80 %. However, the major part of the remaining alpha-Zr(O) and beta-Zr is located in and just above the crucible with a decreasing concentration in the radial direction (fig.7). The maximum value of the mass concentration is about 15% and, due to the hypothesis of stopping the oxidation in blocked meshes, this zircaloy is not oxidized as long as it remains in the crucible.

This material field distribution is dependent on the break criterion chosen and on the thermal transient before loss of integrity. Indeed, the large amount of zircaloy in the central part of the core is linked with its greater heat-up velocity. The allowing time to oxidize zircaloy is consequently shorter than in external rings before cladding ruin.

It is very important to correctly compute this time interval for oxidizing the zircaloy before it is trapped in the crucible, because the amount of this material in the corium governs its evolution during the latter phase of degradation, and especially the kinetics of concrete erosion if vessel rupture occurs.

Concerning the control rods, the degradation begins earlier in the transient and the relocation of Ag-In-Cd just above the water level is emphasized. The most important point is that the flow down of the absorber materials is able to block the channels. Consequently, they are two plugs in each channel : the lower composed of control rod materials and the upper due to fuel rods materials. These two plugs join at the end of the transient near the water level.

Sensitivity Study

In the calculation presented, the mass flow rates in the core are distributed according to flow blockages in the channels. Consequently, when the five central channels are blocked, all the mass flow rate is injected in the most external channel improving its cooling.

In order to quantify the influence of this modeling, a sensitivity study has been done where the mass flow rates are unchanged in the channels despite any blockage in one of them. The debris bed obtained with this sensitivity study has not the same shape and the temperatures distribution is quite different ; the main effect, in this case, is the break of the rods, in the external ring, at the end of the phase 2 (fig 8).

This study highlights the great influence of fluid dynamics in the core behavior and implies to have a multidimensional fluid modeling to correctly compute the cooling and the oxidation of the debris bed.

CONCLUSION

The results obtained, in the TMI-2 accident during the phase 2, are in a rather good agreement with the assumed behavior of the core during this period. The results indicate that the rubble bed forms very early after the beginning of cladding degradation and progresses downwards towards the water level. At the end of this phase, before the momentary reflood of the core, the calculation shows that severe damage has occured in a large part of the core.

A rubble bed, essentially composed of fuel pellet fragments, is supported by a crucible formed, in its upper part, by refrozen materials mainly composed of zircaloy, zirconia and dissolved UO_2 and, in its lower part, by absorber rods materials just above the water level.

However, the rods in the external part of the core are still intact, except the absorber rods which are already molten in their upper part.

The total hydrogen mass released is about 250 kg and the power generated by zircaloy oxidation has induced an acceleration of the core heat-up, while the power injected by stainless steel oxidation and the hydrogen release are much lower (15 kg).

The higher temperatures level reached in this bed implies to have a specific modeling in order to get a good behavior. In addition, heat transfers by radiation from the bed to the upper structures in the core and with the gas must be modeled too.

Moreover, the pronounced axial and radial temperature profiles in the core necessitate a correct modeling of thermalhydraulics which will be handled with at least a two dimensional approach ; a sensitivity study on the distribution logic for mass flow rates in parallel core channels has shown how fluid dynamics play an important role in the core evolution.

Therefore, the next steps in code development will be the introduction of a specific debris bed module and a multidimensional thermalhydraulic module in the reactor vessel in order to improve the application of ICARE2 in large core severe accidents.

Moreover, in order to allow the computation of any kind of severe accidents scenarios in P.W.R, ICARE2 will be coupled with the RCS CATHARE code and with the TRAPF code.

REFERENCES

[1] R.Gonzalez,"Effects of stainless steel claddings on core degradation of a PWR type reactor computed by ICARE2 code" IAEA Workshop on Computer Aided Safety Analysis, Moscow (1990)

[2] C.Gonnier,"In pile investigation at PHEBUS facility on the behavior of PWR fuel bundles in severe accident conditions beyond

the design criteria", Int. Conf. Th. React. Safety,Avignon (1988)

[3] J.C.Micaelli,"CATHARE best-estimate thermalhydraulic code for reactor safety studies", Int. Conf. Th. React. Safety, Avignon (1988)

[4] V.D.Layly,"TRAPF a french stand-alone module for the transport of fission products in a circuit", (CEA tool derived from TRAPMELT), private communication,(1991)

[5] D.J.Osetek,"Results of the four PBF severe fuel damage tests", 15th WRSIM,Washington,(1987)

[6] S.Hagen,"Out of pile bundle experiments on SFD CORA program", KfK 3677, Karlsruhe, (1986)

[7] R.Senemeaud,"Final CSNI TMI-2 Analysis exercise report - French participation", (1989)

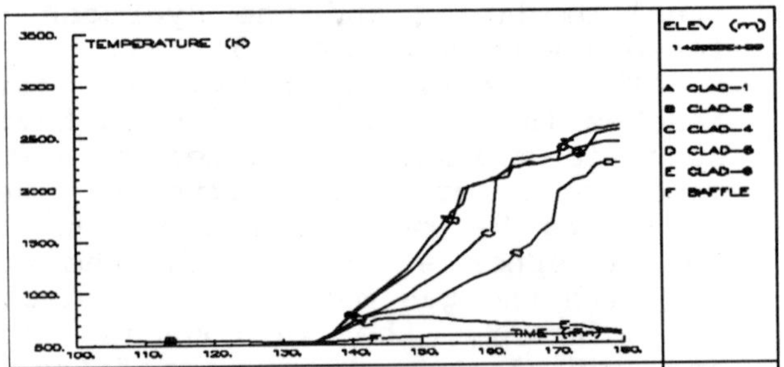

TEMP. EVOLUTION AT ELEVATION 1.46 m (6 RODS + BAFFLE PLATE)
ICARE2 V2 RESULTS FOR TMI2 TRANSIENT (PHASE 2)

fig.1

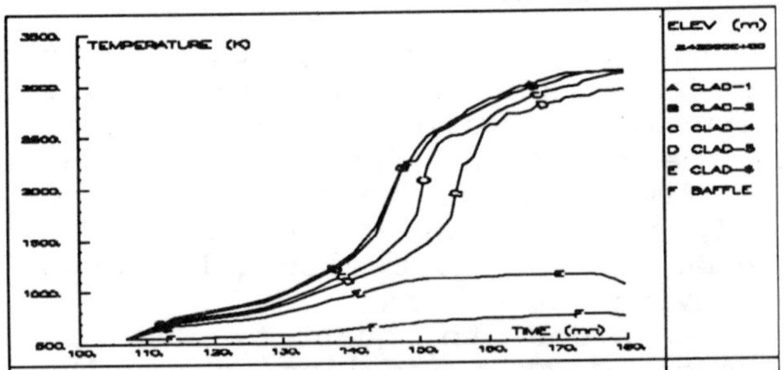

TEMP. EVOLUTION AT ELEVATION 2.42 m (6 RODS + BAFFLE PLATE)
ICARE2 V2 RESULTS FOR TMI2 TRANSIENT (PHASE 2)

fig.2

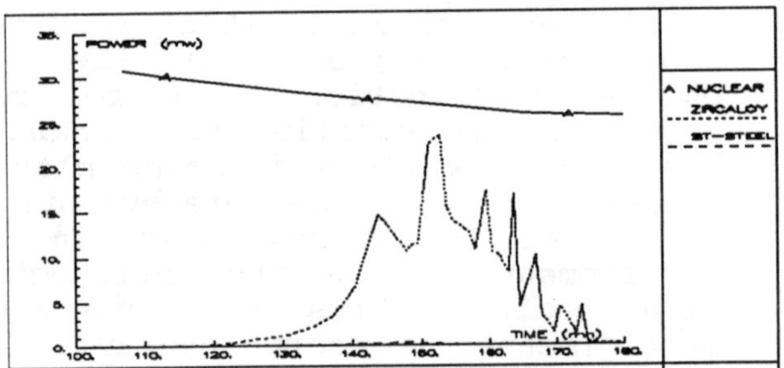

NUCLEAR, ZR AND SS OXIDATION POWERS
ICARE2 V2 RESULTS FOR TMI2 TRANSIENT (PHASE 2)

fig.3

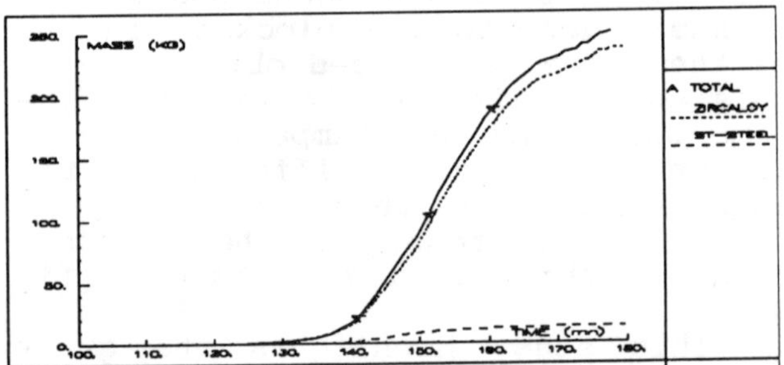

HYDROGEN PRODUCTION DUE TO ZR AND SS OXIDATION
ICARE2 V2 RESULTS FOR TMI2 TRANSIENT (PHASE 2)

fig.4

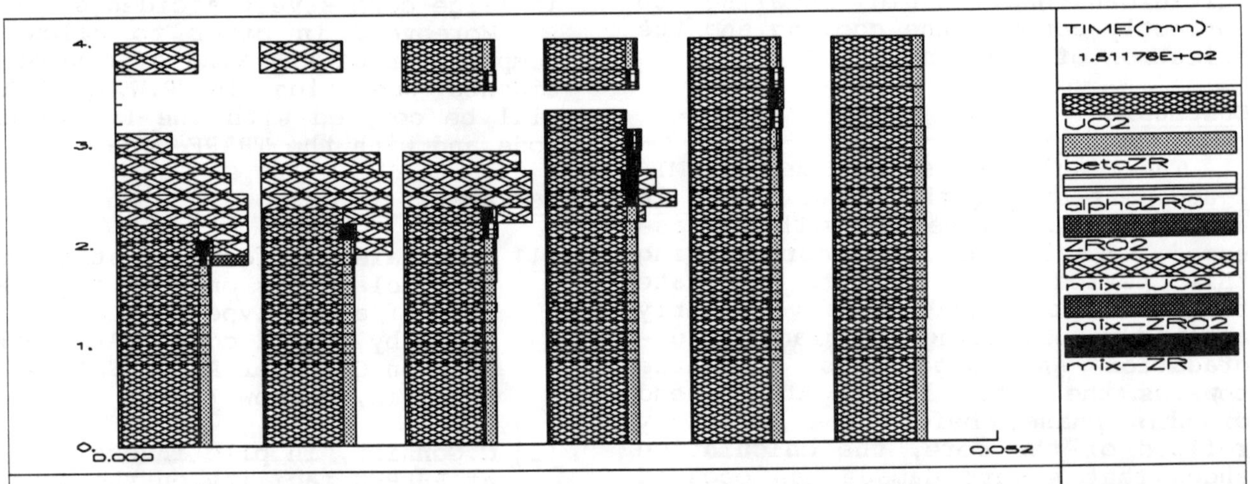

CORE DAMAGE AT TIME 151 mn (FUEL RODS 1 TO 6)
ICARE2 V2 RESULTS FOR TMI2 TRANSIENT (PHASE 2)

fig.5

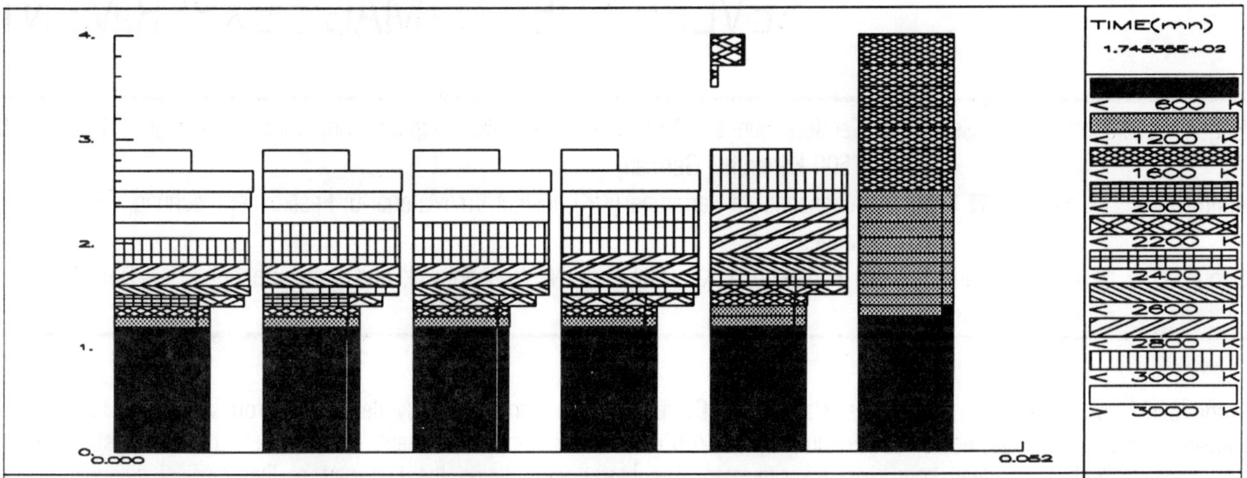

TEMPERATURES FIELD AT TIME 174 mn (FUEL RODS 1 TO 6)
ICARE2 V2 RESULTS FOR TMI2 TRANSIENT (PHASE 2)

fig.6

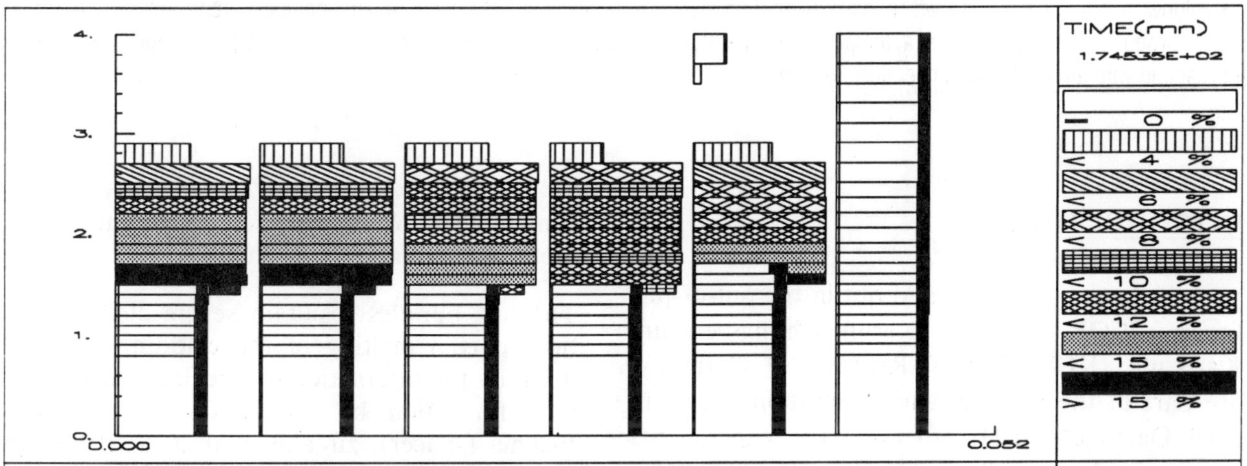

ALPHA-ZR(O) AND BETA-ZR MASS FRACTIONS FIELD AT TIME 174 mn
ICARE2 V2 RESULTS FOR TMI2 TRANSIENT (PHASE 2)

fig.7

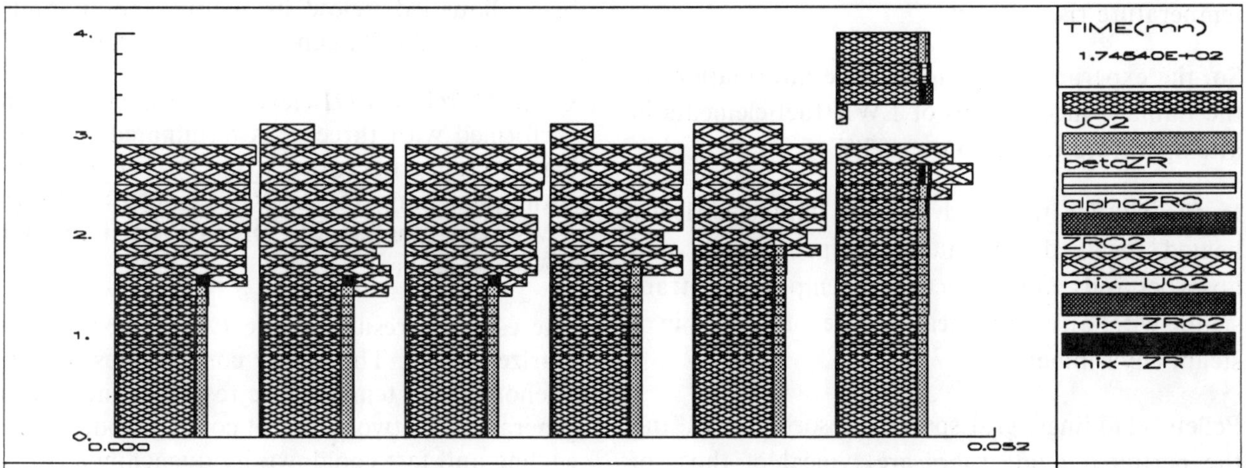

CORE DAMAGE AT TIME 174 mn (FUEL RODS 1 TO 6)
SENSITIVITY STUDY — NO INLET MASS FLOW RATE REDISTRIBUTION

fig.8

INFLUENCE OF REFLOOD IN THE CORA SEVERE FUEL DAMAGE EXPERIMENTS

S. Hagen, F. Siebert and L. Sepold ■ Kernforschungszentrum Karlsruhe (KfK, Hauptabteilung Ingenieurtechnik, P.O. Box 3640, 7500 Karlsruhe, Germany

P. Hoffmann and G. Schanz ■ Kernforschungszentrum Karlsruhe (KfK), Institut für Material-u. Festkörperforschung, P.O. Box 3640, 7500 Karlsruhe, Germany

G. Schumacher ■ Kernforschungszentrum Karlsruhe (KfK), Institut für Neutronenphysik und Reaktortechnik, P.O. Box 3640, 7500 Karlsruhe, Germany

In the CORA experiments on Severe Fuel Damage (SFD) rod bundles with electrically heated fuel rod simulators, full-pellet pins, and absorber rods are subjected to temperatures up to 2400 °C in a steam environment. With these out-of-pile tests small-break LOCA (Loss of Coolant Accident) conditions are simulated. The heatup rate during the first part of the transient amounts to 1 K/s. At about 1100 °C the slow temperature rise is followed by an uncontrolled temperature escalation due to the exothermal heat produced by the zircaloy cladding oxidation. The CORA tests are terminated by power reduction (slow cooldown) or by additional quenching (fast cooldown). The CORA tests which have been terminated by quenching demonstrate an additional, remarkable increase in the hydrogen production combined with a temporary temperature increase, *i.e.*, reflood conditions can trigger an additional zirconium/steam reaction with an additional temperature escalation.

INTRODUCTION

The experiments carried out in the out-of-pile facility "CORA" at the Kernforschungszentrum Karlsruhe (KfK), Federal Republic of Germany, are part of the national and international "Severe Fuel Damage" (SFD) program. The experiments are to simulate small-break LOCA (Loss of Coolant Accident) conditions in a Light-water Reactor (LWR) with the fuel elements exposed to slow temperature rises.

So, the experiments are to provide information on the damage mechanisms of LWR fuel elements in a temperature range up to 2400°C.

In the CORA tests rod bundles with electrically heated fuel rod simulators, full-pellet rods, and absorber rods are subjected to temperature transients with an initial heatup rate of 1 K/s in a steam environment.

Pellets, claddings, grid spacers, absorber rods, and the pertinent guide tubes are typical of those of commercial LWRs with respect to their compositions and radial dimensions.

The test bundles result in severe cladding oxidation, partial melting of the cladding, fuel dissolution by the interaction of zircaloy and UO_2, early melt formation by interactions of zircaloy and Inconel (spacer), zircaloy and steel (PWR absorber), and boron carbide and steel (BWR absorber). Particularly in presence of absorber material, PWR- and BWR-type, the fuel rod cladding is being liquefied below its melting point, and the dissolution of UO_2 can be initiated earlier.

Up to March of 1991 eleven experiments have been performed with three tests terminated by quenching (see test program in Table 1). The experiments with quenching are CORA-12, CORA-13 (both PWR-type tests), and CORA-17 as simulation of BWR conditions.

The essential results of the CORA tests are summarized in (1). This paper concentrates on the the phenomena of temperature response and hydrogen generation for two different cooling modes, i.e. slow cooling and fast cooldown by quenching.

Table 1. *CORA tests performed up to March of 1991*

Test No.	Max. Cladding Tempera-tures	Absorber Material	Other Test Conditions	Date of Test
2	≈ 2000°C	-	UO_2 refer., inconel spacer	Aug. 6, 1987
3	≈ 2400°C	-	UO_2 refer., high temperature	Dec. 3, 1987
5	≈ 2000°C	Ag, In, Cd	PWR-absorber	Febr. 26, 1988
12	≈ 2000°C	Ag, In, Cd	quenching	June 9, 1988
16	≈ 2000°C	B_4C	BWR-absorber	Nov. 24, 1988
15	≈ 2000°C	Ag, In, Cd	rods with internal pressure	March 2, 1989
17	≈ 2000°C	B_4C	quenching	June 29, 1989
9	≈ 2000°C	Ag, In, Cd	10 bar system pressure	Nov. 9, 1989
7	< 2000°C	Ag, In ,Cd	57-rod bundle, slow cooling	Febr. 22, 1990
18	< 2000°C	B_4C	59-rod bundle, slow cooling	June 21, 1990
13	≈ 2200°C	Ag, In, Cd	OECD/ISP; quench initiation at higher temperature	Nov. 15, 1990

TEST SECTION ARRANGEMENT

In the test facility the rod bundle is surrounded by a zircaloy shroud (PWR 1.2 mm, BWR 2.6 mm thick) with a 20-mm ZrO_2 fiber insulation and a high-temperature shield with a fiber insulation. The shroud insulation guarantees a flat radial temperature distribution across the bundle.

The axial temperature profile of the bundle is determined by the cold ends (water supply, ~20°C) and the temperature-dependent electric resistance of the heater. Turbulences by cross flow at the steam inlet (0 mm) have an influence on the axial temperature profile in a way that the bundle is kept relatively cool in the lower part, also during the heatup phase (2).

The test bundle has an overall length of about 2 m with a heated length of 1 m. The decay heat of the fuel rods is simulated by electrical heating using tungsten pins which are centrally located within annular UO_2 pellets of the heater rods.

Two different absorber materials are used within different bundle configurations, i.e. rods with (Ag,In,Cd) are to represent PWR conditions whereas boron carbide absorber rods refer to the experiments simulating BWR fuel elements (see Figure 1).

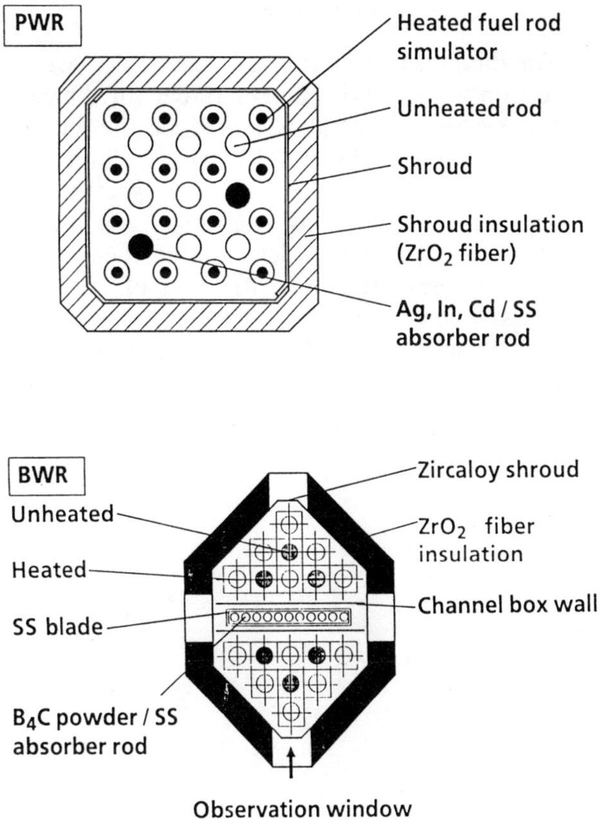

Figure 1. *Bundle arrangements of PWR-type and BWR-type experiments used in the CORA SFD program*

Three grid spacers are inserted in each bundle. In the PWR-type experiments the central spacer is made of Inconel and the upper and lower spacers are made of zircaloy. In the BWR-type experiments zircaloy is used as material for all three spacers.

To study the fragmentation of embrittled fuel element components caused by quenching, the bundle can be flooded with water from the bottom. This is accomplished by an upward movement of a water-filled quench cylinder. During flooding the quench cylinder is being driven hydraulically with an average speed of 1 cm/s. At the onset of quenching the water level is at 220 mm below the bottom of the heated section.

The hydrogen produced by the zirconium - steam reaction is measured at two different positions: at the test section outlet and in the dilution chamber of the off-gas system (Figure 2).

For the measurement at the location above the test section a special dilution chamber (of the gas analysis system) is installed to reduce the gas concentration by a ratio of 1:10 using helium. The gas mixture is transported to the spectrometer via capillary tubing. It is analyzed by a quadrupole mass spectrometer in determining the ion currents that represent the concentrations of the respective gases. From these data the mass production rate of hydrogen is calculated with help of the ratio of the partial pressure of the particular gas and that one of argon (carrier gas) and then multiplied by the argon flow rate through the test bundle. For the quantitative analysis of the hydrogen generation data the measurement system has to be calibrated by means of pretests with calibration gases, i.e. argon with different additions of H_2, H_2O, CO, CO_2, CH_4.

To observe the materials behavior and particularly the relocation during the transient up to ten videoscopes, i.e. video recorders and 35-mm still cameras, are installed at observation windows of the test section.

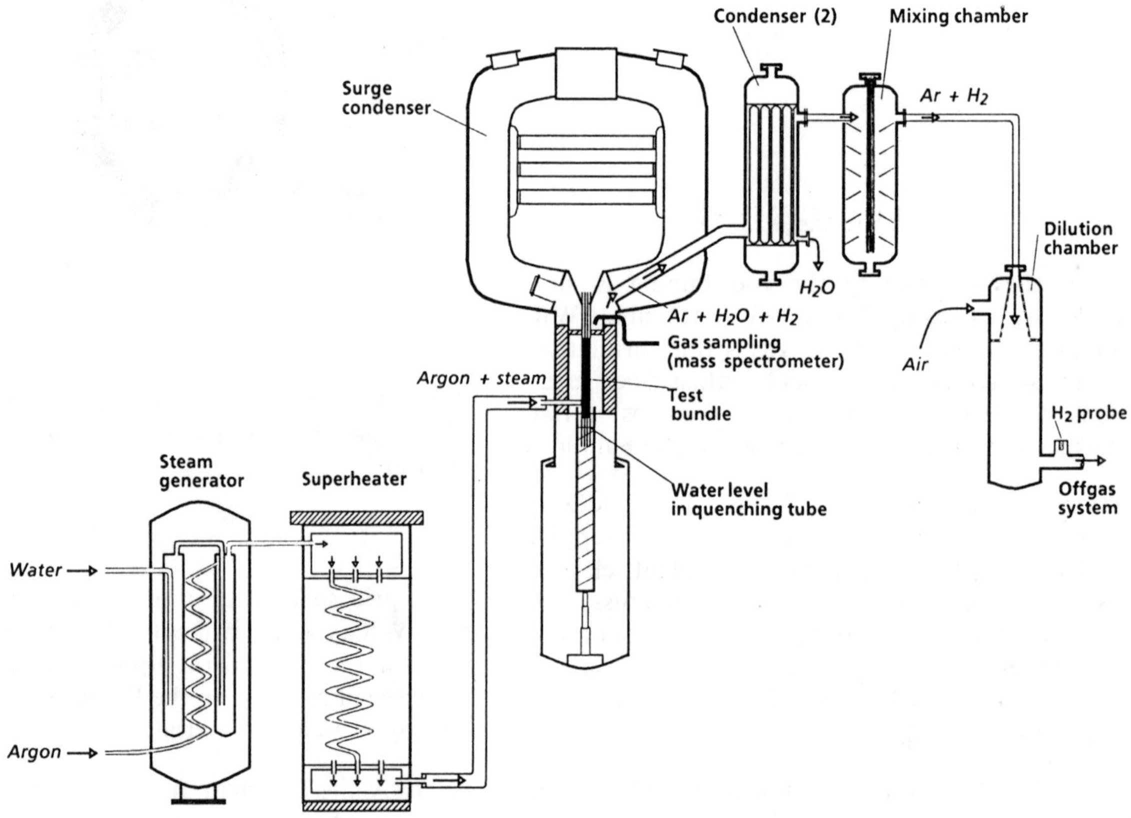

Figure 2. *Simplified flow diagram of the CORA test facility*

TEST CONDUCT

In the CORA experiments the test sequence can be distinguished in the following phases:

- Pre-heating 0000 - 3000 s (argon only)
- Heat-up 3000 - 3900 s (argon + steam)
- Escalation 3900 - 4900 s (argon + steam)

- Cool-down > 4900 s (argon only)

For the first 3000 s, pre-heated argon gas of about 500 °C enters the bundle with a flow rate of 6 - 8 g/s. Between 3000 s and 4700 s the electric power is increased from 6 kW to 28 kW to achieve the initial heatup rate of 1 K/s. At 3300 s within the test a constant flow of superheated steam (2 - 12 g/s) is established in addition to the argon flow.
Between 1000 and 1200°C the slow temperature rise is followed by a temperature escalation, caused by the exothermal zirconium-steam reaction, driving the bundle temperatures to 2000°C and above. The tests are terminated by power reduction (slow cooldown) or by power reduction and quenching with water (fast cooldown). In both cases the steam supply is turned off as well.

OBSERVATIONS OF THE BUNDLE HEATUP

The driving force for the first phase of the heatup in the CORA tests is provided by the electric power input. At a threshold temperature the exothermal reaction heat becomes significant. The threshold value depends on the energy balance within the test rod arrangement, i.e. on the heat losses from the bundle to the environment. In case of high heat losses it is even possible that no temperature escalation at all takes place. With the very effective heat insulation as used in the CORA test section the onset of escalation can be as low as 1100°C.

Sufficiently high steam supply is another prerequisite that a temperature escalation can be achieved.

During the escalation phase of a SFD experiment the electric power (or the decay power in in-pile tests) gains support by the exothermal zirconium - steam reaction:

$$2\,H_2O + Zr = 2H_2 + ZrO_2 + 6700\ J/g\ (Zr)$$

So, the oxidation of the cladding outer surface is a major contributor to reach temperatures of 2000°C and above.
The oxidation of the cladding at the inside does not contribute to the heatup of the fuel rods as long as the cladding is intact and the steam has no access to the inner surface. The unoxidized cladding material, however plays an important role during the flooding phase as will be described later.

The ratio of the chemical reaction energy to the total power was determined to be 30 to 40 % in the CORA experiments (see Table 2).

The electric energy given in Table 2 is the measured total value; it is not corrected for energy losses in the power connection cables that amount to about 10 % of the total power input. The calculated chemical reaction energy is based on the measurement of the total hydrogen generation during the individual CORA experiments. (Thus, energy values are given for those tests only where quantitative H_2 data exist).

Table 2. *Energy release by the Zircaloy oxidation during various CORA tests*

Test	Bundle type	Cool-down mode	Total H_2 production [g]	Oxidation energy [MJ]	Total input of electric energy [MJ]	Percentage of oxidation [%]
CORA-15	PWR	slow	114	17,3	33.1	34
CORA-9	PWR	slow	159	24.2	56.2	30
CORA-7	PWR	slow	114	17.3	33.5	34
CORA-18	BWR	slow	106	16.1	33.1	33
CORA-17	BWR	quenching	128	19.5	28.8	40

In general, the temperature escalation with a heat-up rate of approx. 10 K/s starts at the hot spot of the bundle i.e. in the central or upper region of the CORA test section. From there the oxidation front moves downward. The temperature increase ceases at ~2000°C due to heat losses if no additional electric power is supplied to the bundle.

Figure 3 gives the temperature response of the full-pellet fuel rods during the transients CORA-5 and CORA-16, respectively, both terminated by a slow cooldown. The escalation point (larger temperature gradient) for both experiments is about 1100°C.

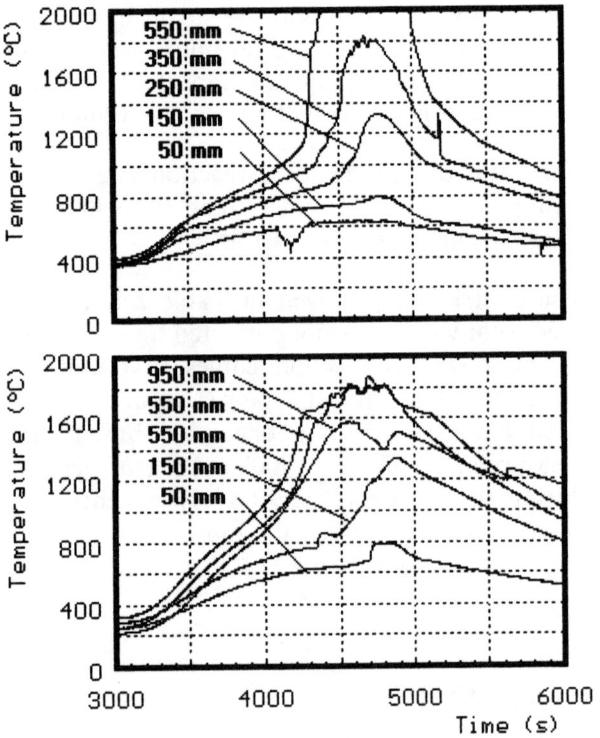

Figure 3. *Temperatures of the unheated (full-pellet) rods in tests CORA-5 (PWR, top) and CORA-16 (BWR), both without quenching*

EFFECT OF BUNDLE QUENCHING

To demonstrate the influence of the cooldown mode on the bundle behavior, the following CORA experiments can be compared: CORA-5/ CORA-12 (PWR-type) and CORA-16/ CORA-17 (BWR-type). CORA-12 and CORA-17 are the tests terminated by power reduction and quenching.

The comparison of the qualitative data on the hydrogen production measured in the off-gas system of the CORA facility between tests CORA-16 (without quenching) and CORA-17 (quenching) in Figure 4 shows a remarkable H_2 peak during the flooding phase of CORA-17.

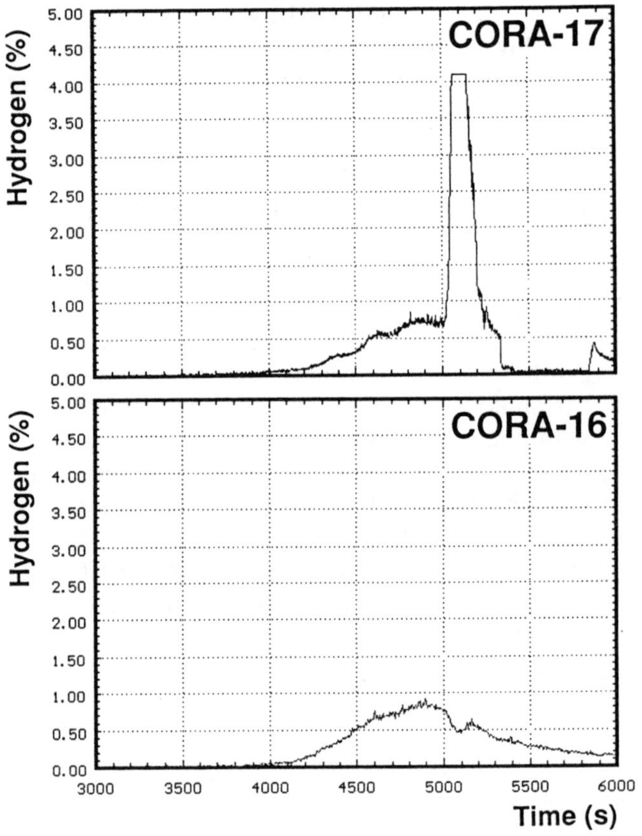

Figure 4. *H_2 production (qualitative data) measured in the off-gas system during BWR-type tests CORA-16 (without quenching) and CORA-17 (with quenching)*

(Note: The hydrogen data are measured in the off-gas system after a dilution by air. The individual curves cannot be compared to each other with respect to the amount of hydrogen because different air volumes were used in the CORA tests for diluting the off-gas. Therefore, all plots on the hydrogen generation in this paper can only be used qualitatively!)

A similar behavior in the hydrogen response is to be seen in test CORA-12. Figure 5 shows the hydrogen production of this test together with the temperature recordings of the full-pellet rods.

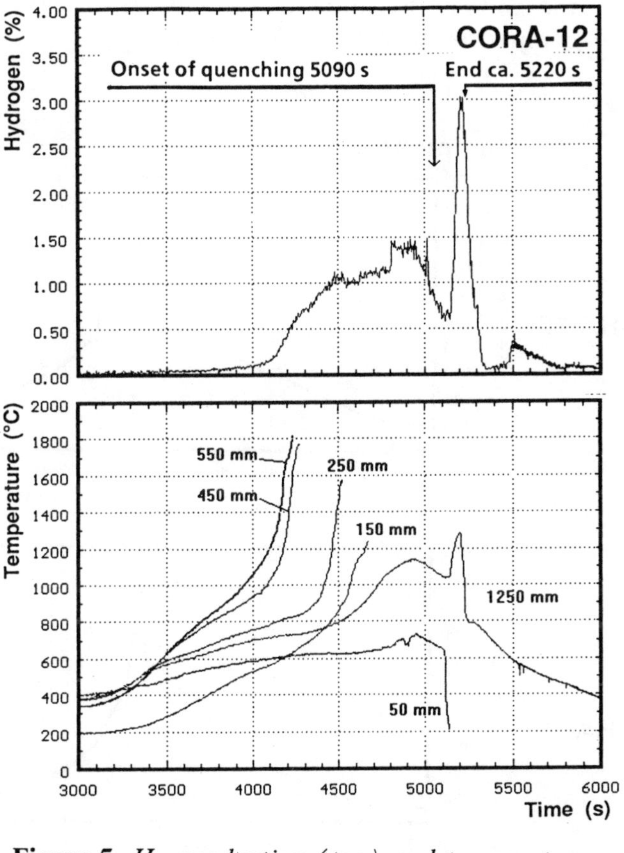

Figure 5. *H_2 production (top) and temperatures of the unheated (full-pellet) fuel rods during quench test CORA-12*

The quench process in CORA-12 starts at 5090 s. The temperature at 50 mm elevation decreases with the rising water level. In the upmost part of the bundle, i.e. at 1250 mm, however, a temporary increase of the temperature is evident together with a strong increase in the hydrogen production. (The thermocouples in the central region had failed in this test).

The temperature escalation during flooding is also demonstrated in the formation of some additional melt. Falling droplets of melt can be observed on video recordings while the quench water front rises.

The temperatures in BWR-type experiment CORA-17 demonstrate a similar behavior (Figure 6). The bottom of the test section remains relatively cool whereas the temperatures in the upper part of the bundle (shown at the 1150-mm elevation of the zircaloy channel box wall because the fuel rod thermocouples had failed in this test) increase as rapidly as the quench water level rises. The hydrogen production rate of test CORA-17 is also depicted in this figure.

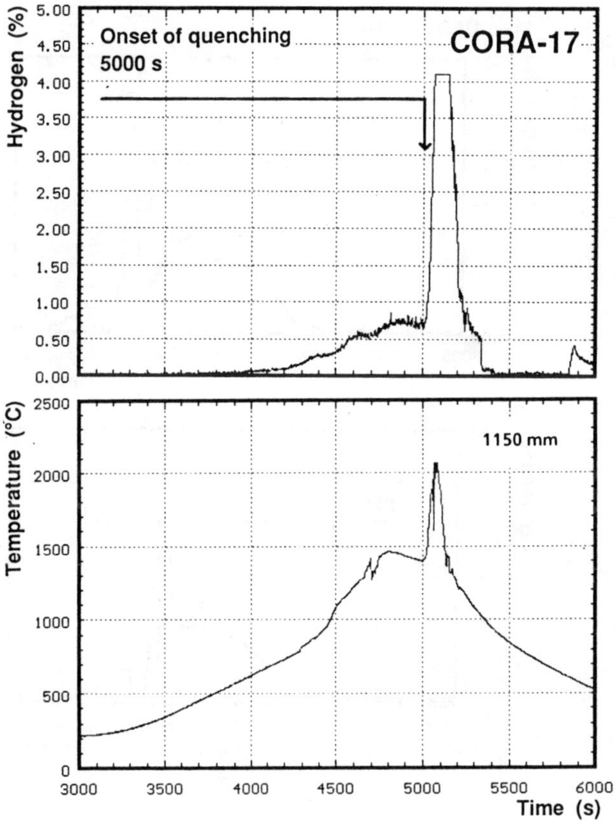

Figure 6. *Temperature of the channel box wall at 1150 mm and H_2 production measured in the off-gas system during BWR-type quench test CORA-17*

The large H_2 peaks in tests CORA-12 and CORA-17 are due to an extremely strong reaction between the saturated steam of the quench water and the zirconium that was not yet completely consumed during the first part of the transient. The major amount of hydrogen is generated during the flooding phase.

The total amount of H_2 measured in CORA-17 amounts to 128 g, 31g of the total were generated between 3000 and 5000 s and 97 g H_2 between 5000 and 6000 s, i.e. during the flooding phase. For comparison: In the LOFT experiment FP-2 the percentage of hydrogen generated during the reflood was 80 % (3).

A hydrogen generation caused by oxidation of the UO_2 as described for PBF experiment SFD-ST (4) was not found, i.e. no hyperstoichiometric fuel was detected in the CORA samples examined so far.

In the third quench experiment, CORA-13 which was selected to become an "international standard problem" (OECD/ISP-31)[1], the quenching of the rod bundle was initiated earlier, i.e. at higher temperatures. In this test the water level was already at the bottom of the heated zone (0 mm) when the electric power was turned off. This conduct led to a hydrogen peak, too. In contrast to the other CORA tests with later flooding, the hydrogen generation trace shows no decrease before the spike at the end of the normal transient, as is shown in comparison with test CORA-12 in Figure 7.

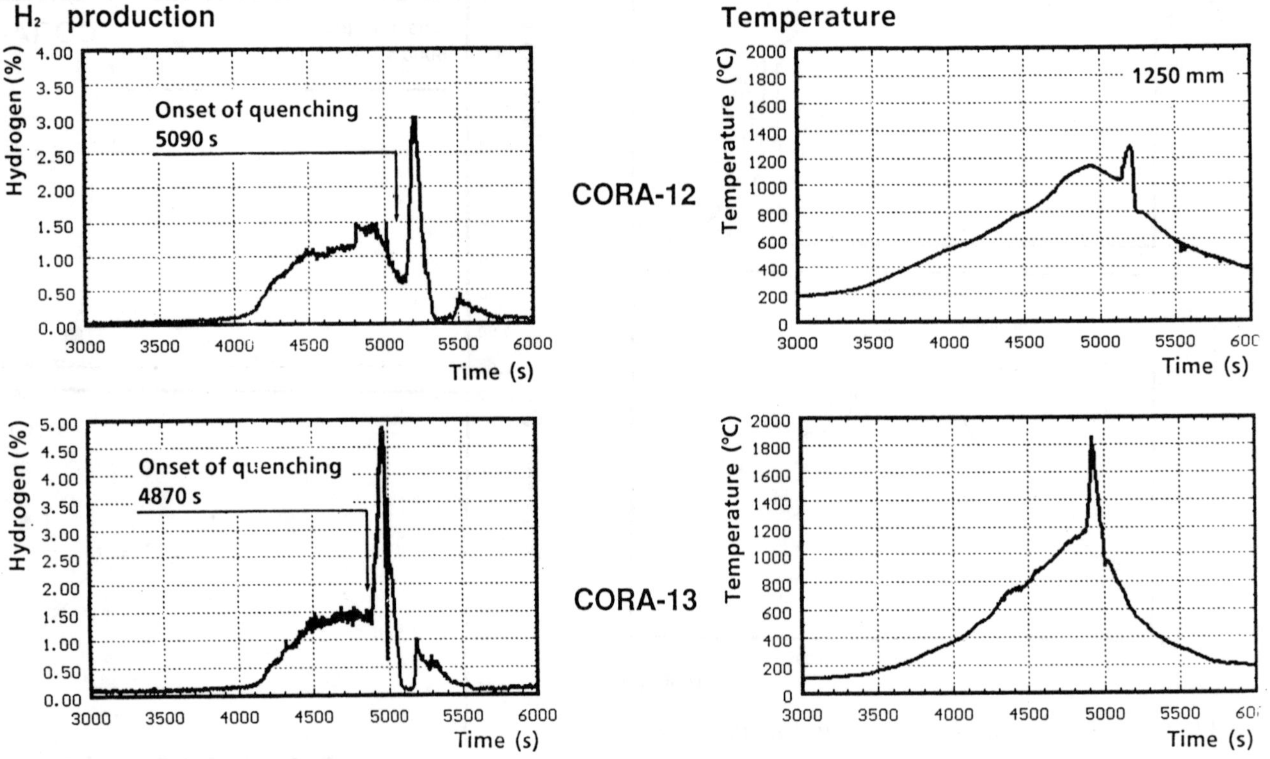

Figure 7. *Qualitative hydrogen data taken at the off-gas system (left) and rod temperatures of tests CORA-13 (quenching at high temperature) and CORA-12 (quenching after power shut-off)*

Similar to LOFT test FP-2 where the upper tie plate made of steel was damaged by temperatures exceeding the melting point as a consequence of the reflood process, the CORA-13 bundle also showed damage in the upmost region where heater rods were destroyed at the end of the experiment.

DISCUSSION OF THE RESULTS

The rate of the temperature rise has a strong impact on the oxidation of the bundle components and the extent of liquid phase formation (1).

---

1 For this reason only a limited information can presently be provided on the results of test CORA-13.

Besides the very strong influence of the heatup rate on the oxidation of the cladding, the steam availability is of interest, too.

There is a difference between the two types of CORA experiments with respect to the steam supply. The steam supply rate in the PWR-type test amounts to 6 g/s whereas 2 g/s are established for CORA experiments of the BWR category (except for the two larger bundles CORA-7 and CORA-18 where the steam rate was doubled).

The percentage of H_2 generated during the quench phase is larger for BWR-type test CORA-17 compared to CORA-12 and CORA-13 of the PWR-type although the zircaloy surface area is almost identical for both categories. The mass of zircaloy, however, differs for both types, i.e. PWR ca. 8 kg, BWR ca. 14 kg, so that more material is available for oxidation of the BWR-type bundle during the quench phase.

In reference (5) is pointed out that the majority of the oxidation occurs below the zircaloy melting point (1760°C) for steam-rich experiments. For steam-starved conditions a larger portion of zircaloy remains unoxidized below 1760°C so that this portion is available for later oxidation.
This could indicate that the experiment CORA-17 - as well as the LOFT experiment FP-2 - had a larger potential for a late oxidation when steam was supplied in excess by the quench water.

Table 3. *Summary of Zircaloy oxidation and hydrogen production obtained from different severe fuel damage tests*

Test	System pressure	Steam flow [a]	Bundle type	Cool-down mode	H_2 production	Total Zr oxidation [b]	Test time at T>1400°C	Fraction of H_2O consumed
	[bar]	[g/s]			[g]	[%]	[s]	[%]
CORA-15	2.2	6	PWR	slow	114	47	~ 1000	17
CORA-9	10	6	PWR	slow	159	66	~ 600	40
CORA-7	2.2	12	PWR	slow	114	23	~ 1200	7
CORA-18	2.2	4	BWR	slow	106	25 [c]	~ 1200	20
CORA-17	2.2	2	BWR	quenching	31 [a]	52 [c]	~ 1000	14 [a]
PBF-SFD-ST	70	16	PWR	fast-quench	172	75	~ 600	16 [d], [a]
PBF-SFD 1-1	70	0,6	PWR	slow	64	28	~ 600	100
PBF-SFD 1-3	70	0,6	PWR	slow	59	26	-	-
PBF-SFD 1-4	70	0,6	PWR	slow	86	38	~ 750	100
LOFT FP-2	11	13	PWR	fast-quench	1000	58	~ 300	46 [a]

[a] During transient, before reflood

[b] Percentage referred to bundle length of 1 m, oxidation of the outer shroud surface neglected (for all CORA tests)

[c] Mass of channel box walls not taken into account for the CORA-BWR experiments

[d] Source: NUREG/CR-5597

The CORA experiments, however, do not confirm any conclusion that steam-starved conditions are a prerequisite for a significant temperature escalation and an additional hydrogen production during a reflood phase.

Table 3 provides a summary of hydrogen production and zircaloy oxidation data evaluated from different experiments.

The fraction of the steam consumed in the CORA experiments was calculated according to (5):

$$H_2O\ Consumption = \frac{m_{H2} \cdot \frac{M_{H_2O}}{M_{H2}}}{\dot{m}_{H_2O} \cdot t}$$

with

- m_{H2} [g] total amount of hydrogen produced
- $\dot{m}_{H2O}$ [g/s] measured steam flow rate before reflood
- t [s] time at temperature (>1400°C)
- M_{H2O}/M_{H2} Ratio of molecular weights of steam/hydrogen

The calculational results, which are average data for the entire length of the bundle and do not reflect local and temporal situations where steam starvation could have existed, demonstrate that the consumption was not complete. This means that all experiments - even those with a steam supply rate of 2 g/s - were run under steam-rich conditions (see Table 3).
There are no results on steam consumption given in this table for the flooding phase, e.g. of CORA-17, because the effective flow through the bundle is unknown. (The flow rate during the quenching is estimated to be in the order of ten times the flow during the transient, i.e. before reflood).

The experimental findings on the oxidation behavior of zircaloy under flooding conditions were not expected. So, an explanation for this phenomenon is needed. The post-test pictures obtained from the bundle reveal the final state only and do not allow to study the conditions during the quench process. Therefore, only a hypothesis for the flooding phenomena can be given at the moment.

Based on the temperature and hydrogen readings it seems clear that an additional oxidation is triggered when fresh, relatively unoxidized zirconium is available for the approaching two-phase (water - steam) flow.

Well before the rising water level wets the hot region of the bundle the turbulent two-phase flow could lead to a heterogeneous temperature distribution, i.e. to inhomogeneous cooling of bundle components and to localized heat generation. The formation of local hot spots could intensify the oxidation of metallic structures. Three possible processes of a re-oxidation are conceivable:

- Metallic residues of cladding oxidize intensively by formation and growth of cracks which penetrate from the oxide scale into the metallic part of the zircaloy. The formation of wedge-shaped fresh oxide supports further splitting into finer cladding fragments.

- Re-solidified metallic melt oxidizes particularly fast when thermal shocks break open oxide scales allowing the oxygen of the coolant to react with the formerly protected, unoxidized material. Consequently, the oxidation could lead to re-melting of the relocated material.

- Molten metallic material is more resistant to thermal shocks. Further relocation and fragmentation of the melt could provide fresh metallic surfaces for a violent oxidation.

This listing indicates that various phenomena could contribute to the enormous heat release observed in the quench experiments.

CONCLUSIONS FROM THE CORA EXPERIMENTS WITH REFLOOD

The results of the quench experiments in the CORA facility demonstrate that core flooding can trigger an uncontrolled temperature escalation regardless if the first heatup of the bundle took place under steam-starved or steam-rich conditions.

The additional temperature excursion is combined with a strong additional hydrogen generation within a very short period of time and with additional relocation of some molten material, i.e. droplets, to lower elevations.

These results which are confirmed by the LOFT FP-2 in-pile experiment are not to lead to the conclusion that in a case of emergency, i.e. during a severe accident a dry reactor core should not be flooded. Indeed, it is unrealistic to conceive the idea that one could define the best period of time for reflooding a dried-out core. However, reactor operators should get to know about the hazard of additionally generated hydrogen during a core reflood.

ACKNOWLEDGMENTS

Various kinds of support of the experimental program and its evaluation are gratefully acknowledged.
The test rods were assembled by Mr. E. Mackert, the test bundles by Messrs. H. Giesmann and Ph. Pfann. The authors would like to thank Messrs. H. Benz, O. Heil, W. Rötzel, and H.J. Röhling for the test preparations and for the test conduct.
Mr. K.P. Wallenfels is acknowledged for arrangement of camera and video systems and for the preparation of the temperature measurement. Messrs. R. Huber, H. Malauschek, and Ms. I. Schub prepared and conducted the on-line measurements of the off-gas samples.
Mr. W. Rötzel's effort in the post-test photography is greatly appreciated. The authors would like to thank Mr. L. Anselment for the sectioning of the bundle and for the preparation of the metallographic samples, Mr. H. Metzger for the investigation of the metallographic samples by optical microscope, and Mr. J. Burbach for the SEM investigations.
The discussions with Mr. W. Hering on the topics of this paper were very helpful.

LITERATURE CITED

1. S. Hagen, L. Sepold, P. Hofmann, G. Schanz, "CORA Experiments on Severe Fuel Damage with and without Absorber Materials", Heat Transfer and Fuel Behavior in Nuclear Reactor Accidents, 26th ASME/AIChE/ANS National Heat Transfer Conference, Philadelphia, 6-9 August 1989, AIChe Symposium Series 289, Vol. 85, pp. 135-140

2. W. Hering, K.Müller, A. Schatz, "Verification of the KESS-III Code System Using CORA Experiments", this conference

3. R.R. Hobbins and G.D. McPherson, "A Summary of Results from the LOFT LP-FP-2 Test and Their Relationship to Other Studies at the Power Burst Facility and of the Three Mile Island Unit 2 Accident", OECD/LOFT Final Event, ISBN 92-64-03339-4, 1991

4. A.W. Cronenberg, "Severe Accident Zircaloy Oxidation/Hydrogen Generation Behavior Noted from In-pile Test Data", Nuclear Technology, Vol 93, 1991

5. A.W. Cronenberg, "In-Vessel Zircaloy Oxidation/Hydrogen Generation Behavior During Severe Accidents", NUREG/CR-5597, 1990

ANALYSIS OF THE SEVERE FUEL DAMAGE TEST PHEBUS B9+ USING ICARE CODE

S. Bourdon ■ Atomic Energy Commission (CEA), IPSN/DRS, Cadarache, France
P. Villalibre ■ Consejo de Seguridad Nuclear, Madrid, Spain
B. Adroguer ■ Atomic Energy Commission (CEA) IPSN/DRS, Cadarache, France
G. Geoffroy ■ Atomic Energy Commission (CEA), DRN/DMT, Saclay, France

The PHEBUS Severe Fuel Damage B9+ has been chosen as OECD/CSNI International Standard Problem 28. This paper describes an analysis of this test performed at Cadarache Nuclear Research Center. The analysis is carried out using a mechanistic fuel behaviour code called ICARE. Calculations focused on predicting the main features of the test : cladding oxidation, mechanical behaviour of the cladding, dissolution of UO_2 by molten Zr and melt relocation. Several fuel dissolution models have been used and UO_2 dissolution is presented along with comparisons to measurements. Agreement is generally good, but some empiricisms were incorporated into the code in order to achieve good agreement. Sensitivity calculations on fuel dissolution and clad failure criteria are discussed and code developments are proposed.

A large programme is presently being conducted in France by the Commissariat à l'Energie Atomique (C.E.A.) to study the behaviour of a PWR fuel bundle during postulated beyond design basis accidents.

The objective of the PHEBUS Severe Fuel Damage (SFD) programme, was to provide data on the state of a severely damaged core. Many phenomena were examined : cladding oxidation and corresponding hydrogen generation ; fuel dissolution by solid and liquid Zr ; zirconia and inconel spacer-grid interaction with cladding Zr ; interaction of absorber materials (Ag - In - Cd alloy) with fuel rods ; relocation of liquefied materials and flow blockage formation. The fission product (FP) release and transport will be studied in the forthcoming PHEBUS FP programme planned for mid 1992. The B9+ test conducted in January 1989 was aimed at studying material interactions and relocation in absence of control rod materials with a nuclear heated bundle having an axially variable cladding oxidation. These phenomena being crucial in assessing the codes, this test has been chosen by the OECD/CSNI for the semi blind International Standard Problem ISP 28, in which only the temperature data have been supplied to participants.

The analytical programme includes the development and the verification of the advanced mechanistic ICARE code to describe with detailed models the main phenomena and processes involved during the progression of a core damage from the initial fuel heat-up to the final relocation and freezing of core materials. The current version V1 of ICARE has been used to calculate the ISP 28 in the framework of a joint French - Spanish PHEBUS Consortium participation. The calculations are compared with the currently available experimental results and the main features of the analysis are discussed.

THE PHEBUS SFD PROGRAMME

Objectives of the programme and of the B9+ Test

The PHEBUS facility enabled the reproduction of an environment as representative as possible of what would exist in a core under conditions beyond design basis accidents. In order to avoid overly complicated integral experiments involving all the aspects of the core degradation, each of the 6 tests of the SFD matrix has been focussed mainly on a few main phenomena occuring during the core degradation [1] [2].

In particular, the objectives of the B9+ test were to study the cladding oxidation, the mechanical behaviour of the cladding with a variable ZrO_2 thickness layer containing molten zircaloy, the dissolution of UO_2 and ZrO_2 by molten zircaloy and the relocation of the melt.

These in pile experiments on the early phase of in vessel core melt degradation up to the ceramic material relocation have significantly increased the international previous data base mainly coming from the PBF SFD tests, the NRU - FLHT tests, the ACRR - DF tests and from the out-of-pile CORA experiments. The PHEBUS-SFD data are intensively used for the development and verification of the mechanistic ICARE code [3].

The facility description

The facility mainly consists of a driver core supplying the test fuel with nuclear power, a SFD loop which produces the thermal hydraulic conditions and a pressurized water loop working as an independent cooling circuit between the SFD loop and the driver core [4]. The SFD loop can carry the fluid flow to the test train through three injection lines (steam, hydrogen, helium). The test train cross-section shows the bundle which consists of 21 unirradiated UO2 fuel rods in a 12.6 mm pitch matrix (Fig. 1).

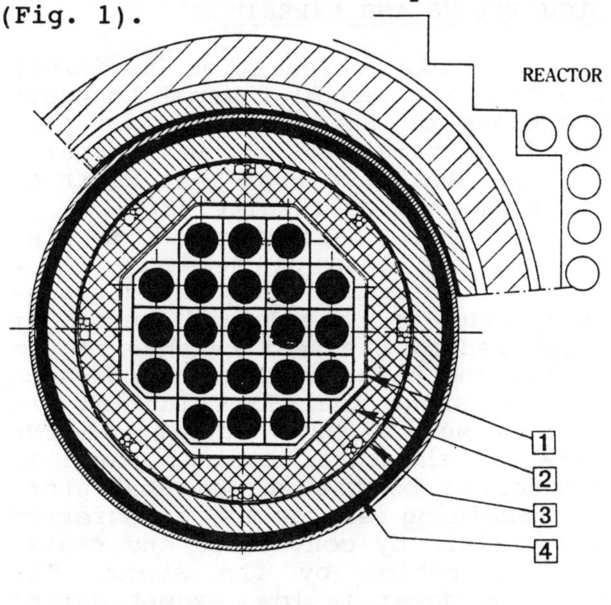

FIG.1: Bundle cross-section
(1) Zr liner, (2) porous ZrO_2, (3) dense ZrO_2, (4) Stainless S. tube

The total fissile height is 0.8 m. Two spacer-grids in inconel 560 mm apart are located on either side of the mid plane.

The insulating shroud of the bundle is a multi-layer structure. The octogonal inner zircaloy liner 0.6 mm thick is surrounded by a thick porous ZrO_2 layer (94 mm ext. diameter and 69.5 mm between faces of the inner octogonal surface), a dense ZrO_2 layer 1 mm thick and an external stainless steel tube 8 mm thick. Different kinds of thermocouples enable the temperature measurements of fuel center, cladding, fluid and shroud at different radial locations and elevations. The hydrogen concentration in the gas mixture at the loop outlet is measured continuously by means of a mass spectrometer. Detailed information on the facility are given in reference [6].

DATA SPECIFICATIONS OF THE B9+ TEST

The experimental scenario has been divided into three parts (Fig. 2). The first oxidizing phase performed with pure steam flow enabled a deep axial profile of ZrO_2 to be obtained in the cladding during a temperature increase by steps up to 1800 K. The second heat-up phase in pure

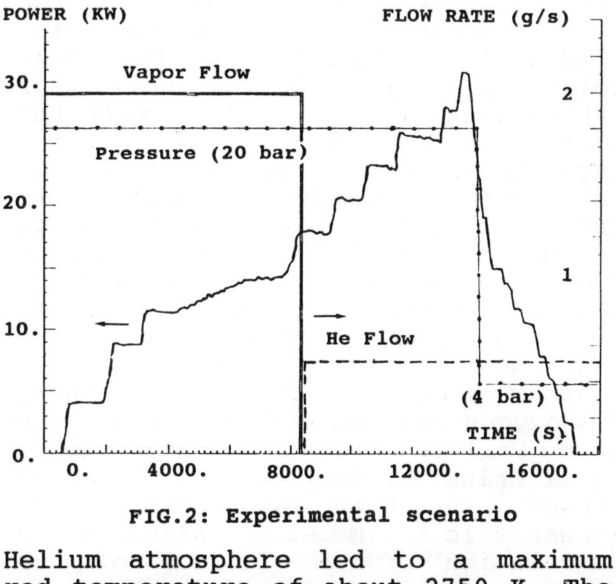

FIG.2: Experimental scenario

Helium atmosphere led to a maximum rod temperature of about 2750 K. The final phase performed with step reductions in nuclear power enabled a slow cooling down in order to keep unchanged the previous bundle geometry. The nuclear power evolution in the bundle versus time and elevation was well defined from measurements and neutronic calculations [6].

The ballooning of the pressurized rods with He was avoided by the low temperature melting of the fusible seal on the rod top plug. The radial external boundary conditions of the insulator shroud was maintained constant at 528 K. The porous ZrO_2 of the shroud was fed with Helium during the oxidation phase in order to limit the variation of its thermal conductivity by H_2 entering from the liner oxidation.

THE ICARE CODE AND THE B9+ MODELLING

Processes considered in the ICARE calculations

The code allows detailed mechanistic descriptions of all the major degradation phenomena occuring during a severe PWR accident. It has been used to prepare and analyse the PHEBUS SFD tests as well as to calculate some other experiments as CORA or PBF and the TMI-2 accident [5]. Thermal-hydraulics in the core with convection and 2D radiative heat transfer, cladding thermal deformations and oxidation, fuel-cladding liquefaction and melting with relocation, control rod degradation and flow blockage formation are all included [5]. Concerning the fuel dissolution, the UO_2 - Solid Zr interaction can be considered if the contact is good and if the temperature is greater than 1473 K. This interaction was not considered here due to the melting of the upper plugs of the rods at 1120 K. Three dissolution models by molten Zr can be chosen and have been used for sensitivity studies ; the Hofmann diffusion model, the convective Kim-Olander model or the pseudo-binary UO_2-oxygen saturated Zr phase diagram from Politis with an instantaneous equilibrium U-Zr-O mixture (model chosen in the base case). No mechanistic model allows a calculation of the failure of the protective ZrO_2 of the cladding and an user-specified criteria is assumed for its dislocation based on a failure temperature and on a lower thickness limit (2300 K and 300 µm in the base case). A new ZrO_2 dissolution model by molten Zr exists based on a Hofmann correlation but has not been used due to some inconsistencies on the calculation of the UO_2 solubility limit from the U-Zr-O phase diagram. No spacer-grid model exists to take into account the Zr-Inconel eutectic formation and the hold-up effect on molten material.

Modelling of the B9+ test

The fluid flow in the bundle has been divided in 3 independent subchannels, each one containing an average rod representing the central rod, the 8 first ring rods and the 12 second ring rods.

The external channel is bounded by the surrounding 5 layers of the shroud with the external surface imposed at the constant temperature of the cooling circuit. Each component and channel is divided axially into 10 equally spaced meshes on the active fuel region (0.8 m). Fuel and material properties are the standard MATPRO ones except the thermal conductivity of the porous ZrO_2 controlling the radial heat losses which has been calculated by a specific model taking into account the temperature, the thermal properties of ZrO_2 and of the filling gas and the porosity of the medium.

COMPARISONS BETWEEN ICARE CALCULATIONS AND EXPERIMENT

The ISP28 calculation performed to define the radial thermal boundary conditions on the liner [7] is presented in this section and compared with the available Post Test destructive Examinations results (P.T.E.).

The main parameter which influences the loss of geometry integrity of the rods during the degradation phase is the thermal behaviour controlled by the heat balance in the bundle. The calculated energy balance (Fig. 5) shows that 75 % (during the steam phase) to 90 % (during the He phase) of the total nuclear power is transferred to the shroud insulator. The remaining energy is transferred to the fluid by convection and radiation absorption by the steam. The oxidation power is low, except during the final escalation (20 %) observed at the hot point (0.6 from the bottom of the fissile length) just before the end of the oxidation phase (8378 s).

Preliminary calculations using the thermal conductivity model of the porous ZrO2 of the shroud led to an overprediction (in the lower part of the bundle) and to an underprediction (in the upper part) of the rod temperatures. These temperatures are very sensitive to the thermal conductivity of the porous ZrO_2 which dominates that of the shroud. This parameter with important uncertainties (porosity, filling gas, 2D geometry) was tuned in sensitivity calculations to better match experimental and calculated temperatures. In these conditions the temperature histories were well predicted (Fig. 3, 4) so the main purpose of the test (the study of durability of the oxidized layer, the dissolution and the melt progression) could be correctly studied. Radial temperature differences are also in agreement with the measured ones during the oxidation phase. During the He phase, thermocouple readings became unreliable due to a too high temperature being reached or due to eutectic formation between rod molten material and thermocouple cladding. Calculated cladding temperature differences are respectively less than 10, 60 and 150 K between central, 1st ring, 2nd ring rods and shroud liner throughout the transient.

The temperature excursion calculated at 8000 s, results initially from a significant power increase (Fig. 2). When the temperature reached 1853 K (ZrO_2 tetragonal-to-cubic phase transition) the enhancement of the oxidation parabolic kinetics of Urbanic-Heidrick (base case) leads to calculate an escalation at the hot level (Fig. 3, 4) as measured in the test. The oxidation power was stopped soon after this increased oxidation by the steam/He flow transition at 8378 s.

The Zr cladding (Fig. 6) is more than half consumed in the upper hot part of the bundle just before the escalation at 8200 s and the ZrO_2 thickness increases quickly up to the steam/He transition leading to a maximum oxidation at 0.6 m of 76 % to 70 % (WT) for the three kinds of rod and 45 % for the liner. Visual observations of the radial cuts of the rods at the upper part of the bundle show near complete cladding oxidation.

The final rapid oxidation behaviour observed at the upper part of the bundle is difficult to calculate for it is very sensitive to the temperature calculated just before the escalation or to the oxidation kinetics correlation chosen. A sensitivity calculation performed with the Prater rate constants led to unrealistic temperature and oxidation escalations before the ones observed in the test and to a total oxidation in the upper part of the bundle.

No oxidation limitation occured due to steam starvation. The good agreement between the calculated and measured total H_2 production ($39\ 10^{-3}$ Kg) confirms the global consistency of the base case calculation concerning the rod thermal and oxidation histories (Fig. 7).

As the temperature rises, the non oxidized Zr melts at the hot zone just after 10500 s and dissolves the UO_2 fuel from 0.3 to 0.7 m. The corresponding dissolution is a strong function of the temperature but also of the oxidation profile (25 % to 75 %) which determines the mass of molten Zr available and its oxygen content. The UO_2 dissolution at one level is limited by its solubility limit in the local U-Zr-O mixture and by the failure of the ZrO_2 scale which governs the flow-down of the mixture and the end of the dissolution.

Figure 8 illustrates the different degradation processes calculated at mid bundle elevation for a first ring rod : the ZrO_2 and Zr-α(O) layers growing, the corresponding Zr-β consumption, the Zr-α(O) melting at 2250 K, the instantaneous UO_2 dissolution corresponding to the pseudo binary phase diagram of Politis and the final cladding failure due to an oxide layer too thin (< 300 μm) when the temperature exceeds the failure threshold value of 2300 K.

The mean dissolution mass fraction calculated with the Politis model is compared with measures from visual examinations of cross-cutting sections of the bundle (Fig. 9). The mean mid-bundle dissolution is underpredicted by the code by nearly a factor of 2. The overestimation in the upper part would be due to an under-prediction of the oxidation which seems to be near complete in the test.

The failure criteria was fulfilled for all the rods between 12000 s and 13700 s at the middle of the bundle. Cross cutting sections of the bundle also show the cladding disapearance for most of the rods in the mid-bundle region from 0.3 to 0.45 m. Figure 6 shows the relocation time of the 1st ring rods at 0.4 m. The oxide shell is too thick to fail at the upper hot levels of the bundle.

The molten U-Zr-O mixture relocates downward along the individual rods in a film process and freezes at the lower nodes 1, 2 and 3 (Fig. 10, 11). This process removes the supply of nuclear power corresponding to the UO_2 dissolution and limits the temperature rise.

The calculated flow blockage profile (the flow blockage is defined to be equal to 37 % at time zero due to undeformed rods) shows a maximum of 50 % at 0.20 m in agreement with the experiment (Fig. 10). The underprediction in the mid-bundle region where cladding disapearance was calculated is mainly due to the non prediction of the local accumulation of mixtures. (As observed in this region on cross-cutting sections). The refrozen material at 0.20 m is calculated as a mixture of UO_2, Zr, ZrO_2 with a respective weight composition of 15 %, 57 % and 28 %.

The final degradation state of the bundle is shown Fig. 11.

DISCUSSION

At the upper part of the bundle between 0.6 and 0.8 m the code seems to underpredict the oxidation (maximum calculated 76 %). At the present time only visual observations of the oxidation are available in this zone. They show a near total oxidation but also local significant UO_2 dissolution on some rods. It is thought that only a partial cladding oxidation in this zone or an internal relocation (between the UO_2 pellets and the slightly deformed cladding) of molten Zr coming from the upper plug zone could explain this local dissolution. So a total oxidation may not be reached at the upper part of the rods. Anyway the tendancy to underestimate the oxidation is consistent with the first PHEBUS SFD B9 test showing that the Urbanic correlation, which gave the better general prediction of the oxidation, tends to underpredict by about 25 % the measured ZrO_2 thicknesses.

Concerning the UO_2 dissolution sensitivity studies have been done with the Hofmann and the Kim-Olander models. These models gave similar results which overpredict the dissolution in the mid-plane by a factor of 3. These results are not consistent with the pseudo binary diagram of Politis who is known to underestimate the UO_2 solubility.

The discrepancy between the three dissolution models might be due to inconsistent data concerning the UO_2 limit of solubility in the mixture. The underprediction of the dissolution by the Politis model can be explained by the value of the calculated maximum fuel temperature which is below the threshold value of 2670 K bounding the low and high UO_2 solubility domains of the Politis diagram. The Kim and Olander model for the same temperature behaviour gives with its own saturation concentration a greater solubility limit. The Hofmann model in the code is not linked with a solubility limit and the dissolution is only limited by the cladding failure criteria or by the freezing of the mixture during the final cooling phase.

Additional calculations have shown the sensitive effect of the cladding failure criteria on the fuel dissolution and relocation. Changing this user input criteria (2300 K and 500 μm) can change significantly the extent of the cladding dislocation area spread on 30 cm around the mid bundle instead of 10 cm with a 300 μm failure thickness. Visual observations of the cross-cutting sections of the bundle show that this second criteria defines a failure area inside the bundle (Fig. 12) in better agreement with the test except on the upper part (node 10) where the slight underprediction of the calculated oxidation (460 μm) leads to a cladding failure not observed in the test. Figures 11 and 12 illustrate the threshold effect of criteria on the degradation.

Besides this, the dissolution of the cladding ZrO_2 shell not considered in these calculations should affect also the UO_2 dissolution and the relocation of U-Zr-O mixtures. The ZrO_2 dissolution model now in development should reduce the fuel liquefaction when both the UO_2 and ZrO_2 dissolution are simultaneously considered. Taking into account these effects, the dissolution models of Hofmann and Kim-Olander would have given better results than those obtained with the present version of the code.

The inconel spacer-grids (nodes 2 and 9) are not modelled in ICARE so the very localized plug observed in the lower part of the bundle at 0.05 m which is mainly due to the spacer-grids relocation cannot be predicted. For the upper grid it is thought that the protective effect of the oxide thick layer had prevented the formation of the low temperature Inconel-Zr eutectic and had avoid the UO_2 dissolution. Limited effects were observed on the cladding with the partially molten lower grid.

CONCLUSION

Calculations of the PHEBUS B9+ test have been done with the ICARE code in the framework of the ISP28 exercise. The complexity of the shroud and of its behaviour at high temperature required some sensitivity calculations on the conductivity of the porous ZrO_2 in order to obtain correct initial thermal conditions for the degradation. In these conditions, ICARE performed a good prediction of the temperature histories of all the components of the test train.

Comparisons with the first available P.T.E. results showed a correct prediction of the oxidation and of the relocalization of molten materials but weaknesses concerning the fuel dissolution by molten zircaloy.

Sensitivity studies showed the effect on the fuel degradation of some threshold phenomena as the escalation of the oxidation near 1853 K or the failure criteria of an oxidized cladding which may change significantly the amount of non oxidized Zr, the duration of the UO_2 dissolution, the extend of the cladding dislocation and the relocation of the molten material. The unmodelled inconel spacer-grids did not affect significantly the rod degradation.

The first P.T.E. data of B9+ appear very valuable concerning the fuel dissolution and the failure criteria for oxidized rods : failure parameters of 2300 K and 500 µm enable to better fit the observed global cladding dislocation.

ICARE code developments are underway with respect of the fuel solubility limits in the U-Zr-O mixture by using a ternary phase diagram deduced from chemical thermodynamic properties and models. The optimization of the simultaneous UO_2-ZrO_2 dissolution model is in progress.

More complete conclusions will be drawn concerning the ICARE ability to predict the degradtion when a revised dissolution model including ZrO2 dissolution is added and when more complete and qualitative P.T.E. data are available.

REFERENCES

[1] B. ADROGUER - C. GONNIER - J.A. MARTINEZ. "The PHEBUS severe fuel damage programme, the facility, the experimental matrix and the analysis of the results." Spanish Nuclear Society magazine - January 1991.

[2] C. GONNIER - G. GEOFFROY - B. ADROGUER. "Phebus Severe Fuel Damage Programme. Main results". A.N.S. Meeting - Portland 21,25 July 1991.

[3] R. GONZALEZ - P. CHATELARD - A. PORRACCHIA. "Effects of stainless steel cladding on the core degradation of a PWR-type reactor computed by ICARE2 code." IAEA workshop - Moscow 14-17 May 1990.

[4] M. BERNE - C. GRANDJEAN - E. SCOTT DE MARTINVILLE - C. GONNIER. "In pile investigations at the PHEBUS facility on the behaviour of PWR-type fuel bundles in severe accidents conditions beyond the design criteria." SFEN - NUCSAFE 88 - Avignon 2,7 October 1988.

[5] R. CHATELARD - R. GONZALEZ - A. PORRACCHIA. "Analysis of the TMI-2 core degradation using ICARE code." - 27th Nat'l Meeting, AIChE Minneapolis (1991).

[6] C. GRANDJEAN - C. GONNIER. "Preliminary Specifications for the International Standard Problem ISP 28". C.E.A. Report. PHEBUS CSD 112/90 - March 1990.

[7] B. ADROGUER. "ISP 28 - Final Specifications for calculations and reported results." CEA Report. PHEBUS CSD 115/90 - Sept. 1990.

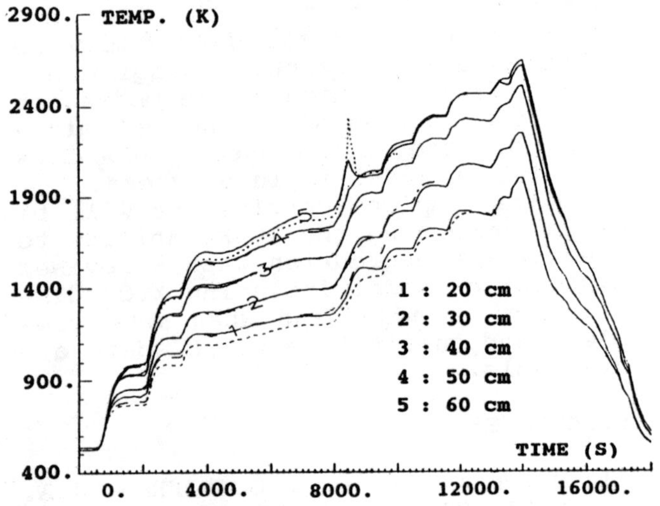

FIG.3: Calculated and measured (--) rod temperatures versus time

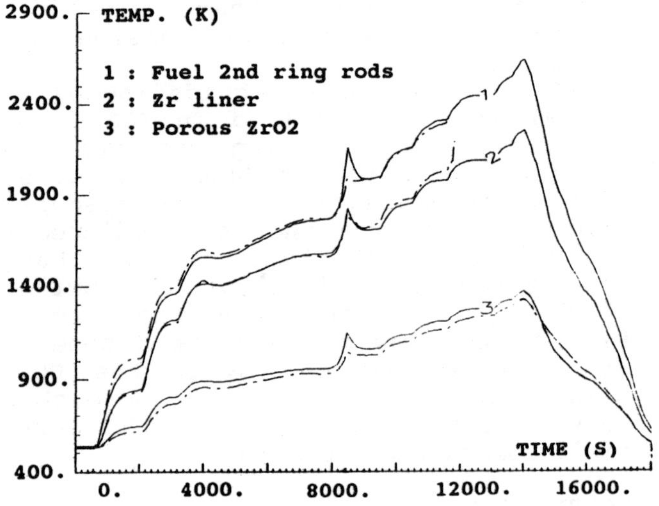

FIG.4: Calculated and measured (...) rod and shroud temperatures at level 70 cm

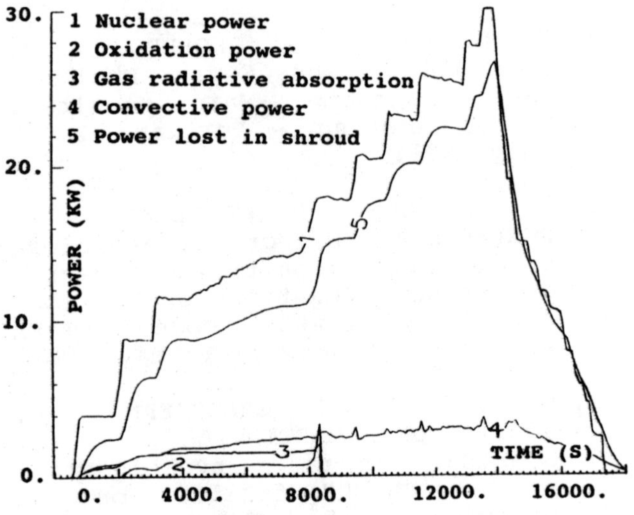

FIG.5: Energy balance in the bundle

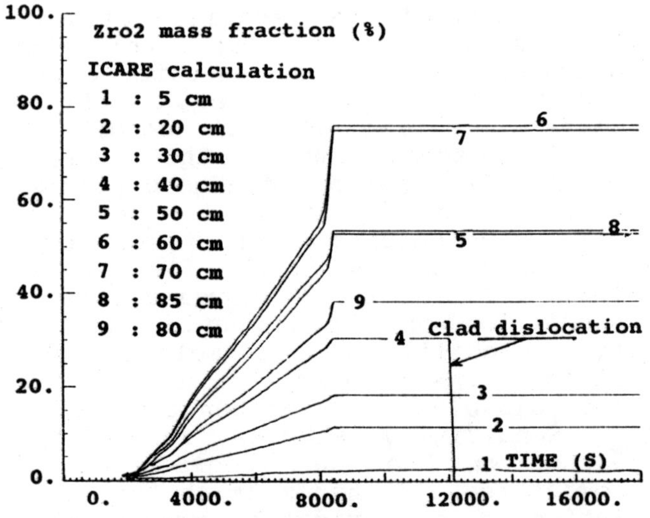

FIG.6: Cladding oxidation versus time at different levels

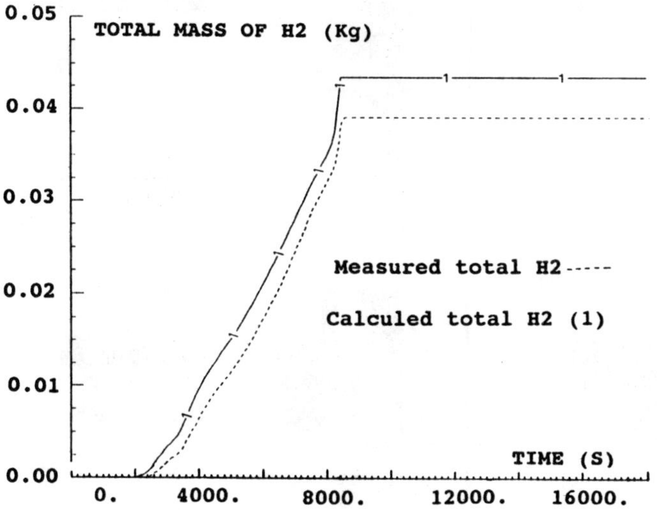

FIG.7: Calculated and measured total H_2 production

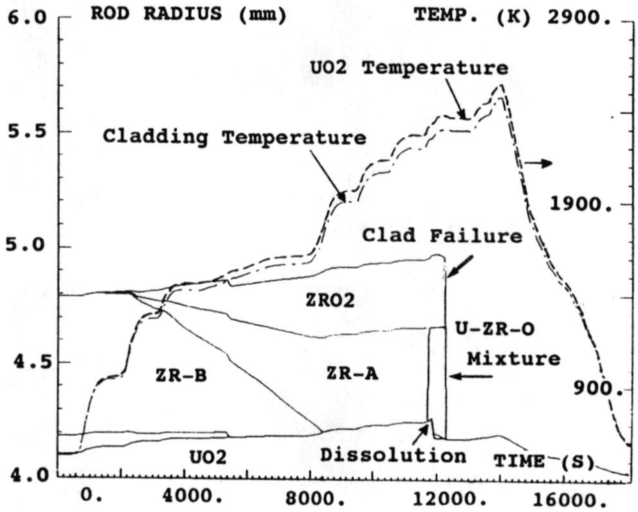

FIG.8: Calculated Temp. and radius of a 1st ring rod versus time

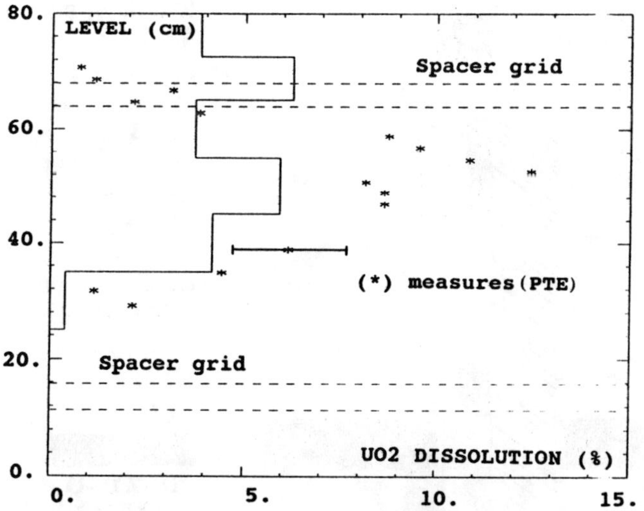

FIG.9: Calculated and measured mean UO_2 dissolution profile at 18000 s

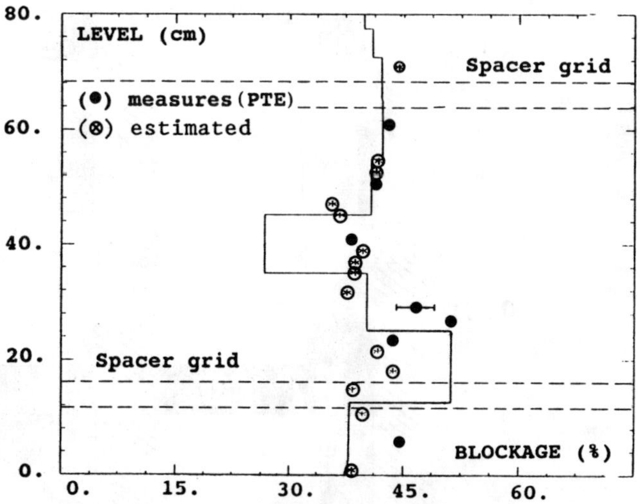

FIG.10: Total flow blockage at 18000 s

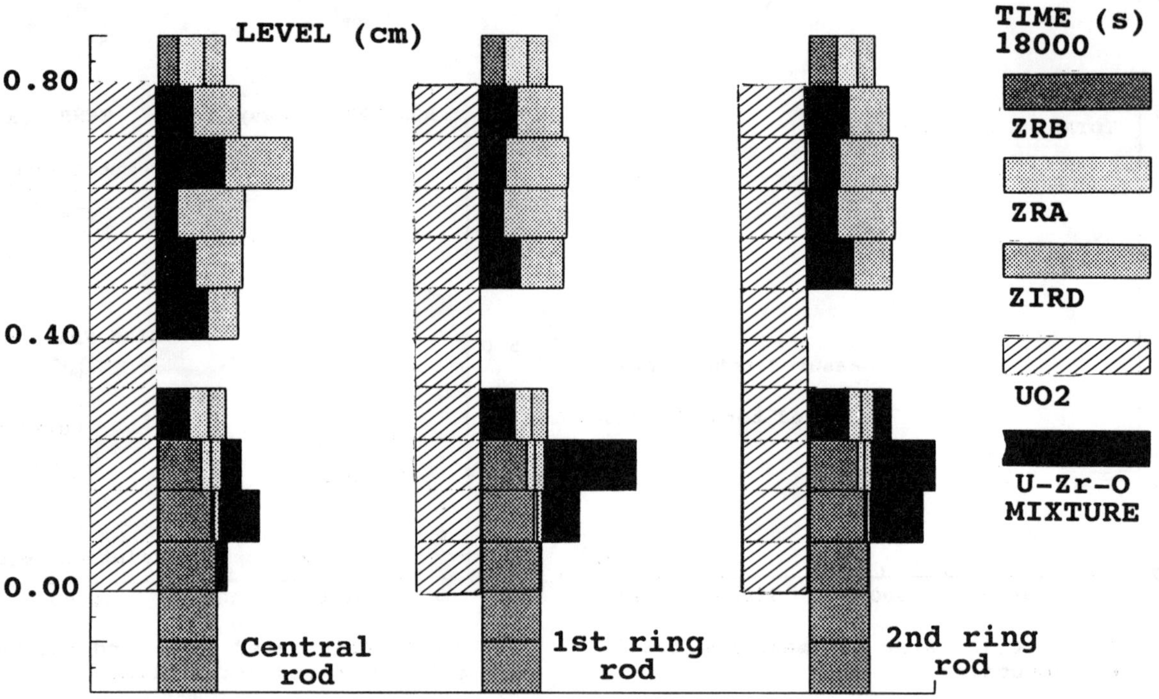

FIG.11: Final picture of the bundle (18000 s) ; Politis model ; Failure criteria (300 μm, T = 2300 K)

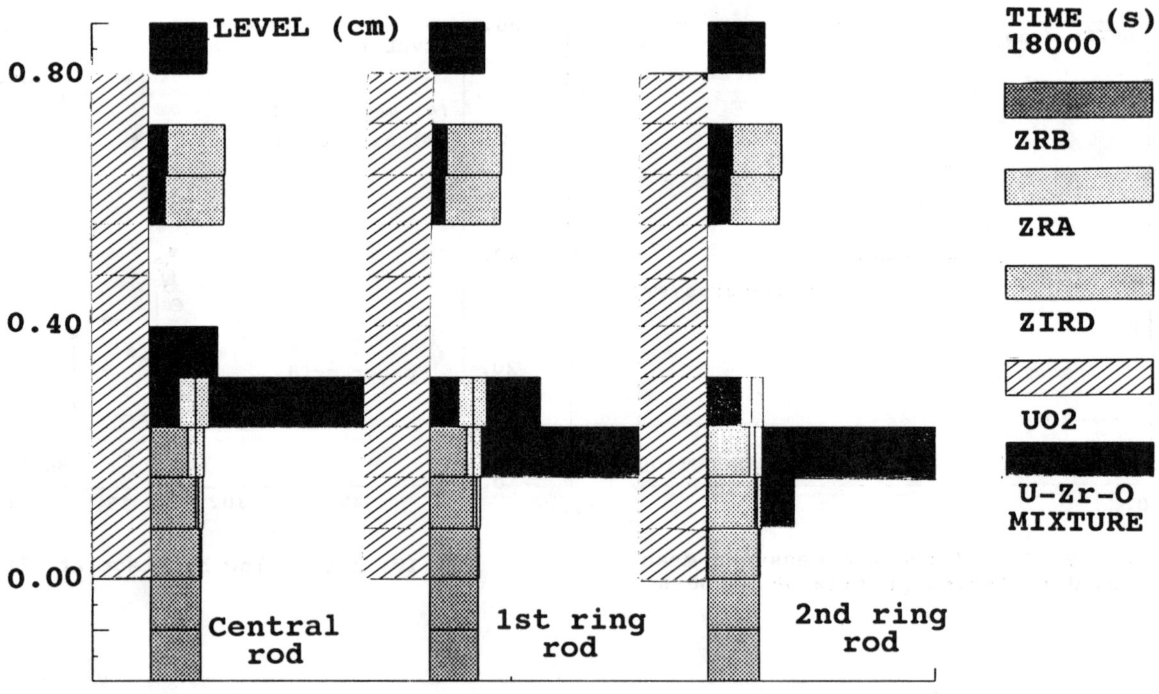

FIG.12: Final picture of the bundle (18000 s) ; Politis model : Failure criteria (500 μm, T = 2300 K)

AN ANALYSIS OF MOLTEN-CORIUM-INDUCED FAILURE OF DRAIN PIPES IN BWR MARK II CONTAINMENTS*

Rusi P. Taleyarkhan ■ Engineering Technology Division, Oak Ridge National Laboratory, Oak Ridge, TN 37831-8057
Michael Z. Podowski ■ Dept. of Nuclear Engineering & Engineering Physics, Rensselaer Polytechnic Institute, Troy, NY 12180-3590

This study has focused on mechanistic simulation and analysis of potential failure modes for inpedestal drywell drain pipes in the Limerick boiling water reactor (BWR) Mark II containment. Physical phenomena related to surface tension breakdown, heatup, melting, ablation, crust formation and failure, and core material relocation into drain pipes with simultaneous melting of pipe walls were modeled and analyzed. The results of analysis have been used to assess the possibility of drain pipe failure and the resultant loss of pressure-suppression capability. Estimates have been made for the timing and amount of molten corium released to the wetwell. The study has revealed that significantly different melt progression sequences can result depending upon the failure characteristics of the frozen metallic crust which forms over the drain cover during the initial stages of debris pour. Another important result is that it can take several days for the molten fuel to ablate the frozen metallic debris layer—if the frozen layer has cooled below 1100 K before fuel attack.

One of the most important aspects of hypothetical core meltdown accidents in BWRs is concerned with the consequences of reactor pressure vessel (RPV) failure and associated corium release into the primary containment. One potentially important mechanism for pressure suppression pool (PSP) bypass is associated with the failure of drain pipes which penetrate the concrete pedestal floor downward into the wetwell volume. Analyses of this issue were conducted by the Nuclear Regulatory Commission's BWR MARK II and MARK III Parametrics Program at the Oak Ridge National Laboratory for the LIMERICK[1] and WNP-2[2] Mark II containments[3]. Due to space limitations, detailed results are presented only for the LIMERICK Mark II containment.

PROBLEM FORMULATION

It is obvious that drain failure characteristics will depend on the specifics of a debris pour history, which may be predicted differently depending upon the code used. Results described later in Section 4 are thus presented parametrically so as to be applicable for various debris pours. However, a focussed analysis is conducted thereafter using a debris pour history predicted by the BWRSAR[4] code for a short-term station blackout accident sequence event, as documented in Ref. 3, and summarized in Table 1. As seen from Table 1, the first debris pour up to 6000 seconds after vessel failure consists largely of molten steel and Zircaloy. Thereafter, a mixture of molten UO_2 and Zircaloy is ejected. Debris bed height increase is evaluated under the implicit assumption that the debris ejected from the reactor vessel instantaneously spreads uniformly over the floor. Possible drain failure scenarios for the Limerick containment were postulated as occuring in 3 stages.

Stage 1 is identified as the situation that occurs at the start of the metallic debris-melt attack on the stainless steel drain cover plate. Therein, two cases are identified. For

*Research sponsored jointly by the U.S. Nuclear Regulatory Commission, Office of Nuclear Regulatory Research under Interagency Agreement DOE 1886-8022-8B and the U.S. Department of Energy under Contract No. DE-AC05-840R21400 with Martin Marietta Energy Systems, Inc.

Table 1. Debris Ejection History For BWR Mark II Drain Failure Analysis

Cavity Radius(m)=3.03

	Zircaloy		Other Metals		Uranium-Dioxide		Debris Bed		Elapsed
Time(min)	Mass (lb)	Height (m)	Mass (lb)	Height (m)	Mass (lb)	Height(m)	Total Ht(m)	Rise Rate (m/s)	Time (s)
0.00							0		
265	0	0	0	0	0	0	0	0.00E+00	
275	3499	0.008	7983	0.016	0	0	0.024	4.06E-05	6.00E+02
285	12610	0.031	28773	0.057	0	0	0.088	1.06E-04	1.20E+03
305	40108	0.097	40820	0.081	0	0	0.178	7.55E-05	2.40E+03
345	40108	0.097	40820	0.081	0	0	0.178	0.00E+00	4.80E+03
365	40108	0.097	40820	0.081	0	0	0.178	0.00E+00	6.00E+03
385	45792	0.111	49444	0.098	14782	0.021	0.231	4.35E-05	7.20E+03
405	84273	0.204	107817	0.214	82745	0.119	0.537	2.56E-04	8.40E+03
885	113869	0.276	352454	0.700	380184	0.547	1.523	3.42E-05	3.72E+04

Case A either the debris internal energy is not high enough to melt the drain cover plate, or, if significant water is present, debris quenching and crust formation occurs. In this case the drain plate remains intact. In Case B the drain plate melts, and relocates downwards into the drain pipe region, and freezes along the cold walls. The aspect of molten metallic debris seeping through the drain plate cover slots after overcoming surface tension forces is also addressed.

Stage 2 is identified as the phase in which the existing Zircaloy-steel layer that may have partially crusted (over the drain cover) is subjected to a stream of molten UO_2 or a eutectic composition of UO_2 and Zircaloy in the second debris pour.

Finally, stage 3 accounts for the probable situation where the molten fuel melts the Zircaloy-steel layer (formed during stage 2) and thereafter attacks the drain pipe region. This is also the situation when significant core-concrete-interaction (CCI) occurs, accompanied by non-condensible gas generation which tends to stir the molten mixture.

The above mentioned scenarios cover most of the important aspects of the core debris attack of the drain pipe region under consideration.

MODELING OF CORIUM-DRAIN INTERACTION

Various mathematical models and related computer programs that have been set up to analyze BWR Mark II debris-drain interactions are described in this section.

Exact Analytical Solution for Simultaneous Melting & Freezing of Semi-Infinite Slabs in Intimate Contact - Neumann Problem

Consider a solid plate in direct contact with hot liquefied debris material. If the melting temperature of the plate is lower than that of the debris, both plate melting and debris solidification (i.e., crust formation) may occur simultaneously. Based upon experiments conducted with simulant materials, Podowski and Lahey[5] observed that the liquefied plate material (of lower density than the debris) would percolate upwards through the cracks in the crust to the debris pool surface, so that the debris pool remains in contact with the solid plate. Assuming that the plate and molten debris are thick (i.e., their thickness is greater than their respective thermal boundary layers), the propagation of plate melting and the buildup of debris crust was approximated by a one-dimensional "semi-infinite" conduction model.

Details regarding the conduction equations, and accompanying boundary conditions can be found in Ref. 5.

The model described above has been coded as the STEF program. This code proved particularly useful for the modeling of stage 2 scenarios as described earlier, where molten fuel (i.e., UO_2 or UO_2-Zr eutectic) attacks a layer of less dense material (i.e., Zircaloy-steel mixture) that was ejected earlier and may have frozen. Interestingly, codes such as CORCON[6] simply assume that the molten fuel debris gets instantaneously relocated to the concrete surface without accounting for the time delay involved. Such was the case in the analysis presented in Ref. 3. On the contrary, the mechanistic model in STEF described above calculates both the amount of fuel crust that would form, and the relative timing associated with the potential fuel-concrete attack.

The 2DKO program[7] for Analysis of Two Dimensional Melting and/or Freezing in Two-Material Structures

The model used in the 2DKO computer code accounts for 2-D phenomena related to simultaneous heat transfer and phase change (melting/freezing) between hot liquids and relatively cold solids. Various applications include the melting of a solid structure when in contact with a continuously delivered melt, melting of a solid plate subjected to an impinging jet of hot liquid, and the melting of a vertical wall in contact with a liquid pool. The 2DKO model has been validated against experimental data, and showed good agreement[7]. This code was considered particularly useful for the present analysis of drain failure, since it allows for transient debris deposition on a plate.

An additional validation of the 2DKO code model was conducted by comparing against results of a one-dimensional (1-D) heat transfer model developed using the HS and CVH modules of the MELCOR[9] code. A sample problem was set up to evaluate the transient thermal response of a two-slab system consisting of molten Zircaloy on a cold steel plate representing the initial stage of debris pour described earlier. For MELCOR calculations, the specific heat capacity of each material was increased by an appropriate amount over a small temperature range near the material's melting temperature, to account for the latent heat of fusion. Transient temperature profiles generated by the two models were in good agreement with one another. Coupled with validation against experimental data, this verification exercise gave confidence in the use of 2DKO for analyzing pertinent aspects of the BWR Mark II inpedestal drain failure problem.

The PLUG program for Analysis of Core Debris Relocation within Drain Pipe

The PLUG computer program was written to evaluate core debris relocation (i.e., plug-type flow) within a drain pipe with simultaneous debris freezing and possible tube melting. The program utilizes mathematical models reported in Ref. 10. Figure 2 provides a schematic representation of the physical configuration for freezing penetration of a moving molten debris fluid in a channel, combined with channel melting. The time period between the passage of the fluid's leading edge and the beginning of wall melting is referred to as the melting delay time, t_m, whereas the time required for the crust buildup at the entrance region to cause complete blockage is designated as t_c. Both t_c and t_m need to be evaluated externally, or known previously. PLUG also calculates thermal and hoop stresses in the frozen debris crust, and compares them to the critical failure stress at the drain pipe melting front. This allows one to obtain an indication of crust collapse, and re-entry of molten core debris into the channel. The model predictions were compared by Best et al.[10] against data obtained from experiments involving molten UO_2 flow from a reservoir into narrow channels. Good agreement was observed. Furthermore, PLUG program evaluations for a sample problem were verified against hand calculations.

DRAIN FAILURE ANALYSIS

The simulation capability developed was used to analyze the BWR Mark II drain pipe regions for their response to attack by core debris. The analysis was conducted in two steps: (1) code calculations over a range of parameters covering the estimated conditions for the Limerick BWR containment, and, (2) application of results of parametric evaluations to analyze the possible sequence of events leading to drain failure. These two steps are described subsequently.

General Parametric Evaluations

The parametric calculations are described in five stages as follows:

Surface Tension Breakdown Evaluations and Analysis. The consequences of surface tension breakdown, and melt entry into the drain pipe through the slots in the drain cover plate for the Limerick containment (Figure 1), were addressed. Assuming negligible contribution from momentum transfer of debris as it settles around the slot, or hole, and ignoring melting or crust formation, a force balance yields, Hc = S x Pw / (rhod x g x Ac), where, Hc is the critical height of molten debris, S is the surface tension of debris material, Pw is the wetted perimeter of the opening, rhod is the density of debris material, g is the gravitational acceleration, and Ac is the cross-sectional area of the opening. Using this relationship it was found that the critical height of molten debris for a variety of conditions is quite often in the several millimeter range.

A thermal analysis using the 2DKO code indicated that it takes less than a second for molten metallic debris to freeze and plug the slots or holes in

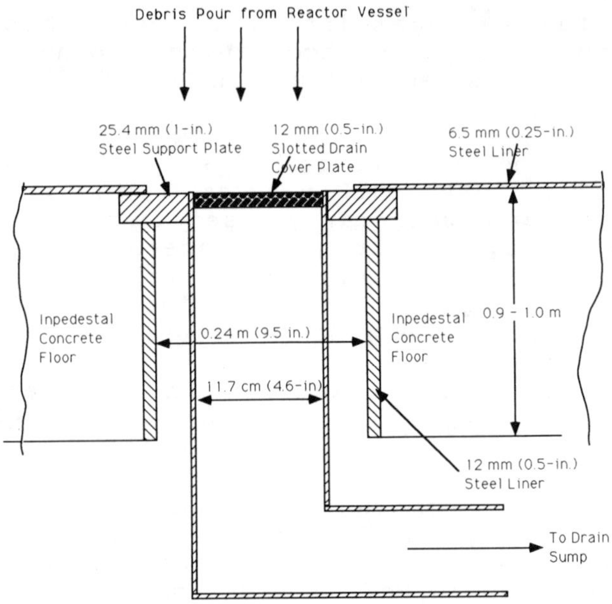

Figure 1. Schematic of Inpedestal Drain Pipe Region for Limerick BWR Mark II Containment

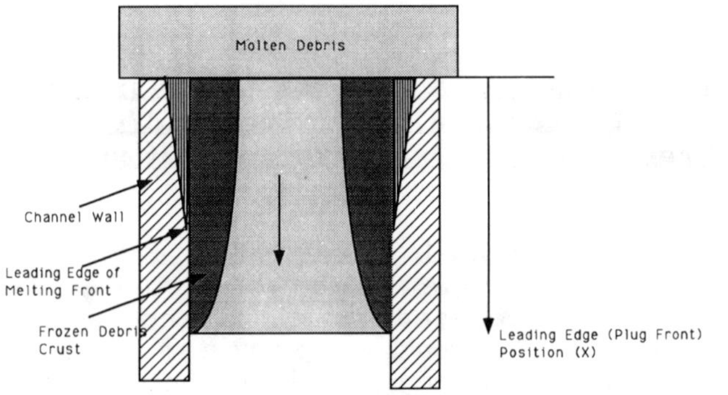

Figure 2. Schematic Representation of Molten Debris Plug Flow Model Characteristics

the drain pipe region, under various boundary conditions. Further, based upon BWRSAR results depicted in Table 1, it would require tens of seconds for several millimeters of molten pool head to form for surface tension breakdown to occur. Hence, for all practical purposes it can be assumed that no significant melt drainage will occur at the start of metallic debris pour.

Falling Debris Parametric Evaluations. The 2DKO program described earlier was used to analyze the response of a stainless steel plate to falling metallic debris as predicted by the BWRSAR code (summarized in Table 1). A test matrix of runs made for this analysis is shown in Table 2. A range of effective heat transfer coefficients were estimated for heat transfer between hot and cold surfaces, as shown in Table 2. It was noted from Ref. 3 that the maximum inpedestal drywell temperature reaches a value between 600 K and 700 K. An upper bound of 600 W/m^2-K was thus calculated for the debris surface-to-drywell effective heat transfer coefficient. Due to minimal structural material in the drain pipe, and the atmosphere therein, and to limit the number of code runs and ensuing analyses, the plate surface-to-ambient heat transfer coefficient was conservatively set to zero. The thickness of the plate was for the most part taken as 12.7 mm (0.5-inch) to correspond with the thickness of the steel cover plate in the Limerick inpedestal. However, a parametric study with different plate thicknesses for selected cases has also been conducted. Based upon BWRSAR predictions for the relative combinations of molten Zircaloy and steel, the material properties of the falling debris were derived by weighting the respective properties of Zircaloy and steel with factors of 0.3 and 0.7 respectively. The calculations were performed until 6000 s after the start of debris pour, after which the 2DKO program is not valid.

Results of the various evaluations are summarized in Table 2. It is seen from Table 2 that the steel plate (i.e., drain cover) melting characteristics are not significantly affected by the choice of debris surface heat transfer coefficient. This indicates that the process of plate melting is dominated primarily by heat conduction. It should be noted that for all cases, the steel plate gets completely molten within 1300 s, and that at the end of 6000 s various melt-crust temperature profiles can result.

<u>Debris Crust Ablation Parametric Analysis with the STEF Program.</u> For generality, ten different cases were analyzed with the STEF program. The results are tabulated in Table 3. The plate (i.e., frozen metallic debris) temperature at infinity (i.e., sufficiently far away from the melting front) was tied to the results obtained from the 2DKO falling debris analysis. This temperature was set at the melting temperature of the plate for the case when the metallic debris surface heat transfer coefficient was set to 0 W/m^2K

Table 2. Results of Falling Metallic Debris Film Analysis with the 2DKO Program

Case	Plate Thickness (in/cm)	Debris Superheat(K)	Tp (K)	Hd (W/m2-K)	tp,f (s)	Fd / Zd % / cm	tp,e (s)	Fd / Zd (% / cm)	Results at 6000 s Falling Debris Temperatures Td,bot (K)	Td,top (K)
1	0.50 / 1.27	0	303	600	847	90 / 5.45	1300	62 / 5.92	1005	592
2	0.50 / 1.27	100	303	600	770	87 / 4.29	1130	54 /5.51	1060	611
3	0.50 / 1.27	0	573	600	730	96 / 4.17	1130	63 / 6.43	1019	597
4	0.50 / 1.27	100	573	600	670	95 / 3.29	997	57 / 4.71	1084	617
5	0.50 / 1.27	0	303	0	847	90 / 5.45	1300	62 / 5.92	1547	1547
6	0.50 / 1.27	100	303	0	770	87 / 4.29	1130	54 /5.51	1547	1547
7	0.50 / 1.27	0	573	0	730	96 / 4.17	1130	63 / 6.43	1547	1547
8	0.50 / 1.27	100	573	0	670	95 / 3.29	997	57 / 4.71	1547	1547
9	0.375 / 0.95	100	303	600	630	98 / 2.82	860	65 / 4.06	1074	617
10	1.0 / 2.54	100	303	600	1420	66 / 6.90	2870	95 / 16.97	1048	597

Notes:
Tp - Initial Temperature of Plate
Hd - Debris Surface Heat Transfer Coefficient
tp,f - time at which plate first starts melting after attack from metallic debris
tp,e - time at which plate is completely melted
Fd - Fraction of debris that is frozen at time tp
Zd - Height of debris crust at time tp
Td,bot - Temperature (of debris) at bottom of debris bed at 6000 s
Td,top - Temperature of debris at top of debris bed at 6000 s

Table 3. STEF PROGRAM EVALUATIONS

Case	Debris Material	Tinf (K)	Plate Material	Tinf (K)	Time to Ablate Plate by .178 m (s)
1	UO2-Zr	2606	Zr-Steel	1820	1120
2	UO2-Zr	2606	Zr-Steel	1123	878700 (244 hrs)
3	UO2-Zr	2606	Zr-Steel	2098	1250
4	UO2-Zr	3113	Zr-Steel	1820	650
5	UO2-Zr	3113	Zr-Steel	1123	2.441e4 (6.8 hrs)
6	UO2-Zr	3113	Zircaloy	2098	550
7	UO2	3113	Zr-Steel	1820	1650
8	UO2	3113	Zr-Steel	1123	No Plate Melting Possible
9	UO2	3113	Zr-Steel	1423	2.441e4 (6.8 hrs)
10	UO2	3113	Zircaloy	2098	1420

(i.e., in the 2DKO evaluations). When the debris surface heat transfer coefficient was taken as 600 W/m^2K, the plate temperature at infinity was set at 1123 K (arrived at by taking the arithmetic average of frozen metallic debris crust temperatures over the height of the bed at the end of 6000 s). Cases 1 through 6 in Table 3 consider situations where the molten debris is a eutectic mixture of UO_2 and Zircaloy, either at the melting temperature of the eutectic or superheated up to the melting temperature of UO_2. As seen from the results, depending upon the degree of plate subcooling, the time it takes to completely ablate the approximately 18 cm of frozen crust could be several (i.e., Case 2). As expected, the shortest duration of time for completely ablating the frozen debris layer plate is when the plate temperature is at its melting temperature. Even in this case (i.e., Case 1), it can take close to 20 minutes. Hence it is evident that a significant time can elapse before the molten fuel debris can reach the concrete basemat, and initiate CCI. Finally, as seen from Table 3, the evaluations show that the amounts of debris crust formation and plate ablation are generally much less when the debris is UO_2 alone, rather than a eutectic mixture of UO_2 and Zircaloy. This is attributed in large measure to the significantly lower thermal conductivity of UO_2 compared to the UO_2-Zr eutectic mixture.

<u>PLUG Program Parametric Evaluations.</u> In order to evaluate the parameters t_m and t_c for PLUG calculations, additional 2DKO program runs were made with the debris bed height fixed at a given value equal to the drain pipe channel radius. The plate thicknesses chosen correspond with channel wall thicknesses for the pipe regions in the Limerick and WNP-2 designs respectively. The debris side heat transfer coefficient was set at 0 W/m^2K due to centerline symmetry. The heat transfer coefficient for the plate (i.e., tube material) interface with the ambient atmosphere was parametrically varied between 5 W/m^2K, and 500 W/m^2K. Results of these evaluations are shown in Table 4.

The PLUG program was next used to evaluate the amount of debris flow through channels. Results obtained for the length of debris penetration, crust buildup, and failure, are summarized in Table 4. As can be seen, a significant amount of debris can exit through the drain pipes before the entrance

Table 4. PLUG PROGRAM EVALUATIONS

Case	Debris Material	Superheat (K)	Tube Wall Thickness (in)	Characteristics Hp (W/m2-K)	tc (s) (see Note 1)	tm(s) (see Note 2)	Xc (m) (see Note 3)	Crust Failure	Maximum Unobstructed Amount of Debris Expelled (kg) - see Note 4
1	Zr-Steel	0	0.25	500	300	No Melting	66	None	About 5500 kg(per drain)
2	Zr-Steel	100	0.25	500	347	No Melting	73	None	About 6000 kg (per drain)
3	Zr-Steel	0	0.25	5	400	70	81	None	About 6800 kg (per drain)
4	Zr-Steel	0	0.50	5	580	284	105	None	About 25,000 kg / drain - Limerick Not Applicable for WNP-2
5	UO2-Zr Eutectic	0	0.25	500	435	>> 435	171	None	About 17,000 kg (per drain)
6	UO2-Zr Eutectic	500	0.25	500	> 600	20	214	Yes	Continuous Pour Established; If not- About 80,000 kg (per drain) for Limerick About 21,000 kg (per drain) for WNP-2
7	UO2-Zr Eutectic	0	0.25	5	> 1000	15	305	Yes	Continuous Pour Established at Rates of: About 1200 kg/s (per drain) for Limerick About 300 kg/s (per drain) for WNP-2
8	UO2-Zr Eutectic	0	0.50	5	600	80	214	Yes	Continuous Pour Established; If not- About 80,000 kg (per drain) for Limerick Not Applicable for WNP-2
9	UO2-Zr Eutectic	500	0.50	5	> 6000	40	> 1030	Yes	Continuous Pour Established at Rates of: About 1200 kg/s (per drain) for Limerick Not Applicable for WNP-2

Notes/Assumptioons:

1) tc - time it takes for channel entrance to get blocked from freezing debris
2) tm - time it takes for tube wall to start melting after first passage of debris front
3) Xc - maximum unobstructed axial debris plug traverse before channel blockage at entrance
4) Mass of debris ejected evaluated as = debris density x cavity area x Xc
5) Height of molten debris pool is assumed to remain constant

blockage condition is met. Crust failure in drain pipes is also indicated in some cases where significant tube wall melting occurs. In these instances a continuous pour of core debris at high discharge rates may also get established.

Limerick Drain Failure Analysis

Drain failure was analyzed in a somewhat focused manner using the results of the parametric evaluations, and the BWRSAR debris pour history given in Table 1. Analysis results are depicted in Figure 3, using a flow chart format.

As seen in Figure 3, upon initial debris attack of the drain pipes, surface tension breakdown may or may not occur depending upon the nature of the pour. Even if breakdown occurs, only negligible drainage will result, as was previously demonstrated.

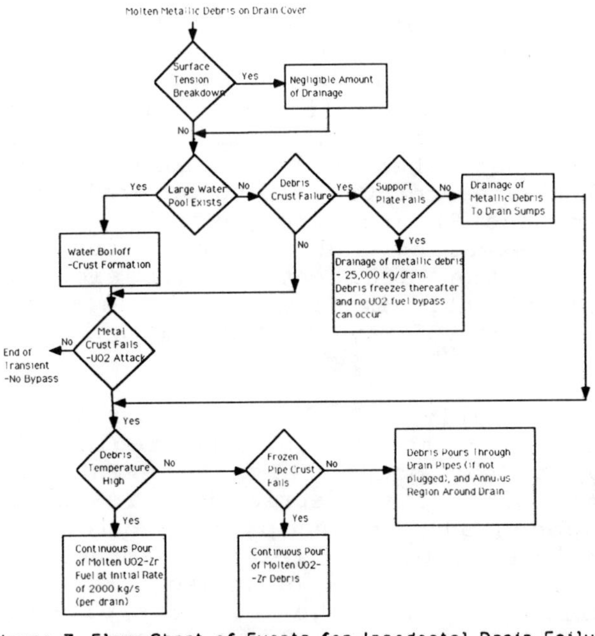

Figure 3. Flow Chart of Events for Inpedestal Drain Failure Analysis in Limerick BWR MARK II Containment

As described below, very different scenarios may take place dependent upon whether or not a significant water pool exists in the inpedestal cavity region.

Significant Water Pool Present in In-Pedestal Cavity Region. In case a significant water pool exists in the inpedestal region, debris freezing from quenching will occur until the water boils off or the metals cool down sufficiently. The degree of subcooling of the metallic debris would clearly depend upon operating conditions. The MELCOR analysis of Ref. 3 indicates that the metallic debris temperature would be lowered to about 1100 K at the end of 6000 s from the start of the debris pour. In any case, it is evident that no significant metallic debris drainage into the drain pipes can occur in this scenario.

Upon ejection of molten UO_2-Zr Eutectic into the inpedestal cavity region, the frozen metallic debris crust may start to ablate. For the likely case where metallic debris has cooled off to about 1100 K, Case 2 (Table 3) results indicate that it may take several days before the entire frozen metallic debris crust layer of about 0.178 m height can be ablated, even if the UO_2-Zr debris does not solidify due to combined heat of radiation and convection to the structural materials in the cavity region. If a UO_2 debris layer is assumed, no metallic crust ablation is predicted (Case 8). From Table 3 it is seen that other relative timings (e.g., 30 minutes for Case 7, to several hours for Case 9) are also possible. These are revealing and important results, since the accident progression is significantly impacted by virtue of the fact that CCI may not occur at all.

Clearly, if the fuel debris does not reach the drain pipe region, no bypass can occur. In the event that the molten fuel debris does reach the inpedestal drain pipe region an indication of the relative amounts of drainage can be obtained from the results of the PLUG program evaluations, Cases 8 & 9 respectively in Table 4. Depending upon the amount of debris superheat, a continuous pour may or may not take place. Significant drainage can be expected in either case. The actual amount of debris expelled may be greater than the amounts indicated in Figure 3, since the simultaneously occurring CCI would tend to remove the entrance blockage, and enlarge drain holes due to radial concrete ablation.

No Significant Water Pool in In-Pedestal Cavity Region. In the absence of a water pool, the 2DKO-based evaluations shown in Table 2 indicate that plate melting can be expected to start in about 10 to 15 minutes from the start of metallic debris pour, at which point in time several inches of debris crust would have formed. Even at the start of plate melting a significant debris crust may have formed. Hence, a qualified conclusion may be drawn that metallic debris drainage will not occur even if the drain cover plate is completely molten. However, a significant amount of concrete degassing may also be occurring simultaneously which may fracture or stir the frozen debris mixture. In case crust failure or bypass do not occur, the 2DKO evaluations (Cases 1 through 4 in Table 2) indicate that at the end of 6000 seconds the entire metallic debris would have crusted based on assumed debris surface-to-ambient heat transfer. The crusted debris temperature would exhibit a temperature gradient ranging from about 600 K at the top of the debris, to about 1000 K at the debris-drain plate region. At the other extreme, where the surface heat transfer characteristics may have been sufficiently degraded (e.g., possibly from the generation of an aerosol cloud layer which blocks radiation heat transfer) only small fractions of the debris would have frozen after 6000 s into the transient. Such a surface condition in unlikely, because no significant CCI can occur during this stage, and because the metallic debris has no significant UO_2 fuel in it. Thus, the formation of dense aerosol clouds above the debris bed to cause adiabatic-type heat transfer surface conditions is not possible.

In the unlikely event of debris crust failure upon drain-cover plate melting, the accident sequence can change significantly. Molten metallic debris above the crust would begin pouring into the drain pipe region. Based upon the 2DKO and PLUG program results (Table 4), it appears that 5000 to 7000 kg (per drain) can be expected to leave the debris bed, and enter the drain sumps as shown in Table 4 for Cases 1 to 3. The metallic melt would freeze before the remaining metallic debris bed above in the concrete pedestal is attacked by the molten fuel debris at the end of 6000 seconds.

Finally from the falling debris analysis for Case 10 (Table 2) it appears that there may be a remote possibility of melting the 25.4 mm (1-inch) thick supporting structure (Figure 1) holding the 0.102 m (4-inch) diameter drain pipe late (i.e., after 2870 seconds) into the transient. From the results of Case 1 through 8 analyses (Table 2), it is evident that this result would hold true even if the debris surface heat transfer conditions were not adiabatic. However, even upon complete melting of the support plate, most of the metallic debris above it would be in a frozen state. Hence, under the assumed conditions, a continuous pour of any significant amount does not appear likely through the opening that may be created. If the support plate were to fail somehow (e.g., from creep rupture or material degradation), and molten metallic debris poured through it, PLUG evaluations (Case 4, Table 4) indicate that up to 25,000 kg (per drain) of debris may pour out before plugging up the channel entrance. If such an event did occur, it would happen about an hour after the first debris pour commenced. Since it does not appear likely that the annulus portion will get plugged up with metallic debris melt, a reasonable pathway does exist for the molten fuel to get ejected out of the inpedestal region after it has ablated the crusted metallic debris.

The progression of the transient after 6000 seconds would essentially remain the same as for the case when a significant water pool exists. However, in the unlikely event the entire drain pipe region (including the annulus) were plugged up with frozen metallic debris, the progression of the molten fuel would be radically different. No fuel can be ejected from the inpedestal region during MCCI for this instance. However, if the annulus region were not filled with frozen debris, the results of melt progression would remain similar to those described previously (i.e., when a significant water pool exists). This indicates that the support plate for the drain pipe in the Limerick containment design can potentially play an important role in preventing early PSP bypass of molten fuel debris.

CONCLUDING REMARKS

Mechanistic models of the combined heat transfer and phase change between either flowing or stationary multi-material corium and the drain pipes have been used in the analysis. These models have been numerically tested and validated against experimental data. Several accident scenarios have been analyzed, including timing, amount and composition of corium released from the RPV onto the containment floor, melt superheat, the effect of water in the in-pedestal cavity region, melt inflow into the drain pipe, pipe heatup and plugging due to crust buildup, etc. in a general manner. The results of these analyses have been used to assess the possibility of drain pipe failure, as well as to estimate the timing and amount of molten corium release to the wetwell. The study has revealed that significantly different melt progression sequences can result depending upon the failure characteristics of the frozen metallic crust which forms over the drain cover during the initial stages of debris pour. Another important result is that it can take several days for the molten fuel to ablate the frozen metallic debris layer - if the frozen layer had cooled below 1100 K.

As a final note it is cautioned that the results presented in this paper are based upon the specific debris pour history predicted using the BWRSAR code, and on the individuals models and assumptions used. Other debris pour histories (predicted by other codes) may cause considerable variations in BWR MARK II containment drain failure characteristics.

LITERATURE CITED

1. "Updated Final Safety Analysis Report," Limerick Generating Station (Units 1 and 2), Philadelphia Electric Company, (1990).

2. "WPPSS Nuclear Project No. 2, Final Safety Analysis Report," Washington Public Power System, (1982).

3. Greene, S. R. et al., "The Response of BWR Mark II Containments to Station Blackout Severe Accident Sequences," NUREG/CR-5565, Oak Ridge National Laboratory (to be published).

4. Hodge, S. A., and Ott, L. J. "Boiling Water Reactor Severe Accident Response (BWRSAR) Code Description and Assessment," NUREG/CR-5318, Oak Ridge National Laboratory (to be published).

5. Podowski, M. Z., and Lahey, R. T. Jr., "Degraded BWR Core Modeling - Physical Simulation of Selected Components," Empire State Electric Energy Research Corporation Final Report EP84-4, (1989).

6. Cole, R. K. Jr., et al., "CORCON-MOD2: A Computer Program for Analysis of Molten-Core Concrete Interactions," SAND84-1246 (NUREG/CR-3920), Sandia National Laboratories, Albuquerque, NM, (1984).

7. Kasprzak, S., Konishi, H. and Podowski, M. Z., "A Method of Analysis of Two Dimensional Melting and/or Freezing in Two-Material Structures," Proceedings of the 24th National Heat Transfer Conference and Exhibition, Pittsburgh, Pennsylvania, 27-33,(1987).

8. Saitoh, T., "An Experimental Study of the Cylindrical and Two-Dimensional Freezing of Water with Varying Wall Temperature," Technology Report of Tohoku University, Volume 41, No. 1, (1976).

9. Summers, R. M. et al., "MELCOR 1.8.0: A Computer Code for Nuclear Reactor Severe Accident Source Term and Risk Assessment Analyses," NUREG/CR-5531, Sandia National Laboratories, (1991).

10. Best, F. R., Wayne, D., and Erdman, C., "A Fuel Freezing Model for Liquid-Metal Fast Breeder Reactor Hypothetical Core Disruptive Accidents," Nuclear Science & Engineering Journal, **89**, pp. 49-60, (1985).

LOWER HEAD FAILURE ANALYSIS[a]

J.L. Rempe, G.L. Thinnes, and C.M. Allison ■ Idaho National Engineering Laboratory, EG&G Idaho, Inc., P.O. Box 1625, Idaho Falls, ID 83415

The U.S. Nuclear Regulatory Commission is sponsoring a lower vessel head research program to investigate plausible modes of reactor vessel failure in order to determine (a) which modes have the greatest likelihood of occurrence during a severe accident and (b) the range of core debris and accident conditions that lead to these failures. Preliminary results, which were obtained with analytic closed-form approximations to assess the important governing parameters in non-dimensional form, illustrate the importance of vessel and tube geometrical parameters, material properties, and external boundary conditions to predict vessel failure.

This paper summarizes preliminary results from an integrated assessment of reactor vessel lower head failure analyses.[1] The ultimate goal of this research is to determine the timing and mode of lower head failure during severe accident conditions. Specific objectives of the analyses are to investigate plausible modes of reactor vessel lower head failure to determine (a) the mode(s) having the greatest likelihood of occurrence and (b) the quantity and composition of corium debris and range of accident conditions that could lead to these failures. All major types of U.S. light water reactor (LWR) vessels are considered, and both high- and low-pressure conditions are addressed for each reactor type.

Potential Lower Head Failure Mechanisms. Several mechanisms have been identified as having the potential to cause reactor vessel lower head failure:

- Melt penetration and failure of a lower head penetration
- Pressurized ejection and rupture of a lower head penetration
- Global creep rupture of the lower head
- Localized effects such as jet impingement induced lower head failure.

Although all four mechanisms are considered in this research project, the initial research summarized in this paper has concentrated on the first three mechanisms.

Plant Design Features that Impact Lower Head Failure. Plant design features can increase the likelihood that vessel failure will be caused by a particular mechanism. Although the lower head failure analyses program is considering representative plant data from each of the U.S. vendors, results for only two plants are discussed within this paper, a Babcock and Wilcox (B&W) PWR and a General

a. Work supported by the U.S. Nuclear Regulatory Commission, Office of Nuclear Regulatory Research, under DOE Contract No. DE-AC07-76ID01570; Ralph Landry, Technical Monitor.

Electric (GE) BWR. For these analyses, it is assumed that all lower head plate materials are SA533B1. Typical lower head thicknesses are also assumed, 21 cm for a GE BWR and 13 cm for a B&W PWR vessel.

Lower head penetrations vary in number, construction, size, and material composition. GE BWR vessels have over 200 penetrations, whereas typical B&W PWR vessels have about 50 penetrations. Instrument guide tube penetrations are typically made by boring holes through the head, inserting a tube through the hole, and welding a nozzle to the tube and vessel inner surface. BWR control rod guide tube penetrations are formed by boring the hole for the guide tube, drilling a countersink for a stub tube that surrounds the guide tube on the vessel inner surface, inserting the guide and stub tubes, and welding the guide tube to the stub tube. To facilitate inserting tubes into the holes in the vessel head, tolerances on the outer diameter of each penetration tube and the diameter of each hole in the vessel head are set in their design. This results in a radial gap, which must be considered in certain penetration failure analyses. The BWR drain line penetration consists of a carbon steel nozzle welded to a hole bored through the vessel lower head. Typically, the instrument guide tubes for PWRs are Inconel and for BWRs are stainless steel. The BWR control rod guide tubes are also stainless steel, and the stub tubes are Inconel.

LOWER HEAD PENETRATION TUBE HEATUP AND FAILURE

Analyses were performed to calculate characteristic tube heatup times, melt penetration velocity, melt penetration distance, and in-vessel and ex-vessel tube failure. Results obtained for predicting the likelihood of in-vessel tube failure, melt penetration distance, and ex-vessel failure are presented for a range of melt compositions (bounded by a metallic melt based upon stainless-steel properties and a ceramic melt based upon UO_2 properties) and tube parameters (based upon the reference PWR and BWR penetrations).

Debris Temperatures Required for Tube Melting. Debris characteristics required to induce tube melting were estimated by applying a heat balance to a penetration tube in contact with corium debris and coolant using the idealized geometry illustrated in Figure 1 to obtain:

$$\frac{\theta_d}{\theta_t} = \frac{c_{pt}\ M_t}{c_{pd}\ M_d} \qquad (1)$$

where c_{pt} and c_{pd} represent the specific heat capacity of the debris and the tube, respectively; M_d and M_t, the mass of the debris or the tube, respectively; and θ_d and θ_t, effective debris and tube temperatures, respectively.

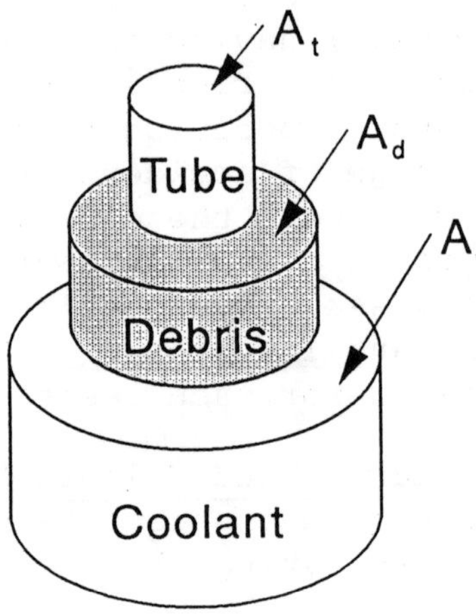

Figure 1. Idealized geometry of penetration surrounded by corium and coolant for predicting debris temperature requirements for tube melting.

Curves in Figure 2 represent Equation (1) evaluated for various debris and tube material combinations. The region above the curves corresponds to cases where the initial debris temperature can induce tube melting. As shown in Figure 2, relatively low debris temperatures may cause tube melting in PWRs that have only lower head penetrations for instrument tubes. For example, in the B&W plant with M_t/M_d ranging between 0.0039 and 0.053, a ceramic debris temperature of 1700 K is required to induce melting of Inconel penetration tubes. In a GE BWR, the ratio of M_t/M_d is significantly higher, ranging between 0.15 and 0.30. Hence, ceramic debris temperatures in excess of 2000 K are required to induce tube melting. The curves in Figure 2 also show that somewhat lower debris temperatures can cause melting if the melt is primarily metallic material or if the penetrations are stainless-steel. Horizontal bands corresponding to θ_d/θ_t for debris temperatures of 3113 K (the melting temperature of UO_2) and 2400 K (an average of melting temperatures for UO_2 stainless steel) are included in Figure 2. As shown in Figure 2, melting is predicted at these debris temperatures for both reactors' penetrations.

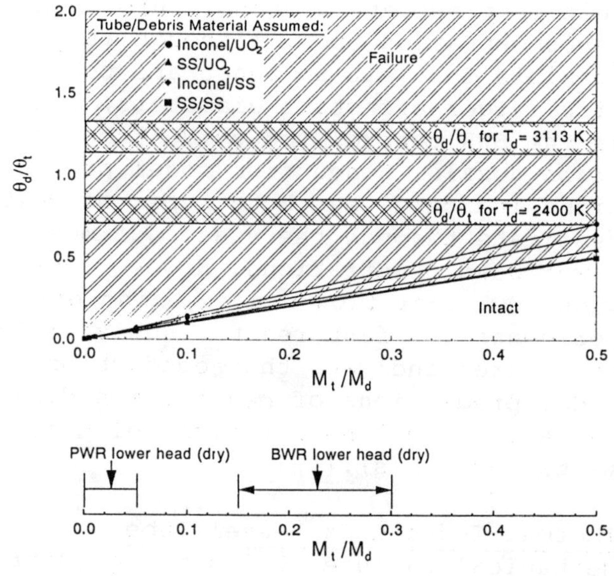

Figure 2. Debris thermal criteria for in-vessel tube failure.

Melt Penetration Distance. An order-of-magnitude estimate of the penetration distance for plug solidification (X_p) is assessed using the bulk freezing model first advanced by Ostensen and Jackson[2,3] and the conduction layer model of Epstein.[4,5] Although recently-developed models more accurately predict melt penetration distances under certain conditions,[6] these two models bound the range of experimentally-observed melt penetrations distances. Figure 3 illustrates the geometry assumed for each model.

The bulk freezing model assumes that turbulence within the flowing melt prevents stable crust formation at the channel wall (see Figure 3).

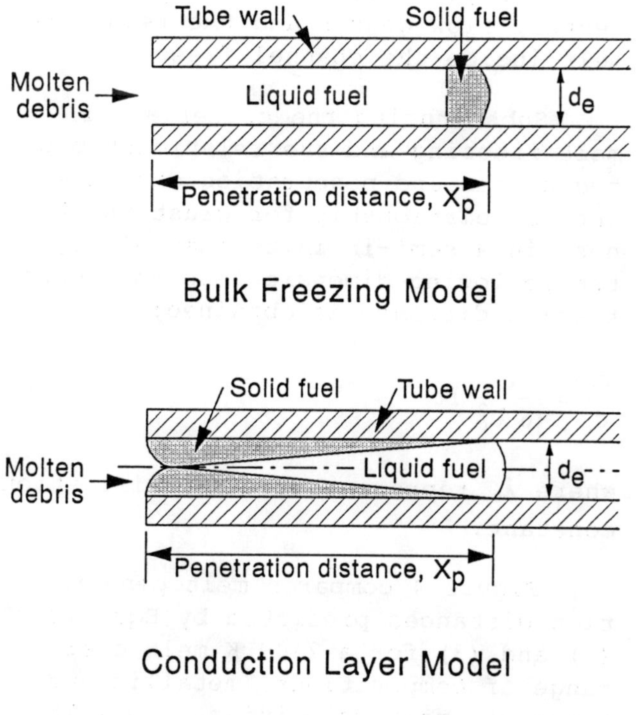

Figure 3. Comparison of bulk-freezing and conduction layer models.

The distance for plug solidification (X_p) in the bulk freezing model is obtained by equating the energy given up by an element of molten debris (of equivalent diameter, d_e, and Peclet number, Pe) upon cooling to a solid state by convective heat transfer from the melt to the wall over the melt penetration distance, X_p.

$$\frac{X_p}{d_e} = 0.4\ Pe^{0.6}\left[\frac{\theta_d}{\Delta T}\right] \tag{2}$$

where

$$\theta_d = \left[\frac{L_d}{c_{pd}} + T_d - T_{mp/d}\right] \text{ and}$$

$$\Delta T = T_d - T_t\ .$$

The conduction layer model assumes that transient freezing is governed by crust buildup at the channel wall, where conduction heat transfer governs melt solidification. Once the frozen layer closes at the channel center, flow ceases and the remaining melt inside the channel freezes.

Substituting the criteria for melt freezing and the relationship between time and penetration distance into a relationship for crust thickness in a semi-infinite channel,[5] the following dimensionless melt penetration distance is obtained:

$$\frac{X_p}{d_e} = \frac{1}{16\,\lambda_c^2}\ Pe \tag{3}$$

where λ_c represents the solidification constant.

Figure 4 compares melt penetration distances predicted by Equations (2) and (3) for a 2400 K melt over a range of compositions (metallic and ceramic), melt velocities, and melt flow diameters. The scales at the bottom of Figure 4 show Peclet numbers

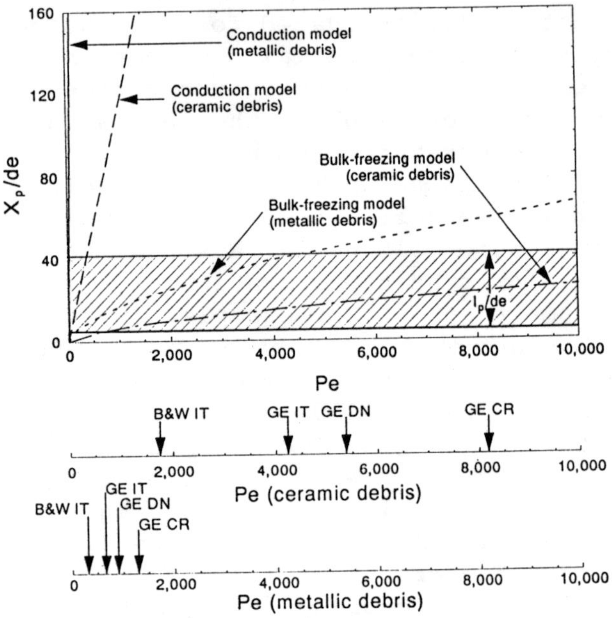

Figure 4. Comparison of penetration lengths predicted by conduction layer and bulk freezing models, assuming a 0.1 m/s velocity melt.

corresponding to ceramic and metallic melts with 0.1 m/s velocity. If the melt follows bulk freezing model predictions, the curves in Figure 4 indicate that it is unlikely for ceramic melt to penetrate a vessel through an instrumentation tube, although it is possible for melt to penetrate tubes with larger effective diameters, such as the BWR control rod guide tube or drain nozzle. Conduction model predictions indicate that both types of melt penetrate through the vessel of most lower head penetrations. However, previous UO_2 fuel flow and freezing experiments for fast reactor subassembly structures indicate that conduction model predictions of penetration distance are too long for high velocity melts (>0.1 m/s).[4,7]

Thermal-Induced Ex-Vessel Tube Mechanical Failure. Using the geometry shown in Figure 5, two methods were applied to assess ex-vessel tube failure potential: (a) a heat balance

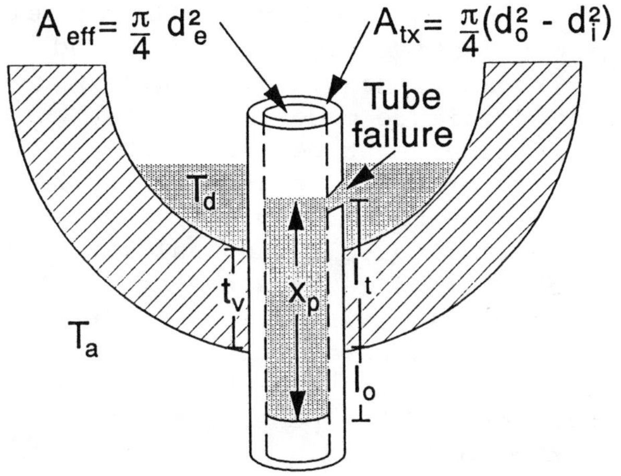

Figure 5. Geometry for analyzing ex-vessel tube failure.

between melt heat loss and melt/tube wall heat transfer and (b) a steady-state heat transfer model, which considers melt decay heat and tube wall radiative heat loss. These methods were examined to define the heat transfer parameters influencing ex-vessel tube thermal-induced mechanical failure.

To determine if tube failure by thermal means is possible, a heat balance was written for the geometry illustrated in Figure 5, where the sensible and latent heat transferred from a molten melt plug is equated to the heat gained by the tube wall.

$$\frac{\theta_d}{\theta_t} = \frac{A_{tx}}{A_{eff}} \frac{\varrho_t c_{pt}}{\varrho_d c_{pd}} \quad (4)$$

In this equation, θ_d is again equal to the effective debris temperature, but θ_t designates the temperature increase between the initial tube temperature and the tube failure temperature, which for ex-vessel failure calculations is assumed to occur when the tube temperature reaches the point at which its ultimate strength reduces to zero (1450 K for Inconel; 1370 K for stainless steel). A_{tx} represents the tube cross-sectional area; A_{eff}, the effective debris flow area; and ϱ_t and ϱ_d, the densities of the tube and the debris, respectively.

Equation (4) has been evaluated for metallic (stainless steel) and ceramic (UO_2) based melt debris (at their solidus temperature) in contact with the reference plant penetrations. Figure 6 presents results obtained by assuming that the initial tube temperature is 550 K. Results illustrate that a higher debris temperature is required to induce tube failure if the debris is metallic and if the tube is Inconel. This phenomena may be understood by noting the higher Inconel failure temperature and the higher UO_2 heat capacitance. Ex-vessel tube failure is also predicted to be more likely for thin-walled penetrations with larger effective diameters for melt flow.

To estimate the steady-state wall temperature for a tube containing debris and its associated decay heat, a heat balance can be applied between the containment at temperature, T_a, and the tube containing debris with ex-vessel length, l_o, equivalent diameter, d_e, and heat flux, $\ddot{q}_d$ to obtain:

$$\frac{\ddot{q}_d}{\epsilon_t \sigma T_a^4} = \frac{1}{d_e/d_o} \left[\frac{T_t^4 - T_a^4}{T_a^4} \right] \quad (5)$$

In this equation, ϵ_t represents tube emissivity; σ, the Stefan Boltzmann constant; and d_e and d_o, debris effective flow diameter and tube outer diameter, respectively.

Equation (5) has been evaluated to construct the failure map shown in Figure 7. As shown in the figure, higher debris heat fluxes are required to fail Inconel tubes, which is due to the higher temperature at which the

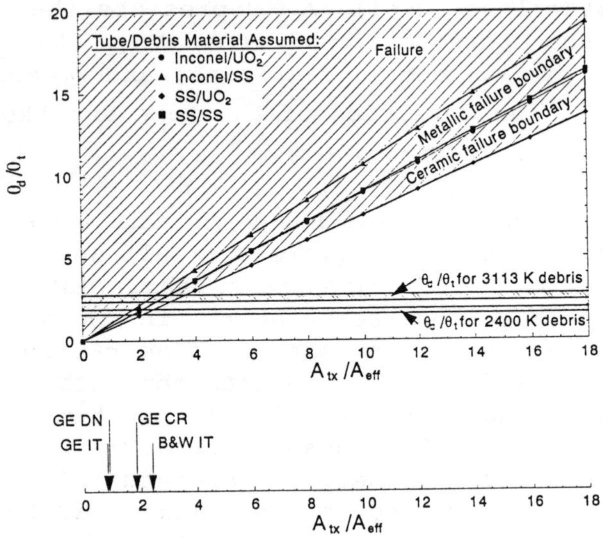

Figure 6. Debris thermal requirements for ex-vessel tube failure using a tube temperature equilibrium analysis.

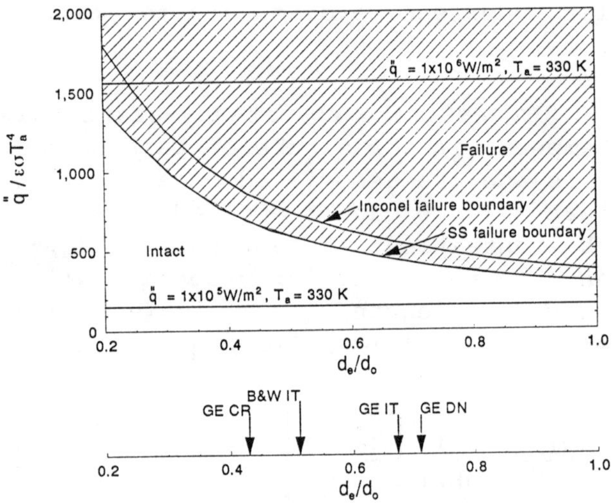

Figure 7. Debris heat flux requirements for ex-vessel tube failure using a steady-state equilibrium analysis.

ultimate strength becomes negligible for Inconel. In addition, Figure 7 illustrates that higher heat fluxes are required to fail thicker walled tubes with smaller outer diameters.

PRESSURE-INDUCED PENETRATION TUBE AND VESSEL HEAD FAILURE

Structural failure of the vessel and penetration tubes is evaluated concurrently, considering three postulated failure modes. The first considers tubes that penetrate the head ejecting from the vessel, while the second considers rupture of the tubes outside the vessel. The analysis uses a mechanics of materials approach to evaluate the plausibility of the two failure modes using structural material properties and characteristic dimensions of the penetration tubes and the lower head. Creep properties of the materials are not considered. The third mechanism considered is global head rupture. Previous lower head severe accident analyses[8,9] in which plastic and creep responses of vessel lower heads were calculated with finite element models, have suggested a simpler approach to calculate vessel rupture based solely upon radial temperature distribution in the vessel wall and the ultimate strength of the head material at temperature. The simplified approach provides a means of comparison between the penetration tube ejection/rupture failure modes and the global head failure mode.

Figure 8 illustrates the generalized geometry of the tube and vessel wall. It is assumed that the tube and attachment weld have melted at the inner surface of the reactor pressure vessel (RPV), and molten debris has entered the tube and been pushed down through the tube by the operating system pressure until sufficient heat is lost through the tube wall to cause some of the debris to freeze. This would cap the tube and provide an additional area on which the internal pressure could act. The only mechanism assumed to resist the ejection is the

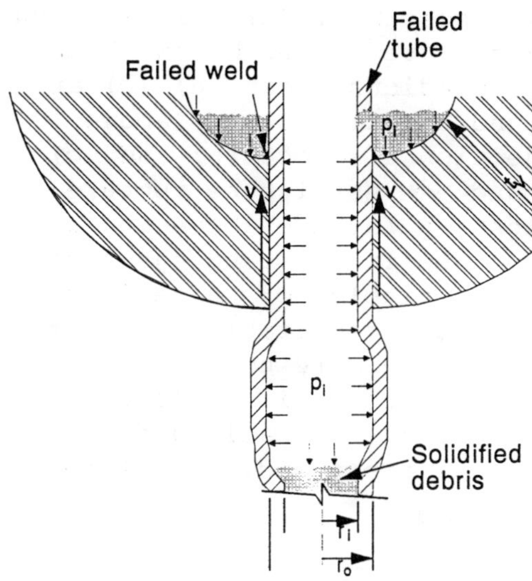

Figure 8. Idealized penetration tube and RPV wall with temperature.

expansion of the tube in the hole, causing a friction force buildup through the vessel wall. The resulting equilibrium equation of these forces provides a means of evaluating ejection, which is achieved when the following inequality is true:

$$p_i \frac{\pi\ d_o^2}{4} > V\ , \qquad (6)$$

where V represents the integral of differential shear, v, across the vessel wall thickness.

The differential shear, v, is a function of internal pressure, hole and tube temperature, hole diameter, tube inner and outer diameter, friction coefficient between the tube and hole, coefficients of thermal expansion (CTE), Poisson's Ratios, and elastic moduli of the head and tube materials. These parameters vary with temperature, which varies radially through the vessel wall and, therefore, causes the frictional shear function to vary radially. No support of the tube outside the vessel wall is assumed to resist nozzle ejection, since instrument guide tubes are quite long and flexible outside the reactor vessel and the control rod guide tubes would have to grow about 3 cm before any external support could be developed. The differential shear force is assumed to be the result of an interference pressure developed between a thick-walled cylinder, the tube, and a surrounding rigid jacket, the vessel wall, over a short increment of tube length. Each increment expands separately because of the local temperature and, for the tube, pressure loads.

Closed form solutions[10] are then used to calculate interface pressures required to force the tube outer diameter to conform to the diameter of the hole in the vessel wall after free expansion. This pressure is limited to the pressure required to cause compression failure of the tube material throughout its wall thickness[11] without any buckling consideration. This limiting interface pressure, p_{LI}, is determined by

$$p_{LI} = \frac{2}{\sqrt{3}} \sigma_u \ln \frac{d_o}{d_i} \qquad (7)$$

where d_i is the tube inner diameter. The ultimate stress, σ_u, is evaluated at the local tube temperature.

Although any spatial temperature distributions for the tube and hole can be modelled (see Figure 8), the calculations for developing the failure maps use a linear distribution for the tube temperatures, T_t, and a distribution representative of steady-state heat transfer out of the head, T_H, for the vessel wall material based upon a common inside vessel wall temperature, which is input to the model.

In addition, the model compares the internal pressure required to rupture the tube with the operating system pressure, p_i. this rupture pressure calculation uses whose material ultimate strength evaluated at the local tube temperature just outside of the vessel lower head.

$$p_i \geq \frac{2}{\sqrt{3}} \sigma_u \ln \frac{d_o}{d_i} \quad . \qquad (8)$$

This comparison determines whether conditions are sufficient to rupture the tube if plugging occurs outside of the vessel wall.

The global rupture model is based upon the following equation:

$$p_i \geq \frac{2}{R_m} \sum_{n=1}^{\infty} (\sigma_u \Delta t_v)_n \quad . \qquad (9)$$

The same vessel lower head wall temperature distribution as used in the tube ejection/rupture model is used in this model. The ultimate strength for SA533B1 as a function of temperature and the radial temperature profile are used to determine the ultimate membrane load capacity of each of 20 increments of the wall thickness. All incremental ultimate load capacities are then summed to develop a total membrane load capacity for the vessel with radius, R_m. This pressure load capacity (the right side of the inequality above) is then compared to the internal pressure, which is an input parameter for the tube ejection/rupture calculation.

Failure maps have been developed for typical incore instrument guide tube penetrations using Equations (6), (8), and (9). Figure 9 illustrates the failure map for a B&W penetration with the maximum gap between the tube outer radius and the vessel wall hole radius. Figure 10 is the failure map for the same penetration with the minimum gap. A similar failure map was generated for the BWR instrument tube penetration, at the maximum radial gap shown in Figure 11. The maps plot tube and global vessel failures in terms of the inner wall temperature and the system pressure.

For the tube failure model three curves are possible: (a) a lower bound ejection failure curve, below which the model predicts ejection; (b) an upper bound ejection curve, above which ejection also occurs; and (c) an upper bound ex-vessel tube rupture curve plotting failure above it. In actuality, the lower bound ejection curve is not viable for these instrument guide tube penetrations. Inner wall temperatures are not hot enough to cause the penetration weld at the vessel inner wall to fail, and ejection would be prevented. Therefore,

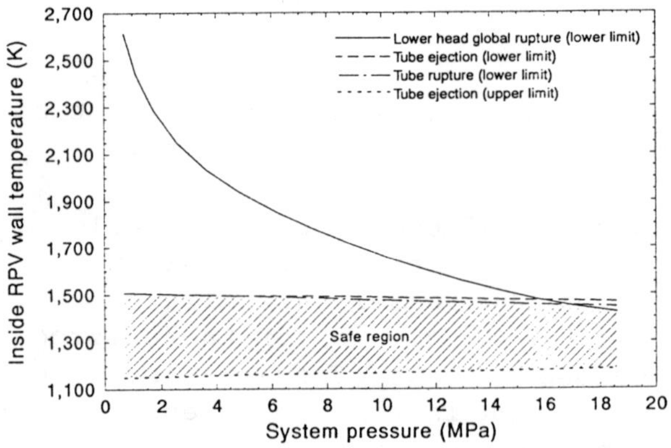

Figure 9. B&W instrument guide tube failure map at maximum radial gap (0.013 cm).

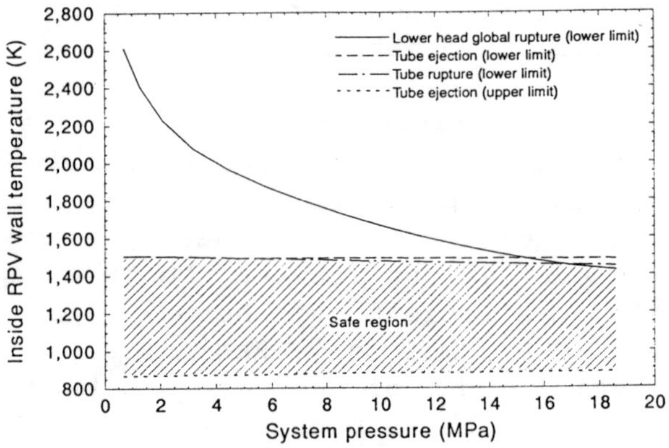

Figure 10. B&W instrument guide tube failure map at minimum radial gap (0.006 cm).and the vessel wall hole radius.

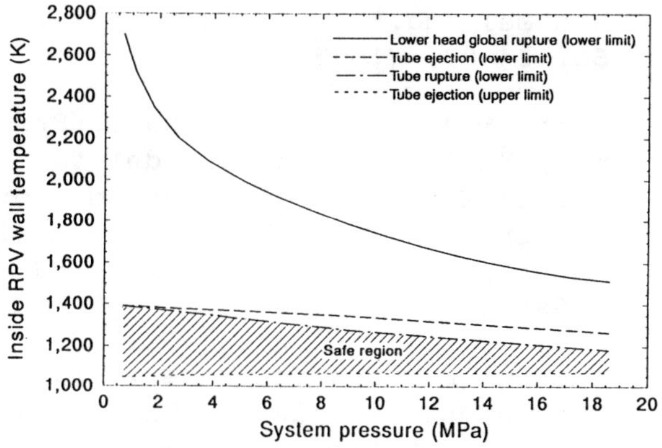

Figure 11. GE instrument guide tube failure map at maximum radial gap (0.025 cm).

the safe regions for these penetration tubes are below the limiting upper bound ejection or rupture curve. The lower head global rupture failure maps are compared with the tube ejection/ rupture failure maps in Figures 9 through 11. Global failure is predicted by this model above the curve. As can be seen, global rupture does not control the safe region in any of the failure maps. However, it must be noted that with creep considerations, the safe region would be reduced because of more limiting tube and global rupture curves.

The most sensitive parameters in these calculations are the lower head inner wall temperature and the tube material ultimate strength, which limits the binding force between the tube and vessel head. The influence of relative difference between mean coefficients of thermal expansion (CTE) of head and tube materials appears not to be as great. The radial tube-hole gap size primarily affects the lower bound ejection curve, which is of significance for the BWR control rod penetrations.

For this analysis, CTE, Young's Modulus, and ultimate strength values, which are temperature dependent, were linearly extrapolated beyond existing test data to zero at higher temperatures. The upper end temperatures on this test data ranged from 920 K to 1260 K. In this model, the ultimate strength of Inconel reduced to zero near 1450 K and that of 304 stainless near 1370 K. Upper bound tube ejection curves and tube rupture curves tended to form in this vicinity.

Comparison of Figures 9 through 11 also indicates that the upper bound on the safe region is always limited by tube rupture at lower pressures. This failure curve is highly dependent upon the ultimate strength of the tube material and the flat temperature profile assumed along the tube length. For more detailed analyses, test data for the more critical material parameters should be extended to temperatures at which the ultimate strength is reduced to about 1 MPa. Otherwise a considerable underprediction of tube and vessel capacity can occur based upon these analyses.

It is not possible to accurately model the phenomenon of creep in the structural response of the lower head at this time because the creep properties for SA533B1, which is the predominant plate material used in both BWRs and PWRs, has not been developed above 922 K. SA533B1 material undergoes a phase transition near 1000 K. This precludes the use of extrapolated Larson-Miller parameters for any simplistic creep rupture analysis and the extrapolation of existing constitutive creep laws for detailed finite element analyses. Idaho National Engineering Laboratory is currently performing creep tests at temperatures above 922 K for SA533B1 material to resolve this problem.

DISCUSSION AND CONCLUSIONS

Preliminary results, which were obtained with analytic closed-form approximations to assess the important governing parameters in non-dimensional form, illustrate the importance of vessel and tube geometrical parameters, material properties, and external boundary conditions on predicting vessel failure. In-vessel tube melting is governed by the tube-to-debris mass ratio within the lower head, where most penetrations are predicted to fail if surrounded by molten debris. Melt penetration distance is dependent upon the effective flow diameter of the tube. Molten debris is predicted to fail a penetration with a larger effective flow diameter, such as a boiling water reactor (BWR) control rod guide tube. Ex-vessel tube failure for depressurized reactor vessels is also dependent upon tube geometry and is predicted to be more likely for BWR penetrations. At somewhat higher pressures (greater than ~0.1 MPa) ex-vessel tube rupture becomes a dominant failure mechanism, although tube ejection dominates failure at lower temperatures.

LITERATURE CITED

1. Rempe, J.L., Thinnes, G.L., and C.M. Allison, *Light Water Reactor Lower Head Failure Analyses*, NUREG/CR-5642 (Draft), EGG-2618 (December 1990).

2. Ostensen, R.W. and J.F. Jackson, "Extended Fuel Motion Study," ANL-RDP-18 (July 1973).

3. Ostensen, R.W. et al., "Fuel Flow and Freezing in the Upper Subassembly Structure Following an LMFBR Disassembly," *Trans. Am. Nucl. Soc.*, **18**, 214 (1974).

4. Epstein, M. et al., "Transient Freezing of Flowing Ceramic Fuel in a Steel Channel," *Nucl. Sci. Engr.*, **61**, 310-323 (1976).

5. Epstein, M., "Heat Conduction in the UO_2-Cladding Composite Body with Simultaneous Solidification and Melting," *Nucl. Sci. Engr.*, **51**, 84-87, (1973).

6. Best, F.R., Wayne, D. and C. Erdman, "A Fuel Freezing Model for Liquid-Metal Fast Breeder Reactor Hypothetical Core Disruptive Accidents", *Nucl. Sci. and Eng.*, **89** (1985).

7. Cheung, F.B. and L. Baker, Jr., "Transient Freezing of Liquids in Tube Flow," *Nucl. Sci. and Eng.*, **60**, 1-9 (1976).

8. Chambers, R.S., *A Finite Element Analysis of a Pressurized Containment Vessel During Core Melt Down*, SAND87-2183 (1987).

9. Anderson, L.E. et al., *Effects of a Hypothetical Core Melt Accident on a PWR Vessel with Top-Entry Instruments*, EPRI, IDCOR Program Technical Report 15.2A (June 1983).

10. Young, W.C., *Formulas for Stress and Strain*, 6th ed., McGraw-Hill, New York (1989).

11. Cook, R.D. and W.C. Young, *Advanced Mechanics of Materials*, Macmillan (1985).

THERMAL ANALYSIS OF A REACTOR LOWER HEAD WITH CORE RELOCATION AND EXTERNAL BOILING HEAT TRANSFER

J.E. O'Brien and G.L. Hawkes ■ Idaho National Engineering Laboratory, EG&G Idaho, Inc., P.O.. Box 1625, Idaho Falls, ID 83415

A thermal analysis has been performed to assess the viability of external water flooding as a cooling strategy to prevent reactor vessel thermal failure in the case of a severe accident with partial core melting and core relocation to the reactor vessel lower head. Vessel wall temperatures and heat fluxes were predicted over a range of decay heating values using a one-dimensional heat conduction model in conjunction with literature correlations for the local heat flux distributions associated with natural convection inside the molten debris pool. A finite-element prediction of flow and heat transfer inside the internally heated molten pool was also obtained. Vessel wall temperature results were highly dependent on the distribution chosen to represent heat transfer along the lower surface of the melt region. The literature correlations, which were obtained for low Rayleigh number laminar natural convection in two-dimensional rather than axisymmetric geometries predicted much higher heat fluxes for the upper melt surface compared to the lower surface. The finite-element computational fluid dynamic predictions, obtained for turbulent natural convection indicated similar heat flux levels for the upper and lower surfaces. Vessel wall temperature predictions were satisfactory when the literature heat flux correlations were used, but inside vessel wall temperatures in excess of the melting point were indicated when the finite-element heat flux distribution was applied.

INTRODUCTION

Under the assumption of a complete loss-of-coolant accident in a PWR with no emergency cooling, core meltdown and core relocation to the reactor vessel lower head are possible. One strategy currently under study to reduce the possibility of reactor vessel failure under these conditions is to introduce water around the outside of the reactor vessel as a heat sink in order to maintain the vessel wall at a temperature low enough to prevent thermal failure.

After core relocation, a molten pool consisting primarily of a mixture of ZrO_2 and UO_2 (corium) is formed in the reactor vessel lower head. A crust of solid material forms around the periphery of the pool, but internal heat generation resulting from the radioactive decay of fission products assures that most of the pool remains molten and, in fact, undergoes significant internal natural convection. The details of this natural convection process are important inasmuch as they provide the local heat flux distribution around the inside surface of the melt. This distribution, in conjunction with the thermal boundary conditions imposed on the outer crust surface, determines the fraction of the total heat dissipation that is transferred through the upper crust to the inside of the reactor vessel (primarily by radiation) and the fraction which must be conducted through the wall of the reactor vessel lower head.

Relatively few experimental studies of natural convection in internally heated fluids have been performed and these have been mostly limited to two-dimensional geometries. The most relevant studies for the purposes of the present analysis are those involving either two-dimensional semi-circular or hemispherical geometries. Jahn and Reineke [1] carried out experiments with the semi-circular trough geometry in a very thin apparatus (raising questions about the two-dimensionality of the flow) using holographic interferometry for observations of isotherms and heat flux. Results of this study included average and local heat transfer distributions for the lower (curved) surface and comparison to numerical predictions in the Rayleigh number range of 10^7 - 10^{11}. A related numerical study of natural convection with internal heat generation in various isothermal two-dimensional enclosures was carried out by Mayinger et al. [2] for a Rayleigh number range of 10^4 - 10^9. Results of that study have been used to formulate correlations for the average heat transfer to the upper and lower surfaces of a semi-circular trough [3]. These correlations have been applied, for lack of better information, to hemispherical geometries and much higher Rayleigh numbers.

An experimental study of heat transfer with internal heat generation in a hemispherical enclosure was performed by Gabor et al. [4]. As with previous studies, this experiment utilized ohmic heating of the fluid to provide the internal heat generation. It is impossible, however, to achieve uniform volumetric heating of the fluid in the hemispherical geometry since the electric flux lines are not parallel. Average heat transfer was measured by performing an energy balance on the cooling water used to maintain the isothermal-wall boundary condition. Results for Nusselt number were found to lie below the correlation of Mayinger et al. [2].

The objective of the present study is to make use of the limited information available on natural convection heat transfer in hemispherical internally heated pools in order to estimate the reactor vessel wall temperatures to be encountered in the case of a severe accident with partial core melting and core relocation to the reactor vessel lower head. Water is assumed to have been introduced in the region outside the reactor vessel to act as a heat sink. Two predictive strategies have been used. The first is a one-dimensional heat conduction model suggested by Epstein and Fauske [3] which makes use of the available natural convection correlations for the heat transfer boundary conditions on the inside surface of the corium crust. This one-dimensional approach is useful when the water level outside the reactor vessel is assumed to be higher than the upper corium surface. If the water level is below the upper corium surface, heat conduction through the vessel wall is highly two-dimensional in the vicinity of the upper corium surface. A finite-element heat conduction model was used to predict vessel wall temperatures for this case.

Numerical predictions of the natural convection fluid flow inside the hemispherical melt were also made using commercial finite-element CFD software. The heat flux distribution predicted for the inside surface of the corium crust will be compared to previous numerical predictions. Implications of the uncertainty in knowledge concerning this local distribution will also be discussed.

PROBLEM DEFINITION

The steady-state thermal analysis presented herein is based on the assumption of a complete reactor vessel loss-of-coolant accident. A 75% core meltdown is assumed with relocation of the molten core material to the vessel lower head. Transient heat conduction effects are not modeled. A schematic representation of the thermal analysis problem addressed in this study is presented in Fig. 1. Uniform volumetric decay heat dissipation is assumed in the displaced core. The calculations have been performed over a range of volumetric heat generation rates centered around 1 MW/m^3, which corresponds to about 0.75% of the core power rating for the displaced volume. A mixture of steam and hydrogen gas at 380 K and 1 atm pressure is assumed to be circulating by free convection inside the reactor vessel. The radiant temperature of the inside of the vessel is also assumed to be 380 K. Final vessel wall temperature results are highly insensitive to the number assumed for vessel internal gas and radiant temperatures up to at least 1000 K. Liquid water has been introduced around the exterior of the reactor vessel. Two cases were considered, corresponding to two different possible water depths. The high

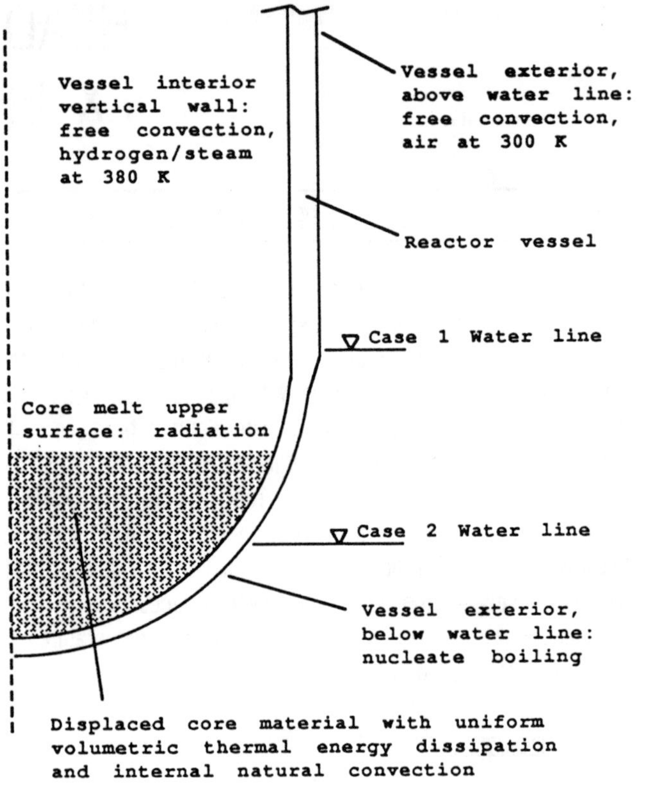

Figure 1. Schematic of lower head thermal analysis model.

water level case assumes a liquid level 0.3048 m (1 ft.) above the transition of the vessel from the bottom head to the cylinder. The low water level case assumes a water level 0.762 m (2.5 ft.) above the bottom of the vessel. Below the water level, external heat transfer is assumed to occur by nucleate pool boiling. The total thermal energy dissipation rate was low enough and the wetted surface area large enough that the critical heat flux could not be achieved. In fact, all heat fluxes of interest were at least a factor of three lower than the pool boiling critical heat flux for water. Consequently, film boiling was not considered in this analysis. Above the water level, heat transfer occurs by free convection to air at 300 K.

The mass of relocated core material was calculated by assuming that 75% of the 69,230 kg of UO_2 in the original core was relocated. In addition, ZrO_2 was added in the amount of 1 mole ZrO_2 per 1.37 mole UO_2. The total mass of the debris bed was therefore 69,217 kg. Thermal properties (density, thermal conductivity, viscosity) of the UO_2/ZrO_2 compound were calculated in accordance with NUREG/CR 5273 [5] for both the solid and liquid phases. The volume (and height in the lower head) of the relocated core materials was determined from the known mass and density values. Since the core materials remain mostly molten, the liquid-phase density (7589 kg/m^3) was used in calculating the melt volume. This volume was such that the low water line was 0.732 m lower than the top surface of the core melt materials. The high water line was 0.963 m above the top surface of the core melt materials. The shape of the melt volume was that of a hemispherical segment, with a height equal to 67% of the lower head inside radius.

ONE-DIMENSIONAL MODEL

Heat conduction through the vessel wall is essentially one-dimensional for the high water level. Therefore, for this case, calculations based on the one-dimensional model suggested in [3] were performed. Required boundary conditions for the model were the heat flux distribution on the inside surface of the corium crust, the heat flux due to nucleate boiling on the external surface of the vessel lower head, and the radiative heat flux at the upper surface of the melt region. Model predictions included the thickness of the solid corium crust, outer crust heat fluxes, and vessel wall temperatures.

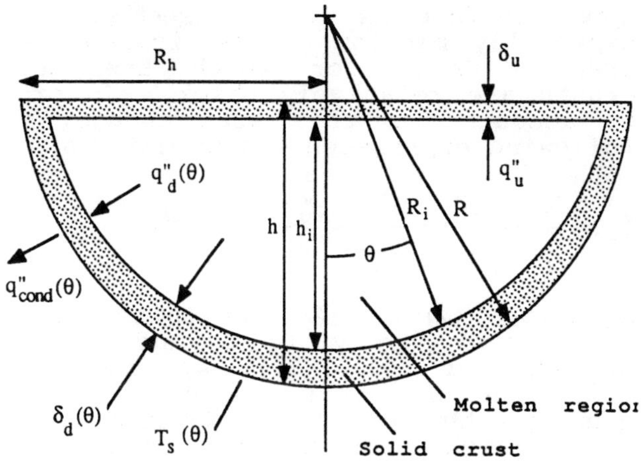

Figure 2. Core melt geometry.

A schematic of the core melt geometry is presented in Fig. 2. It has been prepared to qualitatively show the anticipated local crust thickness distribution, with a relatively thin solid corium crust along the upper surface and a decreasing crust thickness along the lower hemispherical surface for increasing angular distance from the bottom of the vessel.

Formulation of the first one-dimensional model proceeds as follows. For lack of better information, the average heat flux correlations recommended in [3] based on the data of [2] for the semi-circular geometry and relatively low Rayleigh numbers are assumed to apply to the hemispherical-segment geometry of the present problem. Consider first the upper corium surface. An examination of heat conduction through the crust (with volumetric heat generation) requires that

$$\frac{\dot{q}\,\delta_u^2}{2k} + \frac{q''_u\,\delta_u}{k} = T_m - T_s \qquad (1)$$

where q''_u is the heat flux on the inside surface of the upper crust, T_m is the melting point temperature for the corium (assumed to be 2800 K), and T_s is the outer crust surface temperature. From an overall energy balance for the crust,

$$q''_u + \dot{q}\delta_u = q''_{rad} = \sigma\varepsilon(T_s^4 - T_{surr}^4) \qquad (2)$$

where δ_u is the solid crust thickness and natural convection from the upper corium surface has been neglected. Equation (2) assumes a constant area through the thickness of the upper crust. In reality, the area of the upper surface is larger (see Fig. 2), especially for low volumetric heat generation rates for which the solid crusts are relatively thick. In order to account for this area change, the following expression was used for δ_u:

$$\delta_u = \frac{q''_{rad} R_h^2 - q''_u (R_h - \delta_d)^2}{\dot{q}/3 \left[(R_h - \delta_d)^2 + R_h^2 + R_h (R_h - \delta_d)\right]} \quad (2a)$$

Equations (1) and (2) can be solved simultaneously to yield T_s and δ_u. Of course, for Eqn. (2a), δ_d, the lower crust thickness is needed. Consequently, the equations describing the crust thicknesses and heat fluxes for the upper and lower surfaces are coupled. The average heat flux on the inside surface of the upper corium crust, q''_u, is obtained from a correlation based on [2] for a two-dimensional semi-circular geometry:

$$q''_u = \frac{k\Delta T}{R} (0.36\ Ra^{.23}) \quad (3)$$

where the Rayleigh number, Ra, is based on the volumetric heat generation rate:

$$Ra = \frac{g\beta\dot{q}R_m^5}{\alpha\nu k} \quad (4)$$

An average radius, $R_m = (R + h)/2$, has been used for the hemispherical segment characteristic dimension. A corresponding expression may be written for the average heat flux on the inside surface of the lower corium crust:

$$\overline{q''_d} = \frac{k\Delta T}{R} (0.54\ Ra^{.18}) \quad (5)$$

The characteristic melt temperature difference, ΔT, can be obtained from an overall energy balance for the molten portion only, using equations (3) and (5) for the upper and lower heat fluxes. For the hemispherical segment, this energy balance yields

$$\Delta T = \frac{\dot{q}/k \left[h_i/3 \left(3R_i - h_i\right)\right]}{(2R_i - h_i)/R_i \left(.36Ra^{.23} + 1.08Ra^{.18}\right)} \quad (6)$$

where R_i and h_i are the radius and height of the molten region (see Fig. 2), which depend on the magnitude of the volumetric heat generation rate. Use of equation (6) for the characteristic melt ΔT guarantees energy conservation in the molten region. Note that calculation of h_i requires both crust thickness values, indicating once again that the equations for the upper and lower crusts are coupled.

Attention will now be focused on the lower crust. Consideration of radial heat conduction (actually it's not exactly radial for the hemispherical segment) with volumetric heat generation in the lower crust leads to a cubic equation for the inner crust radius:

$$\frac{\dot{q}}{k}\left[\frac{(R^2 - 3R_i^2)}{6} + \frac{R_i^3}{3R}\right] + \frac{q''_d R_i}{k}\left(1 - \frac{R_i}{R}\right) = T_m - T_s \quad (7)$$

where the crust thickness is given by $R - R_i$. This equation is analogous to Eqn. (1) for the plane slab. As it turns out, results obtained using Eqn. (1) were found to be sufficiently close to the results obtained using (7) that (1) was used subsequently. An expression analogous to Eqn. (2) can also be derived for the lower crust:

$$q''_d + \frac{\dot{q}}{6R_i\ h_i}\left[h^2(3R - h) - h_i^2(3R_i - h_i)\right] = q''_{cond}\ \frac{R\ h}{R_i h_i} \quad (8)$$

There are three unknowns in Eqns. (7) and (8): T_s, R_i, and q''_{cond}. An additional equation is available for the temperature drop through the vessel wall thickness:

$$T_s - T_o = \frac{q''_{boil}\ R_o h_o}{(h - R)\ k_v} \ln\left[\frac{h}{R}\ \frac{R_o}{(h - R) + R_o}\right] \quad (9)$$

where R_o is the outer vessel radius. The system of equations is closed by using the Rohsenow [6] pool boiling correlation for the nucleate boiling heat flux, q''_{boil}. The system of nine equations was solved simultaneously using a computer program specially developed for this task. Overall energy conservation was verified using:

$$\dot{q}V = q''_{rad}\left(\pi R_h^2\right) + \overline{q''_{cond}}\ (2\pi Rh) \quad (10)$$

where $\overline{q''_{cond}}$ is the average outer heat flux along the lower crust.

A second one-dimensional model was developed for estimating local heat fluxes and temperatures along the outer corium surface in contact with the inside surface of the reactor vessel lower head. This model made use of the results presented in Fig. 7 of [1] for the local heat flux distribution along the lower (hemispherical) surface of the molten region. Since these results were obtained for a semi-circular trough, care must be taken in applying this distribution to the present hemispherical geometry. In particular, for the semi-circular geometry, the area-weighted average heat flux is the same as the numerical average. For the hemispherical segment, however, the relationship between the area-average heat flux, $\overline{q}''_d$, and the numerical average, $\overline{q}''_N$, is given by

$$\overline{q}''_d = \frac{1}{h} \sum_{i=1}^{N} r_{qi} \overline{q}''_N \left(\Delta h_i\right) \qquad (11)$$

where r_{qi} is the ratio of local to average heat flux provided by the Jahn and Reineke [1] figure (assumed to apply here) and Δh_i is the vertical height of the hemispherical segment associated with each local heat flux value. Since the highest local heat flux values occur near the top of the molten hemispherical region where Δh_i is also the highest, the area-average heat flux is higher than the numerical average. Using 10° angular increments and values of r_{qi} from [1], a value of 0.75 is obtained for $\overline{q}''_N/\overline{q}''_d$ from Eqn. (11). Local heat fluxes for input into the one-dimensional model were then obtained from

$$q''_{di} = .75\ r_{qi}\ \overline{q}''_d \qquad (12)$$

The remainder of the local model was similar to the average model described previously. Upper crust thickness values and lower crust average heat fluxes were supplied to this model from the average model for the volumetric heat generation rate under consideration. Energy closure was verified along the outer crust from

$$\overline{q}''_{cond} = \frac{1}{h} \sum_{i=1}^{N} q''_{cond_i}\, \Delta h_i \qquad (13)$$

FINITE-ELEMENT MODELS

For the low-water-level case, heat conduction in the vessel wall above the water line is highly two-dimensional. A finite element heat conduction model was developed using a commercial program (FIDAP; [7]) to calculate wall temperatures in this region. The finite-element grid used for these calculations included 1888 quadrilateral 4-node elements. Boundary conditions were applied as indicated in Fig. 1. For the purposes of these calculations, the corium pool was replaced by a uniform heat flux boundary condition applied along the inside surface of the vessel wall to a height equal to the top corium surface. Free convection to the gas mixture inside the reactor vessel is assumed above this point.

A finite-element model was also developed for predicting the natural convection flow and heat transfer in the molten corium. This effort was motivated by the fact that the available heat transfer correlations for natural convection with internal heat generation used in the one-dimensional model were applied well outside of their Rayleigh-number range. Furthermore, they were developed for two-dimensional rather than axisymmetric geometries.

The Rayleigh number for natural convection flows with internal heat generation, defined in Eqn. (4), is on the order of 10^{16} for the present molten core problem with $\dot{q}$ = 1 MW/m^3. This Rayleigh number is sufficiently high to raise the question of whether the natural convection in the molten region will be turbulent. The Rayleigh number of Eqn. (4) can be related to the "standard" Rayleigh number by dividing it by the Dammköhler number, Da:

$$Ra_{std} = \frac{g\beta L^3 \Delta T}{\nu \alpha} = Ra \div Da = \left(\frac{\dot{q} g\beta L^5}{\nu \alpha k}\right) \div \left(\frac{\dot{q} L^2}{k \Delta T}\right) \qquad (14)$$

Turbulent natural convection is expected for $Ra_{std} > \sim 10^9$. For the hemispherical segment of molten corium, using R_m as the characteristic dimension, $Ra_{std} = 1.9 \times 10^{10}\ \Delta T$. The characteristic temperature difference, ΔT, can be estimated as a function of $\dot{q}$ using Eqn. (6). For $\dot{q} = 10^6$ W/m^3, $\Delta T \sim 200$ °C. Therefore, $Ra_{std} \sim 3.8 \times 10^{12} \gg 10^9$. Consequently, turbulent natural convection is expected. Of course, Eqn. (6) is not necessarily valid since it was obtained for much lower Rayleigh numbers. Heat transfer coefficients are expected to be much higher for the high Rayleigh number turbulent situation, and therefore characteristic temperature differences will be much lower for a specified $\dot{q}$. But, even if ΔT is only on the order of 10, $Ra_{std} \sim 1.9 \times 10^{11}$, still much greater than 10^9.

The finite-element mesh used for calculating the natural convection flow and heat transfer inside the molten corium pool is shown in Fig. 3. The mesh resolution is highest near the walls, where the gradients are highest. In addition, for turbulent flows, FIDAP employs wall elements in which specialized shape functions (based on "universal" near-wall profiles) are used to accurately capture the sharp variations of the mean flow variables in the near-wall region [7]. A standard (high Reynolds number) k-ε turbulence model was used for predicting eddy viscosity away from the walls. The model included 2820 two-dimensional 4-node isoparametric quadrilateral elements in which velocities and temperatures are approximated using bilinear interpolation. Pressure discretization was accom-

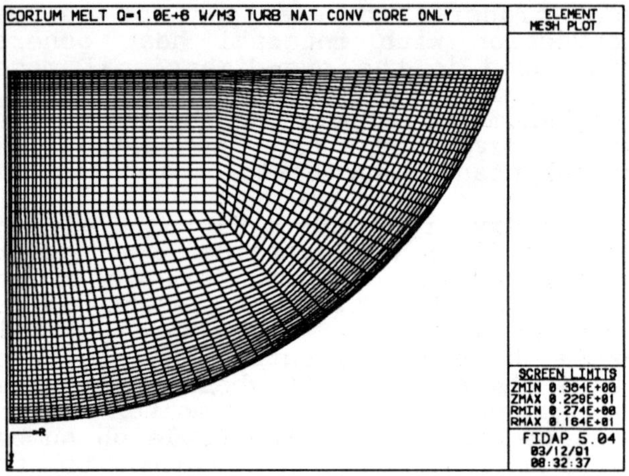

Figure 3. Finite element mesh, molten pool natural convection simulation.

plished by the piecewise-constant discontinuous pressure approximation. The penalty approach was implemented with a penalty parameter value of 10^{-6}. The FEM system of equations was solved using the segregated algorithm. A convergence tolerance of 0.01% was used for the normalized change in the solution vectors.

Solid boundaries at 2800 K were assumed for the top and and side boundaries of the axisymmetric finite-element domain. These boundaries correspond to the inner surface of the solid corium crust. A symmetry boundary is assumed at the centerline. The calculations were performed with $\dot{q}$ = 1 MW/m^3.

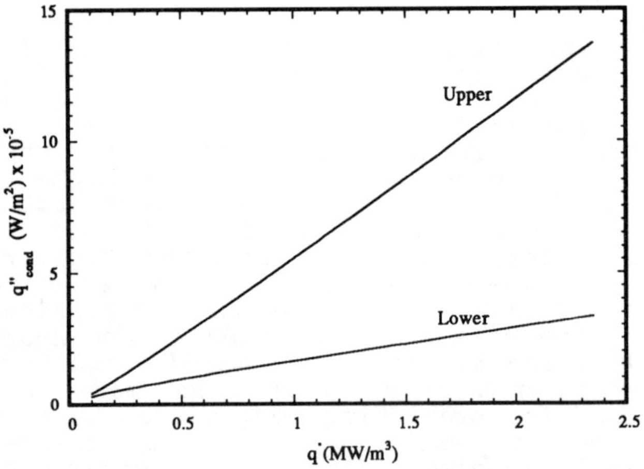

Figure 4. Outer crust average heat fluxes.

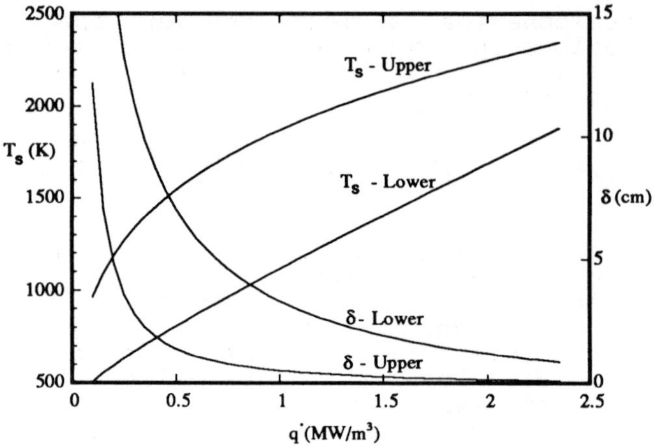

Figure 5. Outer crust average surface temperatures and crust thicknesses.

RESULTS

Results of the one-dimensional model will be presented first. Average heat flux values for the outer surfaces of the upper and lower crusts are presented in Fig. 4 as a function of volumetric heat generation rate. These fluxes are somewhat higher than the corresponding fluxes on the inside crust surfaces since heat generation is occurring through the thickness of the solid crust. Heat flux levels along the upper crust are much higher than along the lower crust, reflecting the use of equations (3) and (5) for the inside crust surface heat fluxes. This allocation of heat transfer in favor of the upper surface is very beneficial in terms of limiting the inside vessel wall temperature. Outer crust temperatures are presented in Fig. 5 as a function of volumetric heat generation rate. The upper crust outside surface is radiating to the inside of the reactor vessel. The lower crust outer surface is in contact with the inside of the reactor vessel. Consequently, concerns about wall thinning and possible creep rupture will be raised when this lower crust outer temperature approaches the vessel melting point, which is around 1700 K. With reference to Fig. 5, this average temperature level is reached only for volumetric heating values in excess of 2 MW/m^3. Of course, there is a very large temperature drop across the vessel wall thickness since the outside of the vessel is in contact with liquid water at atmospheric pressure undergoing nucleate boiling. Average crust thicknesses are also presented in Fig. 5. Lower crust thicknesses are larger than upper crust

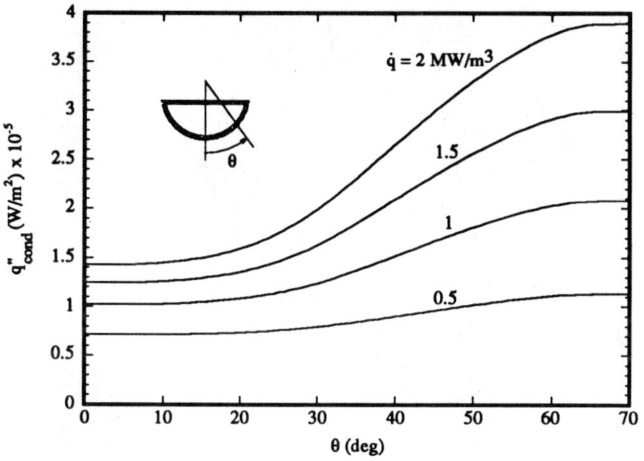

Figure 6. Local heat flux distributions, outer surface of lower crust.

values, again due to the allocation of upper and lower heat fluxes provided by the correlations used.

Outer crust heat flux distributions obtained from local application of the one-dimensional model along the lower surface of the melt region are presented in Fig. 6, with $\dot{q}$ as a parameter. These distributions reflect the adoption of the local data [1] discussed previously. Heat flux values increase monotonically from the bottom of the reactor vessel to a peak value in the corner region near the upper corium crust. Outer crust heat fluxes, q''_{cond}, and inner crust heat fluxes, q''_d, are shown together on Fig 7. for a volumetric heating rate of 1 MW/m^3. The outer heat flux is larger due to internal heat generation through the thickness of the crust. Consequently, the largest

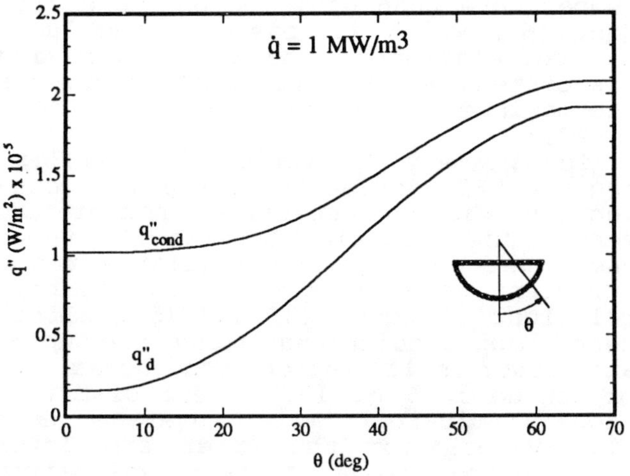

Figure 7. Local heat flux distributions, lower crust.

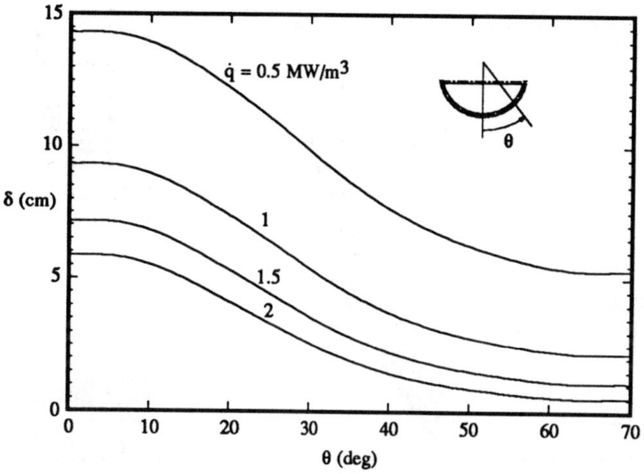

Figure 8. Local lower crust thickness.

difference between the inner and outer heat flux occurs at the bottom of the reactor vessel where the heat fluxes are lowest and the crust thicknesses are greatest. Crust thickness distributions are presented in Fig. 8. Even for the lowest volumetric heating value considered, the crust thickness represents a relatively small fraction of the hemispherical segment radius. Therefore, the geometric departure of the molten region from the basic hemispherical-segment geometry is small.

Local outer crust temperatures predicted from the one-dimensional model are presented in Fig. 9. Maximum temperatures are observed in the corner region near the upper corium crust, as expected. These maximum temperatures are much higher than their counterparts from the average analysis (see Fig. 5). For example, for $\dot{q}$ = 1 MW/m^3 (the

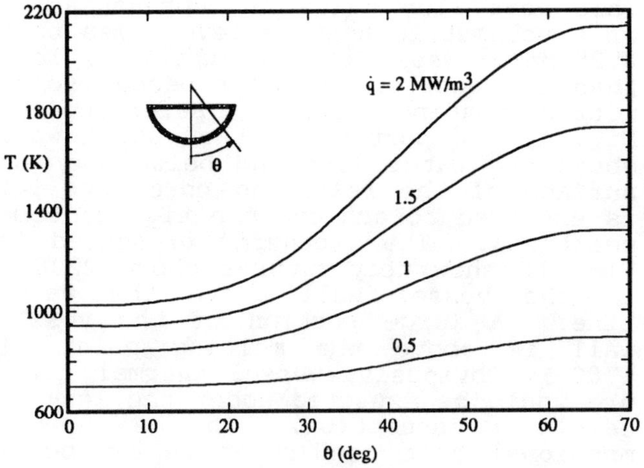

Figure 9. Local temperature distributions, outer surface of lower crust.

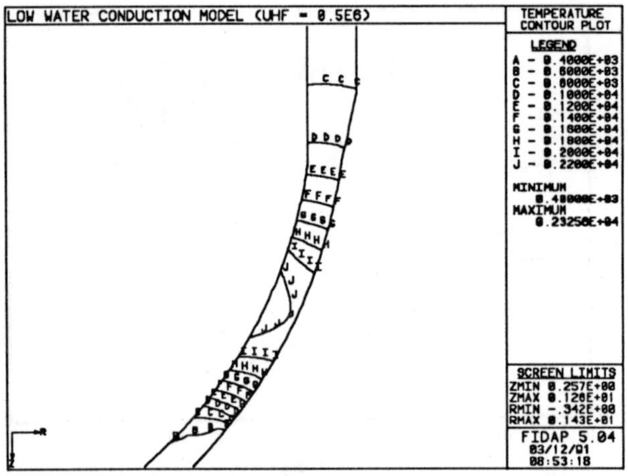

Figure 10. Isotherms in vessel wall, low water level.

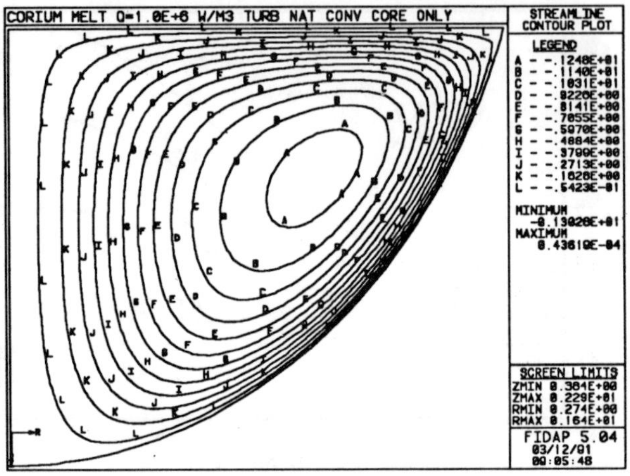

Figure 11. Streamlines, molten pool turbulent natural convection.

approximate expected volumetric heat generation rate for this severe accident scenario) the maximum temperature is 1320 K, whereas the lower-crust temperature obtained from the average analysis was only 1100 K. A maximum vessel inner wall temperature in excess of 1700 K occurs for $\dot{q}$ = 1.5 MW/m^3. Therefore, concerns about possible thermal failure of the vessel wall will be raised for significantly lower values of $\dot{q}$ than were indicated by the average analysis.

Results of the finite-element model constructed to analyze the two-dimensional heat conduction associated with the low-water level case are presented in Fig 10. The results presented in Fig. 10 correspond to a uniform heat flux of 5 x 10^4 W/m^2 applied along the inside surface of the vessel wall to a height equal to the top of the corium. This heat flux value is representative of a volumetric heating level less than 0.25 MW/m^3 (see Fig. 4), which is less than one quarter of the expected volumetric heating level. Referring to Fig. 1, the portion of the vessel wall above the water line and below the top surface of the relocated core material is expected to achieve the highest temperatures. The isotherms presented in Fig. 10 show temperatures above 2200 K in the vessel wall above the water line. A large section of the vessel wall is above the melting point of 1700 K. Obviously, vessel thermal failure would be expected under the low-water-level assumption. In the one-dimensional heat conduction region below the water line, wall temperatures were all below 550 K for this inside-wall heat flux value.

Results of the finite-element turbulent natural convection calculations for the molten corium are presented in Figs. 11 and 12 for $\dot{q}$ = 1 MW/m^3. Streamlines shown in Fig. 11 indicate the presence of a single large convection cell for turbulent natural convection, with upflow near the centerline and downflow along the hemispherical boundary. Laminar calculations presented in [2] predicted multiple convection cells. Laminar calculations with FIDAP also indicated the formation of multiple convection cells. The highest predicted velocities occur in the upward direction about halfway up the centerline, with high velocities also occurring along the lower hemispherical surface.

The maximum predicted temperature in the molten region was only 2812 K. This is a significantly smaller maximum temperature than would be obtained from Eqn. (6), which was based on the laminar correlations. The vigorous mixing associated with turbulence restricts the maximum melt temperature.

The primary motivation for performing the CFD simulation of the melt convection was to determine whether the heat flux allocation between the upper and lower surfaces was similar for the turbulent case as compared to the correlations of Eqns. (3) and (5). Local heat flux predictions along the upper and lower melt region boundaries are presented in Fig. 12. These distributions indicate similar average heat flux values for the upper and lower surfaces whereas, at this volumetric heating level, eqns. (3) and (5) predicted an average upper surface heat

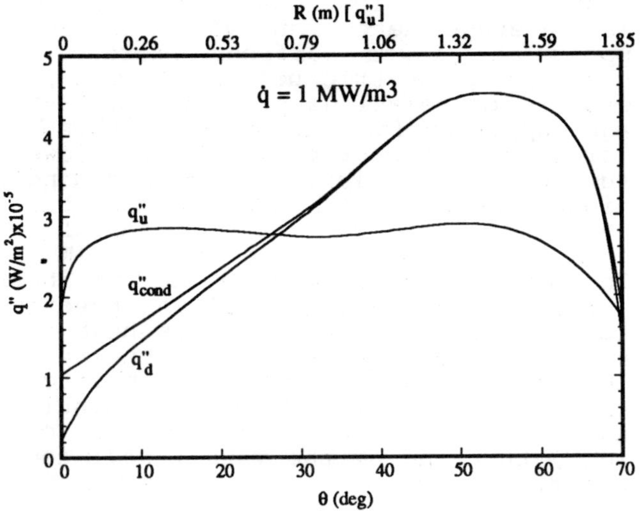

Figure 12. Local heat flux distributions from turbulent natural convection simulation.

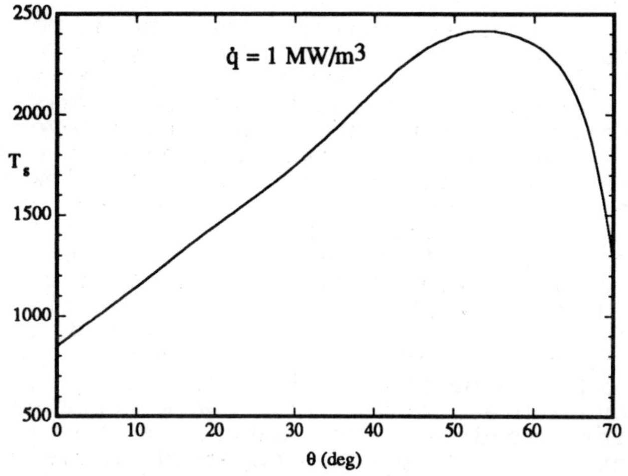

Figure 13. Local lower crust outer surface temperatures using turbulent natural convection simulation.

flux that is larger by a factor of 4 than the lower surface heat flux. This relatively uniform heat flux allocation between the upper and lower surfaces is not unexpected since the mixing associated with vigorous turbulence tends to reduce local heat flux variations compared to laminar or weak turbulent convection. Consequently, according to the turbulent FIDAP simulation, heat flux levels along the lower vessel wall are much larger than those predicted from Eqn. (5). Values for both the inner crust heat flux, q''_d (directly from the FIDAP model) and the outer crust heat flux, q''_{cond} are shown in Fig 12. Values shown in Fig. 12 for q''_{cond} can be directly compared to q''_{cond} values in Fig. 6 (for $\dot{q}$ = 1 MW/m^3). For this volumetric heating level, the peak heat flux shown in Fig. 6 is just over 2 x 10^5 W/m^2. The peak flux in Fig. 12 is over twice as large. The two distributions do have a similar shape, with maximum values occurring near the corner region. The FIDAP simulation predicts a drop in heat flux in the very corner, however.

The lower crust heat flux distribution resulting from the FIDAP turbulent natural convection simulation was provided as input to the locally one-dimensional heat conduction model used previously in order to determine inside vessel wall temperatures. These temperatures are presented in Fig. 13. Outer crust heat fluxes, q''_{cond}, presented in Fig. 12 were also obtained from the one-dimensional model. The peak inside vessel wall temperature resulting from this heat flux distribution is over 2400 K. A large portion of the inner surface of the vessel wall is above 1700 K. Therefore, if the turbulent natural convection heat flux distribution predicted from the FIDAP simulation is correct, thinning of the vessel wall would occur with $\dot{q}$ =1 MW/m^3. To assess possible vessel failure under wall thinning conditions, a stress analysis accounting for the wall temperature distribution and the temperature-dependent material properties must be performed.

Of course, many approximations are included in the finite-element CFD model including a fairly coarse mesh, standard k-ε turbulence model (which may not be at all appropriate for this internally heated natural convection problem), and the use of special wall elements for resolving near-wall behavior. In its favor, the FIDAP prediction does conserve energy on an overall basis and does predict some of the behavior that would be expected for the turbulent situation. Unfortunately, to our knowledge, no other numerical predictions or experiments for turbulent natural convection with a uniformly heated fluid in an isothermal hemispherical enclosure with high Rayleigh numbers exist. This seems to be a research area that requires some additional attention. In the absence of additional research, a reliable prediction of vessel wall temperature distributions under core relocation conditions with external boiling heat transfer cannot be made.

SUMMARY AND CONCLUSIONS

A thermal analysis has been performed to assess the viability of external water flooding as a cooling strategy to prevent reactor vessel thermal failure in the case of a severe accident with partial core melting and relocation to the reactor vessel lower head. Vessel wall temperatures and heat fluxes were predicted over a range of decay heating values using a one-dimensional heat conduction model in conjunction with literature correlations for the local heat flux distributions associated with natural convection inside the molten debris pool. Since the available correlations were obtained for two-dimensional natural convection at much lower-than-expected Rayleigh numbers, a finite-element prediction of flow and heat transfer inside the internally heated molten pool was also obtained for comparison.

Using the literature heat transfer correlations, average heat flux levels along the upper melt surface are much higher than along the lower surface. Accordingly, average inside vessel wall temperatures in excess of 1700 K (approximate vessel melting point) are reached only for volumetric heating values in excess of 2 MW/m^3, which is twice the expected volumetric heating level. Experimental measurements in a two-dimensional semi-circular geometry indicated local heat flux variations along the lower vessel surface which increase monotonically from the bottom of the reactor vessel to a peak value in the corner region near the upper corium crust. Using this local heat flux distribution and a locally applied one-dimensional heat conduction model, maximum predicted temperatures were much higher than their counterparts from the average analysis. A maximum vessel inner wall temperature in excess of 1700 K occurs for $\dot{q}$ = 1.5 MW/m^3.

A finite-element heat conduction analysis was performed to predict vessel wall temperatures for the case in which only partial external flooding occurs, resulting in a water level below the upper surface of the relocated core. For this case, vessel thermal failure would be expected at a volumetric heating level less than one-fourth of the expected level

The finite-element turbulent natural convection simulation predicted the presence of a single large convection cell in the molten region, with upflow near the centerline and downflow along the hemispherical boundary. Predicted heat flux distributions indicated similar average heat flux values for the upper and lower surfaces whereas, at the same volumetric heating level, literature correlations predicted an average upper-surface heat flux that is larger by a factor of 4 than the lower-surface heat flux. Correspondingly higher peak vessel wall temperatures (>2400 K) were therefore predicted for a volumetric heating level of 1 MW/m^3. In fact, a large portion of the inner surface of the vessel wall was predicted to be above 1700 K at this heat generation rate. Therefore, if the turbulent natural convection heat flux distribution predicted from the turbulent natural convection simulation is correct, thinning of the vessel wall could occur.

ACKNOWLEDGEMENT

This work was performed under the auspices of the U.S. Dept. of Energy, DOE Contract No. DE-AC07-76ID01570.

LITERATURE CITED

1. Jahn, M., and Reineke, H. H., "Free Convection Heat Transfer with Internal Heat Sources, Calculations and Measurements," *Proc. 5th Int. Heat Transfer Conf.*, NC 2.8 (1974).

2. Mayinger, F., Jahn, M., Reineke, H. H., and Steinbrenner, V., "Examination of Thermohydraulic Processes and Heat Transfer in a Core Melt," BMFT RS 48/1, Inst. fur Verfahrenstechnik der T.U., Hannover (1976).

3. Epstein, M., and Fauske, H. K., "The Three Mile Island Unit 2 Core Relocation-Heat Transfer and Mechanism," *Nuc. Tech.*, **87**, 1021 (1989).

4. Gabor, J. R., Cassulo, J. C., and Ellison, P. G., "Heat Transfer from Internally Heated Hemispherical Pools," ASME Paper 80-HT-16, ASME/AIChE Natl. Heat Transfer Conference, Orlando., (1980).

5. Hohorst, J.K. et al, "SCDAP/RELAP5/MOD2 Code Manual, Vol. 4: MATPRO-A Library of Material Properties for Light-Water Reactor Accident Analysis," NUREG/CR-5273 (1990).

6. Rohsenow, W. M., "A Method of Correlating Heat Transfer Data for Surface Boiling Liquids," *Trans. ASME*, **74**, 969 (1952).

7. Engelman, M. S., "FIDAP Theoretical Manual," Version 5.0, Fluid Dynamics Intl., Evanston, IL, (1990).

VERIFICATION OF THE KESS CODE SYSTEM USING CORA-EXPERIMENTS

W. Hering, K. Müller, and A. Schatz ■ Institut für Kernenergetik und Energiesysteme, Pfaffenwaldring 31, D-7000 Stuttgart 80, Germany

At the Institut für Kernenergetik und Energiesysteme (IKE) of the University of Stuttgart, Federal Republic of Germany (FRG) the KESS severe fuel damage (SFD) code system is under development, mainly focused on the physical description of the in-vessel phenomena during severe Light Water Reactor (LWR)-accidents. First the included models which describe the main physical processes such as cladding oxidation, mechanical behavior of the fuel rod, eutectic interactions, melt formation, melt release, and material relocation were partly verified using separate effect tests. The verification of the whole code system, however, requires integral tests like the in-pile tests in the French PHEBUS facility at Caderache or the out-of-pile experiments in the CORA facility at the Kernforschungszentrum Karlsruhe (KfK), because the interactions of the physical processes which determine the global bundle behavior have to be considered. To compare the results of the KESS calculation and those of the out-of-pile experiments in the CORA facility an adequate modelling of the specific thermohydraulic conditions is necessary. The results of the calculations are in good agreement with the experimental data over the whole heated length of the test section. This holds in particular for the temperature distribution in the bundle.

INTRODUCTION

At the Institut für Kernenergetik und Energiesysteme of the University of Stuttgart the KESS severe fuel damage code system is under development (1). The main goal is a physical description of the in-vessel phenomena during severe LWR accidents to support accident management measures and development of new reactor concepts. The KESS code is based on a representative concept which describes the physical processes for one rod in detail (micro-level). This detailed description can be considered to be representative for all rods within a given region (macro-level), due to an adequate averaging of the local data, if necessary (1). The modular structure of the code system allows different levels of detail with regard to the description of the physical process (e.g. cladding and shroud oxidation, mechanical behavior of the fuel rod, eutectic interactions, melt formation, melt release, and material relocation (2), so that the most adequate model can be chosen during verification. However, the models which describe the main physical processes such as cladding oxidation, mechanical behavior of the fuel rod, eutectic interactions, melt formation, release, and relocation will be verified using separate effect tests. The verification of the whole code system requires integral tests because the interactions of the single processes have to be considered (3). Besides the in-pile tests in the french PHEBUS facility at Caderache (4). out-of-pile experiments conducted in the CORA facility at the KfK (5). can be used. For code verification using these integral tests detailed analyses have been performed to identify special phenomena resulting from specific test conditions.

ANALYSIS OF THE TEST SECTION

The CORA test section includes the fuel rod bundle with a zircaloy shroud and a high-temperature shield (HTS) for the thermal insulation in radial direction (s. Figure 1).

The standard 25-rod Pressure Water Reactor (PWR)-specific bundle consists of unheated (7 to

9) and heated (16) fuel rod simulators as well as absorber rods (0 to 2). To reduce the radial heat losses, a 20 mm thick ZrO_2 fiber insulation is wrapped around the shroud. Up to ten observatior holes in the shroud allow a detailed documentation of the melt progression and material relocation by video cameras, which are located in the HTS (s. Figure 1) (3).

In test CORA-5 these observation holes were closed by quartz plates, so that no bypass flow could occur in the annular gap (annulus) between the outer surface of the shroud insulation and the inner surface of the HTS.

For temperature measurements thermocouples (WRe5-WRe26) are located in the center holes of the unheated rods as well as on the surfaces of the claddings of the heated rods.

The fluid, composed of argon, as carrier gas, and steam, is injected from the side into the coolant channel through a 27 mm thick tube at the bottom of the heated zone (z = 0.0 m). The inlet flow conditions cause a turbulent zone in the lower third of the bundle. Above 1500 K the exothermal reaction between the steam and the zircaloy (zirconium steam reaction) of the claddings and the shroud leads to a hydrogen production. The release of energy from this reaction causes a temperature escalation which affects melt formation.

The heat source of the out-of-pile experiment is located in the central tungsten pins of the heated rods. Due to the temperature dependency of the specific resistivity of the tungsten, the axial power distribution is temperature dependent. It cannot be measured in the experiment (6).

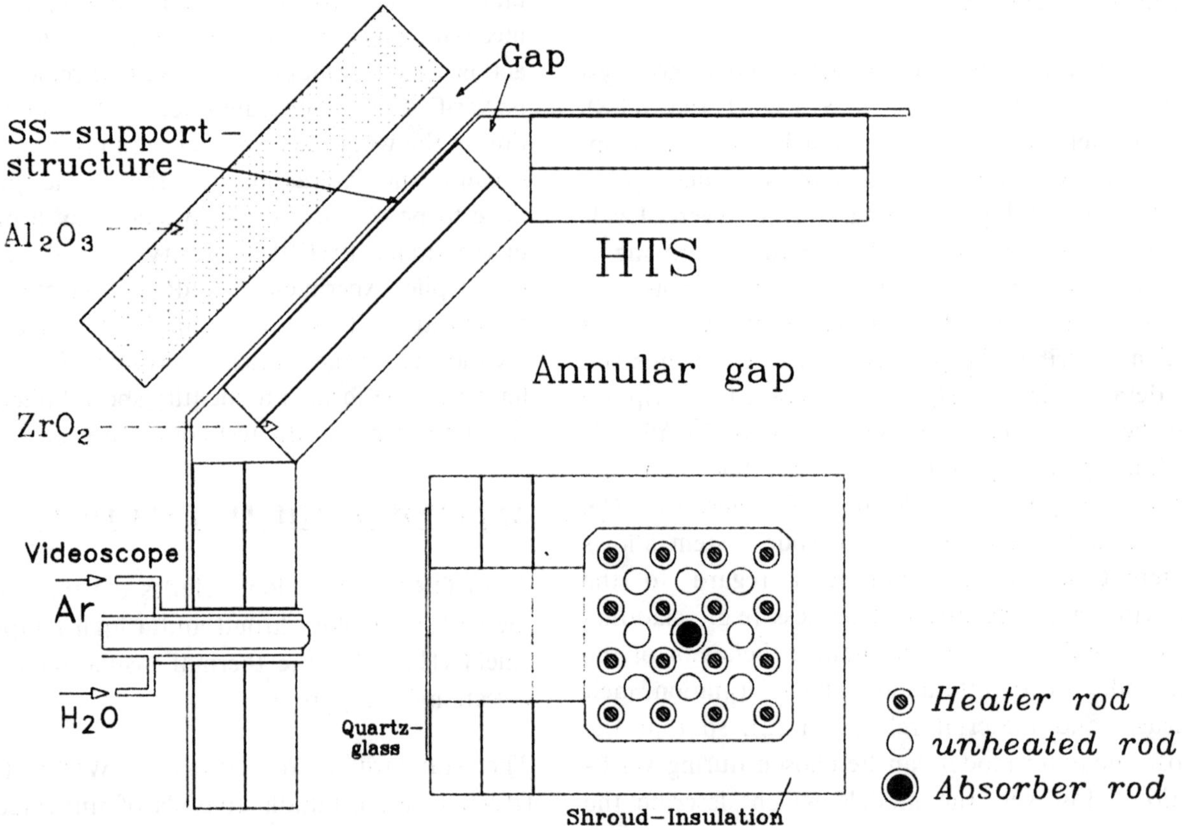

Figure 1. CORA-5 bundle cross section: 16 heated, 9 unheated, and one central absorber rod, surrounded by the shroud with the ZrO_2-insulation, the annular gap, and the HTS including the videoscopes

To recognize such effects and their influence within the whole test section an overview diagram, given in Figure 2, has been developed which combines all informations available during test. The diagram is to support a global test analysis and a comparison between the results of the calculation and the experiment.

In this overview an axial length of 1.2 m of the test section is represented versus time, starting with initial heat up and ending with temperature drop below 1200 K (cool-down phase), so that global as well as local effects can be recognized and analyzed (7).

In the upper part of the diagram the temperatures are given as temperature front lines derived from the thermocouples of the unheated rods (Fuel rod, F, dashed lines). These temperature front lines indicate in vertical direction the axial temperature gradient at any time between 3000 s and 5700 s and in horizontal direction the temperature rise at the twelve elevations given.

The first two lines, the 773 K and 1073 K lines, are also outlined for the heated rods (Simulator, S, dashed-dotted lines) to show the temperature differences measured between heated and unheated rods at low temperatures.

Further, the clad failure derived from the internal pressure history (dotted area) and the damage progression observed as melt release and relocation in the video films (dashed area, (7)) are shown in Figure 2.

The middle part of Figure 2 gives the fluid composition at the bundle entrance. The time delay caused by gas mixing of steam in the carrier gas after 3300 s in the large cavities of the steam generator system ist taken into account. The dashed area represents the volume fraction of hydrogen measured in the off-gas system.

Combining all data of Figure 2, the axial and temporal location of the phenomena observed can be given. The sequence of phenomena, listed in the lower part of the diagram, allows a classification of all PWR-specific CORA tests. Together with the test conditions a selection of tests for verification of the KESS code system can be performed.

ADAPTIONS OF THE KESS CODE SYSTEM

For the calculation using the KESS code system some adaptions had to be made considering the effects listed below:

- Electric heat release, including the temperature-dependent axial power distribution (feedback effect) and the power loss in the extended electrode zones
- Shroud oxidation
- Convective heat transfer between shroud and fluid
- Thermohydraulics with argon as carrier gas, steam and hydrogen
- Turbulent zone in the lower third of the bundle (fluid inlet region)

To simulate the electric heat release, a module HEATROD was developed which calculates the nodal power using the fuel temperature and the total input power measured in the experiment. The heat sources calculated are used as input data for the module HEAT which computes the thermal behavior of the representative rod including all significant phenomena.

The module SHROXI has been included to simulate the oxidation of the shroud.

Furthermore the heat exchange by convection and radiation between the rod and the radial environment including the shroud, the annulus (bypass), and the HTS have been considered in the module UMGEBR.

The thermohydraulic model BOMEL has been extended to handle argon as carrier gas together with steam and the hydrogen produced during the zirconium steam reaction. Moreover, the influence of the fluid injection from the side has been con-

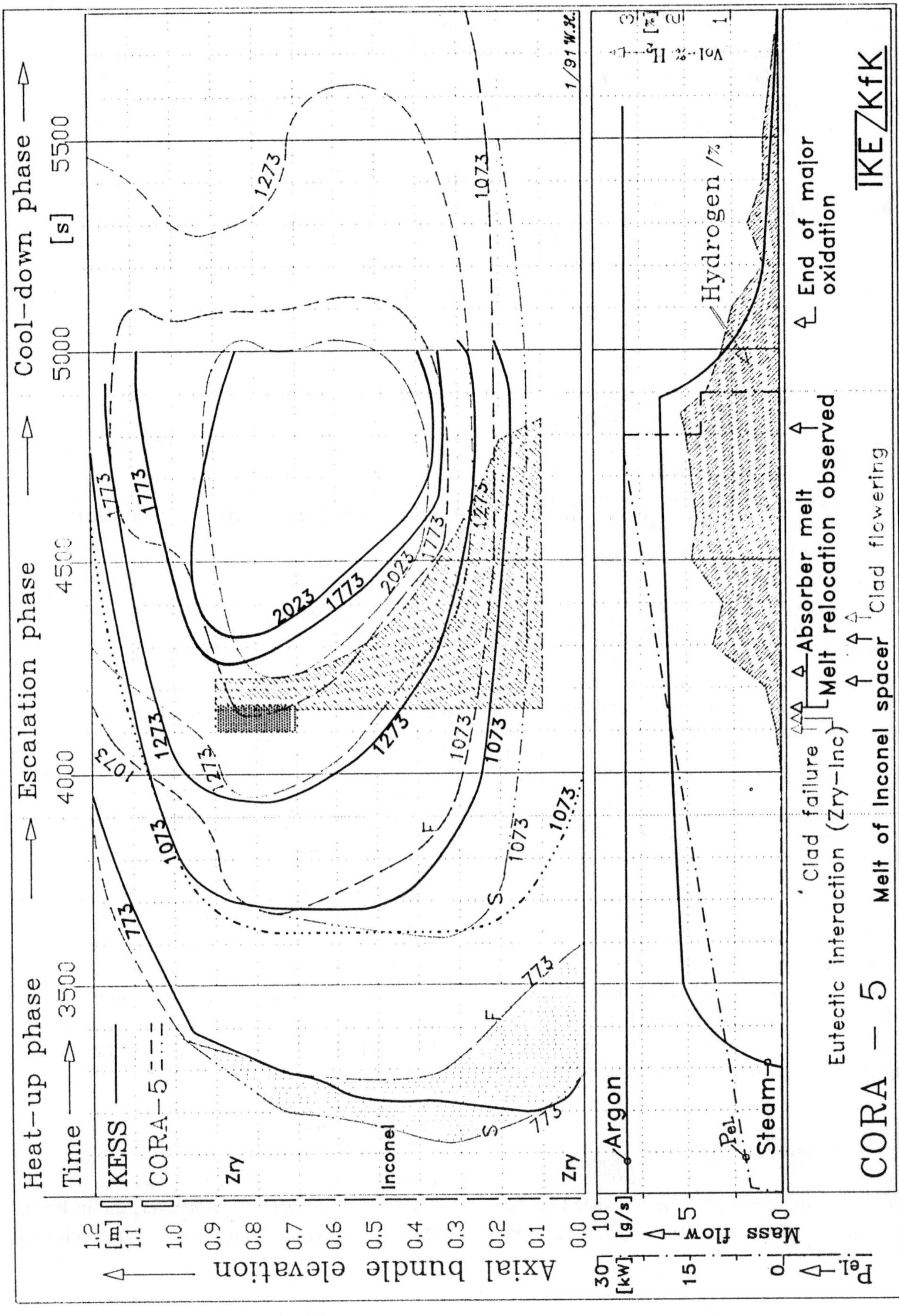

Figure 2. Analysis of test CORA-5: fluid composition (center part), phenomena identified (lower part), and axial temperature profiles (upper part) derived from the experiment (dashed lines) and results of the calculation with the KESS code system (solid lines).

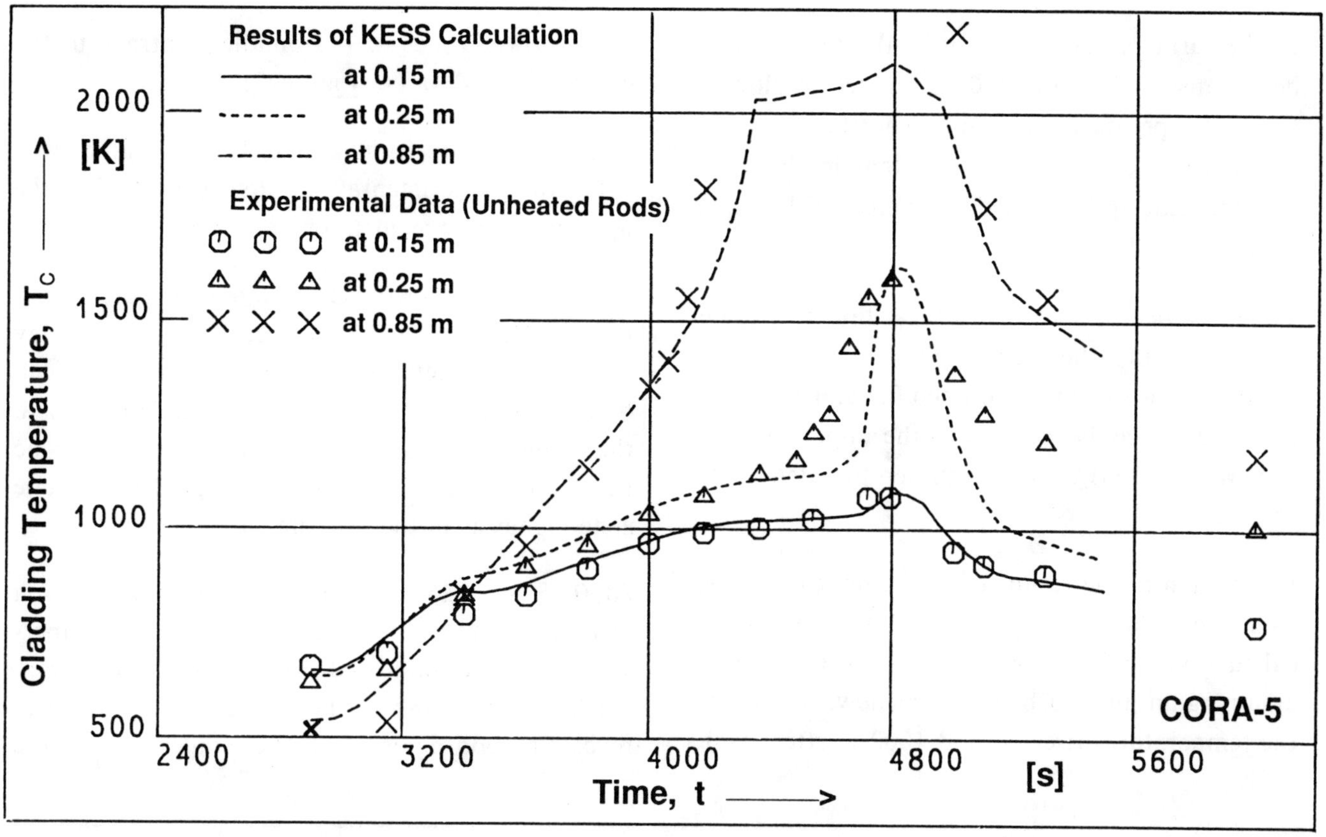

Figure 3. Temperature history of test CORA-5: comparison between the results of the calculation (solid lines) and the CORA measurements (unconnected markers) at 0.15 m, 0.25 m, and 0.85 m bundle elevation

sidered, because the cross flow enhances the heat transfer between fluid and solid structures such as fuel rods and shroud (6). In the CORA-facility this turbulent zone affects the whole bundle because the feed-back of the heating system changes the axial power distribution.

RESULTS OF THE CALCULATIONS

As a first step for code verification calculations were performed using tests CORA-2 (3). and CORA-5, mainly focused on the heat-up and the escalation phase (s. Figure 2).
In CORA-2, a test without absorber rod, the internal fuel rod pressure was high enough to cause ballooning of the claddings which leads to a partial blockage of the coolant channel. The reduced cross section area causes a flow deviation through the observation holes into the annulus (bypass) which influences the bundle behavior significantly. From the experiment only qualitative information can be derived about the flow deviation, whereas the time-dependent behavior remains unknown. Due to these uncertainties in test CORA-2, CORA-5 was chosen for the code verification.

The calculation starts at 2900 s with an axial temperature distribution derived from the experiment. Between 3000 s and 3700 s the increasing steam concentration in the fluid enhances the convective heat transfer. In this time period the influence of the radiative heat transfer between the rods on the cladding temperature is negligible, therefore the 773 K and the 1073 K lines measured show a certain time delay between heated and unheated rods (s. Figure 2). The pertinent temperature front lines obtained by calculation are located between the measured ones, because the behavior of only one representative rod with an averaged power distribution over all rods of the bundle has been considered.

In the turbulent zone (z < 0.3 m), the convective heat transfer is slightly underestimated resulting in higher temperature transients. Therefore the fluid temperature is lower above, so that in the upper half the heat up is delayed compared to the experiment.

The same behavior can be seen at the 1073 K line which indicates the melting point of the absorber material. The deviation between 0.9 and 1.0 m can be explained as heat losses to the upper electrode zone (z > 1.0 m) which are underestimated in this calculation.

In addition to the results of the actual calculations (solid line) the dotted curve gives the results of a calculation which neglects the turbulent zone, assuming an undisturbed laminar flow. In this case the temperatures rises beyond 1000 K after 3900 s in the lower zones of the bundle, contrary to the experimental data (s. Figure 2).

The temperature front line at 1273 K agrees well with the experimental results, except for the uppermost zone as mentioned above.

The comparison of the last two temperature fronts, the 1773 K and the 2023 K line, shows a time delay of about 100 s between 0.3 m and 1.0 m, which will be discussed in detail in the next section. Due to the higher temperature in the upper electrode zone the temperature escalation starts at 0.85 m, one zone higher as in the experiment.

At 0.15 m, 0.25 m, and 0.85 m the comparison given in Figure 3 between the measured temperatures of the unheated fuel rods (unconnected markers) and those of the calculation (solid lines) reveals a good agreement. The temperature at the

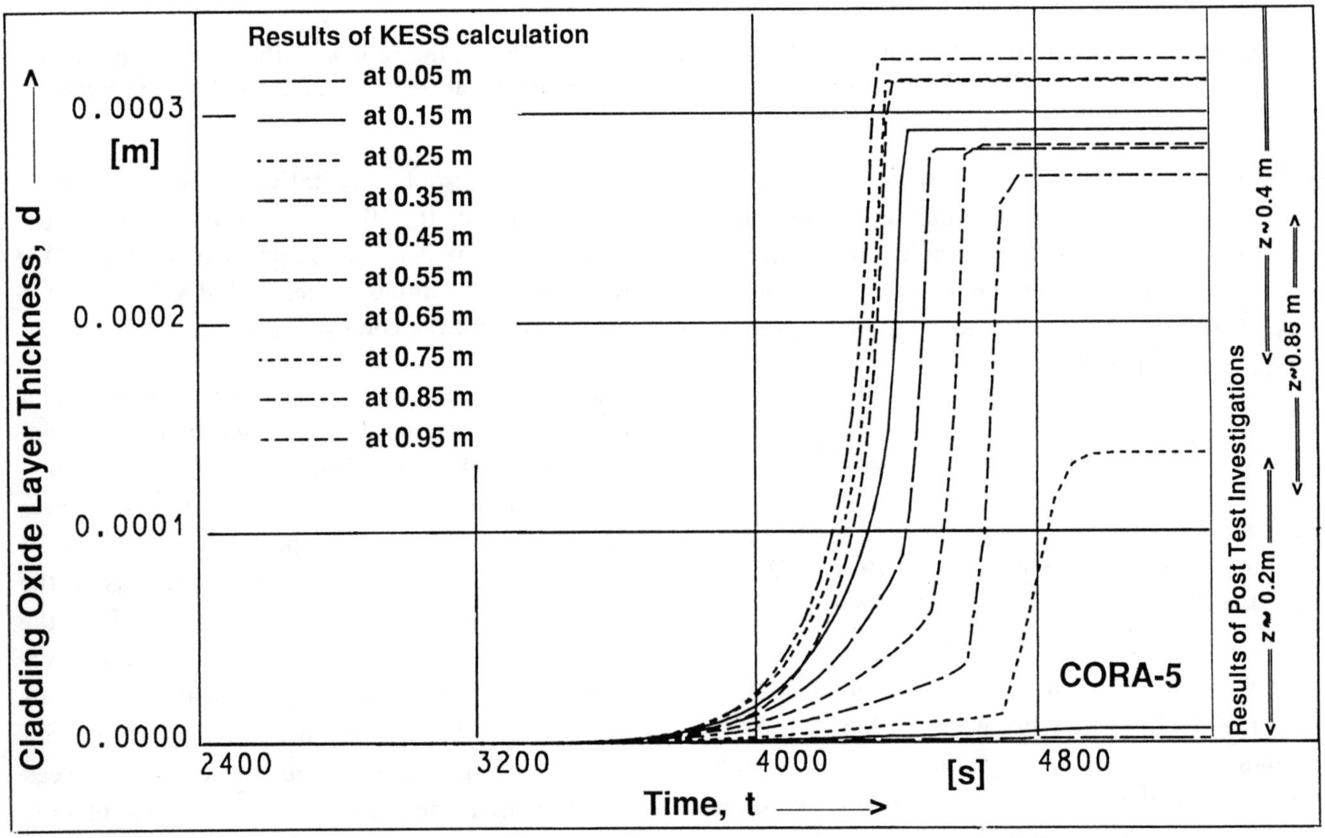

Figure 4. Oxide layer thicknesses (ZrO_2): results of the KESS calculation compared with the data of the post test investigations (right side)

elevation of 0.15 m remains low due to the turbulent flow in this part of the bundle. At 0.25 m the lowest level of solidified melt, the temperature history is dominated by the heat-up due to melt relocation between 4600 s and 4800 s as observed in the experiment. At 0.85 m the temperature escalation due to the zirconium steam reaction is slightly delayed. One possible reason may be that the cladding deformation enlarges the surfaces for oxidation during melt formation and release. Moreover, the temperature measurements may be affected by the failure of the thermocouples, at the melting temperature of β-zircaloy.

In the escalation phase an increase of the hydrogen concentration in the off-gas system was measured after 4000 s (s. middle part of Figure 2), indicating the growth of the oxide scales. For CORA-5 no hydrogen mass flow is available, so that the ZrO_2 layer thicknesses are given in Figure 4. The oxidation in the upper half of the bundle (z > 0.5 m) is stopped by a complete consumption of the zircaloy. No steam starvation was computed at all which was observed in the experiment too (5).

On the right side of Figure 4 the preliminary results of the post-test investigations of CORA-5 are given (9). The arrows indicate the spread of the ZrO_2 layer thicknesses measured by optical microscope at the elevations given. From these measurments the average values were derived to 6.0±1.0 μm at 0.095 m, 54±28 μm at 0.208 m, 323±86 μm at 0.393 m, 309±68 μm at 0.408 m, 424±40 μm at 0.663 m, and 200±57 μm at 0.853 m. No significant difference between heated and unheated rods could be found.

As can be seen, the ZrO_2 layer thicknesses calculated are comparable with the experimental data, except for the upper zones (z > 0.8 m), where higher values were obtained in the calculation.

CONCLUSION

The first step of the verification of the KESS code system has been completed using the integral experiments performed in the CORA facility. CORA-5 has been chosen as a PWR-specific test. A good agreement between calculated and experimental data has been obtained with regard to the melt relocation. The absorber rod was not modelled. The interpretation of the deviations of the temperatures at different bundle elevations show that the adaption of the code to the specific thermal hydraulic conditions of the facility was succesful. A local deviation at the upper end of the heated zone has to be checked further. In the next steps of the KESS code verification the absorber rod model and the unheated rods as well as a more sophisticated relocation model will be applied. Further work has to be done in modelling the low-temperature eutectic reactions leading to early clad failure and melt formation which effects the global bundle behavior, as observed in CORA experiments.

LITERATURE CITED

(1) A. Schatz, M. Bürger, K. Müller, M. Bruder, K.-D. Hocke, E.v. Berg, P. Mayr: KESS a Modular System for the Simulation of Severe LWR Accidents European Simulation Multiconference, Erlangen-Nürnberg, June 10-13, 1990

(2) P. Mayr, M. Bürger, Th. Baas, A. Schatz: Empirische und mechanistische Beschreibung der Zirkon-Wasser-Reaktion im Kern von Leichtwassereaktoren bei hohen Hüllrohrtemperaturen, IKE 2-88, Juli 1989

(3) K. Müller, W. Hering, K. A. Schatz: Verifikation des KESS-III Code Systems anhand des Out-of-pile Experiments CORA-2, IKE 2-91, 1991

(4) E. F. Scott, Ph. Delchambre, P. Hardt: Status of the PHEBUS FP Project, Seminar on the Commision Contribution to Reactor Safety Research, Congress Centre Villa Ponti, Varese, Italy, Nov. 20-24, 1989

(5) S. Hagen, F. Seibert, L. Sepold, P. Hofmann, G. Schanz: Influence of Reflood in the

CORA Severe Fuel Damage Experiments; this conference

(6) W. Hering: Entwicklung eines Simulatormoduls für das CORA-Experiment zum Einsatz in SCDAP, KfK-4493, 1989

(7) W. Hering: Interpretation und Modellierung der Experimentes CORA mit dem erweiterten Kernschmelzcode SCDAP/MOD1, Dissertation IKE, to be published in 1991

(8) W. Hering: Analytical evaluation of CORA experiments using an extended SCDAP version, 26th ASME/AIChE/ANS, Conference on Heat Transfer and Fuel Rod Behavior in Nuclear Accidents, Philadelphia, PA, USA, August 6-9,1989

(9) K. Minato, W. Hering, S. Hagen: Analysis of Cladding Oxidation and Deformation of PWR specific CORA Experiments KfK-4827, to be published in 1991

IGNITION THRESHOLD OF MOLTEN ALUMINUM IN WATER

P. Kuan and B.J. Buescher ■ Idaho National Engineering Laboratory, EG&G Idaho, Inc., P.O. Box 1625, Idaho Falls, ID 83415

Several energetic reactions exist between aluminum and water that may complicate an accident progression in a nuclear reactor with aluminum clad fuel. These are (a) steam explosion, (b) rapid oxidation of aluminum by steam, and (c) spontaneous ignition of molten aluminum in water. This paper focuses on the third reaction, which is likely to occur when molten aluminum is flooded with water. A theoretical treatment is developed that assumes that the ignition is a vapor phase phenomenon, and that the threshold temperature for ignition is the point at which the energy generated from the chemical reaction between aluminum and water is equal to the sum of the energy needed to vaporize the metal and the radiative heat losses from the metal surface. This threshold ignition temperature is determined to be 1684 K, corresponding to a metal vapor pressure of 0.15 mm-Hg. This temperature is in close agreement with the lowest temperature at which spontaneous ignition of aluminum in water has been observed. Similar calculations applied to uranium and zirconium show that their threshold ignition temperatures are 2640 K and 2980 K, respectively. However, for these last two metals, most of the energy generated from the chemical reactions at the ignition threshold would have to come from the diffusion of oxygen into the metal.

Minimum temperatures of energetic reactions between aluminum and water are being reevaluated and updated as one of the tasks to upgrade the safety analysis for the Advanced Test Reactor, a nuclear reactor with aluminum clad fuel at the Idaho National Engineering Laboratory. The purpose of this paper is to re-examine the experimental evidence of the ignition of aluminum in water and to provide a theoretical understanding of the phenomenon.

Apparently, several energetic reactions between aluminum and water exist. First, a steam explosion may occur when molten aluminum is suddenly dispersed upon contact with a solid surface under water (1). The lowest temperature at which an explosion may occur is slightly above the melting point of aluminum (933 K). The explosions are believed to be a phenomenon of rapid vaporization of water caused by the high rate of heat transfer from fragmented aluminum. The process is basically one of thermal-to-mechanical energy conversion. Chemical reactions between aluminum and water are generally minor during these explosions, although some recent experiments have shown that, for some metals, steam explosions can also initiate substantial oxidation.[a]

A second potentially energetic reaction between aluminum and water is the oxidation of aluminum by steam. Such a chemical reaction becomes noticeable at temperatures above 1400 K and increases rapidly as temperature increases, as indicated by experiments performed at Argonne National Laboratory (ANL) (2). Although the reaction can become very rapid above 1800 K, it is not explosive, and the reaction rate is limited by the diffusion of oxygen into the aluminum through a layer of aluminum oxide.

A third reaction between aluminum and water, which is not as well-studied as the previous two, is a chemical reaction accompanied by a rapid rise in pressure, rather than a simple thermal-to-mechanical energy conversion process, as in the case of a pure steam explosion. In tests by Higgins and Schultz (3), molten aluminum at temperatures up to 2553 K was injected into water. Below 1672 K, little oxidation occurred; at 1672 K, the extent of oxidation suddenly jumped to 60%, and a pressure pulse of five atmospheres was also observed. Epstein (4) theorized that the explosive reaction was initiated by the ignition of aluminum in the vapor phase, instead of oxidation by the diffusion of oxygen through a layer of aluminum oxide as observed in the ANL experiments.

In this paper, Epstein's postulate that the ignition of molten metal in water occurs at a metal vapor pressure of 0.15 mm-Hg is reviewed, followed by error estimates for the ignition temperature from existing experimental data. Heat balance calculations are then performed to provide an understanding of the physical basis for the postulate.

EXPERIMENTAL EVIDENCE

Experimental data of the ignition of metal in water exist for aluminum, uranium, and zirconium. The lowest temperatures that spontaneous ignition were observed to occur were 1672 K for aluminum (3), 2773 K for uranium (5), and 2873 K for zirconium (6,7). Epstein claimed that ignition occurs at a

---

a. M. Berman, personal communication.

metal vapor pressure of 0.15 mm-Hg. Based on data published by Hultgren (8), the vapor pressure, p_v, of each metal was fitted to temperature through the least-squares method, and the results are

Aluminum: $\ln(p_v) = 13.624 - 37261/T$ (1a)

Uranium: $\ln(p_v) = 13.435 - 58050/T$ (1b)

Zirconium: $\ln(p_v) = 14.898 - 69742/T$ (1c)

where p_v is in atmospheres and T is in Kelvin.

The calculated temperatures at which the vapor pressures equal 0.15 mm-Hg are 1682 K for aluminum, 2643 K for uranium, and 2977 K for zirconium.

A comparison between the predicted ignition temperatures and the experimental values is shown in Table 1. Assuming that the deviations of experimental values from the predicted values (determined by vapor pressure of 0.15 mm-Hg) are normally distributed with the same mean and standard deviation, the standard error of the mean ignition temperature is 68 K.

Table 1. Predicted and experimental temperatures for ignition of metal in water

Element	Predicted θ_t[a] (K)	Experimental θ_0 (K)	$\theta_t - \theta_0$
Al	1682	1672[b]	10
U	2643	2773[c]	-130
Zr	2977	2873[d]	104

a. Temperature at which vapor pressure equals 0.15 mm-Hg.

b. Lowest temperature at which ignition occurred in Higgins and Schultz's tests (3).

c. Lowest temperature quoted by Epstein from the work of Schnizlein, Baker, and Vogel at which a rise in explosive pressure occurred in their tests (5).

d. Lowest temperature at which a rise in explosive pressure occurred in Argonne experiments (6,7).

IGNITION OF ALUMINUM IN WATER AS A VAPOR PHASE REACTION

Epstein suggested that ignition is a vapor phase phenomenon based on an empirical vapor pressure rule that states that ignition of metals occurs in water at temperatures corresponding to a metal vapor pressure of 0.15 mm-Hg. We postulate that ignition occurs when the heat generated from the reaction exceeds the sum of the heat required to vaporize the metal and the heat losses from the aluminum surface. Such calculations are performed to determine the lowest temperature at which the metal will ignite. To simplify calculations, only radiative heat losses from the metal surface are considered in this paper.

Let N_s be the number density of steam molecules near the aluminum surface, N_{Al} the number density of aluminum atoms in the vapor phase, σ_r the reaction cross-section, and v the RMS one-dimensional relative velocity between aluminum and water molecules. The reaction rate per unit volume near the surface is $N_s N_{Al} \cdot \sigma_r v$, and the reaction rate per unit surface area is $N_s N_{Al} \cdot \sigma_r v \cdot \ell$, where ℓ is the reaction mean free path of an aluminum atom. It is assumed that the aluminum atoms appear in a layer within only one mean free path from the surface. Beyond this layer, the aluminum atoms are assumed to have reacted with the steam molecules to form aluminum oxide. Since, by definition of ℓ, $N_s \sigma_r \ell$ is one, the reaction rate per unit surface area is equal to $N_{Al} v$.

Assuming perfect gas law, the vapor phase number density of aluminum is

$$N_{Al} = p_v/kT \quad (2)$$

and the one-dimensional relative velocity, v, is

$$v = [kT(1/M_{Al} + 1/M_s)]^{1/2} \quad (3)$$

where k is the Boltzmann constant, T is the absolute temperature, M_{Al} is the atomic weight of aluminum, and M_s is the molecular weight of water.

The net energy release rate per unit surface area, E, from the reaction is

$$E = N_{Al} \cdot v \cdot (\Delta H_r - \Delta H_v)/A \quad (4)$$

where ΔH_r and ΔH_v are the heat of reaction and the latent heat of vaporization, respectively, and A is Avogadro's number. The heat of reaction, as quoted by Epstein, is 4199 cal per gram of aluminum reacted, or 4.74×10^5 J/mol. The latent heat of vaporization can be deduced from the vapor pressure curve based on the Clausius-Clayperon equation:

$$\ln(p_v) = \text{constant} - \Delta H_v/RT \quad (5)$$

where R is the gas constant. From Equation (1a), ΔH_v is calculated to be 3.095×10^5 J/mol. Using these numerical values, the energy release rate per unit surface area is

$$E = 4.597 \times 10^{10} \frac{\exp(-37261/T)}{\sqrt{T}} \ (\mathrm{MW/m^2}). \quad (6)$$

The radiative heat loss rate per unit surface area, L_R, for $T >> T_e$ (T_e is steam temperature), is given by

$$L_R \approx \epsilon \sigma_R T^4 \quad (7)$$

where ϵ is the emissivity and σ_R is the Stefan-Boltzmann constant.

Setting Equations (6) and (7) equal, we can solve for the critical surface temperature at which the rate of energy release from oxidation is equal to the rate of radiative energy loss. The critical temperature for aluminum as a function of surface emissivity is given in Table 2. Above the critical temperature, the energy release rate exceeds the energy loss rate, and the reaction escalates exponentially. Below the critical temperature, the reaction is quenched by radiative losses. The vapor pressure of aluminum at the critical temperature for each surface emissivity is also given in Table 2. At $\epsilon = 0.6$, the vapor pressure is equal to 0.154 mm-Hg, almost exactly the same pressure for ignition as conjectured by Epstein. Because of the exponential character of the reaction, the critical temperature at which ignition occurs is relatively insensitive to the assumed surface emissivity. For emissivities between 0.4 and 1.0, the ignition temperature varies from 1645 K to 1734 K. These temperatures are within two standard errors on either side of the experimentally determined ignition temperature.

The analysis above was also applied to uranium and zirconium, but the results were mixed. At an emissivity of 0.6, the calculated ignition temperature for uranium is 3130 K, which is within the experimental error bounds, but the calculated ignition temperature for zirconium is over 4000 K, far above the observed value.

Table 2. Critical aluminum temperature and vapor pressure for which heat generation equals radiative heat loss as functions of surface emissivity

Emissivity	Critical Temperature (K)	Vapor Pressure (mm-Hg)	Heat Generation/ Loss (MW/m^2)
0.4	1645	0.091	0.165
0.6	1684	0.154	0.274
0.8	1711	0.219	0.388
1.0	1734	0.292	0.513

OXIDATION FROM THE DIFFUSION OF OXYGEN INTO METAL

In the previous section, only the oxidation of vaporized metal was considered, and the diffusion of oxygen (from the disassociation of water molecules) into the metal was ignored. For aluminum, the diffusion reaction is negligible at temperatures around 1700 K (2), so the calculations shown above can be justified. For zirconium, the energy release rate from the oxidation of the metal by oxygen diffused into the metal is quite rapid above 2000 K. The reaction rate can be estimated from the Urbanic-Heidrick correlation (9):

$$dW(Zr)/dt = \frac{87.9}{2W(Zr)} \cdot \exp(-16610/T) \qquad (8)$$

where W(Zr) is the mass of zirconium oxidized per unit surface area and T is the absolute temperature. W(Zr) is in kg/m^2, T is in Kelvin, and t is in s.

The energy release rate per square meter, $E_d(Zr)$, as a function of the thickness of zirconium layer before oxidation, δ, is obtained by multiplying dW(Zr)/dt by the heat of reaction (6.52 MJ/kg) and by substituting density (6200 kg/m^3) times δ for W(Zr). The result is

$$E_d(Zr) = 4.622 \times 10^4 \exp(-16610/T)/\delta \qquad (9)$$

where $E_d(Zr)$ is in MW/m^2 and δ is in μm. At 2977 K (zirconium vapor pressure = 0.15 mm-Hg), the energy release rate from the vapor reaction is negligible, but the energy release rate from the diffusion reaction is calculated to be 3.5 MW/m^2 for $\delta = 50$ μm. This is comparable to the radiative heat loss rate at emissivity $\epsilon = 0.8$ and T = 2970 K. Above this critical temperature, the energy release rate from oxidation exceeds the energy loss rate from radiation, so the temperature and the reaction rate escalate.

A similar analysis is applied to uranium. The reaction rate from the diffusion of oxygen into the metal is derived from a correlation given by Wilson et al. (10):

$$dW(U)/dt = \frac{1320}{2W(U)} \cdot \exp(-12582/T) \qquad (10)$$

where W(U) is the moles of oxidized uranium per unit surface area and T is the absolute temperature. W(U) is in mol/m^2, T is in Kelvin, and t is in s.

The energy release rate per square meter, $E_d(U)$, as a function of the thickness of uranium layer before oxidation, δ, is obtained by multiplying dW(U)/dt by the heat of reaction (142 kcal/mol) and by substituting density (18700/0.238 mol/m^3) times δ for W(U). The result is

$$E_d(U) = 4.987 \times 10^3 \exp(-12582/T)/\delta \qquad (11)$$

where $E_d(U)$ is in MW/m^2 and δ is in μm. At 2643 K (uranium vapor pressure = 0.15 mm-Hg), the energy release rate from the vapor reaction is calculated to be 0.12 MW/m^2, and the energy release rate from the diffusion reaction is calculated to be $42.7/\delta$ MW/m^2. The radiative heat loss at an emissivity of 0.8 is 2.2 MW/m^2. If the oxidized layer is less than 17 μm thick, the heat generation rate exceeds the radiative heat loss rate, so the temperature and reaction rate escalate.

Although the calculations shown above support the experimentally determined critical temperatures for zirconium and uranium, the calculated critical temperatures depend on the thickness of an initial oxidized layer. If this theory of ignition is correct, the exact ignition temperature for zirconium and uranium must depend on their oxidation histories. The temperature corresponding to 0.15 mm-Hg is then the minimum temperature at which ignition occurs because the thickness of the oxidized layer must also be less than a certain minimum for the temperature to escalate.

DISCUSSION AND CONCLUSION

The explanation offered in this paper for ignition did not consider the size of the particles. In fact, particle size may play an important role in an ignition. On the one hand, particle size determines the intensity of the reaction. The smaller the particle size, the more intense is the reaction (measured by energy release rate per unit volume) after ignition. On the other hand, it may also affect the ignition temperature itself by changing the heat transfer environment. For example, if the particles are small and densely packed, they may raise the temperature of the heat sink and thus lower the ignition temperature, but they may also limit the availability of water for oxidation and thus raise the ignition temperature. In the extreme case in which a steam explosion occurs, the particles may be so fine that ignition can occur at temperatures not far above the melting point of the metal.[a] However, the experimental determination of ignition, as referenced in this article, was carried out by injecting molten aluminum into water, which mimics a molten core being dropped into a deep pool of water. The results are therefore believed to be applicable to the analysis of accidents in reactors with aluminum clad fuel in the absence of initial steam explosions.

We have offered a theoretical basis for the postulate that metals in water ignite at temperatures corresponding to a unique metal vapor pressure of 0.15 mm-Hg, which seems to fit the experimental values. However, data exist for only three metals; in addition, only one data point exists for each metal. There is, therefore, a critical need for more experimental determinations to reduce the uncertainty of ignition temperatures and to provide more definitive confirmation of the vapor pressure rule than the present data base.

ACKNOWLEDGMENT

The authors would like to thank Drs. R. R. Hobbins and Gus Cronenberg for helpful comments on the paper. This work was supported by the U.S. Department of Energy, under DOE Contract No. DE-AC07-76ID01570.

NOTATION

Roman Letters

A = Avogadro's number
E = energy release rate per unit surface area from vapor phase reaction
E_d = energy release rate per unit surface area from oxygen diffusion reaction
ΔH_r = heat of reaction
ΔH_v = latent heat of vaporization
k = Boltzmann constant
ℓ = reaction mean free path
L_R = radiative heat loss rate per unit surface area
M_{Al} = atomic weight of aluminum
M_s = molecular weight of water
N_{Al} = number density of aluminum molecules
N_s = number density of steam molecules
p_v = vapor pressure
R = gas constant
T = absolute temperature
T_e = steam temperature
v = one-dimensional relative velocity
W(U) = moles of oxidized uranium per unit surface area
W(Zr) = mass of oxidized zirconium per unit surface area

Greek Letters

δ = thickness of metal before oxidation
ϵ = emissivity
θ_0 = experimental temperature
θ_t = predicted temperature
σ_r = reaction cross-section
σ_R = Stefan-Boltzmann constant

a. M. Berman, personal communication.

LITERATURE CITED

1. Wright, R. W., "Dispersal and Pressure Generation by Water-Impact Upon Molten Aluminum," Kinetic Studies of Heterogeneous Water Reactors, STL-372-30 (December 1965).

2. Wilson, R. E., C. Barnes, and L. Baker, "Studies of the Aluminum-Steam Reaction by the Levitation Melting Method," Argonne National Laboratory Chemical Engineering Division Semiannual Report, January-June 1964, ANL-6900 (August 1964).

3. Higgins, H. M. and R. D. Schultz, "Reaction of Metals with Water and Oxidizing Gases at High Temperatures," Aerojet General Corporation, IDO-28000 (1957).

4. Epstein, Leo F., Nuclear Science and Engineering, **10**, 247-253 (1961).

5. Schnizlein, J. G., L. Baker, jr., and R. C. Vogel, "Metal Oxidation Ignition Kinetics and Metal-Water Reaction Studies," Argonne Chemical Engineering Division Summary Report for January, February, and March, 1960, ANL-6145, and "Metal Oxidation and Reaction Kinetics and Metal-Water Reactions," Argonne Chemical Engineering Division Summary Report for April, May, and June, 1960, ANL-6183, Argonne National Laboratory (1960).

6. Vogel, R. C., et al., Argonne Chemical Engineering Division Summary Report for October, November, and December, 1958, ANL-5959, Argonne National Laboratory (1958).

7. Vogel, R. C., et al., Argonne Chemical Engineering Division Summary Report for January, February, and March, 1959, ANL-5996, Argonne National Laboratory (1959).

8. Hultgren, R., et al., Selected Values of the Thermodynamic Properties of the Elements, American Society for Metals (1973).

9. Urbanic, V. F. and T. R. Heidrick, Journal of Nuclear Materials, **75**, 251-261 (1978).

10. Wilson, R. E., C. Barnes, jr., R. Koonz, and L. Baker, jr., Nuclear Science and Engineering, **25**, 109-115 (1966).

ASSESSMENT OF IMPROVED FISSION-PRODUCT TRANSPORT MODELS IN VICTORIA AGAINST ORNL HI AND VI TESTS

P.E. Domagala, J. Rest, and S.A. Zawadzki ■ Materials and Components, Technology Division, Argonne National Laboratory, 9700 South Cass Avenue, Argonne, IL 60439

New models for the release of fission products from fuel have recently been incorporated into a comprehensive code, VICTORIA, for the prediction of radionuclide behavior under severe reactor accident conditions as an alternative to a simple Booth diffusion calculation. It is widely known that the Booth model has severe limitations when used under conditions of changing temperature and power. A new transport model, based on a two-node diffusive-flow formulation, has been implemented in VICTORIA. In addition to the diffusive-flow model, other mechanisms, such as grain growth, grain boundary sweeping, and intergranular bubble behavior, are taken into account. These physically based models focus on fission-product behavior in intact fuel geometries. Whereas VICTORIA is primarily concerned with the behavior of severely degraded fuel geometries, the capability for accurately characterizing fission-product release from intact fuel must be demonstrated before other geometries can be reliably modeled.

Results of VICTORIA simulations are compared with results of in-cell heating tests on irradiated fuel. The fission-product transport models involved in these simulations are assessed and their predictive capability examined.

VICTORIA [1], a comprehensive computer code based on mechanistic models for the prediction of radionuclide behavior under severe reactor accident conditions, is being developed under the sponsorship of the U.S. Nuclear Regulatory Commission, with international participation by Sandia National Laboratories, Argonne National Laboratory, United Kingdom Atomic Energy Agency, Winfrith, Atomic Energy of Canada Limited, Chalk River Nuclear Laboratories, Oak Ridge National Laboratory, and Batelle Columbus Division. VICTORIA contains a comprehensive chemistry package capable of handling 25 elements and 167 chemical species. Chemical speciation, species transport, aerosol production, geometry of the fuel, containment and piping systems, and to some extent, thermalhydraulics are taken into account in the release process.

New models for the Transport of fission products in intact fuel (TRANIF) have recently been incorporated into the VICTORIA code as an alternative to a simple Booth diffusion calculation. It is widely known that the Booth model has severe limitations when used under conditions of changing temperature and power [2]. The TRANIF routines are based on key models that have been successfully implemented in such nuclear-fuel-behavior codes as FASTGRASS [3]. These models, based on the work of Rest [3-5], provide a comprehensive, mechanistic treatment of the release process.

The VICTORIA models for the diffusive flow of fission products, have been replaced by a more realistic model based on the work of Wood and Matthews [2,6]. In this model, the diffusive flow to a spherical boundary is approximated by representing the concentrations of the diffusing species in terms of quadratic functions in two concentric regions. Variational principles are used to calculate the

*Work sponsored by the U.S. Nuclear Regulatory Commission, Accident Evaluation Branch.

radial distribution of the concentration. This approach provides a more accurate solution to the problem while maintaining a reasonable level of computational efficiency.

In addition to the two-node diffusive-flow calculation, mechanisms such as gas bubble nucleation, bubble growth and migration, pore interlinkage, grain growth, and grain boundary sweeping are taken into account. Figure 1 details the implementation of these models in VICTORIA. A full treatment of the previously mentioned phenomena is provided for noble-gas fission products. The resulting releases to the open porosity are calculated by the TRANIF routines and VICTORIA species concentrations are updated accordingly.

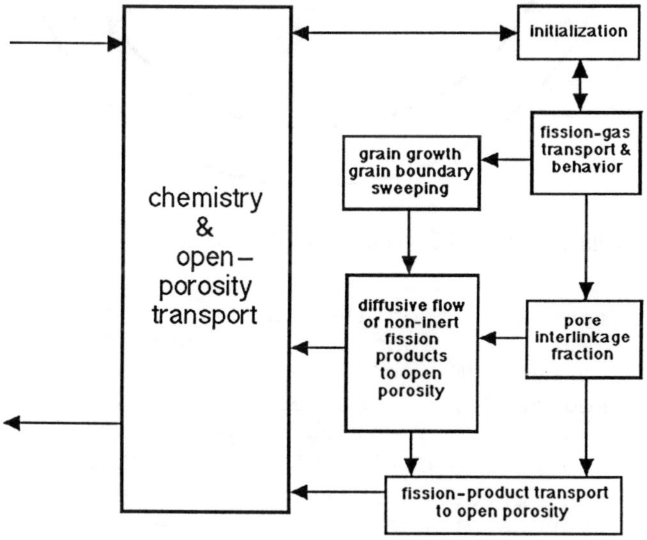

Figure 1. Implementation of TRANIF models in VICTORIA.

The noninert species form a more complex situation. Because these species interact, chemical-equilibrium calculations are required. TRANIF performs an intragranular diffusive-flow calculation based on the two-node solution for fission-product species in the gas phase, as well as a calculation of pore-interconnection probability, grain size, and grain boundary sweeping. Subsequently, solutions for chemical equilibrium and gas phase transport in the open porosity are calculated by the appropriate VICTORIA modules.

VICTORIA is being assessed against the Oak Ridge National Laboratory (ORNL) HI and VI in-cell tests. One objective of these tests is to provide well-characterized results that can be used as a basis for model validation. The HI and VI tests constitute two of the best available sources of data for fission-product release from LWR fuel under heatup accident conditions. In this study, the HI-3 and VI-3 tests are modeled using the VICTORIA code. The primary focus of this effort is the behavior of fission products in the fuel and their subsequent release.

Although VICTORIA is primarily concerned with degraded fuel geometries, the release of fission products from intact fuel is of critical importance. As seen in the HI and VI tests, more than 90% of the more mobile fission products can be released without fuel melting and/or dissolution, and without significant changes in fuel geometry. In addition, the mechanisms that lead to release of fission products from fuel grains in intact fuel are valid also for geometries such as rubble beds. By providing physically based, comprehensive models for fission-product transport in VICTORIA, the entire evolution of the source term during a reactor heatup scenario can be modeled more realistically.

MODELS

Two-Node Diffusive-Flow Model

To calculate the behavior of fission gas in nuclear fuels, it is necessary to calculate the loss of gas by diffusion to the grain boundaries of both gas atoms and gas bubbles. Assuming a spherical grain geometry, and that the only sink for gas atoms is the grain boundary itself, the concentration of gas atoms C_g, within the grain must satisfy the equation

$$\frac{\partial C_g}{\partial t} = \frac{1}{r^2}\frac{\partial}{\partial r}\left(Dr^2\frac{\partial C_g}{\partial r}\right) + K_g, \qquad (1)$$

where D is the gas atom diffusion coefficient, K_g the gas atom generation rate, r the fuel grain radius, and t the time.

Let the concentration of gas atoms in a spherical grain at time t be $C_g^0(r)$. After a small time dt, the concentration becomes $C_g(r)$. Applying the backward Euler approximation, for small dt, Equation (1) can be rewritten as

$$\frac{1}{r^2} = \frac{d}{dr}\left(Dr^2\frac{\partial C_g}{dr}\right) - \frac{C_g}{\delta t} + \frac{C_g^0}{\delta t} + K_g = 0. \qquad (2)$$

Euler's theorem may then be applied to give the variational equivalent of Equation (2):

$$\delta\int_0^{\frac{d_g}{2}} 4\Pi\left[\frac{D}{2}\left(\frac{dC_g}{dr}\right)^2 + \frac{C_g^2}{2\delta t} - \left(\frac{C_g^0}{\delta t} + K_g\right)C_g\right]r^2 dr = 0, \qquad (3)$$

which assumes Dirichlet boundary conditions. An approximate solution to the problem may be obtained by choosing a trial function that satisfies the boundary conditions and minimizes the integral in Equation (3) in terms of the free parameters of the function [6]. Piecewise quadratic functions were chosen because they can be handled more easily and allow an exact representation of Equation (1) for long times.

The grain is divided into two concentric regions of approximately equal volume (see Figure 2). In the

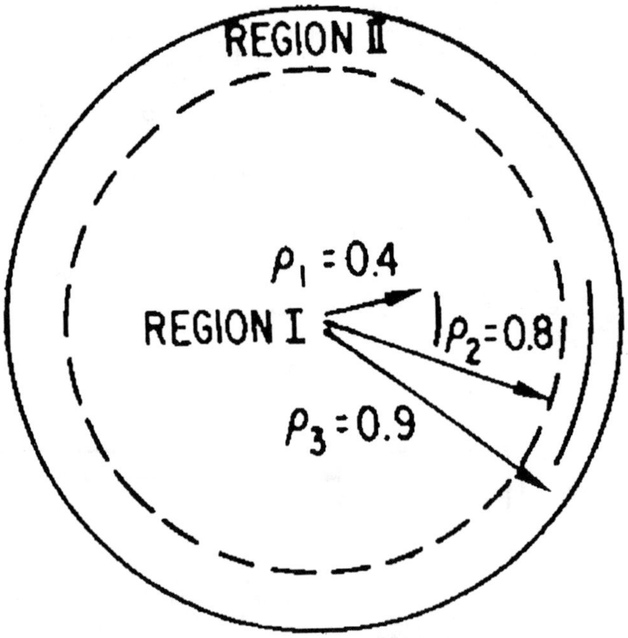

Figure 2. Two-node model of a spherical fuel grain.

central region (Region I), the concentration function is constrained so that $dC_g/dr = 0$ at $r = 0$. In Region II, the concentration function is constrained to a value of $C_g = 0$ at $r = d_g/2$. The two functions are further constrained to be continuous at the boundary between the two regions. This leaves three free parameters, the concentrations C_1, C_2, and C_3 at the radius ratios $r_1 = 0.4$, $r_2 = 0.8$ and $r_3 = 0.9$, where $r = 2r/d_g$. r_1, r_2 and r_3 are defined in Figure 2. Thus, the trial functions are

$$C_g = C_1\left(0.64 - \rho^2\right)/0.48 + C_2\left(\rho^2 - 0.16\right)/0.48 \qquad (4)$$

for Region I, and

$$C_g = 5C_2\left(10\rho^2 - 10 + \rho 9\right) + 10C_3\left(18\rho - 10\rho^2 - 8\right) \quad (5)$$

for Region II. By substituting Eqs. (4) and (5) in Equation (3), and finding the extremum by differentiating with respect to C_1, C_2 and C_3, a set of linear equations is obtained, the coefficients of which can be obtained by direct evaluation.

The flux of gas atoms at the grain surface is

$$J = \frac{6D}{d_g}\frac{\partial C}{\partial r}\bigg|_{r=d_g/2.} \quad (6)$$

Substituting Eqs. (4) and (5) into Equation (6), the gas atom flux to the boundary becomes

$$J = \frac{6D}{d_g^2}\left(-60C_2 + 240C_3\right), \quad (7)$$

where C_1 and C_2 are described [3] by a set of functional dependencies involving D, d_g, and dt.

Grain-Growth and Grain Boundary Sweeping

The model of grain-growth/grain-boundary-sweeping postulates that small, intragranular bubbles and volatile fission-product atoms in the path of a growing grain are swept up by grain-boundary adhesive forces. This action provides a mechanism whereby fission products are collected at the grain faces and edges.

It has been shown [8] that, as a boundary moves, the rate dC_{gb}/dt at which fission products are swept up by the moving boundary is proportional to the rate of change of volume of the grain, i.e.,

$$\frac{dC_{gb}}{dt} = \frac{3eC_I D}{d_{gt}}\frac{d\,d_{gt}}{dt} = \frac{3eC_I D}{d_{gt}}V_{gb} \quad (8)$$

where C_I is the intragranular concentration of a fission product, d_{gt} is the grain diameter at time t, e is a grain-boundary-sweeping efficiency factor, and V_{gb} is the grain-boundary velocity.

The implementation of this theory in the TRANIF routines is based on the FASTGRASS-VFP theory [9]. This model considers two distinct classes of bubbles, those on grain faces and those on edges. A kinetically based model is used to calculate the grain diameter and grain velocity. Changes in fission-product concentrations are then calculated using Equation (8). Based on this calculation and on the face and edge bubble concentrations, a determination is made of whether gas bubbles are caught up and moved along by a moving grain boundary or whether the grain boundary is only temporarily retarded by bubbles and then breaks away.

SIMULATION CONSTRUCTION

Relevant parameters for the HI-3 [10] and VI-3 [11] tests are listed in Table 1. Input decks were constructed for both tests with this data.

It should be noted that the TRANIF routines in VICTORIA require initialization of the fission-gas distribution in the fuel. This information is supplied by running a stand-alone version of TRANIF and generating a restart dump. For evaluation of the TRANIF models, a steady-state irradiation was simulated with the stand-alone TRANIF code. Table 2 lists the relevant parameters for these steady-state simulations. Steady-state irradiations were modeled for both the H. B. Robinson and BR3 fuels as a continuous irradiation at constant temperature and power. Temperatures were brought to ambient levels and power level to zero at the end of

Table 1. Test parameters for the ORNL HI-3 and VI-3 tests.

Parameter	HI-3	VI-3
Fuel rod identification	Rod H-15, bundle B05, H.B. Robinson 2 (PWR)	Rod 1-1002, BR3 reactor,
Belgium		
Irradiation data		
Period	10/71 to 5/74	7/76 to 9/80
Rod ave. heating	23.2 KW/m	25.1 KW/m
Burnup	25,200 MWd/t	42,000 MWd/t
Kr in-pile release	0.35%	10%
Test specimen geometry		
Length	20.3 cm	15.2 cm
Cladding OD	1.072 cm	0.95 cm
Fission-product inventory		
Kr	32.02 mg	36.72 mg
Cs	225.6 mg	231.7 mg
Test temperatures		
Start of test	~300 K	~500 K
Ramp 1, heatup rate	~2 K/s	~0.3 K/s
Hold temperature 1	2273 K	2000 K
Time at temperature	20 min	20 min
Ramp 2 heatup rate		~0.3 K/s
Hold temperature 2		2700 K
Time at temperature		20 min

the steady-state run in one small (~8 min) step to match initial conditions for the in-cell test simulations. Calculated end-of-life gas release for the BR3 irradiation does not fully agree with the reported rod total value because the one-node simulation does not accurately represent systems that contain sizable temperature, chemical, or oxygen partial-pressure gradients.

MODELING RESULTS

Figure 3 shows Kr and Cs release from fuel calculated by VICTORIA with TRANIF for the HI-3 test. While VICTORIA tracks the entire Kr release very well, the predicted Cs release is ~10% over the reported values. This discrepancy can be attributed to the limited treatment of the kinetics of Cs release. For the case of the noble-gas fission products, trapping of the diffusing gas by intra and intergranular gas bubbles are modeled. These types of physical trapping mechanisms have not been included for the non-rare fission products.

Table 2. Summary of parameters used for the steady-state irradiation simulations of test specimens in HI-3 and VI-3

Parameter	HI-3	VI-3
Irradiation time	914.3 day	208.3 days
Linear power	15.7 KW/m	12.2 KW/m
Irradiation temperature	1073 K	1200 K
Grain size	5 mm	8.5 mm
Burnup	2.52 at.%	4.2 at.%
Gas release (Xe, Kr)	~0.24%	~2.0%

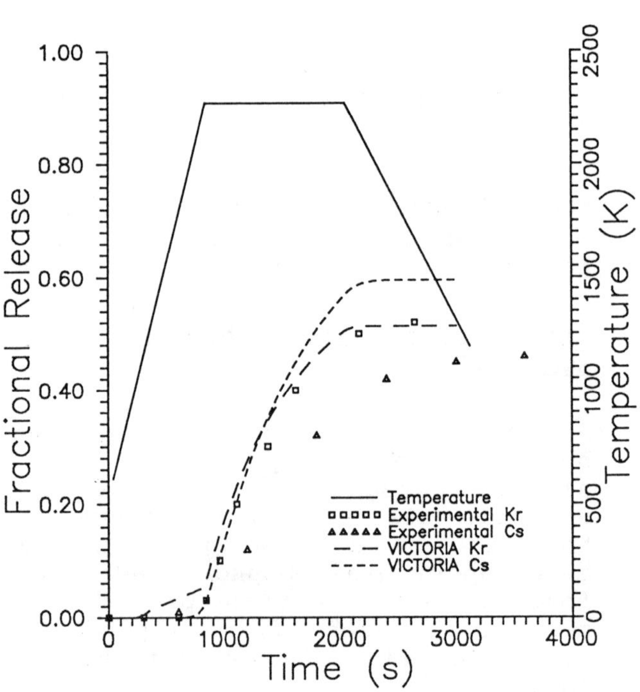

Figure 3. Comparison of Kr and Cs release from fuel in the HI-3 test as calculated by VICTORIA with experimental data.

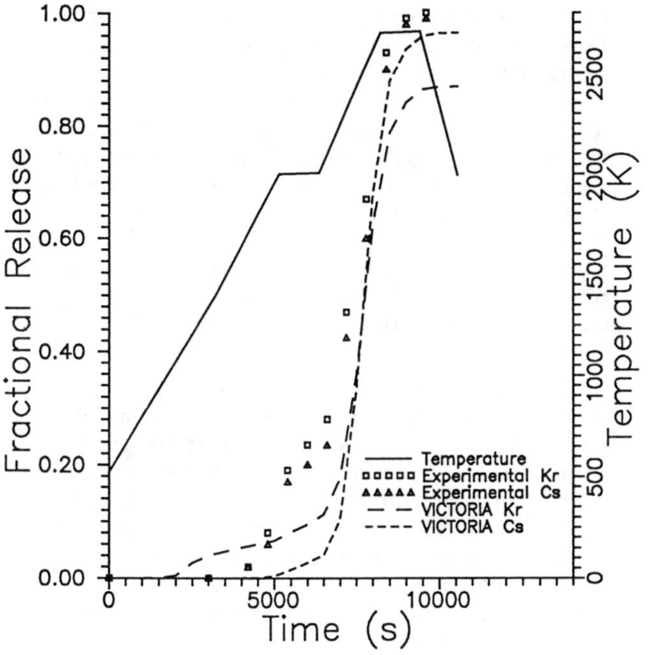

Figure 4. Comparison of Kr and Cs release from fuel in the VI-3 test as calculated by VICTORIA with experimental data.

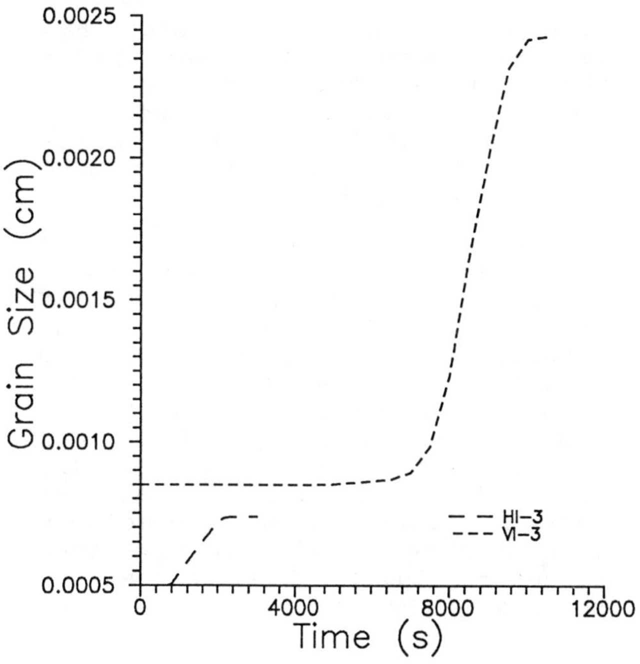

Figure 5. Grain growth as a function of time for the HI-3 and VI-3 test as calculated by VICTORIA.

For the case of the VI-3 test (Figure 4), similar results are obtained. Kr release is fairly well predicted with Cs releases lagging behind early in the test (Kr is initially released from the grain boundaries while Cs has to diffuse through the grain) and then rising above the Kr release due to the lack of intra and intergranular bubble trapping mechanisms.

In both the HI-3 and VI-3 cases, the dominant fission gas release mechanism is found to be grain-growth/grain-boundary-sweeping. As seen in Figure 5, the TRANIF model predicts ~50% and ~250% increases in grain size for the HI-3 and VI-3 tests, respectively. Since a 50% increase was observed in the case of the HI-3 test [8], a 250% increase in grain size for the VI-3 test appears reasonable.

Figure 6 illustrates the importance of the grain-growth/grain-boundary-sweeping mechanism in evaluating fission-product release from fuel. For the case of Kr

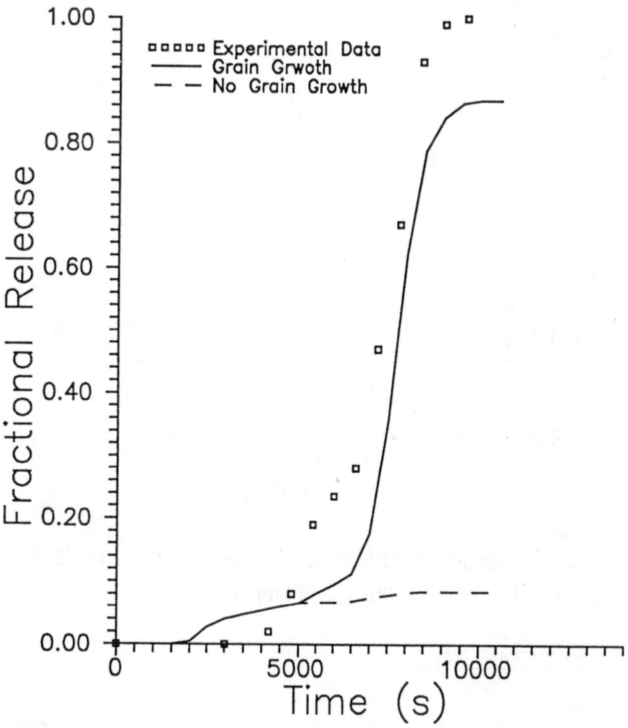

Figure 6. Kr release from fuel for the VI-3 test as calculated by VICTORIA with and without grain growth.

release in the VI-3 test, this mechanism accounts for over 90% of the predicted total release. In fact, this mechanism is responsible for the majority of the intra-granular Kr and Cs release in both the HI-3 and VI-3 tests.

As a further step in the model evaluation, the time step sensitivity of VICTORIA with the TRANIF transport models was assessed. Simulations were run using various values of the time step. In order to eliminate extraneous effects and reduce running time, simulations were run without performing equilibrium chemistry calculations. As seen in Figure 7, no appreciable time step sensitivity at time step values ranging for 0.5 seconds to 4 seconds is observed.

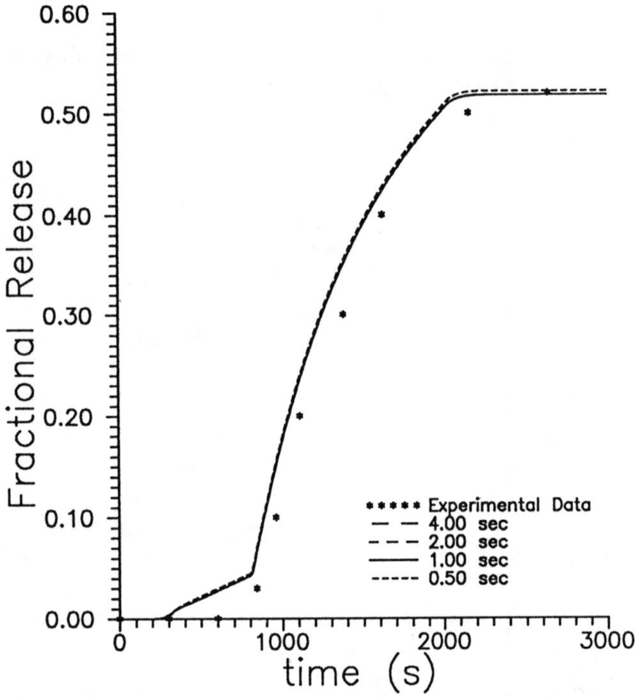

Figure 7. Release of Kr from fuel for the HI-3 test as calculated by VICTORIA using various time step values.

CONCLUSIONS

Based on the results of the HI-3 and VI-3 simulations, the VICTORIA with the TRANIF fission product transport routines are capable of accurately predicting Cs and Kr release from irradiated, intact fuel under simulated reactor heatup-accident conditions. These mechanistic models provide a more accurate and comprehensive characterization of the fission-product-release process by realistically describing the kinetics of the process. For the HI-3 and VI-3 tests, the dominant release mechanism was found to be grain growth/grain boundary sweeping.

VICTORIA with TRANIF provides a physically accurate and computationally sound basis for the modeling of fission-product release from irradiated nuclear fuel under conditions of changing temperature and/or power and form a good basis on which to assess fission-product behavior under conditions of core degradation.

ACKNOWLEDGEMENT

Work sponsored by the U.S. Nuclear Regulatory Commission, Accident Evaluation Branch.

LITERATURE CITED

1. T. J. Heames, D. A. Williams, N. E. Bixler, A. J. Grimely, C. J. Wheatly, N. A. Johns, N. M. Chown, *VICTORIA: A Mechanistic Model of Radionuclide Behavior in the Reactor Coolant System Under Severe Accident Conditions*, NUREG/CR-5545 (SAND90-0756) (1990).

2. M. H. Wood and J. R. Matthews, *J. Nucl. Mater.* **89**, 53-61 (1980).

3. J. Rest, *J. Nucl. Mater.* **120**, 195-212 (1984).

4. J. Rest and A. W. Cronenberg, *J. Nucl. Mater.* **150**, 203-225 (1987).

5. J. Rest, *J. Nucl. Mater.* **168**, 243-261 (1987).

6. J. R. Matthews and M. H. Wood, *Nucl. Eng. Des.* **56**, 439-443 (1980).

7. A. H. Booth, *A Method of Calculating Fission Gas Diffusion from UO_2 Fuel*, Chalk River Nuclear Laboratories Report, CRDC-721 (1957)

8. J. Rest, *Adv. Ceram.* **17**, 223, 1986.

9. J. Rest, *J. Nucl. Mater.*, **131**, 291 (1985).

10. M. F. Osborne, R. A. Lorenz, K. S. Norwood, J. R. Travis, and C. S. Webster, *Data Summary Report for Fission-Product Release Test HI-3*, NUREG/CR-3335 (ORNL/TM-8793) (1984).

11. M. F. Osborne, R. A. Lorenz, J. L. Collins, J. R. Travis, C. S. Webster, H.K. Lee, T. Nakamura, and Y. C. Tong, *Data Summary Report for Fission Product Release Test VI-3*, NUREG/CR-5480 (ORNL/TM-11399) (1990).

ANALYSIS OF LOFT EXPERIMENT LP-FP-2 RESPONSE DURING THE QUENCHING PERIOD*

Ariel Sharon ■ Argonne National Laboratory, 9700 South Cass Avenue, Argonne, IL 60439

The LOFT Experiment LP-FP-2 was terminated by reflooding the core with water after the central portion of the core was badly degraded by melting of fuel rod material. Thermal hydraulic data regarding the core behavior in this period is presented and the differences between the response of the degraded core region and the non-degraded core are discussed.

A model for the degraded core region behavior is presented and its predictions agree well with the data trends. This illustrates that the slow quenching rate of the degraded region in the test could have been controlled by the hydraulic resistance of the degraded channel rather than by thermal limitations.

The potential hydrogen source term that can be generated in-vessel during a severe accident has strong implications on sizing the containment of a power plant and on accident management strategies. This source term can be generated during several phases of a postulated accident.

1. When the core is overheating and melting due to coolant boil-off.
2. When the debris is covered by coolant after a prolonged heatup period.
3. As the debris is falling into the water pool in the lower plenum.

During the first phase, hydrogen generation is typically governed by the cladding oxidation at relatively low temperatures (>1200 K). At temperatures above ~1800 K the oxidation kinetics are fast and steam availability in the coolant channels would typically limit the hydrogen productions.

At temperatures exceeding the cladding melting point, changes in the core geometry may pose an additional limit to hydrogen production, particularly as molten metallic-rich melt is refreezing at colder locations leading to a reduction in the local porosity.

When a degraded core is being flooded with water, large amounts of hydrogen could be generated as large qualities of steam generated by the quenching process can cause oxidation of zircaloy in areas that were previously steam starved.

The third phase in which hydrogen is generated, namely the contact of debris with water in the lower plenum, is dependent to a large extent on the debris composition and its characteristic particle size. Chemical analysis of debris from the TMI-2 accident suggests that the molten debris was already oxidized when it relocated to the lower plenum. It has also been suggested [1,2] that such conditions are rather typical to many postulated accident sequences. Thus, this phase is probably not as an important source of hydrogen production as the in-core quenching phase during core recovery.

One of the main data sources on the degraded core behavior during a recovery attempt is the LOFT experiment LP-FP-2. In this experiment degraded core geometry was obtained in the central fuel module (CFM) which was surrounded by an insulating shroud. The rest of the LOFT core periphery, which was also uncovered in the test, was heated up to a maximum temperature of approximately 1500 K with no significant geometry degradation. A schematic of the core cross section is shown in Figure 1. The rapid heatup inside the CFM was achieved by using highly enriched fuel (9.7%) in the CFM and lower fuel enrichment in the periphery.

* This work sponsored by the DOE ARSAP

The test started with the LOFT facility being at full power (26.8 Mw). At time zero the reactor was scrammed, a cold leg "break" was initialed at 33 seconds and was closed later, before substantial heatup of the core. At 222 seconds a hot leg "break" was initiated, and remained open during the entire transient phase of the test. Based on thermocouple dryout data, the top of the core became uncovered at approximately 662 seconds and at about 1000 seconds thermocouples below the core indicated superheated temperatures. Detailed experimental scenario is provided in Ref. [3].

Inside the central fuel module, cladding thermocouples were placed at three different elevations: 0.25 m (10"), 0.69 m (27"), 1.07 m (42"). Additional thermocouples were placed on guide tubes at elevation of 1.68 m (66"). All elevations are referenced to the bottom of active fuel. The length of the core was 1.68 m, thus the highest guide tube thermocouple was at the top of active fuel. Cladding temperatures exceeding the thermocouple failure temperature of ~2200 K, were reached first at the 1.07 m elevation at ~1540 seconds. Some guide tubes, however, indicated failure temperatures earlier. Failure temperatures were also reached at the 0.69 m elevation but the thermocouples in the 0.25 m elevation reached a maximum of only about 1000 K. At the top of the core (1.68 m) temperatures reached only 1000 to 1500 K prior to the quenching period.

At 1776.0 seconds the hot leg "break" was closed and the primary system was totally isolated. This marks the end of the transient phase of the test. At that time the maximum temperature in the peripheral assemblies reached 1500 K. Using best estimate calculations with MAAP, A. Sharon et al. [4] concluded that at that time core material melting temperatures were reached between elevations of 0.92 to 1.07 m. Some molten material relocated to elevation 0.69 m and lead to the failure of thermocouples at that elevation. Since MAAP assumes that molten core regions are blocked, it was predicted that the CFM become blocked at 1635 seconds. The total amount of hydrogen that was generated at that time was predicted to be 170 grams, and the best estimate was 220 grams.

At 1782.6 seconds the accumulators and safety injection systems were activated leading to flooding of the core. It was noted that the recovery of the CFM was fairly slow and from post test examination it was estimated that the total amount of hydrogen generation was in the range of 550 to 1100 grams [5].

In this paper some interesting thermal-hydraulic analysis of the test data is presented for the quenching period. A simple analytical model is discussed that explains some of the observed data trends.

THERMAL HYDRAULIC DATA DURING THE QUENCHING PERIOD

The pressure response of the primary system and the two accumulators, which were the main source of injection into the reactor vessel, are shown in Fig. 2. Water entered the vessel from 1784 seconds to 1795 seconds and again after 1905 seconds. As the cold (304 K) water from the accumulators entered the core, the pressure decreased from 1.1 Ma to 0.85 Ma in 1788.8 seconds and then sharply increased to 2.5 Ma by 1794 seconds. The net steam generation in the core can be estimated by assuming ideal gas behavior and ignoring gas volume changes due to condensation or evaporation

$$W_{net} = \frac{M_g}{V_g}\frac{dV_{ing}}{dt} - \frac{M_g}{P\gamma}\frac{dP}{dt} \qquad (1)$$

where M_g, V_g, P, γ and t indicate gas mass, gas volume, pressure, isotropic expansion coefficient and time, respectively. Taking derivatives of the pressure and accumulator injection data yield a maximum net steam generation of 7 kg/s at 1791 seconds which corresponds to a gas superficial velocity of 12 m/s or a Kutateladge number of 3.8 based on the total core flow area.

Note that the mass of fuel in the CFM represents only 8.4% of the total core inventory. Thus, in-spite of the high temperatures that were achieved in the CFM compared to the periphery, most of the core thermal energy at the end of the transient phase was stored in the periphery. Therefore, the pressure response reflects the periphery response which would be typical of an overheated core that did not suffer substantial geometry degradation.

Another interesting observation from thermocouples in the periphery is that quenching of the core started only at 1788.8 seconds while injection into the downcomer was initiated at 1784.0. Based on the amount of water injected to the vessel, the core should have been totally covered with water at 1788 seconds, even before any core thermocouple indicated any cooling. This suggests that even with injection directly into the downcomer, most of the water filled the downcomer cold legs and parts of the steam generator before finally entering the core.

Once water entered the core, thermocouples in the periphery cooled at a fairly rapid rate shown in Fig. 3. The initially slower cooling rate followed by a more rapid cooling to a sub-cooled temperature is typical of quenching hot rods that are initially at sub-cooled film boiling. Note that Fig. 3 shows thermocouples from four different elevations in the core. Yet a quench front can not be identified based on the data resolution, and all locations seem to have started to cool down at the same time. Fully quenching of the periphery, between 1792 to 1794, coincides with the end of the rapid pressurization of the primary system.

A totally different quenching behavior is observed inside the CFM. In the bottom, quenching also started at 1788.8 seconds as seen from the lower CFM thermocouple response (0.25 m) in Fig. 4. However, unlike the periphery, cooling at the high cladding temperature preceded at a substantially lower rate. Only at 1792.8 seconds, when the periphery was to a large part already quenched, a more rapid quenching rate is observed followed by another very slow cooling period. Figure 5 shows the response of the same thermocouple for a longer time. It indicates that the cladding reached the saturation temperature only at 1800 seconds and finally cooled down to subcooled temperatures only at 1810 seconds. In the periphery, on the other hand, the cladding cooled very rapidly to saturation and about a second later subcooled temperatures were reached.

At the top of the core all the thermocouples started to heat up at 1788.4 as shown in Fig. 6. Between 1792 and 1794, when the 0.25 m thermocouple indicated rapid cooling, all the thermocouples at the top including those on the upper tie plate (not shown), exceeded their failure temperature of ~2200 K.

At the other two thermocouple elevations, 0.69 and 1.07 m, failure temperatures were reached before the initiation of the quenching process. It is not clear whether the reading of once failed thermocouples can be used for further analysis, but as shown in Fig. 7, many thermocouple readings returned to saturation between 1850 and 1870 seconds. The fairly long quenching period with cooling at the bottom of the CFM and rapid heatup and melting at the top of the CFM is one of the interesting findings from the test with strong implication on accident management strategies regarding the recovery of a badly degraded core. Similar behavior was also observed in some tests in the CORA facility of KFK [6].

The analysis presented in the next section is aimed at trying to understand and quantitatively simulate the observed temperature data trends inside the CFM. In this way the controlling rate processes can be identified

A MODEL FOR THE QUENCHING PROCESS INSIDE THE CFM

Based on the difference in the temperature response between the bottom, non degraded part of the CFM and the periphery, it is postulated that water could not easily enter inside the shroud. In the periphery the cooling rate was determined by boiling heat transfer and two phase instability considerations. On the other hand, inside the shroud the penetration of water, as governed by the hydraulic resistance of the CFM, probably controlled the quenching rate. A schematic description of the postulated quenching process inside the CFM is shown in Fig. 8.

In order to simulate the data trends discussed previously, the CFM is axially nodalized into equal height nodes. A momentum balance between the coolant flow inside the CFM and the quasi-quiescent water column in the periphery yields:

$$\rho_1 g (L - L_{CFM}) = \sum_n \frac{f \, \Delta Z \, W_i^2}{2 D_{h,i} \rho_i A_i^2} \qquad (2)$$

where:

ρ_1 = water density in the periphery
g = acceleration due to gravity
L = height of the liquid in the periphery (referenced in the bottom of vessel)
L_{CFM} = height of the liquid in the CFM (referenced to the bottom of vessel)
f = friction factor (assumed constant)
W_i = gas flow rate in node i
$D_{h,i}$ = hydraulic diameter in node i
ρ_i = gas density in node i
A_i = open flow area in node i
n = number of uncovered nodes.

In solving Equation 2 it should be noted that the mass flow rate can change along the channel due to the consumption of steam and generation of hydrogen. Hence

$$W_i = M_{w,i}\dot{N} \tag{3}$$

where $M_{w,i}$ is the node average molecular weight and $\dot{N}$ is the molar flow rate in the CFM. This flow is constant along the channel and, therefore, from Equation 2, the steam generation in the CFM pool is

$$W_{st} = 18 \left[\frac{2\rho_1 g(L - L_{cfm})}{f\Delta z \sum \frac{M_{w,i}^2}{D_{h,i}\rho_i A_i^2}} \right] \tag{4}$$

The overall heat transfer rate from a saturated CFM pool is therefore

$$Q = W_{st}h_{fg} \tag{5}$$

Assuming that this heat is predominantly due to quenching of the part of the fuel that is covered by water (slow quenching approximation), the rate of change of the water level inside the CFM is then

$$\frac{dL_{CFM}}{dt} = \frac{Q}{\rho_f \ (1 - \varepsilon) A_{CFM} C_F \ (T_n - T_{sat})} \tag{6}$$

where ρ_f and C_f are the fuel material density and specific heat, ε is the porosity of the quenching node, and A_{CFM} is the CFM cross-sectional area inside the shroud.

Equations (4), (5) and (6) are solved for the CFM water level and steam flow rate. Energy balances are performed for nodes above the quenched region to obtain fuel and gas temperatures. Energy generation by hydrogen is also accounted for. Thus, the average densities and molecular weight needed in equation (4) are calculated directly from these energy balances and oxidation rates.

The flow area and hydraulic diameter needed in equation (4) are directly deducted from the local porosity and cladding ballooning deformation which were assumed to be initially equal to that calculated in Ref. [4] at the end of the flooding process. The relationship between the local porosity and the needed geometry parameters are fairly straight forward for slightly deformed geometry and provided in Ref. [7]. For a "crust like" geometry with very low porosity, the hydraulic diameter was calculated as

$$D_h = D_o \left[\frac{\varepsilon(T)}{\varepsilon(T_{sat})} \right]^{1/2} \tag{7}$$

where $\varepsilon(T)$ is the temperature dependent local porosity, T_{sat} is the saturation temperature and D_o is a characteristic diameter of a crack in a quenched core material crust. This value is a model parameter which should typically be of the order of fractions of a millimeter based on observations from in-pile (PBF-SFD) and out-of-pile (CORA) tests.

To facilitate the simulation of the quenching process the model equations (4), (5) and (6) along with the geometry-porosity relations were incorporated into MAAP 3.0 which essentially performed the fuel and gas energy balances, porosity evaluations and the oxidation rate calculations. It should be noted that it was assumed that the porosity of a molten node was set to a low limit of 10% rather than zero originally assumed by MAAP. Post-test examination in Ref. [5] suggest that the local porosity based on the entire CFM cross-section was probably above this value during the test. A porosity value of zero is not suitable since it would prevent any quenching of the CFM.

Following the core nodalization in Ref. [4], twelve axial nodes and two radial rings were used to represent the CFM.

RESULTS AND DISCUSSIONS

Results of the model outlined above are shown in Figure 9, 10 and 11 for the water level, temperature response and hydrogen generation. The initial conditions for the fuel material temperature and mass distribution were taken from Ref. [4] and L_{CFM} was set equal to zero at 1793 seconds. The initial crust porosity was taken as 10% and $D_o = 2.10^{-4}$m.

The calculated water level shows a slow rise followed by a relatively rapid increase in level between 1800 and 1816 caused by the increased porosity due to cooling of the refrozen material. At 1816 seconds, additional melting at the upper half of the bundle is predicted to increase the hydraulic resistance of the channel and again to slow down the water rise. As the consolidated crust like region is quenched at 1850 seconds, the water level quickly covers the core. Temperature calculations (shown in Fig. 10) are in qualitative agreement with the data. The temperature at elevation 0.25 m decreases rather rapidly while the upper portion heats up rapidly to melting. The bundle was finally cooled down at 1860 seconds in agreement with data. Figure 11 indicates that 210 additional grams of hydrogen were generated during the quenching process. This amount is probably much lower than was actually generated during the quenching process. The main parameter that controls the hydrogen generation is the distribution of zircaloy at the beginning of the flooding process. More hydrogen could be predicted using a more mechanistic melt progression model during the heatup and relocation period.

The model is fairly sensitive to the values of the initial porosity distribution, the minimum porosity of molten node and D_o. Faster quenching with reduced hydrogen production is predicted with smaller hydraulic resistance while slower quenching rates with increased hydrogen production is obtained for a combination of model parameters that would increase the hydraulic resistance. Hence, the main contribution of the above analysis is to indicate that hydraulic rather than thermal limitation could have been the controlling rate in quenching the CFM.

The results shown here are generally in agreement with the data trends which indicate that the bottom portions of the assembly would slowly cool down by the penetration of water while the top portion would be heated up and oxidized by the steam. The basic assumption of the model that the hydraulic resistance in the CFM is controlling the quenching rate of the CFM can indeed explain the data trends. To reproduce a better simulation of the data, it is necessary to include a more mechanistic melt progression model that can better predict the porosity and its distribution in the CFM. Also needed are better relationships between porosity and hydraulic resistance.

LITERATURE CITED

1. C. M. Allison, "Perspectives on In-Vessel Behavior as it Affects DCH and DCH Experiments", presented at the DCH Workshop, Annapolis, MO, December 1989.

2. R. R. Hobbins et al., "Review of Experimental Results on LWR Core Melt Progression", Transactions of the 18 Water Reactor Safety Information Meeting, October 1990.

3. J. P. Adams et al., "Quick Look Report on OECD LOFT Experiment LP-FP-2", EG&G report OECD LOFT-T-3804, September 1985.

4. A. Sharon et al., "Simulation of LOFT Experiment LP-FP-2", Using MAAP Version 3.0, FAI report to EPRI, FAI/86-27 June 1986. Also published as an EPRI report with limited distribution.

5. S. M. Jensen et al., "Post Irradiation Examination Data and Analysis of OECD LOFT Fission Product Experiment LP-FP-2", OECD LOFT-T-3810, December 1989.

6. S. Hagen, Presentation in the International CORA Workshop, Karlsruhe, September 1989.

7. A. Sharon, R. E. Henry, C. D. Wu and R. J. Hammersley, "Model for the Recovery of a Badly Degraded Core", Proceedings of ANS Thermal-Hydraulics Division, ANS Winter Annual Meeting, 1989.

Figure 1. Cross-section of the LOFT Core in Test LP-FP-2

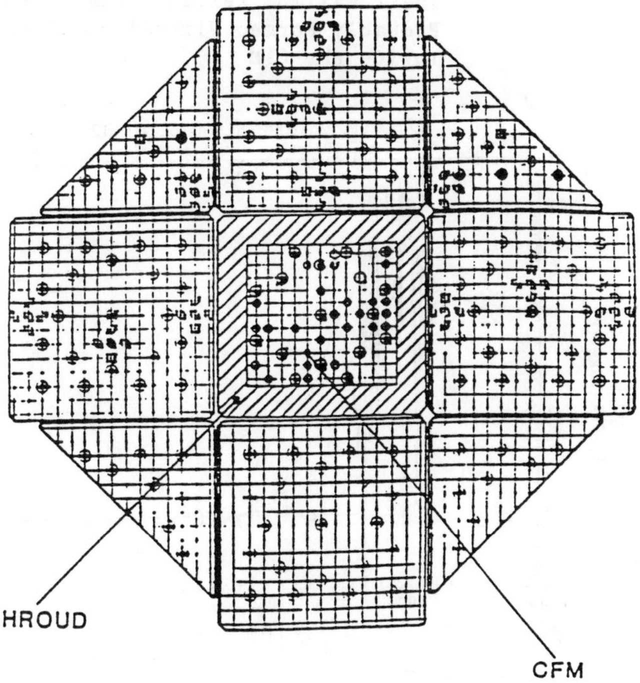

Figure 2. Pressures in the Accumulators and the Primary System During Injection

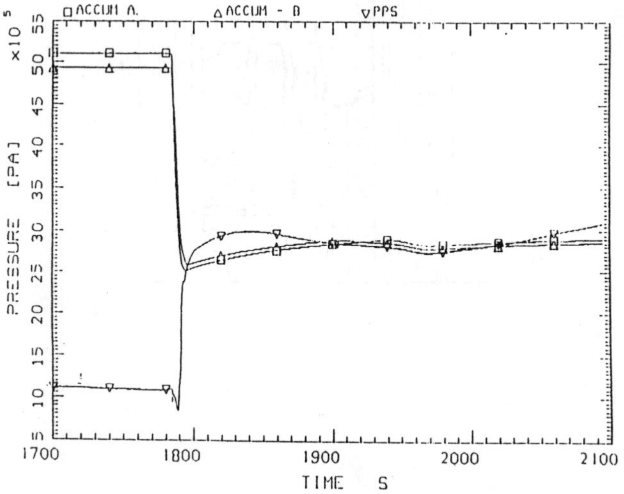

Figure 3. Axial Temperature Distribution in Peripheral Subassembly 2 During the Reflood. The last two digits in the thermocouple identifier represent its elevation in inches from bottom of core.

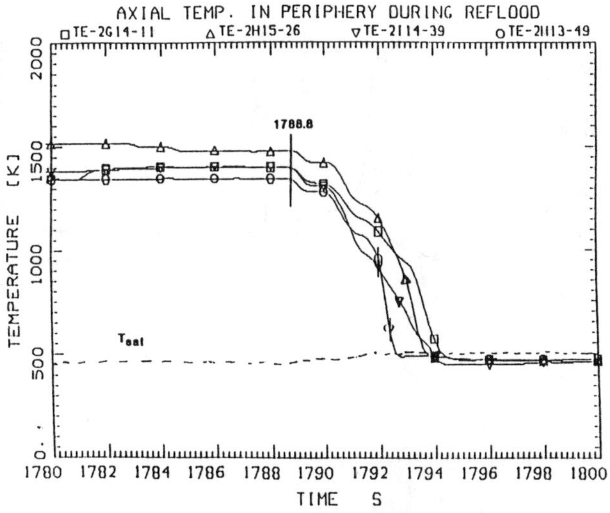

Figure 4. The Response of Typical Thermocouples Inside the CFM and the Periphery

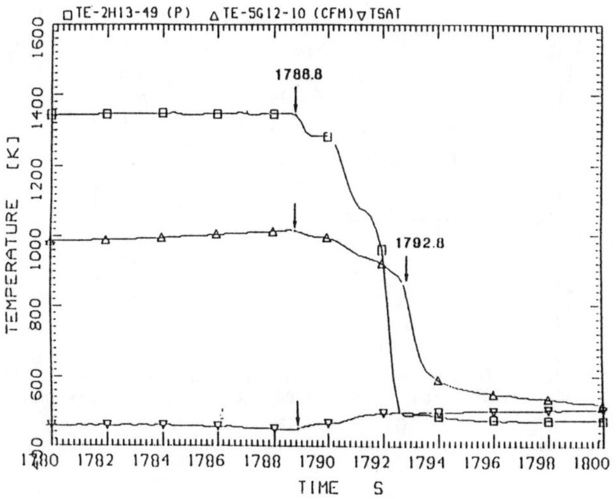

Figure 5. The response of the CFM thermocouple at 10 in. Elevation to the Quenching Process

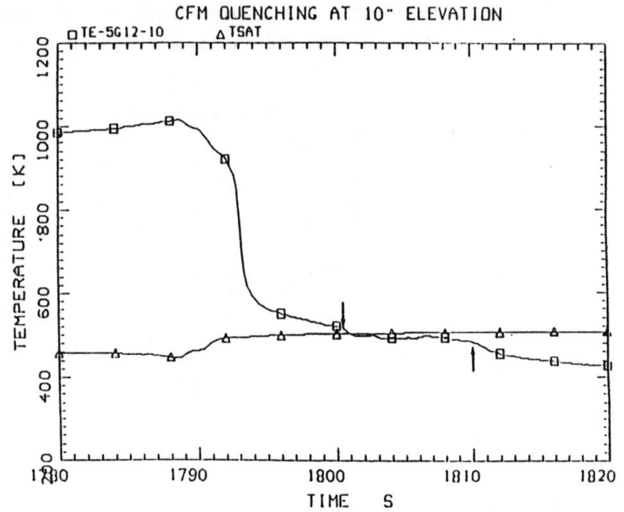

Figure 6. Response of CFM Guide Tube Thermocouples Facing the Four Sides of the Shroud at the 66 in. Elevation (Top of Core).

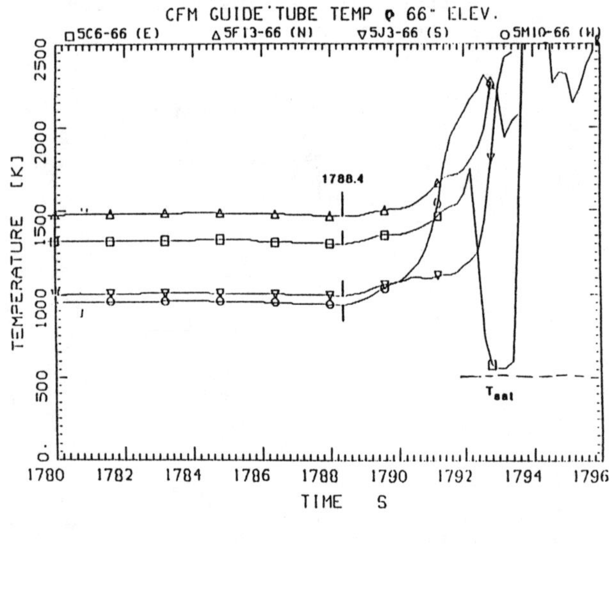

Figure 7. Temperature History During Quenching of the CFM. All Thermocouples Shown Have Reached Their Maximum Reliable Limit and Probably Formed Virtual Junctions Inside the CFM

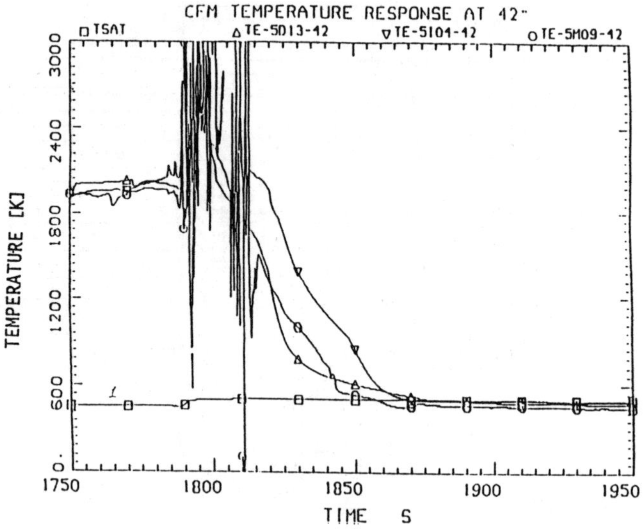

Figure 8. Schematic Description of the Quenching Process Inside the CFM

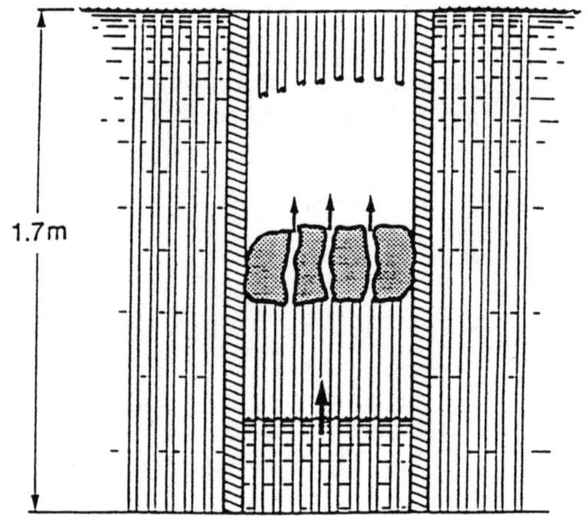

Figure 9. Predicted Water Level during the Quench in LP-FP-2 Experiment

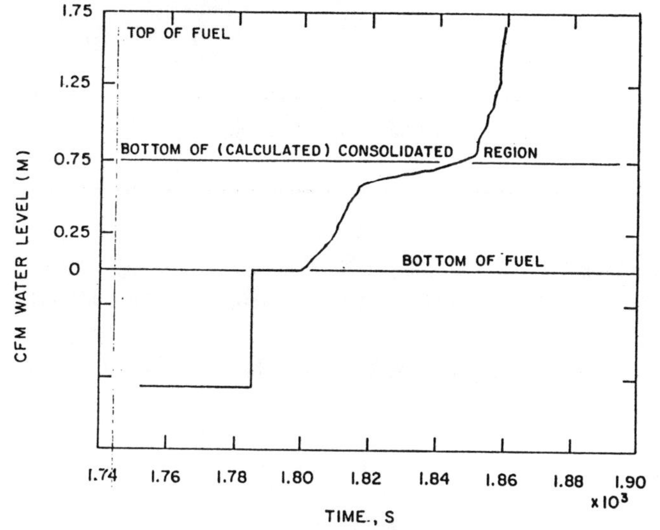

Figure 10. Predicted Temperature Variations at Various Locations During the Quench Period for LP-FP-2 Experiment. Node 2 represents elevation 0.25 m and node 12 elevation 1.68 (top of elevation)

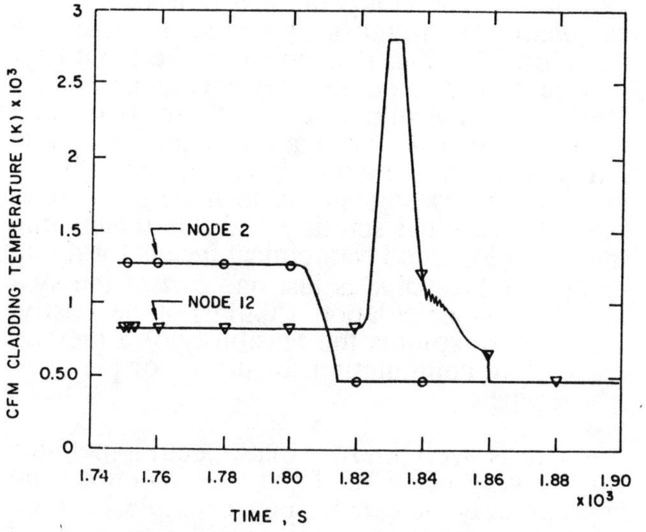

Figure 11. Predicted Hydrogen Generation during the Quench Period

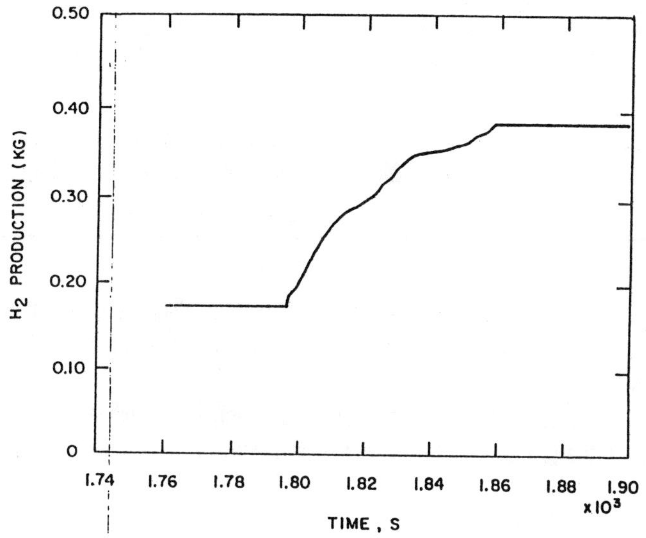

USE OF SPECTRAL METHODS FOR HEAT PIPE ANALYSIS

F. Issacci and I. Catton ■ Mechanical, Aerospace and Nuclear Engineering Dept., University of California, Los Angeles, CA 90024-1597

The cost of numerical studies of heat pipe dynamics is prohibitively expensive both in computational time used and in computer resource allocation. The potential of the spectral methods is explored for the heat pipe analysis to significantly reduce the computational efforts.

Using data sets from the numerical calculations available from past studies, it is shown that a few first terms of Fourier Series, given the proper trial functions, will satisfactorily represent the data sets.

A great deal of effort has gone into calculation of the behavior of the vapor phase of a heat pipe. Work by Bystrov and Goncharov (1983) led to the conclusion that a finite difference solution would be very costly. They found that the time step needed for stable advancement was on the order of 10^{-7} sec or smaller. This led them to a quasi-two-dimensional (2D) approach based on a simple cross-stream direction profile. Subsequent work by others such as Bowman (1987) came to similar conclusions. Bowman now calculates the vapor phase as a 1D process with influence coefficients (Bowman, 1990). As a result most heat pipe analyses assume 1D vapor flow, e.g. Bowman et al. (1990) and Chow and Zhong (1990). The one-dimensional assumption is a direct result of the cost of computation. The use of a 1D model of the vapor phase has, however, been shown to be questionable.

The experimental work by Galaktionov and Trukhanova (1985) shows that the temperature profile is not 1D under steady state conditions. Experimental measurements by Marn et al. (1990) show that when a noncondensible gas is present, its concentration is 2D as is the temperature profile. This is in contradiction to the usual planar front approximation used in heat pipe analysis. The results of 2D transient and steady-state analyses by Tien and Rohani (1974), Faghri et al. (1989) and Jang et al. (1989) show that a 1D model of the vapor flow does not accurately predict the axial heat and mass transfer and pressure drop in a heat pipe. These are important considerations if the heat pipe is operated near its design limits.

The cost of two-dimensional numerical studies has been shown by Bystrov and Goncharov (1983), as mentioned above, Costello et al. (1986), Issacci et al. (1990) and Issacci et al. (1991) to be prohibitively expensive both in computer time used and in computer resource allocation. The computational time needed for calculation of a start-up process, is about two hours on CRAY 2, depending on the input heat flux and heat pipe geometry (Issacci et al., 1991). As a result of the expense associated with computation by numerical methods, one will not do very many. It is our view that a good system design needs to undergo a great deal of analytical scrutiny and this means one must develop good economical means for doing them. The heat pipe is just one part of the system and its calculation should not be costly. This paper explores the feasibility of a method of efficient computation for the vapor phase of a heat pipe.

It is well known that spectral methods can be very efficient if the trial functions are appropriately selected. For example, a one-term approximation will yield fairly good results for flow in a rectangular duct. The

Unnversity of California, Los Angeles, CA

Graetz problem yields good results for the first eigenvalue when a two term approximation is used. At the outset it was not clear how many terms would be needed to properly represent the vapor phase in a heat pipe but data sets from expensive numerical calculations were available from several past studies, Issacci et al. (1990). We will demonstrate the feasibility of a spectral method by showing that a few terms of a Fourier series will satisfactorily represent the data sets. A computer code using a spectral method then remains to be written.

ANALYSIS

A schematic of the vapor flow model used to represent the heat pipe is shown in Figure 1. The spectral method used here is a Fourier analysis of the dependent parameters of the problem. One assumes an approximation for the parameter of interest,

$$\tilde{\rho u} = \sum_{m=1}^{M} \sum_{n=1}^{N} A_{mn}\, f_m(x)\, g_n(y) \tag{1}$$

substitutes into the governing equation

$$L(\tilde{\rho u}) = \varepsilon \tag{2}$$

then weights and integrates and sets the error ε to zero,

$$\int_x \int_y L(\tilde{\rho u})\, f_m(x)\, g_n(y)\, dx\, dy = 0$$

$$m = 1,\ldots,M \qquad n = 1,\ldots,N \tag{3}$$

to obtain an equation(s) for the coefficients A_{mn}.

For multi-dimensional compressible flows, this approach, sometimes called the Galerkin method, requires a great deal of tedious algebra. Before setting out to do this, we decided that the efficacy of the method should be tested. Detailed solutions based on a finite difference procedure were available from an earlier work, Issacci et al. (1991). These results were used to determine how many terms would be needed in the approximation.

The functions chosen to represent the problem parameters must satisfy the proper boundary conditions. It is particularly helpful, although not necessary, to choose orthogonal functions. A series solution for the axial mass flux, ρu, that has the proper symmetry and satisfies the proper boundary conditions is

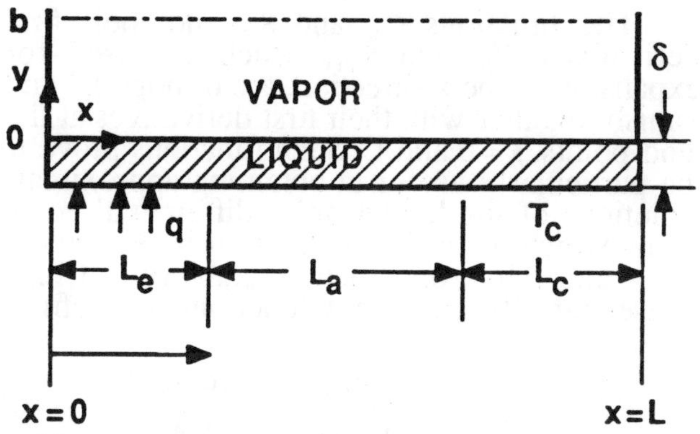

Fig. 1. Vapor flow model in a heat pipe

$$\rho u(x,y,t) = \sum_{m,n} A_{mn}(t)\, C_m\!\left(\lambda_m \frac{x - L/2}{L}\right) \sin \frac{(2n-1)\,\pi y}{2b} + \sum_{m,n} A'_{mn}(t)\, S_m\!\left(\mu_m \frac{x - L/2}{L}\right) \sin \frac{(2n-1)\,\pi y}{2b} \tag{4}$$

The vertical mass flux, ρv, must be selected so that in the steady state limit

$$\frac{\partial \rho v}{\partial y} = \frac{\partial \rho u}{\partial x} \tag{5}$$

as well as satisfy appropriate boundary conditions. A series representation that does so is

$$\rho v(x,y,t) = \sum_{m,n} B_{mn}(t)\, C'_m\!\left(\lambda_m \frac{x-L/2}{L}\right) \cos\frac{(2n-1)\pi y}{2b} + \sum_{m,n} B'_{mn}(t)\, S'_m\!\left(\mu_m \frac{x-L/2}{L}\right) \cos\frac{(2n-1)\pi y}{2b} \tag{6}$$

Scaling x and y,

$$Y = y/b \quad \text{and} \quad X = x/L - 1/2 \tag{7}$$

yields

$$\rho u(X,Y,t) = \sum_{m,n} A_{mn}(t)\, C_m(\lambda_m X) \sin \frac{(2n-1)\pi Y}{2} + \sum_{m,n} A'_{mn}(t)\, S_m(\mu_m X) \sin \frac{(2n-1)\,\pi Y}{2} \tag{8}$$

$$\rho v(X,Y,t) = \sum_{m,n} B_{mn}(t)\, C'_m(\lambda_m X) \cos \frac{(2n-1)\,\pi Y}{2} + \sum_{m,n} B'_{mn}(t)\, S'_m(\mu_m X) \cos \frac{(2n-1)\,\pi Y}{2} \tag{9}$$

The functions C_m and S_m and their first derivatives C'_m and S'_m, which are used for expansion in the x-direction, are orthogonal and vanish together with their first derivatives at the end-points, x = 0 and L. Such functions prove to be particularly useful in obtaining approximate solutions of the higher-order differential equations which arise in problems of hydrodynamics and hydromagnetic stability (Chandrasekhar, 1961). These functions are defined as

$$C_m(\lambda_m X) = \frac{\cosh \lambda_m X}{\cosh \lambda_m/2} - \frac{\cos \lambda_m X}{\cos \lambda_m/2} \quad (10)$$

$$S_m(\mu_m X) = \frac{\sinh \mu_m X}{\sinh \mu_m/2} - \frac{\sin \mu_m X}{\sin \mu_m/2} \quad (11)$$

$$C'_m(\lambda_m X) / \lambda_m = \frac{\sinh \lambda_m X}{\cosh \lambda_m/2} + \frac{\sin \lambda_m X}{\cos \lambda_m/2} \quad (12)$$

$$S'_m(\mu_m X) / \mu_m = \frac{\cosh \mu_m X}{\sinh \mu_m/2} - \frac{\cos \mu_m X}{\sin \mu_m/2} \quad (13)$$

where λ_m and μ_m (m = 1, 2, 3, ...) are the positive roots of the equations

$$\tanh \lambda_m/2 + \tan \lambda_m/2 = 0 \quad (14)$$

$$\coth \mu_m/2 - \cot \mu_m/2 = 0 \quad (15)$$

The orthogonality properties of the normalized functions C_m and S_m are

$$\int_{-1/2}^{1/2} C_m(\lambda_m X)\, C_n(\lambda_n X)\, dX = \delta_{mn}$$

$$\int_{-1/2}^{1/2} S_m(\mu_m X)\, S_n(\mu_n X)\, dX = \delta_{mn} \quad (16)$$

where δ_{mn} is the Dirac delta function, and

$$\int_{-1/2}^{1/2} C_m(X)\, S_n(X)\, dX = 0 \quad (17)$$

The first four roots of equations (14) and (15) are presented in Table 1. For m > 4 the roots are found from

TABLE 1. The roots of the characteristic equations

m	λ_m	μ_m
1	4.7300407449	7.8532046241
2	10.9956078380	14.1371654913
3	17.2787596574	20.4203522456
4	23.5619449020	26.7035375555

$$\lambda_m = (2m - 1/2)\,\pi \ ; \ \mu_m = (2m + 1/2)\,\pi \quad (18)$$

The functions C_m and S_m and their first three derivatives for m = 1, ..., 4 are tabulated by Harris and Reid (1958). Reid and Harris (1958) have also provided a useful sets of integrals involving the C- and the S-functions.

The coefficients in the series representations are found by Fourier analysis of the numerical data obtained by Issacci et al. (1991),

$$A_{mn} = \int_{-1/2}^{1/2}\int_{0}^{1} (\rho u) C_m(\lambda_m X) \sin\frac{(2n-1)\pi}{2} Y dY dX \quad (19)$$

$$A'_{mn} = \int_{-1/2}^{1/2}\int_{0}^{1} (\rho u) S_m(\mu_m X) \sin\frac{(2n-1)\pi}{2} Y dY dX \quad (20)$$

$$B_{mn} = \int_{-1/2}^{1/2}\int_{0}^{1} (\rho v) C'_m(\lambda_m X) \cos\frac{(2n-1)\pi}{2} Y dY dX \quad (21)$$

$$B'_{mn} = \int_{-1/2}^{1/2}\int_{0}^{1} (\rho v) S'_m(\mu_m X) \cos\frac{(2n-1)\pi}{2} Y dY dX \quad (22)$$

RESULTS AND DISCUSSIONS

The numerical results used to calculate the coefficients of the Fourier series, Eqs. (19) to (22), were obtained for a liquid-sodium heat pipe with dimensions, $L_e = L_a = L_c = b$ (see Figure 1).

Spectral methods can be very effective if the first few coefficients are sufficient to represent the data. Figures 2(a) and 2(b) show the first coefficients of a representative calculation. In Figure 2(a) it is shown that the first two coefficients in the y-direction (index -n) are the dominant ones. The same is true for the coefficients in the x-direction (index -m) shown in Figure 2(b).

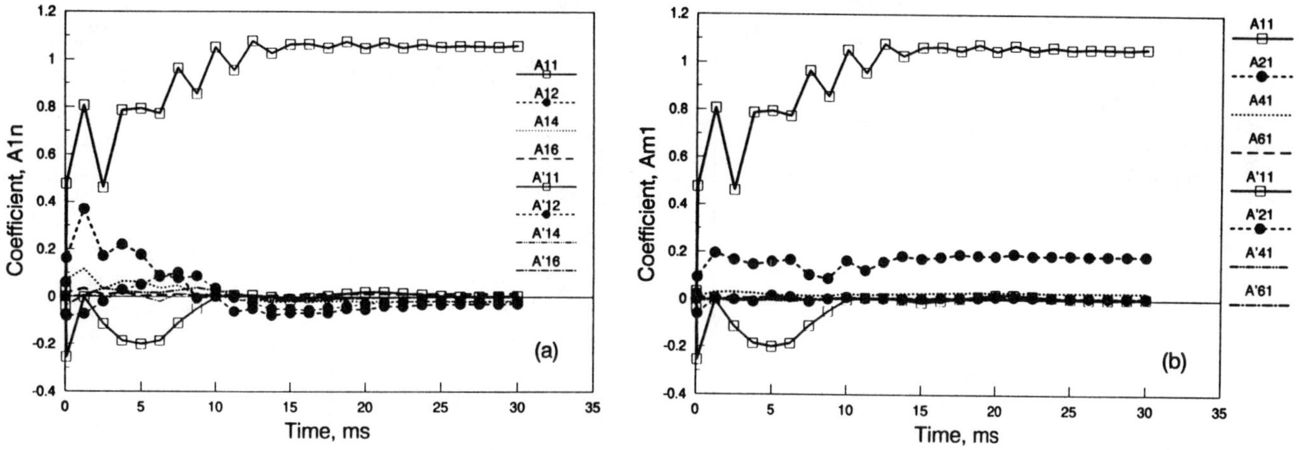

Figure 2. Coefficients A_{1n} and A_{m1} for different n and m

Figures 3 and 4 show the few first coefficients as functions of time for different input heat fluxes for axial and vertical mass fluxes, respectively. The periodic oscillations of these coefficients with time is caused by the multiple wave reflections in the evaporator. Issacci et al. (1991) have explained these oscillations in detail. These figures imply that the mean behavior of the coefficients can be predicted by an exponential asymptote.

One of the major conclusions of previous studies of the dynamic behavior of vapor flow in heat pipes is that, in the complete analysis of heat pipes, the vapor flow can be assumed quasi-steady state. This conclusion is based on the fact that the characteristic time of the vapor flow is significantly smaller than that of the liquid flow. Under assumption of quasi-steady state, the coefficients of the series solutions will only be functions of the input heat flux. Figure 5 shows dependency of the few first coefficients on the input heat flux at steady state. These figures suggest that the coefficients can be represented by a polynomial in terms of log Q, where Q is the input heat flux applied to the evaporator. A general formulation for this dependency is assumed as

$$\text{COEFF} = a_o + a_1 \log Q + a_2 (\log Q)^2 \qquad (23)$$

where COEFF denotes the coefficients A_{mn}, A'_{mn}, B_{mn} and B'_{mn}. The constants of the above equation for the first few coefficients are presented in Table 2.

Figures 6(a) to 6(c) show the axial mass velocity profile at three different cross sections of the heat pipe; middle of the evaporator, adiabatic region and condenser, respectively. The data from the numerical analysis is shown by a solid curve. Series solutions with different number of terms are also shown. In all the figures, it is shown that the calculated data can be easily represented by few first coefficients in the series solutions. The same conclusion can be drawn for the vertical mass flux at the cross section y = 0.75b, shown in Figure 7.

For a steady state or quasi-steady state, a good approximation can be obtained for ρu using a five term approximation, see Figures 2 and 3. The result is

$$\rho u = A_{11}\, C_1(\lambda_1 X) \sin\frac{\pi}{2}Y + A_{12}\, C_1(\lambda_1 X) \sin\frac{3\pi}{2}Y + A'_{11}\, S_1(\mu_1 X) \sin\frac{\pi}{2}Y + A'_{12}\, S_1(\mu_1 X) \sin\frac{3\pi}{2}Y + A_{21}\, C_2(\lambda_2 X) \sin\frac{\pi}{2}Y \qquad (24)$$

Using the coefficients presented in Table 2, ρu at steady state can be represented in terms of the input heat flux as

$$\begin{aligned}\rho u = {} & (0.54 + 0.04 \text{ Log } Q)\, C_1 (\lambda_1 X) \sin \frac{\pi}{2}Y + \\ & (0.67 + 0.02 \text{ Log } Q)\, C_1 (\lambda_1 X) \sin \frac{3\pi}{2}Y + \\ & (0.48 - 0.21 \text{ Log } Q)\, S_1 (\mu_1 X) \sin \frac{\pi}{2}Y + \\ & (-0.7 + 0.29 \text{ Log } Q)\, S_1 (\mu_1 X) \sin \frac{3\pi}{2}Y + \\ & (-0.15 + 0.12 \log Q)\, C_2 (\lambda_2 X) \sin \frac{\pi}{2}Y \end{aligned} \qquad (25)$$

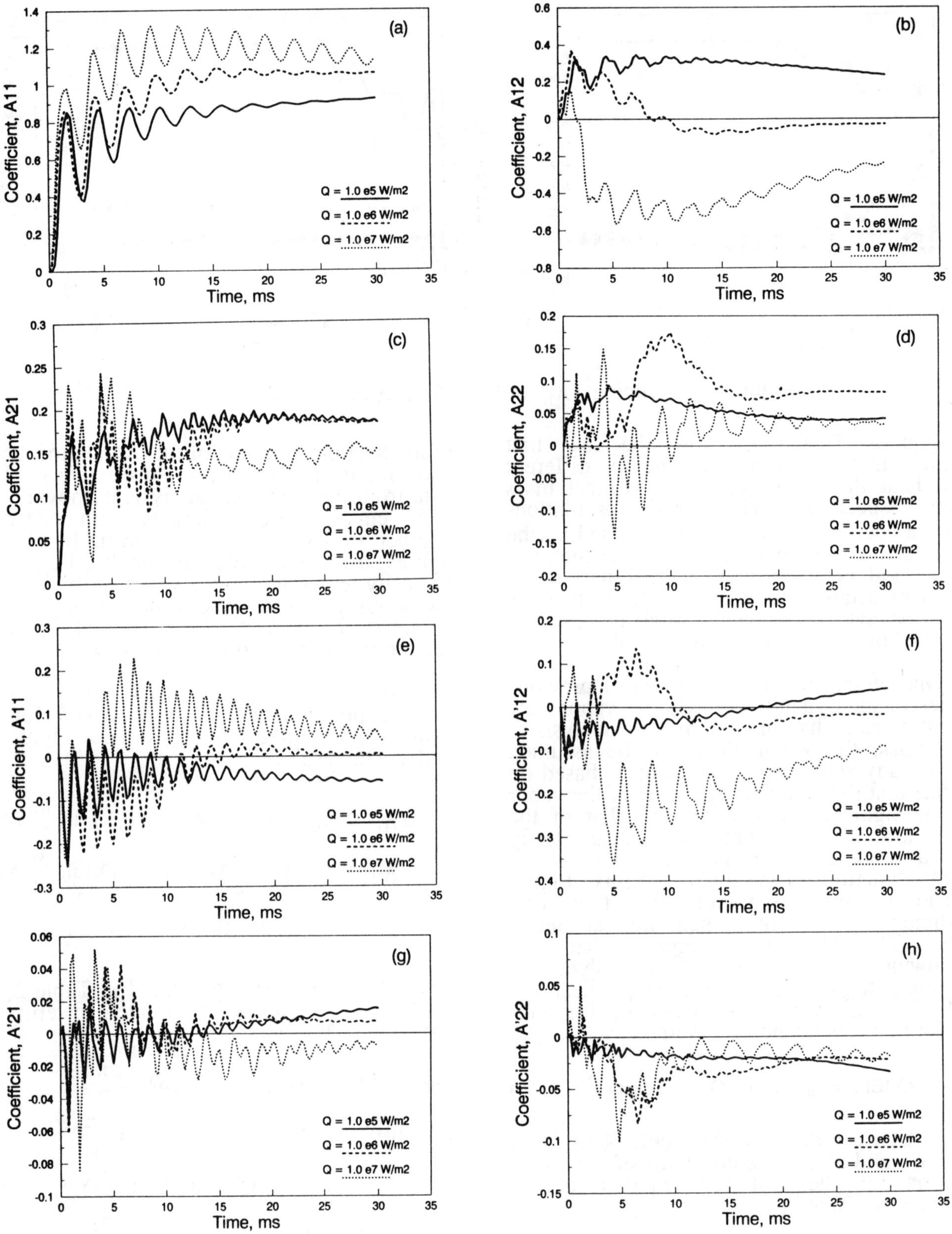

Figure 3. First coefficients for different input heat fluxes; axial mass flux

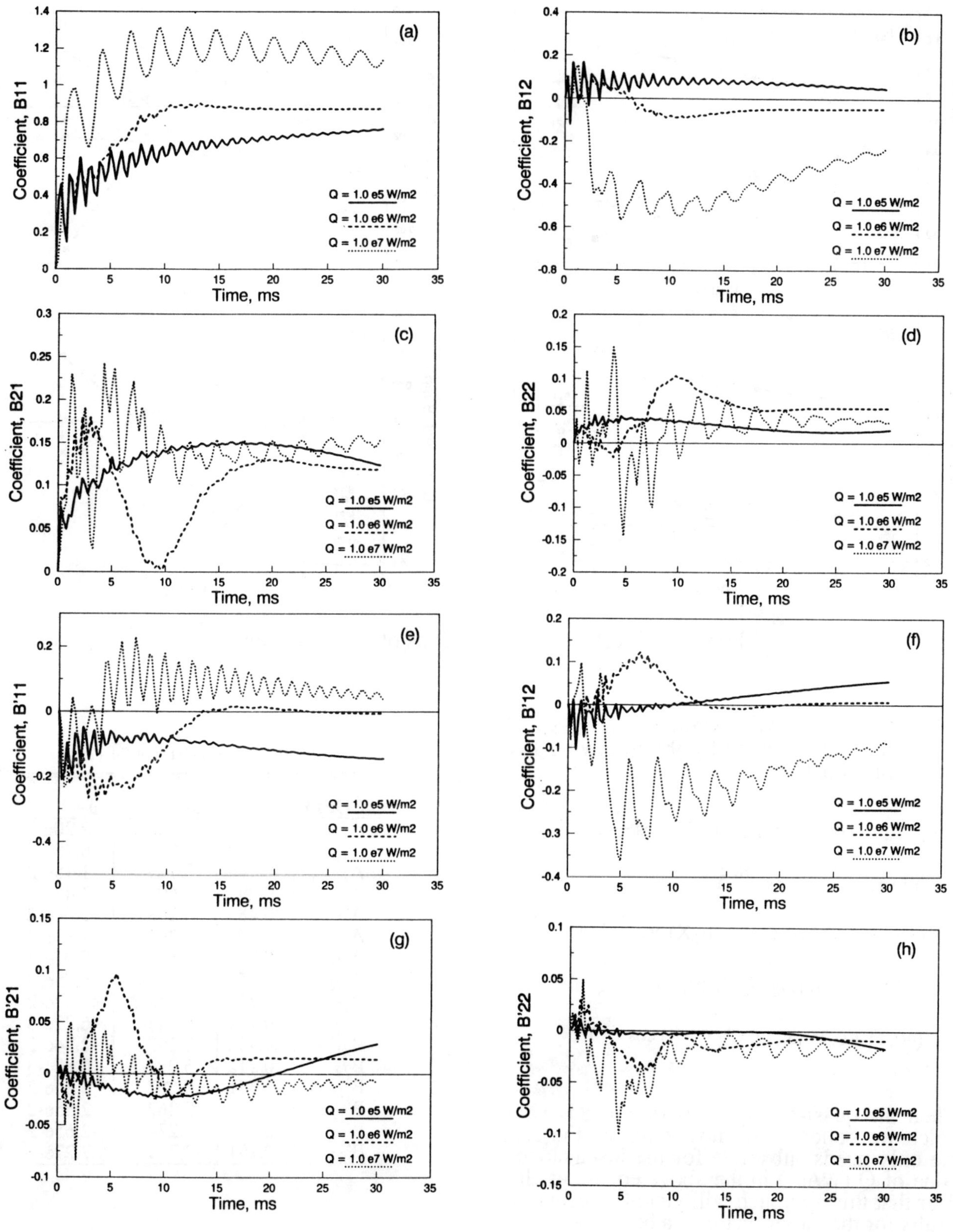

Figure 4. First coefficients for different input heat fluxes; vertical mass flux

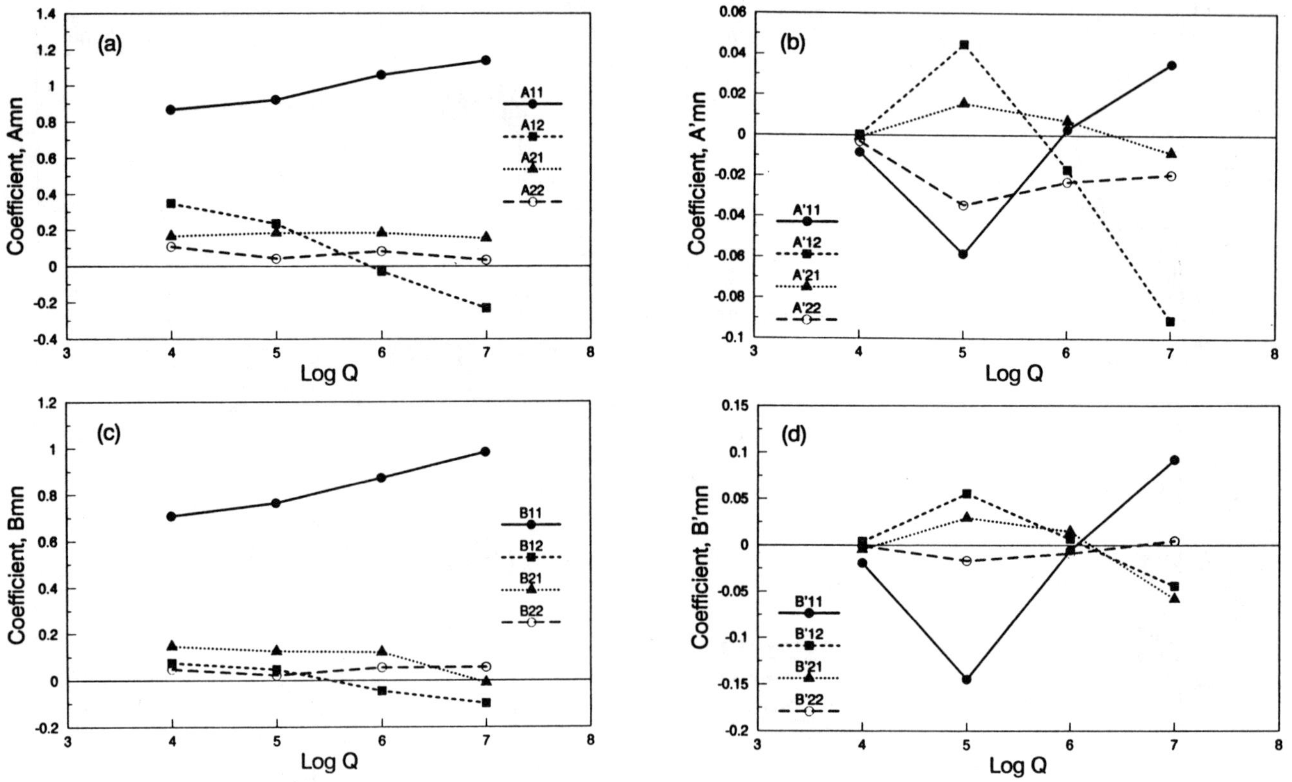

Figure 5. Coefficients as functions of input heat flux

where Q is in W/M^2. Similarly ρv at steady state can be represented by the following five-term approximation

$$\rho v = (0.72 - 0.06 \, \text{Log Q}) \, C'_1 (\lambda_1 X) \cos \frac{\pi}{2} Y + 0.15 \, C'_1 (\lambda_1 X) \cos \frac{3\pi}{2} Y + (1.34 - 0.57 \, \text{Log Q}) \, S'_1 (\mu_1 X) \cos \frac{\pi}{2} Y + (-0.63 + 0.26 \, \text{Log Q}) \, S'_1 (\mu_1 X) \cos \frac{3\pi}{2} Y + (-0.44 + 0.25 \, \log Q) \, C'_2 (\lambda_2 X) \cos \frac{\pi}{2} Y \tag{26}$$

When the quasi-steady approximation is valid, which we believe will most often be the case, one only needs substitute for the instantaneous value of Q (W/m^2) in the above equations. It is clear that this approach will yield fast economic results for the vapor phase of a heat pipe.

TABLE 2. Coefficients of Eq. (23)

COEFF	a_o	a_1	a_2
A_{11}	0.544	4.16e-2	-2.20e-2
A_{12}	0.668	2.09e-2	6.78e-3
A_{21}	-0.145	0.123	-1.15e-2
A_{22}	0.295	-6.69e-2	4.43e-3
A'_{11}	0.485	-0.208	2.06e-2
A'_{12}	-0.692	0.293	-2.97e-2
A'_{21}	-0.211	8.49e-2	-8.01e-3
A'_{22}	0.254	-9.99e-2	8.75e-3
B_{11}	0.716	-5.74e-2	1.37e-2
B_{12}	0.151	6.94e-3	-6.22e-3
B_{21}	-0.438	0.253	-2.74e-2
B_{22}	0.212	-6.98e-2	6.85e-3
B'_{11}	1.341	-0.567	5.59e-2
B'_{12}	-0.629	0.262	-2.55e-2
B'_{21}	-0.678	0.274	-2.65e-2
B'_{22}	0.191	-7.75e-2	7.27e-3

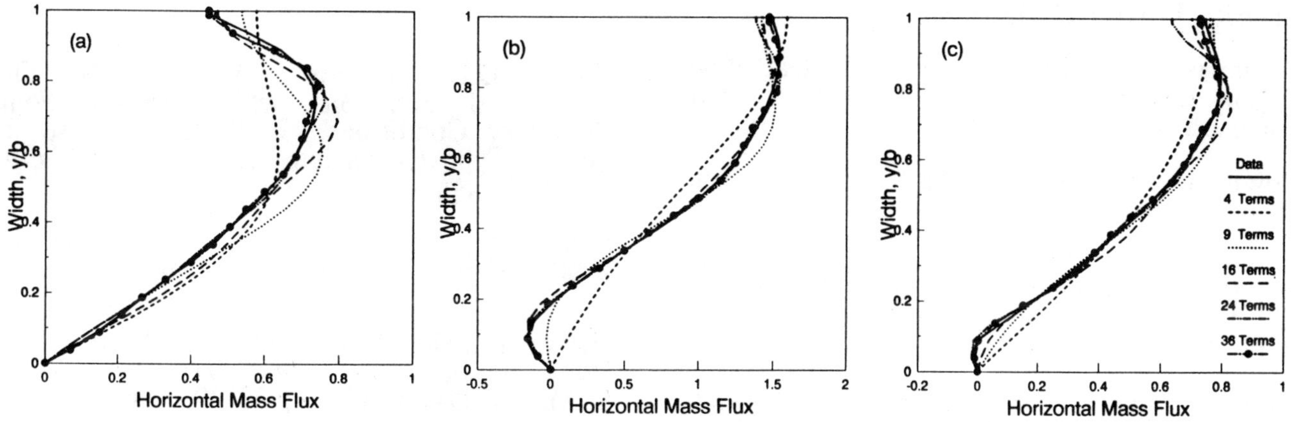

Figure 6. Series solutions compared to the calculated axial mass flux at the middle of (a) evaporator, (b) adiabatic region, (c) condenser

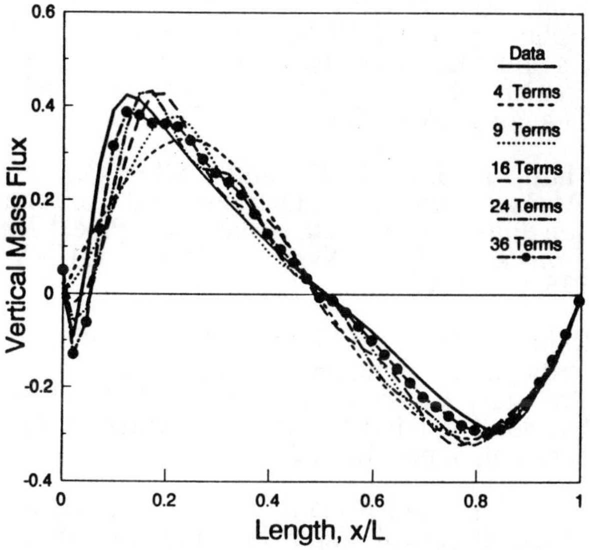

Figure 7. Calculated vertical mass flux

CONCLUSIONS

The potential of the spectral methods has been studied for the heat pipe analysis. It has been shown that the numerical data for the mass flux in different directions can be represented in a series form. The most complicated mass flux profile under steady state condition (including reverse flow) is well represented by first few terms in the series. Interpolation allows one to deal with different heat fluxes.

For a given heat pipe, several numerical solutions can be used to obtain the coefficients in a spectral representation. The spectral representation allows one to represent the dynamic behavior of the heat pipe for use in space thermal models at very little computational cost.

ACKNOWLEDGEMENT

This work was supported by NASA Lewis under Contract No. NAG3-899 and NASA Dryden under Contract No. NCC2-374 Supp. 2.

NOTATION

A_{mn}	coefficients of series solution
b	heat pipe half width
B_{mn}	coefficients of series solution
C_m	beam function
L	heat pipe length
Q	input heat flux (W/msup2)
S_m	beam function
t	time
x	axial coordinate
X	axial coordinate
y	vertical coordinate
Y	vertical coordinate
λ_m	eigenvalues of C_m
μ_m	eigenvalues of S_m
ρu	axial mass flux
ρv	vertical mass flux

Subscripts

a	adiabatic region
c	condenser
e	evaporator

LITERATURE CITED

Bowman, W.J., 1987, "Simulated Heat Pipe Vapor Dynamics," *Ph.D. Dissertation,* Air Force Institute of Technology.

Bowman, W.J., 1990, NASA Workshop on Transient Heat Pipe Analysis.

Bowman, W.J., McClure, W.C., and Towne, M., 1990, "Transient Heat Pipe Modeling, Part 2," *AIAA 28th Aerospace Sciences Meeting,* Reno, Nevada, AIAA Paper No. 90-0061.

Bystrov, P.I., and Goncharov, V.F., 1983, "Starting Dynamics of High Temperature Gas-Filled Heat Pipes," *High Temperature,* Vol. 21, pp 927-936.

Chandrasekhar, S., 1961, *Hydrodynamic and Hydromagnetic Stability,* Dover, New York.

Chow, L., and Zhong, J., 1990, "Analysis of Heat Pipe Start-up from the Frozen State," *AIAA/ASME 5th Joint Thermophysics and Heat Transfer Conf.,* Seattle, WA.

Costello, F.A., Montague, A.F., and Merrigan, M.A., 1986, "Detailed Transient Model of a Liquid-Metal Heat Pipe," *Proc., AIAA/ASME 4th Joint Thermophysics and Heat Transfer Conf.,* Boston, MA.

Faghri, A., Chen, M-M., and Mahefkey, E.T., 1989, "Simultaneous Axial Conduction in the Fluid and the Pipe Wall for Forced Convective Laminar Flow with Blowing and Suction at the Wall," *Int. J. Heat Mass Transfer,* Vol. 32, No. 2, pp 281-288.

Galaktionov, V.V., and Trukhanova, L.P., 1985, "Study of the Process of Heat and Mass Transfer in the Region of the Vapor-Gas Front in a Gas-Regulated Heat Pipe," *J. Eng. Physics,* Vol 48, pp 296-300.

Harris, D.L., and Reid, W.H., 1958, "On Orthogonal Functions which Satisfy Four Boundary Conditions. I. Tables for Use in Fourier-Type Expansion," *Astrophys. J. Supp. Ser,* Vol. 3, pp 429-447.

Issacci, F., Ghoniem, N.M., and Catton, I., 1990, "Vapor Flow Patterns During Start-up Transient in Heat Pipes," *AIAA/ASME 5th Joint Thermophysics and Heat Transfer Conf.,* ASME, HTD-Vol. 135, pp 33-39, Seattle, WA.

Issacci, F., Catton, I., and Ghoniem, N.M., 1991, "Vapor Dynamics of Heat Pipe Start-up," to appear in *J. Heat Transfer.*

Jang, J.H., Faghri, A., Chang, W.S., and Mahefkey, E.T., 1989, "Mathematical Modeling and Analysis of Heat Pipe Start-up from the Frozen State," *Heat Transfer with Phase Change,* ASME HTD-Vol. 114.

Marn, J., Issacci, F., and Catton, I., 1990, "Measurements of Temperature and Concentration Fields in a Rectangular Heat Pipe," *Heat Transfer in Space Systems,* ASME, HTD-Vol. 135, pp 33-39.

Reid, W.H., and Harris, D.L., 1958, "On Orthogonal Functions which Satisfy Four Boundary Conditions. II. Integrals for Use in Fourier-Type Expansion," *Astrophys. J. Supp. Ser,* Vol. 3, pp 448-452.

Tien, C.L., and Rohani, A.R., 1974, "Analysis of the Effects of Vapor Pressure Drop on Heat Pipe Performance," *Int. J. Heat Mass Transfer,* Vol. 17, pp 61-67.

CARBON-CARBON HEAT PIPE DEVELOPMENT FOR SP-100 ADVANCED RADIATOR CONCEPTS

R.D. Rovang ■ Rockwell International, Canoga Park, CA 91303
R.B. Dirling ■ Science Applications International Corp., Santa Ana, CA
R.A. Holzl ■ Delta—G Corp., Sun Valley, CA 91352

Recent work completed in the development of a high-performance, lightweight, carbon-carbon heat pipe is presented. These efforts focused on fabrication of the carbon-carbon tube with integrally woven fins and on the development and demonstration of chemical vapor deposition techniques to protect the inside of the carbon-carbon tube from the potassium working fluid. Approximately 4 meters of carbon-carbon tubing was successfully produced. Characterization tests were performed on portions of these tubes to assess dimensional control, density, burst pressure, and final fiber orientation. Greater than 50 chemical vapor deposition samples were produced during this initial development effort. It was found that a thin (1-to 5-μm) layer of rhenium on the carbon-carbon was needed to obtain a satisfactory coating of niobium. Where this architecture was obtained, highly adhesive, crack-free coatings were produced. Coated tubes were thermally cycled five times to 875 K and showed no degradation.

A program funded by the National Aeronautics and Space Administration–Lewis Research Center (NASA–LeRC) to develop an advanced SP–100 radiator has been undertaken by Rockwell International and its subcontractors; Science Application International Corporation (SAIC), Delta–G, and Thermacore. Overall objectives of this effort are to advance the state of the art and develop the technology base for large, high–temperature radiator subsystems. More specifically, near–term objectives are to demonstrate advancements in radiator design, performance, packaging, weight, cost, life, reliability, and survivability. The basic radiator concept utilizes lightweight, high–performance, carbon–carbon heat pipes, with integrally woven fins, potassium working fluids, and a chemical vapor deposition (CVD) metallic barrier to protect the carbon–carbon tube from the working fluid. In previous work, the major technical feasibility issues associated with this concept were identified, which in general included (1) development of the CVD protective coating, (2) fabrication of the carbon–carbon tube, and (3) techniques to clean, fill, and seal a carbon–carbon heat pipe. This paper presents the progress and results achieved during the recently completed Phase III effort to resolve the first two of these feasibility issues.

During this Phase III effort carbon–carbon tubes with integral fins were successfully produced using a T–300 fiber, an angle interlocking weave architecture, and pitch densification. A barrier coating to protect the inside diameter of these tubes was developed which utilized the CVD of niobium over a thin rhenium interlayer. The rhenium interlayer proved to be critical to the success of this coating technique.

CONCEPT DESCRIPTION

Top–level requirements for this high–temperature, advanced radiator design include the following:

1. No single–point failure modes
2. Capability to reject 2.4 MW of thermal energy with an inlet temperature of 875 K to an effective sink temperature of 250 K
3. Ability to meet all safety and compatibility criteria and fit within two–thirds of the shuttle bay.

To meet these requirements while minimizing weight, a carbon–carbon heat pipe radiator design was selected. The concept utilizes 12 independent lithium loops to transfer heat from the thermoelectric generator to the heat pipe radiators. A tapered manifold is used to contain the transport lithium, the taper being proportioned to heat pipe condenser length (i.e., the width of the manifold is equal to the respective evaporator lengths). The heat pipes use a potassium working fluid and operate well within the predicted entrainment, boiling, sonic, and capillary limits.

The radiator panels utilize carbon–carbon composite heat pipes with a thin metal coating deposited on the inside diameter of the pipe to enhance wettability and prevent material interactions between the base materials and the heat pipe working fluid. The panel is designed to be easily integrated into the current SP-100 radiator configuration and is adaptable to a wide variety of radiator applications.

Table 1 presents the radiator design points for an 875 K thermoelectric system. The indicated net view factor to space of 1.31 is produced by the 17–deg cone

Table 1. Thermoelectric Advanced Radiator Design/Lithium Loop (17–deg Shield Cone Half–Angle)

Thermal Hydraulic	
Lithium loop temperature drop (K)	52.4
Duct to heat pipe temperature drop (K)	40.5
Average heat pipe evaporator flux (W/cm^2)	22.7
Average fin root temperature (K)	808
Lithium flow rate per panel (kg/s)	1.0
Loop pressure drop (kPa)	6
Physical	
Panel area (m^2)	9.28
Number of panels	12.0
Number of redundant panels	1.0
Number of heat pipes per panel	378
Number of redundant heat pipes per panel	34
Heat pipe span across manifold (cm)	4.38
Lithium manifold width (cm)	3.0–17.0
Lithium manifold height (cm)	1.7
Panel mass (kg)	71.2
Heat pipe mass per panel (kg)	42.8
Manifold mass per panel (kg)	22.0
Manifold armor mass per panel (kg)	2.1
Structural mass per panel (kg)	4.3
Heat transport subsystem mass (kg)	107
Total HRSS mass (kg)	961
Geometry	
Net view factor to space	1.31
Effective area for 11 panels (m^2)	134

half–angle. That is, a panel capable of radiating from only one side would have a net (unobstructed) view factor of 1.0. A panel having an unobstructed view to space from both sides has a view factor of 2.0. The view factor from a conical radiator, therefore, falls between these two levels, since the inside cone area has a partially obstructed view of space.

A summary of projected specific masses based on the heat pipes, radiator assemblies, and heat rejection subsystems (HRSSs) is given in Table 2. For this radiator configuration, the specific mass can be improved by increasing radiator area; however, the overall HRSS mass would increase by approximately 2%. The minimum mass system was selected for assessment rather than the minimum specific mass.

The radiator assembly is a deployable, conical design made up of 12 radiator panels. A single panel is shown schematically in Figure 1. One of these panels is totally redundant, each having independent heat transfer loops. There are also 34 redundant heat pipes per panel. Each panel contains 378 heat pipes, 2.5 cm in diameter, ranging from 23 to 73 cm in length, and arranged perpendicular to the panel central axis. A heat pipe cross section is shown in Figure 2.

FEASIBILITY ISSUES

There are two feasibility issues which can be classified as "enabling technologies" that must be resolved to achieve the predicted radiator performance. The first is to produce a carbon–carbon heat pipe tube with integral fins, which will meet both thermal and mechanical requirements while satisfying structural weight goals. The second issue is to develop a coating that protects the carbon–carbon substrate from the 875 K potassium working fluid. Resolution of these two feasibility issues was the objective of the recent Phase III effort.

Table 2. Projected Specific Masses (kg/m^2)

	Physical	Effective
Heat pipes	4.61	3.83
Radiator assembly	7.67	6.37
HRSS	8.63	7.17

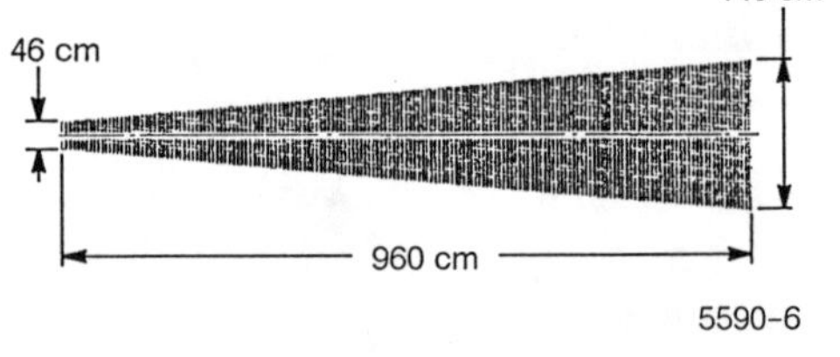

Figure 1. Thermoelectric Panel Design

Carbon–Carbon Heat Pipe Manufacturing

Carbon–carbon heat pipe tube sections were manufactured to demonstrate the feasibility of fabricating thin–wall tubes with integral fins and to provide carbon–carbon material for test, evaluation, and coating/sealing studies.

The reinforcement fiber selected for the fabrication was Amoco T–300 carbon fiber. This polyacrylonitrile (PAN) fiber was selected because it exhibits an appropriate balance between high–thermal conductivity and elastic modulus and ease of handling and weaving. The finned tube was woven from 3,000 filament yarn by Woven Structures, Compton, California. Figure 3 illustrates the angle interlock fiber architecture. Good through–the–thickness thermal conductivity of the tube is achieved by the axial yarn weavers which angle between the tube ID and OD. These axial yarns also increase the interlaminar shear and radial tensile strengths of the tube. The outer layer of hoop (fill) yarns in the tube forms the transverse yarns for the fins, thus ensuring good heat transfer from the tube to the fins.

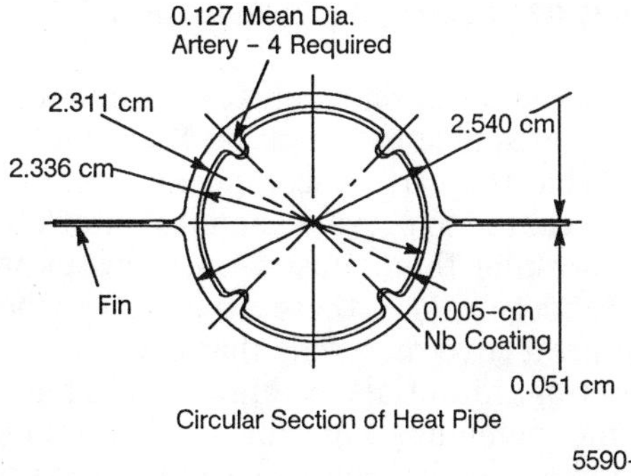

Figure 2. Potassium Heat Pipe Cross Section

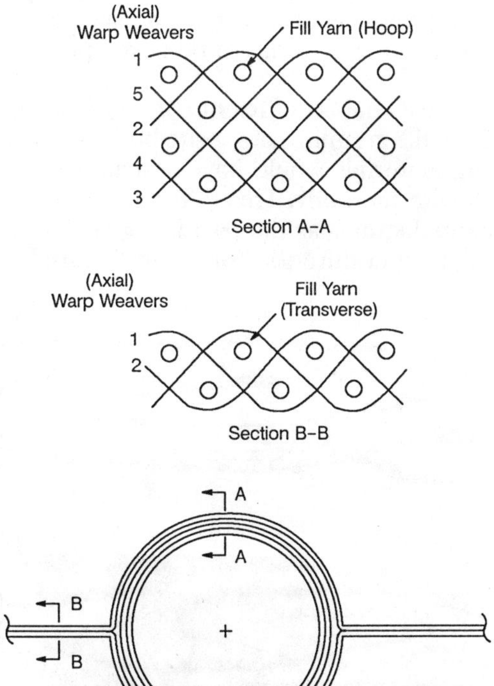

Figure 3. Angle Interlock Weave Architecture

The woven heat pipe preforms were prepregged with an unfilled Novolak/Resole prepreg resin by US Polymeric, Santa Ana, California. The initial resin prepreg of the woven carbon fiber preform allows the otherwise limber preform to be molded into the desired final shape. After curing of the resin, the part is stiff and easily handled. On subsequent pyrolysis of the resin, the remaining carbon matrix maintains the fibers in the desired geometric position and begins the process of filling the fiber bundles and the interstices between bundles with solid carbon. The prepreg resin had a nominal 33% solids content and 8% volatile content. The preform was vacuum bag cured in an aluminum caul plate with a boron nitride coated graphite rod tool used to maintain the desired tube ID during compaction and cure. The autoclave cure was performed at Programmed Composites, Brea, California. After curing, the preforms were inspected for dimensional conformance, and it was found that the fins were close to the desired thickness, whereas the tube wall thickness was less than the nominal specification. Apparently the axial weaver yarns compacted more than expected and future preforms will be redesigned to compensate for this effect.

The cured heat pipe preforms were stacked in graphite restraint fixtures to maintain the desired dimensions and geometry and subjected to a carbon–carbon densification procedure. This consisted of an initial elevated temperature atmospheric pressure carbonization cycle to decompose the cured resin followed by four cycles of repeated coal tar pitch impregnation, carbonization, and graphitization. Allied 15–V pitch was used as the matrix precursor to achieve minimum closed porosity and improved matrix thermal conductivity compared to resin derived matrix material. The graphitization temperature was selected for dimensional stabilization to the maximum coating temperatures, which might be used to seal the inner surface. Figure 4 shows a 0.3–m section of carbon–carbon tube that was produced.

Table 3 presents the as–fabricated properties of the carbon–carbon heat pipe sections. As noted previously, the tube wall thickness was less than the desired 1 mm, and the fin thickness was slightly larger than desired. Although design fin width was 2.5 cm, the fins were actually fabricated at 6.9 cm to allow for additional material for initial coating trials and destructive testing. Bulk density, while typical of 2–D carbon–carbon materials, is somewhat lower than desired and will be increased via additional densification cycles in the future. Tube burst pressure is greater than 4.1 MPa, which was the maximum pressure capability of the testing apparatus used. This greatly exceeds the heat pipe operating pressure. However, higher than average porosity and potentially weakened areas are indicated at the fin tube junction area, and the appropriate fiber architecture in this area should be investigated more fully in the future. The hoop/axial fiber balance measured indicates a lower axial content than desired due to the greater compaction than expected. Future work will increase the axial content to achieve the 2 to 1 hoop to axial fiber volume ratio desired for structural efficiency.

Figure 4. Carbon–Carbon Space Power Radiator Heat Pipe Tube

Table 3. Carbon–Carbon Heat Pipe Tube As-Fabricated Properties

Property	Value
Tube wall thickness (cm)	0.0864
Fin thickness (cm)	0.0584
Fin width (cm)	6.86
Bulk density (g/cc)	1.65
Porosity	0.126
Linear weight (6.86–cm fin) (g/m)	215.0
Linear weight (2.54–cm fin) (g/m)	132.0
Burst pressure (MPa)	>4.14
Fiber volume fractions	
Tube	
Hoop	0.436
Axial	0.124
Fin	
Transverse	0.322
Axial	0.138

Figure 5 presents optical photomicrographs of the carbon–carbon fin and tube cross sections. The traverse of the angle interlock weavers between the ID and OD of the tube section is clearly visible as is the compaction of the hoop yarns. The external surfaces are essentially as-fabricated with only minimal cleanup of loose pitch-derived matrix residue and no machining. Longitudinal splitting of the fill yarns is minimal, although the regions between adjacent yarns shows some porosity. As is typical of most 2–D composites processed using liquid matrix impregnant techniques, no areas of large porosity are evident.

CVD Coating of Carbon–Carbon Tubes

Despite numerous advantages, carbon–carbon has very poor resistance to attack by alkali metals. It was apparent that the carbon–composite heat pipe would have to be coated to prevent chemical attack by the potassium working fluid. Based on previous investigations (Mendelson [1]), there appeared to be several possible protective materials that could be used for a CVD coating. Materials evaluated included niobium, nickel, molybdenum, tungsten, and titanium. These were compared with respect to: solubility in potassium, compatibility with carbon, coating residual stress, carbon–carbon bonding, and coating ductility. Using a "figure of merit" approach, niobium was selected as the primary coating material to be developed.

Because the coefficient of thermal expansion (CTE) of the graphite composite body is low, any of the coating materials would have a significant CTE mismatch with the body. This difference in CTE induces substantial strains as the part is cooled from the processing temperature to room temperature.

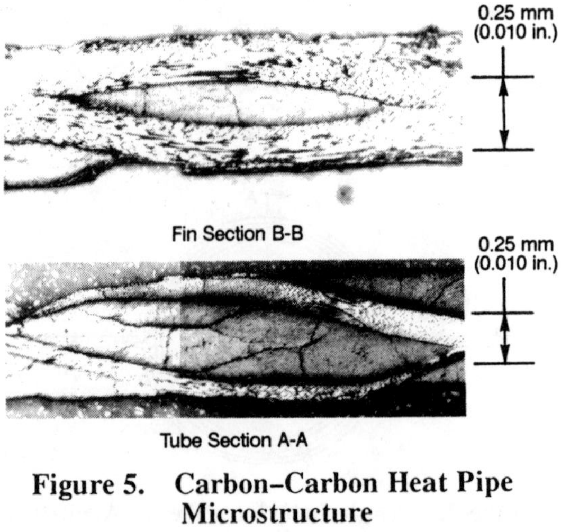

Figure 5. Carbon–Carbon Heat Pipe Microstructure

If it is assumed that the relatively massive and stiff heat pipe body is essentially rigid, the resulting strain induced in the coating upon cooling to room temperature would be 0.0075. If the coating behaved elastically, the resultant tensile stress in the coating (and shear stress at the interface between the coating and the body) would be over 694 MPa (100,000 psi), or greater than a realistically expected ultimate strength for the niobium and, certainly, a very high adhesive strength for the bond between the coating and the substrate.

The selected approach was to use an interlayer between the coating and the body which would perform four functions:

1. Provide some modest gradation in the CTE mismatch
2. Assume a portion of the load from the induced stress
3. Improve the coating adhesion
4. Serve as a carbon barrier to allow the maintenance of ductility in the niobium.

Pure rhenium was selected over the other candidate (niobium carbide) for this interlayer. NbC may have been a more effective carbon barrier, but the rhenium metal was acceptable in this regard and could be expected to have much higher strength and toughness than the brittle ceramic.

The rhenium interlayer was applied at 1100°C, to a thickness of 1 to 5 μm, by pyrolysis of its volatile hexachloride, as follows:

$$ReCl_6(g) \rightarrow Re(s) + 3Cl_2(g)$$

Niobium was then deposited over the rhenium at 1150°C to a thickness of about 100 μm, using hydrogen reduction of niobium pentachloride gas, as follows:

$$2NbCl_5(g) + 5H_2(g) \rightarrow 2Nb(s) + 10HCl(g)$$

The general conditions for these deposition processes were well known. The challenge was to determine whether the coatings could be applied to these carbon–carbon bodies. To provide protection, the coatings had to be defect–free. Crack–free coatings on carbon–carbon, accommodating such a large CTE mismatch, had never been reported.

Satisfactory rhenium deposits were produced from the outset. Considerable early experimentation was required to determine the best gas flows for achieving acceptable, uniform, niobium deposition. At present, 50– to 75–μm niobium coatings of good adhesion to the interlayer and to the body are being made with remarkable, crack–free, integrity. Figure 6 shows a scanning electron micrograph (SEM) of a rhenium/niobium CVD coating after being thermal cycled three times to its nominal operating temperature of 875 K. This sample was taken from the midsection of a 15.2–cm (6–in.) coated tube. The rhenium layer is approximately 5.8–μm (0.23–mils) thick and the niobium thickness is

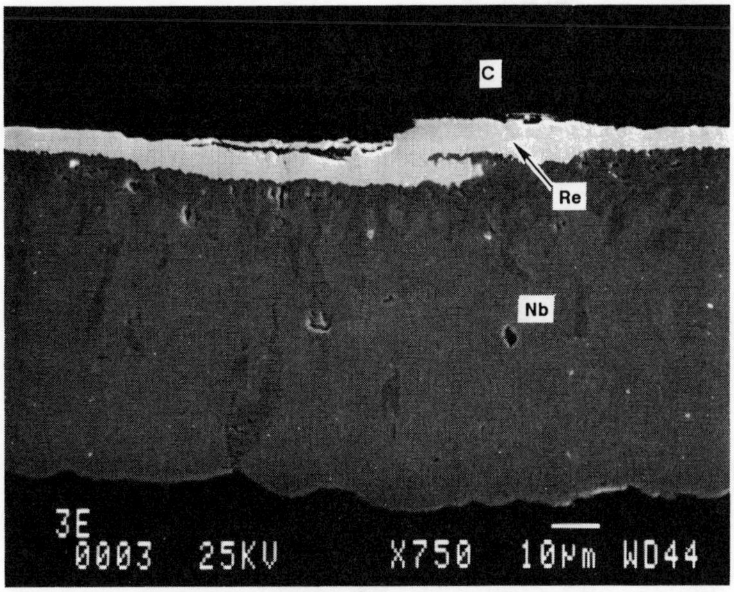

Figure 6. SEM of Re/Nb CVD Coating on Carbon–Carbon Matrix

66–μm (2.6–mils) thick. This SEM shows some penetration of rhenium into the carbon–carbon and excellent bonding to the carbon–carbon and to the niobium.

CONCLUSIONS

Recent development efforts have successfully resolved the two major feasibility issues associated with fabrication of high–temperature, carbon–carbon, potassium working fluid heat pipes. Carbon–carbon tube sections, 61 cm (24 in.) in length with integral fins, were produced. These tubes demonstrated integral fin fabrication techniques, good uniformity, and true axial dimensions. However, three processing improvements (increasing the axial fiber content, additional densification steps to decrease porosity, and modification of the tube/fin junction weave) will be made in subsequent runs. These improved tubes will then be suitable for prototypic heat pipe fabrication.

Niobium CVD coatings with a rhenium interlayer were successfully applied to the inside of the fabricated carbon–carbon tubes. The rhenium interlayer was found to be critical for adhesion and improved ductility in the niobium. Results of these efforts represent a milestone in CVD coating technology and warrant further development as a carbon–carbon protective barrier in support of this advanced radiator concept.

Acknowledgment

The work described in this paper was funded by NASA–LeRC under Contract NAS3–25209.

Literature Cited

1. Mendelson, M. I., "SP–100 Advanced Radiator Program––Phase III Carbon–Carbon Coating Development," Internal Letter, Rocketdyne Division, Rockwell International, 20 December 1989.

TESTING OF ADVANCED CERAMIC FABRIC WICK STRUCTURES AND THEIR USE IN A WATER HEAT PIPE

Z.I. Antoniak, B.J. Webb, J.M. Bates,
M.F. Cooper, and K.A. Pauley ■ Analytic Sciences Department, Pacific Northwest Laboratory[1], Richland, WA 99352

A prototype ceramic fabric/titanium water heat pipe with a ceramic fabric wick has been constructed and tested; it has successfully transported up to 100 W of power at negative tilt angles. Startup and operation both with and against gravity were examined. A fabric emittance of 0.93 was measured with an infrared camera. Ceramic fabric wick testing was begun to aid in the design and construction of an improved prototype heat pipe. Wick testing took place in 1-g; limited testing in 0-g was also initiated. Although analyses of all data are not complete, results to-date suggest that wick performance is better in 0-g than in 1-g.

Both Air Force and National Aeronautics and Space Administration (NASA) future spacecraft thermal management needs span the temperature range from cryogenics to liquid metals. Many of these needs are changing and not well defined and will remain so until goals, technology, and missions converge. Nevertheless, it is certain that high-temperature (> 800 K) and medium-temperature (about 450 K) radiator systems will have to be developed that offer significant improvements over current designs. Earlier papers have discussed prototypical ACF heat pipe design, and reported on preliminary component-level and heat pipe tests [1,2]. This paper discusses follow-on experiments performed in the lower temperature regime as part of a comprehensive advanced ceramic fabric (ACF) heat pipe development program. These experiments encompassed wicking tests using various ceramic fabric samples and heat transfer tests with a 1-m long prototype ACF water heat pipe.

WICKING TESTS

The development of ACF heat pipes requires a thorough understanding of the operation and optimization of various subcomponents. One important area of research is the utilization of ACF materials as wick structures. These structures have been tested by the Pacific Northwest Laboratory (PNL) in both 1-g and 0-g environments.

[1]Pacific Northwest Laboratory is operated for the U.S. Department of Energy by Battelle Memorial Institute under Contract DE-AC06-76RLO 1830.

A typical ACF water heat pipe is constructed with an outer ACF structural sheath, an impermeable metal liner material, and an inner ACF wicking structure. Working fluid pressure seats the thin liner material against the outer ACF layer, and the resulting liner corrugation may be useful as a limited wick primarily in the circumferential direction [1]. ACF arteries may also be added to the heat pipe design for further optimization. A schematic diagram of a heat pipe constructed of ACF materials is illustrated in Figure 1.

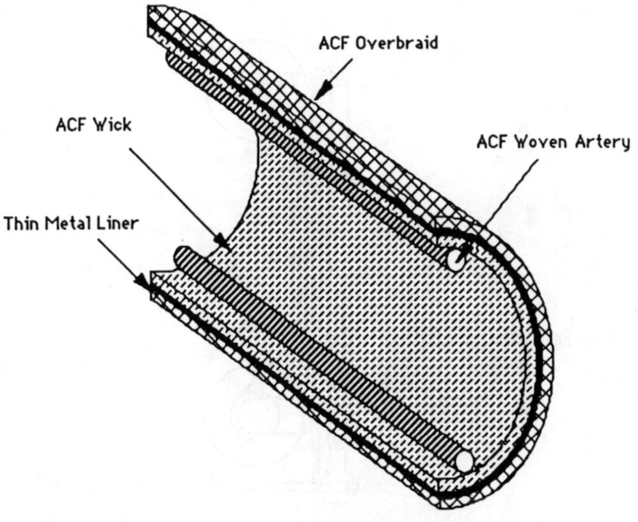

Figure 1. ACF heat pipe cross-section.

A literature review was performed, but few relevant references were found on 0-g wick performance. PNL investigated the feasibility of electroplating samples of Nextel fabric with copper and nickel for possible use as wicks. The nickel coatings showed good adherence to the fabric and were applied in thin layers without rigidizing the fabric. The copper-coated samples did not adhere as well to the fabric and had a tendency to flake off. Increased copper coating thickness improved adherence but only at the expense of rigidizing the fabric. Coated fabric samples were not tested as wicks, but additional work in this area is suggested.

The PNL fluid wicking experiment was intended to procure information necessary for the design of efficient heat pipes for space applications. The primary parameters required are the effective pumping radius and permeability of the ACF wicking materials when operating in 0-g. These parameters can be readily calculated from data on liquid velocity versus distance.[3]

Experimental Apparatus

The 0-g fluid wicking apparatus, shown in Figure 2, is constructed in three parts: a sample ribbon, an actuation device, and a redundant sensor module. Fabric wicking samples are mounted on a 0.381-m wide mylar strip. The mylar strip and samples are initially rolled onto a supply reel. One end of the mylar is fed through the sensing window and attached to a take-up reel. The samples are then moved into the sensing window individually by a hand lever attached to the take-up reel.

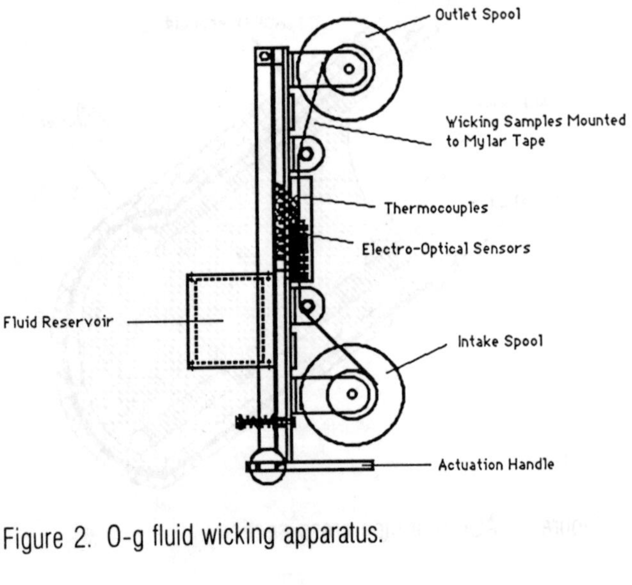

Figure 2. 0-g fluid wicking apparatus.

After a sample is positioned into the sensing window, a hand lever is turned, rotating a cam that brings a liquid reservoir filled with sponge material into contact with the sample. The sample is clamped between the water-filled sponge material protruding from an opening in the reservoir and a plate on the other side of the sample. The experiment for this sample is terminated by simply backing the liquid reservoir away from the sample. A new sample may then be positioned into the window.

Twenty different samples were placed on the mylar carrier strip. These samples consisted of four different types of fabrics (each about 375-μm thick) and one type of sleeving (about 0.003-m in diameter) placed in different geometries. The geometries used were: flat fabric with both sides exposed to air, double layers of flat fabric with both sides exposed to air, and flat fabric stitched to corrugated aluminum foil. Table 1 lists the combination of materials and configurations utilized.

Table 1. Fluid Wicking Sample Material and Configuration

Sample Number	Description
1	Nextel 440 flat fabric/plain weave
2	Astroquartz flat fabric/plain weave
3	Nextel 312 flat fabric/plain weave
4	Nextel 440 tape/woven
5	Nextel 440 sleeving/Nextel 440 thread fill
6	Nextel 440 sleeving/Nextel 312 thread fill
7	Nextel 440 sleeving/Kao-Wool fill
8	Nextel 440 sleeving/empty
9	Same as sample 1 but double layer
10	Same as sample 3 but double layer
11	Same as sample 2 but double layer
12	Filter paper standard
13	Nextel 312 flat fabric/corrugated Al foil
14	Nextel 440 flat fabric/corrugated Al foil
15	Astroquartz flat fabric/corrugated Al foil
16	Nextel 440 tape/corrugated Al foil
17	Same as sample 6 but mounted on Nextel 440 fabric
18	Same as sample 5 but mounted on Nextel 440 fabric
19	Same as sample 7 but mounted on Nextel 440 fabric
20	Same as sample 8 but mounted on Nextel 440 fabric

The samples that used flat fabric were placed over windows cut into the mylar and fastened around the edges of the window. This allowed the wicking to take place in the fabric alone without being disturbed by contact with any other surfaces and allowed the use of the electro-optic sensors described below. The samples were approximately 0.025-m wide and 0.15-m long.

An array of electro-optic sensors was placed along the side of the sensing window. Each of these sensors consisted of an LED and photodetector in one integral unit. The LED and photodetector were mounted on opposite sides of a slot approximately 0.0048-m wide. When a flat fabric sample was rolled into the sensing window, the edge of the fabric was positioned in the slot between the LED and photodetector. When the fabric becomes wet, the transmittance changes, and this can be measured as a change in the resistance across the photodetector. With an array of these sensors along the wicking path, it was possible to measure the time when the wicking front arrived at each sensor. The speed of the wicking front could then be inferred from this data.

In addition to the electro-optic sensors, an array of thermocouples was also attached to the structure that holds the liquid reservoir. When the liquid reservoir is moved into contact with the sample, the thermocouples simultaneously contact the back of the sample. As the wicking front arrived at a thermocouple, the change in temperature from a dry-bulb to a wet-bulb condition was detected.

The third sensing method was a visual record of the experiment taken by a video camera. The samples were lit from the side opposite the camera by a diffuse, fluorescent light source. The wicking front was visible as a small change in the transmittance of the fabric.

<u>Test Results</u>

<u>1-G Testing</u>. Extensive ground testing was conducted on the fluid wicking apparatus both before and after the reduced-gravity experimentation on the NASA KC-135 plane. These tests were performed in both vertical and horizontal orientations.

<u>0-G Testing</u>. Two days of reduced-gravity testing were undertaken with the PNL fluid wicking apparatus on the NASA KC-135 plane. The same set of test samples was flown both days to generate reproducible results.

The only valid data from the 0-g wicking tests was extracted from the video camera record, which was sometimes difficult to interpret. The wicking front could be observed for only the first three samples listed in Table 1. Figure 3 provides preliminary results of the wicking test for these three samples for 1-g vertical and horizontal conditions, and the data obtained during the two days of 0-g flight testing. This data is presented as an average liquid velocity; the error is on the order of 0.001 m/s. Given the paucity and quality of the 0-g data, the desired pumping radius and permeability parameters were not calculated. Nevertheless, it appears that the wicking improves in 0-g; it is theorized that performance in a horizontal 1-g position is lower because of meniscus recession into the wick, and resulting wick effective area reduction.

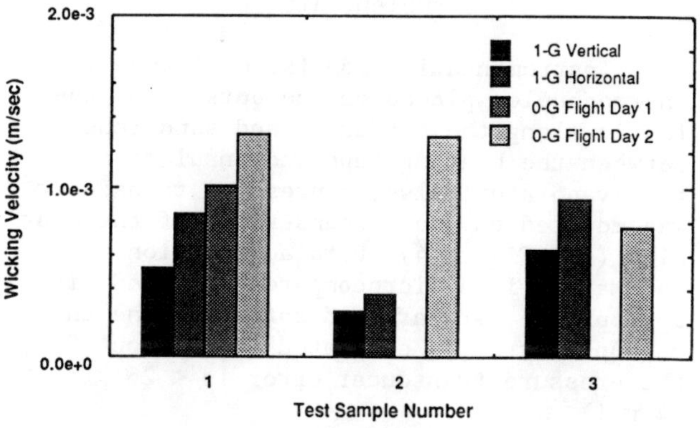

Figure 3. Preliminary results of wicking tests.

The data generated on these flights is limited because of equipment problems during the first day of testing. After about six reduced-gravity parabolas, a power spike destroyed the electro-optical sensor array and shut down the data acquisition computer that contained the acquired data in its volatile memory. The thermocouple technique functioned poorly, mainly because the thermocouple tips had insufficient time to dry between samples. We intend to repeat these experiments, using better and more reliable sensing techniques.

<u>ACF HEAT PIPE TESTING</u>

A non-optimized prototype ACF heat pipe was built using currently-available materials. It was constructed as follows:

- Outer liner — 430-μm (17-mil) thick bi-axially-braided Nextel fabric
- Metal liner — 200-μm (8-mil) titanium alloy
- Wick — ca. 250-μm (10-mil) Nextel, held in place with 100-mesh stainless steel screen
- End fittings — standard Swagelok
- Working fluid — water
- Dimensions — titanium is 0.016-m (5/8-in.) OD x 0.9-m (36-in.) long
- evaporator — 0.229-m (9-in.) (wrapped with electrical heating tape)
- adiabatic section — 0.15 m (6-in.)
- condenser — heat rejection is to ambient air.

Instrumentation consisted of type J thermocouples placed on the outside of the fabric along the condenser and sandwiched between the heating tape and insulation at the evaporator; also, a pressure transducer was mounted on the condenser end of the heat pipe (see Figure 4). Data acquisition was automated via a microcomputer equipped with appropriate hardware and software. The thermocouple error is estimated to be about 2 K; the pressure transducer error is < 25 kPa (4 psi).

Initial performance measurements indicated lower power throughput and reduced condenser temperatures and operating length when the heat pipe was operated against gravity [2]. More recently, we have been able to obtain improved performance at a -5° tilt angle, with near-constant condenser temperature (see Figure 5). This improvement was achieved by further cleaning of the wick, plus the addition of 10 to 15 cc of water after the wick was fully saturated. The heat pipe was then operated with 60 W heat input.

Heat throughput, based on infrared camera scans, is estimated to be nearly the full 60 W. The infrared camera also indicated that surface emittance is about 0.93 (by matching of measured temperatures with the temperatures indicated by the camera), which confirms optical property measurements [1].

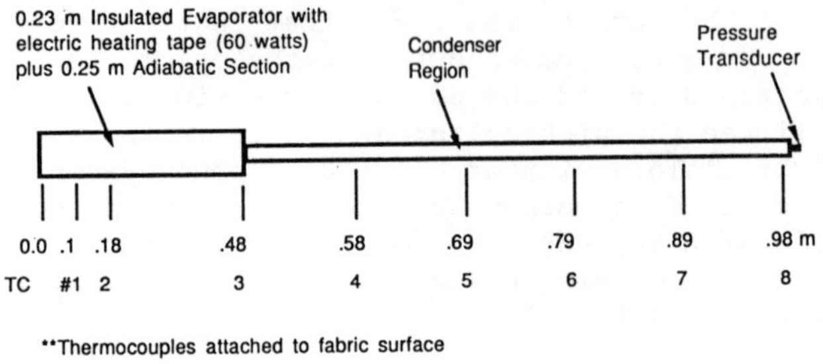

Figure 4. Arrangement of test ACF water heat pipe.

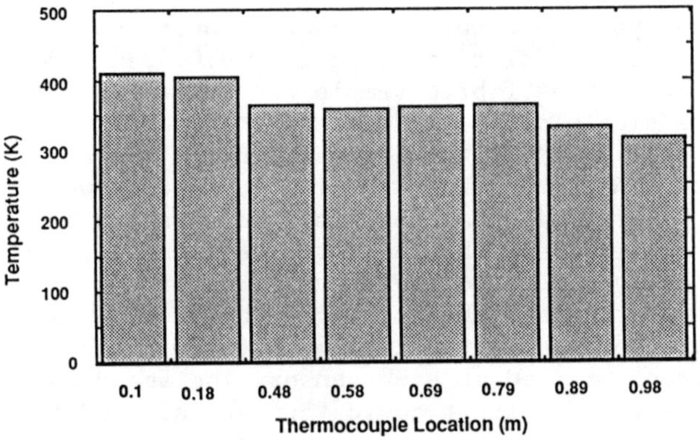

Figure 5. ACF heat pipe axial temperature profile.

Startup at -5° tilt is readily achieved (see Figures 6 and 7). The large temperature differential between the evaporator and condenser sections is due to the temperature drop across the non-optimized fabric and relatively thick titanium liner. Some "droop" is observed at the end of the condenser farthest from the evaporator. This has two causes: in the negative tilt mode, excess liquid pools there, and the pressure transducer at the condenser tip acts as a fin. It is unlikely that the "droop" is caused by non-condensible gases because these were thoroughly purged prior to testing. The thermodynamic (fluid) temperature is very close to the average of the evaporator and condenser temperatures.

We have also been able to maintain constant heat input, from horizontal to a slight negative tilt of - 5 degrees, although temperature at the evaporator does increase by about 12 degrees (see Figure 8). Code predictions with a 1-dimensional model [4] confirm that this is a reasonable performance level.

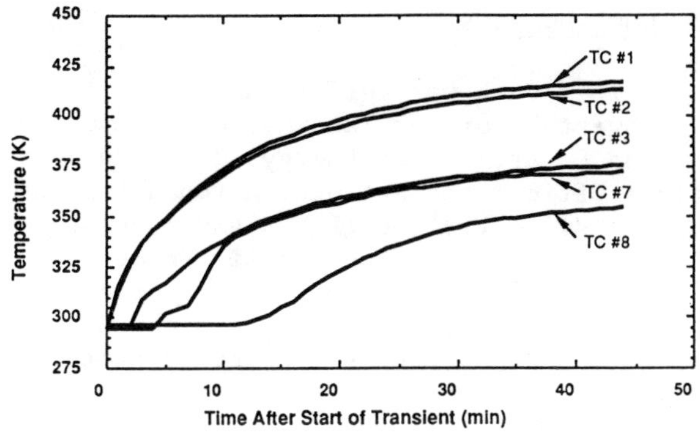

Figure 6. ACF heat pipe startup transient.

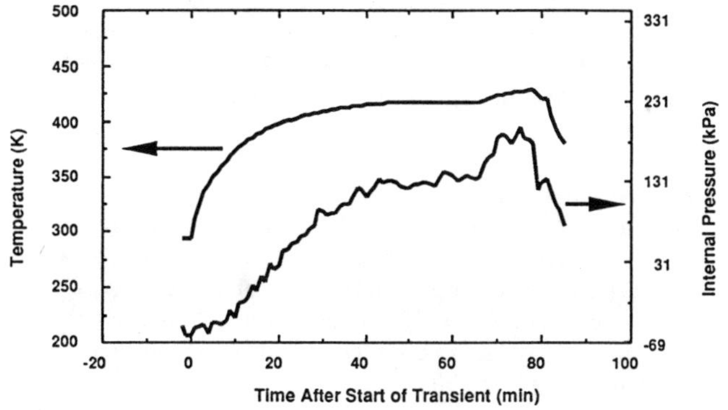

Figure 7. ACF heat pipe transient temperature and pressure.

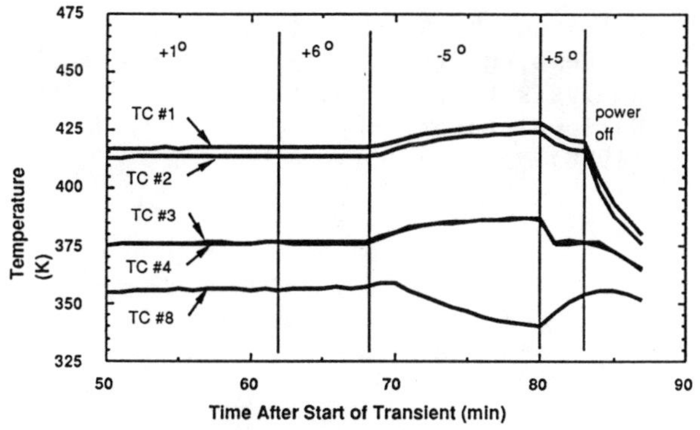

Figure 8. ACF heat pipe test of sensitivity to orientation.

Tests at higher power levels and temperatures representative of a Stirling engine space heat rejection system have taken place. One test, at -5 degrees tilt, reached 114 W and heater temperatures in excess of 550 K (see Figure 9), which destroyed the silicon

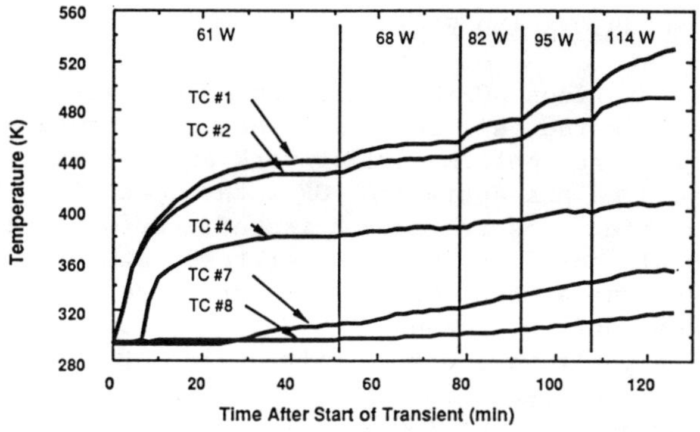

Figure 9. ACF heat pipe test of sensitivity to power.

heating tape. A second high-temperature test, at +3 degrees, was conducted at over 200 W power (see Figure 10).

A larger-diameter (0.0254 m) second water heat pipe with a 100 μm thick stainless steel liner has just been completed. Initial tests parallel the ACF/titanium results. The larger diameter and thinner liner material are unable to sustain atmospheric pressure; when heat input ceases, and temperature and pressure drop, this heat pipe collapses to a nearly-flat geometry at room temperature. Resumption of heat input causes the heat pipe to re-assume its original cylindrical shape; this cycle has been repeated about twenty times with no deleterious consequences. Infrared camera scans taken during startup from a flattened condition indicate that in this startup mode a heat pipe becomes axially isothermal more rapidly than when startup is from a cylindrical geometry. That is, the much-narrowed vapor transport area appears to enhance startup. More data is required to provide insight into this behavior, which has not previously been reported by other heat pipe researchers.

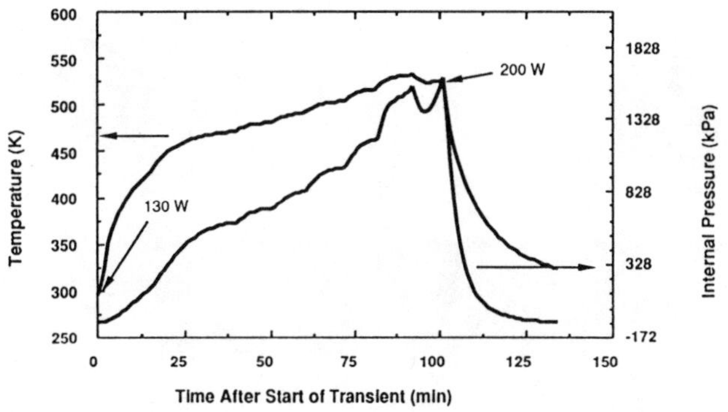

Figure 10. ACF heat pipe test of sensitivity to power.

CONCLUSIONS AND PLANS

Various ACF heat pipe components have been tested; all results to-date are encouraging and confirm earlier work and assumptions. A non-optimized ACF water heat pipe has been constructed and tested. Performance in air to 475 K (water vapor temperature based on a measured 1.60 MPa pressure) is excellent, with about 200 W heat throughput at a positive 3 degree tilt angle. At a negative 5 degree tilt angle, heat throughput is reduced to about 100 W.

The infrared camera also measured a fabric emittance of about 0.93, at a 430 K surface temperature. This agrees well with Nextel fabric optical measurements, considering that the Wien's law emissive peak at 430 K is 6.7 μm. It seems that ACF heat pipes could operate effectively at low temperatures without any coatings; higher temperatures (> 600 K or so) may require high-emittance coatings [1,2].

In the near future, we plan to construct and test an ACF water heat pipe with a 25-38 μm stainless steel or copper liner. Infrared camera and spectrophotometer measurements of fabric optical properties to evaluate radiative heat transfer and the need for coatings will continue. Wicks will be tested, in both 1-g and 0-g, to permit performance optimization. The infrared camera may also prove useful for this task, as it readily detects the temperature change present when a liquid front traverses a wick. Heat pipe tests in a purely radiative heat rejection mode (in a vacuum chamber) are also planned. In the high-temperature regime, we will continue to provide ACF materials to LANL for test on liquid metal heat pipes. Ultimately, an ACF heat pipe test in a space environment will be required; a small possibility exists for realizing this during a FY93 shuttle flight.

ACKNOWLEDGMENTS

This work was performed by Pacific Northwest Laboratory and has been supported by the Department of Energy, Richland Operations Office; the National Aeronautics and Space Administration (Lewis and Johnson Research Centers); and the Air Force (Astronautics and Weapons Laboratories). Pacific Northwest Laboratory is operated for the U. S. Department of Energy by Battelle Memorial Institute.

REFERENCES

1. Antoniak, Z. I., J. M. Bates, and B. J. Webb, "Design of Ceramic Fabric Heat Pipe with Water Working Fluid," Proceedings, 26th National Heat Transfer Conference, S. B. Yilmaz, ed., pp. 230-234, American Institute of Chemical Engineers, New York (1989).

2. Antoniak, Z. I., B. J. Webb, J. M. Bates, and M. F. Cooper, "Construction and Testing of Ceramic Fabric Heat Pipe with Water Working Fluid", Proc. Eighth Symposium on Space Nuclear Power Systems, p. 875, American Institute of Physics, New York (1991).

3. Brennan, P. J., and E. J. Kroliczek, Heat Pipe Design Handbook, pp. 116-169, 280-287, B & K Engineering, Inc., Towson, MD (1979).

4. Woloshun, K. A., M. A. Merrigan, and E. D. Best, HTPIPE: A Steady-State Heat Pipe Analysis Program, LA-11324-M, Los Alamos National Laboratory, Los Alamos, NM (1988).

DROP AND BUBBLE DYNAMICS IN THE U.S. MICROGRAVITY PROGRAM

Taylor G. Wang ■ Vanderbilt University, Nashville, TN 37235

A set of drop and bubble dynamics experiments have been selected for the first United States Microgravity Laboratory (USML) flight in 1992. These experiments will take fullest advantage of what space has to offer; namely, extended microgravity and a containerless processing environment, to study linear and nonlinear wave dynamics, centering dynamics, the rheology of surfaces, electrocapillary phenomena, collision and coalescence, and others. In this talk, the speaker will introduce these experiments and present some of the more recent data obtained in the laboratory in support of the upcoming flight (the author is one of the Principal Investigators to the USML mission).

The availability of an extended zero-g environment provided by space flights might provide a brand new avenue for fluid studies, namely, containerless fluid experiments. For the first time scientists can study the behavior of fluids without the influence of containers.

A set of these experiments has been selected for the United States Microgravity Laboratory (USML) flight in the near future. These experiments will take full advantage of what space has to offer, namely, extended microgravity and a containerless environment, to study linear and nonlinear shape oscillations, bubbles and compound drop dynamics, the rheology of surfaces, collision and coalescence, shapes and instability of rotating spheroids, shapes and instability of deformed drops, etc.

In this paper, a brief introduction will be given to some of these experiments.

LINEAR AND NONLINEAR SHAPE OSCILLATIONS

A liquid drop can be trapped at a stable position in an acoustic standing wave existing in a resonant cavity. The static equilibrium shape of the drop can be controlled by varying the magnitude of the acoustic radiation force, which is proportional (to the second order) to the square of the magnitude of the first order acoustic pressure. In this case, the cavity is of rectangular geometry, and the axis of symmetry is taken as its vertical axis. Depending on the standing acoustic wave or the pressure intensity, a drop may be given the shape of a prolate spheroid, or an oblate spheroid, or a nearly perfect sphere.

Shape oscillations are induced through low-frequency modulation of the acoustic radiation force. The excitation can correspond to either a periodic elongation or compression of the drop at the poles. The restoring force is, in the ideal case, provided only by the interfacial tension which tends to drive the drop back to the equilibrium shape. For low-amplitude vibrations, both modes yield the same results for the resonant frequency, although this is not the case for large displacements (Trinh and Wang, 1980).

The drop oscillation amplitude is monitored using an optical detection technique: a slit, either parallel or perpendicular to the axis of symmetry, is uniformly illuminated, and the shadow of the drop is then centered across the slit, blocking some of the light. As the drop oscillates, the light intensity detected by a photo transistor varies periodically in phase. Up to a reasonably large amplitude, the response is approximately proportional to the drop's deformation.

Normal Mode

The normal mode of drop oscillations of small amplitude for various liquids has been measured as a function of size. The result yields a power law

$$f_2 \sim R^{-1.51}$$

which is very close to Lamb's theory,

$$f_2 \sim R^{-1.50}.$$

Amplitude Dependence

The experimental results are for drops of a silicone oil/CCl_4 mixture immersed in distilled water and for a prolate-biased initial oscillation drive. The percentage of relative change in the decay frequency has been measured as a function of the maximum deformation with the drop in the prolate shape. The measured free decay frequency is inverse linear proportion to deformation, and found to be in good qualitative agreement with the calculation.

Damping Constant Measurements

The damping constant for the fundamental mode has been measured as both a function of drop size of the viscosities of the inner and outer liquids. For viscosities lower than 100 cSt and drop volumes larger than 0.5 cm^3, data was obtained from photographs of oscilloscope traces taken during free decay.

The experimental results on the damping constant for phenetole (1.22 cSt) and for a series of mixtures of silicone oil and CCl_4 has been measured. The increasing values of π_2^{-1} are shown for mixtures with kinematic viscosity equal to 3.25, 6.5, 16.5, 68.6, and 120.4 cSt, in this order.

A linear least-squares fit of the phenetole data yields a power law

$$\pi_2^{-1} \propto R^{-1.4}.$$

Although similar least-squares fits of the data were attempted for all other liquids, the correlation was not good.

Internal Fluid Flow

The motion of the fluid inside the drop can be made visible by the addition of quasi-neutrally buoyant tracer particles. The markers should be small enough to follow the motion of the fluid but large enough to scatter sufficient light for detection. Here we disperse organic dye particles with sizes ranging between 5 and 50 micrometers (μm) in the drop liquid. The particles close to the central midplane containing the symmetry axis are illuminated by a plane light sheet, and the 90° scattered light is photographed while the drop is driven into steady-state oscillations. The exposure usually lasts for a few oscillatory cycles.

Fission Associated with Oscillation

As the oscillation amplitude is greatly increased, the increasingly larger deformation may cause the drop to split. Radiation pressure-induced drop fission was first observed with 2-mm-diameter drops (Marston and Apfel, 1980), and our results confirm that cavitation-free splitting of 1 cm-diameter drops is also possible when the acoustic wavelength is comparable to the drop diameter. A third smaller satellite drop is often formed when the two main droplets split.

BUBBLE AND COMPOUND DROP DYNAMICS

Compound Drop Oscillations

Saffren, Elleman, and Rhim (1981) studied the oscillation frequencies for a concentric three-fluid system. Since ψ is a nonsecular solution of the Laplace equation, it can be expressed in the following forms in each region of the system:

$$\psi(r,t) = \sum_{l,m} [A(l,m;t)r^{l} + B(l,m;t)r^{-(l+1)}]\, Y_{lm}(\theta,\Phi) \quad (1)$$

(shell),

$$\psi_i(r,t) = \sum_{l,m} [A_i(l,m;t)r^{l}]\, Y_{lm}(\theta,\Phi) \quad (2)$$

(core),

$$\psi_i(r,t) = \sum_{l,m} [B_0(l,m;t)r^{-(l+1)}]\, Y_{lm}(\theta,\Phi) \quad (3)$$

(host).

By applying these solutions to the equations of motion with proper boundary conditions, the equations can be reduced to a eigenvalue equation which is essentially the same as a coupled harmonics oscillation. The normal mode frequencies are given by

$$\omega_{\pm}^2 = \frac{K}{J}\,\omega^2, \quad (4)$$

where

$$K = \frac{1}{2}\left(\frac{\sigma m_i}{\tau^3} + \frac{m_0\tau^3}{\sigma}\right) \pm \left[\left(\frac{\frac{1}{2}\sigma m_i}{4\tau^3} - \frac{m_0\tau^3}{4\sigma}\right)^2 + 1\right]^{\frac{1}{2}}, \quad (5)$$

$$J = (1+\Delta\rho_0)(1+\Delta\rho_i)\,\tau^{(2l+1)} - \Delta\rho_i\,\Delta\rho_0\,\tau^{(2l+1)},$$

$$\omega^2 = \left(\frac{\sigma_0\sigma_i}{[R_0R_i]^3}\right)^{\frac{1}{2}} \frac{(l-1)\,l\,(l+1)\,(l+2)}{(2l+1)\,\rho}, \quad (6)$$

and

$$\Delta\rho_i = \frac{(l+1)\,(\rho_i-\rho)}{(2l+1)_\rho},$$

$$\Delta\rho_0 = \frac{l\,(\rho_0-\rho)}{(2l+1)_\rho},$$

$$\sigma = \sqrt{\sigma_0/\sigma_i}$$

and

$$\tau = \sqrt{R_0/R_i}.$$

It is important to note that when the positive square root in Equation (4) corresponds to the normal mode with the higher frequency, the boundary oscillations are in phase. We call this high frequency ("+") mode the "bubble" mode and the lower frequency ("−") mode the "sloshing" mode. The relative boundary displacement of the sloshing mode is out of phase. The relative boundary displacement of these two modes is the bubble mode in which the two boundaries oscillate in phase, and the sloshing mode in which the two boundaries move out of phase. The limiting cases of Equation (4) are shown below.

1. Simple drop (i.e., $R_i = 0$) one gets

$$\omega_l^2 = \frac{\sigma_0 l\,(l+1)\,(l-1)\,(l+2)}{[l\rho_0 + (l+1)\rho]R_0^3} \quad (7)$$

which is Lamb's result.

2. Thin Shell (i.e., $\tau = 1$) then $\omega_{-} = 0$ and

$$\omega_+^2 = \frac{(\sigma_0\,\sigma_i)\,(l-1)\,(l+2)\,l\,(l+1)}{[l\rho_0 + (l+1)\rho]\overline{R}^3}. \quad (8)$$

3. Thick Shell (i.e., $\tau\rightarrow\infty$) then

$$\omega_+^2 = \frac{4\pi\sigma_i\,(l-1)\,l\,(l+1)\,(l+2)}{3V_c[l\rho + (l+1)\rho_i]} \quad (9)$$

and

$$\omega_-^2 = \frac{4\pi\sigma_0\,(l-1)\,l\,(l+1)\,(l+2)}{3V_s[l\rho_0 + (l+1)\,\rho]}. \quad (10)$$

Today there are experimental data on compound drops obtained in an electrostatic chamber which are found to be in good agreement with the experiment.

Centering Mechanism

The physical mechanism governing the centering of a hollow liquid shell in capillary oscillations, which has been observed in experiments, is investigated theoretically. First the shell is assumed to be inviscid and to have a thickness that is much less than its spherical radius. A system of one-dimensional nonlinear equations of motion is derived using a thin-sheet model. From a numerical study the nonlinear effects of the wave are found to cause the core to oscillate slowly relative to the shell, while the center of mass of the whole system remains stationary.

For the $n = 2$ sloshing or bubble mode, if the core is displaced, the lighter side of the shell must oscillate harder to keep the center of mass of the whole shell stationary in the absence of an external force. Therefore, on the average over a wave cycle the lighter side is a region of lower pressure according to the Bernoulli effect, causing the liquid to drift from the heavy side to the light side. Since the force that causes the drift is opposite and arguably also proportional to the displacement, Hooke's law is the logical consequence. Therefore the drift is sinusoidal in time about the center of mass rather than, say, exponential. Exponential behavior can only come from viscosity. The same principle holds for higher modes.

Quantitatively, when the shell in oscillation is non-concentric, with the core displaced upward by Δ, the difference in Bernoulli pressure between the lower and the upper halves of the shell is of the order of $1/2\rho\mathbf{u} \cdot \mathbf{u}$ multiplied by Δ. Since $\mathbf{u}$ in both mode types is of $O(\omega\varepsilon R_s)$ and the shell has a cross-sectional area of $O(R_s D_s)$, an internal force of $O(\rho\omega^2\varepsilon^2 R_s^3 \Delta D_s)$ pushes the liquid upward. The corresponding acceleration of the liquid, which has a mass of $O(\rho D_s R_s^2)$, is of $O(-R_s d^2\Delta/dt^2)$, which is opposite in sign to the displacement, as explained earlier. Using Newton's law and assuming that Δ is proportional to $\exp(-i\Omega t)$ we find that Ω is of $O(\omega\varepsilon)$. This result, if cast into dimensionless form, is consistent with the sloshing mode and the bubble mode.

In one of the experiments (Trinh 1983), the liquid shell is levitated acoustically against gravity without breaking. (Note that without support the shell undergoes a free fall and experiences a zero-gravity condition.) It is interesting to examine the strength of the centering mechanism versus that of gravity. With the shell prevented from falling, the buoyancy effect due to gravity tends to force the core to the top, leading to rupture of the shell. If the shell is in capillary oscillation, the light side at the top must oscillate harder than the heavy side even though there are external forces, otherwise the shell will exert an additional force on itself. Then, on the average over a wave cycle, the lower side has a Bernoulli pressure exceeding that of the upper side of the order of $1/2\rho\mathbf{u} \cdot \mathbf{u}$ times Δ, according to our earlier argument. In a dynamic equilibrium after the slow oscillation subsides, this difference must support the hydrostatic pressure due to gravity between the top and the bottom of the shell of the order of $\rho g R_s$. Since $\mathbf{u}$ is of the order of $\varepsilon\omega R_s$, the equilibrium condition becomes

$$\Delta\varepsilon^2 \sim \frac{g}{\omega^2 R_s}\,, \tag{11}$$

where g is taken as 980 cm s^{-2}. For the typical case we have mentioned $R_s = 0.1$ cm. In order that the shell does not break, Δ should be less than or at most on the order of one, (1) when a reasonable value of ε is used. For the $n = 2$ sloshing mode $\omega = 295$ s^{-1} and $\varepsilon = 0.5$ such that $\Delta \sim 0.5$, which is less than one. For the $n = 2$ bubble mode $\omega = 2400$ s^{-1} and $\varepsilon = 0.05$ such that $\Delta \sim 0.7$, which is still within the safety margin. Therefore the centering mechanism is strong enough to support the liquid shell against the action of gravity.

COLLISION AND COALESCENCE

Pre-Coalescence

In order to understand how a drop mixes with another drop after coalescence, we obtain a simple model of how far a drop, which is initially at rest, can penetrate into the bulk of the same liquid after coalescence with the latter at the latter flat interface with another immiscible liquid. The derivation of the model involved balancing the surface energy available before coalescence, divided by the estimated time taken for total submersion on the one hand, and the estimated rate of viscous dissipation as the drop moves through the bulk on the other. We find that the penetration depth L is proportional to the 5/4th power of the drop diameter d, and is inversely proportional to the square root of the drop viscosity μ:

$$L/d \propto (\rho\sigma d/\mu^2)^{1/4}$$

where ρ is the density of the drop liquid and σ is the surface tension between this liquid and the immiscible host, with the proportionality constant to be determined by experiment. The d dependence is in good agreement with the experimental data.

Mixing Following Coalescence

The problem of mixing following coalescence of drops has not been addressed in detail in the past. Charles and Mason (1960) have studied the partial coalescence of liquid drops at liquid/liquid interfaces, but have concentrated on the formation of satellite drops following coalescence and not on the translation and mixing of coalesced drops. Other studies on coalescence (e.g. Mackay and Mason, 1963) have concentrated on the aspects prior to coalescence; namely, the film drainage problem. Torza and Mason (1970) have studied interactions of drops in shear and electrical fields, and have concentrated on identifying the equilibrium configurations of drops for different surface tension regimes. It appears that one of the main reasons for the lack of controlled experimental investigations on coalescence of drops has been the lack of proper experimental facilities to levitate drops and to facilitate observation of the phenomena.

The current investigation's objective is to study the phenomena of mixing following coalescence and to understand the roles of drop size and drop viscosity. The coalescence of drops of finite size ratio has been conducted in an immiscible system (Anilkumar and Wang, 1990), and the coalescence for very large drop size ratio has been conducted in a vertical column, similar to that used by Charles and Mason (1960).

The coalescence experiments are conducted in an immiscible acoustic levitator with degassed distilled water as the host medium. Silicon oil mixed with bromobenzene is the drop liquid. The mechanism involves levitating the first drop and then introducing the second drop. As the latter slowly seeks the same levitation position, the two drops come into contact with each other and coalesce. The coalescence is delayed until the host liquid between the drops is completely drained. The study of drops of large diameter ratios has been conducted in a vertical channel with water-glycerin mixtures of various viscosities as the drop liquid and silicone oil (DC 200, 100 cs) as the immiscible host through which a drop translates very slowly before coalescing with the bulk liquid at a flat interface.

<u>Finite Size Ratio Drops.</u> The coalescence results in considerable mixing with the smaller drop penetrating the bigger drop as a vortex. The penetration of the smaller drops is basically dictated by the surface tension forces driving the motion, viscous drag resisting the motion, and the diameter ratio of the drop. There is no mixing when drops are of equal size. For higher drop viscosity, the smaller drop is simply lodged in the larger drop, though by the same mechanism. There is no noticeable mixing here.

<u>Very Large Size Ratio.</u> In order to quantify the mixing process, the case of very large size ratio drops has been studied. Here the experimental results can be correlated with respect to the size of the smaller drop and the viscosity of the system. A water-glycerin system has been used as the drop liquid since the surface tension of the mixture is relatively invariant for the viscosity range (1-4 cP) studied. Drops of different sizes generated slowly translate through a viscous oil (DC 200, 100 cs) before coalescing at a flat interface. A laser-induced

fluorescence technique has been used to study the cross section of the flow field. Immediately following coalescence, the drop enters the bulk fluid as a vortex that stretches and translates. Qualitatively this behavior is similar to that observed in finite size ratio drops. The mixing length (drop penetration) measured from the interface is the exponential distance the tip of the vortex translates before stopping.

SHAPES AND STABILITY OF ROTATING SPHEROIDS

From April 29 to May 7, 1985, a Drop Dynamics Module was flown on Spacelab 3. This module incorporated all the acoustic positioning and manipulation capabilities: rotating, oscillating, and shaping of a liquid drop without physical contacts. Drops of water or glycerin-water mixtures of various viscosities were deployed inside an acoustic positioning chamber while orbiting the earth inside the Spacelab module carried in the payload bay of the Space Shuttle. The average level of residual acceleration was around 10^{-3} times the standard earth's gravitational field. The drops were held within a small region inside the chamber by three orthogonal acoustic standing waves generating steady-state forces on the order of 1 dyne, and were driven into rotation by acoustic torque.

The drop volumes varied between 0.5 and 10 cm^3 with viscosities of 1, 10, 20, and 100 cSt, densities between 1.0 and 1.22 g/cm^3, and surface tensions from 55 to 65 dynes/cm.

An initially nonrotating drop was subjected to an acoustically generated torque which caused it to spin up with increasing velocity. The rotational velocity of the liquid was determined through the motion of immiscible tracer particles suspended within the primary drop. The time variation of the shape of the rotating drops was determined by recording the drop profile along three orthogonal views on 16 mm motion picture film.

Data from a typical experimental sequence involved several spin-ups and spin-downs of a 1 cm^3 100 cSt drop. Of particular interest here are both the value of Ω(II), the rotational velocity at bifurcation and the deformation of the drop as a function of rotation speed, since these two characteristics have been precisely determined theoretically.

The most obvious difference is in the location of the bifurcation: Ω(II) was experimentally determined to be 0.47±0.04 while it is calculated to be 0.56. In addition, the experimental rate of change of the relative deformation with rotation rate appears less than that predicted by calculations for the two-lobed shape, while the agreement appears to be quite good for the axisymmetric shape. One might also note that, within experimental uncertainty, there is no clear evidence of the existence of different paths for spin-up and spin-down in the two-lobed regime (although the scatter of the experimental results increases when the data for spin-down are added). These results are representative of the data gathered using several drops of various volumes.

When the deformation of the drop in the two-lobed regime is increased by imposing a larger torque, fission becomes inevitable beyond a certain value. Experimentally, fission generally produces two main drops of equal volume, and a satellite droplet at the center of mass. Fission occurred at a normalized rotational velocity of about 0.21 (±0.04). There is no simple prediction of a critical velocity at fission, but one could compare the data to the velocity calculated for maximum drop deformation (0.26). Calculations also indicate the rotational kinetic and surface energies for a single two-lobed drop and for two separate drops matching the region where the normalized velocity is equal to 0.25.

SHAPES AND INSTABILITY OF DEFORMED DROPS

There have been observations that an intense sound field can break up a liquid drop in levitation by flattening it drastically through the radiation pressure. Using high-speed photography we have observed that for a low-viscosity liquid, at a high sound intensity, it might buckle upward like an umbrella and shatter, or might simply expand horizontally like a sheet and shatter. We believe that at low viscosity, the ripples are capillary waves generated by the parametric instability excited by the membrane vibration, which is driven by the sound pressure. Atomization occurs whenever the membrane becomes so thin that the vibration is sufficiently intense. For any viscosity, the vibration leads to a Bernoulli correction in the static pressure, which is destabilizing. Buckling occurs when an existent equilibrium is unstable to a radial oscillation of the membrane because of the Bernoulli effect. Besides, the radiation stress at the rim of the flattened drop is a suction stress which is also destabilizing, leading to the horizontal expansion and the subsequent break-up.

CONCLUSION

The limited low gravity experiments have yielded some very controversial data. This is precisely why there is a great deal of anticipation of what the next set of space data will bring. Will an extended microgravity and containerless environment provide a truly new avenue for fluid study? Or will it be just another way to look at the same problem? Only with controlled experiments will we find the answer.

REFERENCES

1. E.H. Trinh and T. Wang, "A quantitative study of some non-linear aspects of drop shape oscillations," paper presented at the 100th Meeting of the Acoustical Society of America (1980).

2. P.L. Marston and R. Apfel, "Quadrupole resonance of drops driven by modulated acoustic radiation pressure—Experimental properties," J. Acoust. Soc. Am. **67**, 1499 (1980).

3. M. Saffren, D.D. Elleman, W.K. Rhim, "Normal modes of a compound drop," Proceedings of the 2nd International Colloquium on Drops and Bubbles, Pasadena, CA (1981).

4. E.H. Trinh, "Fluid Dynamical Studies of Free Liquids Using Acoustic Levitation," IEEE Proc in Ultrasonics **2**, 1143 (1983).

5. G.E. Charles and S.G. Mason, "The Mechanism of Partial Coalescence of Liquid Drops at Liquid/Liquid Interfaces," J. Colloid Interface Sci. **15**, 105 (1960).

6. G.D.M. Mackay and S.G. Mason, "The Gravity Approach and Coalescence of Fluid Drops at Liquid Interfaces," Canadian Journal of Chemical Engineering, 203-212 (October 1963).

7. S. Torza and S.G. Mason, "Three-Phase Interactions in Shear and Electrical Fields," J. Colloid Interface Sci. **33**, 67 (1970).

8. A.V. Anilkumar and T.G. Wang, "Drop Coalescence Studies," Proceedings of NASA Workshop on Containerless Experimentation in Microgravity, Pasadena, CA (1990).

ZERO-GRAVITY VOID FRACTION AND PRESSURE DROP IN A BOILING CHANNEL

Masaki Misawa and Samim Anghaie ■ Innovative Nuclear Space Power and Propulsion Institute, College of Engineering, University of Florida, Gainesville, FL 32611

ABSTRACT

The effect of microgravity on the void fraction, slip ratio, and pressure drop in an upward vertical flow with boiling heat transfer is investigated. Two sets of experiments are conducted to measure void fraction and pressure drop in a free falling test module which allows for about 1.5 seconds of testing under 0.02g conditions. Transparent ducts and high resolution photographic method are used to measure void fraction. Image analyzer is used to measure 2-D projection of the void fraction in a plane perpendicular to the viewing direction. In order to calculate 3-D volume averaged void fraction in bubbly and slug flow, bubbles and slugs are assumed to have spherical and rounded cylindrical shapes, respectively. Significant reduction in buoyancy force under microgravity conditions resulted in major changes in two-phase flow distribution and pattern. For low mass fluxes in the range of 30 to 600 kg/m^2/sec, the slip ratio is less than one which is consistent with the measured values of void fraction and also with the measured values of pressure drop along the test section. Microgravity conditions also reduce the bubble detachment rate with subsequent decrease in ebullition and increase in conductive heat transfer rates. This phenomenon in the subcooled boiling region causes the value of the non-equilibrium quality to approach the value of the equilibrium quality, thereby reducing the length of the subcooled boiling region. Under low exit quality and low flow rate conditions similar to those analyzed in this paper, the pressure drop versus mass flux curve induces the regions of negative slope which indicates the likelihood of flow excursion instability.

NOMENCLATURE

D : Tube diameter [m]
G : Mass flux [kg/m^2/sec]
g : Gravitational constant [m/s^2]
j : Volumetric flux [m/s]
u : Average velocity of the phase [m/s]
v : Specific volume [m^3/kg]
x : Thermodynamic equilibrium quality
Greek symbols
α : Volume averaged void fraction
β : Volumetric quality
ρ : Density [kg/m^3]

μ : Viscosity [Pa·sec]
Subscripts
g : Vapor phase
f : Liquid phase

BACKGROUNDS

Results of a few recently performed microgravity two-phase flow and boiling heat transfer experiments[1][2][3] have shown a significant difference in pressure drop and two-phase flow distribution from those of ground-based experiments. This has been primarily attributed to the changes in void fraction and thickening of the liquid film entraining small bubbles. High resolution photographs in a transparent conduit[4] have revealed three distinct flow regimes under microgravity conditions; bubbly, slug, and frothy annular flow. Due to the lack of buoyancy force, the average bubble detachment diameter has been 2 to 4 times larger. The noses and tails of vapor slugs have been observed to be smooth and almost symmetric. In the annular flow regime, smaller nucleation sites have become active. Bubbles generated by fewer nucleation sites have given a frothy look to the liquid film in the slug and the annular flow regimes. The void fraction in the subcooled region has been considerably larger than that of 1-g conditions due to larger bubble size and negative vapor drift velocity.

In most of previous investigaxons, closed loops with boilers and condensers have been used. In these systems, the loop transient time (20-30 sec.)[2] has often been comparable to the duration of low gravity conditions simulated by aircraft low-gravity flights. In addition, the analysis of the effect of reduced gravity on the boiling and condensation has become more even more complicated by their mutual interactions and their impacts on achieving the steady state conditions. This investigation has been solely focused on the analysis of a series of flow boiling tests which have been obtained in an open-loop test facility. The effect of reduced gravity on the two-phase flow pattern, slip ratio, void fraction, and pressure drop have been analyzed.

EXPERIMENTAL FACILITY

Figure 1 shows the schematics of the Freon(R113) two-phase flow boiling loop. The system has been operated at near atmospheric pressures. The 50cm length pyrex test section made of pyrex has been heated uniformly by transparent electrical heaters that covers the inside walls of the 5x5mm square conduit. The transparent electrical heater has been made of Cl_2Sn strips with total resistivity of 308 ohm. This configuration has allowed for the direct observation of the flow pattern change and boiling evolution under gravity conditions changing from 1-g to 0.02g. A high resolution motor driven 35mm still camera has been used to observe flow pattern and void distribution at every 0.25 seconds,

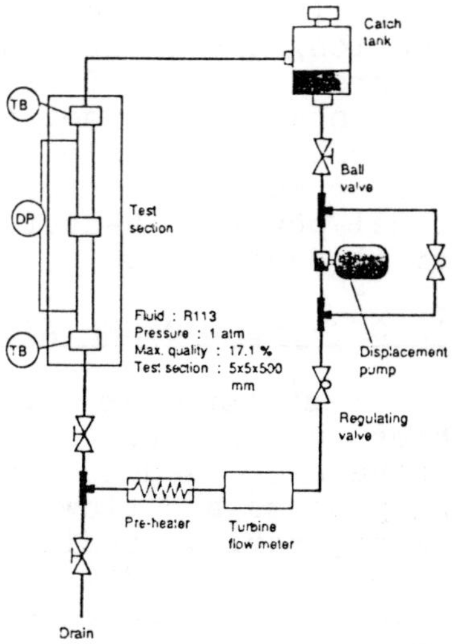

Figure 1. Schematics of R113 two-phase flow boiling loop.

during 1.5 seconds of free-fall test.

The test section has been arranged vertically to minimize the effect of transitional flow pattern change. The working fluid has been pumped in upward direction since the stable flow situation could have easily been obtained with this configuration. The test section for measurement of pressure drop has included a 54 cm length, 4 mm I.D. copper tube. This tube has been heated uniformly by a nichrom coil. A diaphragm type differential pressure transducer has been used to measure inlet-outlet pressure drop.

The mass flow rate has been measured by a turbine flow meter that is insensitive to the changes in gravity. A pre-heater has been used to provide a fine adjustment to the inlet subcooling. The inlet and outlet temperatures were measured by E-type thermocouples placed on the center line of the conduit. The vapor passing through the conduit has been released into the atmosphere through a hole at the top of the reservoir tank, whereas the liquid has been recovered in a reservoir. The pressure drop along the test section, the inlet and outlet temperatures, and the flow rate were recorded in a data logger on real time. An extension cable was connected to the drop module to supply the electrical power to heaters and other electrical devices.

RESULTS OF VISUAL OBSERVATION

Four test conditions with exit qualities ranging from negative to 17.1% have been used to study the boiling process and flow patterns under microgravity conditions. The inlet subcoolings and mass flow rates have been adjusted to observe the three distinct flow patterns, that is, bubbly, slug, and annular flow regimes.

The effective forces acting on the flow have been characterized by three non-dimensional groups; Froude number, Weber number, and Bond number. Under the drop tower microgravity conditions, Froude number has become 50 times larger than that of normal gravity, while Bond number has become 50 times smaller. Thus, flow and void distribution in these tests have been dominated by the inertia and surface tension forces, while the two-phase flow field and bubble motion are predominantly controlled by the gravitational force under normal gravity conditions. Pressure drop and void fraction have been measured at the ending period of the free fall and before the landing of the test module. For all of these test cases, quasi-steady state conditions have been reached after about one second of free fall which is in the order of the flow transient time in the test section. Flow transient time for void fraction measurements has been about 2 to 3 seconds and it has ranged from 1.5 to 10 seconds for the pressure drop measurements. Photographic observation of flow distribution at t=1.0 and 1.25 second has shown similar flow distribution and void fraction. The recorded values of pressure drop measured in the second set of experiments have also shown the similar trend.

The volumetric averaged void fraction has been measured from photographs using an image analyzer. The program IMAGE 1.28P on Macintosh converted the grayscaled 2-dimensional images into a binary mode. By adjusting the threshold, the vapor and liquid have been discriminated into black and white pixels, respectively. The binary images of flow patterns are shown in Figure 2 (a)-(c). IMAGE 1.28p has also been used to count the number of bubbles in a designated area. This information has been used to determine the average bubbles size for each experiment. Using a spherical shape model for bubbles and a rounded cylindrical shape model for slugs, the 2-D area fraction information have been transformed to volume averaged void fraction. Since the observed flow patterns have been bubbly and slug

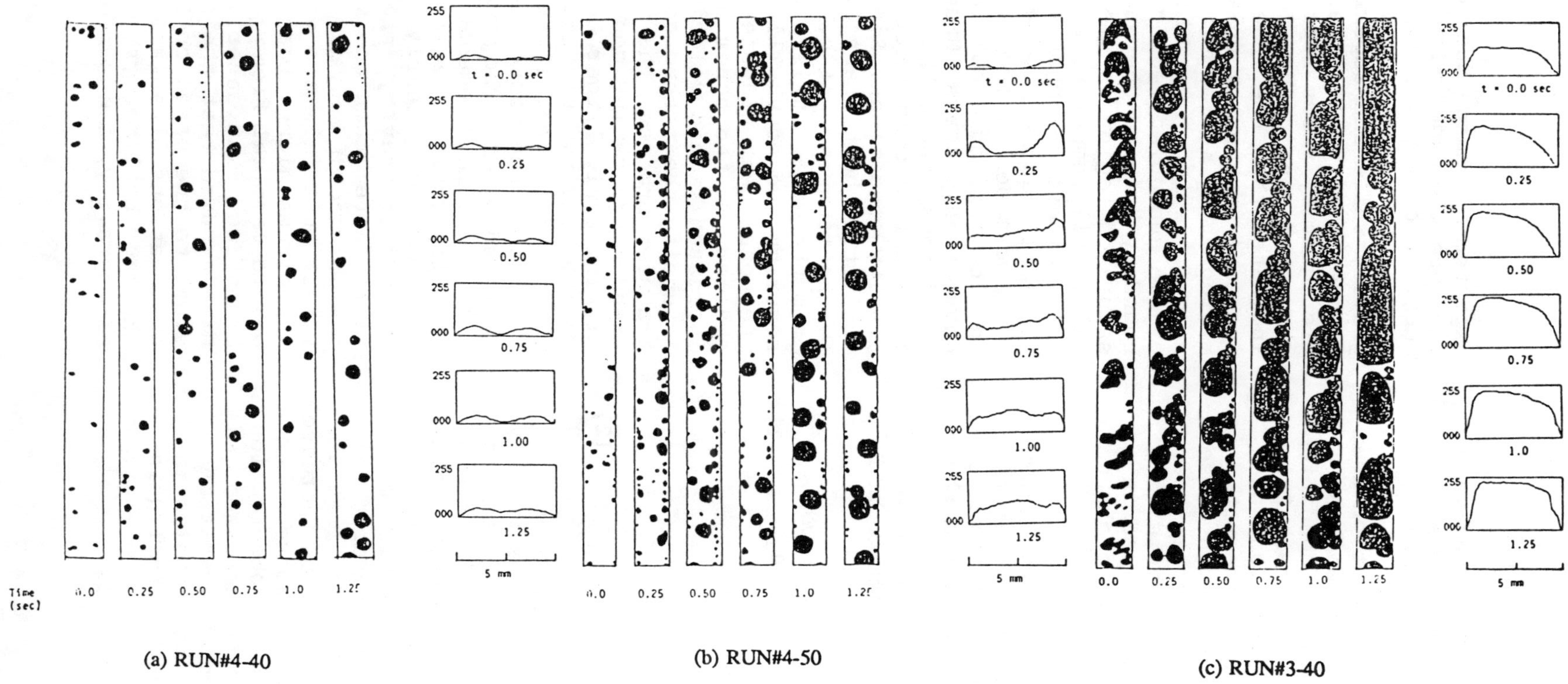

Figure 2. Boiling evolution and radial void distribution for the bubbly and the slug flow under microgravity conditions.

flow, the effect of overlapped vapor phase for inferring the volume averaged void fraction has been insignificant and ignored.

It is noteworthy to mention that the radial void distribution has changed under microgravity conditions. The histograms shown in Figure 2 are the averagexdial distributions of vapor phase scanned and integrated in the axial direction. Although these 2-D histograms have not directly been the representation of the void distribution, they have helped in recognizing the flow pattern change. In the bubbly flow regime under normal-gravity, the bubble distribution has shown a double-hump shape behavior. Under microgravity conditions, the much larger size of bubbles has shifted and merged these peaks at the center.

The total volumetric flux j is related with the void fraction and quality as follows

$$j = j_g + j_f = \frac{Gx}{\rho_g} + \frac{G(1-x)}{\rho_f} \quad (1)$$

$$= u_g \alpha + u_f (1-\alpha)$$

and

$$u_g = (1+C_1)(j_g+j_f) + C_2\sqrt{gD} \quad (2)$$

where C_1 and C_2 are empirically determined constants. As the gravity level decreases, the second term in Eq. (2) becomes smaller. Therefore, to maintain the same volumetric flux of vapor, the vapor velocity u_g decreases as the void fraction α increases. On the other hand, to maintain the same amount of liquid flux, the liquid velocity u_f increases as $(1-\alpha)$ decreases. The above arguments agree with the observation in Figure 2, where the increased void fraction reduces the flow area of liquid considerably and results in further acceleration of the liquid phase.

Heat transfer mechanism is also expected to change due to the change in flow profile under microgravity. In the bubbly flow regime, the photographed region was subcooled under normal gravity while the superficial quality in the corresponding region under microgravity seems approaching to saturated conditions. This fact indicates that the rate of quality increase in the subcooled region under microgravity is larger than that of normal gravity, resulting in shortening of the subcooled region. Since the boundary between the subcooled and saturated region is predicted from the axial wall temperature profile, the wall temperature change during the free-fall has been carefully monitored.

Two modes of heat transfer exist in flow boiling; ebullition and convection process. Under microgravity conditions, the bubbles grow to much larger sizes than those of under normal gravity conditions. Since the ebullition heat transfer is dominated by the frequency of bubble detachment from the wall, it is expected that the contribution of ebullition process to the total heat transfer becomes smaller. On the other hand, the increased liquid velocity seems to intensify the forced convection heat transfer process. The void fraction under microgravity conditions has been significantly larger than that of normal gravity, particularly, in the subcooled region. It should be also noted that the values of void fraction measured under microgravity were larger than those of homogeneous flow in which the slip ratio is equal to 1.0.

This fact clearly indicates that under microgravity, the liquid phase flows faster than the vapor phase. In the absence of buoyancy force, no force exists to accelerate vapor phase to velocities higher than liquid phase in upward vertical flow systems. It should be noticed that other investigators have assumed that the slip ratio is equal to one based on their visual observations of adiabatic flow[5]. In the flow boiling system under microgravity, the bubbles are initially

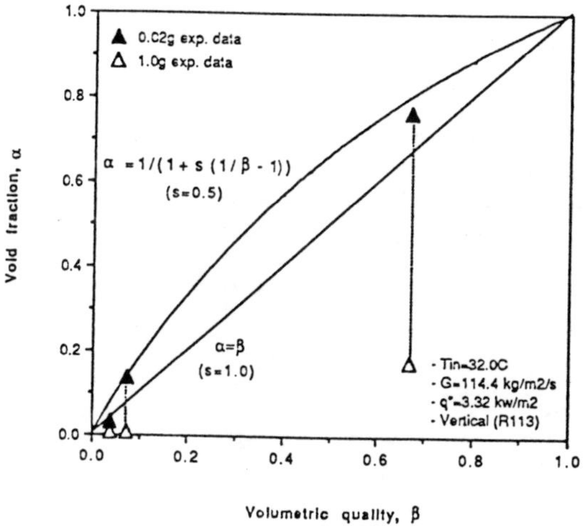

Figure 3. Void fraction α vs. volumetric quality β under microgravity conditions.

formed on the wall with no momentum in the direction of flow. After detachment from the wall, the bubbles are entrained in the liquid stream and move at the same or lower velocity. Therefore, it appears to be reasonable to assume that the average vapor velocity is smaller than that of liquid under microgravity conditions. Figure 3 shows functional dependence between void fraction α and volumetric quality β. The solid curve is a 2nd order polynomial obtained by the least square fitting technique. This curve lies between two curves obtained by assuming the slip ratio is equal to 1.0 and 0.5. At 0.02g, the slip ratio seems to assume a value between 0.5 and 1.0 due to the decrease in the vapor velocity and the increase in the liquid velocity. The slip ratio obtained from the least square fitting to limited number of data obtained in this experiment is 0.74.

PRESSURE DROP

The pressure drop between the inlet and outlet of the test section has been measured by a differential transducer during the free-fall experiments. The test section has been heated uniformly by a nichrom wire tightly wound around the tube. The mass flux has been adjusted to obtain 5 different exit qualities ranging from 2.3% to 34.1%, while the heat flux has been kept constant (6.63kW/m^2). The recorded values of the pressure drop measurements are shown in Figure 4 (a)-(e).

The overshoot at the end of the curve has been caused by the impact on landing. The relatively flat region lasting for nearly one second has indicated the microgravity period during the free fall. The values of pressure drop at t=1.25 seconds after the initiation of free fall have been used for this analysis. The period of free fall was determined by the height of the drop tower. Figure 5 (a)-(e) shows the comparison between the measured values of pressure drops and those obtained by using, first, the two-phase flow homogeneous model and then by using the separated flow model employing the correlation between α and β developed in the previous section. These five curves are not plotted on the same figure due to different inlet temperatures. The frictional component of the pressure drop has been calculated using the Dukler's relation for the two-phase viscosity defined as follows

$$\overline{\mu} = \overline{\rho}\,[x v_g \mu_g + (1-x) v_f \mu_f] \qquad (3)$$

All of the measured values of pressure drop except for the case (e) have been slightly larger than those which have been predicted by the homogeneous model. Using the measured value of s=0.74 from the previous experiments has overestimated the pressure drop. This fact has indicated that the void fraction at microgravity is larger than that of normal gravity under the same operating conditions, which agrees with the measured values of void fraction in a transparent channel.

The predicted pressure drop curves have also shown that the pressure drop decreases as the quality increases. In

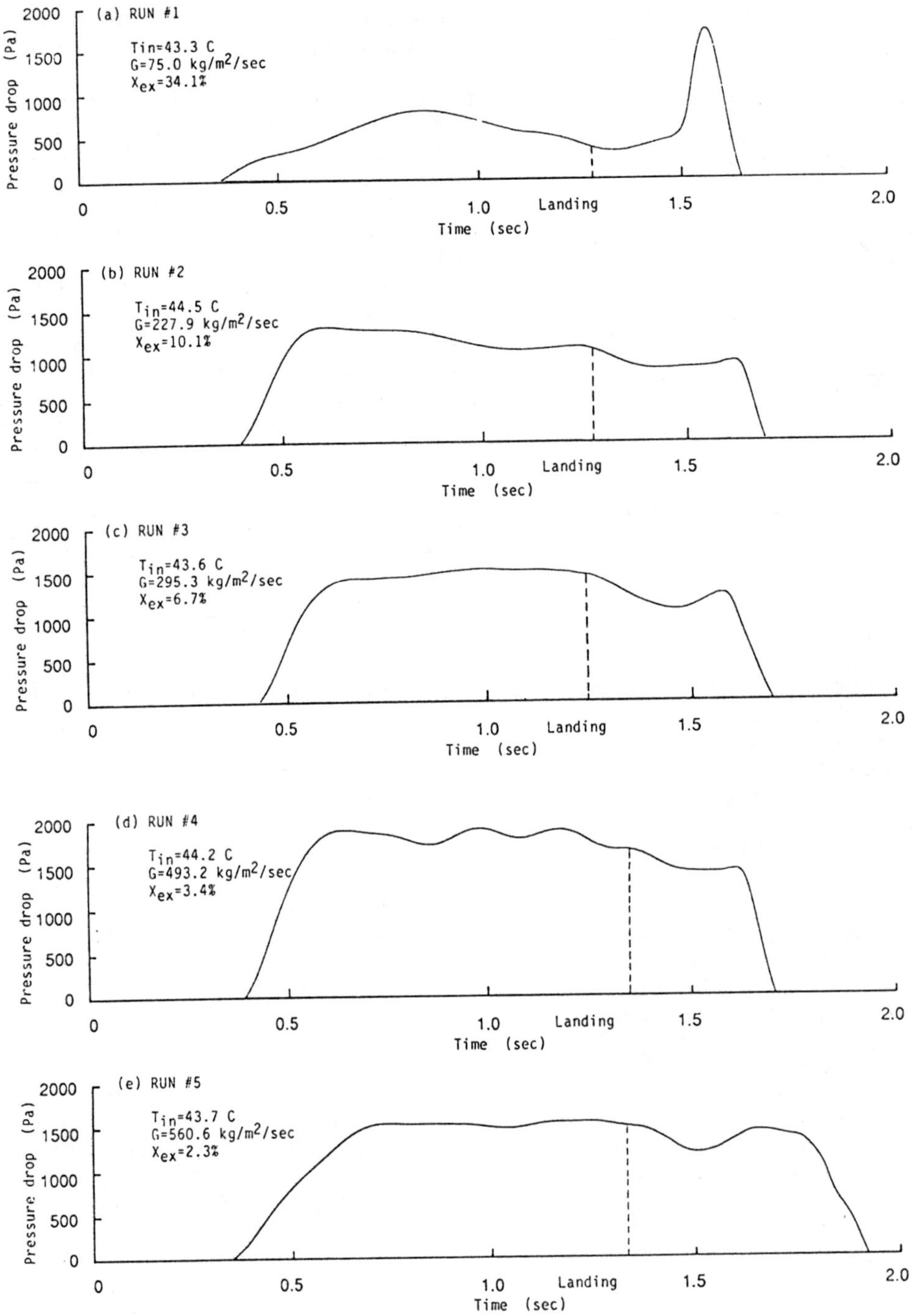

Figure 4. Pressure drop along the test section during the free-fall experiments for 5 different flow conditions with constant heat flux.

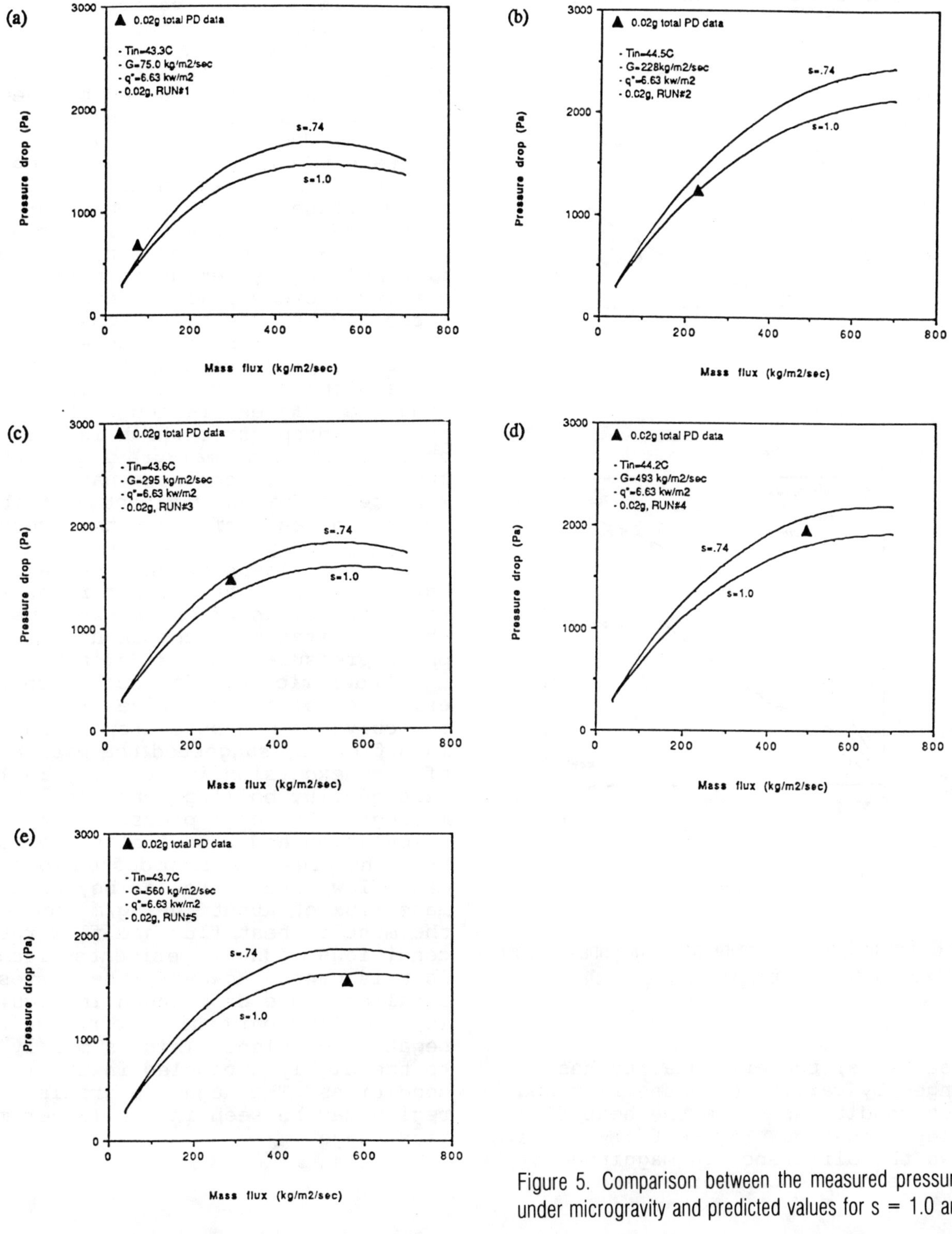

Figure 5. Comparison between the measured pressure drops under microgravity and predicted values for s = 1.0 and 0.74.

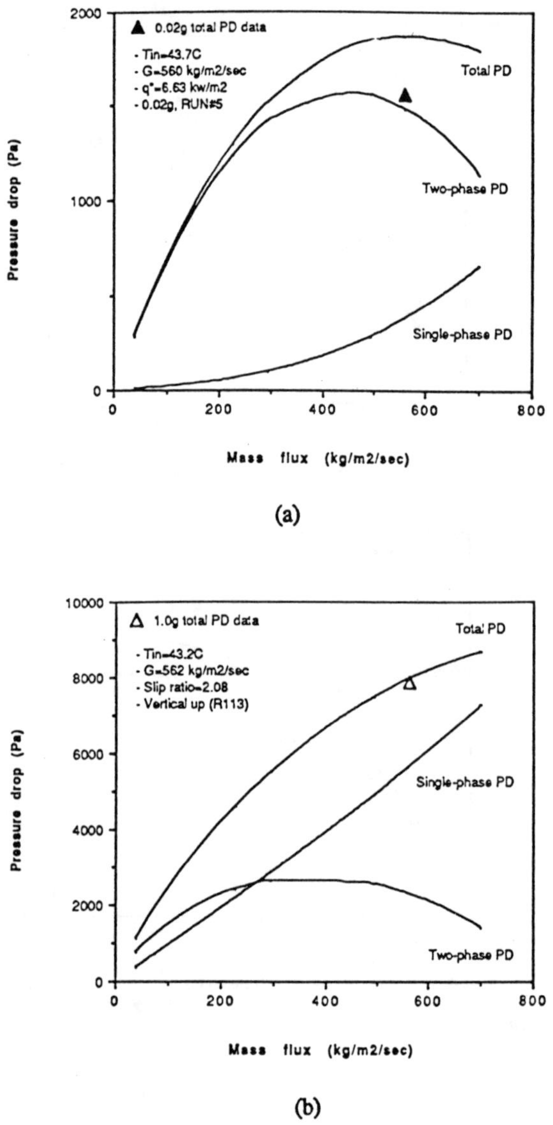

Figure 6. Comparison of pressure drop components under normal- and microgravity conditions.

these tests, the exit quality has changed by varying the mass flux and inlet conditions while the heat flux is kept constant. Figure 6(a)-(b) have shown the difference in magnitude of pressure drop under normal and microgravity conditions. Based on the observation of flow patterns under microgravity conditions, the increase in the two-phase pressure drop seems to be mainly due to the accelerational component which has caused by the significant increase in the void fraction. Under normal-gravity, the increase in the gravitational component of single-phase region is larger than the decrease in the two-phase pressure drop. Therefore, the total pressure drop for a low exit quality vertical flow boiling system under normal-gravity increases monotonously. Under microgravity, the single-phase component of pressure drop exceeds the two-phase pressure drop at a larger value of mass flux, developing a negative gradient region in the total pressure drop curve which is not observed under normal-gravity. Although the frictional component has been reported to be the dominant contributor to the increase of the total pressure drop[3], the present analysis and the results of flow pattern observation have shown that at least for low flow rate conditions of these experiments the accelerational component of two-phase pressure drop is dominant.

Under microgravity conditions, the existence of the negative gradient in the curve of pressure drop versus the mass flux has suggested the possibility of flow excursion instability in low exit quality boiling channels. The microgravity total pressure drop calculation and experiments have shown that the pressure drop decreases as the mass flow rate increases beyond the mass flux of about 500 $kg/m^2/sec$ for the most of heat flux and flow rate conditions of the experiments discussed in this paper. However, the curves of total pressure drop shown in Figure 5 (b) and (d) have exhibited no region of negative gradient. This is mainly due to the highly subcooled inlet conditions. The negative gradient region may be seen in the larger mass flux range.

CONCLUDING REMARKS

The close observation of flow patterns and the measurement of pressure drop under microgravity conditions have revealed the following results.

1. The slip ratio for vertical upward flow in a boiling channel under microgravity is smaller than 1.0. The measured pressure drops under microgravity has been consistently larger than those predicted by the homogeneous model, indicating that the slip ratio should be smaller than 1.0. This has been primarily attributed to the lack of buoyancy force which causes the vapor phase velocity to move slower than the average flow velocity.
2. The significant increase in the void fraction in the subcooled region has suggested that under microgravity conditions, the length of the subcooled region may become shorter.
3. The increased liquid velocity has enhanced the convective component, while the ebullition process has diminished due to the decrease in bubble detachment frequency.
4. The total pressure drop under microgravity has been larger than predicted values by the homogeneous model. The largest contribution has been due to the accelerational pressure drop which has been caused by the void fraction increase.
5. Under microgravity, for the low flow rate conditions of experiments which have been analyzed, the negative gradient of pressure drop versus mass flux in low quality regions has suggested the possibility of flow excursion instability.

Due to the lack of sufficient data points in microgravity experiments, a full quantitative analysis has not been performed. Current research activities have been focused on obtaining data on the void fraction, the wall heat transfer, and pressure drop measurement for a broader range of flow rate and exit quality.

LITERATURE CITED

1. Crowley, C.J. and Sam, R.G. "Microgravity experiments with a simple two-phase thermal system", 8th Symposium on Space Nuclear Power Systems Proceedings, Part 3 pp.1207-1213 (1991)

2. Standley, V.H. and Fairchild, J.F. "Boiling and condensing pumped loop microgravity experiment", 8th Symposium on Space Nuclear Power Systems Proceedings, Part 3 pp.1224-1229 (1991)

3. Lambert, A., Reinarts, T.R., Best, F.R., and Hill, W.S. "KC-135 zero-gravity two-phase flow pressure drop: Experiments and modeling", 8th Symposium on Space Nuclear Power Systems Proceedings, Part 3 pp.1236-1243 (1991)

4. M. Misawa, S. Anghaie, S Kondo, K Furuta, S Enya, and T Yano, "Two-phase flow and boiling evolution under microgravity conditions", The winter annual meeting of ASME, NE-vol.3, pp.19-31 (1989)

5. Dukler, A.E., Fabre, J.A., McQuillen, J.B., and Vernon, R., "Gas liquid flow at microgravity conditions: Flow patterns and their transitions", Int. Symp. on Thermal Problems in Space-Based systems, HTD-vol.83, pp.85-98 (1987)

MICROGRAVITY EXPERIMENTS ON TWO-PHASE FLOW AND HEAT TRANSFER DURING QUENCHING OF A TUBE AND FILLING OF A VESSEL

M. Kawaji and C.J. Westbye ■ Dept. of Chemical Engineering & Applied Chemistry, University of Toronto, Toronto, Ontario M5S 1A4, Canada

B.N. Antar ■ Center for Advanced Space Propulsion, The University of Tennessee Space Institute, Tullahoma, TN 37388

The two-phase flow and boiling heat transfer characteristics have been investigated for quenching of a hot tube and filling of an empty vessel under microgravity on board NASA's KC-135 airplane. Long, continuous and thick filaments and large droplets of liquid were seen to flow in the middle of the tube, unable to rewet the tube wall. The heat transfer rates were much lower than in 1-g under the similar operating conditions and quenching was significantly delayed. In filling a vessel under adiabatic conditions, the liquid was observed to flow along the wall and rarely detach from it. The core of the vessel was primarily voided and no significant masses of liquid were observed to freely float within the vessel.

INTRODUCTION

There are many important processes that occur under microgravity which would involve gas-liquid two-phase flow and boiling/condensation heat transfer phenomena. In refilling of space transfer vehicles with liquid hydrogen/oxygen fuel under microgravity, a transient quenching process is expected involving liquid/vapor two-phase flow and convective boiling heat transfer due to the low boiling points of the fluids and high initial wall temperatures of the refueling lines and receiving tanks. Similar phenomena are also expected in transfer and storage of other cryogenic fluids, emergency cooling of space nuclear power reactors under loss-of-coolant accident conditions and processing of certain materials on the space station.

In the past, only a few studies on two-phase flow under microgravity have been conducted by Heppner et al. (1975), Kachnik et al. (1987), Chen et al. (1988) and Dukler et al. (1988) among others. Many have used the NASA's KC-135 aircraft or Learjet to simulate the zero-g conditions in the study of adiabatic two-phase flow, or boiling and condensation phenomena under steady state conditions. However, the results of those studies will not be directly applicable to the transient, non-adiabatic quenching flow situations described above.

It is known from past studies including our own work on quenching phenomena in 1-g, that the two-phase flow patterns existing under quenching conditions are quite different from those observed under adiabatic, or even boiling but non-quenching conditions (Kawaji et al., 1985). Typically, in quenching of a vertical tube, inverted annular flow and dispersed flow patterns occur with film boiling heat transfer characteristics that are highly dependent on the hydrodynamics (Kawaji and Banerjee, 1987, 1988). In horizontal tubes, the liquid flow is asymmetric about the axis because of stratification due to gravity (Chan and Banerjee, 1981). Since the two-phase flow patterns existing under microgravity conditions are yet relatively unknown but are expected to strongly affect both precursory cooling and quenching rates, it is important to conduct such an experiment to elucidate their behavior under microgravity conditions.

To this end, a series of quenching experiments were performed aboard NASA's KC-135 in June, 1990 to investigate the two-phase flow and convective boiling heat transfer phenomena which occur during rapid cooling and quenching of a hot quartz tube under microgravity. At the same time, the behavior of a liquid-vapor mixture during filling of an initially voided vessel was qualitatively studied using the same experimental apparatus.

EXPERIMENTAL APPARATUS

A compact test loop was constructed incorporating

a transparent, 14mm I.D., 1.2m long quartz tube test section which was heated externally by a nichrome heating tape wound around the tube and connected to an AC power source (Fig. 1). The test loop contained a low boiling point (about 50°C) liquid (R-113) at near atmospheric pressure. A positive displacement pump was used to inject the coolant at a constant rate up to 20 mL/sec for about 20 seconds under microgravity conditions. As the liquid entered the test section it vaporized to give rise to several two-phase flow patterns while cooling the tube at the same time. The flow patterns were recorded using a 35mm still camera, 16mm movie camera and video camcorder located at various axial positions along the tube.

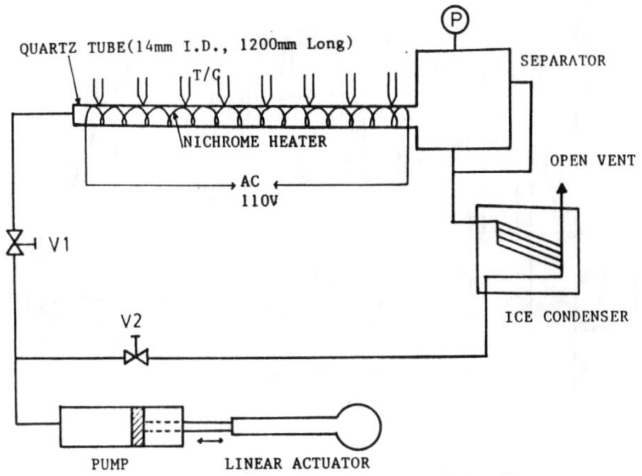

Fig.1 Schematic of the Test Apparatus

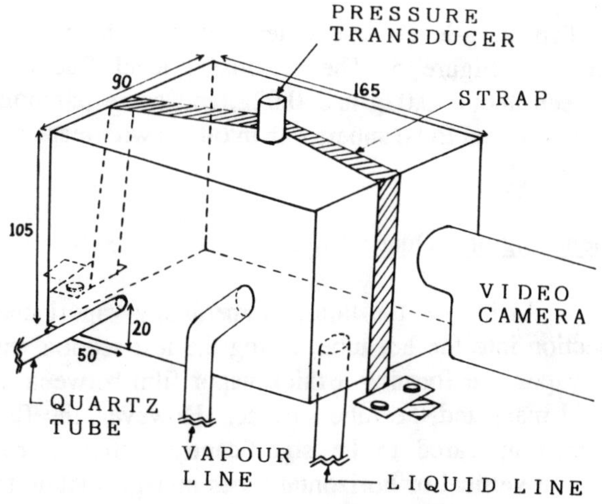

Fig.2 Liquid-Vapor Separator

A vapor-liquid separator, made of a clear lucite vessel (Figure 2) was connected to the exit of the test section. This was intended to function as a separator only in 1-g, however, it was useful to have a large discharge vessel in microgravity experiments also to eliminate potential plugging of the loop between the test section and the condenser with liquid, thereby preventing a build up of back pressure and flow oscillations. A video camera/recorder was installed beside the separator to view the behavior of the liquid/vapor mixture inside. The mixture flowed through the separator and into an ice condenser, made of a series of parallel micro-fin tubes immersed in an ice bath. Here, the vapor was totally condensed and the liquid kept at highly subcooled conditions. A small vent located near the exit of the condenser permitted air displaced by the freon vapor to escape, which was intended to limit the pressure rise in the system. The liquid freon was stored in a reservoir inside the condenser and returned to the pump before the next run.

The temperature response of the test section was recorded with a number of thermocouples externally fixed to the test section with high temperature cement at six axial positions. At each axial location, two thermocouples were located at the top and bottom of the tube to record the azimuthal temperature variation. The wall temperature history provided information on precursory cooling rate and quenching characteristics.

For each run, the test section was initially empty and heated to a desired temperature between 200°C and 320°C. When microgravity condition was reached, the pump was operated and the liquid injected at a specified flow rate. The thermocouple data were collected by a PC based data acquisition system. The 40-Megabyte hard disk functioned without any problem during the microgravity tests.

RESULTS OF 1-G TESTS

The pre-flight experiments were first performed on the ground to obtain reference 1-g data for various injection rates and initial tube wall temperatures. As highly subcooled liquid freon was injected to the test section, it immediately became stratified and vapori to form a vapor film between the liquid and the bottom surface of the tube wall. This vapor film was quite stable and prevented the liquid from contacting and quenching the test section immediately. Heat transfer was in the film boiling mode, which continued until enough heat has been removed from the quartz tube to allow the liquid to come into contact with the wall. Once the liquid was able to rewet the tube wall, heat transfer improved significantly.

At the downstream locations, the wall temperature measured at the bottom of the tube initially remained nearly constant due to the small rates of heat transfer to vapor as shown in Figure 3. When the liquid arrived at the thermocouple location, heat transfer improved and the temperature declined at a faster rate, even though there was no liquid-wall contact. When the quench front arrived and the liquid-solid contact was established, the wall temperature dropped more rapidly producing a characteristic knee in the transient temperature curve.

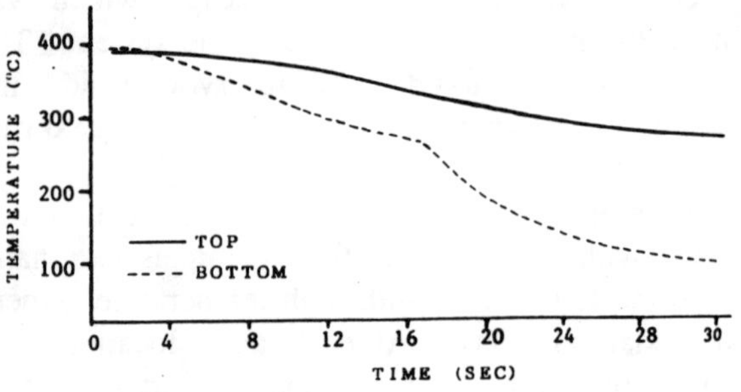

Fig.3 Transient Temperature Response during 1-g Quenching Tests

The top part of the tube wall was cooled at a much slower rate as evident in Figure 3. Quenching was delayed until the tube was filled nearly completely with the liquid, which could take place after the entire bottom part of the tube wall had been quenched. For the vertical orientation in 1-g, more symmetric temperature response is expected around the periphery of the tube (Kawaji et al., 1985).

The pressure transient resulting from the quenching process depended on the rates of vapor production in the test section and condensation in the condenser. Tests with the essentially open system, with air free to flow in and out, showed a very small pressure rise (less than 7 kPa). An increase in the liquid injection rate, which affects the quenching rate, resulted in a greater pressure increase for a given condenser surface area. The pressure rise is an important parameter since a significant back pressure will retard the penetration of liquid and slow down the quenching rate.

RESULTS OF MICROGRAVITY TESTS

The microgravity periods during KC-135 flights were repeatedly provided between the periods of higher gravity as shown in Figure 4, by the plane following a parabolic flight trajectory. The microgravity period lasted for about 20 seconds in each parabola and a total of forty parabolas were conducted on each flight. The higher gravity periods (1.0 - 1.75g) were utilized in this experiment to refill the positive displacement pump and heat up the quartz tube to the desired temperature. The injection of subcooled liquid R-113 was then started as soon as the microgravity period was entered.

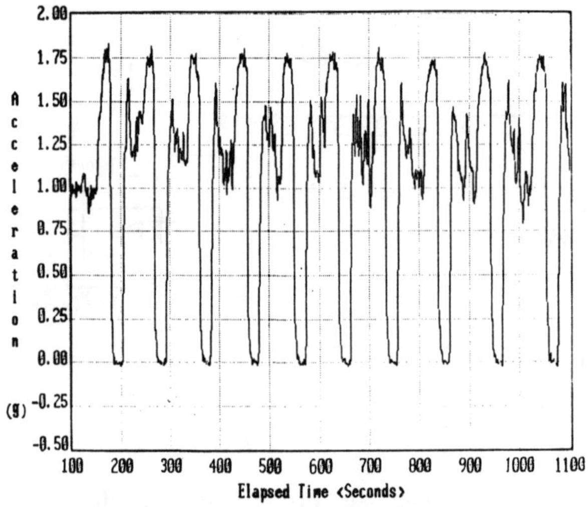

Fig.4 G-levels during KC-135 flights

Typical g-levels encountered during the run are shown in Figure 5. The residual g-level fluctuated between about -0.02g and 0.02g, and some parabolas were observed to be superior than others with respect to g-jitter.

Quenching of a Quartz Tube

As in 1-g, the liquid immediately vapori upon injection into the hot tube, filling the test section with the vapor and forming a thick vapor film between the liquid mass and the tube surface. However, the flow patterns appeared to be significantly different from those in vertical or horizontal tubes in 1-g. In 1-g, the vapor layer is thin and the liquid core is thick in a vertical tube or flows close to the bottom of the tube in a horizontal tube. In microgravity the liquid core was often much thinner, more closely resembling a thick cylindrical liquid filament than that in the traditional inverted annular flow pattern. The thick liquid filament

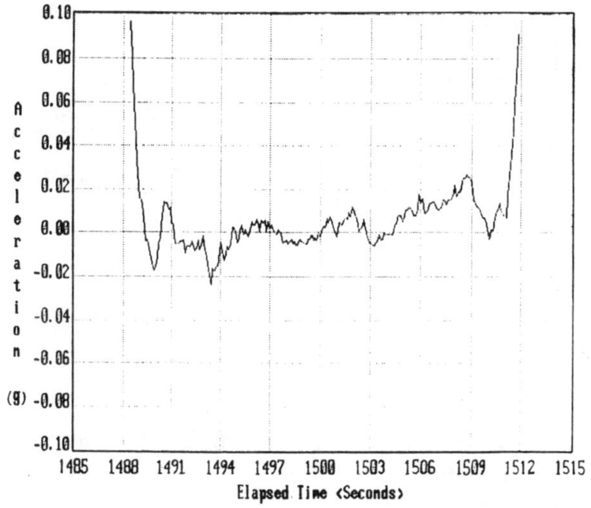

Fig.5 Typical G-level Fluctuations during a Microgravity Period

was observed to flow mostly in the middle of the tube, however, small gravity perturbations affected the distribution of the liquid in the tube. The liquid filament was mostly smooth and continuous, but sometimes bulged at some places nearly filling the tube volume without rewetting the tube surface.

The photographs in Figures 6a to 6d show the flow patterns described above in sequence for the run with the liquid injection speed of about 15 cm/s. Figure 6a shows the flow just after injection began, when the flow was entirely that of vapor at the 35-mm camera location, about 26 cm from the inlet. Shortly afterward, the thick liquid filament appeared as shown in Figure 6b, flowing freely through the thick vapor layer. The filament was not axisymmetric, due to small but fluctuating g-levels. The following pictures in Figures 6c and 6d show the motion of the liquid as the cooling rate, vapor generation rate, vapor velocity and g-level varied continuously with time. At no time in this sequence did the tube become rewetted, although heat was being removed from the tube continuously by film boiling or forced convection to vapor. Had the duration of the microgravity test (20 seconds) been extended, or if the initial wall temperature (about 300°C) been lower, liquid-surface contact would have been obtained and the tube would have quenched at this axial location (26 cm from the inlet). Since the liquid was highly subcooled and came into contact with the tube wall only near the inlet, there was much less bubble entrainment than in 1-g under the same conditions, as demonstrated by the appearance of few bubbles in the liquid filaments.

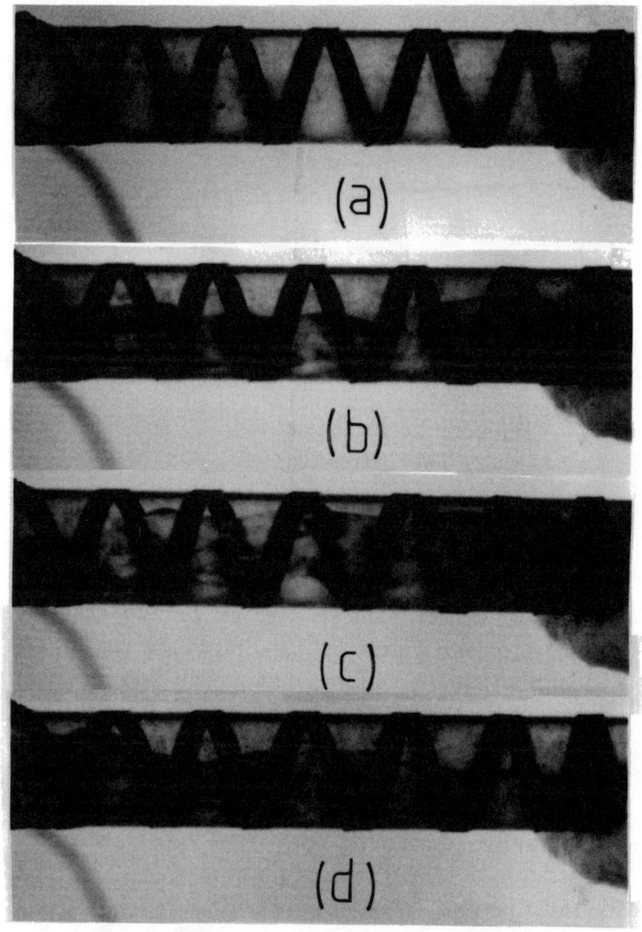

Fig.6 Sequence of Flow Patterns observed during Quenching of a Tube under Microgravity (Run #18: injection speed = 15 cm/s)

Further downstream of the 35mm camera location, the liquid filaments were seen by a video camera located at about 75 cm from the inlet to break up into large droplets due to acceleration and shear by the faster vapor stream. Thus, the two-phase flow exiting the test section and entering the separator was initially a dispersed droplet flow. Although the continued observation of the flow in the separator was hampered by clouding due to white freon vapor, it is conjectured based on the video camera view of the flow at 45 cm from the test section exit that a similar dispersed droplet flow pattern existed throughout the run.

Figure 7 shows the temperature profiles from the top and bottom of the tube at the same axial location, 39 cm from the inlet. The curves are offset because the initial temperature distribution was not uniform, but the exhibited temperature variations are almost identical, as

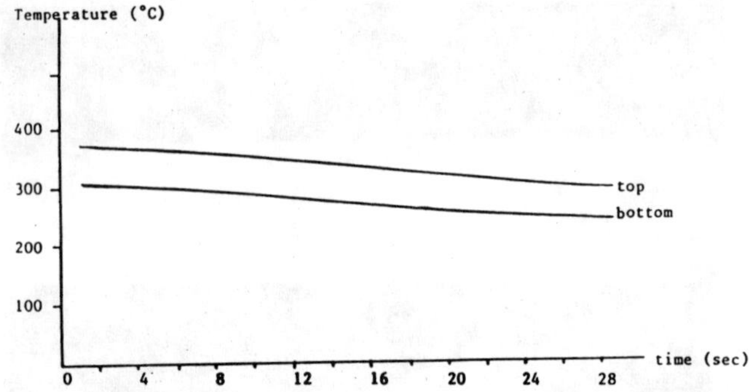

Fig.7 Temperature Profiles at the Top and Bottom of Quartz Tube (Run #18)

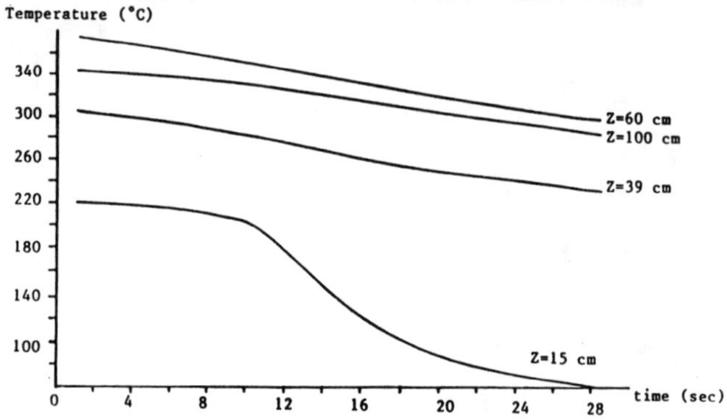

Fig.8 Transient Temperature Profiles at Various Axial Locations (Run #18)

expected due to axial symmetry in the absence of gravity. The rate at which the temperature decreased was much slower than in 1-g, as can be seen by comparison with Figure 3. That reference run in 1-g had similar initial conditions and coolant flow rate, but the rate of heat removal in the film boiling regime was much higher. Quenching occurred within the twenty second period in 1-g and the tube temperature was reduced by about 200°C. On the other hand, in microgravity, quenching did not occur and the temperature decreased by only less than 100°C during the same period.

The transient temperature profiles at various axial locations in microgravity are shown in Figure 8. Again the temperature curves are initially offset because of the non-uniform initial temperature distribution. This resulted from a combination of factors such as irregularities in the spacing of nichrome heating tape and placement of thermocouples with respect to the heater, and the amount of cement used to attach the thermocouples which contributes some heat capacity and contact resistance.

The temperature profile for the thermocouple closest to the inlet (15 cm) showed the normal quenching behavior: temperature slowly decreased until rewetting occurred and heat transfer dramatically improves. The other axial locations did not show this characteristic profile, indicating that the quench front did not propagate substantially down the length of the test section. The equivalent 1-g run showed that the quench front propagates fully from the inlet to the outlet, indicating again that heat transport is impaired under microgravity conditions.

<u>Filling of a Vessel</u>

The fluid behavior in the separator was studied to understand the phenomena involved in filling of an empty vessel under microgravity conditions without the aid of phase separation due to gravity. Several runs were conducted with injection of a single-phase liquid in addition to the two-phase mixture during the flow-boiling experiments. The separator walls were made of 12.5mm thick lucite plates with low thermal conductivity and were neither heated externally or insulated so the filling process can be considered to be nearly adiabatic in the present experiments. Due to the relatively short microgravity periods available, the vessel was far from being filled up at the end of each parabola and the study was therefore limited to the initial phase of the filling process. Actual filling of initially empty and relatively warm vessels with cryogenic fluids such as liquid hydrogen and oxygen will take a longer time and certainly involve film boiling, rewetting, nucleate boiling and convective heat transfer, which will in turn affect the hydrodynamic behavior of liquid and vapor inside the vessel throughout the filling process.

In the cold filling experiments, highly subcooled liquid freon was injected at several different rates (less than 20cm/sec) into the separator. In all cases, the liquid entered the vessel as a continuous liquid jet and travelled across the vessel and reached the opposite wall as shown in Figure 9a. The liquid immediately spread along the wall in all directions, without any significant splash and production of droplets (Figure 9b). The liquid layer spread along the entire surface of the wall opposite to the entrance nozzle, smoothly turned the 90 degree corners and continued to flow along the surfaces of the adjoining four walls. Although the observation of flow inside the vessel became extremely difficult once the wall on the video camera side became covered with the liquid, the vessel was essentially voided in the middle throughout the microgravity period, except for the liquid jet moving from the entrance nozzle to the opposite wall. At the end of an another cold fill run, the liquid remained attached to the vessel wall forming a thick layer with a dynamically moving surface (Figure 9c).

The filling processes during quenching experiments were difficult to observe, as the white cloud of R-113 vapor filled the entire separator vessel soon after the liquid injection into the quartz tube began. There was little liquid carryover in the initial periods and the amount of R-113 vapor produced was excessively large so as to completely block the view thereafter. In order to lessen the initial amount of vapor generation, the quartz tube temperature was lowered and the experiments changed to those of transient flow-boiling. Liquid freon was injected into the quartz tube which

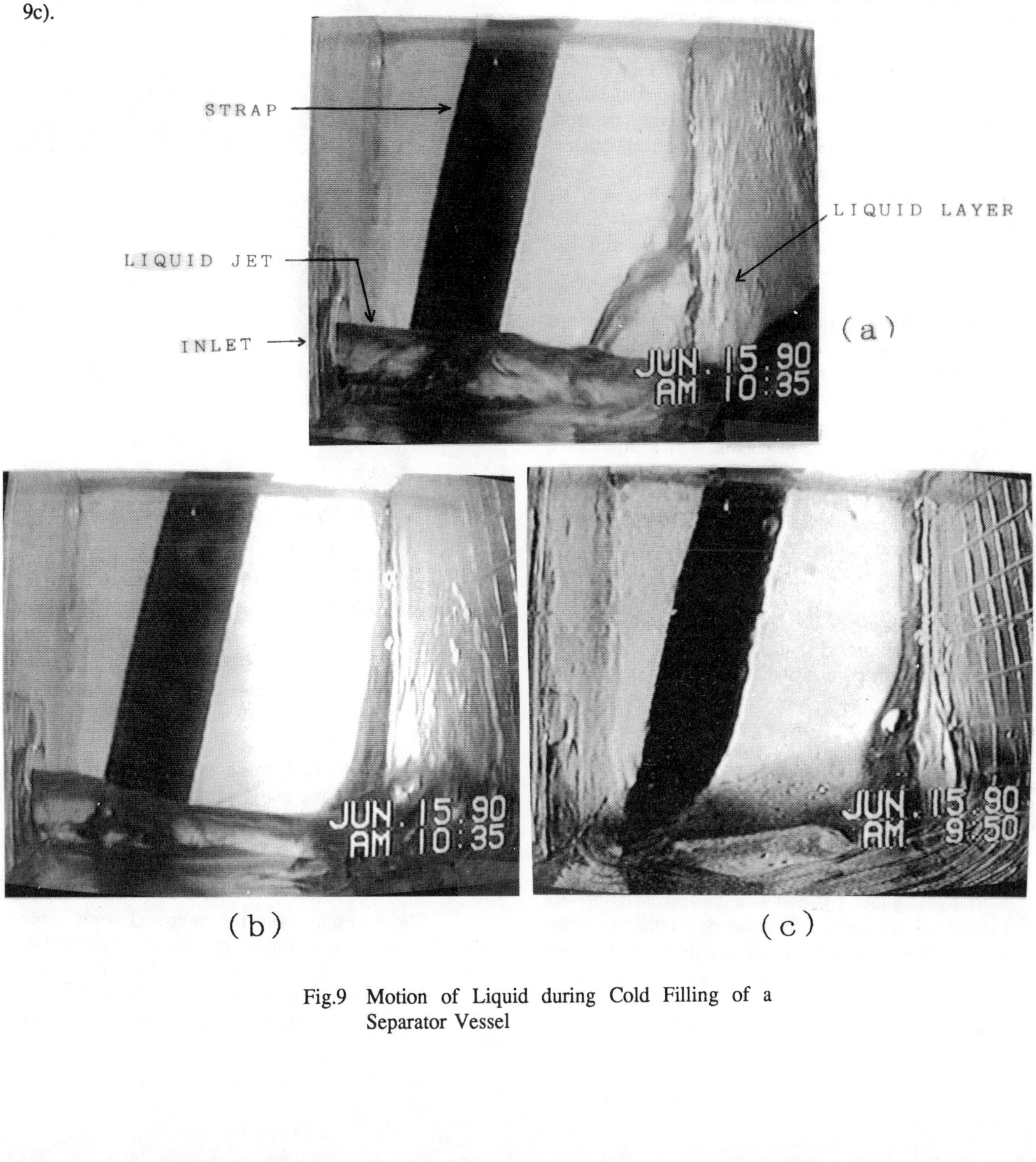

Fig.9 Motion of Liquid during Cold Filling of a Separator Vessel

was continuously heated at a low power level to give a wall temperature of about 60°C. This produced a sufficient amount of vapor within the test section and a jet of vapor-liquid mixture was obtained at the inlet of the separator vessel. Although there was still some fogginess, it was possible to view the motion of the liquid inside the vessel.

In the flow-boiling runs, a jet of vapor-liquid mixture entered the separator vessel at much higher velocities (> 1 m/s) than in the cold fill runs. The jet impinged on the opposite wall but again there was little splash observed and the liquid simply adhered onto the wall to form a liquid layer on the wall (Figure 10a). The liquid layer spread outward in all four directions and flowed along the wall surface leaving the core region of the vessel essentially voided except for the entering two-phase jet (Figure 10b). There were practically no large lumps of liquid detached from the vessel wall and floating elsewhere in the vessel. Similar pictures from other transient flow-boiling runs are shown in Figures 11a-11b.

In the absence of gravity, the initial phase of the vessel filling process was observed to be dominated by the surface tension forces and the liquid tended to adhere to the vessel walls and form layers rather than detached masses. This result suggests a simple venting procedure during filling of a tank with a liquid or a liquid-vapor mixture in which the vapor may be preferentially vented out through a vent nozzle located in the core region of the vessel sufficiently far away from the vessel walls. If the vessel walls are at an elevated temperature relative to the saturation temperature of the fluid and film boiling occurs preventing wetting of the wall surfaces, different fluid behavior is expected and further work would be necessary to determine the liquid behavior in such situations.

(a)

(b)

Fig.10 Motion of Liquid in the Separator Vessel during Injection of a Two-Phase Jet

CONCLUSIONS

The main conclusions from this study are that the two-phase flow patterns encountered in a transient quenching process in microgravity are different from those found under 1-g, and as a consequence, the heat transfer rate is substantially reduced. A thick vapor film exists between the tube surface and the liquid core and without gravity to reduce the thickness of the vapor film, rewetting becomes more difficult and the time required for quenching of a hot surface is greatly extended. In the initial phase of adiabatic filling of vessels under microgravity, the liquid was seen to adhere to and flow along the solid surface leaving the core region of the vessel essentially voided with little detached masses of liquid.

These results will be compared with those obtained in two similar quenching experiments involving R-113 and stainless tube, and liquid nitrogen and quartz tube, to be conducted aboard KC-135 in April, 1991. The information gained here and in future

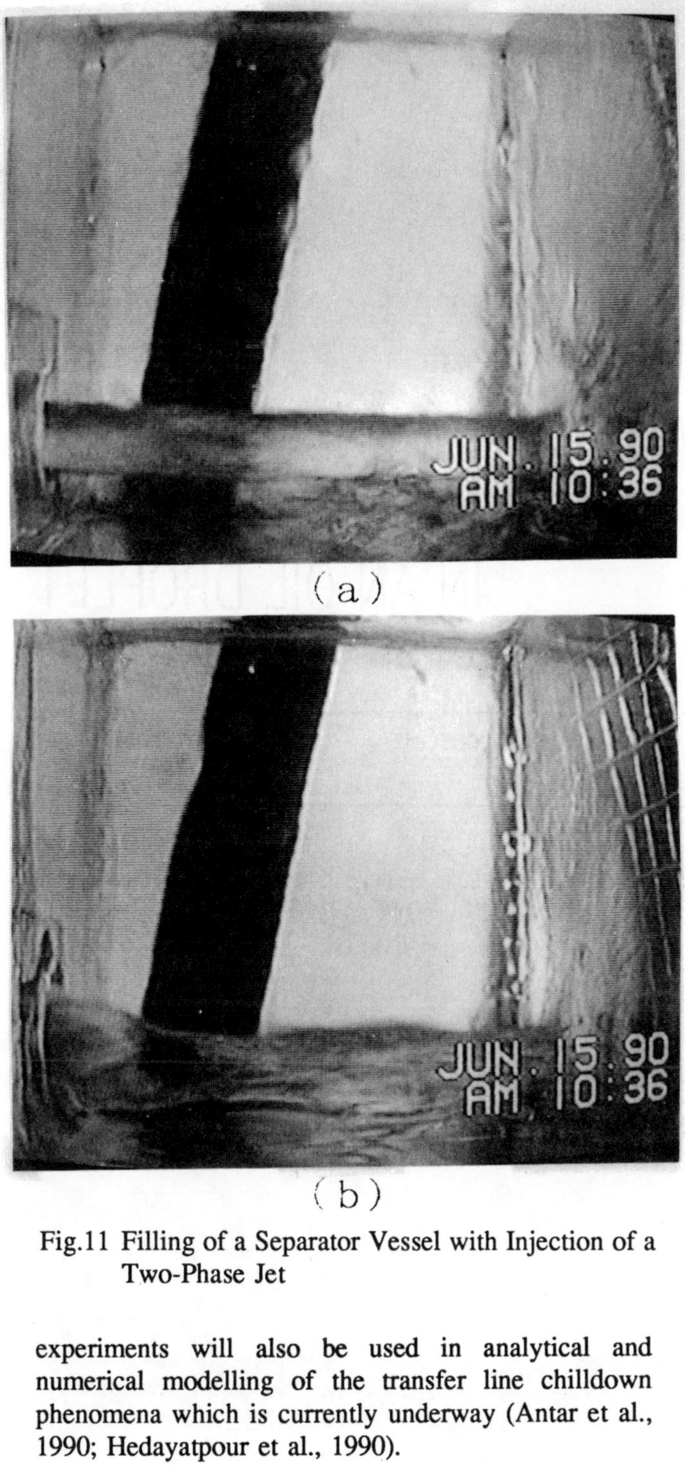

(b)

Fig.11 Filling of a Separator Vessel with Injection of a Two-Phase Jet

experiments will also be used in analytical and numerical modelling of the transfer line chilldown phenomena which is currently underway (Antar et al., 1990; Hedayatpour et al., 1990).

ACKNOWLEDGEMENTS

The authors would like to thank the Canadian Space Agency and Natural Sciences and Engineering Research Council of Canada for providing the financial support and KC-135 flight opportunities for this work.

REFERENCES

Antar, B.N., Collins, F., Hedayatpour, A., and Kawaji, M., 1990 "Transfer Line Chilldown in Low Gravity Environment", AIAA Paper No. 90-2372, AIAA/ASME/SAE/ASEE 26th Joint Propulsion Conference.

Chan, A.M.C. and Banerjee, S., 1981 "Refilling and Rewetting of a Hot Horizontal Tube", ASME J. of Heat Transfer, Vol. 103, pp.281-286.

Chen, I.Y., Downing, R.S., Parish, R. and Keshock, E., 1988 "Reduced Gravity Flight Experiment: Observed Flow Regimes and Pressure Drops of Vapor and Liquid Flow in Adiabatic Piping", AIChE Symposium Series, Vol. 84, No. 263, pp. 203-216.

Dukler, A.E., Fabre, J.A., McQuilen, J.B. and Vernon, R., 1988 "Gas-Liquid Flow at Microgravity Conditions: Flow Patterns and Their Transitions", Int. Journal of Multiphase Flow, Vol. 14, 389.

Hedayatpour, A., Antar, B.N. and Kawaji, M., 1990 "Analytical and Numerical Investigation of Cryogenic Transfer Line Chilldown", AIAA paper No. 90-2373, AIAA/ASME/SAE/ASEE 26th Joint Propulsion Conference.

Heppner, D.B., King, C.D., and Littles, J.W., 1975 "Zero-G Experiments in Two-Phase Fluid Flow Regimes", ASME Paper 75-ENAS-24.

Kachnik, L.J., Lee, D., Best, F., and Faget, N., 1987 "A Microgravity Boiling and Convective Condensation Experiment", Technical Paper 87-WA/HT-12, American Society of Mechanical Engineers.

Kawaji, M. and Banerjee, S., 1987 "Application of a Multifield Model to Reflooding of a Hot Vertical Tube, Part I: Model Structure and Interfacial Phenomena", ASME J. of Heat Transfer, Vol. 109, No. 1, 204-211.

Kawaji, M. and Banerjee, S., 1988 "Application of a Multifield Model to Reflooding of a Hot Vertical Tube, Part II: Analysis of Experimental Results", ASME J. of Heat Transfer, Vol. 110, 3, 710-720.

Kawaji, M., Ng, Y.S., Banerjee, S., and Yadigaroglu, G., 1985 "Reflooding with Steady and Oscillatory Injection", ASME J. of Heat Transfer, Vol. 107, No.3, 670-678.

EXPERIMENTAL DEMONSTRATION OF DIFFUSOCAPILLARY FLOW IN AN OIL DROPLET

D.M. Del Signore, D.L.R. Oliver, and K.J. DeWitt ■ The University of Toledo, Toledo, OH
N. Rashidnia ■ Sverdrup Technologies, Cleveland, OH

The phenomenon of diffusocapillary motion is demonstrated utilizing a vegetable oil droplet suspended in a host medium composed of Dow Corning 200 silicone fluids. The host fluid was prepared in such a way that the average polymer length was varied across the surface of the droplet, thus creating an interfacial tension gradient with concomitant Marangoni flow.

BUBBLE DISTRIBUTION DURING FOAM METAL FORMATION IN A MICROGRAVITY ENVIRONMENT

Kamiel Rezkallah, Darren Mitchell, and Scott Huckerby ■ Mechanical Engineering Department, University of Saskatchewan, Saskatoon, Canada

In a microgravity environment such as that which is simulated during parabolic flights, buoyance forces are virtually eliminated, thereby providing an opportunity to produce "controlled-density" materials (or foam metals) of high porosity. There exists a problem, however, which relates to the behaviour of bubble dispersion and bubble transport in the molten metal, as well as the liquid/gas interface in a microgravity environment. Air bubble distribution in a molten tin was investigated using a specially designed flight apparatus. Pre-prepared samples of different metal/gas ratios were tested on board the NASA KC-135 aircraft in which short durations of near zero-g conditions are experienced (~ 22 seconds). In this paper, the principle of foam metals is briefly explained, the flight hardware is described, and preliminary results of the two-day flight experiment are presented.

A COMPREHENSIVE APPROACH FOR CONSTRUCTING LIQUID-LIQUID FLOW PATTERNS MAP AND CONVERGENCE TO GAS-LIQUID SYSTEMS

Neima Brauner and David Moalem Maron ■ University of Tel-Aviv, Tel-Aviv, Israel

The construction of general two-fluid flow pattern map is presented: The departure from a steady stratified configuration to other bounding flow patterns is analyzed by exploring the stability and well-posedness of the governing equations. In view of the relationships between stability and well-posedness it is shown that the condition for ill-posedness is sufficient to indicate instability while the condition of well-posedness is necessary but insufficient to ensure stability. Of particular interest to flow pattern transition is the *stability and well-posedness map* (SWP), which defines the boundaries of stability and well-posedness of the long waves modes as a function of operational conditions. It is shown that the *integrated frame* of stability and well-posedness considerations comprehensively predict the *complete* stratified/nonstratified transitional boundary in a unified model for horizontal, inclined, various tube sizes, liquid-liquid and gas liquid flows. The transitional boundaries between the other flow patterns encountered in liquid-liquid systems are obtained based on mechanistic models.

The co-flows of two liquid-liquid immiscible phases exhibit a variety of flow patterns, as in two-phase gas-liquid systems. As is well-known, the major effort and attention in the last decades have been focused on developing a wide knowledge on various aspects of gas-liquid flows. With the lack of direct focus on the two-phase mechanisms associated with wide ranges of physical properties,understanding and experience of *liquid-liquid* two-phase flows remains inadequate. This is in no way to imply a lack of interest in the area of liquid-liquid flows. The flows of immiscible fluid mixtures are frequently encountered in the design of a variety of industrial processes and equipments such as liquid-liquid reactors and extraction units, oil transportation lines, etc. A thorough analysis for each of the flow patterns is necessary in view of the different spatial distributions of the phases, which in turn determine the transport rates (momentum, heat and mass). Clearly, each of the flow patterns carries its particular practical interest. For instance, one of the aspects that attracted most attention is the reduction of pressure losses and power requirement in waxy oil pipelines by the addition of a less viscous, immiscible liquid, as a second phase (Wu et al [1], Maron et al [2]).

The authors attempted recently to study *liquid-liquid* systems along two routes; the first by analyzing separately various flow patterns (Brauner & Maron [3], Maron et al [2], Brauner [4]). The separate analyses of stratified, annular or highly viscous core flows for wide ranges of tube size, viscosity and density ratios, show the beneficial regions of operational conditions for each flow pattern.

Moreover, the analyses of the various flow patterns shed some light on the possible mechanisms of transitions to others bounding flow patterns. The second route is by developing the conditions of existence and transitions of the possible configurations (Brauner & Maron [5-7]). It is concluded that the various concepts and results experienced in gas-liquid two-phase flows cannot be readily translated to *liquid-liquid* flow predictions.

The *complete* stratified/nonstratified boundary is established via stability and well-posedness analyses, invoked in parallel on the basic stratified configuration. Stability and well-posedness considerations are shown to constitute *complimentary* predictive frames for stability and departure to all other bounding patterns. Of particular interest are the relations derived between stability and well-posedness, and the characteristics of continuity and dynamic waves. The conditions for marginal stability and well-posedness have been shown to coincide with the conditions for stable kinematic and dynamic waves derived by wave-theory. In general, the conditions of unstable dynamic waves or ill-posedness are sufficient to indicate instability, while the conditions for stable dynamic wave or well-posedness are necessary but insufficient to assume stability (Brauner & Maron [5]).

The main thrust of the present study is in analyzing the various transitional criteria for wide ranges of viscosities, densities and density differential. This is particularly important as the present study is concerned with general liquid-liquid systems. The observed sensitivity to the flow inclination is well-elucidated in view of the present analytical approach. Furthermore, the convergence to the extreme of gas-liquid systems provides an insight on mechanisms of transitions in gas-liquid flows.

STRATIFIED FLOW-TRANSIENT FORMULATION

The analysis of the stability of stratified flow is an initial step towards the development of the transitions conditions to the various bounding flow patterns. Utilizing average one-dimensional two-fluid *transient* formulation, the resulting continuity equations for the two layers and combined momentum equations are (Brauner & Maron [6]):

$$\frac{\partial}{\partial t}(\rho_b A_b) + \frac{\partial}{\partial x}(\rho_b A_b u_b) = 0 \tag{1a}$$

$$\frac{\partial}{\partial t}(\rho_a A_a) + \frac{\partial}{\partial x}(\rho_a A_a u_a) = 0 \tag{1b}$$

$$\left[\rho_b(1-\gamma_b)\frac{u_b}{A_b} + \rho_a(1-\gamma_a)\frac{u_a}{A_a}\right]\frac{dA_b}{dh}\frac{\partial h}{\partial t} + (\rho_b-\rho_a)g\cos\beta\frac{\partial h}{\partial x} + \frac{(\partial P_{ib} - \partial P_{ia})}{\partial x} + \rho_b\frac{\partial u_b}{\partial t} + \rho_b\gamma_b u_b\frac{\partial u_b}{\partial x} - \rho_a\frac{\partial u_a}{\partial t} - \rho_a\gamma_a u_a\frac{\partial u_a}{\partial x} = \Delta f_{ba} \tag{1c}$$

$$\Delta f_{ba} = \frac{-\tau_b S_b}{A_b} \pm \tau_i S_i\left[\frac{1}{A_b} + \frac{1}{A_a}\right] + \tau_a\frac{S_a}{A_a} + (\rho_b-\rho_a)g\sin\beta$$

$$\frac{\partial}{\partial x}(P_{ib}-P_{ia}) = -\frac{\partial}{\partial x}\left\{\sigma\frac{\partial^2 h/\partial x^2}{[1+(\partial h/\partial x)^2]^{3/2}}\right\} \tag{2}$$

where u_a, u_b, h denote here instantaneous local values and γ_a, γ_b are the shape factors which account for a velocity distribution in the two layers. As conventionally used in two-fluid models, the shear stresses τ_a, τ_b, τ_i are expressed in terms of friction factors f_a, f_b, f_i:

$$\tau_a = f_a\rho_a u_a^2/2 \; ; \quad f_a = c_a \, Re_a^{-n_a} \tag{3a}$$

$$\tau_b = f_b\rho_b u_b^2/2 \; ; \quad f_b = c_b \, Re_b^{-n_b} \tag{3b}$$

$$\tau_i = f_i\frac{\rho(u_a - u_b)^2}{2} \; ;$$

$$\rho = \rho_a \quad \text{and} \quad f_i = Bf_a \quad \text{for} \quad u_a > u_b$$

$$\rho = \rho_b \quad \text{and} \quad f_i = Bf_b \quad \text{for} \quad u_b > u_a \tag{3c}$$

Equations (3c) imply that the interfacial shear friction factor is evaluated as equal to that obtained between the faster phase and the pipe wall, augmented by a factor B due to interfacial waviness. For $u_a \simeq u_b$, τ_i is identically zero. Note that the corresponding Reynolds numbers for the two fluids are based on the equivalent hydraulic diameters, defined according to whether the upper layer is the faster one or vice-versa. Thus,

$$D_a = \frac{4A_a}{(S_a + S_i)}\; ; \quad D_b = \frac{4A_b}{S_b} \quad \text{for } u_a > u_b \tag{4a}$$

$$D_a = \frac{4A_a}{S_a}\; ; \quad D_b = \frac{4A_b}{(S_b + S_i)} \quad \text{for } u_a < u_b \tag{4b}$$

$$D_a = \frac{4A_a}{S_a}\; ; \quad D_b = \frac{4A_b}{S_b} \quad \text{for } u_a \simeq u_b \tag{4c}$$

The constants c, n are chosen according to the flow regime in each phase [3]. In distinction to gas-liquid flows (where the gas velocity is of higher order of magnitude and therefore the interface is considered as free surface with respect to the liquid and as a stationary surface with respect to the fast gas phase), in liquid-liquid systems the velocities of the two phases may be of comparable levels and alternatively one phase velocity exceeds the other. Therefore, an 'adjustable definition' of the equivalent hydraulic diameter is part of the solution procedure as implied by Eqs. (4).

Stability and Well-Posedness Analyses

Temporal stability analysis carried out on the linearized form of Eqs. (1) yields the dispersion relation [6]:

$$aC^2 - 2(b_1+ib_2)C + d_1+id_2 = 0 \; ; \qquad C \equiv \omega/k \tag{5a}$$

$$C_{1,2} = \frac{1}{a}(b_1 + ib_2) \pm \frac{1}{a}\left[(b_1 + ib_2)^2 - a(d_1 + id_2)\right]^{1/2} \tag{5b}$$

$$a = \rho_b\frac{A_b'}{A_b} + \rho_a\frac{A_b'}{A_a}\; ; \qquad A_b' = dA_b/dh\; ;$$

$$b_1 = \rho_b\frac{A_b'}{A_b}\gamma_b U_b + \rho_a\frac{A_b'}{A_a}\gamma_a U_a$$

$$b_2 = \frac{1}{2k}\left[\frac{A_b'}{A_b}\frac{\partial\Delta F_{ab}}{\partial U_b} - \frac{A_b'}{A_a}\frac{\partial\Delta F_{ab}}{\partial U_a}\right]$$

$$d_1 = \rho_b\frac{A_b'}{A_b}\gamma_b U_b^2 + \rho_a\frac{A_b'}{A_a}\gamma_a U_a^2 - (\rho_b-\rho_a)g\cos\beta + \sigma k^2$$

$$d_2 = \frac{1}{k}\left[\frac{A_b'}{A_b}U_b\frac{\partial\Delta F_{ab}}{\partial U_b} - \frac{A_b'}{A_a}U_a\frac{\partial\Delta F_{ab}}{\partial U_a} - \frac{\partial\Delta F_{ab}}{\partial H}\right] \tag{5c}$$

Here, C represents the complex propagation velocity of the disturbance, k is the wave number, and $\partial\Delta F_{ab}/\partial(H, U_a, U_b)$ are $\partial f_{ab}/\partial(h, u_a, u_b)$ at steady conditions H, U_a, U_b. The neutral stable wave velocity, C_{rn}, and wave number, k_n (for zero imaginary part for C) are:

$$C_{rn} = \frac{\dfrac{\tilde{A}_b'}{\tilde{A}_b}U_b\dfrac{\partial\Delta F_{ab}}{\partial U_b} - \dfrac{\tilde{A}_b'}{\tilde{A}_a}U_a\dfrac{\partial\Delta F_{ab}}{\partial U_a} - \dfrac{\partial\Delta F_{ab}}{\partial\tilde{H}}}{\dfrac{\tilde{A}_b'}{\tilde{A}_b}\dfrac{\partial\Delta F_{ab}}{\partial U_b} - \dfrac{\tilde{A}_b'}{\tilde{A}_a}\dfrac{\partial\Delta F_{ab}}{\partial U_a}} \tag{6}$$

$$(\frac{\pi}{4})^2 \frac{\tilde{A}'_b}{\tilde{A}_a^3}[(\frac{C_{rn}}{U_a}-1)^2 + (\gamma_a-1)(1-2\frac{C_{rn}}{U_a})]Fr_a^2$$

$$+ (\frac{\pi}{4})^2 \frac{\tilde{A}'_b}{\tilde{A}_b^3}[(\frac{C_{rn}}{U_b}-1)^2 + (\gamma_b-1)(1-2\frac{C_{rn}}{U_b})]Fr_b^2$$

$$- (1 + \tilde{k}_n^2 We_b Fr_b^2) = 0 \qquad (7a)$$

$$Fr_a^2 = \frac{\rho_a}{\Delta\rho}\frac{U_{as}^2}{Dg\cos\beta}\ ; \quad Fr_b^2 = \frac{\rho_b}{\Delta\rho}\frac{U_{bs}^2}{Dg\cos\beta} = \frac{\rho_b}{\rho_a}Fr_a^2/\phi^2\ ;$$

$$\phi = U_{as}/U_{bs}\ ; \quad \tilde{k}_n = k_n D\ ; \quad We_b = \frac{\sigma}{\rho_b D U_{bs}^2} \qquad (7b)$$

where Fr_a, Fr_b, We_b and $\tilde{k}$ represents the Froude, Weber and nondimensional wave numbers and ϕ denotes the ratio of superficial phases velocities. The nondimensional geometrical variables are designated by ~ (area normalized by D^2 and length by D).

It is further of interest to note at this point, that the expression for the wave velocity at neutral stability, Eq. (6) for C_{rn}, has been shown by the authors to be identical to the kinematic wave velocity, C_k, defined as (Wallis [8]):

$$C_k = \left(\frac{\partial U_{bs}}{\partial \epsilon}\right)_{U_{as}+U_{bs},\ \Delta F_{ab}}\ ; \qquad \epsilon = A_b/A \qquad (8)$$

Thus, the neutral stable wave actually represents a continuity wave and its characteristic velocity can be determined either by stability analysis or via the derivative of the liquid flux with respect to its insitu hold-up (concentration). Clearly, both the neutral and continuity waves are based on the steady momentum equation. It has been further shown that the neutral stability condition (Eq. 7a) is identical to the condition for equal kinematic and dynamic wave velocities (Brauner & Maron [5]).

In parallel, the condition underwhich the governing system of Eqs. (1) constitutes a well-posed initial value problem, thus possesses real characteristics, has been shown to be (Brauner & Maron [6]):

$$\tilde{\rho}_b\gamma_b(\gamma_b-1)(\tilde{A}/\tilde{A}_b)^2 + \tilde{\rho}_a\gamma_a(\gamma_a-1)(\tilde{A}/\tilde{A}_a)^2\phi^2$$

$$- (\gamma_b\tilde{A}/\tilde{A}_b - \phi\gamma_a\tilde{A}/\tilde{A}_a)^2 + Fr_{ab}^{-2} + We_{ab}\tilde{k}^2 \geq 0 \qquad (9a)$$

$$\tilde{\rho}_b = 1 + \frac{\rho_b}{\rho_a}\frac{\tilde{A}_a}{\tilde{A}_b}\ ; \ \tilde{\rho}_a = 1 + \frac{\rho_a}{\rho_b}\frac{\tilde{A}_b}{\tilde{A}_a}\ ; \ Fr_{ab}^2 = \frac{\rho_{ab}}{\rho_b}Fr_b^2$$

$$\rho_{ab} = \frac{\tilde{A}'_b}{\tilde{A}_a\tilde{A}_b}\frac{\rho_a\rho_b}{\rho_a/\tilde{A}_a+\rho_b/\tilde{A}_b}\ ; \quad We_{ab} = \frac{\rho_b}{\rho_{ab}}We_b \qquad (9b)$$

The linear stability and well-posedness analyses are carried around a presumed smooth stratified flow configuration. The steady-state solutions are obtained by degenerating the LHS of Eqs. (1) to zero (Brauner & Maron (1989)). Note, however, that in two phase flow problems, the existence of an assumed physical situation, e.g., stratified flow configuration, is not certain under all operational conditions. Therefore, complex characteristics (ill-posed initial-value problem) may not necessarily imply on an incorrect formulation, but may be attributed to physical instability of the assumed flow configuration, whereby transition to a different flow pattern may take place. Inspection of Eq. (9a) points out that for the case of zero-gravity and zero surface tension terms, the basic continuity and momentum equations for plug flow ($\gamma_a=\gamma_b=1$) can not define a well-posed initial value problem. For shape factors γ_a, $\gamma_b > 1$, real characteristics may still be obtained in the absence of gravity and surface tension forces. It has been found that the condition for real characteristics is identical to the condition for stable dynamic waves. In the region where the dynamic waves are unstable, all waves modes are unstable and their evolutions as formulated by the initial value set of Eqs. (1) are ill-posed (Brauner & Maron [5,9]). The relations between stability and well-posedness are further elucidated in view of Figure 2.

Figure 2a represents typical trends of the amplification variation with the wave number as obtained by Eqs. (5) for horizontal liquid-liquid systems, while Figure 2b demonstrates the effects of inclination in downward or upward air-water flows. Also included in Figure 2 the minimal value, k_{rc}, for ensuring real characteristics (Eq. (9)).

For a certain combination of (U_{as},U_{bs}), as in curves (c), a smooth (stable) interfacial structure is maintained for $k>k_n$. In the range of $0 < k < k_n$, where the stabilizing effect of surface tension is insufficient, the disturbances are amplified and hence a wavy interfacial structure develops. For all $k < k_{rc}$, an unstable smooth stratified flow is consistently predicted by both stability and reality of characteristics analyses. However, for any $k_{rc} < k < k_n$, while the governing Equations (1) are well-posed as an initial-value problem, they are still expected to develop a wavy structure. It has been shown that the value of k_{rc} is always within the amplified range, $k_{rc} < k_n$ (Brauner & Moalem Maron [5]). Thus, as long as Eq. (9a) yields a real k_{rc} it corresponds to an amplified mode. The first interpretation for this outcome is that reality of characteristics is only a necessary condition for stability but, clearly, not sufficient. On the other hand, the condition of ill-posedness is sufficient to indicate instability. Clearly, the range of amplified waves, $k < k_n$, varies depending on the (U_{as}, U_{bs}) combinations. For particular combination of (U_{as},U_{bs}) as represented by curve (b), the amplified range almost diminishes (ω_i, $k_n \rightarrow 0$), while for conditions of curve (a), for instance, all wave modes are practically stable. Note, that Figure 2b relates to high pressure system. For the conditions of Figure 2b and in horizontally levelled tube, $\beta=0^o$, a relatively limited range of weakly amplified (unstable) wave modes just appear.

However, all of the growing wave modes are found to be well-posed. As the flow is slightly inclined downward, $\beta=1^o$, all the unstable wave modes disappear and the flow becomes smoothly stable for the entire range of wave numbers. On the other hand, inclining the flow slightly upward, $\beta=-0.5^o$, results in a drastic increase of both the range of amplified wave numbers and the corresponding amplification. Moreover, a range of ill-posed wave modes also appears ($0 < k < k_{rc}$). In view of Figure 2b, the flow inclination is of dramatic effect on the stability of gas-liquid flow in the sense that very slight downward or upward inclination can stabilize or destabilize the stratified flow configuration. However, the impact of a small upward inclination on the stability characteristics is significantly more pronounced. As is shown below, this behaviour forms the basis for elucidating the stratified/nonstratified flow patterns transitions observed in upward and downward inclined flows.

FLOW PATTERNS TRANSITION CRITERION

Stable Stratified Flow Boundaries

Based on the stability and well-posedness analyses, the limiting boundaries for stable and well-posed stratified flows are extracted. These boundaries form the basis to establish what the authors denote as a *'stability and well-posedness map'*, (SWP). The SWP map is then utilized to define general guidelines for constructing a *complete* stratified/nonstratified transitional line to the various bounding flow patterns; slug, pseudo-slugs and annular flows.

Stability and well-posedness map: 'zero neutral stability', 'zero real characteristics' lines. As is shown in Figure 2 depending on the operational conditions, there exists a particular combination (U_{as}, U_{bs}) for which the range of amplified waves diminishes, $k_n \to 0$. In searching for all combinations of (U_{as}, U_{bs}) for which $k_n \to 0$, the so-called *'zero neutral stability' boundary (ZNS)* is obtained. This boundary confines all possible *smooth* stratified flows. The locus of the curve itself represents the departure from *smooth* stratified structure. For any operational set (U_{as}, U_{bs}) beyond the $k_n=0$ boundary, the linear stability analysis predicts exponential growth with time for a finite range of wave numbers, $0 \le k \le k_n$. The disturbances growth in this region may either be damped (due to non-linear effects) and thus end-up with 'stable wavy' stratified flow, or may result in a different flow configuration (due to bridging, for instance.)

In parallel to the 'zero neutral stability' boundary defined by $k_n=0$, a *'zero real characteristics' boundary (ZRC)* is built-up by searching for all combinations of (U_{as}, U_{bs}) which yield by Eq. (9a) real characteristics for $k=k_{rc}=0$. The ZRC curve confines the region for which well-posedness is ensured for all wave modes. The ZNS and ZRC boundaries constitute the *'stability and well-posedness map'* as presented in Figure 3a for horizontal two-fluid system (200 bar, air-water flow). It is of interest to point first some basic observations in view of the typical trends of the structure of the SWP maps.

Generally, the ZRC boundary is composed of two branches, the left one corresponds to $U_b > U_a$ while along the right one $U_a > U_b$. These two branches approach one another at the region of high (actual) velocities of both phases, where the difference of which is just sufficient to balance the gravity term in Eq. (9). The location of this region and its width depend mainly on the density differential. The existence of two branches points out the multiplicity of solutions, which becomes even more complicated along the ZNS boundary, also due to the discontinuities which evolve from laminar/turbulent flow regime transitions in either of the two phases (assumed at Re=1500, [6]). The structure of the SWP map for horizontal air-water system or air-viscous liquids over wide ranges of liquid viscosity and tube diameter has been widely discussed by the authors recently with respect to the stratified/nonstratified transitions in horizontal systems [10].

The ideas and interpretations with regards to horizontal systems prevail basically also in inclined flows, although the limiting ZNS and ZRC boundaries may demonstrate entirely different structures, Figures 3b,c. The first point to note is the significant effect of inclination on the liquid layer velocity relative to the gas velocity, accompanied with drastic variation in the liquid layer thickness. For a given gas and liquid rates, inclining the flow downwards, results in a thinner liquid layer, which is practically insensitive to the gas rate over a wide range. On the other hand, an upward inclination results in a drastic thickening of the liquid layer with high sensitivity to the gas rate. The dramatic changes in the flow geometry, phases velocities and their ratio affect, in view of Eqs. (7) and (9), the ZNS and ZRC locations. For instance, with inclining the flow upward, a large unstable zone extends to relatively low liquid rates over a wide range of gas rates (except at extremely low gas rates, which is out of the practical range). At high gas rates, a limited 'bell-shaped' stable smooth stratified zone still prevails at this particular slight upward inclination. At higher inclincations the resulting variations in the steady stratified solution affect stronger effects on the SWP map.

Construction of the complete stratified/nonstratified transitional boundary. The general implication of the 'zero neutral stability' and 'zero real characteristics' lines, is in defining *three zones*; the area within the ZNS boundary, is well-understood to be the stable *smooth* stratified zone. Beyond the ZRC boundary, the complex characteristics imply that the governing equations of the stratified flow configuration cannot accomodate for the time and space variation associated with a certain range of amplified wave modes. Thus, while the ZNS boundary may represent preliminary transition to a wavy interfacial structure, the ZRC boundary, which is within the wavy unstable region, represents an upper bound for the existence of a wavy stratified configuration, beyond which another flow pattern prevails. It is worth noting, that the stable smooth stratified zone as defined by the ZNS line, is always a subzone, confined within the well-posed region [5].

The *'buffer'* region, which is formed in the range where the ZNS and ZRC lines depart, is characterized by the existence of interfacial disturbances, and as such bears a real potential for flow pattern transition. However, though amplified interfacial waves exist in the *'buffer'* region, the equations which govern the variation of (h, u_a,u_b) in space and time are still well-posed with respect to all of these unstable modes. Whether these disturbances trigger a departure from stratified configuration depends on the relative layers thicknesses. For if $H/D \simeq 0.5 \div 1$, it is likely that the evolution of the interfacial disturbances on the liquid layer will end-up in tube blockage. Thus, when the *entry* from the smooth into the 'buffer' zone (along the ZNS boundary) is associated with a relatively thick liquid layer, above H/D=0.5 line (along the bolded sections of the ZNS in Figures 3a,b) the ZNS line, which in general represents the transition to a disturbed wavy pattern, will also predict the condition for the development of other flow pattern. On the other hand, when the entry to the *'buffer'* zone occurs with relatively thin liquid layer, below $H/D \simeq 0.5$, the ZNS line predicts the development of wavy flow, but it plays no role in predicting transition to other pattern. In this case, the transition to other flow configuration may be 'delayed' and predicted by the ZRC boundary as long as the relative liquid layer thickness, H/D, remains small *in* the 'buffer' zone (along the bolded sections of ZRC in Figures 3). As the relative liquid layer becomes of the order of the conduit radius, $H/D \approx 0.5$, *within* the 'buffer' region, the disturbed interface may trigger flow pattern transition, this time *within* the 'buffer' zone in the vicinity of $H/D \approx 0.5$ line (bolded).

As a corollary, the *entire* transitional boundary for departure from stratified flow is composed of three sections; along ZNS for $H/D > 0.5$, along ZRC for $H/D < 0.5$ and along $H/D \approx 0.5$ in the 'buffer' region.

Annular/Slug Transitional Boundary

The annular pattern is bounded between two limits; one limit relates to break-up of the top-wall film due to the float-up tendency of the lighter core phase *(a)*. In this case, the distruption of the annular configuration ends up in a stratified pattern. This boundary is predicted via the above analysis of the limits of existence of the stratified configuration.

The other limit corresponds to a collapse of the core phase due to growing interfacial waves, which block the core space, leading to slug or bubble flow. The mechanisms of the core disintegration evolves from the inherent Rayleigh instability associated with a moving cylindrical jet, well-known with reference to free-liquid jets. However, in annular flow, further complications due to damping effects of the conduit wall on the eccentric core are to be accounted for. A rather realistic transitional criterion may be established by exploring the critical insitu hold-up which generates favourable conditions for the growing interfacial waves to reach the core center. Favourable conditions for bridging to take place by growing interfacial waves are found to be $A_a \approx A_b$. Recently, Brauner [4] formulated the governing equations for modelling core-annular liquid-liquid systems. A transition line to slug pattern is established by searching for those (U_{as}, U_{bs}) for which the annular model equations fulfill a prescribed criterion on the phases insitu holdup, such as $A_a \simeq A_b \simeq A/2$. The predicted annular/slug boundary is included in Figures 3.

Stratified/Stratified-Dispersed Transition: SD Boundary

Gas-liquid systems are known to stratify even for relatively low gas flow rates, due to the intense float-up of the bubbles and consequent accummulation of gas at the upper tube cross-section. When the density differential is reduced, as in *liquid-liquid* systems, the float-up process becomes moderate and thus the system tends to demonstrate stratified-dispersed flow pattern, whereby, the lighter phase, *a*, remains above phase *b* as swarm of drops. Clearly, under the condition of sufficiently low U_{as}, drops (or bubbles) of phase *a* may be obtained. The lighter phase *a* drops agglomerate at the upper cross-section due to buoyancy (corresponding to $\Delta\rho$) tending to stick together. However, the drops may maintain their shape provided they are small enough, so that the surface tension forces overcome those due to buoyancy, the balance of which yields the critical diameter, d_{cr};

$$\sigma\pi d > \frac{4}{3}\pi\left(\frac{d}{2}\right)^3(\rho_b-\rho_a)g\ ; \quad d_{cr} = \left(\frac{6\sigma}{(\rho_b-\rho_a)g}\right)^{1/2} \qquad (10)$$

For $d > d_{cr}$, bubbles will merge together to form a continuous upper layer. A simple criterion for transition from two continuous stratified layers towards stratified-dispersed upper layer is proposed by requiring that for a given (U_{as},U_{bs}) flow rates, the available upper layer cross sectional area in a stratified configuration, A_a, is smaller than the cross-sectional area of a critical drop-size, A_{cr}. In this case the biggest drop which may occupy the available space, A_a is still smaller than the critical drop size and therefore, swarm of drops or bubbles smaller than d_{cr} may exist within A_a. On the other hand, for $A_a > A_{cr}$ corresponding to larger flow rate of the lighter phase, the available space enables the agglomeration of large drops or bubbles (beyond the critical diameter), which will then merge to form an upper continuous stratified layer. The stratified/stratified-dispersed (SD) boundary is given by:

$$A_a(U_{as},U_{bs})\cos\beta \le A_{cr}\ ; \qquad A_{cr} = \pi d_{cr}^2/4 \qquad (11)$$

Clearly, the SD boundary is of relevance only when it lies within a zone of stable stratification, in which case the stable stratified zone is reduced (area above/left the dotted curves, Figures 3). The stratified-dispersed pattern in (low pressure) gas-liquid systems appears as a very small subzone of the stratified configuration, below the practical flow rates. In liquid-liquid systems (and high pressure gas-liquid systems), extended stratified-dispersed zones appear, at either the top or bottom of the conduit, due to the reduced density differential [7].

Outside the region of stable stratification, other dispersed type of flows may exists [12,13]. However, the exact location of the transitional boundaries are not so well-defined (as in gas-liquid systems), probably due to the low density differentials. Extensive experimental data is required for thorough interpretations. Some aspects concerning the modelling of the various dispersed flow pattern boundaries experienced in liquid-liquid systems have been recently discussed [7,11].

LIQUID-LIQUID FLOW PATTERN MAPS: COMPARISON WITH EXPERIMENTS

Figures 4-6 demonstrate comparisons of the predicted transitional boundaries with horizontal oil-water and gas-liquid systems. Figure 4 corresponds to waxy oil-water system and includes also the upper ZRC branches obtained for varios $\gamma_a \geq 1$ values. Note that since the $H/D \simeq 0.5$ line falls outside the 'buffer' region the complete stratified/nonstratified transitional boundary is determined solely by the ZRC lines. It is interesting to refer at this point to the effects of the shape factor. As Eq. (9) indicates, for γ_a, $\gamma_b > 1$ additional stabilizing terms result. Thus, inclusion of shape factors may extend the region operational conditions for which well-posedness is assured. For the oil/water system of Figure 4, a shape factor greater than one is more likely in laminar (viscous) layer, which is also expected to approach 1 at relatively high rates. Hence, the effect of the oil shape factor, γ_a, is demonstrated with respect to the upper branch along which $U_b > U_a$, whereas along the side branch where $U_a > U_b$, a shape factor of unity may reasonably be assumed for the (high velocity) viscous phase as well. As is shown in Figure 4, the region where core flow has been steadily maintained [14] is indeed located in the predicted region for annular pattern (confined between the AN-SL boundary and the upper ZRC branch with $\gamma_a \simeq 1 \div 1.1$). For the particular physical system of Figure 4, the stratified-dispersed line does not share a common zone with the stratified region (below the upper ZRC branch), thus no stratified-dispersed pattern is expected in this case.

Figure 4 implies that steady core flow is possible at very low water percents as predicted by Moalem Maron et al [2]. A core flow can be maintained up to a maximum water percent, given by the annular/slug boundary, beyond which a transition to slug flow is predicted to take place. The intersection of the annular/slug line with the upper ZRC branch (point A_1 in Figure 4) defines the minimum theoretical oil rate for establishing core flow. Note that the annular/slug transition is presented in Figure 4 as a range obtained for $A_b = A_a/2 \div A_a$ (since $A_b < A_a$ may already lead to bridging by so-formed "areated" slug).

The effects of the oil viscosity on the required water percentage for maintaining steady core flow are described in Figure 5. Again the prediction of the proposed analytical boundary as defined by the ZRC line with $\gamma_a = 1 \div 1.1$ is satisfactory. Both theory and experiment [15] indicate that the required water addition for establishing core flow decreases as the oil viscosity increases.

The general outcome of Figures 4-5 is that the stratified configuration in liquid-liquid oil/water systems may extend beyond the 'zero neutral stability' lines, which stands more fittedly for the evolution of wavy stratified pattern, whereas well-posedness boundaries are more relevant for predicting the departure from stratified pattern.

Figures 6 which relate to gas-liquid systems indicate that in the region of relatively thin liquid layer, $H/D < 0.5$, the experimental data of stratified/nonstratified transition consistently follows the trend of the ZRC boundary. It is to be emphasized that for horizontal flow the ZNS and ZRC boundaries are rather close to each other at this region and the data may seem to follow either of these. In inclined flows, these two boundaries depart significantly. However, the transitional data stick to the ZRC line, although this boundary lies now far deep into the unstable smooth (or stratified wavy) region. This again reinforces the dominant role of the well-posedness boundary in predicting the flow pattern transition in the region of relatively thin liquid layers.

In upwards inclinations (Figure 6d), the ill-posed zone starts penetrating towards lower liquid rates already at $\beta = -0.5^0$. At $\beta = -1^0$ (and steepper upward inclinations), the well-posed region is found to split into two separate regions (Brauner and Maron [9]). The ZNS lines form small bell-shaped zones of stable smooth stratified flow, which shrinks with increasing inclination and does not confine the observed stratified zone. According to the proposed transitional boundary, the role of the ZRC line in predicting transition ends at the intersection with $H/D = 0.5$ line. From this point, the departure from stable stratified configuration is expected first along the $H/D = 0.5$ line in the wavy 'buffer' zone and then along that section of the ZNS line for which $H/D > 0.5$. This is clearly the case in Figures 6a,b. In this case the *entire* stratified/nonstratified boundary is indeed composed of the three sections. However, when almost all the stable region confined by the ZNS line corresponds to $H/D < 0.5$, as in Figure 6c-d, the ZNS line becomes irrelevant in predicting a departure from stratified configuration. In this case, the transitional boundary follows first the ZRC for $H/D < 0.5$ and then the $H/D = 0.5$ line all along in the 'buffer' zone. Note that for the systems of Figures 6, the SD boundary falls outside the predicted stable stratified zone.

It is of further interest to refer to the data which relates to the so-called pseudo-slug pattern (denoted sometimes as large wave stratified flow or wavy annular flow). Pseudo-slugs are characterized by high amplitude roll waves which intensively wet the upper tube walls, without causing the pressure fluctuations typical to slug pattern. Inspection of the data reveals that the regions of pseudo-slugs develop within the herin predicted 'buffer' region around $H/D = 0.5$. Indeed wavy thick layers bear the conditions for large amplitude waves and intensive wetting of the upper wall, hence pseudo-slugs. Clearly, the large amplitude waves are more likely to cause pseudo-slugs in small diameters than in large-diameters [10].

In view of Figures 4-6, it is shown that in spite of the variety of situations which appear at the different two-fluid systems and inclinations, the general guidelines for constructing the stratified/nonstratified transitional boundary are essentially maintained.

THE RELATIVE DESTABILIZING CONTRIBUTIONS OF THE TWO PHASES

The stability equation (7), includes two destabilizing terms which evolve from the mobility of the two-phases and are to be balanced by the gravity and surface tension stabilizing terms. In order to analyze the relative contributions of the two phases destabilizing terms, Eq. (7a) is rearranged in the form:

$$J_a + J_b = 1 + J_\sigma\,; \quad J_\sigma = \frac{\sigma k_n^2}{(\rho_b-\rho_a)g\cos\beta}$$

$$J_a = \frac{\pi^2 \rho_a}{16Dg\cos\beta\Delta\rho}\frac{\tilde{A}'_b}{\tilde{A}_a^3}U_{as}^2[(\frac{C_{rn}}{U_a}-1)^2+(\gamma_a-1)(1-2\frac{C_{rn}}{U_a})]$$

$$J_b = \frac{\pi^2 \rho_b}{16Dg\cos\beta\Delta\rho}\frac{\tilde{A}'_b}{\tilde{A}_b^3}U_{bs}^2[(\frac{C_{rn}}{U_b}-1)^2+(\gamma_b-1)(1-2\frac{C_{rn}}{U_b})] \quad (12)$$

In Eqs. (12) the two phases destabilizing terms, J_a and J_b respectively, represent the combined effects of intertia (as evolve from the LHS of Eqs. (1)) and viscous effects (via C_{rn}, as evolves from the RHS of Eq. (1c)). For liquid-liqud systems where the velocities of the two-phases are comparable, the contributions J_a, J_b terms are well-understood to be both significant. In gas-liquid flows, however, the relative contributions of the gas and liquid destabilizing terms along the transitional boundary are to be *carefully* considered. For the sake of thorough elucidation, Figure 7 presents the liquid destabilizing contribution with respect to the gravity stabilizing term, $J_b(\equiv J_L)$. Clearly, along the ZNS boundary, $(k_n, J_\sigma)\rightarrow 0$, and Eq. (12) reduces to $J_L+J_G=1$. Figure 7b indicates that C_{rn} almost always deviate significantly from the liquid velocity, and thus, $C_{rn}/U_b-1 \neq 0$. Correspondingly, the liquid destabilizing effect as represented by J_L, certainly can not be ignored. Moreover, in view of Figures 6, in the range which corresponds to stratified/intermittent transition, the liquid destabilizing term, in fact, even dominates, $J_L>0.5$. Also to be emphasized that even when $C_{rn}/U_L \approx 1$ (as in air-water system, $\mu_L \approx 1$ c_p), the neglection of the liquid relative contribution may be erroneous over the most of stratified/slug transition boundary, since (C_{rn}/U_L-1) and (C_{rn}/U_G-1) are not the only quantities which determine the relative contributions of the liquid and gas destabilizing effects. This is further demonstrated while following the effects of the liquid viscosity: It is shown in Figure 7 that as the liquid viscosity increases, C_{rn}/U_L may significantly deviate from one and attain high values. However, the liquid relative destabilizing contribution, J_L although proportional to (C_{rn}/U_L-1), decreases with increasing the viscosity and thus the range of liquid dominancy is reduced. This reinforces the statement given by Hanratty [18] that the K-H invicid theory becomes suprisingly more accurate as the liquid viscosity increases.

The liquid phase dominancy along the ZNS boundary, J_L, is continuously reduced as the gas rate increases and $J_G = 1-J_L > 0.5$. Thus, the *complete* transitional stratified/nonstratified boundary in gas-liquid flows, is *'liquid-controlled'*, $J_L >> J_G$ along the stratified/slug transition, and gradually becomes *'gas-controlled'*, $J_G >> J_L$ towards the stratified/annular transition boundary. However, the role of the ZNS boundary in prediction of transition may not extend to high gas rates, and it becomes entirely irrelevant in steep inclination. Thus, for inclined air-water systems, the transition (when determined by linear stability) is practically *'liquid-controlled'*.

It is to be emphasized at this point that previous studies [19-21] related to stratified/slug flow pattern transition, by considering Bernoulli suction effect of the fast gas phase and unjustifiably ignoring totally the liquid destabilizing contribution. As these models are based solely on the gas-contribution, J_G, which shares a minor part along the stratified/slug boundary, a reasonable comparison with experiments required the insertion of augmenting correction constants, which ought to increase with decreasing U_{GS} (greater H/D). The $1/(1-\tilde{H})$ correction (Taitel & Dukler [20]), though yields the larger correction required at lower U_{GS}, is still unphysical, as the other empirical constants; it artificially enhances the negligible destabilizing gas term along the stratified/slug transition, whereas it is practically uneffective along the stratified/annular boundary where H/D <<1.

To sum-up, the relative contribution of the gas or liquid destabilizing terms depends on variety of parameters and operational conditions. In the extremes of definite *'liquid-controlled'* or *'gas-controlled'* zones, empirical constants can be utilized to simplify the general predective transitional criteria, provided that they are employed on the correct dominating contribution.

FURTHER IMPLICATIONS

It is of particular interest to comment briefly on the applications of the proposed analytical approach to systems of small diameters and reduced gravity;

Small diameters systems. Recalling that the longest wave length is of the order of the tube diameter, the stabilizing surface tension term in Eqs. (7,9), is negligible for sufficiently large tubes. Thus, the relevance of the ZNS and ZRC lines defined with respect to infinitely long modes is well-understood. However, as the tube diameter is reduced, the practical long wave limit yields a significant stabilizing contribution which ought to be accounted for. Instead of the *'zero real characteristic'* (ZRC) limit, a *'real characteristic'* (RC) boundary which is based on $k \equiv k_{rc} \simeq 2\pi/D$ in Eq. (9) is established. Clearly, for small tube and a given surface tension, the (RC) line depart from the (ZRC) line, whereby the region of stable stratification expands. This elucidates the delay of stratified/annular transition to higher gas rates observed in small diameter gas-liquid flows. On the other hand, the SD line of Eq. (11) penetrates deeper into the stratified zone, resulting in transition to stratified/dispersed pattern at lower liquid rates. Transitions in small conduits are widely studied by the authors elsewhere [22]).

Reduced gravity systems - microgravity. The expansion of the stratified/dispersed pattern on account of the stratified zone is favoured with increasing the critical dispersed phase diameter, d_{cr}. In view of Eq. (10), large critical diameter $d_{cr}>D$ may be obtained either by reducing the density difference (as in liquid-liquid systems), or by reducing the gravity field as in space conditions. For a prescribed $\Delta\rho=\rho_b-\rho_a$ and tube size, D, there exists a range of gravitation intensity, $0\leq G<G_m$ for which the two fluids may not stratify and practically behaves as "equal densities" or "zero gravity" system. Similarly, Eq. (10) defines a range of density difference $0\leq\Delta\rho<\Delta\rho_m$ for a given D and G, or a maximum tube diameter D_m for a specified $\Delta\rho$ and G, under which one effectively *ignore* the possibility of stratification development. Thus, the range for microgravity ('zero' gravity) systems is defined by either of:

$$G < G_m = \frac{6\sigma}{(\rho_b - \rho_a)D^2}\,; \qquad \Delta\rho < \Delta\rho_m = \frac{6\sigma}{GD^2}\,;$$

$$D < D_m = \left[\frac{6\sigma}{(\rho_b-\rho_a)G}\right]^{1/2} \tag{13}$$

or in terms of Eotwös number, Σ:

$$\Sigma < 1\,; \qquad \Sigma = \frac{6\sigma}{(\rho_b-\rho_a)D^2G} \tag{14}$$

In view of Eqs. (13,14), zero or reduced gravity is not only a matter of prescribing the gravitation intensity. Stated differently, a microgravity system is to be defined with reference to the combination of diameter, density differential and gravity field. Based on physical concepts, a distinction has been made recently by Brauner [11] between 'micrgravity' conditions, where stratification is completely avoided, and finitely 'reduced gravity' systems, which may still demonstrate a stratified configuration. As such, the various flow patterns relevant in space and the transitional boundaries may be defined in terms of 'equivalent systems' under normal gravity.

LITERATURE CITED

1. Wu, H.L., Duijvestijn, P.E.M., Paterno, J. and Guevara, E., *Rev. Tec. Intevep.*, **6** (1), 17-22 (1986).

2. Moalem Maron, D., Brauner, N. and Kruka, V.R., "The Mechanisms of Two Phase Liquid-Liquid Viscous Core Flow", Proc. 6th Miami Int. Symp. Heat Mass Transfer (1990).

3. Brauner, N. and Moalem Maron, D., PCH, *Phy. Chem. Hydro.*, **11** 487-506 (1989).

4. Brauner, N., "Two-Phase Liquid-Liquid Annular Flow , in print, *Int. J. Multi Phase Flow* (1991).

5. Brauner, N. and Moalem Maron, D., "Stability, Well-Psedness and Flow Pattern Transitions in Two-Phase Liquid-Liquid Flows", Proc. 6th Miami Int. Symp. Heat Mass Transfer (1990).

6. Brauner, N. and Moalem Maron, D., "Stability Analysis of Stratified Liquid-Liquid Horizontal Flow", submitted, *Int. J. Multiphase Flow* (1991).

7. Brauner, N. and Moalem Maron, D., "Flow Pattern Transitions in Two Phase Liquid-Liquid Horizontal Tubes", submitted *Int. J. Multiphase Flow* (1991).

8. Wallis, G.B., *One Dimensional Two-Phase Flow: McGraw Hill* (1969).

9. Brauner, N. and Moalem Maron, D., "Analysis of Stratified/Nonstratified Transitional Boundaries in Inclined Gas-Liquid Flows", submitted *Int. J. Multiphase Flow* (1991).

10. Brauner, N. and Moalem Maron, D., "Analysis of Stratified/Nonstratified Transitional Boundaries in Horizontal Gas-Liquid Flows", in print, *Chem. Eng. Sci.* (1991).

11. Brauner, N., *Int. Comm. Heat Mass Transfer*, **17** (3), 271-282 (1990).

12. Russell, T.W.F., Hodgson, G.W. and Govier, G.W., *The Can. J. of Chem. Eng.*, **37**, 9 (1959).

13. Charles, M.E., Govier, G.W. and Hodgson, G.W., *The Can. J. of Chem. Eng.*, **39**, 27-36 (1961).

14. Guevara, E., Zagustin, K., Zubillaga, Y. and J.L. Trallers, J.L., "Core-Annular Flow: The Most Economical Method for the Transportation of Viscous Hydrocarbons", 4th Int. Conf. on Tar-Sands and Bitomans, Egmonton Canada (1988).

15. Oliemans, R.V.A., "The Lubricating-Film Model for Core-Annular Flow", Ph.D. Dissertation Delft University (1986).

16. Andritsos, N., Williams, L. and Hanratty, T.J., *Int. J. Multi phase flow*, **15** (6), 877-892 (1989).

17. Shoham, O., "Flow Pattern Transitions and Characterization in Gas-Liquid Two-Phase Flow in Inclined Pipes". Ph.D. Dissertation, Tel-Aviv Univ. Ramat-Aviv, Israel (1982).

18. Hanratty, T.J., *PCH Phy. Chem. Hydrodynamics*, **9**, 101-114 (1987).

19. Wallis, G.B. and Dobson, J.E., *Int. J. Multiphase Flow*, 1, 173-193 (1973).

20. Taitel, Y. and Dukler, A.E., *AIChE Jl* 22, 47-55 (1976).

21. Mishima, K. and Ishii, M., *Trans. ASME Jl Fluids Engng* 102, 441-445 (1980).

22. Brauner, N. and Moalem-Maron, D., "Identification of the range of small diameter conduits regarding two-phase flow patterns transition" submitted *Int. Comm. Heat Mass Transfer* (1991).

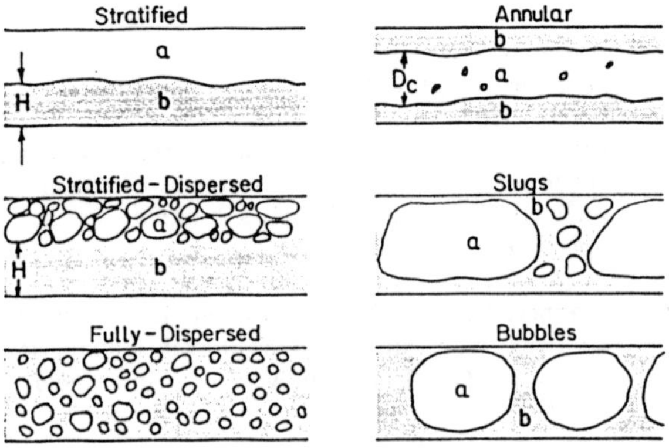

Figure 1. Description of flow patterns in liquid-liquid systems.

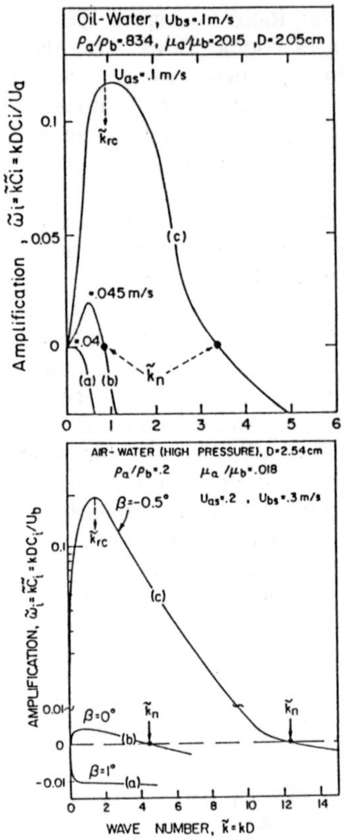

Figure 2. Amplification vs. wave number—effects of flow rates and inclination on the range of unstable modes.

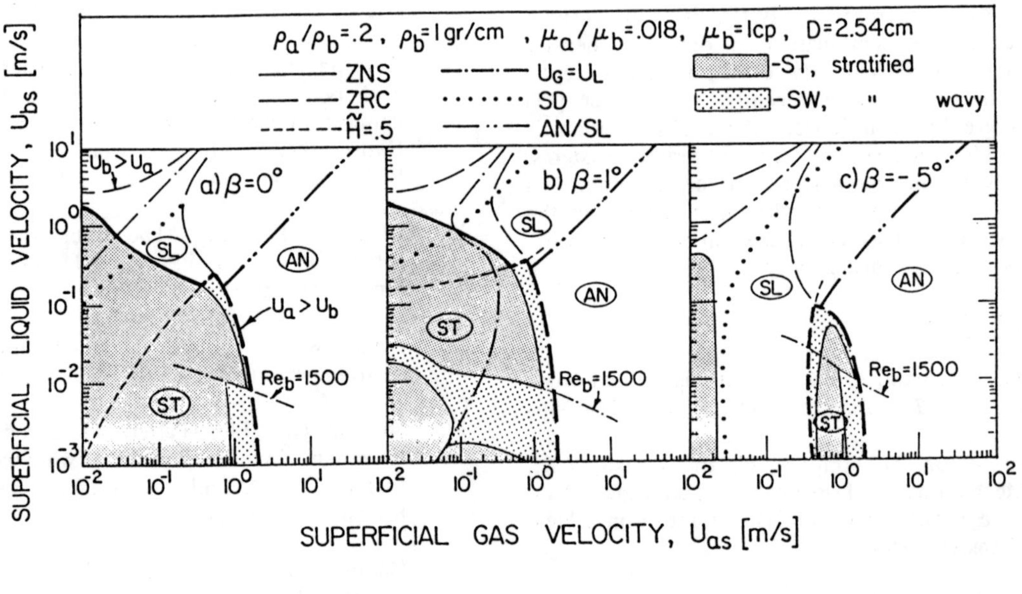

Figure 3. Stability and well-posedness map-construction of the stratified/nonstratified transitional boundary.

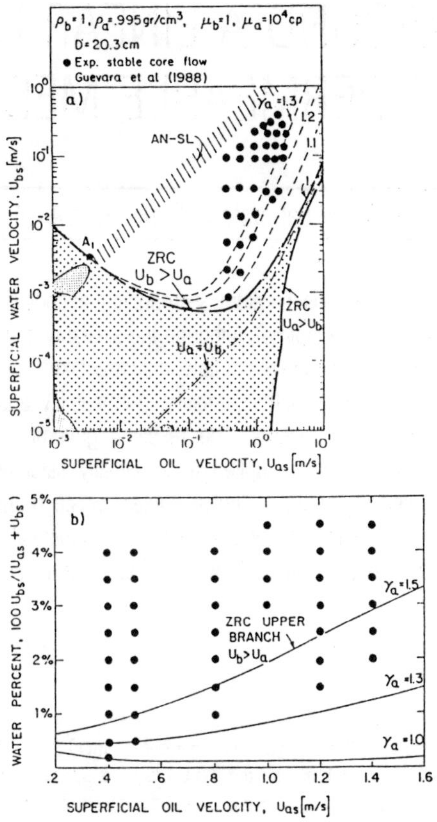

Figure 4. Flow pattern map for a highly viscous oil and water system—comparison with experiment.

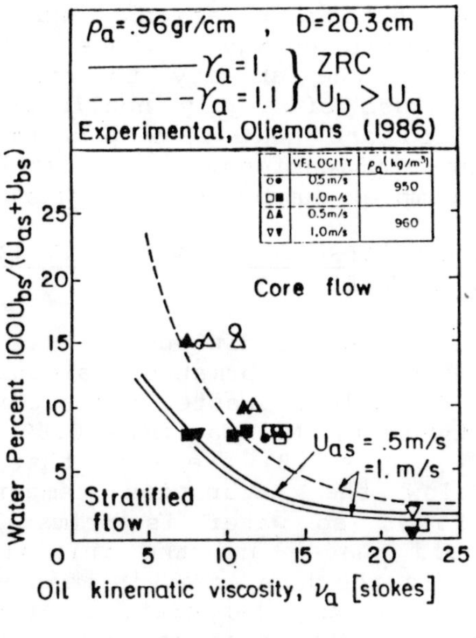

Figure 5. Effect of oil viscosity on the required water percent for stabilization of core flow configuration.

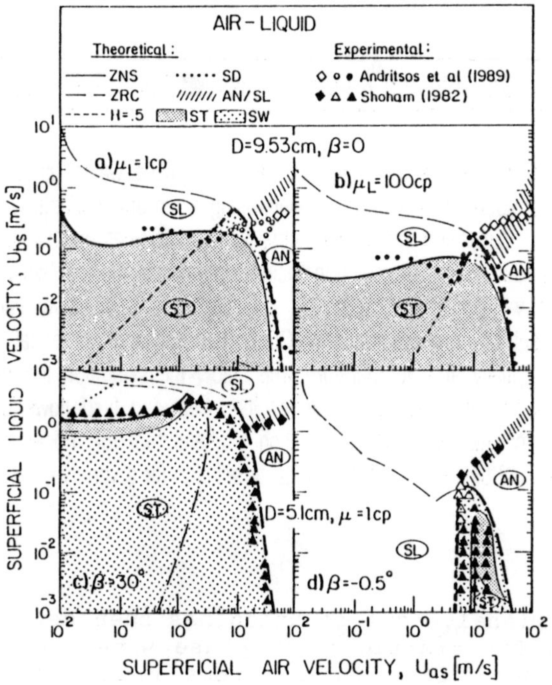

Figure 6. Complete stratified flow boundary based on stability and well-posedness map for gas liquid systems (● ▲ stratified/nonstratified, ○ △ pseudo-slug, ◊ ♦ annular/slug).

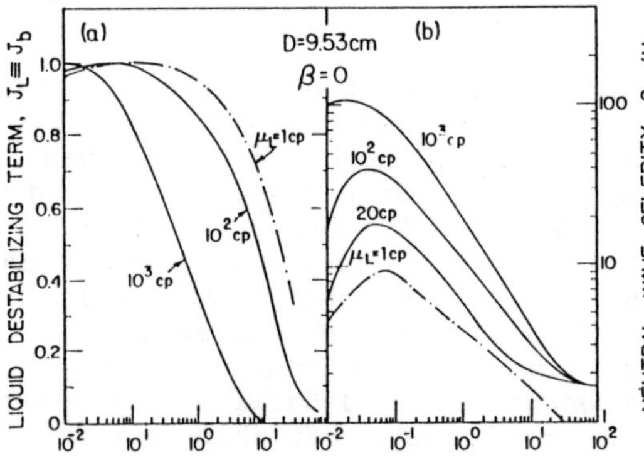

Figure 7. Relative destabilizing contribution of the gas and liquid phases along the ZNS boundary.

A MATHEMATICAL MODEL FOR TWO-PHASE WATER, AIR, AND HEAT FLOW AROUND A LINEAR HEAT SOURCE EMPLACED IN A PERMEABLE MEDIUM

Christine Doughty and Karsten Pruess ■ Earth Sciences Division, Lawrence Berkeley Laboratory, Berkeley, CA 94720

A semianalytical solution for transient two-phase water, air, and heat flow in a uniform porous medium surrounding a constant-strength linear heat source has been developed, using a similarity variable $\eta = r/\sqrt{t}$ (r is radial distance, t is time). Although the similarity transformation requires a simplified radial geometry, all the physical mechanisms involved in two-phase fluid and heat flow may be taken into account in a rigorous way. The solution includes nonlinear thermophysical fluid and material properties, such as relative permeability and capillary pressure variations with saturation, and density and viscosity variations with temperature and pressure. The resulting governing equations form a set of coupled nonlinear ODEs, necessitating numerical integration. The solution has been applied to a partially saturated porous medium initially at a temperature well below the saturation temperature, which is the setting for the potential nuclear waste repository site at Yucca Mountain, Nevada. The resulting heat and fluid flows provide a stringent test of many of the capabilities of numerical simulation models, making the similarity solution a useful tool for model verification. Comparisons to date have shown excellent agreement between the TOUGH2 simulator and the similarity solution for a variety of conditions.

A semianalytical solution has been developed for transient two-phase water, air, and heat flow in a porous medium surrounding a constant-strength linear heat source, using a similarity variable $\eta = r/\sqrt{t}$. While the similarity transformation approach requires a simplified radial geometry, all the complex physical mechanisms involved in coupled two-phase fluid and heat flow can be taken into account in a rigorous way, so the solution may be applied to a variety of problems of current interest.

Application of the similarity solution approach to two-phase fluid and heat flow problems in radial geometry was pioneered in the geothermal well test studies of Grant [1] and O'Sullivan [2]. Grant developed a quasi-analytic approach for the analysis of geothermal wells producing from two-phase reservoirs, using an approximate linearization of the governing equations. O'Sullivan recognized that the similarity variable concept can be used to obtain solutions that include all nonlinearities of the governing equations in a rigorous way.

Our work was motivated by a desire to predict the thermohydrological response to the potential geologic repository for heat-generating high-level nuclear wastes at Yucca Mountain, Nevada, in a partially-saturated, highly-fractured volcanic formation. The geometric simplifications required by the similarity solution preclude detailed predictions of the behavior of the repository as a whole, but the solution can be used to gain insight into the impact of various flow and heat transport parameters on physical conditions near the waste canisters. Other problems that can be studied involve heat sources such as buried power-transmission cables, volcanic dikes, agricultural root-zone heating operations, and underground thermal energy storage systems. Furthermore, the ability to rigorously solve a class of highly nonlinear two-phase fluid and heat flow problems has important applications in the verification of complex numerical simulators.

THERMAL AND HYDROLOGIC CONDITIONS NEAR THE HEAT SOURCE

The ambient conditions in the partially saturated formation around the potential nuclear waste repository at Yucca Mountain, Nevada ($P_0 \approx 0.89$ bars, $T_0 \approx 20°C$, $S_{\ell 0} \approx 0.8$) are such that T is well below the saturation temperature ($T_{sat} \approx 96°C$), so water is primarily in the liquid phase, and the initial heat transfer in the host rock is mainly conductive. As the temperature around a waste package (the heat source) increases to the saturation temperature, evaporation increases and vapor partial pressure becomes appreciable. A convective heat-transfer mechanism known as a heat pipe may contribute to or even dominate heat transfer in this regime. Near the heat source, liquid water vaporizes, causing pressurization and an out-flow of the gas phase, which is composed of noncondensible air and water vapor. The water vapor condenses in cooler more distant regions, depositing its latent heat of vaporization. This creates a nonuniform saturation profile, with liquid saturation increasing with distance from the heat source. The saturation gradient drives the counter-flow of the liquid phase toward the heat source through capillary forces. The liquid then vaporizes again and repeats the cycle. With time the heat pipe moves away from the waste canister, leaving a gas-phase zone in which heat transfer is again conduction-dominated The conditions surrounding a waste package at some time after emplacement are shown schematically in Figure 1.

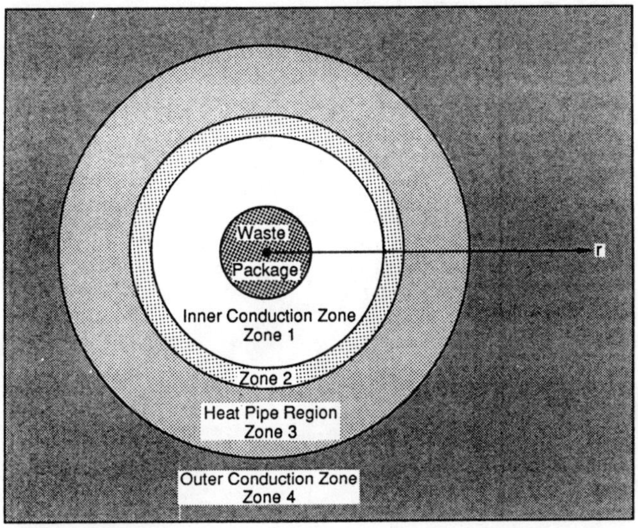

Figure 1. Schematic of the conditions achieved at some time after waste emplacement (not to scale). Water is primarily in the liquid phase in zone 4, because $T < T_{sat}$; two-phase conditions prevail in zones 2 and 3, with $T \approx T_{sat}$; fluid in zone 1 is in the gas phase, with $T > T_{sat}$.

The convective heat transfer of the heat-pipe region is accompanied by small temperature gradients, whereas in the conductive regions temperature gradients are large. Thus, conditions at the waste package may vary greatly depending on the relative importance of conductive and convective heat transfer. If an extensive heat pipe develops, the temperature will remain near the saturation temperature (about 100°C) for long times. Under these conditions, air will be purged from the near-canister region, leaving a gas phase composed purely of water vapor (steam). If heat transfer is primarily conductive, a gas-phase zone, or two-phase zone with only a small amount of immobile liquid, will develop around the waste package. Temperatures may be very high and air is more likely to be present.

The basic requirements for heat-pipe development are (1) the presence of a volatile fluid and (2) a mechanism by which gas-phase fluid can flow away from the heat source and liquid-phase fluid toward it. The groundwater surrounding a geologic nuclear waste repository will be volatile only if the heat-generating capacity of the wastes, coupled with the low thermal conductivity of rock, raises the temperature near the canisters above the saturation temperature of water (T_{sat}) under prevailing pressure conditions. For gas-phase flow away from the heat source to occur, the far-field pressure must be lower than the pressure at the heat source. Liquid-phase flow requires a driving force toward the heat source, which could be gravity, capillary pressure, or a combination of both. Finally, there must be sufficient permeability to both phases to establish the counter-flow necessary for a heat pipe.

For a partially saturated repository environment ambient pressure is near 1 bar and $T_{sat} \approx 100°C$. In contrast, for a deep water-saturated formation fluid pressure at the waste packages (backfill pressure) is likely to be much greater than 1 bar, with the far-field pressure even larger. This makes heat-pipe development unlikely because a) temperatures may never increase as high as T_{sat}; and b) there is no driving force for gas-phase flow away from the heat source when backfill pressure is less than far-field pressure.

METHODOLOGY

The governing equations for two-phase fluid and heat flow for water and air in a porous medium are summarized by Doughty and Pruess [3], and the basic processes represented by the equations are outlined below. A conservation law balances accumulation M and flux Q of each component (water, air, energy), assuming local thermodynamic equilibrium between water, air, and rock. The fluid flux terms include Darcy's law modified for two-phase flow using relative permeability and capillary pressure functions, and binary diffusion between water vapor and air in the gas phase. The inclusion of capillary pressure (which we define as $P_c = P_\ell - P_g$, making P_c a negative quantity) allows the possibility of liquid and gas phase counter-flow. Gas-phase pressure is the sum of air and vapor partial pressures, $P_g = P_a + P_v$. Both the relative permeability and capillary pressure functions are highly nonlinear functions of liquid saturation. The heat balance includes conductive and convective terms with phase-change effects and transport of latent heat. Realistic equations of state for water [4] and air (ideal gas with Henry's law for dissolution in the liquid phase) are used. With these assumptions, the conservation laws form a set of three coupled nonlinear second-order partial differential equations, which are mathematically equivalent to a set of six nonlinear coupled first-order partial differential equations. For single-phase conditions, the six primary dependent variables (unknowns) are temperature T, pressure P, air partial pressure P_a, water flux Q_w, air flux Q_a, and heat flux Q_e. Under two-phase conditions, gas saturation S_g replaces temperature as a primary variable, as temperature can be determined directly from other primary variables through the saturation curve and a vapor pressure lowering correction.

To achieve the symmetry required for the similarity solution approach, the geologic medium is assumed to be uniform and isotropic, the heat source is modeled as an infinitely long cylinder, and gravity is neglected. The geometry of the problem then reduces to radial symmetry, with just two independent variables, radial distance r and time t. If the medium is of infinite extent with uniform initial conditions, and boundary conditions are applied only at $r = 0$ (a line source) and $r = \infty$, and are time-independent, the coupled partial differential equations can be transformed into simpler ordinary differential equations (ODEs) through the use of a similarity variable, $\eta = r/\sqrt{t}$. For example, the conservation equations are

$$\frac{\partial M_m}{\partial t} + \frac{1}{2\pi r}\frac{\partial Q_m}{\partial r} = 0 \qquad (1)$$

where $m = w$ (water), a (air), or e (energy), M is mass or energy per unit volume, and Q is mass or energy flow per unit thickness. After similarity transformation, this becomes

$$-\frac{\eta}{2}\frac{dM_m}{d\eta} + \frac{1}{2\pi\eta}\frac{dQ_m}{d\eta} = 0. \qquad (2)$$

Darcy's Law for liquid flow is

$$Q_\ell = -2\pi r K_\ell \frac{\partial P_\ell}{\partial r} \qquad (3)$$

which becomes

$$Q_\ell = -2\pi\eta K_\ell \frac{dP_\ell}{d\eta}. \qquad (4)$$

Other terms in the flux equations can be similarily transformed. This transformation is known as the Boltzmann transformation in the context of heat-conduction problems, and has been applied by Grant [1] and O'Sullivan [2] to geothermal well test problems.

The resulting ODEs are coupled and nonlinear, so a numerical integration from $\eta=0$ to $\eta=\infty$ is required to solve them. It proves to be convenient to use $z = \ln(\eta)$ as the integration variable. For the linear heat source representing a backfilled or 'closed-hole' nuclear waste canister, the boundary conditions are

$$z = \ln(\eta) = \ln(r/\sqrt{t}\,) = -\infty \qquad (5)$$

($r = 0$ or $t = \infty$)

$Q_w = 0 \qquad Q_a = 0 \qquad Q_e = Q_{e0}$

$P,\ P_a,\ T$ unknown

and

$$z = \ln(\eta) = \ln(r/\sqrt{t}\,) = +\infty \qquad (6)$$

($r = \infty$ or $t = 0$)

$P_g = P_0 \qquad P_a = P_{a0} \qquad S_g = S_{g0}$

$Q_w,\ Q_a,\ Q_e$ unknown.

These boundary conditions constitute a two-point boundary value problem, in which three of the boundary conditions are specified at the $z = -\infty$ limit of integration and three are specified at the $z = +\infty$ limit. The ODEs in z set up a functional dependence between the upper and lower boundary conditions, requiring an iterative process for solution. We use an iterative integration scheme, known as the shooting method [5], which consists of a Newton-Raphson iteration on the unknown boundary conditions $(P, P_a, T) \mid_{z=-\infty}$. For the numerical integration, the limits $z = \pm\infty$ are replaced by finite values z_L and z_U (see below). At the lower limit of the integration, values are guessed for the missing boundary conditions, and the numerical integration of the coupled ODEs is carried out. At the upper limit of the integration, the values of the variables are compared to the specified boundary conditions, and refined estimates of the lower limit boundary conditions are made. This procedure, illustrated in Figure 2, continues until the value of each variable at $z = z_U$ matches the specified boundary conditions P_0, P_{a0}, and S_{g0}. A more detailed discussion of the governing equations and the similarity solution methodology is given in Doughty and Pruess [3,6].

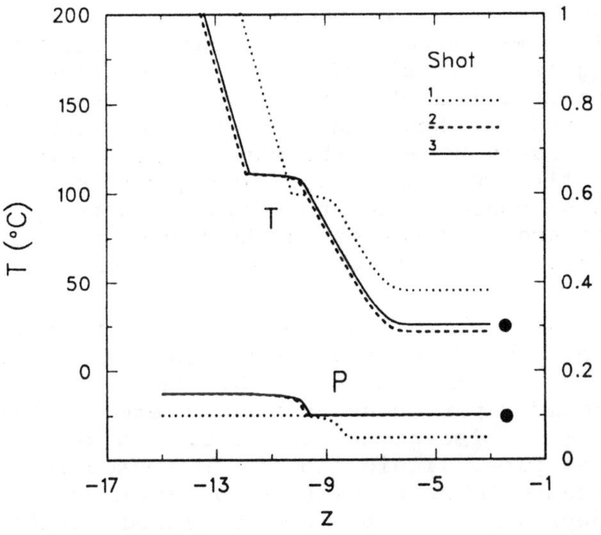

Figure 2. Temperature and pressure profiles illustrating the shooting method for a problem without air. The black dots show the temperature and pressure boundary conditions imposed at z_U.

ILLUSTRATIVE RESULTS

Figure 3 shows an example of the results of the similarity solution. Most of the parameters used to specify this problem, given in Table 1, are typical of conditions expected at the potential nuclear waste repository at Yucca Mountain, Nevada [7]. However, fracture effects are not included and the intrinsic permeability k of the rock matrix is extremely low, so in order to get a stronger heat pipe, intrinsic permeability has been arbitrarily increased by a factor of order 10,000, to 20×10^{-15} m^2, and for consistency capillary pressure strength P_{c0} has been reduced by a factor $\sqrt{10{,}000} = 100$ in comparison to typical Yucca Mountain data. According to the definition of the similarity variable, $z = \ln(r/\sqrt{t})$, Figure 3 represents both a spatial distribution at a given time, with distance from the heat source increasing from left to right, and a time sequence at a given point in space, with time increasing from right to left; most of the discussion of figures that follow is done in terms of a spatial distribution. Figure 3 also shows the results of the similarity solution when no air is present, with other parameters unchanged from Table 1.

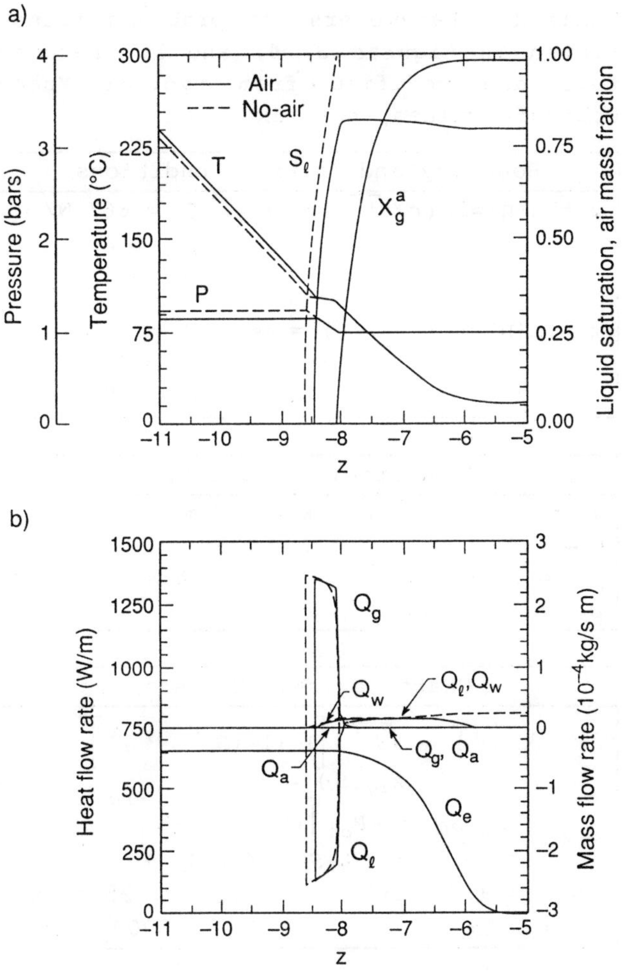

Figure 3. Results of the similarity solution for the with-air and no-air problems described in Table 1. For convenience, air mass fraction $X_g{}^a$ is plotted instead of P_a.

Table 1. Parameters for problems illustrated in Figures 3, 4, and 5. Parameters are modified from typical Yucca Mountain values, see text.

Boundary and Initial Conditions	
$z = \ln(\eta) = \ln(r/\sqrt{t}) = -\infty$	$Q_{e0} = 667$ W/m*
	$Q_{w0} = 0$
	$Q_{a0} = 0$†
$z = \ln(\eta) = \ln(r/\sqrt{t}) = +\infty$	$P_0 = 1$ bar
	$T_0 = 18°C$
	$S_{g0} = 0.2$†

Material Properties	
$k = 20 \times 10^{-15}$ m^2	$\kappa = 2$ W/m K
$\phi = 0.10$	$\tau = 0.5$
$\rho_s = 2550$ kg/m^3	$D^0_{va} = 2.6 \times 10^{-5}$ m^2/s
$c_s = 800$ J/kg K	

Characteristic Curves‡

$$k_{r\ell}(S_\ell) = \sqrt{S^*}\left[1-(1-(S^*)^{1/\lambda})^\lambda\right]^2$$

$$k_{rg}(S_\ell) = 1-k_{r\ell}$$

$$P_c(S_\ell) = -P_{c0}\left[(S^*)^{-1/\lambda}-1\right]^{1-\lambda}$$

$$S^* = (S_\ell - S_{\ell r})/(1-S_{\ell r})$$

$S_{\ell r} = 9.6 \times 10^{-4}$	$P_{c0} = 0.125$ bars
$\lambda = 0.45$	$P_{max} = 5000$ bars

*This thermal power corresponds to high-level nuclear wastes approximately 10 years old.

†In Figure 3, used for with-air problem only.

‡Functional forms of van Genuchten [12]; P_c has been modified slightly so that as $S^* \to 0$, $P_c(S_\ell) \to -P_{max}$ instead of approaching $-\infty$ [3].

Overall the ''with-air'' and ''no-air'' problems show quite similar results. Of course, for the no-air problem, single-phase liquid conditions exist far from the heat source ($S_\ell = 1$ for $z > -8$), whereas the with-air problem is two-phase ($S_\ell = 0.8$), but the pressure and temperature profiles are not very different. The constant pressure and steep linear temperature profiles shown in Figure 3a for $-11 < z < -8.5$ and $-8 < z < -6.5$ indicate conductive regimes, and Figure 3b verifies that mass flow rates are small there. The more gradual temperature decline for $-8.5 < z < -8$ coupled with a large liquid-vapor counter-flow identifies the heat-pipe region. In Figure 3b, Q_g and Q_ℓ are the sum of the air and water components for the gas and liquid phases, respectively. Throughout this paper, the scale on which the mass flows are shown makes air flow vanishingly small, so the mass flow for each phase is essentially water flow (e.g., $Q_g = Q_v$). Note that air mass fraction X_g^a is essentially zero for $z < -8$, so the gas phase there is composed primarily of water vapor (at $z_L = -11$, $X_g^a = 10^{-24}$). The air is purged from the region near the heat source by the vapor flow away from the heat source, as would be any noncondensible gas present in the system. This could have important ramifications for waste-package design with regard to the prevention of corrosion. Extending from the cool end of the heat pipe to the heat-flow front ($-8.2 < z < -6$), there is a small liquid flow away from the heat source, indicating that not all the vapor condensing at the cool end of the heat pipe flows back toward the heat source. The out-flow is necessary because the water vapor forming at the hot end of the heat pipe is much less dense than the liquid water it replaces; it causes the heat pipe to gradually move away from the heat source.

Generally, the presence of air shortens the heat pipe from the cool end ($z = -8$ in Figure 3), by hampering the liquid-vapor counter-flow there. When air is present at the cool end of the heat pipe, liquid flow is decreased there because liquid relative permeability is lower due to the smaller liquid saturation. Vapor flow is decreased because for a given pressure gradient and gas-phase saturation only some of the mass flow is vapor, the rest being air. The heat-pipe shortening with the presence of air is evident from Figure 3, but it occurs at the hot end of the heat pipe ($z = -8.5$ in Figure 3). Because both the with-air and no-air solutions are constrained to match the same pressure and temperature boundary conditions at z_U, any difference in heat-pipe length is shifted toward z_L.

The physical processes associated with the addition of air can be demonstrated more clearly by doing a series of integrations using the no-air boundary conditions for $T(z_L)$ and $P(z_L)$, and sequentially increasing $P_a(z_L)$ from 0 to

10^{-41} to 10^{-26} bars (Figure 4). The boundary conditions at z_U are not matched, but the effect of adding air to shorten the heat pipe is illustrated. Figure 4 shows that as the amount of air present in the system increases, X_g^a increases from zero to one at smaller values of z; because counter-flow occurs only in regions where X_g^a is small, the heat pipe becomes shorter, leading to a greater overall temperature drop; total pressure $P = P_a + P_v$ increases; and S_ℓ decreases because more of the pore space is filled with air. Although the values of $P_a(z_L)$ used in Figure 4 are much too small to be readily distinguished experimentally, with sufficient numerical accuracy they can be used successfully in the similarity solution. This example demonstrates one of the attractive features of the similarity solution: its usual method of employment – taking initial guesses for missing boundary conditions from results of a similar problem – provides insight into the physical effects of various problem parameters.

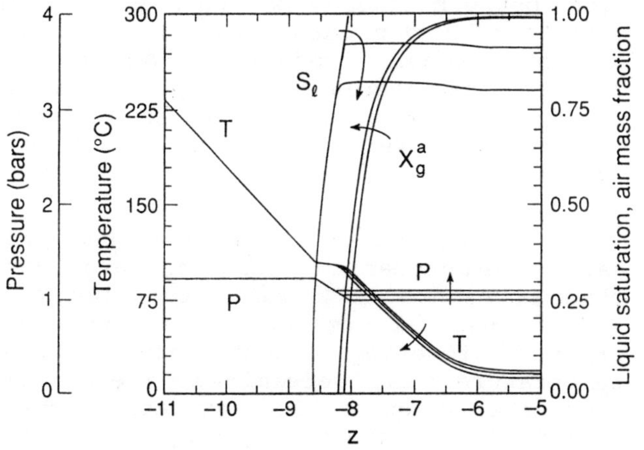

Figure 4. The effect of increasing the amount of air at z_L on the no-air problem shown in Figure 3 and described in Table 1. The arrows show the direction of change as $P_a(z_L)$ is increased from 0 to 10^{-41} to 10^{-26} bars. For comparison, for the with-air problem of Figure 3, $P_a(z_L) = 10^{-24}$ bars.

Two other differences that arise from the presence of air are also illustrated in Figure 3. The pressure decrease that occurs over the length of the heat pipe is larger for the no-air problem because, although the pressure gradient is unchanged, the heat pipe is longer. The pure liquid zone that develops for the no-air problem for $z > -8$ has a much lower compressibility than does the low temperature two-phase zone of the with-air problem, hence the small liquid outflow extends much further from the heat source.

COMPARISON WITH THE NUMERICAL SIMULATOR TOUGH2

Multiphase fluid and heat flow problems that lack some of the symmetries required for application of the similarity variable approach must be solved by numerical simulation, using discretization of the continuous space and time variables. The accuracy and credibility of numerical simulations is a matter of serious concern, because they are subject to space and time discretization errors which are often difficult to quantify. Furthermore, numerical simulations are performed with complex computer programs, and no method is known to directly establish that such programs are free of errors. It is necessary, therefore, to test numerical simulations against known solutions that entail as many complex features as possible. The similarity solution is an ideal tool for such testing, because it gives, within the approximations of the underlying mathematical model, an essentially rigorous solution to a problem that involves the full nonlinear process complexity of transient two-phase fluid and heat flow. The idealizations made pertain only to the flow geometry and to initial and boundary conditions.

We have used the similarity solution to test the numerical simulator TOUGH2 [8,9], which calculates the flow of air and water in gaseous and liquid phases together with heat flow, using the same governing equations and equation of state as does the similarity solution. TOUGH2 employs the integral-finite-difference method to discretize space for one-, two-, or three-dimensional problems that may involve heterogeneous, anisotropic, or fractured/porous media [10]. Figure 5 shows results of TOUGH2 compared with those of the similarity solution for the problem specified in Table 1; the agreement is excellent. A one-dimensional radial mesh with 108 elements was used for the TOUGH2 calculation. The mesh spacing is nonuniform, with finest spacing (0.03 m) used where the similarity solution predicts sharp gradients. A

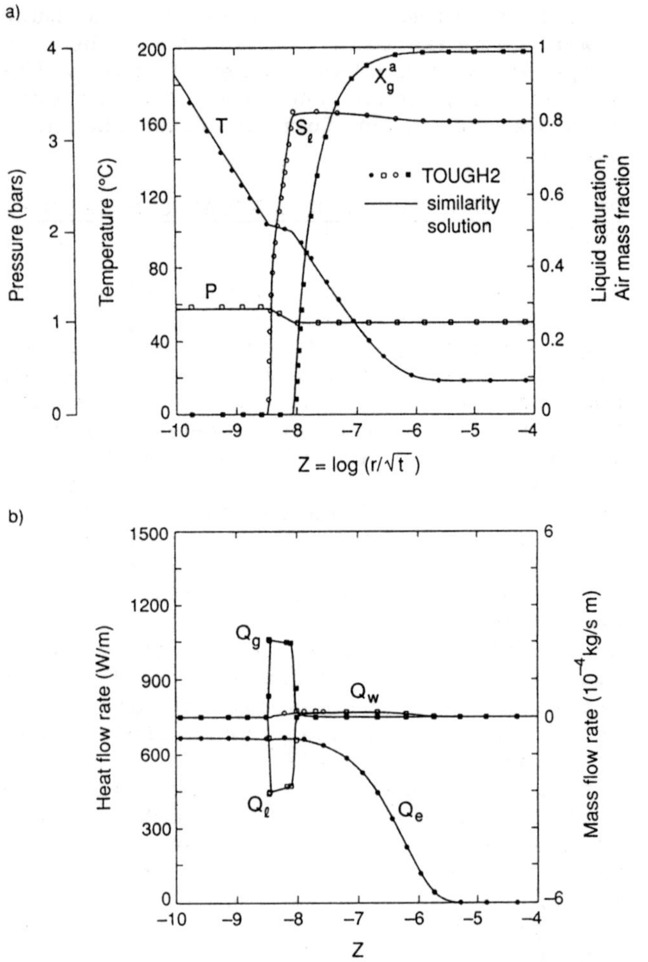

Figure 5. A comparison between results of the similarity solution and the numerical model TOUGH2 for the with-air problem described in Table 1.

heat source is placed in the innermost element ($0 < r < 0.3$ m), and the outermost element ($r = 10{,}000$ m) is at sufficiently large distance to remain at constant temperature, pressure, and saturation for a simulation time of 6.3 years. The TOUGH2 calculation required 800 time steps and 8.5 minutes of CPU time on the Cray X-MP at the National Energy Research Supercomputer Center at Lawrence Livermore National Laboratory.

In contrast, one numerical integration of the similarity solution (using previously-determined correct starting values at z_L) required 228 integration steps and took 2.6 CPU seconds. The shooting procedure requires four integrations per shot, and with reasonable initial guesses, usually converges within three to five shots. The key to successful use of the similarity solution is finding reasonable initial guesses. This may require some preliminary trial integrations, the solution to a similar problem, physical intuition, or a combination of the three. Assuming five trial integrations to arrive at workable initial guesses, and four shots to find the solution, the whole procedure would require 21 shots, or an approximate CPU time of 55 seconds. This sample problem has a relatively short heat pipe, which makes accuracy requirements modest; a numerical integration with only 106 steps, requiring 1.4 CPU seconds, gives correct results also. Conversely, for a long heat pipe, higher accuracy is needed, and computation effort increases correspondingly.

VAPOR PRESSURE LOWERING

In a liquid-vapor system at equilibrium in a porous medium, the vapor pressure depends on the geometry of the vapor/liquid interface, the surface tension between these two phases, and on effects of vapor adsorption on the solid grains. Vapor pressure lowering describes the decrease in vapor pressure that occurs when liquid water occupies the small pore spaces in a geologic medium.

Under two-phase conditions, the primary thermodynamic variables are gas-phase pressure P_g, gas saturation S_g, and air partial pressure P_a. When vapor pressure lowering effects are ignored, temperature is determined iteratively from the steam table saturation curve $P_v = P_{sat}(T)$. Since $P_v = P_g - P_a$, the functional dependence of temperature on the primary variables is $T = T(P_g, P_a)$. Vapor pressure lowering can be modeled using the Kelvin equation, in which the capillary pressure function $P_c(S_\ell)$ accounts for surface tension and interface shape effects. The temperature then depends on saturation as well as vapor pressure, and is given implicitly by

$$P_v = P_{sat}(T) \cdot \exp\left[\frac{P_c(S_\ell)}{\rho_\ell R\,(T+273.15)}\right]. \qquad (7)$$

Thus the functional dependence of temperature on the primary variables is $T = T(P_g, P_a, S_g)$. Vapor pressure lowering can have a large effect; when $S_\ell \rightarrow 1$, $P_c \rightarrow 0$ and $P_v \rightarrow P_{sat}$, but when $S_\ell \rightarrow 0$,

P_v can decline to a few percent of P_{sat} [11].

Figure 6 shows results for two problems which are identical except for the presence or absence of vapor pressure lowering. The primary differences from the problem described in Table 1 are a factor of order 10^4 decrease in intrinsic permeability and a factor of order 10^2 increase in capillary pressure strength P_{c0}, making this problem more representative of the intact rock at Yucca Mountain. The weaker capillary pressure of the Table 1 problem makes vapor pressure lowering effects negligible in that case. For the problems shown in Figure 6, the intrinsic permeability is too small to allow enough fluid flow for a strong heat pipe to develop (Figure 6b), thus heat transfer is primarily conductive, resulting in linear temperature profiles (Figure 6a). Furthermore, the convective gas-phase flow away from the heat source is small compared to the diffusive mass flow between the air and vapor components of the gas phase, resulting in a non-zero air mass fraction X_g^a in the region near the heat source ($-11 < z < -10$ in Figure 6a). If fracture effects were included, a much stronger heat pipe could develop, depending on various assumptions made about fluid flow in the fractures [3].

The capillary pressure function, slightly modified from the van Genuchten [12] formulation [3] has the feature that P_c becomes very large and negative as $S_\ell \to S_{\ell r}$ (see Table 1). Then, from Equation (7), vapor pressure lowering effects become very strong for small S_ℓ, and the two-phase region extends all the way to z_L, as is shown in Figure 6a. Except for the extension of the two-phase zone to small values of z, inclusion of vapor pressure lowering has only minor impact. The slope of the conductive temperature profile for $-11 < z < -10$ decreases slightly for the vapor-pressure-lowering case (Figure 6a), reflecting the increase in thermal conductivity κ with liquid saturation ($\kappa \sim \sqrt{S_\ell}$). Variations in fluid flow rate are less abrupt when vapor pressure lowering is included than when it is not (Figure 6b).

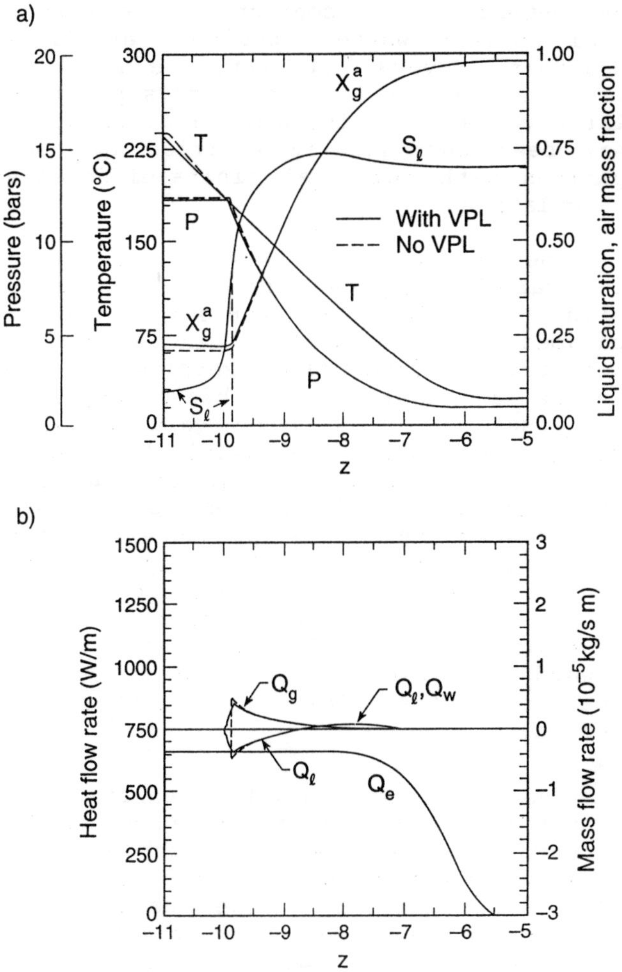

Figure 6. Results of the similarity solution for problems with and without vapor pressure lowering. Note the change of scale for mass flow rate compared to Figure 3.

SUMMARY AND CONCLUSIONS

The mass and energy transport equations for two-phase fluid and heat flow in one-dimensional radial geometry in a homogeneous permeable medium depend on time t and distance r only through the similarity variable $\eta = r/\sqrt{t}$. If initial and boundary conditions can be written as functions of η, then the problem is reduced from a set of partial differential equations to a set of ordinary differential equations. A practically important case for which this is possible is a constant-rate line source at $r = 0$ and uniform initial conditions. These boundary conditions give rise to a two-point boundary value problem which can be solved with an iterative integration scheme. We have applied the similarity variable concept to a problem with a linear heat source emplaced in a partially saturated permeable medium, and have discussed the effect of various parameters in the context of the potential nuclear waste repository at Yucca Mountain, Nevada, although the solution is certainly not limited to this particular application. The similarity solution has been successfully verified by comparison with numerical finite-difference calculations.

One of the strengths of the similarity variable approach is the ability to study the effect of various transport parameters individually, in a simple geometry. Through this exercise one develops a knowledge of basic physical processes, which will be valuable when doing more complicated numerical calculations with realistic geometric detail. The examples in the present paper have illustrated that the presence of air shortens the convective heat-pipe region, thus increasing the temperature at the heat source, and that vapor pressure lowering extends the two-phase region toward the heat source, and makes the onset of counter-flow gradual rather than abrupt.

The similarity-solution results presented here were all calculated using the computer program SIMSOL, which has been fully documented in a laboratory report [13]. In addition to the features described in the present work, the similarity solution can treat an effective continuum representation of a fractured/porous medium, and can include the Klinkenberg approximate treatment of Knudsen diffusion. A detailed description of the similarity solution methodology and results is given in Doughty and Pruess [3,6]. Plans for future work include allowing the inner boundary condition to represent either a heat source or sink, and/or a mass source or sink. This will greatly enhance the applicability of the similarity solution to practical problems, and require only minimal changes to the code. An interesting possible enhancement would be the inclusion of yet another mass component. Components with practical applications include non-condensible gases, (e.g., CO_2), conservative tracers, volatile organic solvents (such as TCE), or non-aqueous phase liquids (hydrocarbon compounds).

ACKNOWLEDGEMENT

The careful review of this work by R. Falta and M. O'Sullivan is greatly appreciated. This work was carried out under U.S. Department of Energy Contract No. DE-AC03-76SF00098, through the Yucca Mountain Project, Sandia National Laboratories, under Document No. SNL 54-1064; numerical computations were supported by the Director, Office of Energy Research, Office of Basic Energy Sciences, Engineering and Geosciences Division, U.S. Department of Energy.

NOTATION

c	specific heat (J/kg K)
D	binary diffusion coefficients (m^2/s) ($D = \tau\phi S_g D_{va}^0 P_0/P$)
D_{va}^0	binary diffusion strength in free gas at $P_0 = 1$ bar (m^2/s)
K	mobility (kg/s mPa) ($K_j = kk_{rj}\rho_j/\mu_j$, $j = \ell, g$)
k	intrinsic permeability (m^2)
k_r	relative permeability
P	pressure (Pa)
P_c	capillary pressure (Pa) ($P_c = P_\ell - P_g$)
Q_e	heat flow rate (W/m)
Q_w, Q_a	mass flow rates (kg/s m) ($Q_m = Q_\ell^m + Q_g^m$, $m = w, a$)
r	radial distance (m)
S	saturation
$S_{\ell r}$	residual liquid saturation
T	temperature (°C)
t	time (s)
X	mass fraction ($X_j^w + X_j^a = 1$, $j = \ell, g$)
z	integration variable ($z = \ln(\eta) = \ln(r/\sqrt{t})$)
η	similarity variable ($\eta = r/\sqrt{t}$)
κ	thermal conductivity (W/m K)
λ	parameter in van Genuchten characteristic curves
μ	dynamic viscosity (Pa s)
ρ	density (kg/m^3)
ϕ	porosity
τ	tortuosity

Subscripts

a	air (also used as a superscript)
c	capillary
e	energy
g	gas phase
L	lower limit of integration

ℓ	liquid phase
r	relative
sat	at saturation (vapor-liquid equilibrium)
s	solid
U	upper limit of integration
v	vapor
w	water (also used as a superscript)
0	boundary condition, reference value

LITERATURE CITED

1. Grant, M. A., ''Quasi-analytic solutions for two-phase flow near a discharging well,'' Report 86, Applied Mathematics Division, Dept. of Scientific and Industrial Research, Wellington, New Zealand (1979).

2. O'Sullivan, M., J., Water Resour. Res. **17**(2), 390-398 (1981).

3. Doughty, Christine and Karsten Pruess, ''A similarity solution for two-phase water, air, and heat flow near a linear heat source in a porous medium,'' Report LBL-30051, Lawrence Berkeley Lab., Berkeley, CA (1991).

4. International Formulation Committee, A formulation of the thermodynamic properties of ordinary water substance, IFC Secretariat, Duesseldorf, Germany (1967).

5. Press, W. H., B. P. Flannery, S. A. Teukolsky, and W. T. Vetterling, Numerical recipes: the art of scientific computing, Cambridge University Press, New Rochelle, New York, Ch. 16 (1986).

6. Doughty, Christine and Karsten Pruess, Int. J. Heat Mass Transfer **33**(6), 1205-1222 (1990).

7. Pruess, Karsten, Y. W. Tsang, and J. S. Y. Wang, Water Resour. Res. **26**(6), 1235-1248 (1990).

8. Pruess, Karsten, ''TOUGH user's guide,'' Report NUREG/CR-4645, Nuclear Regulatory Commission, Washington, D.C. (1987). (Also avail. as Lawrence Berkeley Lab. Report LBL-20700.)

9. Pruess, Karsten, ''TOUGH2—A general-purpose numerical simulator for multiphase fluid and heat flow,'' Report LBL-29400, Lawrence Berkeley Lab., Berkeley, CA (1990).

10. Narasimhan, T. N. and P. A. Witherspoon, Water Resour. Res. **12**(1), 57-64 (1976).

11. Hsieh, C. H. and H. J. Ramey, Jr., Soc. Pet. Eng. J. **23**, 157-167 (1983).

12. van Genuchten, M. Th., Soil Sci. Soc. Am. J. **44**, 892-898 (1980).

13. Doughty, Christine, ''SIMSOL Users Guide,'' Report LBL-28384, Lawrence Berkeley Lab., Berkeley, CA (1991).

STUDY OF CONTINUITY AND PRESSURE WAVES IN CONDENSING TWO-PHASE FLOW IN A PIPELINE

Leksono Mucharam and Michael A. Adewumi ■ The Pennsylvania State University, 202 Mineral Science Building, University Park, PA 16802

The analysis of transient gas condensate flow in pipe is conducted using a one-dimensional compositional two-fluid model. The model is developed using fundamental hydrodynamic equations in which continuity and momentum equations are considered. This model is designed for handling the hydrocarbon fluid flow in pipeline where fluid composition is an important factor in affecting the condensation process. The compositional effect of the fluid is taken into account by coupling a Peng-Robinson equation-of-state-based phase behavior model to the hydrodynamic equations. The resulting system of partial differential equations obtained consists of four dependent variables, namely, pressure, gas volume fraction, gas velocity and liquid velocity. The numerical method of lines is utilized to convert the set of partial differential equations to a system of ordinary differential equations, posed as an initial-value problem (IVP). Integration of the IVP is effected using a versatile ODE solver, in which the Iannelli-Baker's Implicit Runge-Kutta method for stiff problems is implemented. In the analysis using this model, pressure and continuity wave propagation along the pipeline are predicted for various operational procedures and pipeline geometries. The results show the viability of the model in describing the transient behavior of two-phase gas-condensate flow in pipelines, which is very important for natural gas pipeline design considerations.

The increasing role of natural gas as an energy source, especially in the non-transportation sector of an industrial economy, cannot be over-emphasized. This role makes optimal design and operation of gas transmission line, the mainstay of gas transportation, very crucial. A complicating factor is that condensation will invariably take place in these pipelines. Under transient conditions where pressure changes sharply with time, distribution of liquid phase in the pipeline will be greatly affected by the pressure waves propagating along the system. It is therefore of utmost importance to know a priori the point of condensation and quantity of condensate in pipelines. Furthermore, because of the transient nature of the waves, an adequate design of such pipeline must take into account the impact of transients on pipeline design, safety and operation.

Quite often, liquid condensation in pipeline occurs because of the multi component nature of the transmitted natural gas and the inevitable temperature and pressure variation that occur along the pipeline. Several models have been developed and used in oil industries to predict and design the transmission pipelines. Most of them utilize the steady state approach. Adewumi and Bukacek [1] reported that many two-phase flow models give widely conflicting results on predicting pressure distribution along the pipeline system. This makes the design choice even more difficult. Using a steady-state approach, Mucharam and Adewumi [2] developed a compositional model for gas/condensate flow in gas pipelines by coupling the hydrodynamic model with a phase behavior model. Their analysis shows that two-phase gas/condensate flow generally causes a significantly higher pressure drop than the equivalent single-phase flow would, even when the total mass flow rate is the same.

The use of pipelines, in practice, is not limited to transporting natural gas but also for storing the excess gas by using line packing-method. This causes the pipeline gas to experience high pressure while the valve is closed. The gas is then released as needed. Furthermore, the effect of back pressure in a pipeline when a certain flow rate is imposed on the system by accidental or incidental field operation procedure is important to predict. These mean that the transportation of gas in these pipelines becomes transient and therefore requires a transient model for analysis and design. The presence of liquid in the pipeline due to condensation, further complicates the design of such pipeline system. This study attempts to analyze these transients, either deliberately or inadvertently caused, as they may affect gas pipeline design in terms of safety and optimal field operations.

FORMULATION OF THE FLOW MODEL

The two-phase flow model is formulated by coupling two distinct but interrelated components, namely the hydrodynamic description of the condensing two-phase flow in the pipe and the phase behavior description of natural gas system. The resulting model is compositional and hence suitable for studying the flow performance of gas/condensate pipelines. The Peng-Robinson equation of state [3] is used in modeling the phase behavior, and the hydrodynamic formulation is developed by invoking the mass and the momentum conservation laws.

The Hydrodynamic Equations

In formulating the hydrodynamic equations to simulate two-phase gas/condensate flow in pipelines, the modification of the two-fluid model which was formulated and utilized for describing the pneumatic conveying of solid particles in pipe [4] is used in this study. The specific modifications incorporated include compositional effects, the condensing nature of the gas and the associated interfacial mass and momentum transfer between the two phases. This formulation is developed by assuming that each phase, gas or liquid, constitutes a continuum. This assumption enables one to invoke the basic laws of continuum mechanics which, in conjunction with volume averaging technique, produce the governing continuity, and momentum equations. Even though Adewumi and Arastoopour [4] used the integral approach to derive the two-fluid model, the Eulerian approach is also used to derive the governing equations [5].

In developing the two-phase hydrodynamic model for gas/condensate flow in pipelines, four major assumptions have been made. These are :

(1) The amount of liquid is small and the gas flow rate is sufficiently high at which conditions the flow pattern can be assumed to be dispersed. However, both the gas phase and the liquid phase in the system are assumed to obey the laws of continuum mechanics.
(2) The liquid droplets are further assumed to be uniform in size, of diameter d_p. This assumption is particularly relevant to the formulation of the constitutive equations for calculating drag force between liquid droplets and the continuous gas phase.
(3) Flow is assumed to be one-dimensional in the axial direction of the pipeline.
(4) The temperature of the gas and liquid phases are assumed to be equal at every point in the pipeline.

In the continuity equations, the liquid condensation is regarded as a source or a sink term in each phase of the system. The resulting governing equations are presented below.

The Continuity Equations

By considering the mass generation taking place in the system as a result of condensation (or vaporization), the continuity equations for the gas and liquid phases can be expressed respectively as

$$\frac{\partial}{\partial t}(\rho_g\alpha_g)+\frac{\partial}{\partial x}(\rho_g\alpha_g v_g)=\dot{m}_{gL} \tag{1}$$

and

$$\frac{\partial}{\partial t}(\rho_L\alpha_L)+\frac{\partial}{\partial x}(\rho_L\alpha_L v_L)=\dot{m}_{Lg} \tag{2}$$

where $\dot{m}$ is mass rate transfer between the two phases. In this formulation, the mass transfer rate by the phases is assumed to be equal to the mass rate received by the other phase in the system. Therefore, the rate of gas disappearance by condensation equals the rate of liquid appearance, i.e., $\dot{m}_{gL} = -\dot{m}_{Lg}$. Furthermore, $\alpha_g + \alpha_L = 1.0$.

The Momentum Equations

By considering the forces due to friction, drag, gravitation, mass transfer between gas and liquid phases, momentum equations for gas and liquid phases can be written in the differential form as below.
For the gas phase,

$$\frac{\partial}{\partial t}(\alpha_g\rho_g v_g)+\frac{\partial}{\partial x}(\rho_g\alpha_g v_g^2)=\begin{array}{l}-\frac{\partial p}{\partial x}-F_{dg}-F_{wg}\\ -F_{mg}-\alpha_g\rho_g\, g\sin\theta\end{array} \tag{3}$$

For the liquid phase,

$$\frac{\partial}{\partial t}(\alpha_L\rho_L v_L)+\frac{\partial}{\partial x}(\rho_L\alpha_L v_L^2)=\begin{array}{l}-F_{dL}-F_{wL}\\ -F_{mL}-\alpha_L\rho_L\, g\sin\theta\end{array} \tag{4}$$

The Energy Equation

In this model, temperature changes along the pipeline is taken into account by using the Coulter and Bardon [6] equation which has been modified in this study. While the Coulter and Bardon equation was developed for steady state flow condition and horizontal pipe, the modified equation can be used for both horizontal and inclined pipeline system. Using the macroscopic form of the energy balance equation

$$\Delta H+\frac{\Delta v^2}{2\,g_c}+\frac{g}{g_c}\Delta h = Q-W_s \tag{5}$$

as the starting point and employing some simple algebraic manipulation, we can derive the expression for temperature profile as:

$$T(x)=\begin{array}{l}T_s+(T_o-T_s)\,e^{-a_1x}+\left(1-e^{-a_1x}\right)\\ \left[\frac{\eta}{a_1}\left(\frac{dP}{dx}\right)-a_o\sin\alpha\right]\end{array} \tag{6}$$

where

$$a_o = \frac{W_t\, g}{g_c J\,\pi\, d\, U\, M}\ ; \qquad a_1 = \frac{\pi\, d\, U\, M}{W_t\, C_p}$$

Wall Friction Forces In this study, the frictional force between each fluid phase and the pipe wall is determined using the volume-fraction-modified form of the single phase model. Furthermore, the mixture properties in the single phase model are replaced with the appropriate phasic properties. For gas and liquid system, the frictional forces are written, respectively, as follows:

$$F_{wg} = \frac{2\,\alpha_g\, f_g\, \rho_g\, v_g\, |v_g|}{d} \tag{7}$$

$$F_{wL} = \frac{2\,\alpha_L\, f_L\, \rho_L\, v_L\, |v_L|}{d} \tag{8}$$

where f_g and f_L, are Darcy friction factors for gas and liquid phases respectively. These parameters are calculated using Chen's equation [7].

Interfacial Drag Force In hydrocarbon fluid transportation, it is common for gas and liquid phases to exist together in the same system with liquid phases having density relatively higher than the gas phase. This will cause slip between gas and liquid phase to occur in the system where one phase attempts to move faster than the other phase. As a result, an interfacial drag force develops between gas and liquid surface when one phase tries to resist the faster moving phase. The magnitude of this force depends on the shape of gas-liquid interface, slip velocity and the phasic properties.

The model used to represent the interfacial drag force in this study is the one presented by Hughes et al. [8]. Because of the high-velocity nature of the flow and relatively small amount of liquid present, it is assumed that the system operates in the dispersed flow regime.

$$F_{dg} = \frac{C_D\, \rho_g (v_g - v_L)|v_g - v_L|\, A_t}{8 g_c} \tag{9}$$

where A_t, the total surface area of the droplets per unit volume, is given by

$$A_t = \frac{6\,\alpha_l}{d_p} \tag{10}$$

The drag coefficient, C_D, and the two-phase Reynolds number, N_{Re}, are defined as:

$$C_D = \frac{24}{N_{Re}} B_1 + B_2 \quad ; \qquad N_{Re} = \frac{\rho_g |v_g - v_L|\, d_p}{\mu_g}$$

where

$$B_1 = 1.0 + (N_{Re})^{0.687} \quad ; \quad B_2 = \frac{0.42}{\left(1.0 + 42500\, N_{Re}^{-1.16}\right)}$$

The maximum diameter of the droplet is predicted by using the Weber number criterion, i.e.,

$$(d_p)_{max} = \frac{W_e\, g_c\, \sigma}{\rho_g (v_g - v_L)^2} \tag{11}$$

The suggested values of the critical Weber number that give the maximum stable droplet diameter range from 8 to 20. The average droplet diameter according to Ulke [9] is given by

$$\bar{d}_p = 0.06147\,(d_p)_{max} \tag{12}$$

Moeck [10] recommends 13 for the critical Weber number. However, Ulke [9] suggested that the critical number should be set at the upper limit for droplet size in steady state flow condition. The critical Weber number used in this model is 14.

Mass Transfer Effect Force Since condensation is of primary concern in this study, the effect of momentum transfer between two phases due to mass transfer becomes very important. By assuming that the gas and liquid phases are in local equilibrium at all times in the segment of pipeline, the rate of mass transfer can be determined from the phase equilibrium calculations by flashing the system at the pressure and temperature conditions in the pipeline segment concerned. From the result of the phase equilibrium calculations, the rate of mass transfer between the phases is calculated using the equation below.

$$\dot{m}_{Lg} = \frac{4[\,(m_L)^n - (m_L)^o\,]\, G_m}{\pi\, D^2\, x_D\, \Delta L} \tag{13}$$

Where superscripts o and n denote the pipe segment's inlet and outlet conditions.

When gas and liquid velocities are different (as they almost always are), the mass transfer from one phase to another could affect the momentum force in both phases. This mass transfer could have effect either on the velocity or the mass change of both phases. It is assumed in this study that the effect of the mass transfer on both phases is equal. In this model, the mass transfer effect force equation is given by

$$F_{mg} = \frac{\dot{m}_{Lg}\,(v_g - v_L)}{g_c} \tag{14}$$

In the system of four partial differential equations described by Equations (1) to (4), there are fifteen unknown variables, four of which are considered the primary dependent variables that will be solved. In order for the system of Equations to be closed, eleven other relationships are required. These relationships consist of four constraining equations, four constitutive equations and three variables determined from equations of state. The four constitutive equations relate the wall friction forces, F_{wg} and F_{wL}, interfacial drag force, F_{dg}, and the mass transfer force, F_{mg} to the primary dependent variables and they are given by equations (7), (8), (13) and (14). The following constraints are necessary.

$$F_{mg}+F_{mL}=0 \quad ; \quad F_{dg}+F_{dL}=0$$

The Phase Behavior Model

Due to the high dependence of the thermo-physical properties of the gas/gas condensate on pressure and temperature, it is quite important to use a compositional model to simulate the two-phase gas/condensate flow in pipelines. In this study, a two-phase equilibrium phase behavior model is developed using the Peng-Robinson Equation of State.

Density

The gas and liquid densities are estimated, respectively, by

$$\rho_g = \frac{M_g P}{Z_g R T} \quad ; \quad \rho_L = \frac{M_L P}{Z_L R T}$$

where Z_g and Z_L are the compressibility factors of gas and liquid phases respectively that are obtained from the Peng-Robinson equation of state. The M_g and M_L are the molecular weights of gas and liquid respectively that are obtained from phase equilibrium calculations.

Mass Fraction

The gas and liquid mass fractions are estimated, respectively, by

$$f_{mg} = \frac{M_g V}{M_g V + M_L L} \quad ; \quad f_{mL} = \frac{M_L L}{M_g V + M_L L}$$

where V and L are mole fraction of gas and liquid phase respectively.

Specific Heat

Specific heat, C_p, is defined as enthalpy change with respect to temperature at constant pressure. Mathematically, it can be expressed as

$$C_p = \left[\frac{\partial H}{\partial T}\right]_p \tag{15}$$

Joule-Thompson Coefficient

The Joule-Thompson coefficient, η, is an important thermodynamic parameter necessary for estimating heat flow. Mathematically, it is defined as

$$\eta = \left[\frac{\partial T}{\partial P}\right]_H = \frac{\left[\frac{\partial H}{\partial P}\right]_T}{C_p}$$

By substituting Equation into Equation , the Joule-Thompson coefficient of a phase can be estimated from the Peng-Robinson equation of state using a numerical approach.

THE NUMERICAL TECHNIQUE

The governing equations used in this model are expressed in Equations (1) to (4). This set of equations is a system of partial differential equations with time, t and distance x, as independent variables, and it is classified as a system of first order hyperbolic partial differential equations. The stiffness characteristics of the hyperbolic differential equations has been recognized and investigated, in detail, by several investigators (e.g. Tiwari [11]). In mathematical terms, when the eigenvalues of a system of differential equations describing physical processes have a large variation, then the system is referred to as a stiff system. The degree of stiffness is measured by the stiffness ratio which is the ratio of the largest eigenvalue to the smallest one.

In this study, the numerical method of lines is used to solve the system of partial differential equations. This numerical technique has been investigated and used intensively in the Soviet Union since 1934, and due to its flexibility and versatility this method has gained wide acceptance as a viable approach to the integration of partial differential equations. However, only limited theoretical analysis of the numerical method of lines has been performed. This method has been successfully applied to multiphase flow problems [12].

The implementation of the method of lines to the governing equations of the transient flow model basically involves converting the system of partial equations into a system of ordinary differential equations by discretizing the equations in the spatial domain and leaving the time domain continuous. The resulting set of ordinary differential equations can then be integrated with an appropriate ordinary differential equation solver.

The hydrodynamic parameters which are selected as the primary variables for transient model are pressure, gas volume fraction, gas velocity, and liquid velocity. These variables and the physical properties of the fluid vary with time as well as space in this model. Equations (1)-(4) describe the distribution of these variables. When the governing equations are discretized in the spatial domain, the following semi-discretized equations (in note i) result.

The continuity equation for gas-phase becomes

$$\frac{d}{dt}(\alpha_g\rho_g)_i = \left[-\nabla_x(\alpha_g\rho_g v_g)\right]_i + (\dot{m}_{gL})_i \tag{16}$$

Similarly, continuity equation for liquid-phase is

$$\frac{d}{dt}(\alpha_L\rho_L)_i = \left[-\nabla_x(\alpha_L\rho_L v_L)\right]_i + (\dot{m}_{Lg})_i \tag{17}$$

The momentum equation for gas-phase becomes

$$\frac{d}{dt}(\alpha_g\rho_g v_g)_i = \begin{matrix}\left[-(\alpha_g)\nabla_x p\right]_i - \left[\nabla_x(\alpha_g\rho_g v_g^2)\right]_i \\ -(F_{gg}+F_{dg}+F_{wg}+F_{mg})_i\end{matrix} \tag{18}$$

Similarly, momentum equation for liquid-phase is

$$\frac{d}{dt}(\alpha_L\rho_L v_L)_i = \begin{matrix}\left[-(\alpha_l)\nabla_x p\right]_i - \left[\nabla_x(\alpha_L\rho_L v_L^2)\right]_i \\ -(F_{gL}+F_{dL}+F_{wL}+F_{mL})_i\end{matrix} \tag{19}$$

where ∇_x is the difference equivalent partial differential operator with respect to the flow in x-axis direction. Equations (16), (17), (18) and (19) constitute a set of ordinary differential equations for transient model in this study. Based on the selected primary variables for the transient system, the left-hand sides of the equations can be expanded into the non-conservative form, can be expressed in the matrix notation as:

$$\begin{bmatrix} b_{11} & b_{12} & b_{13} & b_{14} \\ b_{21} & b_{22} & b_{23} & b_{24} \\ b_{31} & b_{32} & b_{33} & b_{34} \\ b_{41} & b_{42} & b_{43} & b_{44} \end{bmatrix}_i \frac{d}{dt}\begin{bmatrix} U_1 \\ U_2 \\ U_3 \\ U_4 \end{bmatrix}_i = \begin{bmatrix} E_1 \\ E_2 \\ E_3 \\ E_4 \end{bmatrix}_i \tag{20}$$

where U_1, U_2, U_3, and U_4 are the primary variables which represent pressure, gas holdup, gas velocity and liquid velocity, respectively. These primary variables are arranged in that order to avoid singularity in matrix B. The components of matrix B and vector matrix E are defined as shown in Mucharam [5]. Using matrix representation, equation (20) can be written in the compact form:

$$[B]\frac{d\vec{U}}{dt} - \vec{E}(U) = \vec{0} \tag{21}$$

The resulting set of ordinary differential equations can be solved or integrated by any appropriate ordinary differential equation integrator. In this model, a stiff and stable implicit Runge-Kutta method formulated by Iannelli and Baker [13] is used to solve Equation (21). The semi-discretization scheme used produces a stiff system of non-linear ordinary differential equations which must be solved using a stiffly stable solution method.

In this model, the system of ordinary differential equations represented by Equation is solved numerically using the implicit Runge-Kutta method developed by Iannelli and Baker [13]. The generated matrix is a square-banded matrix B [(m x n) x (m x n)] and a column matrix E [(m x n) x 1], where m is number of given grid points and n is number of primary dependent variables. The Jacobian matrix generated in the implicit Runge-Kutta method is the derivative of all components in column matrix E with respect to the primary dependent variables in the system of equations. At position i, the Jacobian Matrix, $\left[\frac{\partial E}{\partial U}\right]_i$ can be written as

$$\left[\frac{\partial E}{\partial U}\right]_i = \begin{bmatrix} \frac{\partial E_1}{\partial p} & \frac{\partial E_1}{\partial \alpha_g} & \frac{\partial E_1}{\partial v_g} & \frac{\partial E_1}{\partial v_l} \\ \frac{\partial E_2}{\partial p} & \frac{\partial E_2}{\partial \alpha_g} & \frac{\partial E_2}{\partial v_g} & \frac{\partial E_2}{\partial v_l} \\ \frac{\partial E_3}{\partial p} & \frac{\partial E_3}{\partial \alpha_g} & \frac{\partial E_3}{\partial v_g} & \frac{\partial E_3}{\partial v_l} \\ \frac{\partial E_4}{\partial p} & \frac{\partial E_4}{\partial \alpha_g} & \frac{\partial E_4}{\partial v_g} & \frac{\partial E_4}{\partial v_l} \end{bmatrix}_i \tag{22}$$

In solving the system of partial differential equations formulated, both boundary and initial conditions must be imposed. Forward finite difference scheme is used in the system when the condition is given at the outlet of the system, otherwise, backward finite scheme is used in the system when the condition is given at the inlet of the system.

DISCUSSION OF THE TEST RESULTS

Case Studies

Transient flow phenomena do occur in gas transmission pipeline system due to some operational activities in the field such as flow rate adjustments, pipeline networks, and field instrument failures. Three cases are tested to evaluate the transient effect on pressure profiles along the pipeline system.

The pipeline system used in the test has horizontal orientation with the length of 5000 feet. One end of the pipeline is designated as the inlet at which the gas initially enters the pipeline, and the other end is called as outlet where the gas comes out from the pipeline system. The first case of the test is to evaluate the effect of transient phenomena on pressure profiles along the pipeline system when the valve located at the outlet is suddenly closed such that no flow comes out from the outlet or the fluid velocity at the outlet is zero. The second case of the test is quite similar to the first case. However, the valve at the outlet is partially closed such that the different constant flow rates can be assigned as the boundary condition. The resulting transient effects on pressure profiles along the pipeline are then studied. Finally, the third case of the test is to evaluate the effect of transient flow on the pressure profiles along the pipeline system when different size of pipeline diameter is used in the test performed similar to the first and the second cases. In addition, attempts were also made to evaluate the effect of some numerical parameters on the predicted results. In particular, effect of time step size and grid size on predicted results were studied. Parametric studies were also conducted to evaluate the effects of inlet gas flow rate and liquid holdup on the pressure transient behavior.

Discussion

The transported gas composition and its properties used in the test runs are presented in Table 1. This fluid data is obtained from the gas field, and for the purpose of this application, the composition of the gas has been lumped into four pseudo-components. The result of the lumping is presented in Table 2.

In this test, several cases of transient phenomena were made in a horizontal pipeline system by imposing different flow rates at the outlet end of the pipeline for a certain gas flow rates at the inlet. This experiment is very important, especially in analyzing the effect of back pressure in a pipeline when a certain flow rate is imposed on the system by accidental or incidental operational procedures. Furthermore, by knowing the transient phenomena in the pipeline, considerations for safety can be given in pipeline design, especially in the case of high gas flow rate where the pressure build-up caused by transient flow can be significant. For this particular test, a 5000-foot-long horizontal pipeline with constant diameter was considered as the test section of the pipeline. In order to examine the effects of transient flow in pipeline, several gas flow rates and pipe diameters were used in the test such that the importance of various transient effects caused by flow rates and pipeline geometry can be evaluated. Figure 1 shows the back pressure waves along the pipeline system caused by a sudden valve closure at the end of a pipeline where the velocity suddenly becomes zero. The sharp change in pressure caused by back pressure wave can be seen from the plot, and this pressure wave moves backward along the pipeline as the time increases. From the plot, the maximum pressure buildup can also be seen when a sudden closure in pipeline is takes place. For the pipeline design purposes, the predictions of these transient phenomena are very important to give an optimum safety factor for gas transmission pipelines. Similar runs are made for different inlet gas velocities. It is found that for lower inlet gas velocities, the maximum pressure buildup due to back pressure is lower than those for higher inlet gas velocities. Figure 2 shows the effect of back pressure on two different pipeline diameters at the same amount of gas flow rate. From this experiment, it is found that the pressure wave in the pipeline with diameter of 14 inches is relatively higher than that of the pipeline with diameter of 26.75 inches. This phenomenon is logical because compression due to back pressure will increase density more sharply in the smaller-diameter pipe because of low storativity. This difference causes higher pressure buildup in smaller diameter pipeline when the same gas flow rate is imposed on both pipelines.

The plot of pressure wave caused by back pressure at various inlet gas flow rate is presented in Figure 3. It shows that pressure buildup caused by higher inlet gas flow rate is higher than that of lower inlet gas flow rate. This is reasonable since more gas means higher density and hence lower compressibility.

The test section is discretized into 20 grid blocks and the time step used in this calculation is 0.005 second. From this experiment, it is also found that degree of stiffness of the pressure wave increases as the gas velocity increases. For higher flow rates, smaller grid size and time step size are recommended in the numerical calculations in order to have better accuracy. In order to evaluate the effect of numerical schemes on the prediction of the pressure profile, production runs were made using different grid sizes and time step sizes for the same case. The plots presented in Figure 4 shows the effect of different grid size on the predicted pressure transient caused by back pressure in the pipeline. In this analysis, 20 grids and 40 grids were used in the numerical calculations. Both curves are very close to each other and they have the same trend in oscillation caused by pressure wave. Figure 5 shows the predicted pressure profile in the pipeline at the elapsed times of 2.030 sec, after valve closure, with results presented for three different time step sizes of 0.002, 0.005, and 001 second. The results show that although pressure oscillation is relatively higher for bigger time step size, however overall predictions using these three time step sizes have relatively small deviations. This shows that the numerical scheme used is quite stable.

The transient effect of flow rate changes or pipe closure is felt at different points in the pipeline at different times with the point closer to the valve experiencing the effect the earliest. This phenomenon can be seen clearly from the plot of predicted pressure versus elapsed time at different position in the pipeline system shown in Figure 6 and Figure 7 where the shock fronts of the predicted pressure indicate the time when back pressure is felt at that point.

Another back pressure analysis on the transient model was also conducted at the same pipeline geometry by imposing various gas flow rates at the outlet while maintaining a constant pressure at the inlet of pipeline system. The purpose of this test is to analyze the effect of back pressure caused by gas flow rate changes at the outlet of the pipeline system. This experiment is commonly found in field practice where the flow change in the field could take place intentionally or accidentally. The results of this test is presented in Figure 8 where three different gas velocities, 0, 2 and 4 ft/s are imposed at the outlet with the initial gas velocity of 8 ft/s. It shows that the pressure buildup caused by back pressure increases as the imposed gas flow rate at the outlet decreases. The smaller gas flow rate imposed at the outlet the bigger steepness of the pressure wave obtained. This causes the problem to be more difficult to solve numerically.

The effect of liquid phase volume on the back pressure is also examined using the transient model. As expected, the pressure buildup caused by the higher liquid holdup is higher than that caused by the lower liquid holdup. This phenomenon is obvious since the the overall density of fluid system increases as the liquid holdup increases, so that the overall compressibility of the higher liquid holdup is lower than that of the lower one. This analysis shows that the correct information about the percentage of liquid holdup in the system would be a significant contribution for the pipeline design where safety and optimal field operations are the main concern.

CONCLUSIONS

One-dimensional compositional two-phase flow model has been developed in this study using the basic hydrodynamic formulation, to describe the problem of gas condensate flow in pipelines. Numerical methods are implemented in ODE solvers for solving the system of ordinary differential equations arising from the transient models.

The transient model has demonstrated its capability to predict the shock waves which arise in a two-phase transmission line when a sudden change of the flow rate or the pressure is imposed on the system either accidentally or by design. The use of back pressure test to evaluate the transient model is accomplished. From this test, the maximum back pressure effect due to flow rate or pressure changes on the system is evaluated by using parametric studies. This parameter is very important for designing a safe pipeline.

From the numerical tests conducted on the transient model, it is found that the grid size used in the model has little effect on the results predicted, which is more than the effect of the time step size. The size of time step used in the model is determined by the coefficient ratio between two consecutive variables in the specified grid point system. The larger the coefficient ratio in the system the smaller time step required in the calculation.

Further study for improving this model should include the effect of various flow patterns when the amount of liquid phase and the hydrodynamic behavior change significantly along the pipeline system. For more complex pipeline system, implementation of a two-dimensional model would improve the capability of the model to predict two-phase flow performance in a gas-condensate transmission system.

NOTATION

a = first (energy) parameter of EOS
A_p = cross-sectional area of pipe, ft^2
b = second (volume) parameter of EOS
c = third (volume) parameter of EOS
C_D = drag coefficient
C_p = specific heat, BTU/°F–lb-mole
d_p = droplet diameter, ft
d = internal pipe diameter, ft
f_g = Fanning friction factor of gas phase
f_L = Fanning friction factor of liquid phase
F_{dg} = interfacial drag force on gas, lb_f/ft^3
F_{dL} = interfacial drag force on liquid, lb_f/ft^3
F_{gg} = gravitational force due to gas phase, lb_f/ft^3
F_{gL} = gravitational force due to liquid phase, lb_f/ft^3
F_{mg} = force due to phase change from gas to liquid, lb_f/ft^3
F_{mL} = force due to phase change from liquid to gas, lb_f/ft^3
F_{gg} = gravitational force due to gas phase, lb_f/ft^3
F_{gL} = gravitational force due to liquid phase, lb_f/ft^3
F_{wg} = wall friction force of gas phase, lb_f/ft^3
F_{wL} = wall friction force of liquid phase, lb_f/ft^3
g = gravity acceleration, ft/s^2
g_c = conversion factor, 32.174 $lb_m.ft/lb_f.sec$
G_m = total mass flow rate, lb_m/s
H = enthalpy, BTU/lb
K = slope value of Runge-Kutta method
L = pipeline length
m_g = gas mass fraction
m_L = liquid mass fraction
M = molecular weight, lb_m/lb–mole
$\dot{m}_{gL}$ = mass rate of phase change from gas to liquid, $lb_m/s{\cdot}ft^3$
$\dot{m}_{Lg}$ = mass rate of phase change from liquid to gas, $lb_m/s{\cdot}ft^3$
$\underline{p}$ = pressure, psia
p = average pressure, psia
R = universal gas constant
Re = Reynolds number
t = time, s
T = absolute temperature, R
U = overall heat transfer coefficient, $BTU/hr.ft^{2}\cdot{}^{\circ}F$
v = velocity, ft/s
V = molar volume, ft^3/Lb-mole
We = Weber number.
x = distance, ft
z_i = mole fraction of component i in feed

Greek Letters

α = in situ volume fraction
Δ = incremental distance, ft
ε = pipeline roughness, ft
η = Joule-Thompson coefficient
κ = equilibrium ratio
μ = viscosity, cp
ρ = density, lb_m/ft^3
σ = surface tension, lb_f/ft
ω_i = acentric factor of component i
θ = angle
ξ = tolerance

Subscripts

c = critical condition
g = gas
L = liquid
o = inlet or upstream condition
s = surrounding
1 = downstream

Superscripts

n = new time level
o = old time level

LITERATURE CITED

1. Adewumi, M. A. and Bukacek, R. F., Journal of Pipelines, Vol. 5, p. 1, (1985).

2. Adewumi, M.A. and Mucharam, L., SPEPE J., p. 85, (Feb. 1990).

3. Peng, D. and Robinson, D. B., Ind. Eng. Chem. Fundam., Vol. 15, No. 1, p. 59, (1976).

4. Adewumi, M. A. and Arastoopour, H., The 1st World Congress on Particle Technology, Nuremberg (F.R.G.) (April 1986)

5. Mucharam, L.: "One-Dimensional Compositional Modeling of Gas and Condensate Flow in Pipelines," PhD Thesis, Penn State U., (1990).

6. Coulter, D. M. and Bardon, M. F., Oil and Gas J., p. 107, (February 1979).

7. Chen, N. H., Ind. Eng. Chem. Fundam., Vol. 18, No. 3, p. 296, (1979).

8. Hughes, E.D., Lyczkowski, R. W. and McFadden, J. H., EPRI, NP-143, Vol. 2, (February 1976).

9. Ulke, A. (Eds : Veziroglu, T. N. and Bergles, A. E.), Multi-Phase Flow and Heat Transfer III. Part A. Fundamentals, Elsevier, p. 59, (1984).

10. Moeck, E. D., AECL, 3650, (1970).

11. Tiwari, S. N., Abolhassani, J. S., and Smith R. E., 4th Int. Conf. On App. Numerical Modeling, Tainan, p. 759, (Dec. 1984).

12. Adewumi, M. A. and Arastoopour, H., AIChE Symposium Series, no. 257, volume 83, (1987).

13. Iannelli, G. S. and Baker, A. J., 26th Aero. Sci. Mtg, (Jan. 1988)

Table 1. Composition and physical properties of the Iranian gas.

Comp.	Zi	Tci	Pci	MWi	ω_i	Vci
H2O	0.0007	1165.14	3207.90	18.06	0.344	0.9162
CO2	0.0249	548.16	1070.16	44.00	0.225	1.4820
N2	0.0007	195.76	492.45	28.00	0.040	1.4031
C1	0.8485	343.68	667.38	16.00	0.008	1.5608
C2	0.0572	550.32	708.54	30.07	0.098	2.3333
C3	0.0265	666.24	615.93	44.10	0.152	3.2004
iC4	0.0049	735.18	529.20	58.12	0.176	4.1463
nC4	0.0105	765.96	551.25	58.12	0.193	4.0201
iC5	0.0041	829.32	490.98	72.00	0.227	4.8242
nC5	0.0046	845.88	489.51	72.00	0.251	4.7927
nC6	0.0041	868.00	419.00	76.20	0.260	4.8800
nC7	0.0033	1150.00	410.00	112.40	0.280	5.3000
nC8	0.0028	1180.00	395.00	133.00	0.295	5.5000
nC9	0.0020	1070.54	331.80	128.26	0.444	8.7729
nC10	0.0010	1111.60	304.40	142.29	0.490	9.6612
nC11	0.0006	1149.84	285.18	156.18	0.535	10.4053
nC12	0.0003	1184.94	264.60	170.34	0.562	10.3785
nC13	0.0002	1216.45	249.90	184.37	0.623	12.2972
nC15	0.0003	1272.61	220.50	212.42	0.706	13.8738

Table 2. Assigned composition and properties of the lumped system.

Comp.	Zi	Tci	Pci	MWi	ω_i	Vci
C1	0.8485	343.68	667.38	16.00	0.008	1.5608
HYP1	0.1127	620.04	633.60	40.91	0.143	3.1364
HYP2	0.0125	1135.50	425.18	120.15	0.408	8.2219
HYP3	0.0263	575.04	1160.82	30.89	0.223	1.4657

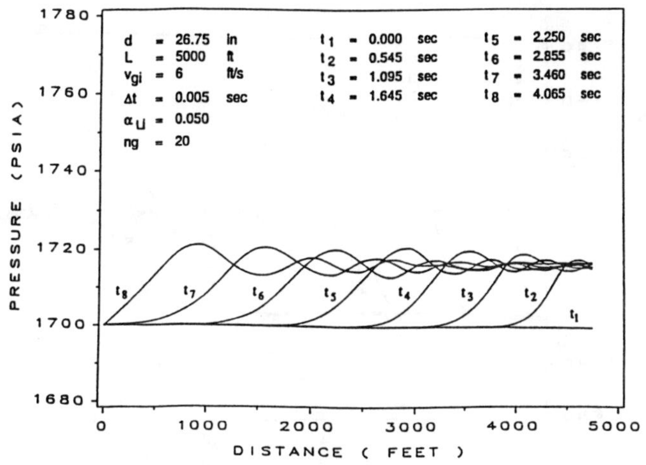

Figure 1. Pressure wave movement in the pipeline for 26.75 in. pipe diameter and 6 ft/s inlet velocity.

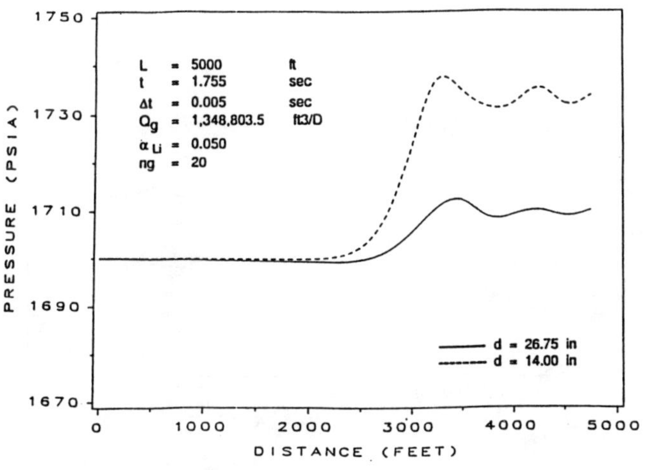

Figure 2. Pressure buildup in pipeline caused by back pressure in 14 and 26.75 in. pipeline diameters at elapsed time of 1.755 sec.

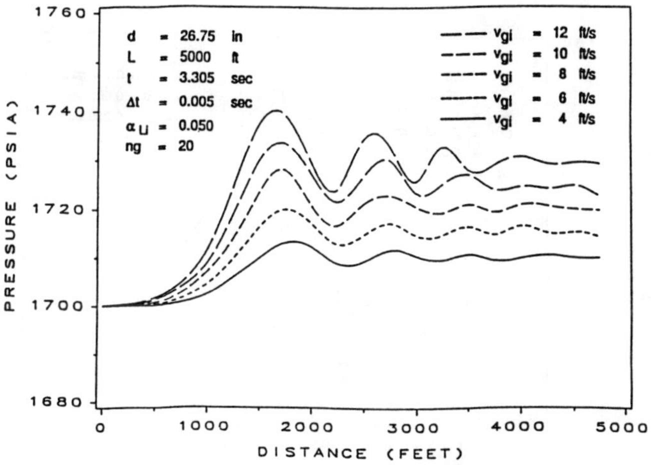

Figure 3. Pressure buildup in pipeline caused by back pressure at various inlet gas flow rates.

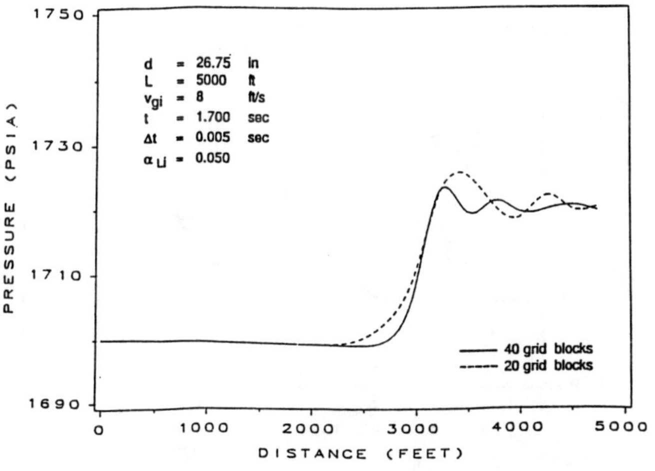

Figure 4. Predicted pressure profiles in the pipeline using two different grid blocks, 20 and 40, at elapsed time of 1.7 sec.

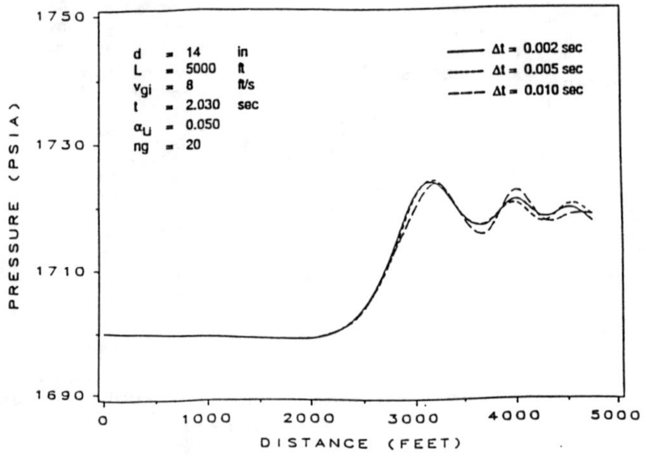

Figure 5. Predicted pressure profile in the pipeline using three different time steps, 0.002, 0.005 and 0.01, at elapsed time of 2.03 sec.

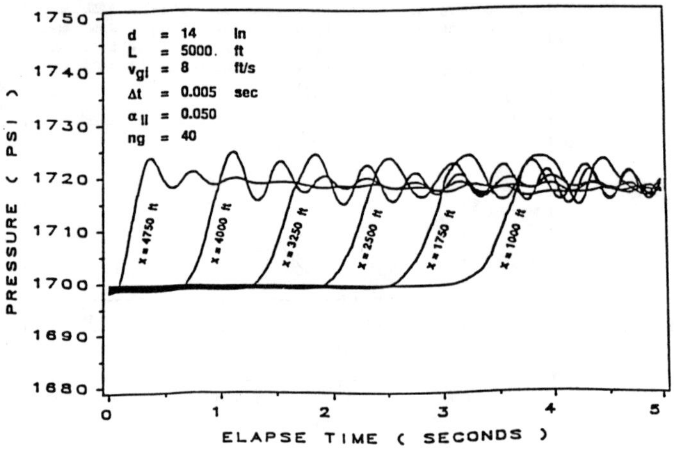

Figure 7. Predicted pressure history at six different positions in the pipeline.

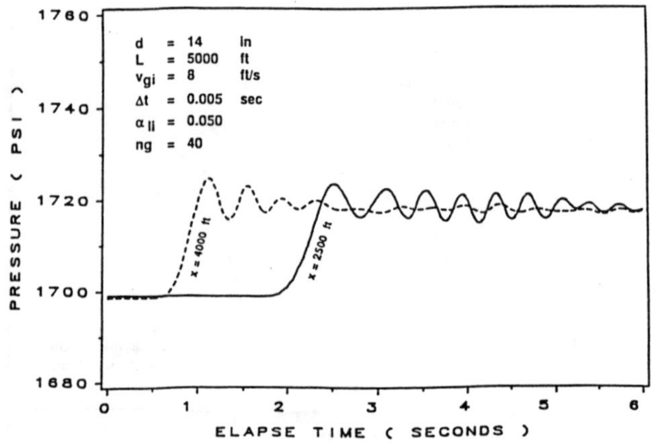

Figure 6. Predicted pressure history at two different positions in the pipeline.

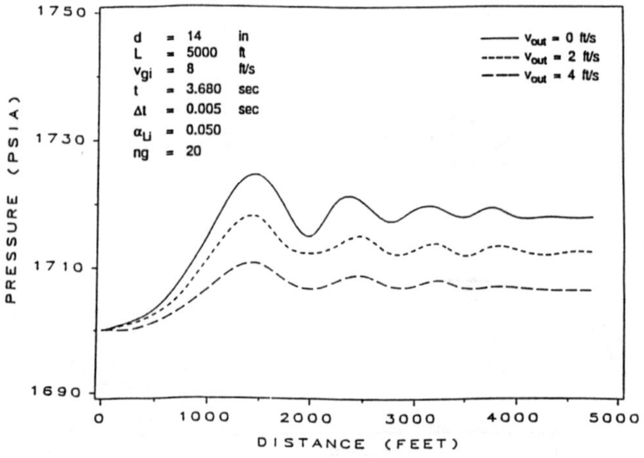

Figure 8. Predicted pressure profiles along the pipeline as the result of three different gas flow rate adjustments at the outlet.

ANALYSIS OF WELLBORE HYDRAULICS IN AIR AND MIST DRILLING USING A MULTIPHASE FLOW MODEL

Shifeng Tian and Michael A. Adewumi ■ The Pennsylvania State University, 202 Mineral Science Building
University Park, PA 16802

A fundamental wellbore hydraulics model is developed to simulate the pneumatic transport of solid cuttings in the drill-string/wellbore annulus. The model incorporates most of the parameters involved in air and/or mist drilling and forms the basis for a predictive tool for the optimal lifting velocity, an essential ingredient in the optimal design of the air and/or mist drilling program. Extensive parametric analysis of the system is performed to explore the predictability of some of the phenomena associated with air and/or mist drilling. The model is demonstrably capable of predicting pressure drop profile, velocity profiles, concentration distributions in the annulus and the associated phenomena under various simulated drilling conditions. Model prediction shows very good agreement with experimental data.

Drilling with air or gas in place of mud as a circulating fluid has several advantages including significantly higher penetration rate, longer bit life, substantial savings in cost and rig-up time, minimization of damage to wellbore, and elimination of some of the problems commonly associated with mud drilling such as lost circulation, etc. These advantages have made significant difference in many field cases [1-11]. The benefits associated with air drilling in comparison with conventional mud drilling are well documented and its potential is even more far reaching if this technology can be effectively utilized. Even though the technology of air drilling is not new, its hydraulics are radically different from those of conventional mud drilling. While extensive technical information is available for mud drilling wellbore hydraulics, the same can not be said of air drilling.

Fundamental technical information on air drilling hydraulics is scanty, even though the technology is widely used in the Eastern part of the United States (for drilling oil and gas wells) and in the western United States for Geothermal wells. In fact, reliable estimates claim that air drilling technique could be effectively used for more than 30 % of the wells drilled in the United States, thus resulting in tremendous costs and time savings. The major inhibition to its acceptance and widespread use is mainly the absence of systematic predictive and design procedure. Rule of thumb and empiricism is still the dominant method used by engineers to design an air drilling program. The lack of a systematic design procedure has prompted several investigators to examine this problem [13, 15, 16]. More recently, this problem has been re-visited by researchers in an attempt to answer many unanswered questions, to foster a greater understanding of the physics involved and to create a rational basis for the development of fundamental predictive strategy [14, 17, 18, 19, 20, 21].

Another inhibition to the wider use of air drilling is the difficuty for air drilling to deal with large formation water invasion. However, mist drilling has been successfully used to handle such cases for up to 100-200 barrels per hour of water invasion [6, 10]. Mist drilling basically converts the invasion water into small water droplets, which are dispersed in the flowing air stream. This is usually achieved by injecting a chemical, usually a surfactant (e.g. soap) into the inlet air stream. This reduces the air-water interfacial tension, thus enhancing the process of mist formation. There is even less technical information about mist drilling available in the literature. In this work, we modified the multiphase hydrodynamic model, which has been used to simulate dry air drilling wellbore hydraulics [19, 20, 21], to investigate mist drilling hydraulics.

Air or mist, as the circulating fluid, performs all the traditional functions of drilling mud, particularly hole cleaning. The importance of hole cleaning cannot be overestimated in determining drilling efficiency. In order for air and/or mist, or any other drilling fluid, to be effective in performing this task, it must be circulated in adequate quantity. On the other hand, circulating more fluid than is required will result in unnecessary additional compression or pump load. The key to realizing a cost reduction is to determine optimal fluid volumetric flow rate. It has since been demonstrated, by experimental measurements and field observations, that the optimal cuttings lifting velocity is not constant but rather dictated by operating and lithological parameters [15, 18]. This predicament and the need to evolve fundamental physics-based predictive/design model for wellbore hydraulics in air and mist drilling motivated this work. Multiphase hydrodynamic approach is used to describe wellbore hydraulics associated with air and mist drilling.

THE HYDRODYNAMIC APPROACH

In a complete departure from the approaches utilized hitherto, the authors [19, 20, 21] applied the theory of multiphase hydrodynamics to simulate air drilling wellbore hydraulics. The results obtained are very encouraging. These analyses reveal the existence of some interesting phenomena which should be very important to the design of an optimal air drilling program. A brief recapitulation of the salient elements of the basic framework for the hydrodynamic description of air and mist drilling wellbore hydraulics is called for in order to provide a focal point for the ensuing discussion.

The Hydrodynamic Variables

The main characteristic that differentiates air and mist drilling from conventional drilling is that air or mist is used as the circulating fliud respectively instead of mud. One major purpose of this circulating fluid is the transportation of the solid "cuttings" from the bottom of the hole to the surface thereby cleaning the hole. In the case of mist drilling, the invading water needs to be removed in addition to the cuttings. The basic practical design variables for air and mist drilling program include optimal air volumetric flow rate and minimum power requirements. The independent variables which characterize the transport of solid particles in the wellbore are: (1) The physical properties of the drill cuttings, including size and size distribution, shape and density. (2) Thermo-physical properties of the transporting fluid such as density and viscosity. (3) Penetration rate, as this will determine the feed rate of solid cuttings or solid loading. (4) The rate of formation water invasion, which determines the quantity of liquid that needs to be removed from the wellbore. (5) The geometrical configuration of the wellbore/drill-string annulus and size of the various components. (6) Fluid transport velocity.

The basic design parameters which are dictated by the above independent variables include: (1) Optimal transport velocity (2) Associated pressure loss and/or power requirement. In addition to these however, one should possess the ability to predict choking phenomenon because of its importance in this system. In order to take all these variables into consideration, a hydrodynamic approach is proven to be viable.

Discussion of the Model

Formulation of the generalized hydrodynamic model using the multiphase approach hinges on the assumption that each phase can be described as a continuum. Once this assumption is admissible, the basic laws of continuum mechanics can be invoked. These will yield local conservation equations which can be averaged using one of the averaging techniques to yield a system of governing conservation equations. Details of these have been presented variously in the literature [22, 23]. Specific application and formulation for wellbore hydraulics in air drilling have been presented earlier by the authors. Its extension and application to mist drilling will be presented here. The assumptions factored into the development of this wellbore hydraulics model are enumerated below:

1. Solid particles of uniform size and density constitute a phase.

2. Invading water forms uniform size of droplets and hence constitutes a phase.

3. The system is isothermal and there is no interphase mass transfer.

4. The particulate and the liquid phases are assumed to be incompressible while the gas phase behavior is describable by the ideal gas equation of state.

5. Pressure gradient is assumed to exist in the gas phase only.

6. The tangential and radial components of the velocity vector are assumed to be negligible. For this reason, pressure is assumed to vary only in the axial direction.

The first and second assumptions enable us to define one gas phase, one liquid phase and several distinct particulate phases. Associated with each phase is a set of hydrodynamic parameters such as density, phasial velocity and volumetric fraction. In addition, the third assumption enables us to eliminate the energy equation thereby reducing the number of partial differential equations defining our problem. The assumption of incompressible solid and liquid phases need no justification while the use of ideal gas equation of state to describe the gas phase is based on the factor that the pressure in the annulus is not very high. Since the fifth assumption assigns all pressure gradient to the gas phase, then only one pressure is assigned to the system. The assumption of negligible radial and tangential components of the velocity field further simplifies our system of equations since the terms involving these velocities are dropped. The resulting governing equations are summarized below:

Continuity Equations

$$\frac{\partial[\alpha_g \rho_g V_g]}{\partial z} = 0 \quad (1)$$

$$\frac{\partial[\alpha_l \rho_l V_l]}{\partial z} = 0 \quad (2)$$

$$\frac{\partial[\alpha_{si} \rho_{si} V_{si}]}{\partial z} = 0 \quad i=1,2,...,n \quad (3)$$

Momentum Equations

$$\frac{\partial[\alpha_g \rho_g V_g^2]}{\partial z} = \begin{array}{l} -\frac{\partial P}{\partial z} + \frac{\mu_g}{r}\frac{\partial}{\partial r}\left[r\alpha_g \frac{\partial V_g}{\partial r}\right] \\ -F_{Dgl} - \sum_{i=1}^{n} F_{Dgsi} - \alpha_g \rho_g g \end{array} \quad (4)$$

$$\frac{\partial[\alpha_l \rho_l V_l^2]}{\partial z} = \begin{array}{l} \frac{\mu_l}{r}\frac{\partial}{\partial r}\left[r\alpha_l \frac{\partial V_l}{\partial r}\right] \\ +F_{Dgl} - \sum_{i=1}^{n} F_{Dlsi} - \alpha_l \rho_l g \end{array} \quad (5)$$

$$\frac{\partial[\alpha_{si} \rho_{si} V_{si}^2]}{\partial z} = \begin{array}{l} \frac{\mu_{si}}{r}\frac{\partial}{\partial r}\left[r\alpha_{si} \frac{\partial V_{si}}{\partial r}\right] \\ +F_{Dgsi} + F_{Dlsi} - \alpha_{si} \rho_{si} g \quad i=1,2,...,n \end{array} \quad (6)$$

where,

$$\rho_g = \frac{PM_g}{RT} \quad (7)$$

$$\alpha_g + \alpha_l + \sum_{i=1}^{n} \alpha_{si} = 1 \quad (8)$$

The following expression is commonly used for the drag forces [23].

$$F_{Dgj} = \frac{3}{4}\frac{C_{Dgj}\rho_g(V_g - V_j)\,|V_g - V_j|\,\alpha_j \alpha_g^{-2.67}}{d_j} \quad (9)$$

for j=l,s1,s2,.....,sn

where,

$$C_{Dgj} = \begin{cases} \frac{24}{N_{Re_{gj}}}\left(1.0 + 0.15 N_{Re_{gj}}^{0.687}\right) & N_{Re_{gj}} < 1000 \\ 0.44 & N_{Re_{gj}} \geq 1000 \end{cases} \quad (10)$$

and,

$$N_{Re_{gj}} = \frac{\rho_g (V_g - V_j) d_j}{\mu_g} \quad \text{for } j=l,s1,s2,...,sn \quad (11)$$

In dry air drilling, even when the drilling speed is very high, the solid loading into the system is very low. The concentrations of the cuttings in the wellbore are very dilute. In the mist drilling case however, since water comes into the wellbore, the presence of cuttings together with the droplets may render the system less dilute, hence the hydrodynamic interactions between cuttings and droplets should not be ignored. The following equations are used to estimate the drag forces between the cuttings and the droplets.

$$F_{Dlsi} = \frac{3}{4}\frac{C_{Dlsi}\rho_l(V_l - V_{si})\,|V_l - V_{si}|\,\alpha_{si}\alpha_l \alpha_g^{-2.67}}{d_{si}} \quad (12)$$

for i=l,s1,s2,...,sn

where,

$$C_{Dlsi} = \begin{cases} \frac{24}{N_{Re_{lsi}}}\left(1.0 + 0.15 N_{Re_{lsi}}^{0.687}\right) & N_{Re_{lsi}} < 1000 \\ 0.44 & N_{Re_{lsi}} \geq 1000 \end{cases} \quad (13)$$

and,

$$N_{Re_{lsi}} = \frac{\rho_l (V_l - V_{si}) d_{si}}{\mu_l} \quad \text{for } i=1,2,.....,n \quad (14)$$

The Weber number is commonly used as a criterion for determining the stable droplet diameter. The relationship between Weber number and the maximum diameter of a stable droplet is given as:

$$d_l = \frac{W_e \sigma}{\rho_g (V_g - V_l)^2} \quad (15)$$

Hinze [26] and Isshiki [27] suggested the following expression to determine the Weber number.

$$W_e = 12\left[1 + \left(\frac{\mu_l^2}{\rho_l d_l \sigma}\right)^{0.36}\right] \quad (16)$$

As Equations (15) and (16) indicate, if the fluid properties, namely viscosity and density are fixed, the droplet diameter is determined by the surface tension of water, σ. This explains why chemical (e.g. surfactant) is injected into the system in the case of mist drilling. The droplet diameter is determined by the type and quantity of the chemical. This implies that the droplet diameter can be operationally controlled. In conducting parametric studies, the effect of droplet size was investigated.

Initial and Boundary Conditions

Clarification of the terminology "initial conditions" needs to be made here since the governing equations being solved are non-transient by definition. This is necessitated by the numerical method of lines (MOL) used for integrating these equations. This method is conventionally formulated for Initial-Value-Problems (IVP) such that the spatial variables are discretized while the time is left continuous. We apply MOL to our problem by treating the axial coordinate, z time-like while discretizing the radial coordinate, r. In so doing, the inlet conditions (at the bottom of the well) become the initial conditions, hence the terminology being used here. Details of this approach have been presented by Adewumi and Arastoopour [23]. Boundary conditions are then imposed in the radial coordinate.

Boundary Conditions. Equations (1)-(6) constitute the conservative form of the system of partial differential equations describing the system under consideration. By algebraic manipulation, these equations can be transformed into the non-conservative form in the primary dependent variables, namely V_g, V_l, V_{s1}, V_{s2}, ... , V_{sn}, α_l, α_{s1}, α_{s2}, ... , α_{sn} and P. The one remaining variable α_g is obtained from Equation (8). The following boundary conditions and initial conditions are imposed. At $r = R_{DP}$ and R_w, we impose the no-slip or small constant-slip conditions for the gas, liquid as well as the particulate phases. Additional boundary conditions are obtained by assuming that the gas phase occupies the very first thin layer of fluid adjacent to the wall thus making the gas void fraction at the wall close to unity, hence the liquid and solid fractions are close to zero.

Initial Conditions. Values of V_{go}, V_{lo}, V_{sio} (i = 1, 2, ... , n), α_l, α_{si} (i = 1, 2, ... , n) (hence α_{go}) and P_o are specified at the pipe inlet. In air drilling, the parameters usually monitored are the drilling rate, the volumetric air input and pressure drop (indirectly). Since the knowledge of gas and solid velocities, pressure and void fraction at the system inlet (i.e. at the bottom of the hole) is necessary to impose the required inlet conditions, these variables must be estimated from the available pieces of information. From the drilling rate, the volume of rock (solid) generated per unit time can be estimated; using this and knowing the average density of the rock, its mass flow rate can be estimated. A quick estimate of the bottomhole pressure can be made from the drill pipe pressure and an estimate of pressure loss in the drill string and bit orifices. Similarly, the gas velocity at the bottom of the hole can be estimated from the volumetric flow rate. This process provides the necessary inlet conditions to be imposed on our system of equations to define the problem. Thus, V_{go} and P_o can be inferred; Either V_{sio} or α_{sio} must be guessed for the other to be estimated from the value of W_{sio}. This procedure is necessitated by the lack of information since these parameters are not usually measured in drilling operations. However, parametric studies conducted using various values of V_{sio} (while α_{sio} are adjusted, within reasonable bounds, to meet the specified W_{sio}) show no significant effect on pressure drop and velocity profile, hence no erroneous results are expected from guessing these values. The value of α_l needs to be determined from the water invasion rate, which is dependent on the formation drilled. One of the purposes of this work is to study the effects of different rates of water invasion.

Numerical Solution

The method of lines was used in solving the above set of partial differential equations. This technique transforms a set of partial differential equations into a system of coupled ordinary differential equations by discretizing the dependent variables in all but one dependent variable. The resulting set of ordinary differential equations is then integrated by an appropriate technique. For a detailed discussion of the basic principle and mechanics of this method, see Adewumi and Arastoopour [23]. Numerical method of lines is implemented in FORSIM VI package [28]. This package was used in integrating the above set of quasi-linear partial differential equations.

DISCUSSION OF RESULTS

Figure 1 shows the geometrical configuration of the wellbore/drill string annulus. The entrance of the air-liquid-solid flow into the annulus is located at the bottom of the well just above the drill bit. The system chosen for simulation is typical of the South western part of West Virginia. Table 1 contains all the pertinent system description parameters used as input data into the model. The information is obtained from field operators in this area. Except where so indicated in any figure, all pertinent parameters are the same as those stated in Table 1.

The solid particles, the liquid and the transporting medium (air) are fed into the bottom of the system at predetermined rates. The basic dependent variables being predicted by the model are Pressure distributions, velocity distributions (solid, liquid and gas) , the solid and liquid concentration distributions and segregation behavior along the wellbore. All these need to be monitored for good understanding of the wellbore hydraulics. In the practical sense, the two parameters of interest to the operator are the pressure loss resulting from the solid transport and the quantity of solid transported. Both are easily deduced from the basic hydrodynamic parameters predicted by the model.

Dry Air Drilling

The effects of air flow rate on dry air drilling wellbore hydraulics have been extensively studied. Experimental results [14, 17, 18] confirmed that an optimal air velocity, which would yield the minimum total pressure loss in the annulus, does indeed exist in the transportation of rock particles in an annulus. The multiphase hydrodynamic model is able to predict this optimal air velocity for variaous air drilling conditions.

Figure 2 is a typical result from the hydrodynamic model and is illustrative of the relationship between total pressure drop in the wellbore and the air velocity at the bottom of the well. This shape of the curve is characteristic of vertical particulate transport. It goes through a minimum point, but is not symmetrical about it; rather, it is skewed such that the left leg has a higher average absolute slope than the right leg. The meeting point between the two legs represents the air flow rate that would yield minimum pressure loss in the wellbore annulus. This is why the corresponding air velocity at this point is referred to as the optimal lift velocity.

This optimal point also marks a boundary between two distinct regimes of air-solid behavior in the wellbore annulus. At a higher air flow rate than the optimal one, i.e., the right leg, the system tends to be very dilute and fast flowing. The particles attain high velocity and the wall frictional forces, which depend on absolute phase velocity, are dominant compared with the particulate phase gravitational force. As the air flow rate is decreased, particles have higher residence times in the system, which means that the solid particle inventory increases. This results in a greater particle concentration in the wellbore annulus and hence much larger gravitational forces which eventually dominate frictional forces. Any further air velocity reduction will cause a dramatic increase in particle inventory and concentration with a disproportionate increase in pressure loss in the wellbore.

A plot of corresponding total solid volume fraction as a function of air velocity is also presented in figure 2. It vividly demonstrates the above discussion. As the air velocity is decreased from a relatively high level, the solid concentration in the annulus increases gradually. However, after the optimal point is reached, further reduction of air velocity results in a very sharp increase in solid concentration coupled with a dramatic increase in pressure loss.

Extensive parametric study reveals that the total pressure loss in the annulus is significantly affected by gas flow rate, inlet pressure, drilling speed and solid density, while optimal air velocity is a strong function of cutting size. Large cuttings require higher optimal air velocity to lift.

In reality, the size distribution of drill cuttings should be close to being continuous. The hydrodynamic model provides a way to deal with different particle sizes. However, consideration of more solid phases involve extensive computer time. A more practical way to deal with the non-uniform size system is by using an average size to represent a narrow cut of particle size ditribution or, in other words, to discretize the continuous particle size distribution. For illustration purposes, here we present some results for a binary-sized system, which means that all the cuttings are classified as either coarse or fine particles. Therefor, the model used consists of Equations (1) and (14) [and n is set equal to 2] except thoes involving liquid and all liquid related terms are dropped.

All the curves in Figure 3 have the same characteristic shape as discussed earlier. The curves for 100% coarse particles and 100% fine particles provide the boundaries that envelope the curves for all fine-coarse mixture runs. It is remarkable to note that all the curves intersect at about the same point, which is referred to as the "critical point". For air velocities higher than the critical velocity, the pressure drop associated with the fine-particle system is higher than that of the coarse-particle system. For air velocities below it, the curves are reversed. One of the most striking characteristics of this system depicted in Figure 3 is the effect of the presence of coarse particles on the system behavior, particularly in the region below the critical point. In this region, the pressure drop curve for 50% coarse system is very close to the curve for 100% coarse rather than in between of the 100% coarse and 100% fine curves. Even the curve for 10% coarse is much closer to 100%-coarse curve than the 100%-fine cueve. As for the region above the critical point, the curves behave according to normal expectation. Observe in this figure that the 50-50% mixture curve lies almost half way in between the 100%-fine curve and the 100%-coarse curve. Other curves are in similar proportion.

Figure 4 shows two plots of the predicted optimal velocity as a function of the volume percentage of coarse particles present in the mixture. The fraction of coarse in the mixture is varied between 0% (uniform-sized fine-particle system) to 100% (uniform-sized coarse-particle system). The curves show a very sharp increase in the optimal velocity of the system as the fraction of coarse increases from zero. The rate of increase decreases rapidly as the fraction of coarse increases beyond this initial range. Thereafter, little change is experienced. What this depicts is that the coarse particles, even when present in very small fractions, determines the behavior of the system in terms of the optimal lift velocity.

Supon [14] invesigated air drilling wellbore hydraulics with a laboratory-scale model. The system description parameters are contained in Table 2. Supon measured optimal air velocity as a function of the percentage of coarse particles, from 0% to 100%. Figure 5 shows the comparison of the predicted results using the hydrodynamic model with Supon's experimental data. Considering the scatter in these data, the model's prediction is very good. Quite remarkable is the quanlity of prediction at lower values of coarse particle concentration where the change in optimal velocity with respect to the fraction of coarse particles is most rapid.

The dominant behavior of the coarse particle in the coarse-fine particle mixture discussed above is quite important, since it implies that the common practice of using some average particle size to characterize a non-uniform size system would be erroneous. A comparative analysis of the solid volume fraction distribution would help to understand this behavior.

Three cases studied are shown in Figure 6. Two of them are uniformly sized system (100% fine and 100% coarse) while the third is a 50% fine plus 50% coarse by volume mixture. As expected, at the same lift velocity, the coarse-particle system has more solid in the system than the fine-particle system, while the mixture system in between. This is mainly due to the factor that the bigger particles have higher residence time in the wellbore than the fine particles. At high lift velocity, there is virtually no difference between the three systems in terms of their concentration. This is due to the high drag forces exerted on the solid particles thus resulting in effective hole cleaning by the circulating air. As air velocity is decreased, however, particles will start to accumulate in the wellbore due to reduced drag force. While this air velocity may still be sufficient to lift the fine particles, it is not for the coarse particles. This causes a very sharp increase in particle concentration of the uniform coarse system at low air velocity as indicated in Figure 6. Quite striking though is the result for the 50-50% mixture. In this case, the model predicts that the overall solid distribution pattern follows very closely that of the 100% coarse particles. It is logical that this behavior would cause the pressure drop profile of the 50-50% mixture to be very close to that of the 100% coarse system as indicated in Figure 3.

Figure 7 shows the relative contributions of the coarse particles and the fine particles to the overall solid fraction in the 50% coarse-50% fine mixture system. In the region of low air velocities, the contribution of the coarse particles to the total solid fraction is dominant, since the fine particles can be cleaned out very quickly even at lower air velocities while the coarse particles accumulates in the wellbore. At higher velocities, on the other hand, both coarse and fine particles can be effectively cleaned, thus both would attain low concentration in the wellbore. The relative contributions of the coarse particles and the fine particles are comparable, even though the coarse particles contribution is still slightly larger.

The above analyses are for binary-sized particle systems. Ternary-sized particle systems are also investigated. These results reaffirm that the previous finding from the binary-sized particle system can be extended to a multi-sized particle system. In a multi-sized particle system, it is the coarsest particles, even when present in very small quantities, are the ones that dictate the hydraulics of the system.

Mist Drilling

In some areas, formation water invasion into wellbore during drilling is inevitable. The multiphase hydrodynamic model outlined in Equations (1)-(14) is used to simulate mist drilling. For demonstration purposes, here we only consider one particle size, so that all the solid particles constitute one phase and are described by a single set of hydrodynamic variables. From the previous discussion, this particle size is not simply an average of the cutting size distribution, rather it should be a weighted average size which would reflect the dominant effects of the coarse particles in the system.

The invading formation water is transported together with the drill cuttings by the air. in order to mantain the water in the system as a mist, some chemical must be injected into the system through drill pipe. The purpose of the injected chemical is to lower the water-air interfacial tention and thus enhance the breakeage of the invaded water into small droplets. It needs to be pointed out that although the terminology of "mist" drilling is commonly used in the industry (and hence adhered to in this paper), the droplet size is not necessary within the range normally classified as mist in the chemical engineering literature.

Figure 8 compares the predicted pressure drop profiles from dry air drilling and mist drilling. All the other conditions for both cases are the same, except that a constant flow rate of 27 bbl/hr water is imposed on the system in mist drilling case. The assumed stable droplet size in this case is 0.228 mm. The optimal air velocities in both cases are very close to each other. This implies that smaller droplets, like fine paticles, have little effect on the optimal air velocity of the system. Additional water, as expected, results in higher pressure loss when the air velocity is above the optimal point. However, when air velocity is below the optimal point, the two curves get closer to each other as the air velocity decreases. If the air velocity is further reduced, the two curves may cross each other at some low air velocity. This means that at high air velocity the invading water simply constitutes more load to be transported, while at lower air velocity the mist in the system may help to increase the cuttings-carrying capacity of the air. The case of a higher water invading rate (Q_l = 50 bbl/hr) is shown in Figure 9. More water causes higher pressure loss or power requirement. The effect of water invasion rate on optimal air velocity is very insignificant.

Figure 10 shows the volume fraction profiles corresponding to the case with Q_l = 27 bbl/hr. At higher air velocity, the particle volume fractions in dry air drilling and mist drilling are basicaly the same. The liquid fraction is much higher than that of the particles. The interesting thing is that when the air velocity is below the optimal point, the particle fraction profiles in air and mist drilling start to increase very sharply, but at the same time deviate from each other, with the particle fraction in dry air drilling being much higher than that in mist drilling. The increase in the liquid fraction is not so sharp. It indicates that because of the smaller diameter and lighter density of the liquid, the droplets attain a higher velocity than the particles do. At higher air velocity, both particles and droplets can be transported by the air and the higher liquid velocity does not help to further increase the carrying capacity of the air. At low air velocity, however, the drag force resulting from the velocity difference between the droplets and the particles does help to increase the cuttings-carrying capacity of the air. This explains why at low air velocity the pressure loss in mist drilling is not significantly higher than that in dry air drilling as indicated in Figure 8.

As mentioned earlier, the droplet size in mist drilling is dependent on the type and amount of the chemical injected into the wellbore. Figure 11 compares two mist drilling cases with different droplet sizes. The liquid incursion flow rates as well as all the other conditions in both cases are the same. The pressure losses in the larger droplet diameter case are much higher than that with small droplets at low air velocities. However, at the high end of the air velocity, the two curves cross to each other. As discussed earlier, at low air velocity, the gravity force is the dominant one with respect to the pressure loss. Smaller droplets attain higher velocity at the same air velocity, hence a small concentration in the system, which in turn results in a lower pressure loss. On the other hand, at high air velocity, friction and drag forces are the dominant forces in the system. Smaller droplets, taken as a whole, have a larger total surface area, which in turn causes higher friction loss, since the liquid mass flow rates in both cases are the same.

Figure 12 shows the corresponding velocity profiles of particles and droplets. All the curves show a linear relationship between particle (and droplet) velocity and the air velocity. In the smaller droplet size case, the droplet velocity is higher than the particle velocity. The reverse is the case for the larger droplets. This results in a positive drag force between the droplets and the particles in the smaller droplet case and a negative one in the larger droplet case. Hence the particle velocity in the smaller droplet case is higher than that in the larger droplet case. It needs to be pointed out that even though the particle velocity difference between the two cases is very small compared with the particle velocity values at high air velocity, it is very siginificant at low air velocity. This explains the difference between the particle volume fraction for the two cases as indicated in Figure 13 at low air velocity. The volume fraction profiles shown in Figure 15 reflects the same concepts as discussed earlier. Smaller droplets helps to increase the cutting carrying capacity of the air at low air velocities, thereby resulting in a lower particle concentration in the system.

CONCLUSIONS

A multiphase hydrodynamic model has been used to study the wellbore hydraulics of air and/or mist drilling. One of the most important findings of this work is that the practice of using some average particle size to characterize non-uniform size system could be very erroneous since even a small degree of non-uniformity in particle size appears to affect the system behavior quite drastically. Several phenomena were identified through production runs using this model. Comparison of the predicted results with experimental data has demonstrated the descriptive and predictive capabilities of the hydrodynamic approach. Several specific conclusions are also inevitable from this study and the most salient ones are enumerated below.

1. There exists an optimal air velocity which would yield the minimum pressure loss during the transportation of drill cuttings. Such optimal velocity is predictable using the model presented here. When the lift velocity is decreased below the optimal velocity, pressure loss increases very sharply. It is not advisable to operate at lift velocity below the optimal velocity.

2. Size of drill cuttings is critical to the value of the optimal air velocity. In a multi-sized particle system, the value of the optimal lift velocity is essentially determined by the size of the coarsest particles even when these are present only in very small fraction.

3. Whereas at low values of coarse particle fractions, the effect of coarse particle fraction on the optimal air velocity is very sharp, at higher values, the effect is less drastic. Predicted optimal velocity, as a function of the binary size distribution compares very well with the available experimental data.

4. When large quantity groundwater invasion occurs during drilling, additional power is needed to lift the water with solid particles. Conversion of the invading water into small droplets by injecting some chemical into the system helps to reduce the power requirement. However, at high air velocity, the effects of the droplet size on the power requirement become less significant.

NOTATION

C_D = drag coefficient
D = well depth
d_s = particle diameter
d_l = droplet diameter

F_D = drag force term
g = gravitational acceleration
M_g = gas molecular weight
n = number of particulate phases
P = pressure
r = radial coordinate
R = gas constant
R_{DP} = drill pipe radius
R_w = radius of the well
T = absolute temperature
V = velocity
W_e = Weber number
W_s = solid mass flow rate
z = axial coordinate

Greek Symbols

α = volume fraction
μ = viscosity
ρ = density
σ = surface tension of water

Subscripts

g = air or gas phase
l = liquid phase
si = i-th solid phase

LITERATURE CITED

1. Wilson, G. E., World Oil, p. 128, (August 1951).

2. Nicolson, K. M., Drill. & Prod. Pract., p. 300, (1953).

3. Jackson, G., Can. Min. & Met. Bulletin, p. 16, (Jan. 1961).

4. Toombs, W. E., Drill. & Prod. Pract., p. 7, (1962).

5. D'Agostino, J. D., World Oil, p. 91, (June 1974).

6. Hook, R. A., L. W. Cooper, and E. W. Payne, World Oil, p. 95, (April 1977).

7. Hook, R. A., L. W. Cooper, and E. W. Payne, Oil and Gas J., p. 86, (June 20,1977).

8. Bowen, R. C. and W. G. Parkhouse, Pet. Eng. Int., p. 106, (Sept. 1978).

9. Parkhouse, W. G. and H. A. Teesdale, Pet. Eng. Int., p. 120, (March 1982).

10. Parkhouse, W. G. and H. A. Teesdale, Oil and Gas J., p. 104, (Sept. 24, 1984).

11. Sheffield, J. S. and J. J. Sitzman, Oil and Gas J., p. 133, (Sept. 9, 1985).

12. Angel, R. R., Trans. AIME, 201, p. 325, (1957).

13. Mitchell, R. F., J. Pet. Tech., p. 2120, (1983).

14. Supon, S. B., An Experimental Study of the Annulus Pressure Drop in a Simulated Air Drilling Operation, Ph.D Diss., Penn State U., (1989).

15. Machado, C. J. and C. U. Ikoku, J. Pet. Tech., p. 2645, (1982).

16. Sharma, M. P. and D. V. Chowdry, S.E. Conf. on Theoretical & App. Mech., p. 205, (May 1984).

17. Supon S. B., N. Fairchild, and M. A. Adewumi, Proc. of the SPE Eastern Reg. Mtg., p. 119, (1987).

18. Supon, S. B. and M. A. Adewumi, Proc. of the SPE Eastern Reg. Mtg., p. 263, (Oct. 1989).

19. Adewumi, M. A. and S. Tian, Proc. of the SPE Eastern Reg. Mtg., p. 253, (1989).

20. Adewumi, M. A. and S. Tian, Proc. of the SPE Eastern Reg. Mtg., p. 151, (1990).

21. Tian, S. and M. A. Adewumi, Proc. of The CIM/SPE Int. Tech. Mtg, Calgary, Canada, p. 12-1 (June 1990).

22. Soo, S. L., Particulates and Continuum: Multiphase Fluid Dynamics, Hemisphere Publishing Corporation (1989).

23. Adewumi, M. A. and H. Arastoopour, AIChE Symp. Series, No. 257, Vol. 83, p. 199, (1987).

24. Adewumi, M. A. and H. Arastoopour, J. Powder Technology, Vol. 48, p. 67, (1986).

25. Adewumi, M. A. and H. Arastoopour, the Int. J. Fluid Mech., 3 (1), p. 53, (1990).

26. Hinze, J. O., App. Sci. Res., Vol. A1, p. 275, (1948).

27. Isshiki, M. N., Report 35, Transportation Technical Research Institute, Tokyo, Japan (July 1959).

28. Carver, M.B., D.G. Stewart, J.M. Blair and W.N. Selander, The FORSIM VI Simulation Paackage for the Automated Solution of Arbitrarily Defined Partial and/or Ordinary Differential Equation Systems, AECL-5821, (February 1978).

Table 1. Air and mist drilling system parameters.

Radius of the Drill Pipe	= 0.04445 m ($3\frac{1}{2}$ -*inch DP*)
Radius of the Wellbore	= 0.135 m ($10\frac{5}{8}$-*inch hole*)
Depth of the Well Simulated	= 5.0 m
Density of Cuttings	= 2650 kg/m^3
Cuttings Diameter (coarse)	= 0.911 mm
Cuttings Diameters (fine)	= 0.228 mm
Solid Viscosity	= 0.001 Pa.s
Air Viscosity	= 0.000018 Pa.s
Solid Mass Flow Rate	= 1.54 kg/s
Drilling Rate of Penetration	= 36.58 m/hr. (2 ft/min)
Bottomhole Pressure	= $5x10^5$ Pa
Inlet Solid Velocity	= 0.285 m/s
Inlet Solid Volume Fraction	= 0.04
Water Invasion Rate •	= 4.29 m^3/hr (27 bbl/hr)
Droplet diameter •	= 0.228 mm
Water Viscosity •	= 0.001 Pa.s
Inlet Droplet Velocity •	= 0.285 m/s
Inlet Liquid Volume Fraction •	= 0.08

• Mist drilling only.

Table 2. Supon's experimental system parameters.

Radius of the Drill Pipe	= 0.0254 m (1-*inch DP*)
Radius of the Wellbore	= 0.0445 m ($1\frac{3}{4}$-*inch hole*)
Depth of the Well Simulated	= 7.3152 m (24 *ft*)
Density of cuttings	= 2650 kg/m^3
Cuttings diameter (coarse)	= 0.983 mm
Cuttings diameter (fine)	= 0.254 mm
Solid Viscosity	= 0.0002 Pa.s
Air Viscosity	= 0.000018 Pa.s
Average Solid Mass Flow Rate	= $6.24x10^{-3}$ kg/s
Bottomhole Pressure	= $2.5x10^5$ Pa
Inlet Solid Velocity	= 0.285 m/s
Inlet Void Fraction	= 0.998

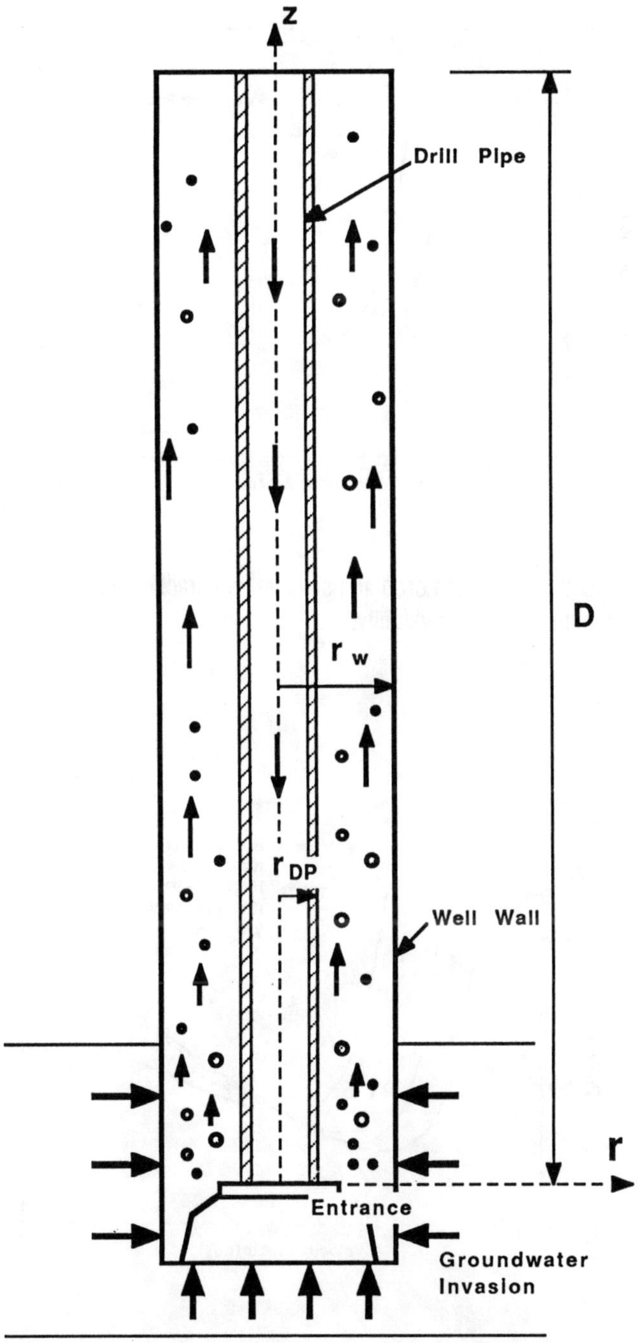

Figure 1. Schematic representation of the drill string/wellbore annulus in air and/or mist drilling.

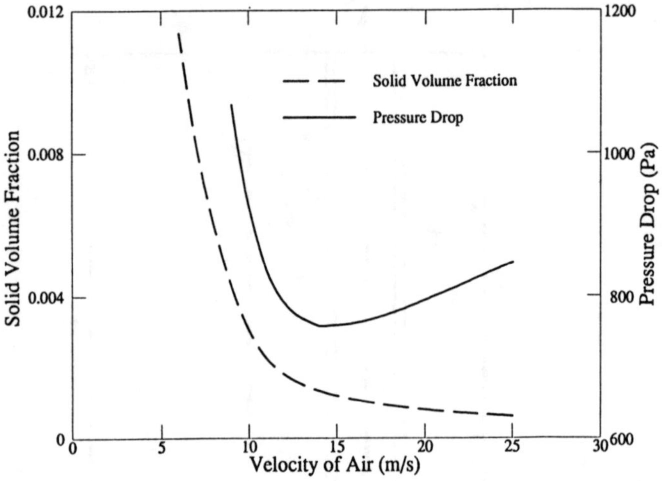

Figure 2. Pressure drop and solid volume fraction distributions as a function of air velocity.

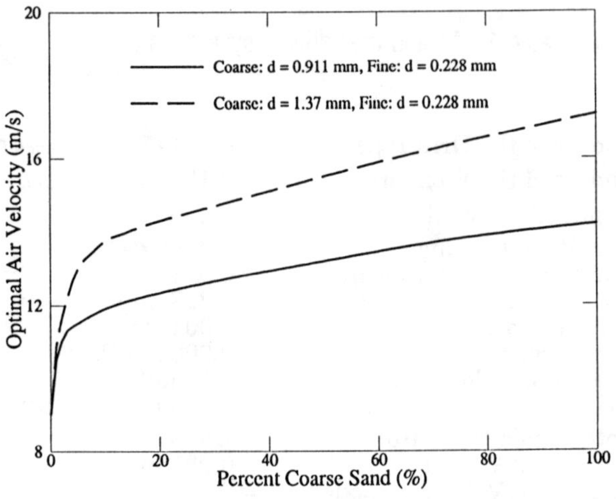

Figure 4. Effect of particle size distribution on optimal air velocity.

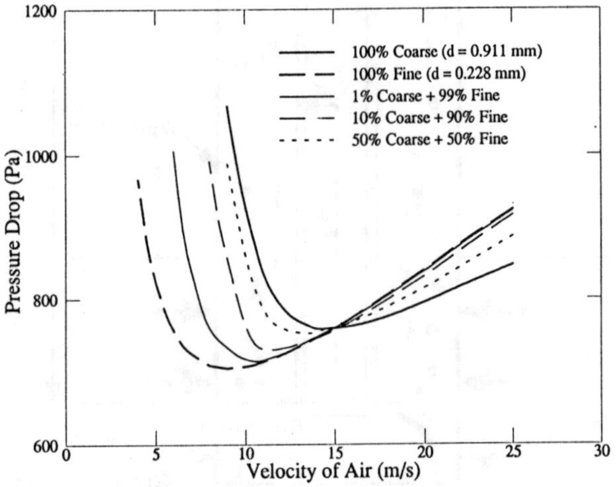

Figure 3. Effect of air velocity on pressure drop and optimal lift velocity for various coarse to fine ratio.

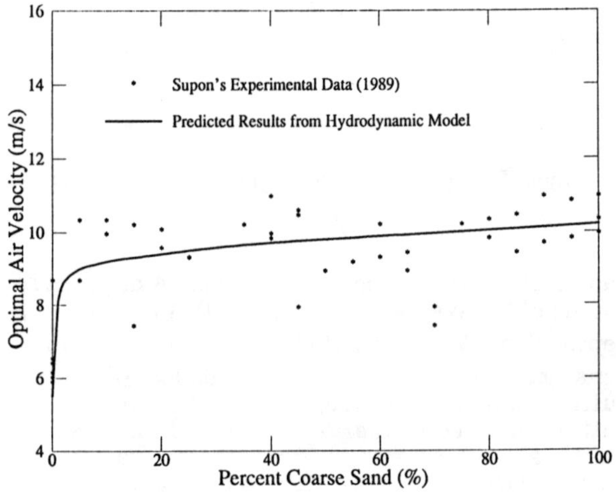

Figure 5. Comparison of predicted results with Supon's experimental data.

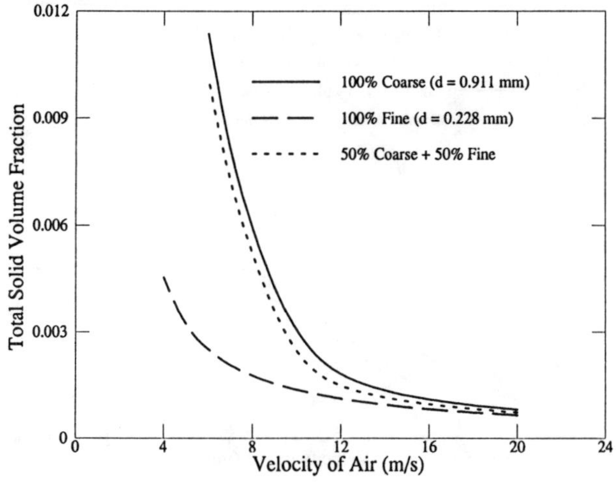

Figure 6. Effect of air velocity on total solid volume fraction.

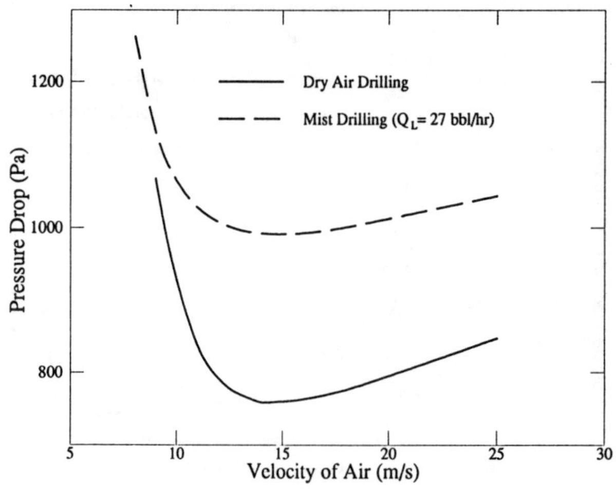

Figure 8. Comparison of pressure loss between dry air drilling and mist drilling.

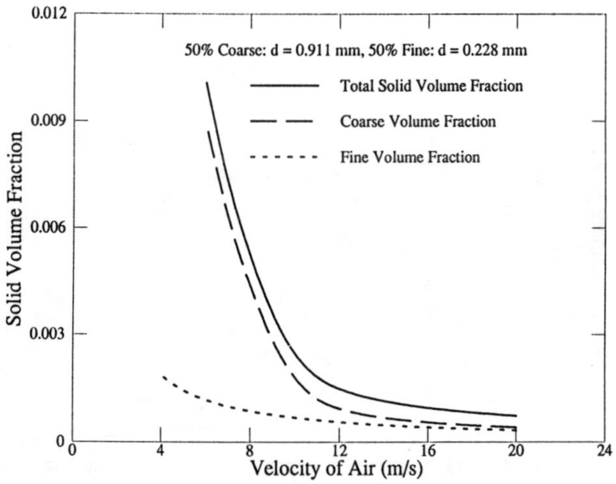

Figure 7. Relative contributions of the fine and coarse particles to the total solid volume fraction.

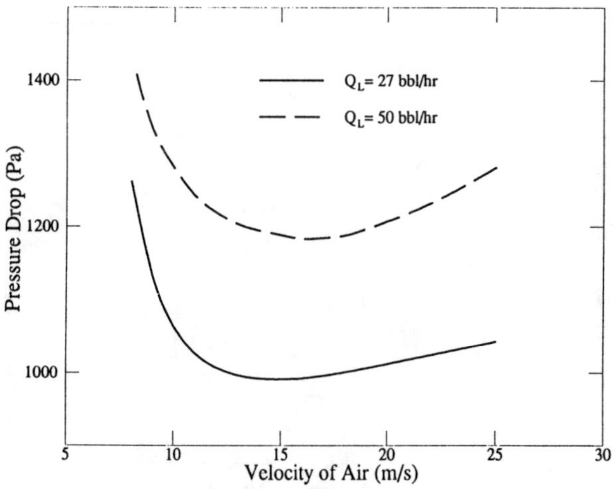

Figure 9. Effect of water invasion rate on pressure loss.

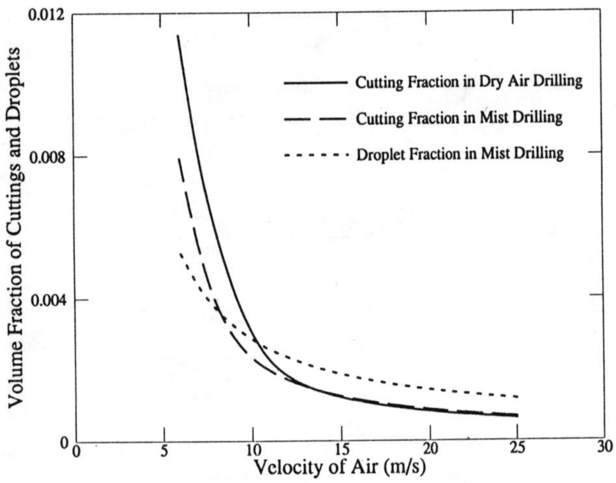

Figure 10. Comparison of solid particle concentrations in air drilling and mist drilling.

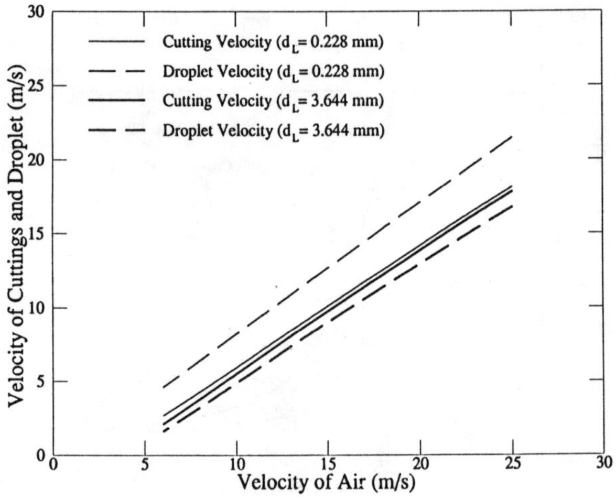

Figure 12. Velocities of particles and droplets as functions of air velocity.

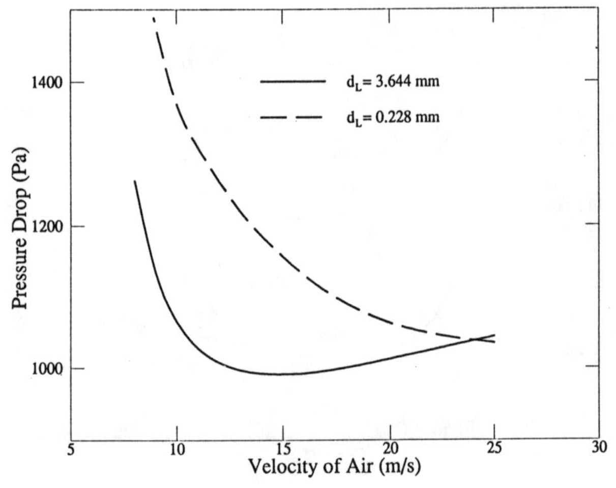

Figure 11. Effect of droplet diameter on pressure loss and optimal air velocity.

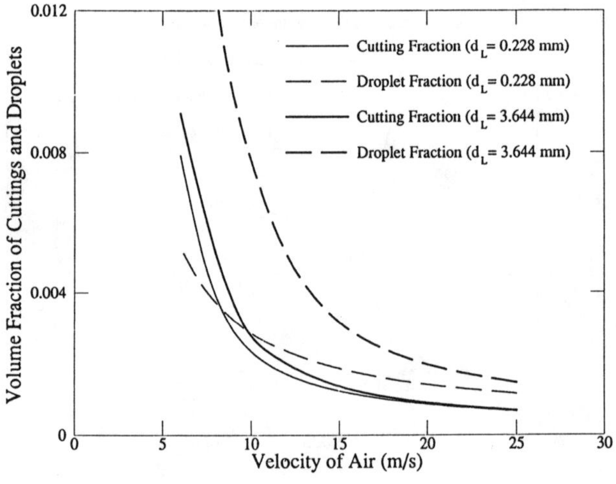

Figure 13. Volume fractions of particles and droplets as functions of air velocity.

A GENERALIZED METHOD FOR CORRELATION OF ADIABATIC TWO-PHASE FLOW FRICTIONAL PRESSURE DROP DATA

N.T. Obot[1], M.W. Wambsganss, D.M. France[2] and
J.A. Jendrzejczyk ■ Materials and Components Technology Division, Argonne National Laboratory, Argonne, IL 60439

A generalized method, based on an analogy with the frictional law of corresponding states for single phase flows, is proposed for the correlation of two-phase flow frictional pressure drop data. It is validated using air-water data obtained on narrow rectangular channels and a circular tube for values of total mass flux (G) in the 50-2000 kg/m^{2}s range. As with single phase flow, the results indicate that the pressure drop for air-water mixtures can be predicted from the proposed correlations provided the critical quality (or superficial gas Reynolds number) and the critical pressure gradient for transition from bubble/plug-to-slug flow are known. The fact that these critical parameters are observed to occur for values of quality (x) less than 0.01 for $50 \leq G \leq 2000$ kg/m^{2}s makes it clear that flowrate and pressure drop data for the $x < 0.01$ range are needed in order to apply the proposed method, just as data for laminar and transitional flow are required for the treatment of the single phase counterpart.

INTRODUCTION

Extensive studies of frictional pressure drop for gas-liquid flows have been made for nearly a half-century. Despite the widespread literature, two-phase pressure drop prediction methods generally have not incorporated fundamental physical phenomena and consequently the resulting correlations include inherent error. Such correlations are usually validated and optimized over small parameteric ranges, but they are not generally applicable without this validation. In principle, there is no reason that two-phase flow pressure drop data should not be correlated in terms of friction factor and Reynolds number defined either using the gas or mixture properties. With the latter, such a treatment has been frustrated by the fundamental difficulty that the physical properties of two-phase mixtures, such as density and viscosity, cannot be determined reliably.

Although it is customary to analyze two-phase flow pressure drop data using the concept of friction multipliers, first introduced by Martinelli and co-workers [1,2], it is almost of general knowledge that the errors associated with this method can be significant. For example, the Lockhart-Martinelli curve overpredicts gas-liquid stratified flow data by more that 50% [3,4]. Also, for immiscible liquid-liquid flows, the deviations are greater than those reported for gas-liquid flows, and the actual pressure gradients are less than the predicted values by about 200% [5].

The difficulty in reconciling the Lockhart-Martinelli method with the various experimental data is the result of a number of factors, of which those dealing with the flow mechanisms will be discussed here. Lockhart and Martinelli postulated, based simply on their data trend, that "a tentative value of 1000" for gas or liquid Reynolds number (Re_G or Re_L) is the upper limit for the existence of laminar isothermal two-phase flow, while the "rough preliminary criteria" for the transition region were reported to be between 1000 and 2000 for Re_G or Re_L. To date, no refinements to these values have been made and little is known about what constitutes laminar or transitional two-phase flow. Also, whether these boundaries vary with duct geometry, total mass flux, or flow direction (i.e. horizontal, vertical or downflow), or the extent to which they may be affected by liquid properties (e.g., viscosity and surface tension), has rarely been acknowledged in the literature. It follows that whatever corrections may be needed to account for these factors are presently unknown.

Chisholm [6,7] developed a Lockhart-Martinelli type correlation with a coefficient that is dependent on, among other things, the types of flow (laminar or turbulent flow of the liquid and gas). Wambsganss *et al.* [8] demonstrated that the correlation is not particularly accurate over a large range of total mass flux (G), with minimum errors on the order of ±40%. In that study, the Chisholm correlation was modified to optimize predictions over a specific range of interest. This useful result does not represent a general approach for predicting pressure drop in an arbitrary two-phase or two-component flow.

Friedel [9] developed an empirical correlation

[1]Also, Clarkson University, Potsdam, NY 13699-5705.

[2]Also, University of Illinois at Chicago, Chicago, IL 60680.

using some 25,000 data points. The terms in the correlation include the Froude and Weber numbers, gas and liquid viscosities, friction factors and specific volume as well as the quality. As documented by Wambsganss *et al.* [8], the errors are unacceptable for $G < 200$ kg/m^{2}s and are on the order of ±40% at high G values. Another problem with this correlation is that it provides no insight into the physical nature of the flow.

The objective of this study was to develop a general approach for correlating gas-liquid pressure drop data based on physical phenomena and with a minimum of empiricism. In particular, the single-phase flow method of Obot [10] and Obot *et al.* [11] was tested with two-phase flow data to form the basis of such a general approach. The results of this development effort are presented in this paper.

ANALYSIS

Single Phase Flow.

In the papers by Obot and co-workers [10,11], it was demonstrated that, at the same reduced conditions, data for single phase friction factor in smooth passages are completely generalized. Specifically, data obtained on smooth circular and noncircular passages were reconciled by accounting for the inseparably connected effects of transition to turbulence and the length scale needed to reduce the data to nondimensional form using the following relations:

$$Re_m = \psi_R Re_a \quad , \quad f_m = \psi_f f_a \tag{1}$$

where $\psi_R = Re_{c,r}/Re_{c,a}$, $\psi_f = f_{c,r}/f_{c,a}$, $Re_{c,r}$ and $f_{c,r}$ are the reference values of the critical Reynolds number and friction factor at the onset of transition to turbulence, respectively, while the corresponding values for any geometric or flow condition of interest are denoted by $Re_{c,a}$ and $f_{c,a}$. In equation (1), Re_a and f_a refer to the Reynolds number and friction factor data for which the critical parameters are $Re_{c,a}$ and $f_{c,a}$. As established therein, Re_m is the similarity parameter in that two channel flows having the same Re_m are dynamically similar to one another and, in the case of the smooth passages, must have almost identical values of f_m.

Two-Phase Flow.

The concept of two-phase flow friction factors is not new, as previous researchers [12-14] have reported pressure drop data in terms of friction factor defined using gas or liquid conditions. The difficulties with the friction factor concept are two-fold; first, the gas or liquid friction factor varies markedly with the total mass flux (G) and duct geometry; and second, no generally valid approach has been proposed to account for these effects.

Of the possible definitions, the most useful form, insofar as the treatment of data for fixed G values is concerned, is the two-phase friction factor (f_G) based on superficial gas properties and it is given by:

$$f_G = [(\Delta P_{TP}/L)\rho_G D_h A^2]/2\dot{m}_G^2 \tag{2}$$

where $\Delta P_{TP}/L$ = the measured two-phase flow pressure drop over a length L, D_h = the hydraulic diameter, A = cross-sectional flow area and $\dot{m}_G$ is the gas mass flowrate. The simplest parameter that could possibly be used to correlate f_G data is the quality (x) or the superficial gas Reynolds number (Re_G); the latter is given by equation (3):

$$Re_G = \dot{m}_G D_h / A\mu_G \tag{3}$$

To calculate f_G using adiabatic air-water data, a constant value of 1.2 kg/m^3 was used for ρ_G, while $\mu_G = 1.85 \times 10^{-5}$ Pas was the basis for all calculations of Re_G. For a given G, the superficial liquid Reynolds number barely changes with quality (x) for low values of x, decreases only moderately with x for intermediate quality, but drops off sharply with increasing x for high quality. These considerations prompted the definition of the two-phase friction factors in terms of the gas phase mass flowrate.

For two-phase flows, it is well-known that the pressure gradient versus quality curve changes slope at the transition between some, but not all, flow patterns [15,16]. For example, the results of Wambsganss *et al.* [8] indicated a pronounced discontinuity at the transition from bubble or plug to slug flow. However, no discernible change in slope was detected for the change from slug to annular flow which is known to be more gradual and not as abrupt.

In this paper, an analogy is drawn between the laminar-to-turbulent transition of single phase flow and the bubble/plug-to-slug flow pattern transition for two-phase flow. Bubble/plug flow occurs at low mass qualities, the flow is relatively quiescent, and, at least in the case of horizontal flow, an analogy with single-phase laminar flow can be argued. Slug and annular flow, on the other hand, are more chaotic and certainly non-homogeneous; here an analogy with single-phase turbulent flow can also be argued.

With the analogies established, the appropriate forms of equation (1) can then be applied to the two-phase flow data using the transition from bubble/plug-to-slug flow pattern as the critical point for the definition of the critical parameters, in particular, the critical quality and the critical pressure drop. Accordingly, the applicable equations for two-phase or two-component flows are given by:

$$Re_m = (Re_{c,r}/Re_{G,ac})Re_{G,a} = \psi_{R,G} Re_{G,a} \tag{4}$$

$$X_m = (x_{c,r}/x_{c,a})x_a = \psi_x x_a \tag{5}$$

$$f_m = (f_{c,r}/f_{G,ac})f_{G,a} = \psi_{f,G} f_{G,a} \tag{6}$$

In equations (4)-(6), the subscript c, r refers to the critical parameters for the reference condition. When the two-phase friction factor and Reynolds number are calculated using equations (2) and (3), the two reference critical parameters ($Re_{c,r}$ and $f_{c,r}$) in equations (4) and (6) can be the single phase data or those for a particular mass flux (G) of two-phase flow. The choice of the reduced quality X_m as the similarity parameter instead of Re_m dictates the selection of the critical quality for a particular G as the reference value, but the reference critical friction factor can be that for the particular G or the single phase value. The denominator for each reference parameter is the two-phase critical value for any G value under consideration, while the remaining data for any such G are denoted by $Re_{G,a}$, x_a and $f_{G,a}$.

There are several observations. The first is that, with the single phase data for $Re_{c,r}$ and $f_{c,r}$ as the reference, Re_m or f_m is independent of the length scale used to reduce the flowrate and pressure drop data to nondimensional form. Thus, as with the treatment of single phase flow friction, two-phase flow gas Reynolds number and friction factor can be calculated using the hydraulic diameter or any arbitrarily selected length scale. But, to reconcile experimental data obtained on ducts of various shapes or for a range of mass flux (G) values, it is necessary to invoke the corresponding states relations, equations (4)-(6). Also, when values of the Reynolds number ($Re_{G,a}$) and the two-phase friction factor ($f_{G,a}$) are computed using constant values of air viscosity (μ_G) and density (ρ_G), respectively, it is clear that Re_m and f_m are independent of these values for μ_G and ρ_G with the single phase critical parameters as the reference.

EXPERIMENTAL DATA

The data used to calculate the results presented subsequently were obtained on the Argonne two-phase flow facility with narrow rectangular channels and a circular tube having an internal diameter of 6.35 mm. The three rectangular configurations are characterized by (H,A,D_h) = (19.05,6,5.44), (3.16,0.17,5.44) and (9.52,6,2.72), where H = channel height, A = aspect ratio (H/W, W is the channel width) and D_h = the hydraulic diameter; the dimensions for H and D_h are in mm. The description of the apparatus and the test procedures are detailed in Wambsganss *et al.* [8] and will not be repeated here. The data used in the analysis were obtained as part of the study of Wambsganss *et al* [8,17]. Whereas tests with the three rectangular arrangements covered a wide range of mass flux (G) values, data for the circular tube were obtained only for G = 500 kg/m^{2}s and these were intended to establish whether the general features of the profile for two-phase flow pressure gradient versus quality were the same as those for the rectangular channels.

EXPERIMENTAL RESULTS

Two-phase Flow Critical Parameters

We begin this presentation and discussion of the critical parameters with a comparison of the be-

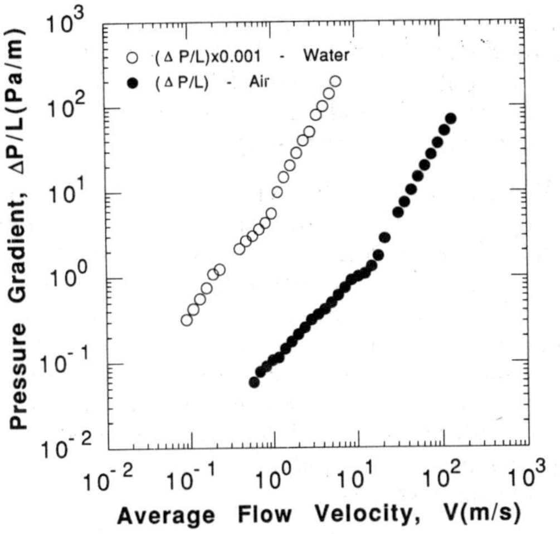

Fig. 1: Single phase pressure gradient vs. average flow velocity.

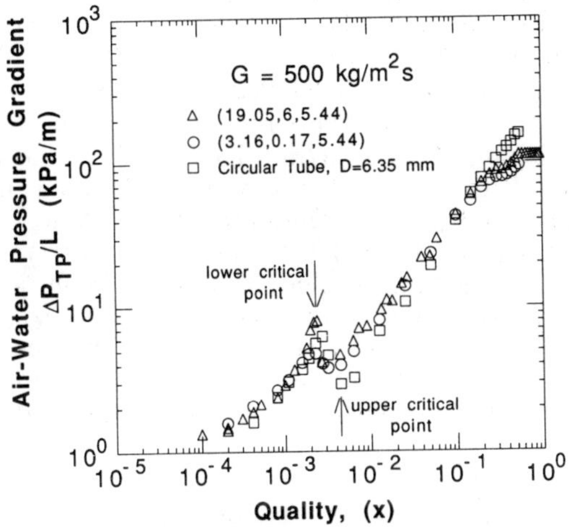

Fig. 2: Two-phase flow pressure gradient vs. quality.

havior for single phase and two-phase flow pressure gradients with increasing flowrate. For the former, it is well known that there are basically three types of flow (i.e., laminar, transitional and turbulent flow), and the lower and the upper critical points correspond to the onset of transition to turbulence and fully turbulent flow, respectively. As can be inferred from the (9.52,6,2.72) results of Fig. 1, a plot of the single phase pressure gradient ($\Delta P/L$) versus the average flow velocity (V) for air and water, the critical points are exhibited by inflection points.

For the air-water mixture, Fig. 2 shows that, as with single phase flow, there are also three regions; $\Delta P_{TP}/L$ initially increases with quality (x), attains a maximum and a minimum at some values of x, and then increases with increasing x. These features are in evidence for plots of pressure gradient versus the superficial gas velocity (U_{GS}) though, due to space limitations, such plots are not shown here. There is another similarity between single phase and two-phase flow behavior, that is, the extent of the intermediate region of Fig. 2 over which the pressure gradient decreases with x is very narrow, paralleling the extent of the single phase transition region. The characteristic features of Fig. 2 were also documented for capillary tubes [15].

With regard to the type of flow in each of the three regions of Fig. 2, little is known. The physical picture is further complicated by the fact that, unlike the single phase counterpart, there are no established criteria for laminar, transitional or turbulent two-phase flow. According to Ozawa *et al.*[15], the intermediate zone of decreasing pressure gradients is the region over which there is a change from capillary bubbly flow to slug flow. This is consistent with the results of the flow pattern studies for each of the three rectangular geometries [8,17].

Figure 2 shows that there are small differences between the three sets of data for $G = 500$ kg/m^{2}s. This is a reflection of the moderate differences in the critical parameters. For instance, the lower critical (maximum) point ($\Delta P_{TP,c}/L$,x_c) are (5.22,0.0018), (4.78,0.00238) and (6.26,0.00276) for (19.05,6,5.44), (3.16,0.17,5.44) and the circular tube, respectively, with $\Delta P_{TP,c}/L$ in kPa/m. Expressed as the two-phase critical friction factor and gas Reynolds number ($f_{G,c}$,$Re_{G,c}$), the values are (in the same order) (21.03,264.7), (10.98,350.4) and (12.52,473.7). The trends on Fig. 2 imply that, as with single phase flow [10,11], two-phase flows having the same $f_{G,c}$ and $Re_{G,c}$ are dynamically similar to one another and, hence, will exhibit similar behavior at the same reduced conditions, either with respect to flow regime or pressure gradient. This observation, which is central to the generalized treatment in this study, prompted a thorough analysis of the data to determine the effects of flow conditions and duct geometric details on the critical parameters.

Determination of Critical Points. Since a typical $\Delta P_{TP}/L$ versus x (or superficial gas velocity,U_{GS}) profile is characterized by two critical points, the lower and the upper critical points (Fig. 2), it is sufficient to consider the lower values, insofar as the application of the corresponding states equations, (4)-(6), is concerned. These values, denoted hereafter as $Re_{G,c}$, x_c and $f_{G,c}$ were deduced from the pressure drop data for all channel configurations. Also, for the single phase counterpart, it was established that the lower critical point, which corresponds to the onset of transition to turbulence, is exhibited by a sharp rise in the root-mean-square (rms) value of the wall pressure fluctuations [18]. Since the general shapes and behavior of the rms pressure profiles with increasing quality x for the $0 < x \leq 0.05$ range do not differ materially from those for single phase flow, the values of the critical quality were also deduced from the rms pressure data from which $Re_{G,c}$ values were calculated. The rms pressure data are given in the original reports [8,17].

Trends for Critical Variables, x_c, $Re_{G,c}$ and $f_{G,c}$. Figures 3 and 4 respectively are plots of x_c and $R_{G,c}$ versus G. The upper plot of each figure gives the appropriate values determined from the pressure gradient data for the three channel arrangements, while a comparison of the latter set with those deduced from the rms pressure data is given in each of the lower plots for the (19.05,6,5.44) configuration. The

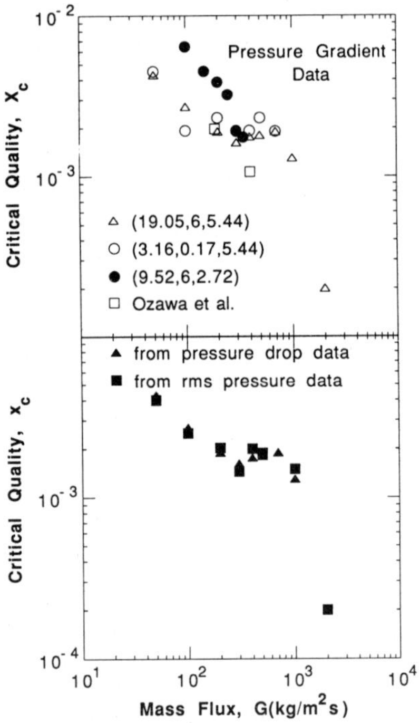

Fig. 3: Critical quality vs. G.

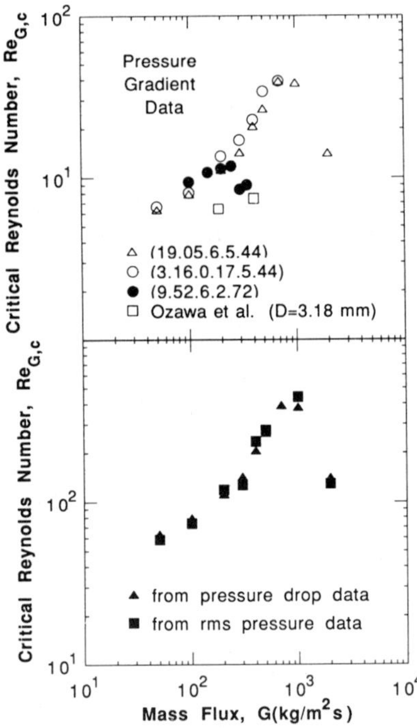

Fig. 4: Critical gas Reynolds number vs. G.

two data points determined from the pressure drop data of Ozawa *et al.* [15] are also given in the upper plots of Figs. 3 and 4.

In Fig. 3, the critical quality decreases initially with G, is almost independent of mass flux for 200 $< G <$ 1000 kg/m^{2}s, but it drops off sharply for the G value for which the liquid Reynolds number at zero quality corresponds to turbulent flow. The lower and the upper values of the critical Reynolds number with air or water flowing alone (Fig. 1) were 2666 and 3770, respectively. Single phase friction factors were obtained only with (19.05,6,5.44) and (9.52,6,2.72) channel geometries, and the differences between the two sets of critical Reynolds number values were no more than 25. The lowest value of x_c is obtained with $G = 2000$ kg/m^{2}s which corresponds to a liquid Reynolds number of 10886 at zero quality.

Since the gas Reynolds number is given by $Re_G = xGD_h/\mu_G$, it follows from the trend on Fig. 3 that $Re_{G,c}$ must increase initially with increasing G (Fig. 4). As with the behavior of x_c, there is a marked drop in $Re_{G,c}$ when the liquid Reynolds number at zero quality is in turbulent flow. Whether $Re_{G,c}$ represents the critical Reynolds number below which the flow is laminar is difficult to ascertain from the results of this study. Suffice it to state that these values of $Re_{G,c}$ are considerably lower than the value of 1000 suggested by Lockhart and Martinelli [1]. Also, the results suggest that, if indeed there exists "laminar" two-phase flow, then the the critical gas Reynolds number for transition to turbulence can be expected to depend upon the total mass flux and the geometric details of the flow channel.

From Figs. 3 and 4 it is evident that there are some differences between the three sets of data for the rectangular channel, notably at small values of G for which there are detectable effects of channel geometry on the flow patterns [17]. The closeness with which the values of x_c or $Re_{G,c}$ deduced from the rms pressure data, using the single phase criterion for transition to turbulence, follow those determined from the pressure gradients appears to suggest that laminar two-phase flow may exist for Re_G (or x) $\leq Re_{G,c}$ (or x_c). Further studies are needed before definite conclusions can be made.

Finally, attention is directed to Fig. 5 which shows the variation of the two-phase critical friction factor ($f_{G,c}$) with the total mass flux, G. For a given G, it is emphasized that each value of $f_{G,c}$ is associated with a value of x_c (Fig. 3) or $Re_{G,c}$ (Fig. 4). Fig. 5 shows that $f_{G,c}$ increases with increasing G; the rise with G is very gradual when values of the liquid Reynolds unmber at zero quality are in laminar flow. For low G values, the effect of channel geometry is significant, paralleling the behavior noted already for $Re_{G,c}$ or x_c. For example, the values of $f_{G,c}$ for the (9.52,6,2.72) are nearly one-half of those with the (19.05,6,5.44) configuration for $50 \leq G \leq 250$ kg/m^{2}s. For air or water flowing alone, the critical friction factor was 0.00773.

Two-phase Flow Friction Factors

The results for two-phase flow friction factor (f_G) are presented on Figs. 6-10. For Fig. 6, which was prepared using the data of Fig. 2, the plots to the left and the right respectively are those of f_G versus the gas Reynolds number (Re_G) and the quality, x. Figs. 7-9 are plots of f_G versus Re_G for the three rectangular channels and various values of mass flux (G). The alternative representations of the same friction factor data as f_G versus x are given on Fig. 10 where, beginning from top to bottom, the f_G data are scaled by 100, 1 and 0.01 for (19.05,6,5.44), (3.16,0.17,5.44) and (9.52,6,2.72), respectively. For clarity and due to the amount of information, the G value associated with each data set is not shown in Fig. 10, but the symbols are the same as those for the corresponding plots of Figs. 7-9. The format on Fig. 10 is intended to provide a clearer perspective for viewing the trend for each channel. It is noted that the appropriate data for $G = 500$ kg/m^{2}s for the results of Fig. 6 are also included in some of the figures.

These figures show that the well pronounced maximum and minimum, observed on the typical plot of Fig. 2, are exhibited by only a slight downward shift in plots of f_G vs. Re_G or x. Figs. 7-9 show that the logarithmic plot of f_G versus Re_G is represented by a series of nearly straight lines with a strong parametric dependence on G and the slopes of these lines are almost insensitive to variations in mass flux for $Re_G < 10^4$. For f_G versus x (Fig. 10), it is observed that , although the slopes of the lines are almost independent of the mass flux for $x < 0.3$, the sensitivity of f_G to variations in G for a given x is less pronounced than that for a specified Re_G. This is a reflection of two factors: first, the critical quality is not a strong function of G (Fig. 3); and second, quality is a more appropriate parameter than Re_G for characterizing f_G because the liquid and gas flowrates are accounted for in its definition.

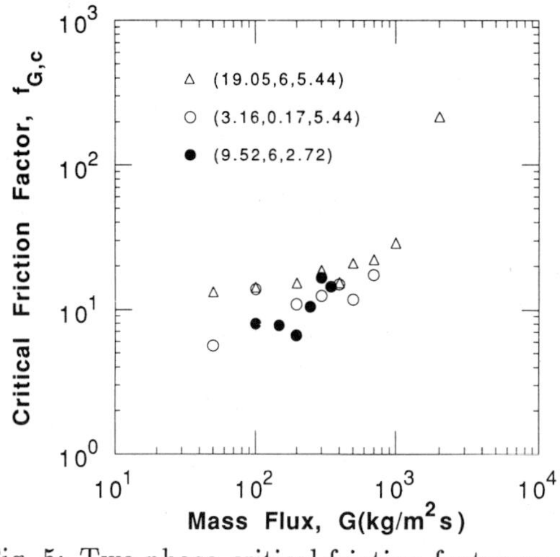

Fig. 5: Two-phase critical friction factor vs. G.

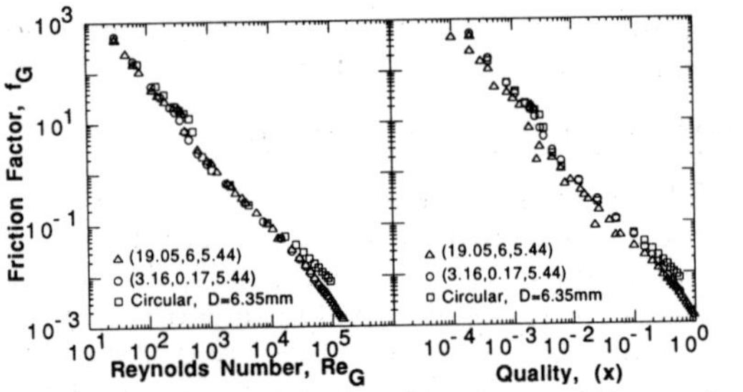

Fig. 6: Two-phase friction factor vs. Re_G and x.

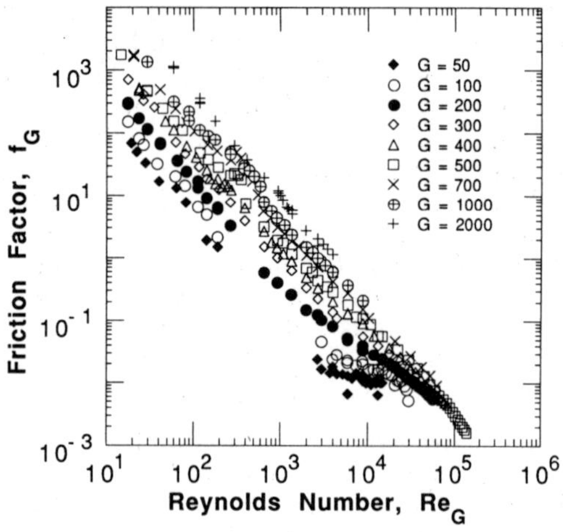

Fig. 7: Plot of f_G vs. Re_G for (19.05,6,5.44).

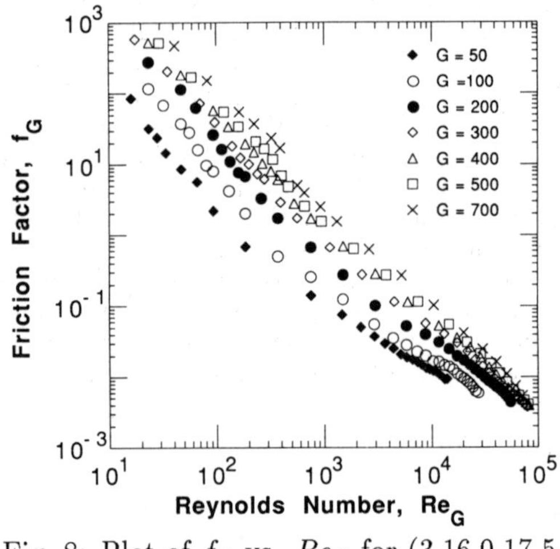

Fig. 8: Plot of f_G vs. Re_G for (3.16,0.17,5.44).

Using the results of Fig. 6, it is established that f_G varies roughly as the -1.35 and -1.33 power of Re_G and x for $10 < Re_G < 10^5$ and $10^{-4} < x \leq 0.6$, respectively. In other words, the two-phase pressure drop, ΔP_{TP}, varies roughly as the 0.65 power of the superficial gas velocity (U_{GS}). Thus, for a given mass flux (G), the decrease of f_G with Re_G is more pronounced, while ΔP_{TP} rises less rapidly with U_{GS}, than when the gas or liquid is flowing alone.

Based on plots of (ϕ^2 - 1) *vs.* quality x (where ϕ is a friction multiplier), Beattie [19] established three distinct regions for pressure drop data: a low

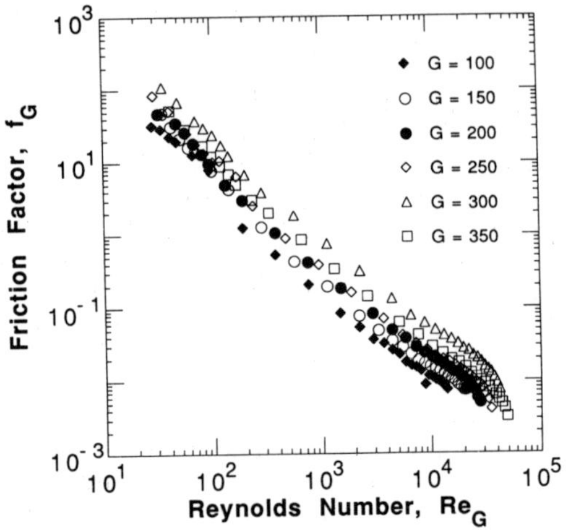

Fig. 9: Plot of f_G vs. Re_G for (9.52,6,2.72).

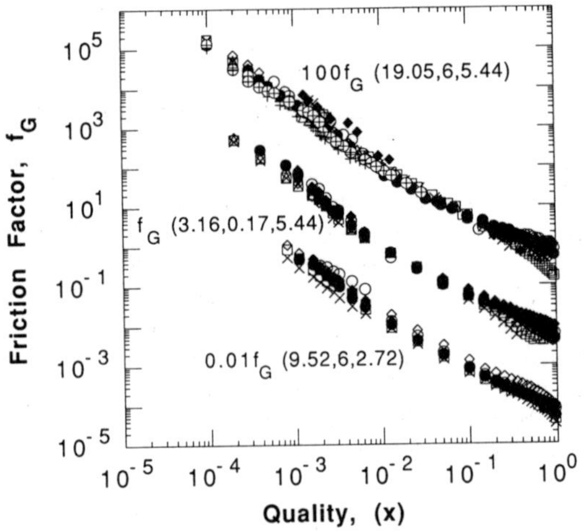

Fig. 10: Two-phase friction factor vs. quality.

quality region, a narrow intermediate zone, followed by a high quality region. For both the low and high quality regions, (ϕ^2 - 1) was found to vary as the 7/6 power of x, and this exponent was independent of mass flux at low and high qualities, but varied with mass flux in the intermediate zone. Although the pressure drop behavior at low and high qualities is confirmed by the present analysis, Figs. 6-10 show that, when the results are presented as f_G *vs.* Re_G or x, the intermediate zone is virtually nonexistent.

Figure 10 shows that, even without the corrections to account for the differences in the critical values for two-phase friction factor and quality, the three sets of data can be represented by a mean curve with errors that are, for the most part, under ±40%. In fact, for the (19.05,6,5.44) configuration, the results of which were compared to the Chisholm [6,7] and Friedel [9] correlations by Wambsganss *et al* [8], with large deviations as noted in the introductory section, the magnitudes of the spread around a mean curve are even smaller than those obtained with each of the above correlations. Similarly, the collapse of the (3.16,0.17,5.44) data is excellent, while the spread on the lowest plot for (9.52,6,2.72) is as good as can be expected due to the variation of the critical parameters with mass flux for this range of G. The results on this figure suggest that it is possible to treat adiabatic two-phase frictional pressure drop data without the introduction of the usual dependence on the type of flow (laminar or turbulent flow of the liquid and gas). Strangely enough, although adiabatic two-phase flow pressure drop has been studied for nearly four decades, straightforward analysis of the type presented herein for variation with x and constant G has not been considered.

It is evident that the effect of G on f_G for each plot of Fig. 10 is more pronounced over the $x \leq 0.005$ region than for the $0.005 < x \leq 0.3$ range. This is not the result of uncertainties in the measurements. The speculation is that this is a two-phase flow analog of the single phase situation where effects of geometric/flow conditions on friction factor are more pronounced in laminar than turbulent flow. This is another indication that laminar two-phase flow may exist for $x < 0.005$.

Beyond a certain quality, typically between 0.4 and 0.6 depending on G, the pressure gradient barely increases with increasing quality. This is reflected in the presence of inflection points followed by a further decrease of f_G with increasing Re_G (or x) for $G \leq$ 500 kg/m^{2}s. For G = 700, 1000 and 2000 kg/m^{2}s, the upper limits to the range of quality covered in the experiments were 0.35, 0.013 and 0.0067 respectively; hence, the profiles for these G values do not exhibit the above behavior.

Validation of the Proposed Method

As stated already, to apply equations (4)-(6) to two-phase flow data, there are two options. The single phase critical values can be selected as the reference, and for air or water flowing alone, the values are 2666 and 0.00773 for $Re_{c,r}$ and $f_{c,r}$, respectively. With these numbers as the reference, it can be determined from the results on Figs. 3 and 5 that $\psi_{R,G}$ and $\psi_{f,G}$ vary between 5 and 50 and, 0.00005 and 0.0015, respectively. Alternatively, the two-phase critical values for a particular G can be selected as the reference. In this case, and using the (19.05,6,5.44) channel critical parameters for G = 500 kg/m^{2}s as the reference (*i.e.*, $Re_{G,c}$ = 264.65, x_c = 0.0018 and $f_{G,c}$ = 21.03), the scaling factor, $\psi_{R,G}$, varies between 0.5 and 5, while $\psi_{f,G}$ assumes values between 0.1 and 5. For ψ_x, the values are between 0.3 and 9. For any particular G and channel geometry, the appropriate values for $\psi_{R,G}$, ψ_x and $\psi_{f,G}$ are easily calculated from the results on Figs. 3-5.

These two alternative representations of the same experimental data are illustrated in Figs. 11 and 12. In Fig. 11, the upper and the lower set of data are for the situations where the G = 500 kg/m^{2}s two-phase critical data for the (19.05,6,5.44) configuration and the single phase critical values were selected as the reference, respectively. For Fig. 12, the reference value for the abscissa (x_c = 0.0018) is that for the (19.05,6,5.44) channel and the two-phase critical friction factor ($f_{G,c}$ = 21.03) for this channel is also used as the reference for the upper plot. The single phase critical friction factor is the reference for the lower plot. Each profile on Figs. 11 and 12 was prepared using all of the data on Figs. 7-9 and the G = 500 kg/m^{2}s circular tube data of Fig. 6. For the sake of clarity, and due to the large amount of experimental data (a total of 720 data points) and parametric information, a uniform symbol is used for all data. For each value of G, the critical parameters used in the calculations correspond to those in Figs. 3-5.

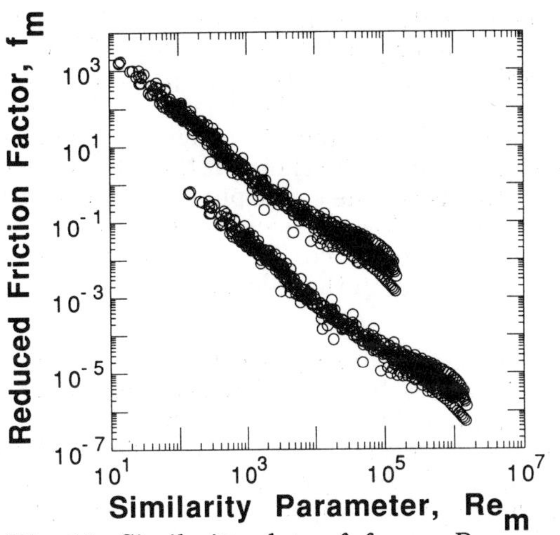

Fig. 11: Similarity plots of f_m vs. Re_m.

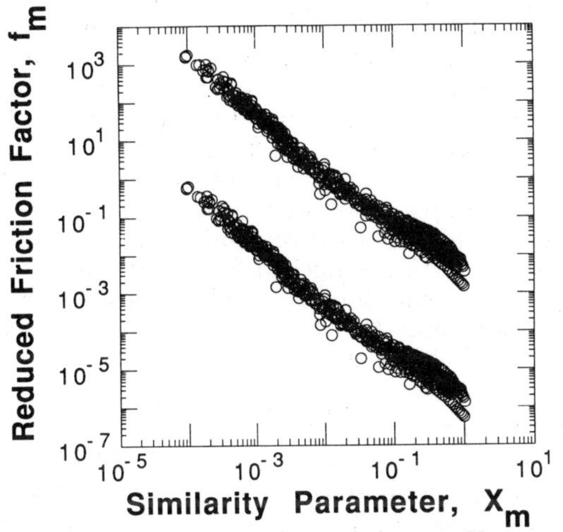

Fig. 12: Similarity plots of f_m vs. X_m.

Despite the marked variation of f_G with G (Figs. 7-9), the reduced data of Fig. 11 are closely approximated by single curves. With X_m as the similarity parameter (Fig. 12), the general behavior of these profiles is exactly the same as on Fig. 11 and there are little detectable differences between the two figures. Given the ±3% uncertainty associated with the determination of the two-phase critical parameters from the original data and, the ±2% and ±5% uncertainties in the flowrate and pressure drop measurements respectively, the collapse of the data is as good as can be expected. Although the general trend is not modified by the choice of the reference critical values, the use of the single phase data gives profiles that fall well below those obtained with a particular G data as the reference. This is a direct consequence of the differences in the calculated values of $\psi_{f,G}$ and $\psi_{R,G}$.

The fact that two-phase flow pressure drop data for various G values and flow channels are satisfactorily treated via the corresponding states method gives rise to several observations. The first is that the effects of duct geometry, mass flux and, possibly, the physical properties of the gas-liquid mixture manifest themselves through variations in the critical values for quality (or gas Reynolds number) and two-phase friction factor. These differences in the critical parameters are largely responsible for the disagreement among the results obtained by various other investigators. Also, the availability of two-phase pressure drop and flowrate data for the low quality region, inclusive of the intermediate zone of decreasing pressure gradients, is central to the application of the proposed method, just as laminar and transition data are needed for the treatment of the single phase counterpart. And, further, the analysis establishes that Re_m or X_m is the similarity parameter for adiabatic two-phase flow friction.

Correlation of Pressure Drop Data

From Figs. 11 and 12 it is evident that the variation of f_m with Re_m or X_m is not linear over the entire Re_m or X_m range. Several empirical relations for f_m as a function of Re_m as well as X_m were developed using the results of Figs. 11 and 12. Of these, the most accurate predictive equations are given by:

$$f_m = \exp[A(Re_m)] \quad ; f_m = \exp[B(X_m)] \qquad (1)$$

where:

$$A(Re_m) = a_o + a_1 \ln Re_m + a_2(\ln Re_m)^2 + a_3(\ln Re_m)^3 \qquad (2)$$

$$B(X_m) = b_o + b_1 \ln X_m + b_2(\ln X_m)^2 + b_3(\ln X_m)^3 \qquad (3)$$

The numerical values of the constants, a_0, a_1, a_2 and a_3; b_0, b_1, b_2 and b_3, along with the ranges of validity and the reference critical values are given in Table 1. For each empirical correlation, the calculated values (f_m, cal) are within ±20% and ±30% of the experimental data (f_m, act) for about 85% and 90% of the 720 data points, respectively. This is illustrated in the typical scatter plot in Fig. 13, where the largest deviations are observed in the narrow $0.008 < X_m < 0.03$ range. Although refinements to these correlations may be necessary as more data become available, the degree of concurrence established here is a remarkable one in the difficult area of two-phase flow, where the development of reliable pressure drop correlations based on physical phenomenon has received somewhat meager treatment.

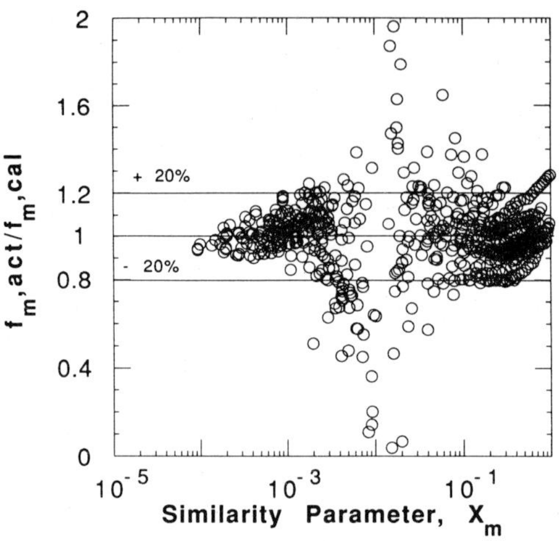

Fig. 13: Typical scatter plot.

A study of the calculated values of $f_{m,act}/f_{m,cal}$ was made to determine whether the trend for $0.008 < X_m < 0.03$ was associated with some physical phenomena. A total of 34 out of 720 data points gave deviations greater than 50%; 20, 8, 4, and 2 data points were for the (19.05,6,5.44), (3.16,0.17,5.44) (9.52,6,2.72) rectangular geometry and the circular tube, respectively. For the most part, these data points are located in the vicinity of the minimum pressure gradient. It is likely that this small region may be the seat of marked pressure fluctuations, in which case, a number of pressure drop readings at a given quality would need to be averaged to obtain representative mean values.

Although only one set of circular tube data (G = 500 kg/m^{2}s) is common to Figs. 11 and 12, there is no reason that the reduced circular tube data for other G values should not follow the mean curves obtained from these results. Support for this observation comes from the results in Fig. 6 from which it can be established that, as f_m *vs.* Re_m or *vs.* X_m, the three sets of data are indistinguishable. It is therefore reasonable to suppose that the present correlations are equally applicable to circular tubes.

DISCUSSION

There are indications that the critical parameters are also in evidence for two-phase flow in vertical channels [13,20]. These studies which involved use of tube diameters on the order 25.4 mm appear to suggest that the critical parameters do exist for passages with hydraulic diameters that are greater than those considered herein. The implication is that, although the present analysis is made for horizontal passages having small hydraulic diameter, for which reliable data were available, the method is applicable to vertical flows as well as to larger diameter tubes. Also, the speculation is that the method should work for immiscible liquid-liquid flows because there are indications that these flows are also characterized by critical parameters [5].

The practical significance of the present method relates to the calculation of frictional pressure drop for adiabatic gas-liquid flows which, from the definitions of f_m and Re_m (or X_m), is quite straightforward provided that the two-phase critical parameters are known. From an experimental point of view, the fact that emerges is that pressure drop measurements need not be made over a wide quality range since knowledge of the critical parameters should afford calculations of the two-phase pressure drop from the equations developed in this study.

Finally, the difficulty in preparing similarity plots using previously published data is explained by the fact that measurements were made for conditions that precluded determination of the critical values for quality and pressure drop. For the few studies that provided low quality data, see for example Ozawa *et al.* [15], the analysis was complicated by the fact that the mass flux (G) was not held fixed with increasing quality, x. This proper procedure of acquiring two-phase pressure drop data for fixed G values was overlooked in most previous studies. Since the friction multiplier increases with increasing G for a given x, it may indeed be argued that the magnitudes of the differences between experiments and the Lockhart-Martinelli correlation are intimately associated with this experimental problem.

CONCLUDING REMARKS

It is established that a typical profile of two-phase pressure gradient, $\Delta P_{TP}/L$, vs. quality x (or superficial gas velocity, U_{GS}) exhibits a maximum and a minimum over a certain range of x (or U_{GS}). The values of the critical quality (x_c) are largest for mass flux (G) less than 100 kg/m^{2}s, are less sensitive to variations in G for $200 < G < 1000$ kg/m^{2}s, but drop off sharply for the G values for which the liquid Reynolds number at zero quality (Re_{LO}) correspond to fully turbulent flow. Expressed as the critical gas Reynolds number ($Re_{G,c}$) based on the superficial velocity, $Re_{G,c}$ increases with increasing G, attains its highest value for Re_{LO} value within the transition region, and then decreases with increasing G. The two-phase critical friction factor ($f_{G,c}$), based on the local peak pressure gradient and gas conditions, increases with increasing G.

The logarithmic plot of the two-phase flow friction factor (f_G) versus the gas Reynolds number (Re_G) is represented by a series of nearly straight lines with strong parametric dependence on the mass flux (G), and the slopes of these lines are almost independent of G for $Re_G \leq 10^4$. By contrast, when the results are presented as f_G versus quality (x), the deviations of the data from the mean curve through all data for the rectangular channels and the circular tube are, for the most part, under $\pm 40\%$.

By applying the corresponding states equations, the extensive air-water data for rectangular channels and a circular tube are closely approximated by single profiles, establishing that pressure drop data for two-phase flow can be treated in exactly the same way as the single phase counterpart. Unlike the latter which is characterized by one similarity parameter (Re_m), it is established that there are two similarity parameters(Re_m and X_m) for adiabatic gas-liquid mixtures, either of which can be used to generalize the pressure drop data.

ACKNOWLEDGEMENTS

This research was funded by the U.S. Department of Energy, Office of Conservation and Renewable Energy, Division of Advanced Industrial Concepts, and represents a U.S. contribution to Annex I of the International Energy Agency (IEA) Program on Research and Development in Heat Transfer and Heat Exchangers. The authors wish to thank Messrs E. B. Esen and K. H. Snell for help with the preparation of the illustrations and the typing of the manuscript.

LITERATURE CITED

[1] Lockhart, R.W. and R.C. Martinelli, *Chem. Eng. Prog.*, **45**, pp. 39-48 (1949).

[2] Martinelli, R.C. and D.B. Nelson, *Trans. ASME*, **70**, pp. 695-702 (1948).

[3] Hoogendoorn, C.J., *Chem. Eng. Sci.*, **9**, pp. 205-217 (1959).

[4] Govier, G.W. and M.M. Omer, *Can. J. Chem. Eng.*, **40**, pp. 93-104 (1962).

[5] Charles, M.E. and L.U. Lilleleht, *Can. J. Chem. Eng.*, **44**, pp. 47-49 (1966).

[6] Chisholm, D., *Int. J. Heat Mass Transfer*, **10**, pp. 1767-1778 (1967).

[7] Chisholm, D., *Int. J. Heat Mass Transfer*, **16**, pp. 347-358 (1973).

[8] Wambsganss, M.W., J.A. Jendrzejczyk, D.M. France and N.T. Obot, "Two-Phase Flow Patterns and Frictional Pressure Gradients in a Small, Horizontal Rectangular Channel," Argonne National Laboratory Report ANL-90/19 (1990).

[9] Friedel, M., "Improved Friction Pressure Drop Correlation for Horizontal and Vertical Two-Phase Pipe Flow," Paper 2., European Two-Phase Flow Group Meeting, Ispra, Italy (1979).

[10] Obot, N.T., *Trans. ASME J. Fluids Eng.*, **110**,

pp. 431-440 (1988).

[11] Obot, N.T., E.B. Esen and T.J. Rabas, *Int. J. Heat Mass Transfer*, **33**, pp. 2133-2143 (1990).

[12] Bergelin, O.P. and C. Gazely, Jr., "Co-Current Gas-Liquid Flow. 1-Flow in Horizontal Tubes," Heat Transfer and Fluid Mechanics Institute, Berkeley, California (published by ASME), pp. 5-18 (1949).

[13] Govier, G.W. and W.L. Short, *Can. J. Chem. Eng.*, **36**, pp. 195-202 (1958).

[14] Beattie, D.R.H., "Friction Factors and Regime Transitions in High Pressure Steam-Water Flows" ASME Paper 75-WA/HT-4, ASME Winter Annual Meeting, Houston (1975).

[15] Ozawa, M., K. Akagawa, T. Sakaguchi, T. Tsukahara and T. Fujii, *Bulletin JSME*, **22**, pp. 1763-1770 (1979).

[16] Chisholm, D., Two-phase Flow in Pipelines and Heat Exchangers, Longmans Inc, New York, 137 (1983).

[17] Wambsganss, M.W., J.A. Jendrzejczyk, D.M. France and N.T. Obot, "Two-Phase Flow Patterns and Frictional Pressure Gradients in a Small, Rectangular Channel: A Comparison Between Two Horizontal Orientations," Argonne National Laboratory Report ANL-90/46 (1990).

[18] Obot, N.T., J.A. Jendrzejczyk and M.W. Wambsganss, "Transition to Turbulence in Flow Passages: Part 1. Direct Determination of Its Onset," To be published in *Trans. ASME J. Fluids Eng.*

[19] Beattie, D.R.H., "Two-Phase Pressure Losses-Flow Regime Effects and Associated Phenomena," Australian Atomic Energy Comm., AAEC/TM 589 (1971).

[20] Oshinowo, T. and M.E. Charles, *Can. J. Chem. Eng.*, **52**, pp. 438-448 (1974).

TABLE 1.
Values of Constants for Correlations.

a_o	a_1	a_2	a_3	Validity Range	Reference ($Re_{c,r}$, $f_{c,r}$) or ($x_{c,r}$,$f_{c,r}$)
12.606	-1.842	0.005	0.002	$10 \leq Re_m \leq 10^5$	(264.65, 21.03) - Fig. 11 (upper)
8.628	-1.712	-0.023	-0.003	$10^2 \leq Re_m \leq 10^6$	(2666,0.00773) - Fig. 11 (lower)
b_o	b_1	b_2	b_3		
-5.089	-0.872	0.071	0.002	$10^{-4} \leq X_m \leq 1$	(0.0018,21.03) - Fig. 12 (upper)
-12.936	-0.838	0.079	0.002	$10^{-4} \leq X_m \leq 1$	(0.0018,0.00773) - Fig. 12 (lower)

DEVELOPMENT OF AN ISOKINETIC PROBE FOR MEASUREMENTS OF SOLID MASS FLUXES IN CIRCULATING FLUIDIZED BEDS

Blaine Herb, Suisheng Dou, Kemal Tuzla,
and John C. Chen ■ Institute for Thermal-Fluid Engineering and Science, Lehigh University, 111 Research Drive, Bethlehem, PA 18015-4791

A solid sampling technique was developed and used successfully to measure the local time-average solids mass flux in circulating fluidized beds (CFB's). Net flux measurements resulted from the difference between upward and downward flux components, sampled under conditions which introduced the least possible disturbance to the flow. Sample results were obtained in a pilot-scale CFB riser of 15 cm I.D. and 10.8 m height. FCC catalyst particles with Sauter mean diameter of 68 μm were used in the tests. The results showed the local time-average solids mass flux to vary with radial position and indicated an upflowing solid suspension in the core and downflowing solids in an annulus adjacent to the riser wall.

INTRODUCTION

Many industrial processes have evolved in which solid particles are contacted with a gas to achieve chemical or physical change within the gas or solid phase. Examples of these processes include solid particle drying, coal combustion, fluid catalytic reactions, and calcining reactions. Depending on the relative velocity between the gas and the solid, gas and particle properties, and equipment size and geometry, several different vertical gas-solid flow regimes result. At superficial gas velocities above the particle terminal velocity and at relatively high solids mass fluxes, an entrained suspension of solids forms which is known as a CFB.

Knowledge of the solids distribution and flow behavior in the CFB riser is key to successful design and operation of CFB systems. The solids distribution governs the pressure drop occurring along the CFB riser and is directly related to the mean solids residence time within the riser. The solids concentration determines the gas-solid interfacial area per unit volume of the mixture, which is expected to directly affect gas-solid reaction and mass transfer rates. Furthermore, heat transfer coefficients measured at the riser wall and on submerged surfaces show strong dependence on the solid concentrations in the vicinity of the heat transfer surface, [1]. Solid mass flux measurements are important to describing the solid flow behavior in CFB's. In addition, erosion rates of exposed surfaces are expected to tie closely with the solid flow behavior along these surfaces.

Few measurements of solid mass fluxes in CFB's have been reported in the literature. Van Breugal et al. [2], using an isokinetic sampling probe in a 30 cm riser operating under dense-phase flow conditions, reported a parabolic solids flux profile with downward solid flow along the riser wall. Using a non-isokinetic sampling probe, Bierl et al. [3] also measured strong radial variations in the solid mass flux. Combining these measurements with information on radial variations of time-average solids concentration obtained from x-ray measurements, they concluded that the CFB consisted of a dilute, upflowing solids core surrounded by a dense, downflowing solids annulus.

More recently, Montceaux et al. [4] measured particle mass fluxes in the fully-developed region of the CFB in a 15 cm I.D. riser using fine catalyst particles. For area-average solid volume fractions below 0.05, normalized solids flux profiles at a given gas velocity were found to be independent of solids mass flux. At higher solids concentrations, the normalized solids flux profiles were no longer independent of total solid mass flow. The experimental findings supported their theoretical analysis which predicted the existence of similar solid flux profiles below some critical solid concentration.

The objective of this paper is to describe the experimental sampling technique used to measure time-average solid mass fluxes in a CFB and to present sample solid mass flux measurements taken in a 15 cm I.D. CFB riser.

EXPERIMENT

Experiments were carried out in two different CFB systems: a lab-scale CFB with a 5 cm I.D. riser with a height of 2.7 m and a pilot-scale CFB system, shown in Figure 1, having a cylindrical riser with an inside diameter of 15 cm and a height of 10.8 m. As Figure 1 shows, primary air and secondary air were introduced below and above a sintered metal distributor, respectively. Under normal test conditions, a dense bed was established between the distributor and the secondary air injection elevation. Above the secondary air injection point, the circulating fluidized bed was established and existed until the smooth right-angle exit to the cyclone.

Axial positions within the riser were measured relative to the secondary air inlet location, above which the total air flow was established. Particles captured by the cyclone were returned through a downcomer leg which provided a particle storage hopper and standpipe combination for the measurement of solid recirculation rate. The return particles were horizontally transported to the CFB riser by means of secondary air flow.

Thirty-three pressure taps were installed along the height of the riser. Each pressure tap line was purged at a constant rate to prevent particles from entering and blocking the pressure tap lines. The pressure tap lines were manifolded to six differential pressure transducers to allow the pressure drop to be measured simultaneously across six segments of the riser.

In a typical experiment, the desired superficial velocity was set by adjustment of both primary and secondary air flows and particles were charged to the riser at the desired rate by adjusting

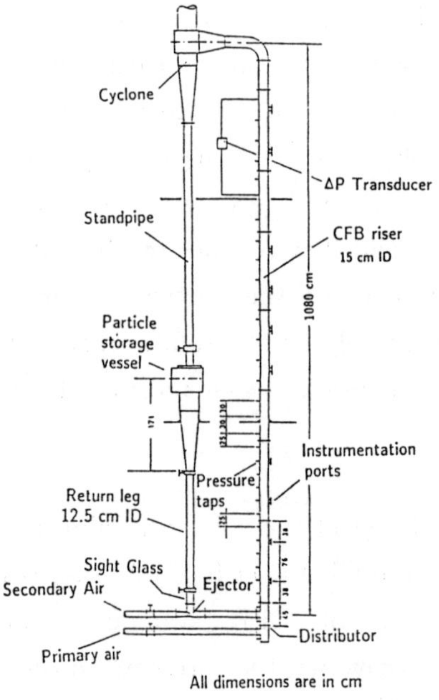

Figure 1: Schematic of CFB Test Loop

the solids feed valve. The solid recirculation rate (total solid mass flux in the riser) was measured by temporarily closing the butterfly valve above the particle storage vessel and measuring the time to accumulate a packed bed volume of 0.0032 m^3 in the transparent standpipe. A separate measurement of the apparent packed bed density of the test particles was used to calculate the mass of solids collected in the transparent standpipe. Due to the surge volume provided by the particle storage vessel, the normal feed rate of returned solids continued without interruption during this measurement procedure.

The results presented in this paper were obtained for 87 μm FCC particles with a terminal velocity in ambient air of 0.27 m/s. Data were obtained for superficial gas velocities ranging from 1-3 m/s and solid recirculation rates in the range of 10-50 kg/m^2s in the CFB riser.

SAMPLING PROBE AND MEASUREMENT METHOD

To measure the axial component of the time-averaged solids mass flux locally within the CFB, a sampling apparatus and technique were developed. The sampling probe was designed to mount in the instrumentation ports and to traverse the riser I.D. Figure 2 sketches the design of the sampling probe tip. To ensure the accuracy of the upward solid flux measurements, the probe was designed for isokinetic sampling. With the probe tip facing upstream, isokinetic sampling requires that the local gas velocity aspirated through the probe tip equals the local gas velocity that would exist at this point if the probe were not present. If the aspiration velocity is below the isokinetic velocity, the gas flow is deflected around the probe tip along with fine particles carried by the gas which causes the solid sample collected to be too small. If the aspiration velocity exceeds the isokinetic velocity, the collected solid sample will be too large since excess gas and solids are drawn into the probe tip. Under isokinetic sampling conditions, the static pressures measured inside and outside the probe tip are expected to be equal. Thus, inside and outside pressure taps made of 1.6 mm stainless tubing are located 25 mm from the probe tip to detect the isokinetic sampling condition.

Since gas must be drawn through the probe tip in order to collect a solid sample, isokinetic sampling is only possible in the direction of the local gas flow. Isokinetic sampling of the downward solid flux component along the riser wall is not expected to be possible if the gas flow is upward.

The tube through which the two-phase flow is drawn is made of 5 mm I.D. stainless tubing (1.6 x 10^{-5} m^2 area). The tip is machined to a 10° knife edge to introduce the least possible disturbance to the gas-solid flow.

Figure 3 provides a schematic drawing of the sampling system used in conjunction with the sampling probe. The four-way valve may be switched in two different positions: one to sample the gas-solids mixture through the probe and the

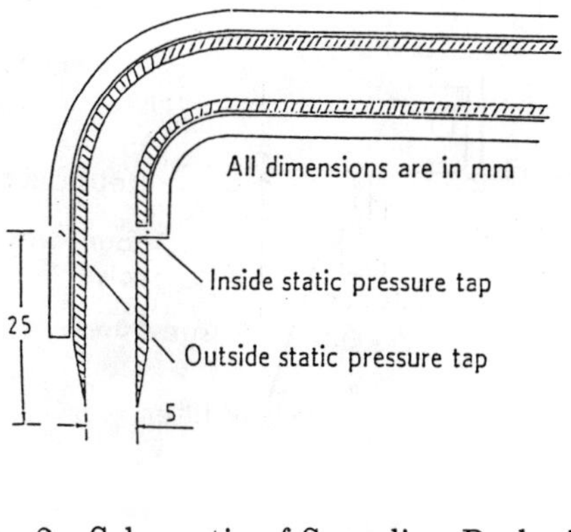

Figure 2: Schematic of Sampling Probe Tip

other to purge the probe tip with air. In the sampling position, the gas-solid flow drawn through the probe tip and led to the sample collection vessel where the solids are disengaged and settle to the bottom of the collection vessel. The sampled gas flows through a porous filter, one of two rotameters where the gas flow is monitored, and continues to flow into an evacuated vessel. With the four-way valve in the purge position, a stream of purge air flows in the reverse direction through the sample jar and continues out through the probe tip to prevent particles from entering the probe. The horizontal run of tubing leading from the probe tip to the sample jar was minimized to prevent solid particles from settling out and blocking the probe.

To detect the isokinetic condition, the pressure tap lines were lead to an inclined manometer capable of resolving pressure differences as small as 0.25 mm of water. When the pressure taps were not being used to measure the difference in static pressures, the manometer was bypassed to prevent pressure oscillations from blowing out the merian fluid and a purge flow was introduced through the pressure tap lines to prevent solids from penetrating and blocking them.

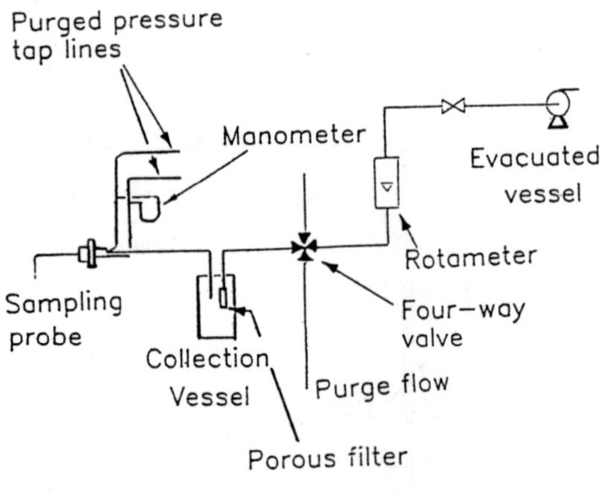

Figure 3: Solids Sampling Apparatus

Figure 4 show the results of a test of the probe's ability to measure the local gas velocity in single-phase gas flow. With the probe oriented along the riser centerline and facing upstream, the local gas velocity was measured isokinetically within ± 0.2 m/s maximum uncertainty. By blocking the sample line, the local gas velocity was calculated within ±0.2 m/s maximum uncertainty from a measurement of the pressure rise at the probe tip due to stagnation of the fluid, using a standard pitot tube equation:

$$V_{pitot} = \sqrt{C\,(2\,\Delta P\,/\,\rho_f)} \tag{1}$$

where

V_{pitot} = Local gas velocity
ΔP = Stagnation pressure rise
ρ_f = Fluid density
C = Constant = 0.98 - 1.0

The excellent agreement between the local gas velocity measurements obtained using the two measurement methods indicates that the probe was capable of detecting the isokinetic condition.

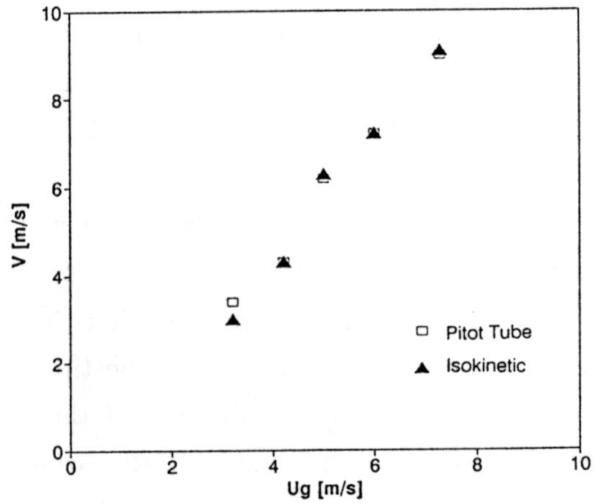

Figure 4: Single-Phase Qualification of Isokinetic Sampling

With the probe facing upstream, the following procedure was used to obtain an isokinetic solids flux measurement. Initially, the probe and pressure taps were purged and the manometer was bypassed. Then, the system was switched into the sampling mode. With the manometer bypass valve and the pressure tap purge valves temporarily closed, the aspiration rate through the probe was adjusted to zero the pressure difference. Sampling was stopped by switching to the purge mode and the sampling vessel was removed, emptied, and replaced. Then, the system was switched to the sampling mode while timing the collection of solids. After a solid sample of approximately 100 grams had accumulated in the collection vessel, the system was switched to the purge mode, the sampling time was recorded, and the collected solid sample was weighed. The solid mass flux through the probe tip was obtained from the following expression

$$G_{s,loc} = M_s/(A_p t_c) \qquad (2)$$

where

$G_{s,loc}$	= Time-average local solid mass flux	kg/m^2s
M_s	= Mass of solids collected	kg
A_p	= Collection area of probe tip	m^2
t_c	= Collection time	s

To qualify the isokinetic solid sampling technique for two-phase, a dilute gas-solid flow was generated in which the solids were observed to flow uniformly upwards through a transparent 5 cm I.D. acrylic riser with no downward solid flow along the wall. With the probe positioned to measure the upward solid mass flux along the riser centerline, solid mass fluxes of 276 μm sand particles were measured at different aspiration velocities while maintaining the solid recirculation rate at approximately 13 kg/m^2s. Figure 5a plots the mass flux normalized with respect to the solid recirculation rate (measured during the sampling interval). as a function of the aspiration velocity. For this dilute-phase flow condition, a uniform solid mass flux profile was expected, having a normalized mass flux close to unity. As Figure 5a indicates, the upward flux was accurately measured over a range of aspiration velocities centered about the isokinetic point where the local flux at the centerline was found to approximately equal the solid recirculation rate measured in the return leg of the CFB. Figure 5b plots the pressure difference (P_{out}-P_{in}) measured as a function of aspiration velocity and indicates that an accurate solid mass flux measurement can be attained with aspiration rates corresponding to probe pressure difference readings ranging from approximately -1 to +1 mm H_2O. A complete radial traverse indicated a relatively flat mass flux profile at this operating condition as was expected.

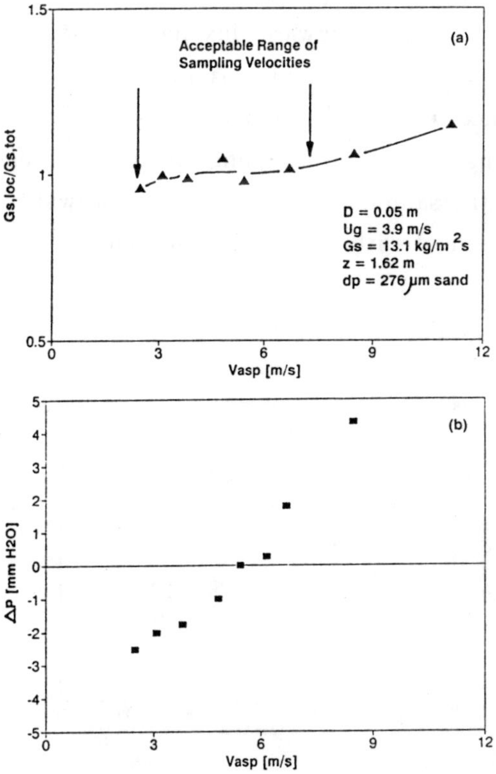

Figure 5: Qualification of Isokinetic Solid Flux Measurements

At higher solid flow rates (increased solid holdup), a downward flow of solids along the riser wall was observed. Therefore, to obtain an accurate measurement of the local time-averaged net solids mass flux, the solids fluxes in both the upstream and downstream directions were measured at a given position, and the difference between the two measurements (upward-downwards flux) yielded the net solids mass flux. A sensitivity study revealed that the downward component of the solids flux was sensitive to the aspiration velocity in the region near the wall where the downflow of solids was more significant. The increase in the measured downward solids flux with increasing aspiration velocity, shown in Figure 6, was expected to result from drawing excess gas and low velocity solids through the probe tip. At the minimum aspiration velocity necessary to continually withdraw solids without blocking the sampling line, the pressure inside the probe tip was lower than the outside pressure. The downward flux measurements were taken at the minimum aspiration velocity to introduce the least disturbance to the gas flow while sampling solids. This sampling condition approaches isokinetic sampling and yields a downward flux measurement with best possible accuracy.

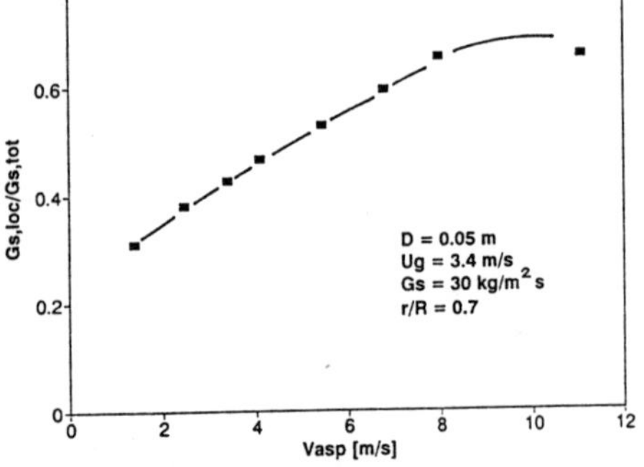

Figure 6: Sensitivity of Net Solids Flux Near the Wall

The uncertainty in the measured local mass flux due to the uncertainties in the collection time, sample mass measurement, and area of the probe tip was estimated to be less than 10% of the measurement. Added uncertainty in the net solids flux measured in the vicinity of the riser wall resulted from the uncertainty in measuring the downward component of the mass flux at this location.

RESULTS AND DISCUSSION

The Net Solids Mass Flux Profile

The net solids mass flux profiles are constructed from separate time-average profiles obtained for both the upward and downward solid mass fluxes. Assuming that the upflowing and downflowing solids flux measurements included only solids that would have passed through the probes sampling area in the upward or downward direction, respectively, in the probes absence, the time-average net solids flux is given by the difference between the time-average up and downflowing solids flux measurements [2]. Two measurements of both the up and downflowing solids fluxes were obtained at eight radial locations (32 flux measurements in total) to determine the solids flux profile at a given operating condition and probe elevation.

Figure 7 shows sample upward and downward solid fluxes plotted against normalized radial position in the riser. The profiles were measured at an elevation of 1.5 m in the 15 cm I.D. riser. FCC particles having a mean diameter of 87 μm were flowing at 20 kg/m^{2}s with a superficial air velocity of 2.4 m/s. The normalized radius is +1 at the inner tube wall next to the test port, 0 at the riser centerline, and -1 at the tube wall directly opposite from the test port. The normalized upward solid flux profile is parabolic and the maximum solids flux at the riser centerline attains a value close to three times larger than the cross-sectional average solid flux. The downward solid flux profile is also parabolic in shape. Within the central core of the riser the downward solid flux is relatively small in comparison to the upward flux; however, it is important to note that a downward flow of solids persists into the central core of the riser. As the wall is approached, the magnitude of the downward solids flux increases sharply and exceeds the magnitude of the upward solids flux near the wall. This suggests the predominance of downward flowing solids in an annular region in the vicinity of the riser wall. The net upward solids flux profile is obtained by adding the upward and downward components of the solid mass flux.

Effect of Radial Position

Figure 8 shows sample measurements of both local, time-average solids mass flux (triangles) and local, time-average solids concentration (circles) plotted against normalized radial position. Solid concentrations were obtained by time-averaging dynamic solid volume fraction measurements obtained using a capacitance probe technique [5]. The data were obtained at an elevation 1.5 m above the secondary air inlet using 87 μm FCC particles flowing at 20 kg/m^{2}s with a superficial gas velocity of 2.4 m/s. The net local solids flux measurements are represented by the solid triangles. The cross-sectional average net solids mass flux obtained by integrating the profile over the riser cross section (16 kg/m^{2}s) agrees reasonably well with the independent total solids flux measurement (20 kg/m^{2}s). As in figure 7, the net solid flux profile is parabolic with net upflow in the riser core and net solids downflow along the riser wall. The mass flux profile decreases gradually in the core region and then drops sharply upon entering the downflowing annulus. It is important to note that the magnitude of the local solids flux deviates significantly from the cross-sectional average value. In the center of the riser, the local flux is more than twice the magnitude of the average net solid upflow. Near the riser wall, the downward flux is comparable to the average net solid upflow (20 kg/m^{2}s), but oppositely directed.

For this operating condition, the annulus, defined by the presence of net solids downflow, is approximately 1 cm thick and occupies 27% of the riser cross-sectional area. Based on an integration

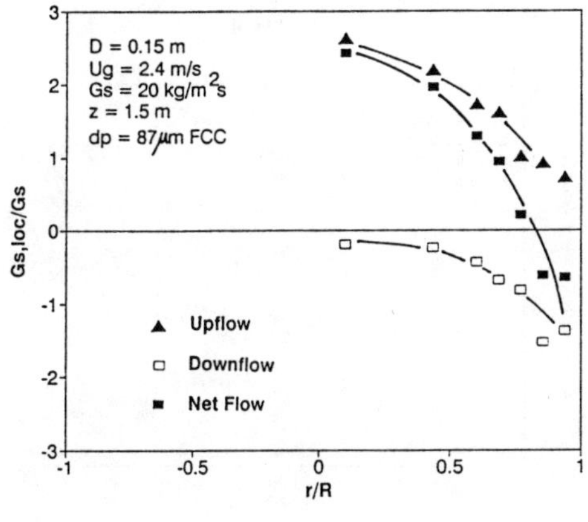

Figure 7: Upward and Downward Solid Mass Fluxes

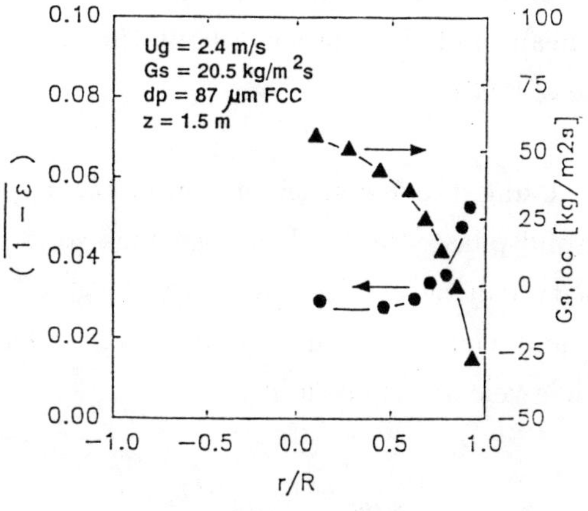

Figure 8: Radial Variation of Solid Concentration and Mass Flux

of the net solid flux profile across the core, the upflow through the core at this operating condition exceeds the measured net solid throughput by 12%. The mass of solids flowing up through the core per unit mass of solids flowing down through the annulus is given by

$$M_c/M_a = (M_c/M_t)/(1 - M_c/M_t) \tag{3}$$

and attains a value of 9 for this case. Since negligible solid downflow is expected at the exit to the riser, the mass flow of solids through the core is expected to equal the solid throughput at the riser exit. From the measurement location (a=1.5 m) to the riser exit, the average rate of change of the net solid upflow through the core with elevation is 0.005 kg/m/s. This quantity represents the net rate of solid transfer from the upflowing core to the downflowing annulus. Since mass is conserved, this is also the rate of increase in solid downflow in the annulus with decreasing elevation in this section of the riser.

As Figure 8 also illustrates, the solid concentration (solid circles) clearly increases from the riser centerline toward the wall. The radial concentration profile is also flatter in the riser core ($r/R < 0.7$) and undergoes a sharp increase near the riser wall. The local solid fraction also shows significant radial variations about the area-average value of 0.037.

Using the equation of continuity applied to the solid phase, the local time-average solid particle velocity can be obtained from the local solid mass flux and concentration measurements. The solid particle velocity is given by

$$V_{s,loc} = G_{s,loc}/(\rho_s \, \epsilon_s) \tag{4}$$

where

$V_{s,loc}$ =	Local time-average particle velocity	m/s
$G_{s,loc}$ =	Local net time-average solid flux	kg/m^2s
ρ_s =	Solid particle density	kg/m^3
ϵ_s =	Local time-average solid volume fraction	(-)

Figure 9a plots the local time-average solid particle velocity calculated using the data in Figure 8. The shape of the solid velocity profile is similar to the solid mass flux profile. Solid particles in the riser centerline experience a net upward flow, with calculated average particle velocities exceeding 1 m/s. As the riser wall is approached, the time-average particle velocity decreases, first gradually and then more rapidly. Close to the riser wall, the FCC particles flow downward with an estimated

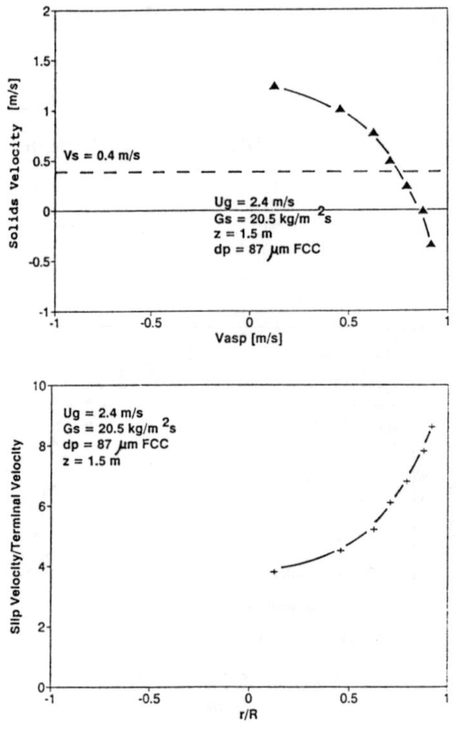

Figure 9: Radial Variation of Solid Particle and Slip Velocities

speed of 0.4 m/s. Hartge et al. [6] have directly measured solid particle velocity profiles in a 40 cm I.D. riser using fiber optic probes and obtained similar profiles. The dashed line drawn at 0.4 m/s represents the value of the mean solid particle velocity defined by:

$$V_s = G_s / (\rho_s \langle \epsilon_s \rangle) \quad (5)$$

In Figure 9b, an estimate of the local particle slip velocity is plotted. The local gas velocity is assumed to be fairly uniform and is approximated by $U_g / \langle 1 - \epsilon_s \rangle$. This plot indicates that the local slip velocity increases from the centerline toward the riser wall. For this operating condition, the estimated slip velocity varies from four to eight times the particle terminal velocity (U_t=0.3 m/s). Obviously, the relative velocity between the gas and the solid is a strong function of radial position.

The above information on the radial variations of the solids flow and concentration behavior provides supporting evidence for a core-annular flow structure within the CFB riser. The core region is dominated by a relatively dilute suspension of solids that flows upward at relatively high speeds. The detection of a downward solids flux component in the core does suggest the slight presence of downflowing solids in this region. Thus, the core may be comprised of disperse upflowing solids with intermittent downflowing solid particle clusters. The annular region in the vicinity of the wall is not a thin layer but is hundreds of particle diameters in thickness. This annulus is characterized by a dense gas-solid mixture with a high tendency to flow downward. The detection of an upward solids flux component in this region suggests the occasional presence of upflowing solids. High speed video photographs taken through a transparent section of the riser revealed a persistent downflow of strands of particles along the riser wall which was occasionally interrupted and temporarily replaced by a disperse upflowing mixture.

CONCLUSIONS

The sampling method used in this study was shown to provide accurate solid mass flux measurements in CFB's. Measurements of the time-average local solid mass fluxes in CFB systems provide clear evidence of a core-annular flow structure in CFB risers. The central core region is dominated by a dilute upflowing suspension of solids flowing at high velocities. The presence of a downward solid flux component in the core region suggests that particle clusters flow intermittently downward within this region. The annulus, having a thickness on the order of centimeters, is occupied by a dense, downflowing solid suspension having relatively low velocity. The detection of an upward solid flux component and video photographs indicate that the downflowing solid strands may be temporarily interrupted by a disperse upflowing mixture. A fundamental understanding of the nonuniform solid flow behavior in the CFB systems is critical to successful modeling of industrial CFB processes.

ACKNOWLEDGMENT

This work was supported by Air Products & Chemicals, Inc. and Lehigh University.

NOTATION

A_p	Probe Sampling Area	m^2
d_p	Sauter Mean Particle Diameter	μm
G_s	Total Solid Circulation Rate	kg/m^2
$G_{s,loc}$	Net Local Solids Mass Flux	kg/m^2s
M_a	Solid Flow Down Through Annulus	kg/s
M_c	Solid Flow Up Through Core	kg/s
M_t	Total Solid Flow Rate	kg/s

t_c	Sample Collection Time	s
U_g	Superficial Gas Velocity	m/s
U_t	Terminal Velocity	m/s
V_{pitot}	Pitot Tube Velocity	m/s
V_s	Mean Solids Velocity	m/s
$V_{s,loc}$	Local Solids Velocity	m/s
z	Axial Distance Above the Secondary Air Inlet	m

Greek Characters

ϵ_s	Solid Volume Fraction	-
$<\epsilon_s>$	Cross-sectional Average Solid Volume Fraction	-
Δ P	Difference Between Inside and and Outside Pressures	Pa
ρ_s	Solid Particle Density	kg/m^3

LITERATURE CITED

1. Dou, S., Herb, B., Tuzla, K., Chen, J. C., "Heat Transfer Between an Immersed Tube and a Circulating Fluidized Bed", presented at AIChE Annual Meeting in San Francisco., (1990).

2. Van Breugal, J.W., Stein, J.M., and de Vries, R.J., "Isokinetic Sampling in a Dense Gas-Solids Stream", Proc. Instn. Mech. Engrs., 184, 18-23, (1970).

3. Bierl, T.W., Gajdos, L.J., McIver, A.E. and McGovern, J.J., D.O.E. Report No. EE2449-11, (1980).

4. Montceaux, L., Azzi, M., Molodtsof, Y. and Large, J.F., "Overall and Local Characterization of Flow Regimes in a Circulating Fluidized Bed", in Circulating Fluidized Bed Technology, pp. 185-201, Ed. P. Basu, Pergamon, (1986).

5. Herb, B., Tuzla, K., Chen, J.C., "Distribution of Solid Concentrations in Circulating Fluidized Beds", Fluidization VI, Engineering Foundation, New York, 65-72, (1989).

6. Hartge, E.U., Li, Y., and Werther, J., "Analysis of the Local Structure of the Two-Phase Flow in a Fast Fluidized Bed", Circulating Fluidized Bed Technology II, 153- 160, (1988).

7. Rhodes, M. J., Laussman, P., Villian, F. and Geldart, D., "Measurement of Radial and Axial Solids Flux Variations in the Riser of a Circulating Fluidized Bed", Circulating Fluidized Bed Technology II, (1988).

CORRECTION OF ERRORS IN INTRINSIC THERMOCOUPLE SIGNALS RECORDED DURING QUENCHING

James E. Park, Kenneth W. Childs,
and Gerard M. Ludtka ■ Martin Marietta Energy Systems, Inc., P.O. Box 2003, Oak Ridge, TN 37831-7039
William Chu ■ Fluid Dynamics, Inc., Evanston, IL 60201

The error in a surface temperature measured using an intrinsic thermocouple is quantified by numerically simulating the quenching of a hot (500°C) nickel cylinder into cold (20°C) water. A two-dimensional (axisymmetric) analysis indicates an error of about 40°C that occurs within 0.1 s of immersion and persists for almost 1.7 s. A three-dimensional simulation of the region near the separated lead wires predicts an error of similar magnitude. To correct for the lead conduction, a parameter estimation algorithm suggested by J.V. Beck is being developed to determine the governing surface heat flux - surface temperature curve. In this paper, the error analysis is described, the parameter estimation algorithm is presented, and a preliminary one-dimensional implementation of the algorithm is used to process the data from a quenching experiment.

The surface temperature of a solid object undergoing rapid cooling (quenching) may be measured using sensors attached directly to the surface or mounted within the object behind the cooled surface. Each approach has problems that must be addressed.

The problems appear as errors in the indicated temperatures. Instrumentation errors may occur due to electronic noise, calibration, truncation during data recording, etc. Conversion of the recorded voltages into temperatures can introduce additional errors.

This paper presents numerical estimates of the particular error introduced by heat conduction through the leads of surface mounted thermocouples (TCs) during rapid transient cooling (i.e., quenching). It is shown that conduction through the lead wires during boiling causes a significant depression of the surface temperature in the immediate vicinity of the attachment point(s).

Because of the depression in the local temperature, the problem of extracting the undisturbed surface temperature from the measured value is multidimensional. A parameter-estimation technique to extract the surface heat flux and the undisturbed surface temperature from the signal from a surface-mounted TC is described. A one-dimensional antecedent of the algorithm is used to smooth data from a quenching experiment.

Embedded Sensors

If the sensor is mounted behind the surface, at best it responds to the temperature at the point behind the surface. The surface temperature during the quenching operation must be extracted from the data using Inverse Heat Conduction techniques (see Beck et al., for a review).[1] Generally, the further behind the cooled surface the sensor is located, the more difficult the analysis and the less accurate the results. Also, sensors may introduce errors into the measurements by altering the heat flow near the sensing element. This variation in heat flow can occur due to incompatible materials, lack of good thermal contact between the sensor and the test piece, or impurities (such as brazing flux) introduced during installation of the sensor. Beck[2,3] has described the analysis techniques appropriate for this configuration. Attia and Kops[4] analyzed the steady-state problem of combined error due to below-surface installation and losses through exposed leads.

Surface-Mounted Sensors

If sensors are attached to the surface of the object being cooled, a different set of problems must be faced. Chemical reaction between the cooling fluid and the sensor is possible. In some cases, a reliable thermal contact between the surface and the sensor cannot be maintained. Attachment by mechanical means (screws, straps, adhesives) introduces additional mass and possibly electrically dissimilar materials to modify the voltage output from the sensor. If the lead wires are not carefully installed, mechanical stress due to differential thermal contraction during cooling can lead to failure of the sensor before the quench has been completed. Spot welding is often the method selected to fasten a TC to a surface.

Heat loss through the lead wires is a source of error in measurements made using surface-mounted TCs. The wires act as fins, creating preferential heat flow paths; the enhanced heat flow lowers the temperature of the bimetallic junction more rapidly than the surface temperature not disturbed by the lead. Thus, the temperature indicated by a surface TC may be different from the true (undisturbed) surface temperature at any time during the quench.

Beaded TCs. A beaded thermocouple consists of two leads of dissimilar metals that are joined into a bead at the sensing junction by spot welding or brazing. The bead is attached to the surface for which the temperature is sought. The mass of the junction bead contributes to measurement errors because of its thermal inertia; the temperature of the bead differs from the surface temperature during temperature transients such as quenching.

The effect of the heat loss through TC wires has been studied for the past three decades. For a surface-mounted sensor, emphasis has been on the calculation of the response of an attached TC to sudden changes in the temperature of the plate. An idealized configuration is studied; the lead wires and the bead are represented by a cylinder attached normal to a flat surface. This problem was treated by Henning and Parker[5] and Keltner and Bickle.[6] Both teams presented analytic solutions to approximations to the problem. Gat et al.[7] added the treatment of temperature-dependent transport properties using two numerical methods. Keltner and Beck[8] applied their unsteady surface element (USE) method to this problem. A later paper (Keltner and Beck)[9] considered the added complexities of contact resistance between the wire and plate and lateral convection (with constant heat transfer coefficient) from the wire. A simplified evaluation technique followed (Beck et al.).[10] Litkouhi and Beck[11] examined the two-dimensional character of the problem. The solutions presented in those papers are limited to a single lead wire, either insulated or subject to a constant surface heat transfer coefficient.

Unfortunately, if quenching of hot metal with a cold liquid leads to boiling, the surface heat flux on the part is highly nonlinear (Moreaux and Beck; Park and Ludtka).[12,13] That is, the surface heat flux is a strong function of surface temperature and the temperature of the quenching liquid (Figure 1); thus the linear transform-based techniques used in the foregoing references are not applicable.

Intrinsic TCs. The effect of heat conduction through the leads on the indicated temperature can be reduced by using an intrinsic TC. In this design, the leads are individually spot-welded to the surface, usually one or two wire diameters apart (Kollie).[14] The apparent electrical junction is located in the base material, somewhere between the spot welds. The temperature at the virtual junction is nearer to the temperature of the undisturbed surface than the depressed temperature at the base of the lead wires.

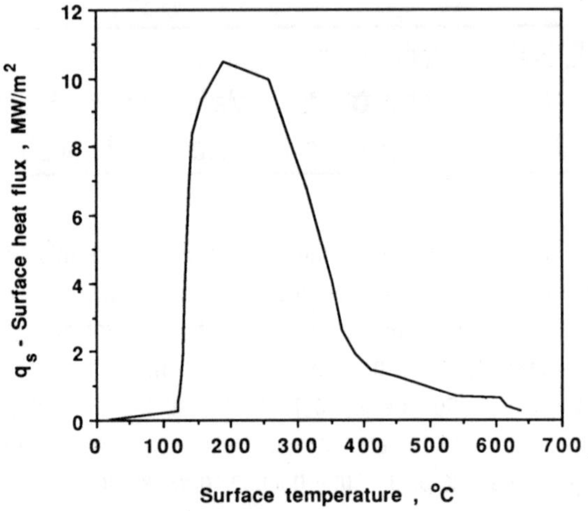

Figure 1. Surface heat flux vs surface temperature.

This paper presents numerical estimates of the error due to conduction through the leads of a beaded and an intrinsic TC. A parameter-estimation technique is proposed to extract the surface heat flux and the undisturbed surface temperature from the signal measured by an intrinsic TC. A one-dimensional antecedent of the algorithm is applied to data from a quenching experiment.

Use of Model to Estimate the Measurement Error

Two problems are of interest: First, two-dimensional heat flow in the plate near the cylinder representing a beaded TC is calculated. Second, the three-dimensional heat flow in the plate near the attachment points for the two leads that constitute an intrinsic TC is calculated. In each case, heat flow is also calculated in the wires. The temperatures at the TC junction and at an undisturbed surface location are calculated as a function of time for each configuration. The difference between the calculated temperatures at the two locations is the error introduced by lead conduction. Temperatures were calculated using the HEATING6 thermal analysis code (Elrod).[15]

The numerical calculations simulate cooling of a cylinder made from nickel tubing. This is part of an apparatus used by Park and Ludtka[13] for a series of quenching experiments. The wall thickness of the cylinder is 9.5 mm (3/8 in.). The assumption is made that the cylinder has a large enough diameter that the wall can be considered a flat plate.

In the problem of interest (see the description of the experiment below), the interior of the cylinder is isolated from the quenching medium (water) so that the interior surface is assumed to be adiabatic. From preliminary transient conduction analyses, it was determined that the heat flow through the plate remained one-dimensional during the quench to within about 1 mm of the cooled surface. Thus, the adiabatic boundary condition 9.5 mm below the cooled surface is replaced with the one-dimensional temperature solution at a depth of only 1.08 mm. This value is an integer multiple of one wire radius. This device eliminated the necessity of meshing almost 90% of the nickel plate.

The properties used for nickel and the thermocouple wires (Chromel and Alumel) are assumed to be constant and are given in Table 1.

Table 1. Physical Properties for Wires and for Nickel

	Nickel	Chromel	Alumel
Thermal conductivity W/(m·K)	83	19.2	29.7
Thermal diffusivity, m^2/s	19.84×10^{-6}	4.91×10^{-6}	6.59×10^{-6}

The initial temperature for the entire plate and wire was set at 500°C. At the centerline and at the one-dimensional boundaries far from the centerline, the heat flux was zero. At the bottom boundary, the time-dependent temperature was prescribed. On the surface of the wire and the upper surface of the plate, the heat flux was a specified function of the local surface temperature as determined by Park and Ludtka.[13] The weld between the wire(s) and the plate was assumed to be perfect (no contact resistance).

<u>**Beaded TC - Two-Dimensional Model**</u>

Far from the wire, the heat flow to the surface is one-dimensional. At that point, the calculated surface temperature should be the same as that calculated using the one-dimensional model described above. By running several calculations using different radial extents for the meshes, the one-dimensional surface was located 4.36 mm (about 69 wire radii) from the centerline of the wire. The radius of the wire is 0.0025 in. (0.0635 mm). Figure 2 shows the more coarse of two finite-difference meshes used.

For the two-dimensional model, the chromel and alumel wires are presumed to act as a single wire having a diameter and thermal properties as suggested by Attia and Kops,[4] that cylindrical wire is assumed to be attached normal to the nickel plate, and the problem is axisymmetric (r-z).

<u>Results</u>. The results are shown in Figure 3. Solutions for three cases are presented. Each case was solved with the minimum cell width of one wire radius and then resolved using a mesh formed by dividing each cell width into two cells. Two of the cases use constant values for the surface

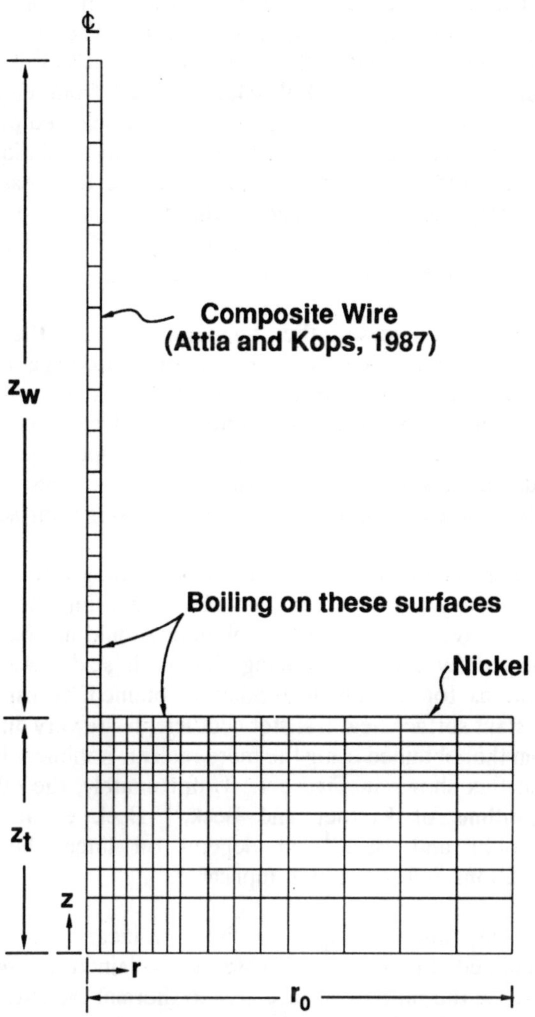

Figure 2. Finite-difference mesh used to determine temperature field in the vicinity of the beaded TC junction.

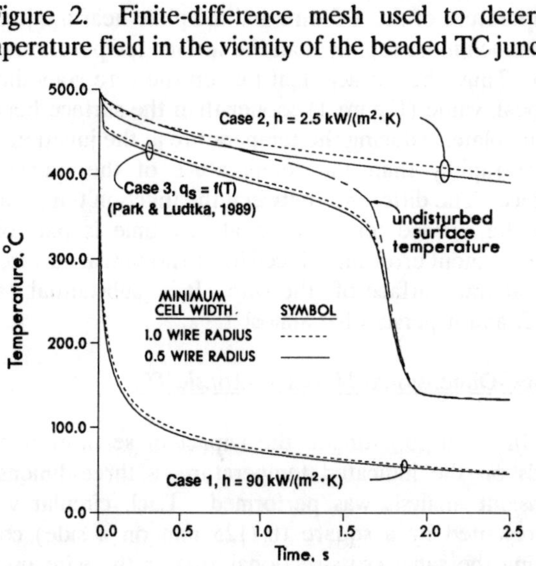

Figure 3. Indicated and undisturbed surface temperatures calculated using two-dimensional axisymmetric model with various surface boundary conditions.

heat transfer coefficient (h) as boundary conditions at the cooled surface of the plate and the wire. The values of h were chosen to match the maximum [90 kW/(m^2·K)] and minimum [2.5kW/(M^2·K)] values derived from Figure 1. The constant-h boundary conditions were selected because they are linear and thus can be used with the techniques of Keltner and Beck,[9] for example. For the third case, the boundary condition applied at the cooled surface of the plate and the wire is the nonlinear surface heat flux appropriate for transient boiling (Figure 1).

For each case, the temperatures for the node representing the bead temperature are plotted against time in Figure 3. The node on the centerline at the interface between the plate and the wire was selected to be the junction node. For the third case, the temperature of the undisturbed surface far from the TC junction is also shown. This is the temperature to be measured using a surface TC!

For each of the cases, the solution using a finer mesh differs only slightly from that obtained using the coarse mesh. The character of the solutions should not be changed by further reducing the mesh size. Also, the solutions for the linear problems obtained by using the constant surface heat transfer coefficients are very different from that obtained using the more realistic nonlinear boiling heat flux shown in Figure 1. Unfortunately, the efficient algorithms of Keltner and Beck,[8,9] Beck et al.,[10] and Litkouhi and Beck[11] all depend on linear boundary conditions and so are not applicable.

The unique shape of the nonlinear temperature calculated for the third case is explained as follows. Because the ratio of surface area to thermal capacity of the wire is high relative to the same ratio for the plate, the temperature of the TC wire and the plate near the junction falls much more rapidly than the surface temperature of the slab. Thus, the surface heat flux on the wire goes through its peak value (Figure 1) sooner than the surface heat flux of the plate, reducing the temperature at the junction much more rapidly than the temperature of the undisturbed surface. The difference between the junction temperatures and the surface temperature at any time is part of the measurement error introduced by the nonlinear surface heat flux at the surface of the wire. It is substantial, about 40°C, and it persists for almost 1.8 s.

Three-Dimensional Model - Intrinsic TC

In order to estimate the impact of separation of the leads on the indicated temperature, a three-dimensional transient analysis was performed. Each circular wire is represented by a square (0.1125 mm on a side) column having the same cross-sectional area as the wire used for the two-dimensional calculation. The wires are two "equivalent diameters" apart. The surface heat flux shown in Figure 1 is applied to the slab surface and, using the surface area for the cylindrical wire, to the wire surfaces.

Description. A view of the model is shown in Figure 4. The detail for the TC wires and the fine zoning used in the vicinity of the wire-to-plate junctions can be seen. The one-dimensional transient solution at a depth of 1.08 mm was used as the bottom boundary condition.

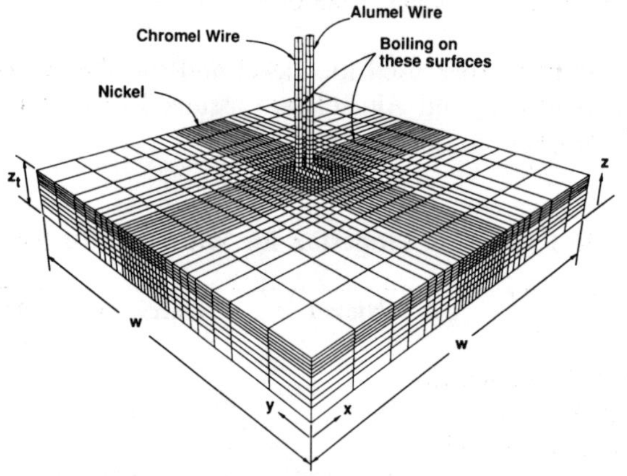

Figure 4. Finite-difference model used to determine three-dimensional temperature field near lead wires for an intrinsic TC.

Results. The problem was run out to t = 0.06 s, requiring 1000 time steps. This time span covers the initial rapid drop in temperature seen in Figure 2 for the third case. The cost of the calculation precluded extension of the simulation to longer times. Figure 5 shows the undisturbed surface temperature, the temperatures at the junctions between each wire and the plate, and the temperature midway between the vertical portion of the two lead wires. The latter point was arbitrarily selected as the virtual junction for the intrinsic thermocouple. The undisturbed temperature is actually three curves – one curve at the surface of the plate on each of two outer edges and one curve at a corner of the surface.

The steep descents occurring between 0.02 and 0.03 s in the temperatures from the two wire junctions correspond to the heat flux on the surface of each wire passing through its peak. These curves have slightly different shapes because of the difference in thermal properties (Table 1). In each case, the error is about 40°C, which is similar to the two-dimensional results.

The selection of a node on the plate surface midway between the two leads to represent the junction of the intrinsic TC was arbitrary; the actual temperature that corresponds to the TC output and the physical point at which that temperature occurs are not known. The temperatures calculated along the lead-plate intersections at a particular time (not presented) show a distinct gradient, with the lowest temperatures at the point where the lead leaves the plate surface. To determine the temperature

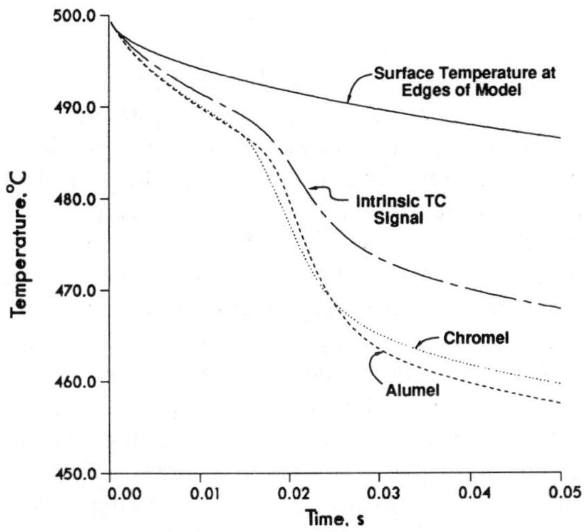

Figure 5. True surface temperature and temperatures calculated near the intrinsic surface TC 3-d problem.

indicated by the TC during the simulation, the actual voltage generated at the dissimilar metal interfaces at each time must be calculated by integrating the appropriate path functions of the Seebeck coefficients and the local temperature gradients (Pollock).[16] That voltage could then be translated into an indicated temperature using a calibration function (Park et al.).[17]

Summary of TC Lead Error Analysis

Lead losses bias the temperature indicated by a surface-mounted TC. For a quench from 500°C, the error for a type-K TC attached to a nickel plate occurs immediately on immersion of the hot part and persists for almost 2 s. The magnitude of the error for a beaded TC is apparently about 40°C. An intrinsic TC using leads separated by 1-2 wire diameters has an error of similar magnitude.

Parameter Estimation

The purpose of the experiment mentioned above (Park and Ludtka)[13] is to determine the boiling curve for quenching – that is, a curve of surface heat flux vs surface temperature that allows calculated temperatures for the apparent TC junction to match the measured values is desired. Since the heat flow near the TC attachment is multidimensional and distorts the temperature indicated by the TC, a multidimensional analysis of some type is required to interpret data recorded from surface TCs.

Two-dimensional numerical inverse heat conduction algorithms, such as those developed by Bass, Drake, and Ott,[18] and Osman and Beck,[19] assume that the surface temperature at any time is a smooth function (e.g., a superposition of sine waves or low-order polynomials) of distance along the cooled surface. Thus, these methods do not consider the local disturbance that the sensor introduces into the temperature field being measured.

Technique Suggested by J. V. Beck

A quenching experiment (Park and Ludtka)[13] had resulted in highly oscillatory TC signals that required smoothing either before or as a part of an inverse heat conduction analysis to determine the surface heat flux curve (Park, Chu, and Ludtka).[20] J. V. Beck[21] suggested that the heat flux curve be determined using the techniques of parameter estimation (Beck and Arnold).[22] Using parameter estimation together with a one-dimensional heat conduction analysis, a smooth boiling curve could be economically extracted by sequentially determining less than fifty parameters rather than several thousand unknown surface heat flux values (one for each temperature measurement). The temperature data would be automatically smoothed by the conduction analysis.

More specifically, the unknown heat-flux curve is approximated by straight-line segments. The surface heat flux in each segment is determined from the local surface temperature by interpolating between the end points of the segment

$$q_s = Q_k + (T_s - T_k)(Q_{k+1} - Q_k)/(T_{k+1} - T_k), \qquad k = 1, 2, ..., K \quad T_{k+1} \le T_s \le T_k \tag{1}$$

Starting at the highest surface temperature (k = 1), a conduction analysis is done to calculate a temperature curve at the TC junction over that increment of the data. The unknown parameter to be estimated is Q_{k+1}, the surface heat flux at the low-temperature (T_{k+1}) end of the straight-line segment. This basic procedure is repeated for each segment of the surface heat-flux–surface-temperature curve, with the temperature solution in the plate and wire for each segment of the curve serving as the initial condition for the next segment.

The method developed to extract a surface heat flux curve from signals generated by a surface-mounted TC is described below. Starting with Beck's suggestion, a technique has been devised that treats the problem of high-frequency oscillations in the data as well as the two-dimensional distortion of the local temperature field due to conduction along the TC lead wires.

Model

Because of the lead conduction, the temperature field near the TC attachment point is two- or three-dimensional even if the overall quenching problem can be considered to be one-dimensional. Using a two-dimensional axisymmetric analysis, the lead wire pair is modeled as a single, composite wire (Attia and Kops).[4] Heat flow through the nickel plate and through the composite wire attached normal to the surface of the plate is governed by the transient conduction equation

$$\rho c \frac{\partial T}{\partial t} = \frac{\kappa}{r} \frac{\partial}{\partial r} \left(r \frac{\partial T}{\partial r} \right) + \kappa \frac{\partial^2 T}{\partial z^2} \tag{2}$$

Initial and Boundary Conditions

The initial temperature is specified. For the normal quenching experiment, the initial temperature is uniform. The inner boundary of the nickel plate is assumed to be adiabatic. Since the heat flow is presumed to be one-dimensional at some distance ($r = r_o$) from the centerline, that boundary is also adiabatic. The heat flux is also zero at the symmetry axis. Thus, the boundary conditions are

$$\begin{aligned} q_s &= -\kappa \frac{\partial T}{\partial z} (r, 0) = 0 \\ q_s &= -\kappa \frac{\partial T}{\partial r} \left(r_0, z \right) = -\kappa \frac{\partial T}{\partial r} (0, z) = 0 \end{aligned} \tag{3}$$

The heat flux on the surface of the wire ($r = r_w$) and the upper surface of the nickel plate ($z = z_t$) is given by Eq. (1).

Method

Overview. As suggested by Beck, the unknown heat flux curve is approximated by straight-line segments, Eq. (1). The boiling curve is the same on the surface of the wire and on the surface of the nickel. In order to start the process, the heat flux is assumed to be constant in the first segment, $Q_2 = Q_1$. An alternative would be to specify Q_1.

If one tries to determine each parameter Q_{k+1} in turn without considering coupling with Q_{k+2}, the value at the end of the next boiling-curve segment, the surface heat flux curve develops oscillations that eventually cause the process to fail. To avoid those oscillations, each pair of adjacent line segments is considered in turn. Thus, for k = 1, 2, ... , K-1, the parameters Q_{k+1} and Q_{k+2} are determined such that the sum

$$S_k = \sum_{i=L_k}^{L_{k+2}} \left(Y_i - T_i \right)^2 / D_k \tag{4}$$

is minimized. The sum, S_k, is the mean-square deviation per point between the TC temperature calculated by the model, T_i, and the measured value, Y_i. The number of points included in the sum, D_k, is given by

$$D_k = L_{k+2} - L_k + 1 \tag{5}$$

For the axisymmetric case, the TC junction is assumed to be at r = 0, $z = z_t$. L_k, L_{k+1}, and L_{k+2} are the indices corresponding to the times at which the minimum calculated surface temperature drops below T_k, T_{k+1}, and T_{k+2} respectively. Minimization of S_k is an iterative process. D_k will vary during the iterative minimization of the sum because the calculated temperature depends on the value of the parameters used, dropping faster if the surface heat flux is raised. The values of L_k and Q_k are known from the solution for the previous line segment. After the sum S_k has been minimized, the values of Q_{k+1} and L_{k+1} are accepted; Q_{k+2} and L_{k+2} are used as starting guesses to analyze the next pair of line segments. Fitting two segments for each minimization in Eq. (5) suppresses the tendency for the solution curve to oscillate.

Parameter Estimation

In the iterative process, one starts with estimates for Q_{k+1} and Q_{k+2} for the kth line segment and attempts to correct those values of Q to give a lower value of S. The process is repeated until S is constant within some tolerance. The present algorithm is detailed in the appendix.

Equations (2) and (A10) were approximated using an explicit, conservative finite-difference technique.

Summary of Experiment

The experiments and the preliminary conclusions have been described fully in Park and Ludtka.[13]

Apparatus. The apparatus (Figure 6) consisted of a heavily instrumented hollow nickel cylinder, a quench tank fabricated from a 200-L (55-gal) drum, a large vacuum furnace, and a Keithley high-speed data acquisition system.

The cylinder (304.8-mm long × 152.4-mm dia) was fabricated from a length of commercial nickel tubing, with welded caps enclosing the two ends (the top and bottom in the quenching orientation). Eighteen type-K TCs were attached to the cylinder. Ten were intrinsic TCs spot welded to the outer surface; eight were brazed into radial holes drilled into the wall of the cylinder from the inside; lead wires from these were contained within the cylinder. Leads from all the TCs were routed to the top of the cylinder, through the stainless steel pipe attached to the cylinder. From the top of the pipe, the leads were routed to the Keithley A/D recorder.

Procedure. The nickel test piece was heated in the vacuum furnace (operated at a pressure of less than 1.3×10^{-3} Pa) to a temperature of 810°C and held until the oven

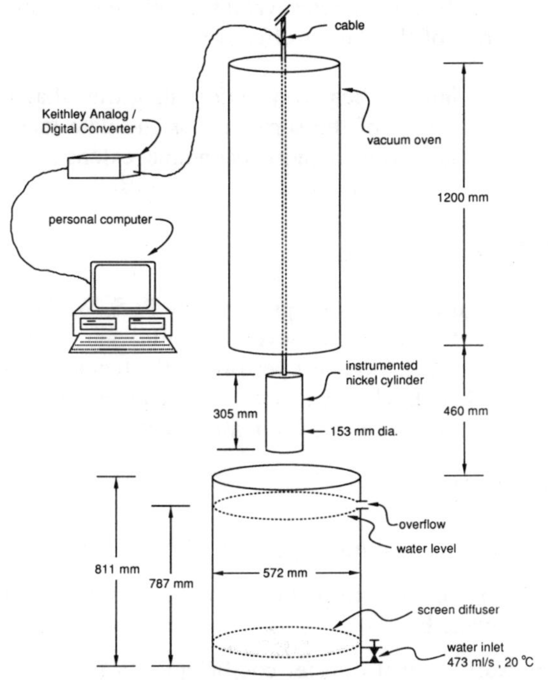

Figure 6. Experimental quenching facility.

controller indicated a uniform, stable temperature. Water inlet temperature for the experiment was 20°C. The cylinder was lowered from the oven to a position about 25 mm above the surface of the water; the data recording system was then activated, and the cylinder was dropped into the water. Signals from sixteen of the TC were digitized and recorded at 4-ms intervals for 40 s. The transient TC outputs recorded by the Keithley acquisition system were analyzed using a CRAY X/MP supercomputer.

Sample Data. Temperatures extracted from signals generated by a surface-mounted intrinsic TC mounted 186 mm (7.33 in.) above the bottom of the cylinder are shown on Figure 7. Because the recorder was started after the test piece left the oven, the initial recorded temperature is about 735°C instead of the 810°C recorded in the interior of the oven. The signal contains a great deal of "noise" which has been attributed to formation and collapse or shedding of steam bubbles on the hot surface (Park and Ludtka).[13] Before 12.5 s, the boiling is characterized as stable film boiling; at about 12.5 s, the surface film becomes unstable and collapses. Nucleate boiling then dominates, decaying into natural convection as the surface temperature drops below 100°C at about 27 s.

Model Parameters. To date, the parameter estimation algorithm has been implemented using one independent variable, the direction (z) normal to the surface of the plate. The entire thickness of the nickel plate (9.5 mm) was discretized using twenty computational cells. The smallest cells (near the cooled surface) were 0.9 mm wide; the width of cells further from that surface increased gradually.

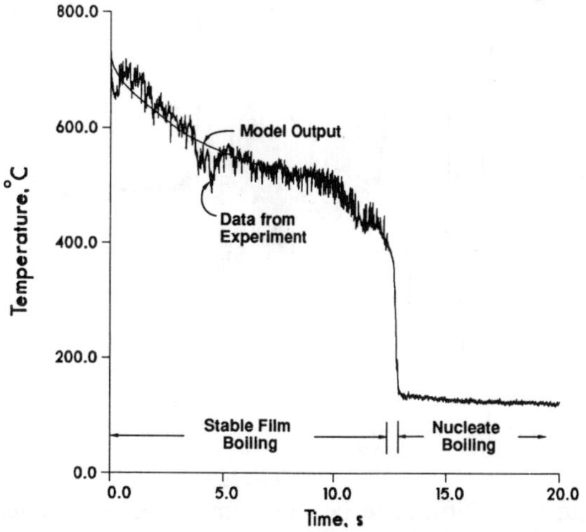

Figure 7. Temperature (°C) measured from a surface-mounted intrinsic TC during quenching and smoothed (one-dimensional) values using parameter estimation.

Because the calculation was one-dimensional, the wire was not included in the model. The present implementation of the algorithm is explicit, with a time of 1×10^{-4} s. Thus, forty time steps were used to span one data point; 2×10^5 time steps were required to span 20 s of data (5000 points). Depending on the number of iterations used to determine each value of Q_k, each problem solution required as much as 1200 s of processor time.

Preliminary Calculated Results - Comparison with Data. The calculated surface temperature at the TC is also shown in Figure 7. Several observations are noted. First, the calculated temperature is indeed smoothed; essentially no trace of the high-frequency oscillation remains. Second, the model result appears to capture the trend of the data. However, major upswings in the recorded signal that appear at about 1.5 s and 4.0 s are not captured by the algorithm. Third, the smoothing appears to be satisfactory in the nucleate boiling regime.

Figure 8 shows details of the difference between the smoothed and the measured temperatures. Except for the regions noted in the previous paragraph, the differences are roughly centered about 0. The differences are markedly smaller in the nucleate boiling regimes.

The boiling curve that produced the calculated surface temperatures shown in Figure 7 is shown in Figure 9. The solid line is the initial guess for the curve; the dashed line resulted from the calculation. Clearly, the initial estimates for the Q_k were not satisfactory in the region of peak heat flux. To repeat, the surface heat flux shown in Figure 9 is for the one-dimensional problem. This curve is not a correction for fin cooling from the intrinsic TC lead wire.

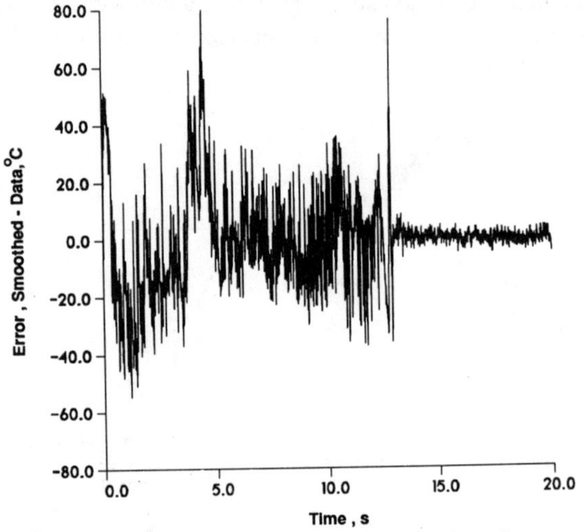

Figure 8. Difference between smoothed and measured temperatures (°C) from quenching experiment.

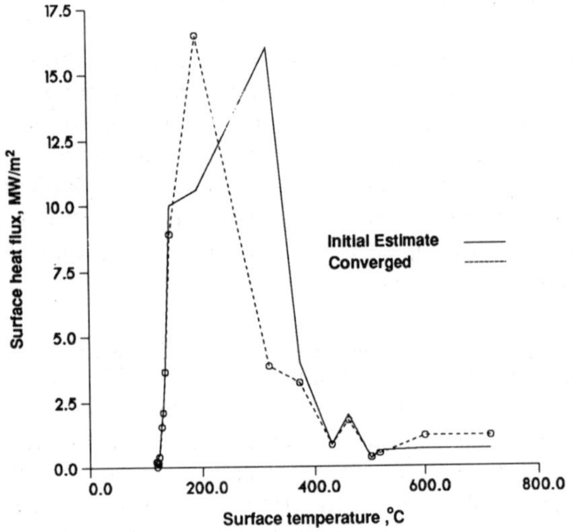

Figure 9. Initial and final values for parameters, Q_k, determined by parameter estimation algorithm.

CONCLUSIONS AND RECOMMENDATIONS

Although the comparison shown in Figure 7 is encouraging, the algorithm presented in this paper has been demonstrated for only one dimension. Some specific conclusions and directions for future development follow.

1. Two- and three-dimensional model studies indicate that surface temperatures measured using an intrinsic TC may be in error as much as 40°C due to lead heat loss.

2. The oscillations in the data recorded during quenching are smoothed using the proposed algorithm. The smoothed curve appears to follow the major trends in the data.

3. The parameter estimation scheme used to correct the two parameter values determined over each span of the data is effective.

4. An implicit heat conduction algorithm that allows time steps of the same size as the data increment would permit a more economic extension to the two-dimensional problem of interest.

ACKNOWLEDGMENTS

The figures have been prepared by R. S. Wallace; A. I. Rose word processed several drafts and the final manuscript; C. H. Shappert performed the technical editing. Professor J. V. Beck of Michigan State University motivated the use of the Parameter Estimation technique.

NOTATION

c = specific heat for metal, Eqs. (3) and (A10), J/(kg·K)

D_k = number of data points in the kth and k + first data spans, Eq. (5)

h = surface heat transfer coefficient, kW/(m^2·K)

K = number of points in the surface heat flux - surface temperature table

L_k = index of last data point in kth span of data

n = direction normal to the surface, Eq. (A11)-(A13)

Q_k = parameter for high-temperature end of interpolating function for kth span of surface heat flux vs surface temperature curve, Eq. (1), W/m^2

q_s = surface heat flux, W/m^2

r = radial coordinate, m

r_o = outer radius of mesh, m

r_w = radius of wire, m

S_k = average mean-squares deviation per data point for the kth span of the data, objective function to be minimized by parameter estimation, Eq. (4)

t = time variable, s

T_i = calculated temperature corresponding to ith data point, Eq. (4), °C

T_k = high-end temperature of interpolating function for kth span of surface heat flux vs surface temperature curve, Eq. (1), °C

T_s = surface temperature, °C

X_{ij} = sensitivity coefficient relating change of ith calculated temperature to jth parameter, Eq. (A4), (°C·m^2)/W

Y_i = ith data point, Eq. (4), °C

z = axial coordinate, m

z_t = plate thickness, m

Greek letters

ρ = density of metal, kg/m^3

κ = thermal conductivity of metal, W/(m·K)

APPENDIX: PARAMETER ESTIMATION USING SENSITIVITY COEFFICIENTS

An overview of parameter estimation techniques is available in Beck and Arnold.[22] The implementation for the present problem is presented below.

The minimum of S_k [Eq. (4)] corresponds to the total variation of S_k being zero,

$$\delta S_k = \frac{\partial S_k}{\partial Q_{k+1}} \delta Q_{k+1} + \frac{\partial S_k}{\partial Q_{k+2}} \delta Q_{k+2} = 0 \qquad \text{(A1)}$$

Since the variations in the Q may not be zero, the partial derivatives must each vanish identically at the optimum,

$$\frac{\partial S_k}{\partial Q_{k+1}} = -\frac{2}{D_k} \sum_{i=L_k}^{L_{k+2}} (Y_i - T_i) \frac{\partial T_i}{\partial Q_{k+1}} = 0 \qquad \text{(A2)}$$

and

$$\frac{\partial S_k}{\partial Q_{k+2}} = -\frac{2}{D_k} \sum_{i=L_{k+1}+1}^{L_{k+2}} (Y_i - T_i) \frac{\partial T_i}{\partial Q_{k+2}} = 0 \qquad \text{(A3)}$$

The dependence of L_{k+1} and L_{k+2} on the parameters has been neglected in Eqs. (A2) and (A3). The partial derivatives in Eqs. (A2) and (A3) are the sensitivity coefficients,

$$X_{i,j} = \frac{\partial T_i}{\partial Q_j} \qquad \text{(A4)}$$

The change in the calculated temperature from iteration n to iteration n + 1 is related to the change in the parameters by Eq. (A5),

$$T_i^{n+1} = T_i^n + X_{i,k+1}\,\delta Q_{k+1} + X_{i,k+2}\,\delta Q_{k+2} \qquad \text{(A5)}$$

Substituting Eqs. (A4) and (A5) into (A2) and (A3) yields a pair of linear equations for the change in the parameters.

$$\sum_{i=L_k}^{L_{k+2}} X_{i,k+1}^2\,\delta Q_{k+1} + \sum_{i=L_{k+1}+1}^{L_{k+2}} X_{i,k+1} X_{i,k+2}\,\delta Q_{k+2} = \sum_{i=L_k}^{L_{k+2}} (Y_i - T_i^n) X_{i,k+1} \qquad \text{(A6)}$$

$$\sum_{i=L_{k+1}+1}^{L_{k+2}} X_{i,k+1} X_{i,k+2}\,\delta Q_{k+1} + \sum_{i=L_{k+1}+1}^{L_{k+2}} X_{i,k+2}^2\,\delta Q_{k+2} = \sum_{i=L_{k+1}+1}^{L_{k+2}} (Y_i - T_i^n) X_{i,k+2} \qquad \text{(A7)}$$

These equations are easily solved for δQ_{k+1} and δQ_{k+2}. Finally, the estimates for the parameters are updated using Eqs. (A8) and (A9).

$$Q_{k+1}^{n+1} = Q_{k+1}^n + \delta Q_{k+1} \qquad \text{(A8)}$$

$$Q_{k+2}^{n+1} = Q_{k+2}^n + \delta Q_{k+2} \qquad \text{(A9)}$$

Sensitivity Equation. The sensitivity coefficients are conveniently calculated using a method of Beck, Litkouhi, and St. Clair.[23] Equation (2) is differentiated with respect to each parameter, the order of differentiation is reversed, and the definition of the sensitivity coefficients, Eq. (A4), is applied, giving

$$\rho c \frac{\partial X_j}{\partial t} = \frac{\kappa}{r} \frac{\partial}{\partial r}\left(r \frac{\partial X_j}{\partial r}\right) + \kappa \frac{\partial^2 X_j}{\partial z^2}, \; j = k+1,\, k+2 \qquad \text{(A10)}$$

The initial conditions for Eqs. (A10) are $X_j = 0$. The boundary conditions are derived by differentiating Eqs. (3) and (1) with respect to the parameters. Thus, the normal derivative of X_j is zero on the adiabatic boundaries. On the quenched surfaces, the boundary condition for X_2 (= X_1) is

$$\frac{\partial q_s}{\partial Q_2} = \frac{\partial}{\partial Q_2}\left(-\kappa \frac{\partial T}{\partial n}\right) = -\kappa \frac{\partial X_2}{\partial n} = -\kappa \frac{\partial X_3}{\partial n} = 1 \qquad \text{(A11)}$$

For j > 2, the boundary conditions are

for $k > 1$,

$$-\kappa \frac{\partial X_{k+1}}{\partial n} = \begin{cases} (T_s - T_k)/(T_{k+1} - T_k)\,, & T_{k+1} < T_s < T_k \\ (T_{k+2} - T_s)/(T_{k+2} - T_{k+1})\,, & T_{k+2} < T_s < T_{k+1} \end{cases} \qquad \text{(A12)}$$

and

$$-\kappa\frac{\partial X_{k+2}}{\partial n} = \begin{cases} (T_s - T_{k+1})/(T_{k+2} - T_{k+1}) \ , \ T_{k+2} < T_s < T_{k+1} \\ (T_{k+3} - T_s)/(T_{k+3} - T_{k+2}) \ , \ T_{k+3} < T_s < T_{k+2} \end{cases} \quad \text{(A13)}$$

The similarity of Eq. (A10) to Eq. (2) allows simplification of the computer code used to solve the equations. Equation (A10) is solved using the same finite-difference technique, code, and mesh used to solve Eq. (2). In fact, through careful programming, the geometric factors that appear in the finite-difference approximations need be calculated only once per time step and applied to advance the temperature solution T and the sensitivity coefficient solutions X_j. The particular values X_{ij} are the values at the mesh location selected for the temperature solution T_i.

LITERATURE CITED

1. Beck, J. V., et al., *Inverse Heat Conduction - Ill-Posed Problems*, John Wiley & Sons, New York, 1985.

2. Beck, James V., *J. Heat Transfer*, 124 (May, 1962).

3. Beck, James V., *Nuc. Eng. Des.*, **7**, 9 (1968).

4. Attia, M. H. and L. Kops, "Distortion in Thermal Field Around Inserted Thermocouples in Experimental Interfacial Studies - Part II: Effect of the Heat Flow Through the Thermocouple," ASME paper 87-WA/Prod-2 (1987).

5. Henning, C. D. and R. Parker, ASME *J. Heat Transf.*, 146 (May, 1967).

6. Keltner, N. R. and L. W. Bickle, "Intrinsic Thermocouple Measurement Errors," ASME-AIChE Heat Transfer Conference, paper No. 76-HT-65, August 9-11, 1976.

7. Gat, Uri, et al., *Int. J. Heat Mass Transf.*, **18**, 1337 (1975).

8. Keltner, N. R. and J. V. Beck, *J. Heat Transf.*, **103**, 759 (November, 1981).

9. Keltner, N. R. and J. V. Beck, *J. Heat Transf.*, **105**, 312 (May, 1983).

10. Beck, James V, et al., *AIAA J.*, **22**, 9, 1328 (1984).

11. Litkouhi, B. and J. V. Beck, *AIAA J.*, **23**, 10, 1609 (October, 1985).

12. Moreaux, F. and G. Beck, "Influence of Organic Polymer Addition on the Stability of Film Boiling in the Case of Quenching in Subcooled Water," pp. 2067-2071 in *Proceedings of the Eighth International Heat Transfer Conference*, Hemisphere Publishing Corp., Washington, D.C., (1984).

13. Park, J. E. and G. M. Ludtka, "Calculation of Surface Heat Transfer Parameters During Quenching from Surface Thermocouple Signals," pp. 33-42 in ASME HTD - Vol. 119, *Heat Transfer in High Energy/High Heat Flux Applications* (1989).

14. Kollie, Thomas G., Oak Ridge National Laboratory, private communication with Gerard M. Ludtka, Oak Ridge Y-12 Plant (1986).

15. Elrod, D. C., et al., "HEATING6: A Multidimensional Heat Conduction Analysis with the Finite-Difference Formulation," NUREG/CR-0200, ORNL/NUREG/CSD-2/V2, Oak Ridge Gaseous Diffusion Plant, Oak Ridge, Tenn. (October 1981).

16. Pollock, D. D., *The Theory and Properties of Thermocouple Elements*, ASTM Special Technology Publication 492 (1971).

17. Park, James E., et al., "Quenching Heat Flux Calculated Using Information From Deeply Buried Thermocouples," ASME HTD-Vol. 112, *Heat Transfer Measurement, Analysis, and Flow Visualization*, Book No. H00504 (1989).

18. Bass, B. R., et al., "ORMDIM: A Finite Element Formulation for Two-Dimensional Nonlinear Inverse Heat Conduction Analysis," NUREG/CR-1709, ORNL/NUREG/CSD/TM-17, U.S. Nuclear Regulatory Commission, Washington, D.C. (Dec., 1980).

19. Osman, A. M. and J. V. Beck, *J. of Thermophys. and Heat Transf.*, **3**, 2, 146 (1989).

20. Park, J. E., et al., "Smoothing Strategies for Surface Temperatures Measured During Vigorous Transient Boiling," pp. 91-99 in ASME HTD-Vol. 129, *General Papers: Phase Change and Convective Heat Transfer*, (1990).

21. Beck, James V., Michigan State University, private communication with James E. Park, Martin Marietta Energy Systems, Inc. (June 15, 1987).

22. Beck, James V. and Kenneth J. Arnold, *Parameter Estimation in Engineering and Science*, John Wiley & Sons, New York, 1977.

23. Beck, J. V., B. Litkouhi, and C. R. St. Clair, Jr., *Numerical Heat Transf.*, **5**, 275 (1982).

FEASIBILITY EXPERIMENTS ON USING CONDENSATION IN SYSTEMS WITH ELEVATED HEAT EXCHANGERS TO ENHANCE PASSIVE SAFETY

R.D. Coffield and J.E. Schmidt ■ Westinghouse Electric Corporation, Advanced Energy Systems, P.O. Box 158, Madison, PA 15663

A series of tests were performed to demonstrate that the condensation heat transfer occurring in elevated heat exchangers relative to a lower situated boiling core can be an extremely effective mode of passive heat removal under loss of coolant accident (LOCA) conditions. The application of the concept is to a low pressure heavy water production reactor design. Under emergency conditions, a continuous supply of water to the core is assured by coolant flow from a large elevated tank (1363 m^3) of borated light water (BLWT) early in the transient. At later stages of the accident, gravity drain of system leakage floods the reactor coolant system in the lower containment and then flows into overflow sumps back into the vessel to maintain core cooling. The condensation heat transfer mode is not needed prior to draining of the BLWT which would take over one hour into the transient. The tests show that several times this reactor decay power level can be removed by heat exchanger condensation even with the upper hot leg piping and heat exchanger tubes having been filled with air. Testing also shows the concept is feasible for the case of a hot leg break where steam excapes to the containment in parallel to that flowing to the heat exchangers. The paper provides sufficient information to support further, more detailed testing, to fully quantify this passive means of decay heat removal. Although the tests were performed for a heavy water concept, the same condensation heat transfer mechanisms exist in PWRs for similar conditions particularly in those plants which utilize straight-through steam generator designs.

This paper describes analyses and experiments that have been performed for a heavy water production reactor design to help characterize shutdown heat removal through the primary heat exchangers under loss of coolant accident (LOCA) conditions. The basic plant configuration is shown in Figure 1. The system is configured for natural circulation flow in both the reactor and coolant flow loops. Many of the Figure 1 features are dedicated to mitigation of LOCA damage.

A sketch of relative positions between the primary heat exchangers and the reactor vessel are shown by Figure 2. If an LOCA were postulated to occur, the following progression of events would occur in the overall reactor system:

- o Makeup water will be supplied by the helium gas pressure driven pressurizer.
- o Active nonsafety-grade pumps in the Reactor Coolant Treatment System (CTS) will supply additional D_2O makeup to that from the pressurizer.
- o If the makeup supply is not adequate, active nonsafety-grade pumps in the Residual Heat Removal System (RRS) will be activated to pump back heavy water leakage.
- o If the break flow exceeds the capacity of the RRS pumps, then a gravity feed of borated light water from a large elevated water storage tank passively supplies makeup to the cooling system.
- o If the earlier actions do not supply sufficient makeup water, a coolant boiling phase is reached where heat is passively removed through both: 1) condensation heat transfer in the primary heat exchangers with the circulating water depositing the heat into the cooling tower basin which acts as the ultimate heat sink or, 2) through condensation heat transfer to inside of the steel containment shell with the ultimate heat sink being natural convection air flow on the outside of the containment shell.

This report focuses on the inherent heat removal capability in the primary heat exchangers (PHXs) via condensation in the presence of steam/air mixtures (condensable/noncondensable gases).

J. E. Schmidt, Westinghouse Electric Corporation, Madison, Pennsylvania and R. D. Coffield who is now with Westinghouse Bettis Atomic Power Laboratory, West Mifflin, Pennsylvania.

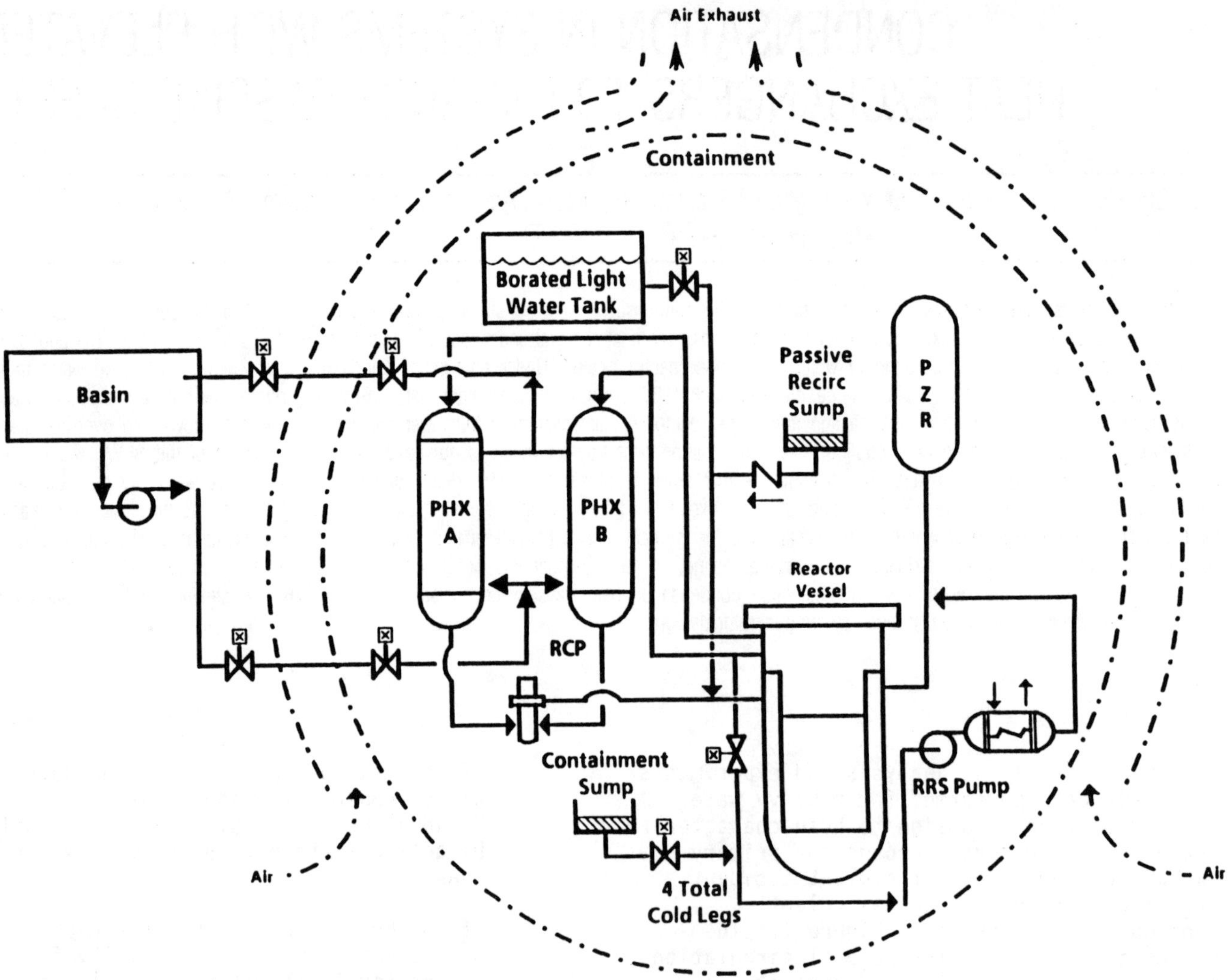

Figure 1: Passive Containment Cooling Concept for a Low Pressure Heavy Water Reactor

For both large and small breaks in the reactor coolant system, the core will remain cool for the period that the borated light water tank (BLWT) supplies water, and no significant steaming will occur. However, after the BLWT is emptied, the system will gradually heat up and core steaming will start. Scoping calculations performed for this stage of the accident indicate that core steaming has the capability of removing extremely large quantities of heat by condensation heat transfer in the PHXs. For the applicable Reynolds number, Reference 1 recommends the heat transfer coefficient correlation (where film refers to the film of condensed water):

$$h_{avg} = 1.60 \left[\frac{k_f^3 \rho_f^2 g h_{fg}}{\mu_f L \ (T_s - T_w)} \right]^{1/4} \tag{1}$$

with

k_f = film thermal conductivity

ρ_f = film density

μ_f = film viscosity

h_{fg} = latent heat of vaporization

L = tube length

T_s = saturation temperature

T_w = wall temperature

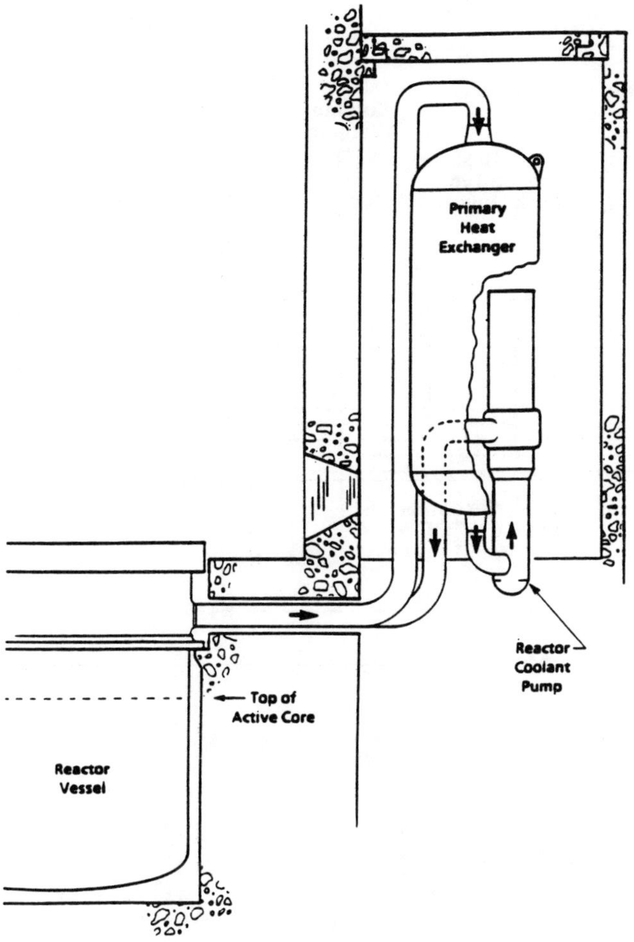

Figure 2: Relative Position of Reactor Vessel Nozzles to Reactor Coolant System

Application of this correlation indicates that the condensation heat transfer coefficient would be over 3.41 kW/m^2-°C (600 BTU/hr-ft^2-°F). When combined with an average circulating water temperature of 66°C (150°F) and over 46,450 m^2 (500,000 ft^2) of tube surface area in all eight heat exchangers, many times more heat rejection capability than required is provided. For example, a maximum rate of steam flow to the heat exchangers can be estimated by assuming choked steam flow in all intact hot legs. This would result in a potential heat load to the heat exchangers of about seven percent full power, which is indicative of initial shutdown (BLWT drainage will take several hours even for large LOCA conditions and decay heat would then be less than one percent full power). Once the steam reaches the heat exchanger inlet, velocities of less than 15.24 m/s (50 ft/s) are predicted at the inlet to the tubes, and with the complete condensation that would occur, water films of 0.0254 cm (0.010 in) would result in a condensate velocity of less than 0.061 m/s (0.2 ft/s). The Equation (1) prediction does not consider the effects of noncondensable gases on decreasing the condensation heat transfer coefficients. However, as will be described, even if the heat transfer coefficient were decreased by a factor of ten due to the presence of noncondensable gases, more than adequate heat removal capability would still exist.

TESTING PERFORMED TO CHARACTERIZE PRIMARY HEAT EXCHANGER CONDENSATION

To experimentally model the phenomena of condensing steam in the primary heat exchangers for the prototype, the test configuration shown by Figure 3 was developed. The position of the heat exchanger tube sheet relative to the vessel outlet nozzle is 2.74 m (9 ft) in the model which is approximately the same dimension as in the prototype. However, the test heat exchanger tube length is only 3.05 m (10 ft) (see Figure 4) as compared to over 7.63 m (25 ft) in the plant; and thus, the relative driving

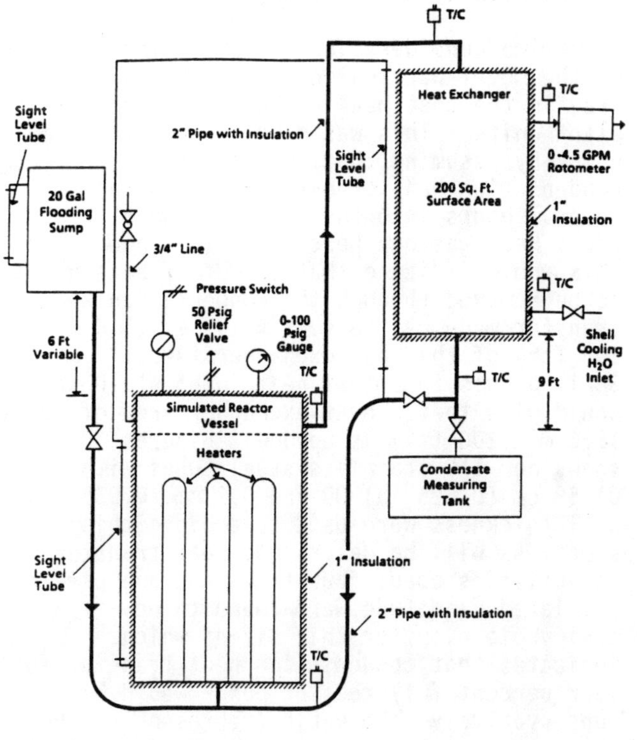

Figure 3: Heat Exchanger Steam Condensation Test

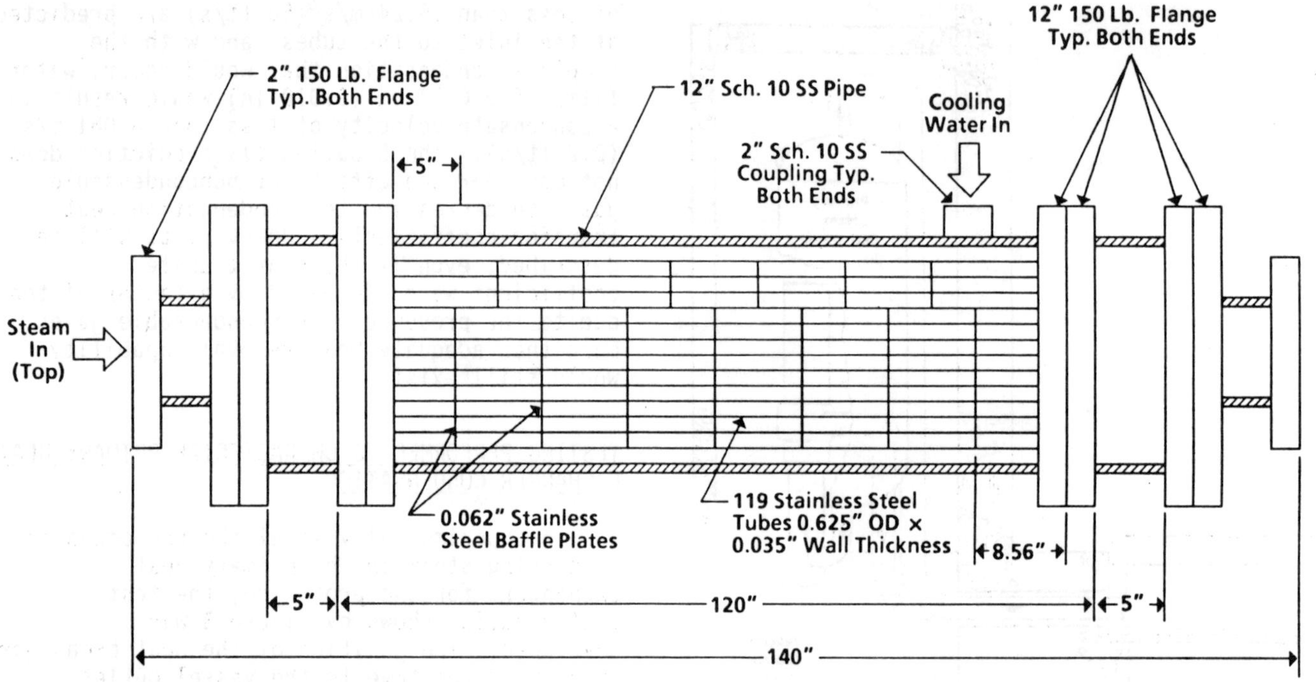

Figure 4: Condensation Test Heat Exchanger

head of condensate in the cold leg versus steam rising in the hot leg is conservatively represented by the test model.

Prime consideration in developing the size of the model was scaling the heat transfer area in the test heat exchanger to that of the plant units. This was done for a reference point by assuming that the time of condensation heat transfer starting was several hours into the transient when the decay heat was one percent of full power. This would indicate that 27 MWt of heat needs to be removed through the condensation heat transfer mode by the 4.65 x 10^4 m^2 (5.0 x 10^5 ft^2) of the heat exchanger tubes. Scaling this to a test heat input of 10 kW would give that a heat exchanger area of about 18.6 m^2 (200 ft^2) is needed. Figure 4 shows how 119 stainless steel tubes with a 01.59 cm (0.625 in) OD and 0.9 mm (0.035 in) wall thickness were used to achieve this area. As will be described, heat transfer capabilities about four times the one percent simulated magnitude were found to be achievable even for this sizing which indicates that condensation heat transfer for four percent full reactor power would be conservatively achievable (representing the reactor decay heat load which would exist at considerably less than one hour into the transient).

The circulating water loop was modeled as shown in Figure 3 with a prototypic counterflow configuration. For all tests, flow rates representative of natural circulation were conservatively modeled with flow rates in the range of one to two percent full flow (i.e., for the temperature increases that would develop, flows of over twice these magnitudes would develop in the circulating water loop).

The electrical heaters were designed to supply over 40 kW of power with a very low heat flux so that there would be no concern relative to critical heat flux. In fact, for reactor vessel water levels higher than the outlet nozzle (as was the case for all these tests), over a factor of ten margin exists.

Figures 5 and 6 show the test hardware located in position for the testing.

TEST RESULTS

The test procedure was to allow the simulated reactor core to boil at various power levels from 10 to 50 kW, and then measure (through the heat exchanger cooling water ΔT) the condensation performance of the heat exchanger. High power levels were used to heat the water to within a few degrees of saturation temperature and then the power

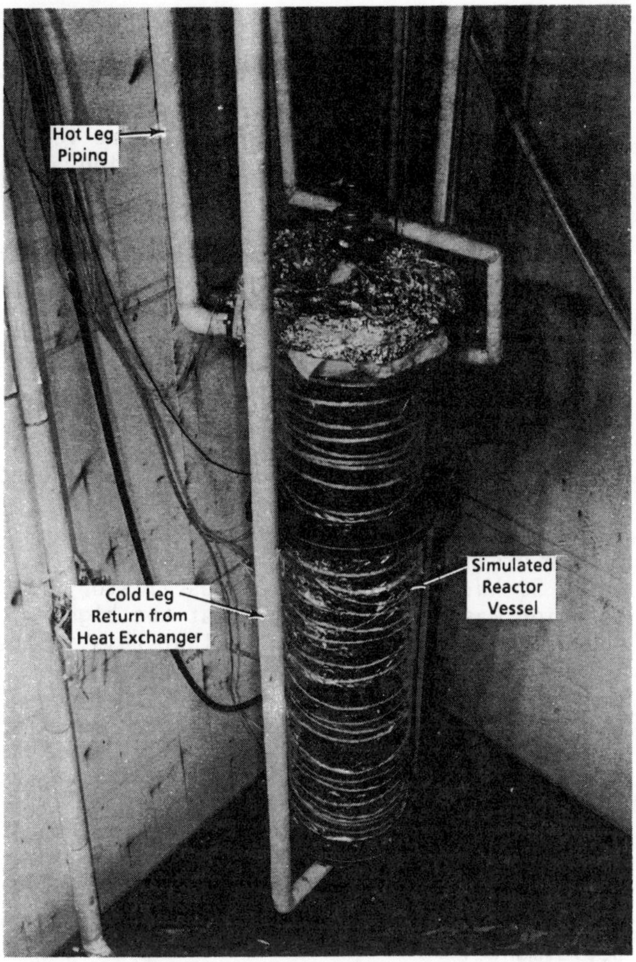

Figure 5: Simulated Reactor Vessel

was decreased to that for the the particular test (at least half an hour prior to boiling initiation). At the initiation of all tests, the hot leg piping at the inlet of the heat exchanger was opened (through a top vent line) to insure that the elevated hot leg and the inside of the heat exchanger tubes were filled with noncondensable gas (air).

The condensation heat transfer in the heat exchanger was found to effectively remove all the steam heat (heat of vaporization) and subcool the condensate for power levels below 40 kW. This would correspond to prototype reactor power levels a few minutes into the transient which significantly exceeds the required heat removal of about one percent full power described earlier. A general characteristic of all the tests was that significant boiling would initiate when the heater and upper portions of the vessel reached 110°C (230°F) which was the saturation temperature near the top of the heater region for the hydrostatic pressure that existed. As the hot fluid from the heater region accelerates and rises under buoyancy, its saturation temperature decreases which promotes an intense build-up of steam to the hot leg riser. Once steam was climbing to the heat exchanger through the hot leg piping and condensing in the heat exchanger, a positive thermal driving head would develop between the hot and cold legs which would cause a flow of the cold condensate back into the reactor vessel. This flow would cool the lower reactor vessel region and gradually cool the heater and upper vessel regions sufficiently to result in the boiling process eventually stopping. This terminated the heat exchanger condensation process until boiling resumed. For all tests, steam continued to be generated for long periods even though the water in the bottom third of the vessel had been cooled by over 56°C (100°F). As

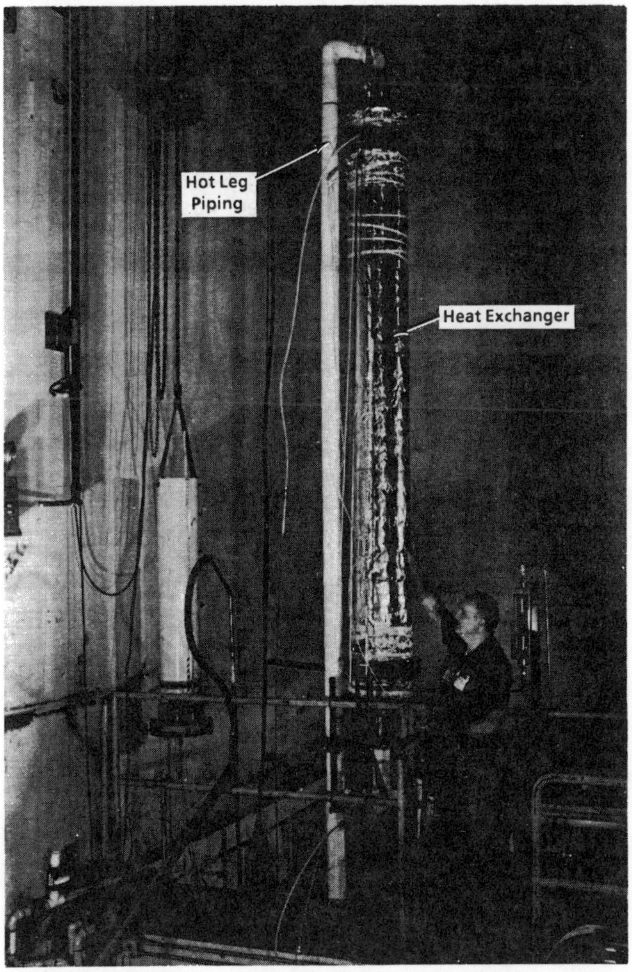

Figure 6: Test Heat Exchanger

would be anticipated, it took longer times at the higher power levels for this cooling from the cold condensate return to stop the boiling process. Minimum times that boiling was sustained were about fifteen minutes (for lower powers near 10 kW) and maximum times were in the two to four hour range (near the 40 kW level).

In all cases, the cooling water ΔT indicated power input from the condensation heat transfer to be in the range of 10 to 20 percent higher than the actual power input. This was due to the superheated liquid and steam passing through somewhat cooler water to near the top of the simulated reactor vessel. This created a non-thermal equilibrium flow of steam into the hot leg that was partially out of phase with the heater power (i.e., in addition to heater power, the depth and degree of stratification and superheat of the upper vessel liquid region would impact rate of vapor flow up the hot leg piping to the heat exchanger).

For power levels approaching 50 kW, it was found that the rate of heat transfer exceeded the capability of the heat exchanger condensation coolant flow of 1.7 x 10^{-2} m^3/s (4.5 gpm). The cooling water would boil since the cooling system was not designed for this magnitude of heat input which is well beyond the level of concern in the scaled system.

Table 1 shows typical data recorded for the tests.

A high degree of confidence exists for the calorimetric heat balance obtained from the heat exchanger cooling water (HX CW) since both the flow and mixed mean outlet temperature were measured by several independent methods. For the coolant outlet temperature, the temperature was measured after the flow passed through a highly turbulent mixing chamber at two separate positions in series on the outlet line, and the temperatures at both locations were very close (within one percent). In addition, the labyrinth flow path on the shell side of the heat exchanger (see Figure 4) promoted a high degree of coolant water mixing within the heat exchanger. For the coolant flow measurement, the flow was intermittently collected and weighed to confirm the rotometer flow measurements (see Figure 3). The two methods of flow measurement have results within a few percent of each other.

The boiling/condensation process was found to approach and sustain the steady state durations indicated by Table 1 with very little oscillatory behavior for test powers below 30 kW. However, a more oscillatory behavior was found to develop at the higher power levels. It is believed that this is due to the condensation location intermittently varying among the 119 tubes. For a period, the tubes with maximum condensation occurring would have a pressure increase relative to other tubes with less flow (in parallel flow) because of the larger build-up of condensate. The other tubes which had been less effective would then become more dominate resulting in some oscillatory behavior which did not detrimentally impact the overall heat transfer performance.

For all tests, the water level in the simulated reactor vessel was maintained higher than the outlet nozzle. This assured over a factor of ten margin to heater rod critical heat flux but allowed boiling on the upper portions of the heaters.

Assessments the heat transfer rates for the data given in Table 1 indicated that

Table 1: Typical Data for Boiling/Condensation Heat Transfer Tests

Test Power (kW)	Calorimetric Heat Balance from HX CW (kW)	Duration of Boiling/Condensation (hrs)
10.0	11.1	0.25
20.0	20.9	2.0
30.0	32.7	2.0
35.0	39.0	4.0
40.0	44.0	4.0
45.0	49.0	2.0
50.0	Exceeds CW Capacity	---

sufficiently high magnitudes of heat transfer coefficients existed to meet the passive heat removal requirements. As described earlier, even if the noncondensable gas (air) were to decrease the high condensation heat transfer coefficient of Equation (1) by an order of magnitude, more than adequate rates of heat transfer would still exist which was substantiated by the tests. In the next phase of this investigation, computer codes (e.g., TRAC and/or RELAP) which can evaluate the various phenomenon occurring in the heat exchanger plus the interactions relative to the integrated system will be used to help quantify the spatial distribution of heat transfer in the heat exchanger. For example, the testing of this current phase seems to indicate that most of the condensation takes place over the top several feet of the tubes with the condensate subcooling to the cooling water inlet temperature as it flows through the remainder of the tube length. A combination of the computer code analytical modeling plus test results from the next more detailed testing phase should enable preliminary design analyses to be performed with a high degree of confidence.

The correlation of Equation (1) is only applicable for the case of no noncondensable gases present. It was introduced earlier to demonstrate how much margin was available to accommodate the presence of noncondensable gases. Although these tests have demonstrated ample heat transfer capability for the desired HWR safety function, additional testing is needed to develop a correlation which accounts for the presence of noncondensable gases.

In another test series, a separate line to the main hot leg pipe was opened allowing steam to blow down to the outside atmosphere in parallel with the steam condensing in the heat exchangers. This line was sized to have one seventh the flow area of the hot leg pipe in order to simulate the rupture of one out of eight hot leg pipes in the prototype system. This parallel line was opened while stable condensation was occurring in the heat exchanger and, for all cases, no effect on the heat exchanger condensation process was found. This separate line, in addition to the main hot leg pipe, discharged from the simulated reactor vessel outlet plenum. Thus, when steam was being generated, a large inventory of superheated water existed (in the upper vessel region) which provided sufficient steam inventory for both parallel lines. In the reactor prototype, assuming that only one eighth the core decay power is being lost from the hot leg break would indicate that a low containment pressure increase of less than 0.34 bar (5 psi) would be needed to develop a sufficient ΔT between the condensing steam (escaping through the break and condensing on the inside of containment) and the free convection air on the outside of containment to remove the heat. The same computer code model, as used to predict the Table 1 type data, will be used to evaluate these tests.

CONCLUSIONS

Several key conclusions can be derived from the condensation heat transfer tests that were performed:

- o Condensation heat transfer in elevated heat exchangers relative to boiling occurring in a lower positioned reactor vessel is an extremely effective mode of heat removal. In the case of the reference design discussed, this mode of heat removal was found to be over four percent of full power for a system believed to be reasonably well scaled as far as the most significant controlling parameters. This effective heat removal was found to occur even with air inside the hot leg piping and heat exchanger tubes.

- o For hot leg piping rupture simulation, the release of steam out of a parallel line with proportionately smaller diameter than the other main line feeding the heat exchanger (simulating rupture of one out of eight hot legs) did not detrimentally impact the heat exchanger condensation heat transfer process. (Note that loss of water inventory is not a concern because the water from the 1363 m^3 BLWT maintains sufficient water into the vessel downcomer initially and then after flooding the lower containment, the overflow sump feed makes up for losses out of the parallel flow path simulating the failed pipe.)

Based on these tests, it is recommended that designers of heavy water production reactors give strong consideration of this decay heat removal method as they develop passive heat removal systems.

LITERATURE CITED

1. McAdams, H.W., Heat Transmission, 3rd Edition, McGraw-Hill Book Company, New York, 1954.

HEAT TRANSFER AND FRICTION FACTOR BEHAVIOR IN THE MIXED CONVECTION REGIME FOR AIR UP-FLOW IN A HEATED VERTICAL PIPE

G. Fu, M.J. Driscoll, N.E. Todreas, and S. Yesilyurt ■ Massachusetts Institute of Technology, Cambridge, MA 02139

An experimental investigation of heat and momentum transfer in upward air flow in a vertical riser under mixed convection conditions is presented. Such risers are specified in advanced nuclear power reactor concepts for passive removal of decay heat. The measured Nusselt numbers showed decreases of up to 50% due to large property changes and progression into the mixed convection region. They are correlated for the range $Re \geq 5000$, $1\times10^7 \leq Gr_q \leq 2.5\times10^8$ by application of a Reynolds Number-dependant multiplier to the forced convection Dittus-Boelter correlation. The friction factor results, which exhibit larger errors due to constraints on the length of the test section, suggest a nominal 10% increase above comparable isothermal values for the Buoyancy Number range < 0.1 with density computed at the axially-averaged bulk air temperature.

The Modular High Temperature Gas-Cooled Reactor (MHTGR) is one of the next generation power reactors currently being developed in the United States under Department of Energy sponsorship. A major emphasis in the development of these reactors is improvement of safety margins by use of passive safety systems. A feature of the MHTGR design, shared in principle by the annular passage outside the uninsulated guard vessel of the liquid metal cooled PRISM reactor design, is the use of an uninsulated steel reactor vessel and an array of riser tubes in the reactor cavity to provide a passive decay heat removal path from the core to the ambient air as illustrated in Figure 1.

The design of this Reactor Cavity Cooling System (RCCS) requires attention, not only to its engineering aspects, but also to the fundamental heat transfer and fluid flow processes involved. This requires characterization of the heat transfer coefficient and the friction factor inside the riser tube in the operating regime, which involves a condition intermediate between forced and free convection, i.e., mixed convection—a regime not yet well understood. Accordingly, the objective of the present work was to determine the heat transfer coefficient and friction factor inside the riser under its projected operating conditions.

GENERAL CHARACTERIZATION OF HEATED DUCT FLOW

A useful characterization of riser performance can be developed by examining the momentum balance for upward flow in a vertical heated duct.

G. Fu is presently at
S3 Technologies, Columbia, MD 21045.

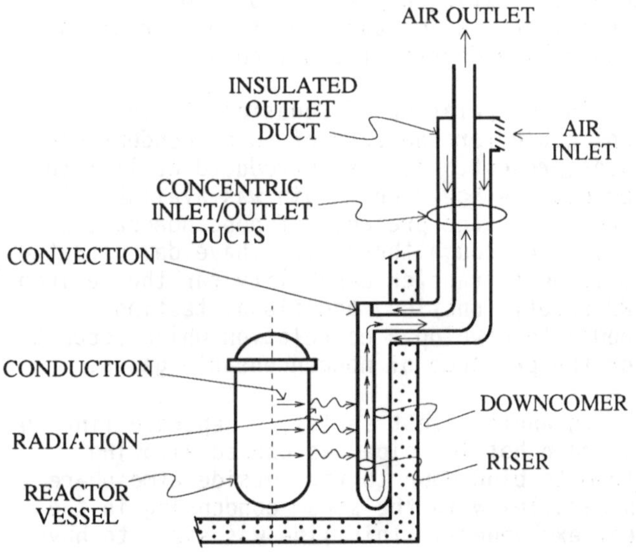

Figure 1: Reactor Cavity Cooling System (RCCS) (from Reference [1]).

$$\frac{\partial(\rho w^2)}{\partial z} = -\frac{dp}{dz} + \frac{1}{r}\frac{\partial}{\partial r}\left(r(\mu + \mu_t)\frac{\partial w}{\partial r}\right) + \rho_o \beta g (T - T_o) \quad (1)$$

where p is the pressure defect $= p_{actual} - p_{hydrostatic}$.

Averaging over the cross section area, A_c, and assuming dp/dz is constant across this cross section, one gets

$$\left(\frac{dp}{dz}\right)_{acc} = -\left(\frac{dp}{dz}\right)_t + \left(\frac{dp}{dz}\right)_b + \left(\frac{dp}{dz}\right)_f , \quad (2)$$

where

$$\left(\frac{dp}{dz}\right)_{acc} = \frac{1}{A_c}\int_A \frac{\partial(\rho w^2)}{\partial z}dA \; , \left(\frac{dp}{dz}\right)_t = \frac{1}{A_c}\int_A \left(\frac{dp}{dz}\right)dA \; ,$$

$$\left(\frac{dp}{dz}\right)_b = \frac{1}{A_c}\int_A \rho_o\beta g(T - T_o)\, dA \; ,$$

and

$$\left(\frac{dp}{dz}\right)_f = \frac{1}{A_c}\int_A \frac{1}{\rho}\frac{\partial}{\partial r}\left(r(\mu + \mu_t)\frac{\partial w}{\partial r}\right)dA$$

$$= -\frac{1}{A_c}(2\pi R)\,\mu_w\left(\frac{\partial w}{\partial y}\right)_w = -\frac{2\pi R}{A_c}\tau_w \; ,$$

since $(\mu_t)_R = 0$ and $\left(\frac{\partial w}{\partial r}\right)_R = -\left(\frac{\partial w}{\partial y}\right)_w$.

Assuming that the fluid properties are constant (and taken at reference temperature, T_o), except for density in the buoyancy term, that the flow is fully developed (the acceleration term is neglected) and applying the continuity condition, integration of Eq. 1 over the length of the duct, L, yields:

$$0 = \Delta p_t - \frac{2\pi R}{A_c}\tau_w L + \beta\rho_o g\int_o^L \Delta\bar{T}dz \; , \qquad (3)$$

where

$$\Delta p_t = p_{in} - p_{out} \text{ and } \Delta\bar{T} = \frac{1}{A_c}\int_A (T - T_o)\, dA .$$

Now applying the definition of the skin friction factor and the Blasius solution,

$$\tau_w = c_f\frac{1}{2}\rho_o w_o^2 = \frac{f}{8}\rho_o w_o^2 = 0.316\text{Re}^{-1/4}\,\frac{\rho_o w_o^2}{8} ,$$

and evaluating $\Delta\bar{T}$ from the energy equation as

$$\Delta\bar{T} = \frac{q''\pi Dz}{c_p\dot{m}} = \frac{4q''}{c_p\rho_o w_o}\left(\frac{z}{D}\right) ,$$

Eq. 3 can be written in the following dimensionless form:

$$\Delta\bar{p}_t = -\frac{2\text{Gr}_q\left(\frac{L}{D}\right)^2}{\text{Re}^3\text{Pr}} + \frac{0.158\left(\frac{L}{D}\right)}{\text{Re}^{1/4}} , \qquad (4)$$

where $\Delta\bar{p}_t$ is Δp_t normalized by $\rho_o w_o^2$.

Eq. 4 can be used to demonstrate several possible riser configurations. If both ends of the duct are open to ambient air, $\Delta\bar{p}_t = 0$ and the riser creates its own flow, thereby operating in what can be termed the "pure" natural convection regime. In this condition the operating Re number from Eq. 4 is

$$\text{Re} = \left[\frac{\text{Gr}_q\left(\frac{L}{D}\right)}{0.039\ \text{Pr}}\right]^{4/11} . \qquad (5)$$

Our heated riser of length, L, can be modified by the addition of a pump or a throttling valve. If the duct flow is "pumped", $\Delta\bar{p}_t > 0$, and the Re number will be greater than that of Eq. 5. On the other hand, if the duct is throttled by a valve, $\Delta\bar{p}_t < 0$, the Re number will be less than that of Eq. 5.

The heated riser can also be pumped by an unheated chimney of length, L_c. In this case, Eq. 4 becomes:

$$\Delta\bar{p}_t = 0 = -\frac{2\text{Gr}_q}{\text{Re}^3\text{Pr}}\left(\frac{L}{D}\right)\left(\frac{L + 2L_c}{D}\right) + \frac{0.158}{\text{Re}^{1/4}}\left(\frac{L + L_c + L_u}{D}\right) , \qquad (6)$$

which accounts for the buoyant and friction components of the pressure change in the chimney. Figure 2 illustrates these modes of riser operation.

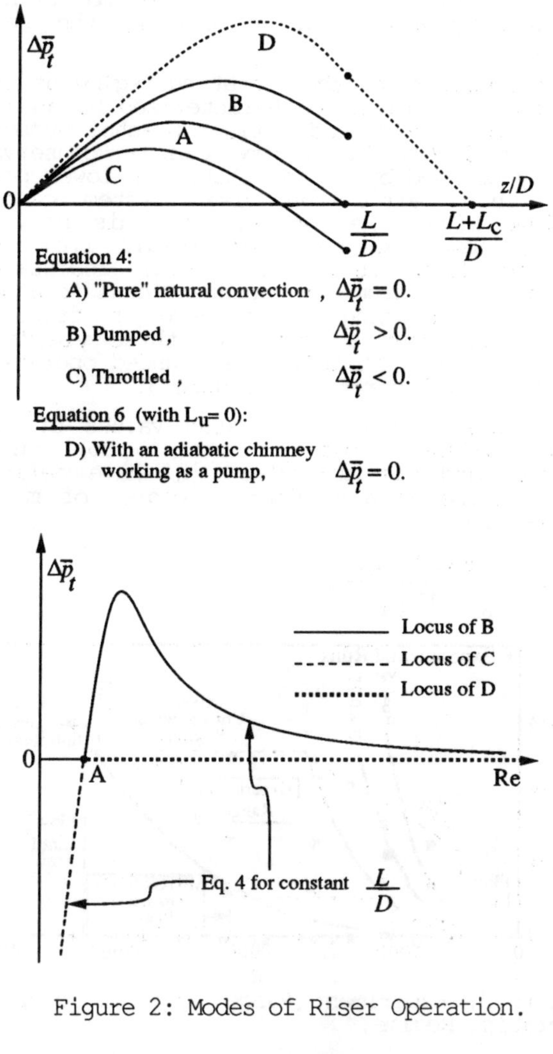

Figure 2: Modes of Riser Operation.

The reactor cooling system of Figure 1 utilizes an unheated inlet section, an unheated chimney section and a heated riser whose performance, which is governed by Eq. 6, will be considered next.

PERFORMANCE RANGE OF THE REACTOR RISER TUBE

The RCCS is designed as the ultimate barrier to prevent reactor core and vessel damage for severe accidents. The loss of coolant accident in which the reactor system can be either depressurized or pressurized is one of the most severe accidents in terms of the maximum core and reactor vessel temperature attained. In a loss of coolant accident, the decay heat generated inside the core will all be either deposited in the graphite moderator or transported to the ambient air through the RCCS. To predict the riser operation conditions, namely mass flow rate and outlet air temperature, one must know the decay heat generation rate as a function of the time and what portion of that decay heat is deposited in the graphite moderator. We will use the data at about 100 hours after the reactor has been shut down as the reference case, to approximate the maximum core, reactor vessel, riser wall temperature, riser outlet air and air flow rate, although these maxima do not in reality all occur at exactly the same time.

The extreme of the operating regime of the reactor system can be characterized by an inlet Re of 2.2×10^4 and a riser Grashof number, Gr_q, of 5×10^7. The regime can be conservatively bounded by noting that, for lower powers, this self-driven riser system cannot achieve higher Re, while, for this maximum power, lower Re can only result from increased flow resistance. Such resistance could arise from an unintentional channel blockage. Assuming an increase in friction factor of about a factor of three yields a lower Re bound of 10^4. This bounded operating regime is illustrated on Figure 3.

Figure 3 also illustrates values of the dimensionless buoyancy number, Bo, which Jackson and his co-workers [2] have defined and utilize as a measure of degree of mixed convection:

$$Bo = \frac{8 \times 10^4 Gr_q}{Re^{3.425} Pr^{0.8}}.$$

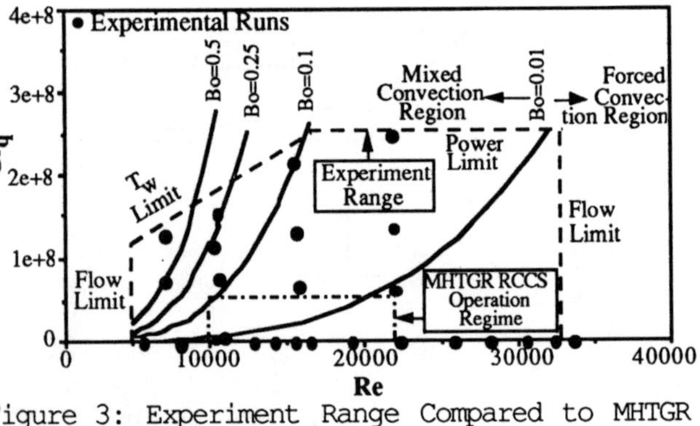

Figure 3: Experiment Range Compared to MHTGR Operating Regime.

Taking the value of Bo = 0.01 as a conservative criterion for onset of significant mixed convection effects, we observe that a portion of the riser operating regime may be in mixed convection. This is of design significance, since heat transfer coefficients and friction factors can vary significantly from forced convection values under mixed convection conditions. The determination of actual riser operating conditions within this operating regime requires simultaneous solution of the conservation equations (Eq. 6 is the relevant momentum equation), with assumptions for the heat transfer coefficients and fraction factor.

We next assess the relevant mixed convection correlations and define the experimental conditions for which new design data is needed.

RELEVANT CORRELATIONS IN THE LITERATURE

A thorough review of the mixed convection literature has been made by Fu [3]. Correlations for Nusselt number proposed by Jackson and co-workers [2], Petukhov and Polyakov [4] and Connor and Carr [5] have been assessed. All reflect the fact that in an aiding mixed convection flow, two types of velocity profiles can occur. As shown in Figure 4, for the Type I profile, the velocity near the wall has increased due to the buoyancy force, but the maximum flow velocity is still at the pipe center (lines 1 to 3 in Figure 4). For the Type II profile, the buoyancy effect is so large that the maximum velocity is no longer at the center of the pipe, but at some location between the wall and the center (line 4 in Figure 4).

Figure 4 illustrates the mechanism for heat transfer behavior proposed by Hall and Jackson [6]. Specifically, as the buoyancy effect increases, the shear stress near the

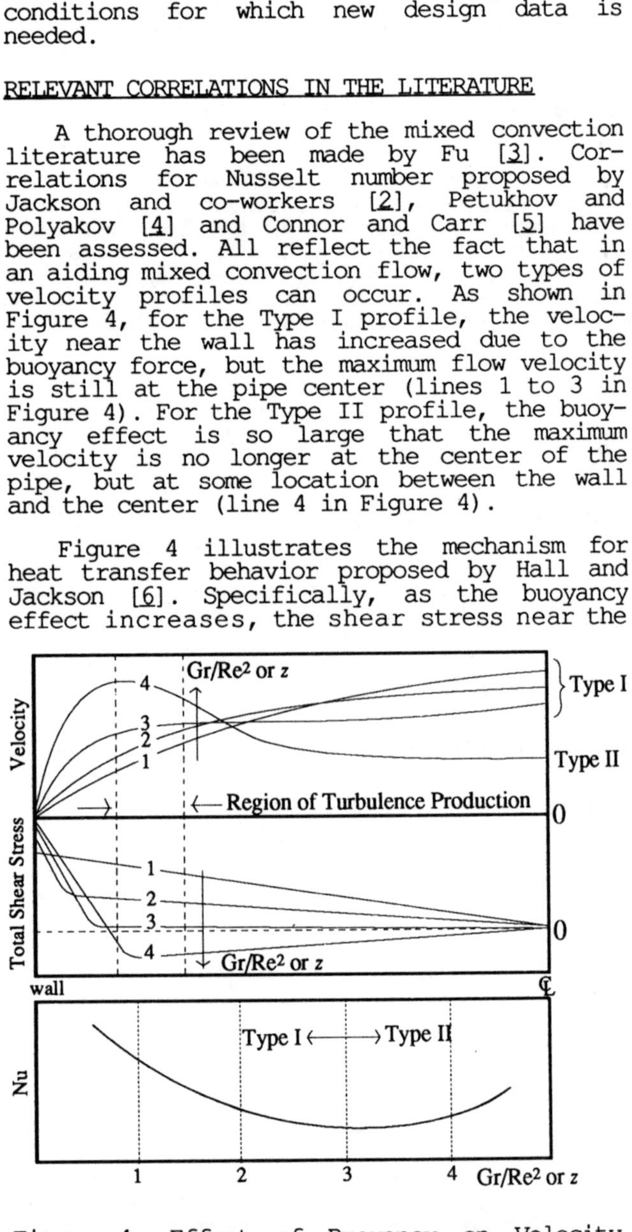

Figure 4: Effect of Buoyancy on Velocity, Shear Stress and Heat Transfer for Aiding Turbulent Flow (from Symolon [14]).

wall decreases and the Nusselt number decreases until a point where turbulence production is close to zero. At that point laminarization occurs and the Nusselt number reaches its minimum value. Further increase of the buoyancy effect will make the velocity profile change from Type I to Type II and the turbulent shear stress become negative. However, the absolute value of this turbulent shear stress production starts increasing, causing the heat transfer rate to also start improving from its minimum value.

We select Jackson's correlation as revised in [7] for focus since it is based on the most comprehensive study currently available. The correlation is

$$\frac{Nu}{Nu_{fc}} = \left(\left| 1 - \frac{Bo}{\left(\frac{Nu}{Nu_{fc}}\right)^2} \right| \right)^{0.46}. \qquad (7)$$

Note that the term in the brackets on the right hand side of Eq. 7 has been taken as the absolute value, which makes the correlation capable of describing both Type I and Type II flows. The correlation was also compared by Jackson with available air experiment results by Steiner [8] and Carr, et al. [9] in the fully developed region.

Compared to the heat transfer behavior in turbulent mixed convection, the friction factor has received much less attention. Neither a satisfactory theory nor a modest data set exists for friction factor in mixed convection. A key reason for this is the difficulty in measuring the friction factor either from the velocity gradient at the wall or from direct friction pressure drop measurements.

Petukhov and Polyakov [4] employed a theory similar to the Reynold's analogy principle in which they predict similar behavior between the Nusselt number and the friction factor in a turbulent aiding flow. We present their correlation since it is recently published and presumably is based on the extensive long term effort in this field of the authors. The correlation for calculating the friction factors in a turbulent aiding flow is

$$f = \left[\frac{1 + 0.83e^2}{1.82\log_{10}\left(\frac{Re}{8}\right) + 0.076e^2E^{-0.25}} \right]^2 \qquad (8)$$

where

$$e = \frac{Gr_q x10^3}{Pr{\cdot}Re^{2.75}} \text{ and } E = \frac{Gr_q}{Pr{\cdot}Re^4}.$$

Their forced convection friction factor correlation is

$$f = \left[1.82 \log_{10}\left(\frac{Re}{8}\right)\right]^{-2} \qquad (9)$$

In this assessment we also present heat transfer and friction results from a companion MIT project which has developed a code, CONDOR, for mixed convection analysis based on the k-ε model [10].

These literature results for both the Nusselt number and friction factor show deleterious trends for the mixed convection regime. Specifically, for modest Bo (0.1 to 1.0) laminarization occurs and the Nusselt number decreases to values which approach 50% of the comparable forced convection values. For friction factor, while the evidence is less extensive, after a modest decrease for conditions below Bo ~ 0.3, friction factors can increase to levels a factor of three above comparable forced convection values for Bo reaching values around five.

EXPERIMENTAL DESIGN AND OPERATING PROCEDURES

The uncertainties in mixed convection performance suggested that experiments should be performed to bracket the expected regime of reactor riser operation. An experiment was designed for this purpose, which covers the experimental range illustrated in Figure 3. The experiment also achieves at least the same air temperature rise, about 140°C, as the design case experiences, so that air property change effects will be reflected in our test results. The experimental constraints limiting operation of the current apparatus are:

- Lower Re of 5000 due to error in measured air flow rate using the standard ASME orifice.
- Upper Re of 33000 due to the maximum Δp measurable by the pressure transducer.
- Maximum wall temperature of 400°C due to limitation of the thermocouple insulation.
- Maximum power of 4000 W, consistent with the wall temperature limitation imposed by the thermocouples and the available laboratory power supply.

The design of the experiment is a compromise among the need to represent key features of the real system, the desire to create a readily interpretable experiment, and practical restrictions on cost and size. The overall apparatus is a once-through system with ambient air as the flow medium. The "riser" in the present investigation is a 63.5 mm ID commercial seamless cylindrical carbon steel pipe (instead of a rectangular slot, as in the present MHTGR design). It has a 8.6 m (135 L/D) heated section, which is heated uniformly by helically-wound wrap-around electrical heating tapes at a pitch of 70 mm, and a 1.7 m (27 L/D) inlet unheated section incorporating a flow straightener, see Figure 5. The total height is approximately half that of a real riser. The pipe is insulated by an inner layer of 12.7 mm thick fiber glass insulation, a middle layer of 63.5 mm thick thermo-12 calcium silicate insulation and an outer layer of 25.4 mm thick fiber glass insulation. The insulation is covered by a PVC pipe to prevent rain damage, since the pipe is located out-of-doors. The heater power is provided by a 208 V single phase power line. A variac is used to control the input heating power in the range of 0 to 240 V and 0 to 17 A, hence 0 to 4 kW. Air flow is provided by a centrifugal blower. The

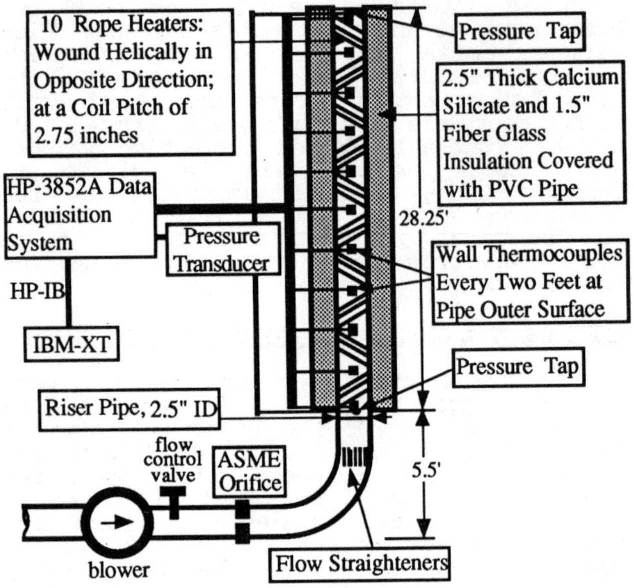

Figure 5: Schematic of Riser Mockup (not to scale).

flow rate is controlled and adjusted using an in-line control valve.

The directly measured parameters are air flow rate, total pressure drop, heating power (from current and voltage) and temperatures. The air flow rate is measured using an ASME standard orifice [11]. A pitot-tube is used as a secondary monitor. The total pressure drop is measured between two wall taps using a commercial differential pressure transducer manufactured by MKS Instrument, Inc., based on the measurement of electric capacity. This pressure transducer has a full scale range of 133 Pa; the minimum measurable differential pressure drop depends upon the system stability. One of the pressure taps is located at the beginning of the heated section and the other is six inches from the end of the heated section. The heater power measurement is simple: an ammeter and voltmeter pair gives P=I*V. However, the actual power input to the flowing air is the difference between total input power and heat loss through the insulation. This total heat loss has been determined by running the experiment with the entrance and exit blocked and insulated. There are fourteen T-type thermocouples imbedded in the outer wall of the pipe to monitor the wall axial temperature profile. The glass insulation on the thermocouples restricts the maximum wall temperature to 400°C. The thermocouples start where heating begins; successive thermocouples are positioned every two feet thereafter, each rotated 90° circumferentially from the preceding one.

The temperatures and pressure drop are collected using a computerized data acquisition system, the Hewlett Packard HP-3852A. All instruments used in this experiment were calibrated as discussed by Fu [3].

The experimental runs are rather simple to conduct. Heater power is adjusted to the level designated using a variac, and air flow is adjusted using the in-line valve. The wall temperatures and the pressure drops are monitored at an IBM-XT computer using self-programmed software through the data acquisition system HP-3852A. The steady state situation is considered adequately approached when wall temperature drift with time is less than 1°C per 30 minutes.

The major parameters we are seeking are the friction factor and the Nusselt number. Since the measured parameters are the air flow rate, the wall and inlet air temperatures and the total pressure drop, the Nusselt number and friction factor must be calculated from these measured parameters.

Nusselt Number

Since the wall heat flux can be found from the difference between the total heat input and the heat loss through the insulation, the local Nusselt number can be obtained easily as

$$\mathrm{Nu} = \frac{(q''_{\mathrm{nom}} - q''_{\mathrm{loss}})D}{k(T_w - T_a)} \, . \tag{10}$$

The air stream bulk temperature, T_a, is calculated from the energy balance as follows:

$$T_a(z) = \frac{1}{\dot{m}c_p}\int_0^z (q''_{\mathrm{nom}} - q''_{\mathrm{loss}})\,dz + T_{\mathrm{in}} \, . \tag{11}$$

This Nu number will be compared with the forced convection Nu number calculated using the Dittus-Boelter correlation

$$\mathrm{Nu} = 0.023\,(\mathrm{Re})^{0.8}(\mathrm{Pr})^{0.4} \, , \tag{12}$$

where Re will be evaluated at the local air bulk temperature and Pr is taken as constant, 0.7, since it is virtually unchanged in the temperature range of the experiment.

Friction Factor

In the current experiment the friction factor is determined from measurement of the total pressure drop. Therefore, our results will unavoidably combine developing and fully developed contributions, yielding a channel-averaged friction factor. Define this channel-average friction factor as

$$\overline{f_1} = \int_0^L \frac{f}{\rho}dz \Bigg/ \int_0^L \frac{dz}{\rho} \, . \tag{13}$$

From the relation

$$\Delta p_f = \int_0^L \frac{fG^2}{2D\rho}\,dz \, , \tag{14}$$

for constant G one has

$$\overline{f_1} = 2D\,\Delta p_f \Bigg/ \left(G^2\int_0^L \frac{dz}{\rho}\right) . \tag{15}$$

Since the total pressure Δp_t was measured instead of Δp_f, Eq. 15 becomes

$$\overline{f}_1 = 2D(\Delta p_t + \Delta p_b - \Delta p_{acc}) \Bigg/ \left(G^2 \int_0^L \frac{dz}{\rho}\right) . \quad (16)$$

The buoyancy head, Δp_b, can be evaluated as follows:

$$\Delta p_b = \int_0^L [\rho_{in}(T_{in}) - \rho(T_a)]\, g\, dz \ . \quad (17)$$

The air density, r, can be expressed by treating air as a perfect gas, as:

$$\rho = \frac{353.12}{T(K)}, \ \text{kg/m}^3 . \quad (18)$$

The acceleration pressure drop is calculated as follows:

$$\Delta p_{acc} = G^2 \left[\frac{1}{\rho_e(T_e)} - \frac{1}{\rho_{in}(T_{in})}\right] . \quad (19)$$

In Eq. 13 the definition of the friction factor involves integrals. The fluid density as a function of location is not known before the flow rate is calculated. However, the flow rate depends on the friction factor. This makes the Eq. 13 definition difficult to use in engineering applications.

Consequently, in engineering applications the channel average friction factor is typically defined in terms of a reference density r_o, such that Eq. 16 becomes:

$$\overline{f}_2 = \frac{2\rho_o D(\Delta p_t + \Delta p_b - \Delta p_{acc})}{\overline{L}G^2} \quad (20)$$

Note that $\overline{f}_2$ is very similar to the apparent friction factor in an isothermal flow. The evaluation of the reference density in Eq. 20 will be discussed subsequently, when the experimental results are presented.

EXPERIMENTAL RESULTS

The friction factors and the Nusselt numbers are the major results sought. The wall temperatures are also of interest. In addition, the changes of the dimensionless numbers, Re, Gr and Bo, along the channel height will be presented, since their behavior is related to the local Nusselt number behavior. Table 1 summarizes the experimental conditions investigated, which are illustrated in Figure 3. Error bars for all experimental data have been assessed at the 1 σ level [3] and are shown for data which encompass the experimental range on the figures which follow.

Re, Gr and Bo Number

The use of air as the flow medium in the experiments leads to significant temperature changes over the test length. Since gas properties are much more sensitive to temperature change than are those of liquids, the property changes of the flowing air must be considered in interpreting the experimental results. For example the air density will decrease about 40% when the temperature increases from the ambient temperature, 20°C, to 200°C, a temperature a number of runs have reached. These property changes will affect the dimensionless numbers. Typically for these conditions, the Re decreases at almost a constant rate: for example, from 10,000 at channel inlet to a little below 7,000 at the channel exit, a 30% decrease. The Gr_q exhibits a rapid decrease in the first half of the channel, then decreases at a much slower rate. The channel exit Gr_q is less than 4% of the channel inlet Gr_q number. The rate of decrease of the Bo number is between those of the Re and Gr_q numbers; overall, an 85% decrease takes place.

Table 1 Experimental Conditions for Heated Test Runs

Run	Symbol (Fig. 10)	Bo (Inlet/Exit)	T_{air} (°C) (Inlet/Exit)	T_w (°C) (Inlet/Exit)*	Re (Inlet/Exit)	Gr_q (Inlet/Exit)
1	⊞	0.0092/0.0051	18.2/68.7	53.8/93.6	21800/19600	6.22E7/2.41E7
2	△	0.031/0.014	17.3/88.3	60.2/115.1	15600/13500	6.71E7/1.84E7
3	⊡	0.037/0.0080	14.2/195.5	136.6/270.9	21800/15500	2.49E8/1.64E7
4	○	0.020/0.0073	16.8/120.5	91.5/174.4	21900/17800	1.37E8/2.46E7
5	●	0.060/0.017	16.0/148.1	98.5/196.5	15700/12100	1.32E8/1.52E7
6	+	0.10/0.017	21.5/258.1	169.9/332.3	15400/10100	2.14E8/0.79E7
7	◇	0.15/0.042	25.8/153.3	107.0/204.8	10200/8000	7.80E7/0.91E7
8	■	0.23/0.045	26.6/212.4	151.5/286.5	10200/7300	1.15E8/0.69E7
9	□	0.29/0.045	22.6/249.7	173.8/336.6	10400/6900	1.54E8/0.58E7
10	◆	0.50/0.13	17.4/133.1	111.4/242.3	7000/5600	7.03E7/0.79E7
11	▣	0.87/0.13	12.9/216.8	167.9/254.8	7200/4900	1.30E8/0.52E7

* Inlet at 2 feet from start of heated length and exit at 2 feet from end of heated length.

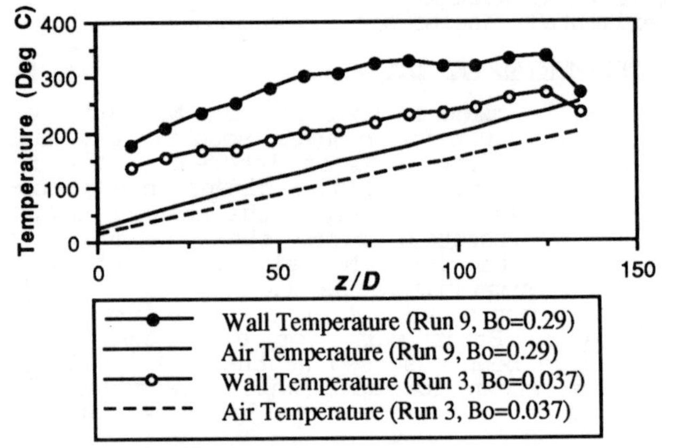

Figure 6: Wall Temperature for Low and High Bo Number Cases.

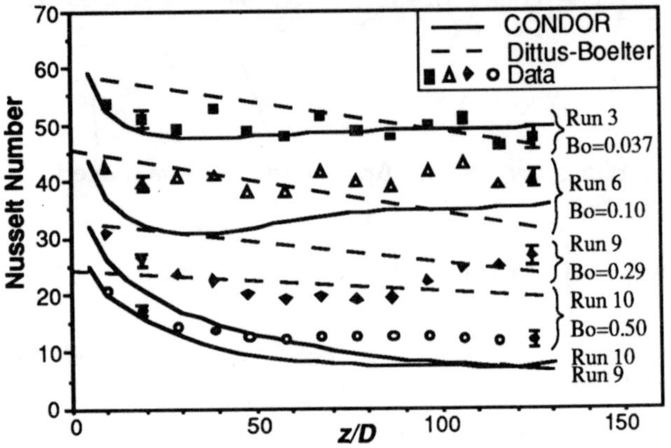

Figure 7: Heat Transfer Behavior for the Range of Bo Number Tested.

Wall Temperatures

Two typical wall temperature profiles are shown in Figure 6. They represent low and high inlet Bo number cases. The air stream temperatures computed from the energy balance are also shown in the figure. The wall temperature variation is very close to linear for the inlet Bo equal to 0.037 case, whereas for the high (0.29) Bo case, the buoyancy effects cause a noticeable distortion in the wall temperature over the entire duct length. This phenomena of wall temperature distortion has also been observed by other investigators, namely Argonne National Laboratory investigators, and Miyamoto, as reported by Sohn, et al. [12], and Tanaka, et al. [13].

Nusselt Numbers

Local Nusselt number results are given as functions of axial location and local Bo number in Figure 7. They span the inlet Bo number range investigated and represent different Nusselt number variation trends. Dittus-Boelter correlation and CONDOR predictions, based on the local bulk temperature, are also shown for each case. The inlet Bo number of Run 3 is close to 0.037 and hence the data are considered to be in the forced convection regime. The measured Nusselt number is very close to the calculated values from the Dittus-Boelter correlation for large z/D. The CONDOR results indicate the extent of the developing region (to $z/D \approx 30$). Run 6 with an inlet Bo number of about 0.10 is considered to be just inside the mixed convection regime. The results show that the measured Nusselt numbers do not yet exhibit an obvious decrease. There is no obvious reason to expect that the data and the CONDOR results should be higher than the Dittus-Boelter correlation at the larger z/D locations. For Bo increased to 0.29 as in Run 9, the effect of the buoyancy force is reflected by a gradual decrease in Nusselt number from the inlet until the middle of the channel and then a gradual recovery. Note that the associated CONDOR predictions do not recover as z/D increases. The suspected reason for this behavior is presented in the next section. At the channel exit, the Nusselt number reaches or exceeds the forced convection values. For further increase of Bo number, the Nusselt numbers stop recovering from their minimum values, as shown by Run 10 (Bo = 0.50). The Nusselt results are flat at about 70% to 50% of the forced convection Nusselt number values, depending on how large the buoyancy force is. It should be pointed out that at high Reynolds number the measured Nu numbers exhibit ripple - periodic undulations every three or four measured points. The ripples disappear at lower Re numbers. The ripples in Nusselt number are due to corresponding ripples in the wall temperature distributions. However, careful examination of the manner in which the heater coils were wrapped, and the thermocouple locations, does not show a direct connection with the observed ripples.

Friction Factors

The measured isothermal friction factors (at no heat input, i.e., cold run conditions), while within typical error range of forced convection friction factors, were consistently higher than the nominal values by about 3% to 5%. This may be attributed to flow development (i.e., "entrance") effects. The contribution of the terms Δp_b and Δp_{acc} to the error compared to other factors (see Eq. 16 and Eq. 20) varies with the Reynolds number. For Re = 15,000 they are small, at Re = 10,000 they are comparable, whereas at Re = 7,000 they are the major contributors.

The non-isothermal (mixed convection) friction factor measured in the experiments is shown in Figure 8 as a function of inlet Bo numbers. Both $\bar{f}_1$ and $\bar{f}_2$ are shown in the figure, where $\bar{f}_2$ is plotted at $\rho_o = \rho_{in}$, and $\rho_o = \bar{\rho}$, the density evaluated at the average bulk temperature, $\bar{T}_b = 0.5\ (T_{in} + T_e)$. As can be seen from the figure, $\bar{f}_2$ at $\rho_o = \bar{\rho}$ is very close to $\bar{f}_1$. It is clear from the figure that the density has a large effect on the magnitude of the friction factor. In fact, $\bar{f}_1$ decreases with Bo number while $\bar{f}_2$ with $\rho_o = \rho_{in}$ increases with Bo number.

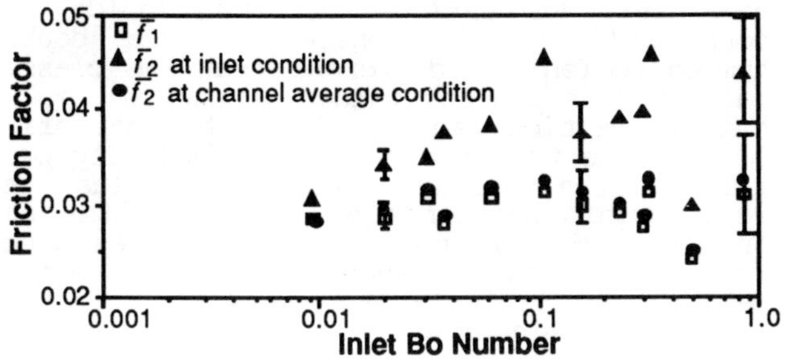

Figure 8: Experimental Friction Factor Results Interpreted with Various Air Density Assumptions.

ASSESSMENT OF PREDICTIVE CORRELATIONS USING THE EXPERIMENTAL RESULTS

We next assess and recommend predictive approaches for Nusselt number and friction factor for design applications using our experimental results.

Nusselt Number

For Nusselt number we consider the Jackson correlation and the k-ε model in the CONDOR code. Figure 9 presents the Jackson correlation and the CONDOR predictions for three experimental cases of Figure 7, which span the Bo range of interest. Note that due to the large property changes along the duct, each experimental case appears as a trajectory of local conditions, with Bo decreasing from inlet to outlet. Also, recall that the Jackson correlation is applicable to fully developed flow and was compared only with data that had small property changes. On the other hand, the CONDOR predictions and the experimental results presented here represent developing, local property dependent behavior.

Consequently, in comparing CONDOR results with the Jackson correlation, only the outlet portion (i.e., lower Bo region) of the CONDOR trajectory is appropriate. Figure 9 illustrates that the predicted trajectories of Runs 9 and 10 suggest disagreement with Jackson regarding the minimum Nusselt number and the Bo (~0.10 vs 0.25) at which occurs. However, if the disagreement is compensated by shifting the CONDOR results to the higher Bo number, CONDOR predicted trajectories approach the Nusselt values of the Jackson

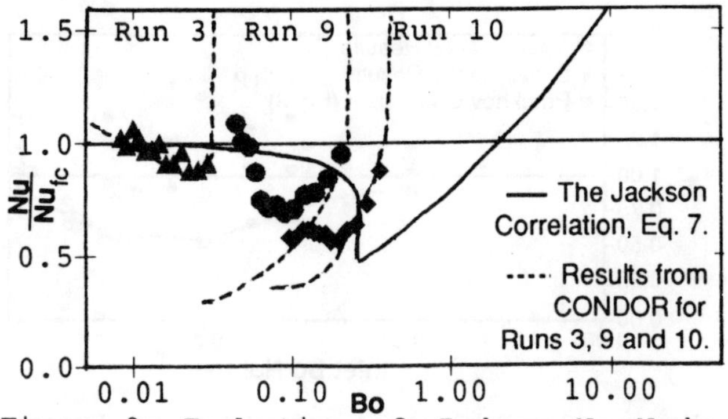

Figure 9: Evaluation of Jackson Nu Number Correlation Against Test Results and CONDOR.

correlation. Nevertheless, the trajectory of each run will still undershoot the lowest Nusselt value (Nu/Nu_{fc} = 0.5) of the correlation, reflecting the fact that the code does not allow the velocity field to transition from a Type II profile (maximum velocity close to wall) to Type I profile (maximum velocity at pipe center). This occurs even though the local Bo decreases to the necessary range based on the local condition transition hypothesis, consistent with our interpretation of the Jackson correlation in terms of local conditions.

Figure 9 also includes the experimental data for these three cases. We observe three characteristics:

1. The local conditions are fairly well represented by the Jackson correlation in the low mixed convection regime (Run 3).
2. The Nusselt number does appear to recover after a precipitous decrease in the moderate mixed convection region (Run 9 and to some extent Run 10).
3. The local Bo value for recovery (~0.1)is lower than that predicted (0.25) by the Jackson correlation (Run 9) and reasonably represented by the CONDOR prediction (Run 10).

For design use both the Jackson correlation and the CONDOR code have current limitations — large local condition variations and apparent error in the predicted Bo number for minimum Nusselt number for the Jackson correlation — and for the CONDOR code the inability to switch back to a Type I velocity profile (and the corresponding Nu (Bo) behavior), once a Type II velocity profile has been established. While design needs are typically for low Bo values, i.e., less than 0.1 where both approaches are reasonably satisfactory, we prefer to present a modified Dittus-Boelter correlation for design use. Our modification accommodates the observed decrease in Nusselt number (up to about 50%) due to the large property changes, and the progression into the mixed convection region, by a multiplier. The proposed correlation is:

$$Nu = Nu_{DB} \,/\, [1 + (4200/Re)^3] \tag{21}$$

for $Re \geq 5000$; $1\text{x}10^7 \leq Gr_q \leq 2.5\text{x}10^8$,

which is plotted with all the data measured in this study, in Figure 10. Observe that all the data effectively lies within about ±30% of this proposed correlation.

The applicability of this correlation to an actual riser case where the axial heat flux is expected to be higher near the reactor core midplane than at the top and bottom is an important question. The correlation has been designed to apply for local conditions. CONDOR k-ε computer modeling has confirmed that one can deal with variable $q'(z)$ using this methodology. Indeed, $q'(z)$ was not exactly constant in the subject experiments because of non-uniform (and different) heat losses at the ends of the test section. Therefore, we conclude this correlation can

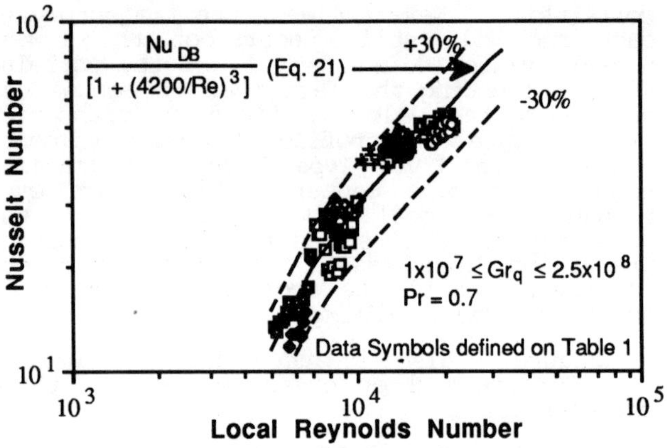

Figure 10: Proposed Correlation for Heat Transfer Results.

be used for actual risers with axially varying heat flux input.

Friction Factor

Figure 11 compares Petukhov's correlation, presented in Eq. 8, with experimental results of $\bar{f}_1$ and $\bar{f}_2$ with $\rho_o = \rho_{in}$. For each case the friction factor has been normalized by the isothermal friction factor from Petukhov's forced convection correlation, Eq. 9. The experimental results for $\bar{f}_1$ are generally larger than that predicted by Petukhov. Further, they do not exhibit the recovery and increasing trend predicted by Petukhov for Bo numbers greater than about 0.1. A friction factor calculation based on channel average density is most appropriate since, as Figure 8 demonstrates, it closely approximates $\bar{f}_1$. However, use of a channel average density involves an iteration process in the flow rate and heat transfer calculations, because the down-stream air temperatures are not known when the calculations start. Nevertheless, since designers generally can closely estimate the channel average temperature, use of $\bar{f}_2$ with $\rho_o = \bar{\rho}$ is recommended. From Figure 11 (since $\bar{f}_1 \cong \bar{f}_2$ with $\rho_o = \bar{\rho}$) we recommend use of

$$f/f_o = 1.1 \text{ for Bo} < 0.1.$$

For larger Bo, sufficient results are not available to make a definitive recommendation. However, $f/f_o = 1.1$ for Bo up to 1.0 appears justifiable and conservative, pending completion of further experiments.

Summary

In forced convection and in the strongly established mixed convection region, the friction factor and the heat transfer coefficient both increase, albeit with different ratios of proportionality. A transition region occurs, however, where these parameters are inversely proportional. Our k-ε computer code computations show that in this transition region the radial velocity profile changes from its conventional turbulent forced convection profile into a "double humped" profile dominated by near-wall buoyancy. Hence, while the Reynolds analogy parameter, Nu/RePrf = constant, holds for both profiles, the constant in the double humped region" is different from the classical Reynolds analogy value. Consequently, in this transition region, where the similarity between the temperature and the velocity profiles in the wall layer no longer holds, the relationship between the friction factor and Nusselt number is not clear. This area needs further explanation and experimental confirmation, particularly more accurate friction factor determination.

CONCLUSIONS AND RECOMMENDATIONS

1) The measured Nusselt numbers are in qualitative agreement with the Jackson correlation which we consider the best among published correlations. The present experiments provide an opportunity to examine the local Nusselt number behavior along the flow direction under mixed convection conditions, since the air properties change substantially. The Jackson correlation is also in good agreement with the experimental results regarding the extent of the heat transfer impairment, both showing a maximum decrease in heat transfer of about 50%, but it does not replicate well the variation of Nusselt number with Bo number. This bound is, however, very useful in RCCS integral heat removal calculations since the limiting "worst case" is of most concern from a reactor safety point of view.

2) Therefore, a better Nusselt number correlation or even a better dimensionless parameter group, instead of Bo, should be developed in the future. For present purposes the most useful approach would appear to be to present the data on a modified Dittus-Boelter Nu versus Re plot, as in Figure 10, where systematic differences from the Dittus-Boelter correlation are readily evident and have been empirically correlated. The recommended heat transfer correlation represented in this figure is Eq. 21, where all parameters are based on local properties:

$$Nu = Nu_{DB} / [1 + (4200/Re)^3] \qquad (21)$$

for Re $\geq$ 5000; $1 \times 10^7 \leq Gr_q \leq 2.5 \times 10^8$.

3) The friction factor behavior under the present experiment's conditions depends on which properties are used in evaluating the friction factor, because the air temperature and its properties change substantially along

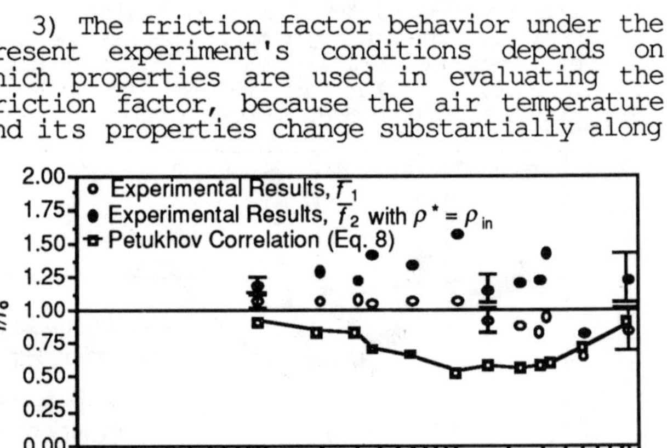

Figure 11: Evaluation of Petukhov Friction Factor Correlation with Tested Results.

the channel. For the condition where the inlet temperature is used as the reference temperature, a maximum friction factor increase of 50% over isothermal correlation values is inferred. For the condition where the channel average bulk temperature is used, an initial increase (to Bo = 0.1), followed by a decrease in the friction factor, is evident from the experiments. The reason for this difference is the much larger effect of density variations on the friction factor evaluated from the friction pressure drop measurements, compared with effect of viscosity variation on the friction factor obtained from empirical correlations. The results do not support the recommendation of the Petukhov correlation (Eq. 8). However, a value of f/f_o = 1.1 for Bo < 0.1 is recommended, where density is computed at the axially-averaged bulk air temperature.

ACKNOWLEDGMENT

This study was sponsored by the DOE and conducted under the technical direction of Bechtel National, Inc. The authors are grateful for the technical assistance provided during the course of this work by Dr. S. Ghose of Bechtel and others at GA Technologies, Inc. and Oak Ridge National Laboratory.

NOMENCLATURE

English Letters

A -- area
c_f -- skin friction factor
c_p -- heat capacity
D -- pipe diameter
f -- friction factor
G -- mass flow velocity
g -- gravitational acceleration
h -- heat transfer coefficient
I -- current
k -- thermal conductivity
L -- length
$\dot{m}$ -- mass flow rate
p -- pressure
P -- power
q'' -- heat flux
R -- pipe radius
r -- radial direction
T -- temperature
V -- voltage
w -- velocity
z -- flow direction

Dimensionless Number

$$\mathrm{Bo} = \frac{8\times10^4 \mathrm{Gr_q}}{\mathrm{Re}^{3.425}\mathrm{Pr}^{0.8}}$$ -- Buoyancy number

$$\mathrm{Gr_q} = \frac{g\rho^2\beta q'' D^4}{k\mu^2}$$ -- Grashof number

$$\mathrm{Nu} = \frac{hD}{k}$$ -- Nusselt number

$$\mathrm{Pr} = \frac{c_p\mu}{k}$$ -- Prandtl number

$$\mathrm{Re} = \frac{\rho w D}{\mu}$$ -- Reynolds number

Greek Letters

β -- thermal expansion factor
μ -- viscosity
μ_t -- turbulent viscosity
ρ -- density
τ -- shear stress

Subscripts

a -- air
acc -- acceleration
b -- buoyancy
c -- cross section, chimney
DB -- Dittus-Boelter
e -- exit
f -- friction
fc -- forced convection
in -- inlet
loss -- heat loss
nom -- nominal
o -- reference condition
out -- outlet
t -- total,turbulent
u -- unheated
w -- wall

REFERENCES

1. Ghose, S.K. & A.S. Shenoy, "Natural Circulation Air Cooling System for MHTGR," *IECEC*, August 1988.
2. Cotton, M.A. & J.D. Jackson, ASME HTD-Vol.84, pp. 43-50, 1987.
3. Fu, G., Sc.D. Thesis, MIT Nuclear Engineering Dept., January 1991.
4. Petukhov, B.S. & A.F. Polyakov, *Heat Transfer in Turbulent Mixed Convection*, Hemisphere Publishing Co., 1988
5. Connor, M.A. & A.D. Carr, *Proc. 6th Int'l Heat Transfer Conf.*, Vol. 1, pp.43-48, Toronto, Canada, August 1978.
6. Hall, W.B. & J.D. Jackson, ASME 69-HT-55, 1969.
7. Jackson, J.D., M.A. Cotton & B.P. Axcell, *Int'l J. of Heat and Fluid Flow*, Vol. 10, No. 1, pp.2-15, March 1989.
8. Steiner, A., *J. Fluid Mech.*, Vol. 47, part 3, pp.503-512, 1971.
9. Carr, A.D., M.A. Connor & H.O. Buhr, *J. Heat Transfer*, pp. 445-452, November 1973.
10. Yesilyurt, S., SM Thesis, MIT Nuclear Engineering Dept., February 1991.
11. Leary, W.A. & D.H. Tsai, "Metering of Gases by Means of the ASME Square-Edged Orifice with Flange Taps," Sloan Lab for Auto and Aircraft Engines, MIT, July 1951.
12. Sohn, D.Y., F.B. Cheung, A Hunsbedt & E.L. Gluekler, "Decay Heat Removal by Buoyancy-Induced Turbulent Natural Convection in a Vertical Parallel-Plate Channel," ASME Winter Annual Meeting, Chicago, 1988.
13. Tanaka, H., S. Maruyama & S. Hatano, *Int'l. J. Heat Mass Transfer,* pp.165-174, Vol. 30, No.1, 1987.
14. Symolon, P.D., Ph.D. Thesis, MIT Mech. Engineering Dept., August 1982.

TURBULENT HEAT TRANSFER IN THE DEVELOPING AND FULLY DEVELOPED REGIONS OF A VERTICAL CHANNEL

D.Y. Sohn ■ Medium Engine Development Department, Daewoo Motor Co., Ltd., Inchon, Korea
F.B. Cheung ■ Department of Mechanical Engineering, The Pennsylvania State University, University Park, PA 16802

Turbulent heat transfer in a vertical parallel-plate channel heated both symmetrically and asymetrically is studied using an implicit finite-difference method. The problem is formulated by applying the conservation laws with the turbulent motion being described by a low-Reynolds-number k-D model. A variable-grid pattern is employed in the numerical computation along with the use of a combined inner and outer iteration procedure. The inner iteration treats the problem of velocity-pressure coupling utilizing a predictor-corrector approach weighted by an under-relaxation factor, whereas the outer iteration minimizes the error associated with the lagging technique. Numerical calculations have been performed over wide ranges of Grashof and Graetz numbers covering both the developing and fully developed flow regions. Based upon the numerical solutions, correlations are developed for the induced flow rate and the local Nusselt numbers in different flow regions. The correlations obtained in this study can be conveniently used to predict the maximum wall temperature and the induced flow rate in both performance and design calculations of a reactor-vessel air cooling system.

INTRODUCTION

In recent years, the process of turbulent natural convection in a vertical parallel-plate channel has received considerable attention mainly because of its relevance to decay heat removal from advanced reactors. A typical example is the so-called reactor-vessel air cooling system (RVACS) that utilizes the means of natural convection to reject heat from the vessel wall to the ambient air. One characteristic feature of the RVACS is that the air flow is induced entirely by buoyancy as a result of decay heating of the vessel. As such, both the air flow rate and the convective heat transfer coefficient are unknown quantities that need to be determined in the design and performance calculations.

Miyamoto *et al.* [1] were the first to measure the turbulence quantities and the temperature distribution in a vertical parallel-plate channel. They obtained a correlation for the heat transfer coefficient in the fully developed flow region. Pedersen *et al.* [2] carried out a similar set of experiments but with much higher wall heat fluxes and larger physical scales to simulate the performance of a full-scale segment of the RVACS. They measured the induced air flow rate and axial variations of the wall temperatures. Recently, Cheung and Sohn [3] performed a theoretical study of the problem utilizing a low-Reynolds-number k-D turbulence model developed by To and Humphrey [4]. Correlating equations for flow and heat transfer were presented for a symmetrically heated channel under fully developed flow conditions.

In this study, the work of Cheung and Sohn [3] is extended to the case of a symmetrically as well as an asymmetrically heated channel, and correlating equations are obtained numerically for both the developing and fully developed flow regions.

MATHEMATICAL FORMULATION

The conservation equations for the mean turbulent flow field induced in the channel can be obtained by time-averaging the instantaneous variables which are expressed as the sum of the mean components, denoted by an overbar, and the fluctuation components, denoted by a prime, *i.e.*,

$$T = \overline{T}+T', \quad u = \overline{u}+u', \quad v = \overline{v}+v', \quad \rho = \overline{\rho}+\rho', \quad \text{etc.} \tag{1}$$

After substituting these variables into the conservation equations, time-averaging, and applying the order-of-magnitude analysis, the following governing equations are obtained:

Continuity Equation:

$$\frac{\partial(\overline{\rho u})}{\partial x} + \frac{\partial(\overline{\rho v})}{\partial y} + \frac{\partial(\overline{\rho' v'})}{\partial y} = 0 \tag{2}$$

Momentum Equation:

$$\overline{\rho u}\frac{\partial \overline{u}}{\partial x} + \overline{\rho v}\frac{\partial \overline{u}}{\partial y} + \overline{\rho' v'}\frac{\partial \overline{u}}{\partial y}$$

$$= -\frac{d\overline{P}_d}{dx} + (\rho_\infty - \overline{\rho})g + \frac{\partial}{\partial y}\left[(\mu + \mu_t)\frac{\partial \overline{u}}{\partial y}\right] \quad (3)$$

Energy Equation:

$$\overline{\rho}C_p\overline{u}\frac{\partial \overline{T}}{\partial x} + \overline{\rho}C_p\overline{v}\frac{\partial \overline{T}}{\partial y} + C_p\overline{\rho' v'}\frac{\partial \overline{T}}{\partial y} = \frac{\partial}{\partial y}\left[\left(\lambda + \frac{\mu_t C_p}{Pr_t}\right)\frac{\partial \overline{T}}{\partial y}\right] \quad (4)$$

where $\overline{P}_d$ is the difference between the local pressure inside the channel and the ambient pressure at the same elevation x, *i.e.*, $\overline{P}_d = \overline{P}(x) - P_\infty(x)$. Note that the density of air is a function of temperature. Thus the ambient density ρ_∞ can vary in the axial direction, *i.e.*, $\rho_\infty = \rho_\infty(x)$, depending on the temperature stratification a_s, in the ambient region [5].

The initial and boundary conditions for the velocity and temperature fields are

$$x = 0: \quad \overline{P}_d = -\frac{1}{2}\overline{\rho}_o U_o^2, \quad \overline{T} = T_o, \overline{u} = U_o \quad (5a)$$

$$y = 0: \quad \overline{u} = \overline{v} = 0, \quad -\lambda\frac{\partial \overline{T}}{\partial y} = q''_{h,c} \quad (5b)$$

$$y = H: \quad \overline{u} = \overline{v} = 0, \quad \lambda\frac{\partial \overline{T}}{\partial y} = q''_{c,c} \quad (5c)$$

$$x = L: \quad \overline{P}_d = 0 \quad (5d)$$

where the convective heat fluxes $q''_{h,c}$ and $q''_{c,c}$ can be evaluated by considering the process of surface radiation between the channel walls.

The methods of To and Humphrey [4] is adopted in the present work using a two-equation k-D turbulence model which includes the effect of density fluctuation. The isotropic dissipation of turbulent kinetic energy is defined as

$$\epsilon = \nu\overline{\frac{\partial u_i' \partial u_i'}{\partial x_j \partial x_j}} = \nu D \quad (6)$$

In the variable property field, ϵ is not an independent variable. Therefore, it is more convenient to solve the D-equation first and then multiply the result by ν to obtain ϵ.

$$\mu_t = C_\mu \cdot F_\mu \cdot \overline{\rho}^2 k^2/(\mu D) \quad (7)$$

$$\overline{\rho}\overline{u}\frac{\partial k}{\partial x} + \overline{\rho}\overline{v}\frac{\partial k}{\partial y} = \frac{\partial}{\partial y}\left[\left(\mu + \frac{\mu_t}{Pr_k}\right)\frac{\partial k}{\partial y}\right] + \mu_t\left(\frac{\partial \overline{u}}{\partial y}\right)^2 - \mu D - \beta\frac{\mu_t}{Pr_t}\frac{\partial \overline{T}}{\partial x}\left(\overline{u}\frac{\partial \overline{u}}{\partial x} + \overline{v}\frac{\partial \overline{u}}{\partial y} + g\right) \quad (8)$$

$$\overline{\rho}\overline{u}\frac{\partial D}{\partial x} + \overline{\rho}\overline{v}\frac{\partial D}{\partial y} = \frac{\partial}{\partial y}\left[\left(\mu + \frac{\mu_t}{Pr_\epsilon}\right)\frac{\partial D}{\partial y}\right] + C_1 \cdot F_1 \cdot \frac{D}{k}\mu_t\left(\frac{\partial \overline{u}}{\partial y}\right)^2 - C_2 \cdot F_2 \cdot \mu\frac{D^2}{k} - C_3 \cdot \beta\frac{D\,\mu_t}{k\,Pr_t}\frac{\partial \overline{T}}{\partial x}\left(\overline{u}\frac{\partial \overline{u}}{\partial x} + \overline{v}\frac{\partial \overline{u}}{\partial y} + g\right) \quad (9)$$

where $C_1 = C_3 = 1.44$, $C_2 = 1.92$, $C_\mu = 0.09$, $Pr_k = 1.0$, and $Pr_\epsilon = 1.3$. The boundary conditions are [4]:

$$y = 0: \quad k = 0: \quad D = 2\left(\frac{\partial k^{1/2}}{\partial y}\right)^2 \quad (10a)$$

$$y = H: \quad k = 0: \quad D = 2\left(\frac{\partial k^{1/2}}{\partial y}\right)^2 \quad (10b)$$

From dimensional considerations of the above governing equations, it can be shown that the induced flow rate in terms of the Reynolds number, Re, and the Nusselt numbers at the hotter and colder walls, Nu_h and Nu_c, respectively, are functions of the Grashof number, Gr, thermal loading factor, Ω_H, heat flux ratio, γ_H, aspect ratio α, and the ambient temperature gradient a_s. These parameters are defined as

$$Re = \frac{\rho U_o D_h}{\mu}, \quad Nu_h = \frac{h_h D_h}{\lambda}, \quad Nu_c = \frac{h_c D_h}{\lambda_o} \quad (11)$$

$$Gr = \frac{g\beta q''_w D_h^4}{\lambda \nu^2}, \quad \Omega_H = \frac{\beta q''_w D_h}{\lambda}, \quad \gamma_H = \frac{q''_{c,c}}{q_{h,c}}, \quad \alpha = \frac{L}{D_h} \quad (12)$$

where $q''_w = q''_{c,c} + q''_{h,c}$ is the total wall heat flux. The objective of this study is to determine the functional dependence of Re, Nu_h, and Nu_c.

NUMERICAL SOLUTION METHOD

The governing equations are solved numerically by an implicit finite-difference method. A variable-grid pattern [6] is employed in the numerical computation for which 70% of the grids are distributed in the near wall regions. The computer program so developed consists of two iteration loops. The inner loop treats the problem of velocity-pressure coupling whereas the outer loop minimizes the error associated with the lagging technique. Note that the coupling between the velocity and pressure calculation is an important factor because the present flow field is self-adjusted.

The calculated pressure at the outlet is the only reference point that can be used to examine the induced flow rate at the inlet. Hence a predictor-corrector procedure weighted by an under-relaxation factor is adopted. The under-relaxation is introduced since the present velocity profile is considerably skewed. This iteration process is different from the method of Patankar and Spalding [7] as illustrated in Raithby and Schneider [8].

To reduce the error associated with the lagging technique, the coefficients and source terms at the $(i+1)$th station are first determined using the upstream values obtained at the (i)th station, and the new values of flow and turbulence quantities are determined subsequently by the Thomas algorithm for tirdiagonal matrix. Once the axial velocity profile, $(U_{i+1,j})^k$, is known, the transverse velocity, $(V_{i+1,j})^k$, is computed explicitly from the continuity equation. Utilizing the solution just obtained at the (k)th iteration step, the coefficients and source terms are then updated, and the new values for k, D, and T are calculated. This procedure is repeated until a prescribed convergence criterion is satisfied. A detailed description of the numerical approach is given in [5].

RESULTS AND DISCUSSION

In the numerical computation, more than 230 cases covering the ranges of $1 \times 10^8 \leq Gr \leq 1 \times 10^{14}$ and $2000 \leq Re \leq 2.5 \times 10^5$ for $\Omega_H = 1$, 10, 100, and 1000, $a_s = 0$, 0.5, 0.75, and 0.8, $\gamma_H = 0$, 0.25, 0.5, 0.75, and 1, and $T_o = 293K$, are performed. Based upon the results obtained for these cases, correlating equations are derived. All results are successfully correlated with the reliability of $R^2 \approx 99\%$ where R^2 is the coefficient of multiple determination.

(a) Correlation for the Induced Flow Rate

The induced flow rate is expressed in a dimensionless form in terms of the Reynolds number. Previous studies of natural convection [1,9] correlated (Re/Gr) as a function of $(Gr \cdot Pr/\alpha)$. In the present study, the induced flow rate is also a function of the ambient temperature stratification, the heat flux ratio, and the thermal loading factor. If $a_s = 1$, no flow will be induced in the channel. For the case of symmetric heating, a multiple-linear-regression analysis of the present numerical data gives

$$\frac{Re}{Gr} = 0.593(1-a_s)^{0.546}\Omega_H^{-0.059}$$

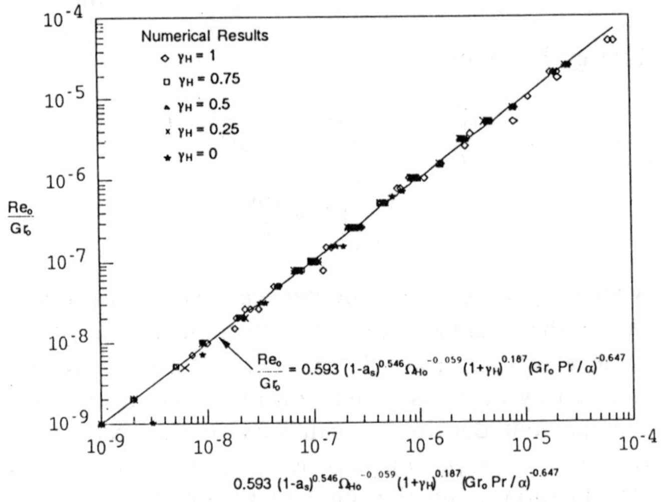

Figure 1. Correlation of the induced flow rate based upon the numerical results.

$$(1+\gamma_H)^{0.187}\left(\frac{Gr \cdot Pr}{\alpha}\right)^{-0.647} \quad (13)$$

The fitted curve and the numerical data are plotted in Figure 1.

(b) Correlations for the Local Nusselt Numbers in the Developing Region

Nelson and Wood [9] provided the local Nusselt number correlation for the laminar case with symmetric heating. Their result is given by

$$Nu = 0.536\left(\frac{x}{Re\ Pr\ D_h}\right)^{1/2} = 0.536(x^+)^{-1/2} \quad (14)$$

where x^+ is $(1/Gz)$, and Gz represents the Graetz number. Figure 2 summarizes the present results obtained for the symmetric heating case. In the inlet region, the local Nusselt numbers are well correlated by equation (14). In the developing region, the local Nusselt numbers do not vary according to the laminar behavior. Rather, they can be correlated by a different linear function of x^+ in the log-log plot. The resulting equation is

$$\begin{aligned} Nu_h &= Nu_c \\ &= 0.0282\ Re^{0.676}\ \eta^{-0.0394}\ \Omega_H^{-0.0601}\ x^{+-0.278} \quad (15) \end{aligned}$$

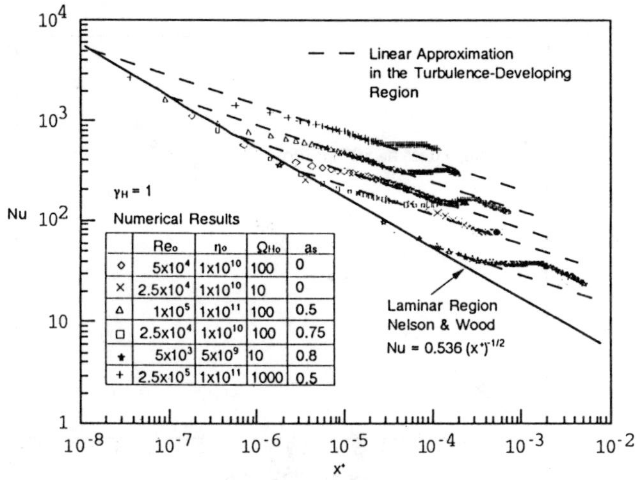

Figure 2. The local Nusselt number variation in the turbulence-developing region.

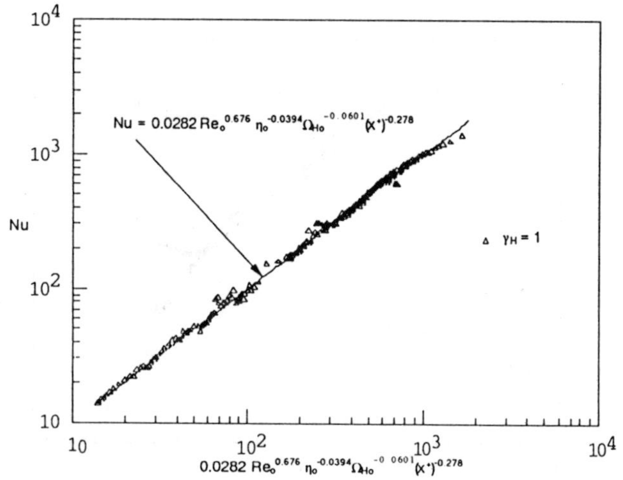

Figure 3. Correlation of the local Nusselt number in the turbulence-developing region based upon the numberical results.

The fitted curve and the numerical data are plotted in Figure 3.

For the asymmetric heating case, equation (14) needs to be modified by considering the thermal loading factor since it characterizes the wall heating condition. The correlating equations at the hotter wall and the colder wall are found to be

$$Nu_h = 0.0282\ Re^{0.676}\ \eta^{-0.0394} \left(\frac{2\Omega_H}{(1+\gamma_H)}\right)^{-0.0601} x^{+-0.278} \tag{16}$$

$$Nu_c = 0.0282\ Re^{0.676}\ \eta^{-0.0394} \left(\frac{2\gamma_H \Omega_H}{(1+\gamma_H)}\right)^{-0.0601} x^{+-0.278} \tag{17}$$

The above equations are well applied to the case of $\gamma_H = 0.25$ as shown in Figure 4.

(c) Correlations for the Local Nusselt Numbers in the Fully Developed Region

The wall temperature variation in the fully-developed region is nearly linear in spite of the effect of property variation, and its rate of change in the axial direction is the same as the rate of change of the bulk temperature. The bulk temperature variation is given by

$$\theta_b = \left(\frac{2\eta\Omega_H}{Re\ Pr}\right) \cdot X = 2 \cdot \Omega_H \cdot x^+ \tag{18}$$

Therefore the wall temperature variations in the fully-developed region can be written as

$$\theta_{h,FD} = (\theta_{h,FD} - \theta_b)_{\text{const}} + 2 \cdot \Omega_H \cdot x^+ \tag{19a}$$

$$\theta_{c,FD} = (\theta_{c,FD} - \theta_b)_{\text{const}} + 2 \cdot \Omega_H \cdot x^+ \tag{19b}$$

The corresponding local Nusselt number variations are expressed as follows by dividing the above equations by $\Omega_H/(1+\gamma_H)$ and $\Omega_H \cdot \gamma_H/(1+\gamma_H)$, respectively:

$$1/Nu_{h,FD} = 1/[Nu_{b,h}]_{FD} + 2 \cdot (1+\gamma_H) \cdot x^+ \tag{20a}$$

$$1/Nu_{c,FD} = 1/[Nu_{b,c}]_{FD} + 2 \cdot [(1+\gamma_H)/\gamma_H] \cdot x^+ \tag{20b}$$

where $[Nu_b]_{FD}$ is the Nusselt number based on the bulk temperature in the fully-developed region. Once the correlation for $[Nu_b]_{FD}$ is obtained, the local Nusselt number variations can be easily determined. The fully-developed Nusselt numbers, $[Nu_b]_{FD}$, can be well correlated by the following equations as shown in Figure 5.

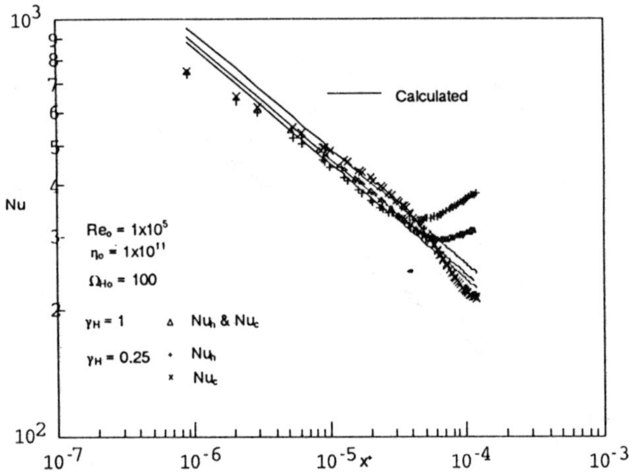

Figure 4. Application of the correlation for the local Nusselt number in the turbulence-developing region for the asymmetric heating case.

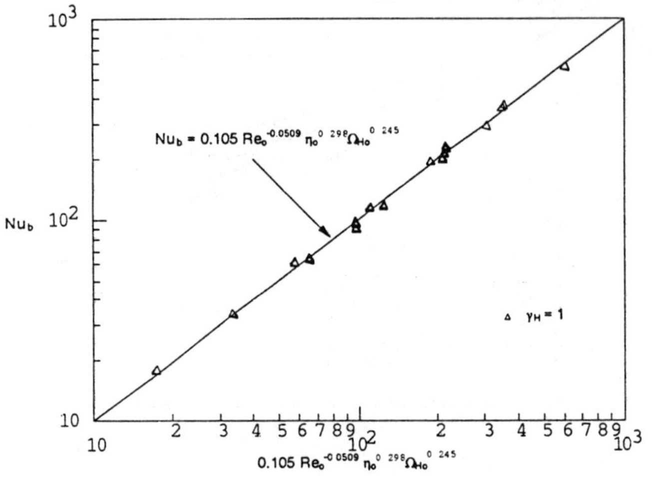

Figure 5. Correlation of the fully-developed Nusselt number, Nu_b, based upon the numerical results.

$$[Nu_{b,h}]_{FD} = 0.105\ Re^{-0.509}\ \eta^{0.298}\left(\frac{2\Omega_H}{(1+\gamma_H)}\right)^{0.245} \quad (21)$$

$$[Nu_{b,c}]_{FD} = 0.105\ Re^{-0.509}\ \eta^{0.298}\left(\frac{2\gamma_H\Omega_H}{(1+\gamma_H)}\right)^{0.245} \quad (22)$$

With equations (13) to (22), the maximum wall temperature can be predicted rather easily [5].

CONCLUDING REMARKS

Numerical computations have been conducted for various Grashof numbers, Graetz numbers, thermal loading factors, heat flux ratios, and ambient temperature gradients. Based on the numerical results, correlating equations have been obtained for the flow and heat transfer characteristics in the developing and fully developed regions. These correlations can be conveniently used to predict the maximum wall temperature and the induced flow rate in both performance and design calculations. Experimental data, however, are needed to validate the numerical correlations.

REFERENCES

1. Miyamoto, M., Katoh, Y., Kurima, J., and Sasaki, H., *Proc. 8th Int. Heat Transfer Conf.,* **4**, 1593, (1986).
2. Pedersen, D. *et al., AIChE Symp. Ser., #257,* **83**, 145, (1987).
3. Cheung, F.B. and Sohn, D.Y., *Proc. Nineth Int. Heat Transfer Conference,* **4**, 1412, (1990).
4. To, W.M. and Humphrey, J.A.C., *Int. J. Heat Mass Transfer,* **29**, 573, (1986).
5. Sohn, D.Y., "A Numerical Study of Turbulent Natural Convection in a Vertical Parallel-Plate Channel with Symmetric and Asymmetric Heating," Ph.D. Dissertation, Pennsylvania State University, (December 1989).
6. Anderson, D.A., Tannehill, J.C., and Pletcher, R.H., Computational Fluid Mechanics and Heat Transfer, Hemisphere, New York, (1984).
7. Patankar, S.V. and Spalding, D.B., *Int. J. Heat Mass Transfer,* **15**, 1787, (1972).
8. Raithby, G.D. and Schneider, G.E., *Numerical Heat Transfer,* **2**, 417, (1979).
9. Nelson, D.J. and Wood, B.D., *Proc. Eighth Int. Heat Transfer Conference,* **4**, 1587, (1986).

NOMENCLATURE

A_s Ambient temperature gradient
C_p Specific heat

D	Modified turbulence dissipation
D_h	Hydraulic diameter
$F's$	Wall functions
g	Gravity
Gr	Grashof number
H	Channel width
h	Heat transfer coefficient
k	Turbulent kinetic energy
L	Channel height
N_u	Nusselt number
P	Pressure
Pr	Prandtl number
q''_w	Wall heat flux
Re	Reynolds number
T	Temperature
U_o	Induced velocity at inlet
u	Axial velocity
v	Transverse velocity
x	Axial coordinate
y	Transverse coordinate

GREEK SYMBOLS

α	Aspect ratio
β	Thermal expansion coefficient
ϵ	Dissipation of turbulent kinetic energy
γ_H	Heat flux ratio
η	Gravity factor
λ	Thermal conductivity
μ	Dynamic viscosity
ν	Kinematic viscosity
ρ	Density
θ	Dimensionless temperature, $\beta(\overline{T} - T_o)$
Ω_H	Thermal loading factor

SUBSCRIPTS

b	Bulk condition
c	Colder wall
D	Dissipation field
FD	Fully developed region
h	hotter wall
k	Turbulent kinetic energy field
o	Inlet condition
t	Turbulence quantity
∞	Ambient condition

SUPERSCRIPTS

$'$	Fluctuating quantity
$-$	Mean quantity
k	Iteration step

CAVITY-TO-CAVITY INTERACTION IN NUCLEATE BOILING: THE EFFECT OF HEAT CONDUCTION WITHIN THE HEATER

Kemal O. Pasamehmetoglu and Ralph A. Nelson ■ Nuclear Engineering and Technology Division, Los Alamos National Laboratory, Los Alamos, NM 87545

This paper presents a numerical study aimed at analyzing bubble behavior as a function of site location. The effects of site distribution on the nucleate boiling curve are examined. Simple local-instantaneous models that mimic the bubble behavior on the boiling surface were implemented into a three-dimensional finite control volume conduction code. For a given site density, sample cases were run for uniform and nonuniform site distribution.

Our results indicate considerable deviation from linearized theories that always assume a uniform distribution. It is shown that bubble emission frequency is a strong function of site location. Consequently, the bubble flux density is shown to deviate from a simple periodic behavior with increasing nonuniformity in site distribution. This study further indicates that a uniform site distribution results in miniumum area- and time-averaged surface superheat and minimum temperature variations on the boiling surface. As the distribution becomes less uniform, average surface temperature and surface temperature variations along the boiling surface increase.

INTRODUCTION

Nucleate boiling is a highly desirable and thermally efficient mode of thermal energy transport that yields high heat fluxes at reasonable surface temperatures. Therefore, nucleate boiling phenomena have a wide range of applications.

In the last 60 yr, nucleate boiling has been the subject of numerous studies that have yielded both empirical correlations and theoretical models. Unfortunately, the state of the art suggests that almost every experiment calls for a different correlation.

There are very few studies in the literature where bubble-to-bubble or cavity-to-cavity interactions between a pair of adjacent sites are considered. The effect of multiple neighboring cavities on bubble behavior and, ultimately, on the boiling curve is not fully understood yet. The present study is aimed at further understanding of one piece of the puzzle, that is, the thermal interaction between cavities through heat conduction within the solid. Theoretically, the nucleation sites may interact through a number of mechanisms: (a) sonic waves generated during the early bubble growth and departure stages, which is postulated to promote nucleation; (b) hydrodynamic interaction because of liquid motion, pressure changes, and convective heating during the bubble growth and departure, which may promote or inhibit nucleation; (c) seeding, which is only valid for cavities in the very near vicinity of a growing bubble and which may promote intermittent activations; and (d) thermal interaction through heat diffusion within the heated solid, which may promote or inhibit nucleation. The available literature is reviewed briefly in the next few paragraphs.

Based on an analysis of high-speed motion pictures, Henley and Hummel [1] concluded that nucleation of a bubble at one site inhibited nucleation at all surrounding sites for a brief period thereafter. They also observed a strong tendency for bubble formation at the surrounding sites following the departure of the central bubble. Sonic interaction between the growing bubbles was postulated to be responsible for this behavior.

Chekanov [2] experimentally studied the interaction of two adjacent cavities using boiling water at atmospheric pressure. Chekanov found that the time elapsed between the growth of the two bubbles followed a gamma distribution. For a distance between cavities less than 3 departure diameters, Chekanov observed a "repulsive" interaction, when the growth of a bubble at one site inhibited the growth of the bubble at the adjacent site. At distances greater than 3 bubble diameters, the interaction was postulated to be "attractive," such that the growth of one bubble enhanced the growth at the adjacent site. At distances much larger than bubble diameter there were no interactions. Chekanov postulated that the results could be explained by sonic interactions.

Calka and Judd [3] boiled dichloromethane on a coated glass surface and observed bubble growth from adjacent sites. They found that the time elapsed between the growth of two adjacent bubbles followed a Gamma distribution, where the shape parameter vs the separating distance curve was in qualitative agreement with Chekanov's plot. Although they were quantitative deviations between the studies of Chekanov and Calka and Judd, the disagreement between the two studies is mostly in the interpretation of the results in defining attractive and repulsive interactions. For further discussion, the readers are referred to the study of Calka and Judd [3]. The mechanism for attractive interaction was postulated to be the seeding mechanism, where the growing bubble in one cavity activates a surrounding cavity intermittently through vapor seeding, based on the observations of Judd and Lavdas [4].

More recently, Judd [5] provided some more data using dichloromethane boiling over a glass surface. The results were in general agreement with the results of Calka and Judd.

As indicated by Sultan and Judd [6], the bubble flux density distribution over a heating surface is also an indirect indication for nucleation phenomenon and possible interactions. Their study (Sultan and Judd, [7]) indicated that sites are distributed in cluster forms over the heating surface, which suggests that bubble formation may promote activation in the neighboring sites, contrary to Chekanov's postulate. To examine this phenomenon further, Sultan and Judd [6] obtained data using boiling water at atmospheric pressure, heated by a copper block 6 in. (15.24 cm) in diameter, reduced to 2 in. (5.08 cm) at the boiling surface. Using a bubble detection probe, they obtained the time elapsed between the bubble formation at two adjacent sites. The data show that the elapsed time remains constant for saturated boiling at distances greater than ~5 mm. For distances shorter than this critical separation distance, the elapsed time decreases with decreasing distance. This critical separation distance appears to decrease with increasing subcooling. Sultan and Judd [6] recommended that this behavior could be explained through radial heat conduction within the heater. They postulated that the superheated liquid near the bubble base will act as an instantaneous ring source following the bubble departure.

It is conceivable that the time-dependent heat-transfer coefficient under a waiting and growing bubble acts as a transient source and sink term, which alters the thermal behavior in the vicinity. This is the essence of thermal site interaction. However, compared with this scenario, the model proposed by Sultan and Judd seems highly exaggerated. Furthermore, the model of Sultan and Judd is independent of the adjacent bubble history, as well as of the existence of other cavities.

In summary, these experimental data possibly include the effects of all of the interaction mechanisms cited in the introduction. However, all the mechanisms are not necessarily important within the same data set.

The present study is aimed at investigating the thermal interactions within the heater only. Our current closure relationships do not permit us to look into other types of interactions, such as sonic and hydrodynamic interactions. The study is expected to isolate the thermal interaction problem and examine the consequences of such interaction on the boiling curve.

In the next section, we will describe our computer model specifically developed to analyze the heater conduction effects on cavity-to-cavity interaction in nucleate boiling.

MODEL DESCRIPTION

Heater Analysis of Nucleation Dynamics (HANDY) is a three-dimensional, implicit-explicit, finite-volume, conduction-heat-transfer computer program that we are developing to analyze the nucleate boiling process. The finite control volumes are octagonal and square cylinders. Figure 1 illustrates the horizontal mesh structure. Each octagonal cell on the heater surface ($z = 0$) may contain a nucleation site (not necessarily active). The cell size in the horizontal direction is typically on the order of a bubble departure diameter (~1 mm), whereas the cell height can be made very small, depending on the heater thickness. The time-step size is on the order of a fraction of a millisecond because the phenomena of interest such as bubble waiting time and growth time are typically several milliseconds. Under these restrictions, the axial cell size limits the program stability. Thus, the code is made implicit-explicit in the axial direction with a degree of implicitness 0.5, which provides an unconditional stability in the axial direction. The solution scheme is fully explicit in the horizontal direction because the cell and the time-step sizes of interest do not violate the stability criterion. Furthermore, this scheme provides an easier coupling between octagonal- and square-cell temperature fields and results in a tridiagonal matrix that yields a computationally more efficient solution than a fully implicit case. Further details on the HANDY code can be found in Pasamehmetoglu and Nelson [8].

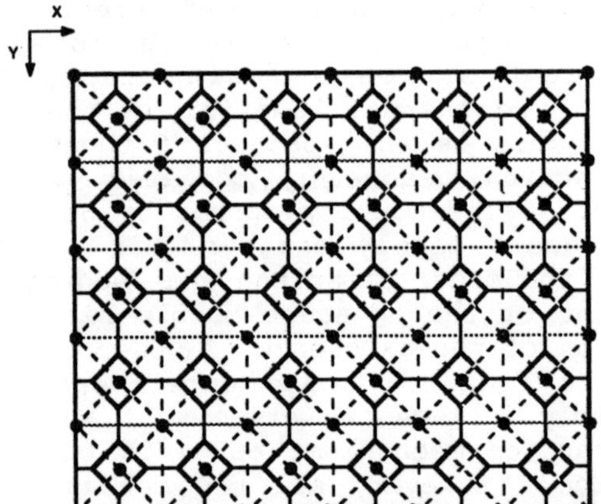

Fig. 1. Horizontal mesh structure of HANDY.

The boiling surface is characterized by a local (cell-averaged) and instantaneous (averaged over the time-step size) heat-transfer coefficient that satisfies the required boundary condition for the conduction solution. The coupling is time-wise explicit for the surface temperature dependence of the heat transfer coefficient. Version 1 of HANDY is only applicable to saturated water boiling at atmospheric pressure. The bubble behavior is simulated using literature-based models, which were reformulated (when necessary) to yield local-instantaneous heat transfer coefficients. Symmetry conditions supply other needed boundary conditions.

The current computer model is applicable to the discrete bubble region of nucleate boiling curve. There are numerous cavities on the boiling surface, some of which are activated and generate bubbles. Underneath the growing bubble and within its vicinity, latent heat transport and enhanced convection are important. In the region away from these bubbles, the surface is cooled by natural convection to single-phase liquid. Likewise, the heat-transfer mechanisms change locally as a function of time because of various stages in bubble dynamics. Thus, the computer model requires various closure relationships that will provide the heat-transfer coefficients as a function of location and time. In the following subsections, these closure relationships are summarized. These models are chosen among various models and correlations available in the literature. They are chosen because (a) they are widely accepted in the boiling heat transfer community and (b) when combined and used within HANDY, they provide the expected qualitative and quantitative boiling curve behavior. For brevity, extensive assessment and discussion of each model is not included in this paper.

Natural Convection Model. The surface cells without active cavities are assumed to be cooled by natural convection. For these cells, the well-known heat-transfer correlation given by

$$Nu = C_{NC}\ Ra^{1/3} \quad (1)$$

is used (Holman, [9]) to predict the extrapolated thermal layer thickness under natural convection conditions. The advantage of this correlation is that when it is written in terms of the heat-transfer coefficient, h_{NC}, it is independent of characteristic length. The proportionality constant C_{NC} currently is set to 0.15. This is a commonly accepted magnitude for isothermal surfaces, even though values as high as 0.6 have been reported in the literature. Thus, knowing the value of the heat-transfer coefficient, the extrapolated thermal layer thickness can be estimated as

$$\delta_{T,NC} = \frac{k_l}{h_{NC}} \quad . \quad (2)$$

Then, the local instantaneous heat-transfer coefficient is obtained through one-dimensional transient conduction solution within the liquid layer. The governing equation becomes

$$\frac{\partial T_l}{\partial t} = \alpha_l \frac{\partial^2 T_l}{\partial z^2} \ , \quad (3)$$

with the following initial and boundary conditions:

$$@\ t = 0, \qquad T_l = T_{sat}$$
$$@\ z = 0, \qquad T_l = T_w\ (x,y,t)$$
$$@\ z = \delta_{T,NC'}, \qquad T_l = T_{sat}$$

In the code, the solution to this transient problem is obtained numerically using the thermal thickness in user-defined axial cells. The solution scheme is fully implicit; however, the transient boundary temperature, $T_w\ (x,y,t)$, is the old time-temperature from the heater solution.

Cavity Activation Criterion. The user can define the desired number of cavities (limited to one cavity per surface cell) and cavity radii in the octagonal surface cells. The criterion used for the activation of this cavities is determined based on the theory developed by Han and Griffith [10]. Based on a maximum extrapolated thermal layer thickness, δ_{NC}, the range of cavities that can be activated in saturated liquids may be obtained from

$$R_{c,max \atop c,min} = \frac{\delta_{T,NC}}{3}\left(1 \pm \sqrt{1 - \frac{6A}{\delta_{T,NC}\ \Delta T_{w,NC}}}\right) \quad (4)$$

where

$$A = \frac{2\sigma T_{sat}}{\rho_g h_{fg}} \quad . \quad (5)$$

The cavities with radii between $R_{c,max}$ and $R_{c,min}$ will be activated. However, study of Shoukri [11] show that, on natural surfaces, cavities with sizes closer to the lower range are present and will be activated.

Bubble Departure Criterion. Currently, a simple bubble departure criterion is used such that a fixed departure diameter determines the departure condition. We used the correlation of Cole and Rohsenow [12], which is given by

$$D_b = C_B \sqrt{\frac{\sigma}{g\ \Delta\rho}} \left(\frac{\rho_l C_{p,l} T_{sat}}{\rho_g h_{fg}}\right)^{5/4} \quad (6)$$

where $C_B = 1.5 \times 10^{-4}$ is recommended for water. For the calculations reported in this paper, we used $C_B = 1.25 \times 10^{-4}$, which yields a bubble departure diameter of 2 mm for saturated water at atmospheric pressure. This number was more suitable for number density estimates needed for the present calculations.

Bubble Waiting Time Models. According to Hsu [13] and Han and Griffith [10], the bubble from an active cavity will grow if the liquid temperature at the bubble cap location reaches the required superheat. For a hemispherical nucleus emerging from a cavity, the required superheat can be calculated from

$$\Delta T_b = \frac{A}{R_c} \quad . \quad (7)$$

According to Han and Griffith, the isotherm tangent to the bubble cap will be located at a vertical location, $z_b = 1.5\ R_c$, relative to the heating surface, a short lateral distance away from the nucleus.

After bubble departure, bulk liquid comes into contact with the heater surface, and the thermal boundary layer starts developing. The development of the thermal boundary layer, δ_T, may be obtained from the solution of the transient conduction equation in the liquid given by Equation (2). However, if z_b is less than the nodal spacing, Δz_1, for the numerical solution of Equation (2) (quite often true because the cavity radii are typically on the order of μm), then an approximate approach is used as

$$\delta_T(t) = \sqrt{\pi \alpha_1 t} \quad . \tag{8}$$

This is the well-known solution for extrapolated thermal layer thickness in a semi-infinite solid for a constant boundary temperature. The instantaneous temperature at location z_b is calculated through a linear interpolation where the quasi-steady assumption is invoked by using the instantaneous wall temperature. If the calculated temperature exceeds ΔT_b from Equation (7), the bubble starts growing.

During the waiting time, except during the first time-step size following bubble departure, the convective heat-transfer coefficient is computed from the numerical solution in the liquid field if $z_b > \Delta z_1$. Otherwise, the convective heat-transfer coefficient is calculated as

$$h = \frac{k_1}{\delta_T} \quad . \tag{9}$$

For the first time-step size following bubble departure, a quench heat-transfer coefficient is required. The instantaneous contact temperature for two semi-infinite solids at uniform temperature is given by

$$\Delta T_c = \Delta T_w \left(1 + \sqrt{\frac{k_1 C_{p,1} \rho_1}{k_H C_{p,H} \rho_H}} \right) \quad . \tag{10}$$

In our problem, the heater has a certain temperature distribution in the axial and lateral directions. Knowing the contact temperature, the finite control volume equations are solved to obtain a value of h_1 that would yield a cell temperature equal to the contact temperature. Similarly, a heat-transfer coefficient, h_2, that would result in a liquid contact temperature to that given by Equation (10) within one time-step size is calculated. The maximum of the two magnitudes, h_1 and h_2, is chosen as the quench heat-transfer coefficient.

Bubble Growth Time Models. Isobaric and isothermal growth equations are considered for the bubble growth rate. For the isobaric growth, the modified Rayleigh solution provided by VanStralen *et al.* [14] is used. This growth equation is given by

$$R_1(t) = 0.8165 t \sqrt{\frac{\rho_g h_{fg} \Delta T_{w,o} e^{-\sqrt{t/\tau_d}}}{\rho_1 T_{sat}}} \quad , \tag{11}$$

where $\Delta T_{w,o}$ is the wall superheat at the time bubble starts growing and τ_d is set to 10 ms when used within this equation. For the isothermal growth, the bubble is assumed to grow due to microlayer evaporation. Assuming conduction across the microlayer of average thickness $\delta_o/2$, the energy equation yields

$$\frac{dR_2}{dt} = \frac{k_1 \Delta T_w(t)}{\rho_g h_{fg} \delta_o(t)} \quad , \tag{12}$$

where the initial microlayer thickness is formulated as (Olander and Watts, [15])

$$\delta_o = C_M \frac{\rho_g h_{fg}}{\rho_1 C_{p,1} \Delta T_w} \sqrt{Pr_1 \, R_2(t)} \quad . \tag{13}$$

Note that this formulation yields $R_2 \sim t^{1/2}$, which typically characterizes the asymptotic bubble growth. For these calculations, we set $C_M = 1.6$. For the isothermal growth, we neglected the effect of relaxation microlayer suggested by VanStralen *et al.* [14]. Lee and Nydahl's study [16] indicates that the liquid isotherms do not surround the bubble during asymptotic growth. The isotherms fold within a thin layer near the bubble base. Thus, the contribution of relaxation microlayer would be minimal. The isothermal and isobaric growth equations are combined using the approach suggested by VanStralen *et al.* [14] as

$$R(t+\Delta t) = R(t) + \frac{d}{dt}\left[\frac{R_1(t)\, R_2(t)}{R_1(t) + R_2(t)}\right] \Delta t \quad . \tag{14}$$

The bubble is assumed to depart when the radius exceeds the departure radius discussed earlier.

The heat-transfer coefficient during bubble growth is calculated through quasi-steady conduction across the microlayer.

Underneath the growing bubble, the average microlayer thickness is assumed to be $\delta_o/2$, whereas the average extrapolated thermal layer thickness is set to δ_o outside the growing bubble. Thus the cell-averaged heat-transfer coefficient is given by

$$h(t) = \frac{k_l}{\delta_o R_d^2}\left[2R^2 + \left(R_d^2 - R^2\right)\right] \quad . \tag{15}$$

Further details and discussion of these models can be found in the report by Pasamehmetoglu and Nelson [8]. Using the HANDY code, which contains the above-described models, we looked at a sample problem for nucleation dynamics. The sample problem is described in the following section.

SAMPLE PROBLEM DESCRIPTION

As a sample problem, we looked at the boiling of saturated water at atmospheric pressure. A 1-mm-thick copper plate was chosen as the heater material. The heater was supplied with a constant heat flux boundary condition of 70 kW/m^2 at the lower surface. If we look at Gaertner's [17]) and Sultan and Judd's [8] boiling curve, the expected average surface temperature is between 10 and 12°C for these conditions. We assumed a number density, N/A = 61,695 active sites per m^2, which was held constant for all cases studied. Note that Gaertner's [17] number density data yield about 46,000 sites/m^2, Sultan's [18] data yield 85,000 sites/m^2,and Shoukri's [11] data yield 155,000 sites/m^2. Thus, the number used in this study is between Sultan's and Geartner's measurements. The present study concentrates on the effects of site location. To isolate the problem, the additional effects of cavity-size variations were eliminated and will be considered in the future. All the cavities were assumed to be identical and to have cavity radii of 3.5 μm.

The diameter of the octagonal cells was set equal to the bubble departure diameter. The numerical parameters used for the current calculations are given in Table 1. Using the same numerical parameters, we ran five sample cases by changing the nucleation site locations. Figure 2 illustrates all five cases. Note that for a constant number density, Case 1 corresponds to a uniform site distribution, whereas Case 5 is the most nonuniform one among the five cases considered. The results of the calculations for these cases are given in the following section.

TABLE 1
NUMERICAL PARAMETERS USED IN THE CURRENT CALCULATIONS

Δz (m)	1.0×10^{-4}
Δx (m)	2.013×10^{-3}
Δz_1 (m)	3.734×10^{-5}
Δt (s)	5.0×10^{-4}

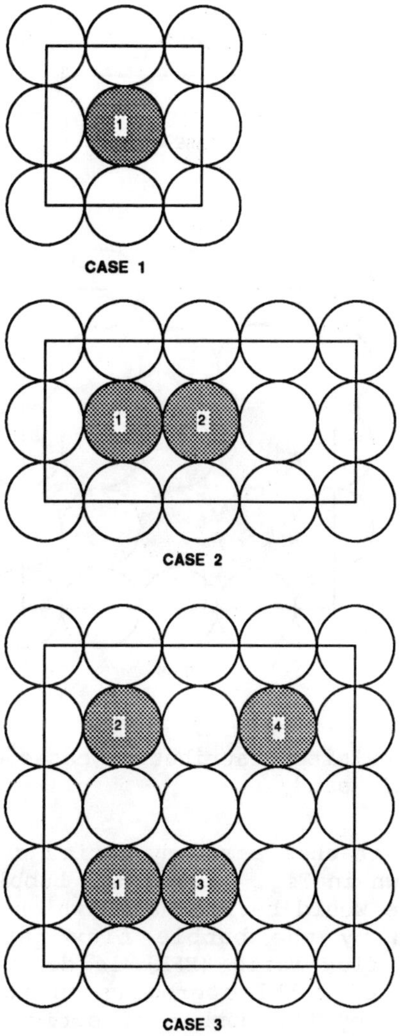

Fig. 2.
Various site distribution cases examined in this study.

RESULTS

Case 1 corresponds to a uniform cavity distribution over the heating surface for a given number density. In this case, the

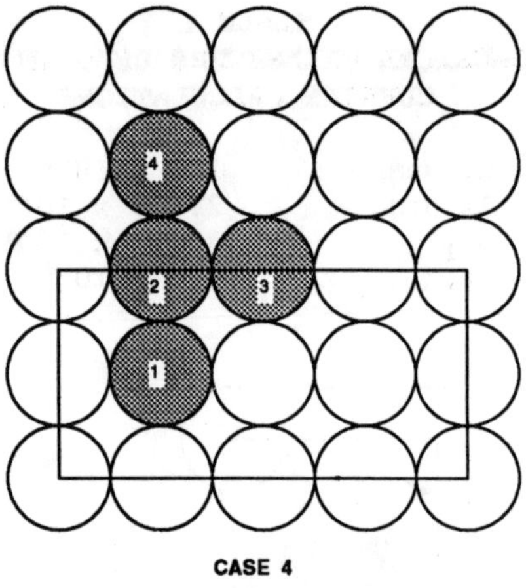

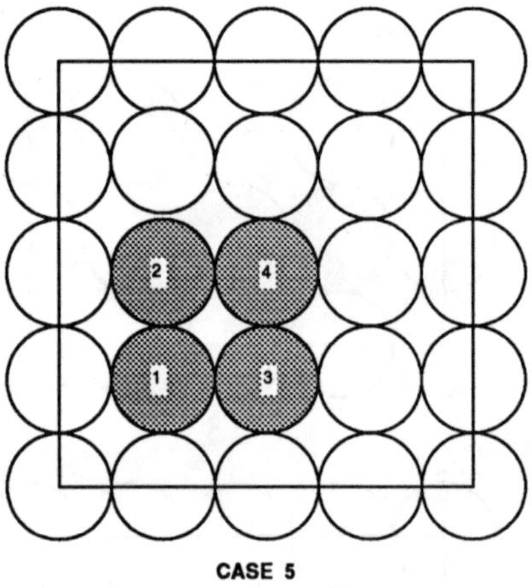

Fig. 2. (Cont)
Various site distribution cases examined in this study.

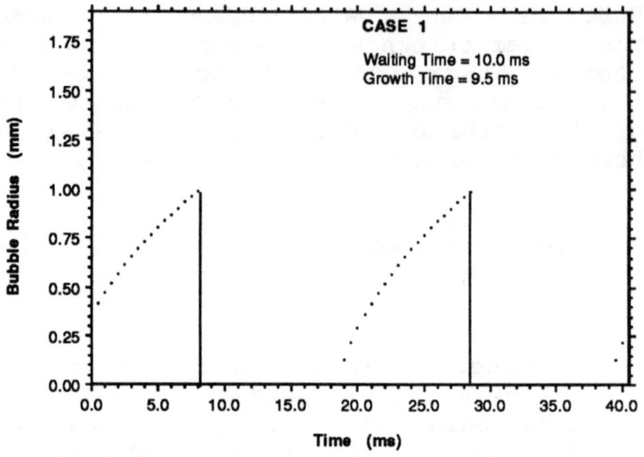

Fig. 3.
Bubble behavior for Case 1.

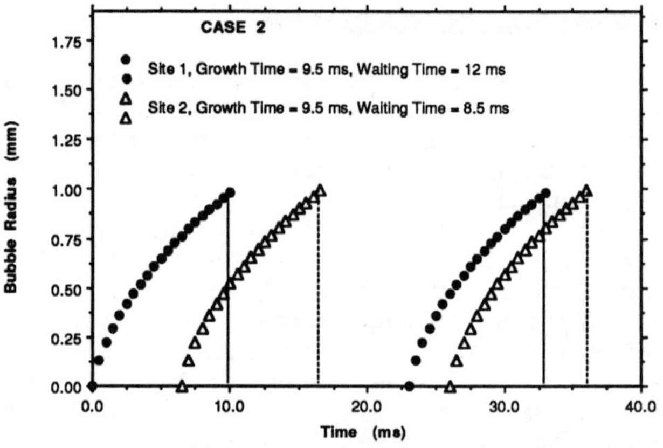

Fig. 4.
Bubble behavior for Case 2.

bubble reaches a steady-period of 19.5 ms, as shown in Fig. 3. All the bubbles on the surface would be in phase, which is illustrated by the bubble flux ratio. The bubble flux ratio (BFR) is defined as the number of bubbles growing on the surface divided by the number of active cavities. For a uniform distribution, BFR is periodic and is either one or zero.

Case 2 corresponds to a small deviation from a uniform distribution as a pair of adjacent cavities have a separating distance equal to 1 bubble diameter. In this case, site number 1 generates bubbles with a frequency of 46.5 Hz, whereas site number 2 generates bubbles with a frequency of 55.6 Hz, as shown in Fig. 4. The behavior of individual bubbles is expected to be periodic.

The numerical calculations indicate a small deviation from periodic behavior, as the waiting time for bubble 2 occasionally decreases from 8.5 ms to 8 ms. This is possibly a consequence of the numerical integration using 0.5-ms time-step sizes and is not caused by a system nonlinearity. However, the boiling surface behavior has a higher order periodicity. One way of characterizing the surface behavior would be again to look at BFR. It can be shown that BFR has a higher order periodicity where the patterns will repeat themselves every 387 ms. Thus, the period for the BFR increases by nearly 20 times as compared with Case 1.

Case 3 represents a combination of Cases 1 and 2. In this case, all four bubbles are out of phase, as shown in Fig. 5. However, bubbles generated from sites 2 and

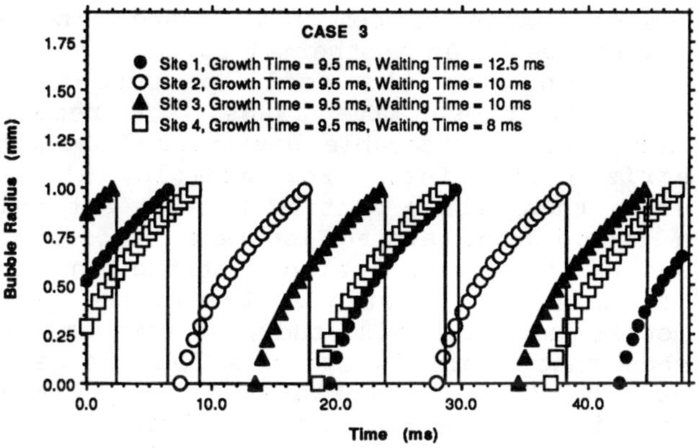

Fig. 5.
Bubble behavior for Case 3.

3 are out of phase but have the same period of 19.5 ms. Bubbles emitted from site 1 have the longest period (22 ms), whereas site 4 generates bubbles with period of 17.5 ms. Figure 6 illustrates the BFR for this case over a time frame of 100 ms. The BFR patterns are expected to repeat themselves with a period of 7.5 s, which is 385 times the period of uniform distribution (Case 1).

Cases 4 and 5 increase the cavity distribution nonuniformity even further. For Case 4, sites 1 and 4 are symmetric and generate bubbles with a frequency of 50 Hz. Bubbles emitted by site 3 have the highest frequency (60.6 Hz), whereas bubbles generated by site 2 have the lowest frequency (40 Hz). Finally, Case 5 has the highest degree of nonuniformity among all five cases that we considered. Sites 2 and 3 are symmetric and have a bubble emission frequency of 50 Hz. Sites 1 and 4 emit bubbles with frequencies of 40.8 Hz and 58.8 Hz, respectively. The BFR for Cases 4 and 5 have a pattern repeat period of 8.25 s and 8.33 s, respectively. Figures similar to those for Cases 1 through 3 can be generated for these cases as well. For brevity, they are not included in the paper. Thus, it can be concluded that the heater surface characteristics, characterized by the BFR in this study, become more and more nonlinear as the nonuniformity in cavity distribution increases. It would be interesting to see the effects of this behavior on the boiling curve.

Figures 7 and 8 illustrate the area-averaged instantaneous surface superheat, instantaneous maximum surface superheat, and instantaneous minimum surface superheat for Cases 1 and 5, respectively. The surface-averaged wall superheat and the maximum wall superheat increase as the site distribution becomes less uniform (as we go from Case 1 to Case 5). On the other hand, increasing the nonuniformity decreases the minimum surface superheat, thus, creating a larger temperature scatter on the boiling surface. Time- and area-averaged temperatures typically are used for plotting the boiling curve. We obtained the time-averaged values by averaging the data over the last second of the 2.5-s run, and these are reported in Table 2. Note that for Cases 3 through 4, bubble flux repeat periods are much larger than a second; thus, time-averaging done over a second does not really represent an appropriate average. However, the obtained values is expected to be within ±0.05°C of the exact value.

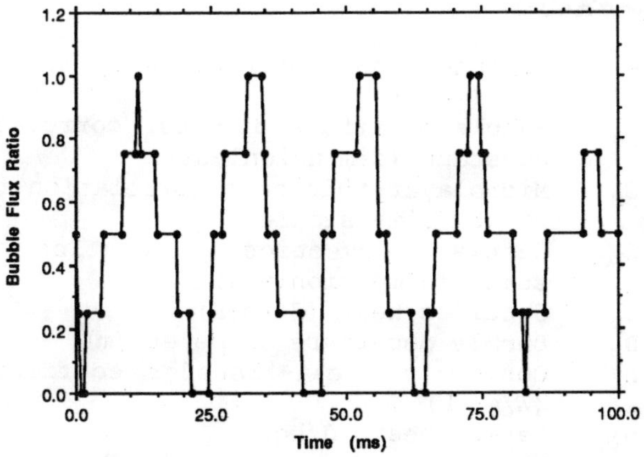

Fig. 6.
Bubble flux ration for Case 3.

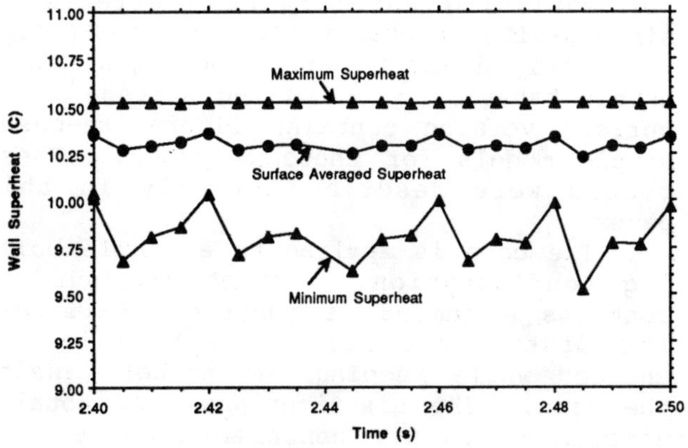

Fig. 7.
Wall superheat behavior for Case 1.

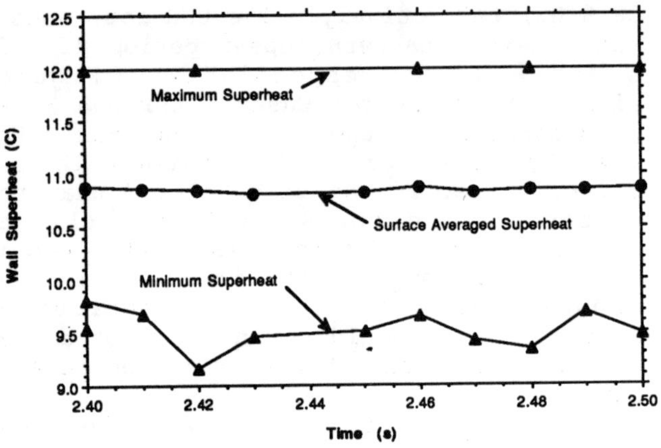

Fig. 8.
Wall superheat behavior for Case 5.

TABLE 2
TIME-AVERAGED WALL SUPERHEATS

Case	Time and Area Average Wall Superheat	Time Average Minimum Superheat	Time Averaged Maximum Superheat
1	10.30	10.52	9.82
2	10.54	11.11	9.66
3	10.68	11.36	9.50
4	10.78	11.84	9.50
5	10.85	11.98	9.50

SUMMARY AND CONCLUSIONS

In this paper, we studied the effect of the distribution of active sites on the bubble behavior and its consequences on the boiling curve. A three-dimensional conduction computer program is being developed for analyzing the nonlinear effects of site-to-site interactions. The HANDY computer program has various models that mimic bubble behavior on a boiling surface. The current version contains literature-based simple models for these purposes. These models were described briefly in this paper.

The code is applied to a simple boiling configuration in which the surface contains a number of identical cavities. The distribution of these cavities is changed while keeping the number density the same. The distributions from totally uniform to highly nonuniform cases were considered in five separate cases. Using symmetry boundary conditions, the total number of active cavities was kept at maximum four. All cases are run using a copper heater, which is commonly treated in the literature as an isothermal surface.

Even with the small perturbations tried in these five cases, the results indicated considerable deviation from linearized theories. For example, it was shown that bubble waiting time is not only a function of average surface temperature and cavity size but also a strong function of the cavity location. Within the parametric range of this study, we were able to show that there is up to a 55% change in waiting time. Similarly, the periodic behavior of the BFR becomes highly complex if the site distribution deviates from uniform. It also was shown that the site distribution has a noticeable influence on the time- and area-averaged wall temperatures. This is the quantity typically used in plotting the boiling curve. The average temperature shows an increase as the site distribution becomes increasingly nonuniform. Likewise, a nonuniform distribution promotes a larger span between the minimum and maximum surface temperatures.

Note that the present model represents a small perturbation from linearized models. Nevertheless, it presents strong indications that the actual nucleate boiling phenomenon is highly nonlinear. The effects of cavity-size variations are observed to enhance the nonlinear behavior and will be included in the future studies. As the next step in our modeling efforts, we will incorporate a more dynamic bubble growth model into HANDY. Also, the effects of hydrodynamic interactions are being incorporated into the code. Techniques to identify and analyze nonlinear and chaotic behavior also will be integrated into the interpretation of the results in an effort to better understand the behavior.

NOTATION

A	Superheat parameter given by Equation (7) (m-°C)
C_B	Bubble departure diameter correlation constant (dimensionless)
C_M	Microlayer thickness correlation constant (dimensionless)
C_{NC}	Natural convection correlation constant (dimensionless)
C_p	Specific heat (J/kg-K)
D_b	Bubble departure diameter (m)
h	Convective heat-transfer coefficient (W/m^2-K)
h_{fg}	Latent heat (J/kg)
k	Thermal conductivity (W/m-K)
Nu	Nusselt number (dimensionless)
Pr	Prandtl number (dimensionless)

R	Bubble radius (m)
R_1	Bubble radius as a result of isobaric growth (m)
R_2	Bubble radius as a result of isothermal growth (m)
Ra	Rayleigh number (dimensionless)
R_b	Bubble departure radius (m)
R_c	Cavity radius (m)
s	Separation distance between adjacent sites (m)
T	Temperature (°C)
T_{sat}	Saturation temperature (°C)
t	Time (s)
t_{end}	Problem end time (s)
z	Axial coordinate (m)
z_b	Elevation of the isotherm passing through nucleus cap (m)
α	Thermal diffusivity (m^2/s)
Δt	Time-step size (s)
ΔT_b	Superheat at bubble cap (°C)
ΔT_c	Superheat at contact (°C)
ΔT_w	Wall superheat (°C)
$\Delta T_{w,o}$	Wall superheat at time bubble starts to grow (°C)
Δx	Octagonal cell diameter (m)
Δz	Axial grid spacing within the heater (m)
Δz_1	Axial grid spacing within the liquid layer (m)
$\Delta \rho$	Density difference between liquid and vapor phases (kg/m^3)
δ_T	Extrapolated thermal layer thickness (m)
ρ	Density (kg/m^3)
σ	Surface tension (N/m)
τ	Elapsed time between bubble growth at adjacent sites (s)
τ_d	Bubble growth time (s)

Subscripts

g	Vapor
H	Heater
l	Liquid
max	Maximum
min	Minimum
NC	Natural convection

LITERATURE CITED

1. Henley, J. J. and Hummel, R. L., *Industrial and Engineering Chemistry, Fundamentals* **6**,(4), 603-606 (1967).

2. Chekanov, V. V., translated from *Teplofizika Vysokikh Temperatur* **15**(1), 121-128 (1977).

3. Calka, A., and Judd, R. L., *Int. J. Heat Mass Transfer* **28**, 2331-2342 (1985).

4. Judd, R. L., and Lavdas,C. H., *Trans. ASME, J. Heat Transfer* **102**, 461-464 (1980).

5. Judd, R. L., *Trans. ASME, J. Heat Transfer* **110**, 475-478 (1988).

6. Sultan, M. and Judd, R. L., *Trans. ASME, J. Heat Transfer* **105**, 3-9 (1983).

7. Sultan, M. and Judd, R. L., *Trans. ASME, J. Heat Transfer* **100**, 56-62 (1978).

8. Pasamehmetoglu, K. O. and Nelson, R. A., Los Alamos National Laboratory report LA-UR-91-977 (1991).

9. Holman, J. P., *Heat Transfer*, 5th edition (McGraw-Hill, New York) (1981).

10. Han, C-Y and Griffith, P., *Int. J. Heat Mass Transfer* **8**, 887-904 (1965).

11. Shoukri, M., Master Thesis, McMaster University, Hamilton, Canada (1974).

12. Cole, R., and Rohsenow, W. M., *Chem. Eng. Prog. Symp. Series* **92**(65), 211-213 (1969).

13. Hsu, Y. Y., *Trans. ASME, J. Heat Transfer* **84**, 207-216 (1962).

14. VanStralen, S. J. D., Sohal, M. S., Cole, R., and Sluyter, W. M., *Int. J. Heat Mass Transfer* **18**, 453-467 (1975).

15. Olander, R. R. and Watts, R. G., *Trans. ASME, J. Heat Transfer*, Vol. 91, pp. 178-180 (1969).

16. Lee, R. C. and Nydahl, J. E., Trans. ASME, J. Heat Transfer **111**, 474-479 (1989).

17. Gaertner, R. F., *Trans. ASME, J. Heat Transfer* **87**, 17-29 (1965).

18. Sultan, M., Master Thesis, Mc Master University, Hamilton, Canada (1977).

KUOSHENG BWR/6 CHANNEL FLOW STABILITY ANALYSIS

J.R. Wang, Chunkuan Shih, B.S. Pei, C. Pan,
and H.P. Chou ■ Department of Nuclear Engineering, National Tsing Hua University, Hsinchu, Taiwan
M.L. Hou ■ Taiwan Power Company, 242, Sec. 3, Roosevelt Road, Taipei, Taiwan

The present study aiming at systematically investigating BWR stability using RETRANO2/MOD3 code has been undertaken by National Tsing Hua University and Taiwan Power Company. The primary objective of this paper is to determine the channel flow stability map under natural circulation conditions for Kuosheng Nuclear Power Plant (KNPP). Mean-while, two transient loci of KNPP recirculation pump trip (RPT) with decreasing feedwater enthalpy are also drawn on the stability map to assess whether similar power oscillation phenomena as LaSalle-2 event may occur at KNPP. The result indicates that both transient loci have enough stable margins to unstable boundary and reveals that KNPP's operations will be stable under such RPT with decreasing feedwater enthalpy transients.

Boiling flow in Boiling Water Reactors (BWRs) is susceptible to hydrodynamic instability, which may cause flow oscillations. Sustained flow oscillations could induce boiling crisis, disturb control systems, or cause mechanical damage. Recently, instability phenomena were observed in several BWR plants[1-4] and had renewed interest in the area of BWR instability. A joint R&D program aiming at systematically investigating BWR stability analysis using RETRANO2/MOD3 code has been undertaken by National Tsing Hua University and Taiwan Power Company. This paper emphasizes on the channel hydrodynamic stability analysis of Kuosheng Nuclear Power Plant (KNPP), which has two units of GE designed BWR/6 with rated core thermal power of 2894 MW and rated core flow of 10645 kg/sec (23472 lb/sec).

The approach to investigate the flow stability of KNPP consists of two steps. The first step is to establish the stability map under natural circulation conditions (at 100% rod line) which is the condition expected to be more susceptible to instability than the normal flow condition. Parametric studies in inlet orifice form loss coefficient, fuel channel nodalization and time step size are included in this step. The second step is to analyze the dual recirculation pump trip (RPT) transients with isolated feedwater heater and plot the loci on the stability map. The purpose of this study is to determine the KNPP channel flow stability map, meanwhile, to assess whether similar power oscillation phenomena as LaSalle-2 event may occur at KNPP.

DESCRIPTION OF KNPP RETRANO2/MOD3 MODEL

Analytical tools for BWR stability prediction are essential in the verification of the adequacy of stability margins. The time domain computer code, RETRANO2/MOD3[5] has been applied for that purpose and has demonstrated excellent predictions for the BWR stability analysis [6-10]. The present study uses the RETRANO2/MOD3 code to analyze KNPP flow instability. Two different KNPP RETRANO2/MOD3 models are established in which the channel model is used to determine the flow stability map and the total plant model is used for dual RPT transient analyses.

KNPP RETRANO2/MOD3 Channel Model

The channel model simulates the active fuel (12 volumes) and the unheated core inlet and outlet regions of the fuel bundle with standard RETRAN control volumes and junctions, as shown in Fig. 1. The unheated core inlet region includes the fuel assembly inlet orifice and the fuel lower tie plate. The unheated core outlet region includes the upper tie plate. The lower plenum and upper plenum are treated as time dependent volumes

(TDVs) which allow flexibility in the specification of boundary conditions and the imposed perturbations. The actual geometry of KNPP fuel assembly is accurately modeled by the data derived directly from fuel assembly specification[11].

KNPP RETRAN02/MOD3 Total Plant Model

Fig. 2 shows the total plant nodalization of KNPP RETRAN02/MOD3 model, which has been verified against a series of power tests to assure the adequacy of the model[12]. This nodalization, applied to a wide range of transient events, is essentially the same as those used by other utilities for BWR transient analysis. In this model, a total of 29 volumes, 43 junctions and 4 heat conductors are contained.

In the fuel region, 3 fluid volumes, connected by 3 heat conductors, are used to simulate the behavior of coolant within the fuel channels. Results of the core nodalization study [12] in several transient cases indicate that there is no apparent advantage in using twelve, instead of three, volumes to model the fuel region. The bypass region is separately represented by a single fluid volume connected with a conductor. The point kinetics is used to calculate the reactivity feedback effect and dynamic slip model is adopted for core thermal hydraulic calculations. The steam dome, steam dryer, separator, upper downcomer and standpipes are each represented by a single fluid volume. In the present modeling, the bubble rise model is employed for the separator, upper downcomer and dryer regions. Two recirculation loops are each modeled with 3 volumes: the recirculation pump (RP) with suction pipng, the discharge piping, and the generation-2 jet pump. Each of the RPs is simulated with the RETRAN pump model. To simplify the RETRAN simulation, the four main steam lines are lumped into a single equivalent line, which is then divided into 4 fluid volumes before the steam header. Seven safety relief valves(SRVs) with different setpoints are modeled between the first volume and the suppression pool. Only one main steam isolation valve (MSIV) is used to simulate the isolation between inside and outside of the containment. Following the steam header, two flow streams are simulated, one to turbine via a turbine control volume (TCV), the other to the condenser via a bypass valve (BPV). Each SRV and BPV is modeled by a critical flow junction with isoenthalpic expansion option to calculate the chocked flow rate. The suppression pool, condenser and turbine are each simulated by a large "sink" volume. A single feedwater fill junction is used to represent the feedwater lines. The amount of feedwater injected into the vessel and the feedwater temperature are simulated by the control system.

ESTABLISHMENT OF KNPP FLOW STABILITY MAP

The method to establish KNPP flow stability map is developed based on an instability plane plotted with nondimensional core inlet subcooling versus nondimensional power to flow ratio, which is the same as that of the Ishii-Zuber instability plane [13]. On this plane, an instability threshold is determined by the following procedure [10].

1) To establish a steady state condition that corresponding to natural circulation flow and a power level corresponding to a rod pattern for 100% power at rated circulation for a given channel power, a channel pressure drop and an inlet subcooling. Establishment of initial conditions requires some iteration in which a trial channel flow rate must be input and then performs a steady-state RETRAN calculation with the loss coefficients of the spacers and the tie plates specified and the loss coefficients of the inlet orifice unspecified. The trial channel flow rate is changed and the steady-state analysis repeated until the RETRAN calculated inlet orifice loss coefficient converges to its known value.

2) The channel power, which is given explicitly by input data, is gradually increased keeping the inlet subcooling constant. The lower plenum pressure and enthalpy, and upper plenum pressure are also kept constant during calculation.

3) The instability threshold is determined by the point of onset of instability from the oscillation characteristic of the core inlet flow as illustrated in Fig. 3. Due to the small increasing rate of channel power, it is believed that this approach will not introduce significant uncertainty in the determination of threshold power.

4) By repeating the above procedure for various core inlet subcoolings, the stability map is obtained.

Before determining the stability map, the sensitivity on channel nodalization and time step size has been studied. Results of the channel nodalization study indicate that the instability threshold of a fine mesh (44 volumes) channel modeling is about the same as that of base modeling (12 volumes). However, the computing time for the fine mesh modeling is ten times longer. Similar results are also observed for the study of time step size, where 5 ms and 10 ms are adopted. Finally, a value of 10 ms was selected for the time step size, because the calculated results are essentially unchanged. Several parametric studies have been investigated and showed that the inlet orifice loss coefficient , K, is one of the dominant parameters for instability analysis [14]. In this study, instability boundary by using different K (0.337, 10.23 and 23.85) is determined and shown in Fig. 4. It shows that the larger K increases the ratio of single-phase pressure drop and makes the channel more stable as is expected. The results also show that instability boundary of the larger K (23.85) is irregular which is under further investigation now.

LOCI OF KNPP DUAL RPT TRANSIENT ANALYSIS ON STABILITY MAP

It is well recognized that channel hydrodynamic stability analysis presents the most conservative stability margins. As a result, the inclusion of the loci for two transient analyses of KNPP RPT with isolated feedwater heater (with or without reactor scram) on channel flow stability map should be on the pessimistic side of prediction. The plant initial conditions in the RETRAN analysis of both transients are presented in Table I. The plant model described in section 2 is used. Table II shows the comparison of the event sequences of both transient analyses. The transients are initiated at 1 second by tripping dual RPs, which in turn cause the coast down of the recirculation flow as well as core flow. A reduction in core flow results in an increase in core void fraction and, hence, decrease in core power due to negative void reactivity feedback effect. The feedwater enthalpy is assumed to decrease linearly with a rate of 1Btu/lb/sec for the case of RPT with scram (Case 1) and of 2Btu/lb/sec for the case of RPT without scram(Case 2). Fig. 5 and 6 show the power behaviors and core flow responses of both transients, respectively. At the initial stage of the transients, a decrease in coolant temperature causes a reduction in the void content of the core that is equivalent to a positive reactivity insertion. Due to the swelling effect of the increased voids, an increase in water level results in a high-level turbine trip signal and scrams the reactor at 30 second. If the reactor scram signals are bypassed, the core power in Case 2 is increased to 90% at 100 second. The loci of the calculated transients in Cases 1 and 2 are also shown on the stability map in Fig. 7. The figure indicates that both transient loci have enough stable margin to unstable boundary. It reveals that KNPP's operations will be stable under such RPT with decreasing feedwater enthalpy transients.

CONCLUSIONS

In this study, a plant specific RETRAN02/MOD3 model for the KNPP stability analysis has been developed including: (1) establishment of KNPP flow stability map and (2) loci of KNPP dual RPT transient analysis on stability map. Based on calculated results, the following conclusions can be drawn:

a) The KNPP channel flow stability map under natural circulation condition is determined. The instability threshold channel power levels under various inlet subcooling conditions are determined by the results from RETRAN time domain analysis. These calculations, in turn, allow us to establish the traditional stability map by informations such as system pressure, flow rates, degree of inlet subcooling, and channel power at the point where flow instability starts.

b) Strong dependence of the channel inlet orifice loss coefficient has been identified to show that accurate modeling of this particular parameter is of vital importance in the analysis. Flow oscillations can be easily demonstrated if one choose to use a smaller inlet loss coefficient.

c) Two transient loci of KNPP RPT (with or without reactor scram) are performed to assess whether similar power oscillation phenomena as LaSalle-2 event may occur at KNPP. It indicates that both transient loci have enough stable margin to unstable boundary and reveals that KNPP's operations will be stable under such RPT with decreasing feedwater enthalpy transients.

ACKNOWLEDGMENT

The authors would like to express their gratitude for the financial support of Taiwan Power Company and many valuable comments provided by the reviewers.

REFERENCES

[1] NRC Information Notice No.88-39: LaSalle Unit 2 Loss of Recirculation Pumps with Power Oscillation Event.
[2] S.A. Sandoz and S.F. Chen, "Vermont Yankee Stability Tests during Cycle 8", Trans. of ANS, 45 (1983)
[3] Araya, F., et al., "Parametric Study on Neutron Flux Oscillation Event at LaSalle 2 with RETRAN Code", Proc. ANS 1989 National Heat Transfer Conf., Philadelphia, HTC-VOL. 4 (1989)
[4] Takigawa, Y., et al., "Caorso Limit Cycle Oscillation Analysis with Three-Dimensional Transient Code TOSDYN-2", Nucl. Tech., Vol. 79 (1987)
[5] C.E. Peterson, et al., "RETRAN02 A Program for Transient Thermal-Hydraulic Analysis of Complex Fluid Flow Systems", EPRI NP-1850-CCMA-A (1984)
[6] K. Hornyik, "BWR Stability Analysis with RETRAN", Proc. 2nd Int. RETRAN Conf., EPRI NP-2494-SR (1982)
[7] J.M. Sorensen, et al., "BWR Stability Calculations with RETRAN", Proc. 3rd Int. RETRAN Conf., EPRI NP-3803-SR (1985)
[8] J.M. Sorensen, et al., "The Reactor Analysis Support Package (RASP), BWR Stability Analysis", EPRI NP-4498 (1988)
[9] S.J. Peng, et al., "BWR Core Stability Analysis with RETRAN", Proc. 5th Int. RETRAN Conf., EPRI NP-5781-SR (1988)
[10] Araya, F., et al., "Summary of RETRAN Calculations on LaSalle 2 Neutron Flux Oscillation Event Performed at JARI", Proc. 1990 Int. Workshop on BWR Stability.
[11] "Final Safety Analysis Report of Kuosheng Nuclear Power Station Unit 1 and 2", Taiwan Power Company.
[12] J.R. Wang, et al., "RETRAN02/MOD3 Analysis of Kuosheng BWR/6 Startup Tests", Proc. 4th Int. Topical Meeting on Nuclear Reactor Thermal Hydraulics, Germany, Oct. 1989.
[13] M. Ishii and N. Zuber, "Thermally Induced Flow Instabilities in Two Phase Mixtures", 4th Int. Heat Trans. Conf., B5.11, 1970.
[14] J.R. Wang, et al., "Investigation of Chinshan BWR/4 Stability Analysis", Proc. 1990 Int. Workshop on BWR Stability. Long Islands, N. Y. USA.

Table I. Initial Conditions of KNPP RPT Transients Analyses

Core Power	2894.0 MWt (100%)
Steam Dome Pressure	1040.0 psia (7.16 MPa)
Core Flow	23472 lb/sec (10645 kg/sec)
Steam Flow	3459 lb/sec (1568.7 kg/sec)
Feedwater Flow	3459 lb/sec (1568.7 kg/sec)
Feedwater Temperature	420.0°F (215.5°C)
Narrow Range Water Level	2.95 ft (90.0 cm)

Table II. Event Sequences of KNPP RPT Transients Analyses

Events (sec)	Case 1 (with scram)	Case 2 (w/o scram)
Initial Manual Trip on Dual RP	1.0	1.0
Loss of FW Heater (FW Enthalpy Decrease)	1.0 (1Btu/lb/sec)	1.0 (2Btu/lb/sec)
Core Power and Flow Start Dropping	1.36	1.36
Water Level Reach L8	29.8	-
T/B Trip Due to High Water Level (L8)	29.9	-
Reactor Scram	30.0	-
End of Analysis (Core Power)	100.0 (5%)	100.0 (90%)

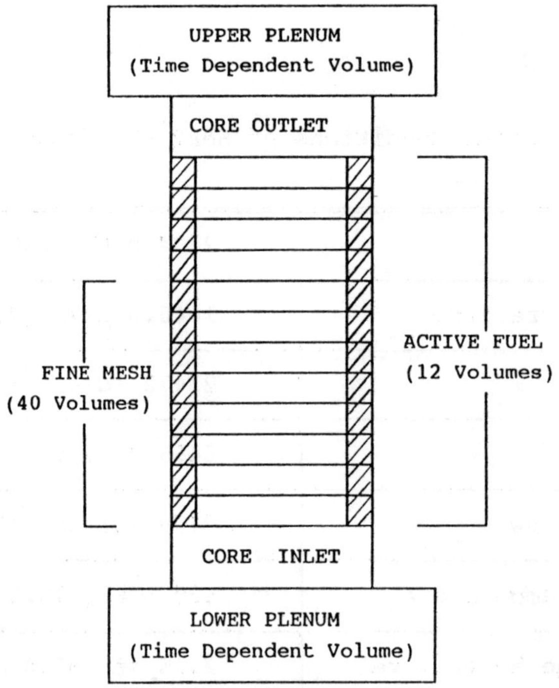

Figure 1. RETRAN model for KNPP channel stability analysis.

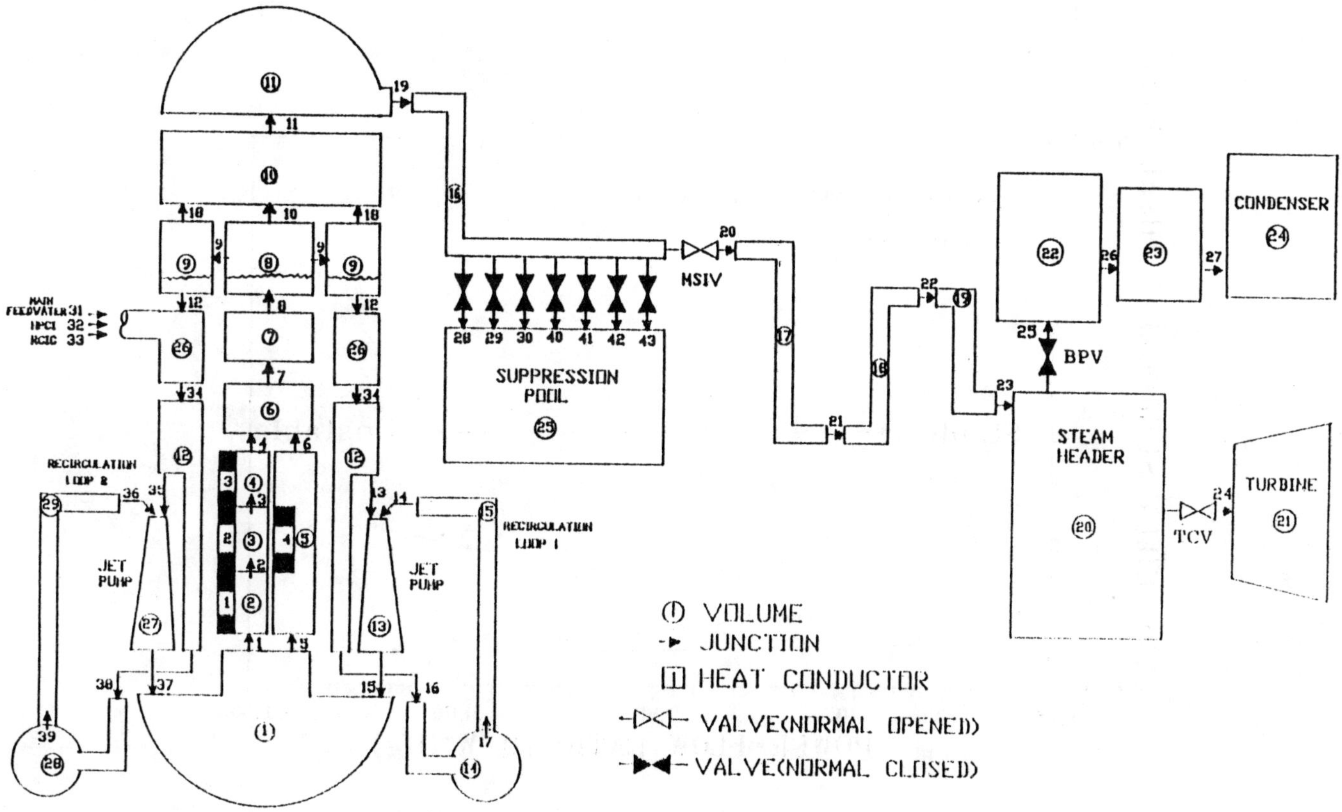

Figure 2. Total plant nodalization of KNPP RETRAN model.

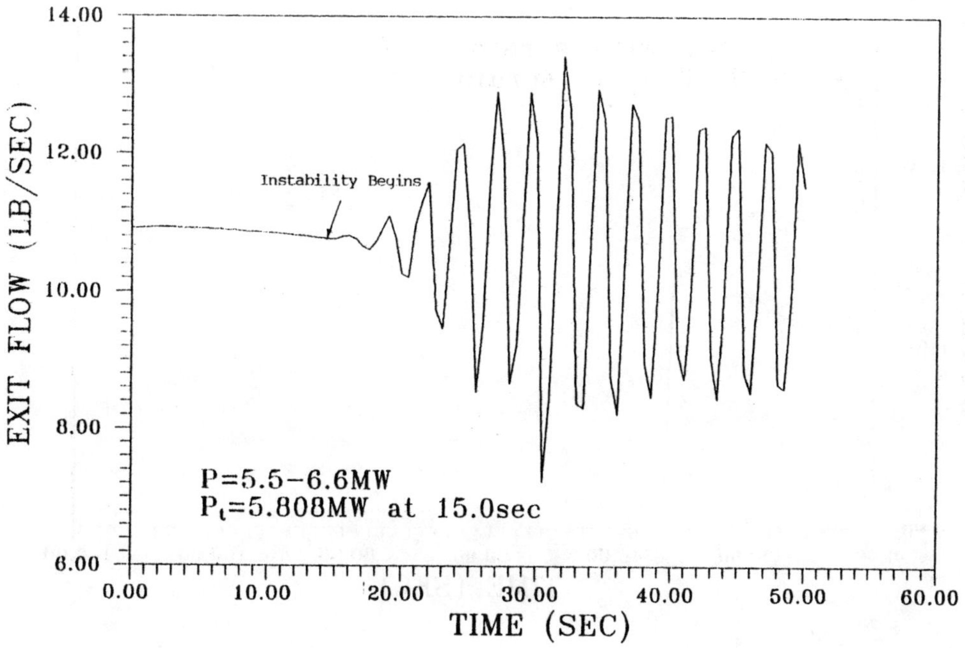

Figure 3. Onset of instability in KNPP under natural circulation conditions.

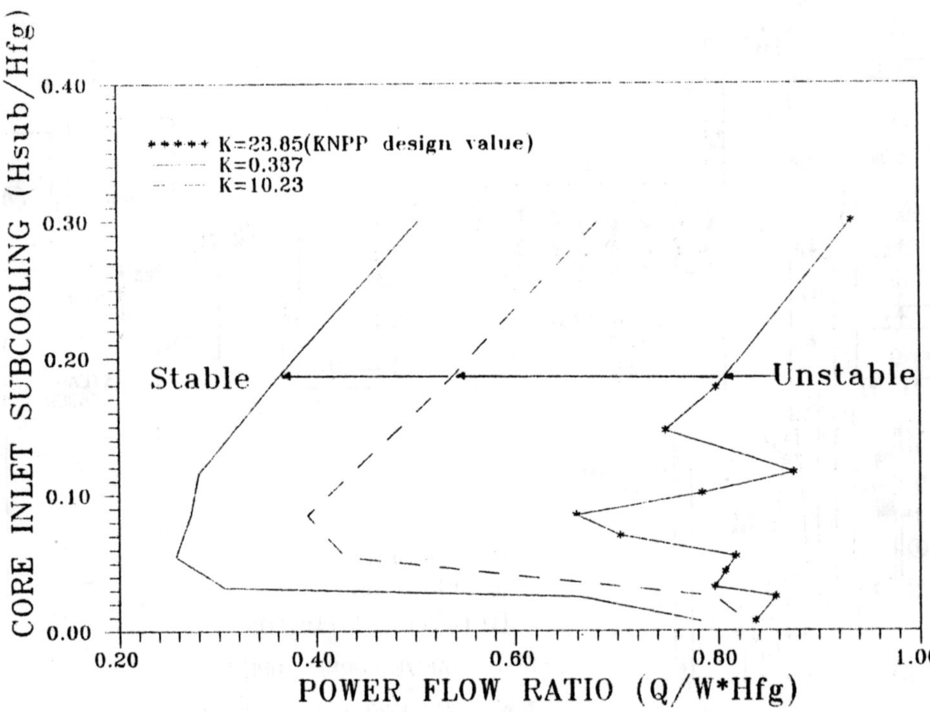

Figure 4. Effect of channel inlet orifice loss coefficient on instability boundary.

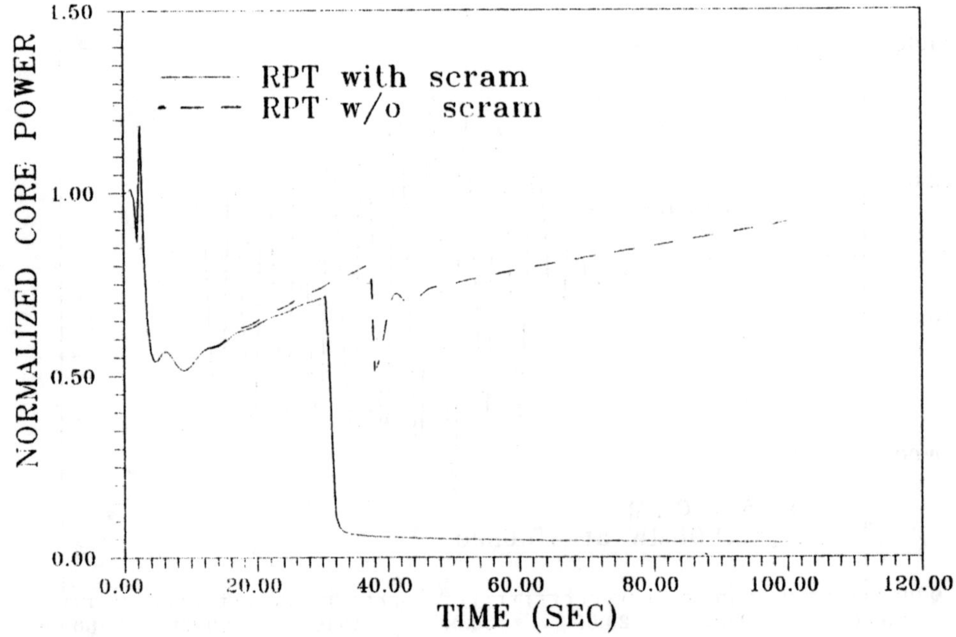

Figure 5. Core power histories for KNPP RPT with decreasing feedwater enthalpy transients analyses.

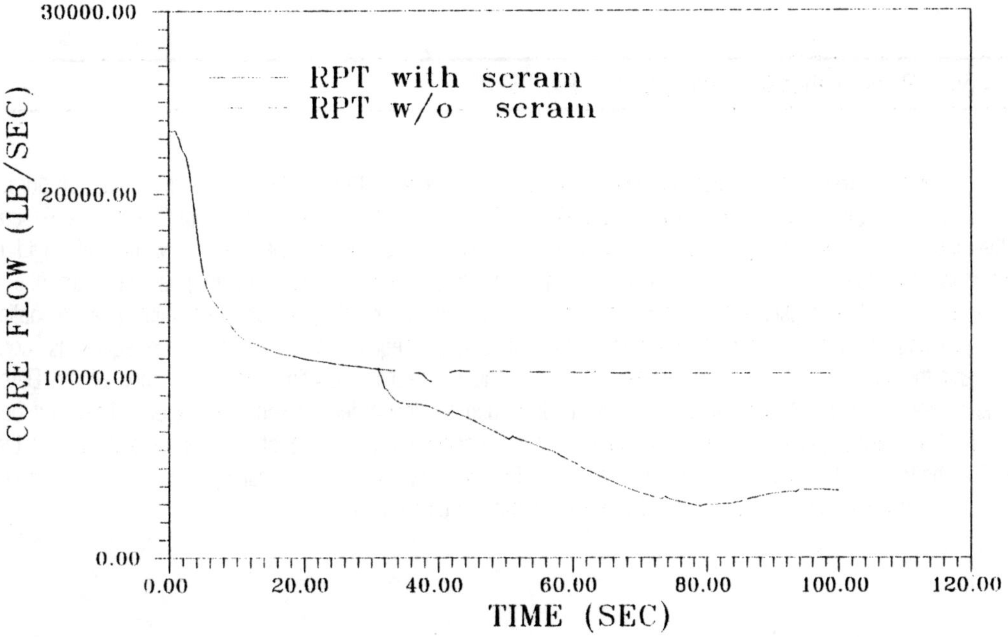

Figure 6. Core flow histories for KNPP RPT with decreasing feedwater enthalpy transients analyses.

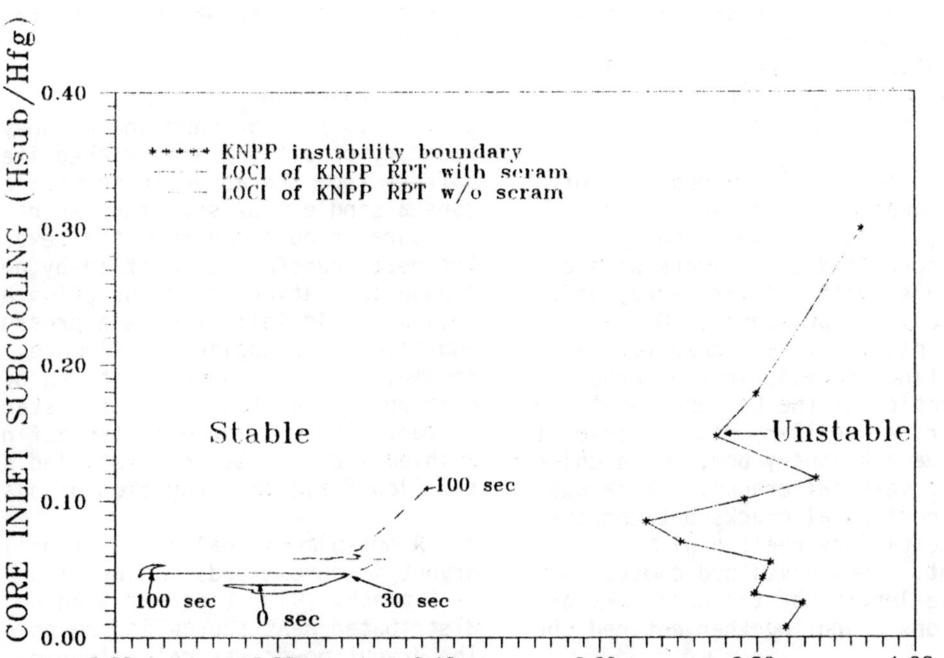

Figure 7. Loci of calculated KNPP RPT transients on stability map.

THERMAL HYDRAULICS OF STEAM GENERATOR SLUDGE

Asim Ulke ■ Bettis Atomic Power Laboratory, West Mifflin, PA 15122

A one-dimensional model of two-phase flow in porous media has been developed. The model considers a slab of porous medium on a horizontal nonporous surface, both under water. Heat is transferred from a vertical surface. The vapor, which is generated, flows vertically upward and the liquid is replenished by a counterflow. Under steady state conditions the mass flow rate of liquid equals that of vapor at every point along the flow path. The liquid saturation (ratio of liquid volume to pore volume) is maximum (the full mobile saturation limit) at the top surface and decreases with distance into the porous medium reaching a minimum value (the immobile saturation limit). At this point the liquid and the vapor velocities approach zero. The solution requires a minimum of five equations; continuity and momentum for the liquid and the vapor phases and one energy equation. The momentum equations are extensions of the Darcy equation to the inertial flow regime in porous media. All five equations can be described as first order differential equations. They can be integrated with respect to distance into the sludge pile using the appropriate boundary conditions at the top of the porous slab and at the transition point to "dryout" or at the bottom of the slab whichever comes first. The conservation of chemical species was incorporated into the above system of equations to assess concentrating factors and likelihood of material corrosion.

The sludge deposits on top of the tubesheet in commercial steam generators create regions of high susceptibility to tube corrosion. It is believed that the corrosion occurs when the sludge deposit becomes too deep for the liquid to penetrate freely. This leads to liquid deficient regions with high chemical concentrations on the tube surface.

Thermal hydraulics in sludge deposits of commercial steam generators was studied by Eckert, et al., [1] and Baum and Greaney, [2]. Eckert, et al., conducted experiments with a single tube and also with a square array of 9 tubes at relatively low pressures, 100 to 500 kPa (15-72.5 psia). Visual observations showed three distinct regions in the porous bed: A dryout region at the bottom containing superheated vapor; a capillary region above it characterized by a stationary bed; and a chimney region containing vertical cracks. There was also a layer of horizontal cracks between the chimney and the capillary regions with negligible height. They developed closed form solutions for the lengths of the capillary and the chimney regions which together defined the wetted length.

Baum and Greaney performed tests with a single tube at high pressures, 5.5 to 7.6 MPa (800-1100 psia). Porous material was packed in an annular gap around the tube. The materials which were tested included samples from two commercial plants. The wet region was assumed to consist of two regions; a mixed flow region on top and a divided flow region below it. An iterative solution was obtained for the combined regions.

The solution presented in this paper is similar to that of Baum and Greaney in that the boundary conditions are matched iteratively. It differs from the above in several aspects: It uses a single flow structure where the vapor pressure is obtained from the Leverett function; the heat transfer is modified by porosity and liquid saturation (fraction of pore volume filled by liquid); it uses a pressure gradient equation which applies to flow beyond the Darcy regime; it allows specifications of porosity and permeability as functions of distance into the sludge. These allow a better definition of the problem and results in a detailed solution of the flow field as a function of distance.

A one-dimensional model of boiling and dryout in porous beds was previously presented by Lipinski [9]. Lipinski used uniformly distributed heat source in the porous domain. The dryout power was determined by maximizing the heat flux with respect to the saturation at the top of the bed. This model differs from that of Lipinski [9] in the form of heating and in the solution tecnique to calculate the dryout location.

ANALYTICAL MODEL

The primary objective of the analytical model presented in this paper is the determination of the wetted length. The wetted region is described by a single set of equations.

Figures 1 and 2 show the geometry and the flow field considered in this work. The conservation equations can be written for a differential element of height dx in the wetted region. The continuity for the liquid yields

$$\frac{dV_{s\ell}}{dx} = -\lambda_{\ell g}/\rho_\ell \qquad (1)$$

where $\lambda_{\ell g}$ is the vaporization rate per unit volume. $V_{s\ell}$ is the superficial liquid velocity or "filter" velocity found by dividing the liquid volumetric flow rate by the cross sectional area. The continuity for the vapor can be written similarly

$$\frac{dV_{sg}}{dx} = \lambda_{\ell g}/\rho_g \qquad (2)$$

The velocities of the liquid and the vapor phases can be related to the pressure gradients by the Kozeny-Carman equation, Chu, et al., [3], Lipinski [9] and Chung and Cotton [10].

$$\frac{dP_\ell}{dx} = -\frac{\mu_\ell}{K_{r\ell}k}V_{s\ell} - \frac{\rho_\ell}{n_{r\ell}n}V_{s\ell}^2 + \rho_\ell g \qquad (3)$$

$$\frac{dp_g}{dx} = -\frac{\mu_g}{K_{rg}k}V_{sg} - \frac{\rho_g}{n_{rg}n}V_{sg}^2 + \rho_g g \qquad (4)$$

where k is the laminar permeability given by

$$k = \frac{\varepsilon^3 d_p^2}{150(1-\varepsilon)^2} \qquad (5)$$

and n is the turbulent permeability obtained from the Ergun's model, Ergun [4].

$$n = \frac{\varepsilon^3 d_p}{1.75\,(1-\varepsilon)} \qquad (6)$$

The porosity is represented by ε and the particle diameter by d_p.

Dybbs and Edwards [12] have shown that Ergun's model correlates a set of experimental data to within the experimental error.

The relative permeabilities were calculated using the forms developed by Verma, et al., [5].

$$K_{r\ell} = S_e^3 \qquad (7)$$

$$K_{rg} = 1.2984 - 1.9832\,S_e + 0.7432\,S_e^2 \qquad (8)$$

where S_e is the normalized saturation given by

$$S_e = \frac{S_\ell - S_{\ell p}}{S_m - S_{\ell p}} \qquad (9)$$

The values of the immobile saturation $S_{\ell p}$ and the full mobile saturation S_m are 0.2 and 0.895 respectively, based on the experimental results of Verma, et al., [5].

The turbulent relative permeabilities, $n_{r\ell}$ and n_{rg} were assumed to have the same values as $K_{r\ell}$ and K_{rg} respectively, as in chung and Cotton [10].

The capillary pressure P_c is given by

$$P_c = P_g - P_\ell \qquad (10)$$

The following correlation of the capillary pressure versus liquid saturation data of Leverett [6] is given by Udell [11].

$$f = 1.417(1-S) - 2.120(1-S)^2 + 1.263(1-S)^3 \qquad (11)$$

where f is the dimensionless pressure defined by

$$f = \frac{P_c}{\sigma}\left(\frac{k}{\varepsilon}\right)^{1/2} \qquad (12)$$

The dimensionless saturation S is defined similar to S_e of Verma, et al., given by Equation (9) above.

$$S = \frac{S_\ell - S_{\ell p}}{1.0 - S_{\ell p}} \qquad (13)$$

The evaporation rate per unit volume is given by

$$\lambda_{\ell g} = \frac{1}{A_c h_{fg}}\frac{d\dot{q}}{dx} \qquad (14)$$

For a triangular pitched tube array the flow area is given by

$$A_c = \cos(30) \times \text{Pitch}^2 - \frac{\pi D_o^2}{4} \quad (15)$$

The heat transfer rate per unit length can be obtained from

$$\frac{d\dot{q}}{dx} = U_o\ (T_p - T_s)\ \pi\ D_o \quad (16)$$

In calculating the overall thermal conductance, only the fraction of the tube surface in contact with liquid was considered to be significant.

$$U_o = 1/\left(\frac{D_o}{D_i\ h_i} + \frac{D_o\ \ell n\ D_o/D_i}{2\ k_t} + \frac{1}{h_o S_\ell \varepsilon}\right) \quad (17)$$

The primary side forced convection heat transfer coefficient was obtained using the Dittus-Boelter correlation.

$$h_i = \frac{k_{p\ell}}{D_i}\ 0.023\ (R_e)^{0.8} Pr^{0.4} \quad (18)$$

The secondary side boiling heat transfer coefficient was calculated by Thom's correlation, [7] which is recommended by Rohsenow [14] for low quality boiling.

$$h_o = 1970\ e^{P_s/4344}\ (T_w - T_s) \quad (19)$$

Defining a conductance coefficient for the primary side

$$U_p = 1\ /\ \left(\frac{D_o}{D_i\ h_i} + \frac{D_o\ \ell n\ D_o/D_i}{2\ K_t}\right) \quad (20)$$

the tube outside wall temperature can be found iteratively using

$$T_w = T_p - q''/U_p \quad (21)$$

where

$$q'' = U_o\ (T_p - T_s) \quad (22)$$

The concentration of chemical species can be calculated from a simple mass balance. The molecular diffusion and the chemical species mass carried by the vapor phase were neglected.

$$\frac{dC}{dx} = C\ \lambda_{\ell g}/\rho_\ell\ V_{s\ell} \quad (23)$$

where, C is the ratio of the local to the bulk chemical concentration.

The boundary conditions at the lower boundary of the wet region complete the definition of the problem, which are $S = 0$ and $V_{s\ell} = 0$.

METHOD OF SOLUTION

Equations (1), (2), (3), (4), (16) and (23) form a set of first order differential equations. They can be integrated with respect to the independent variable X, the distance from the top of the porous region. In this work a fourth order Runga-Kutta procedure was used for starting the solution. The solution was continued using the Adams-Bashford quadrature formula with a corrector, both of which are fourth order, Kunz [13].

The integration requires initial values for the six variables $V_{s\ell}$, V_{sg}, P_ℓ, P_g, $\dot{q}$ and c at x = 0. The downward superficial liquid velocity $V_{s\ell}$ is initially assumed and calculated iteratively. The upward vapor velocity V_{sg} is calculated by the continuity condition that the mass flow rates of liquid and vapor are equal at any given location. The liquid pressure P_ℓ is given by the bulk conditions. The vapor pressure P_g is calculated from Equations (10), (11), (12) and (13) with $S_\ell = 0.895$, the mobile saturation limit. The heat transfer rate $\dot{q}$ is calculated using Equations (16) through (22). The concentration of chemical species is equal to the bulk value. As the integration proceeds downward, one of three events can happen: 1-The liquid depletion occurs with $V_{s\ell} \to 0$, but normalized saturation $S > 0$. The solution is restarted with a higher initial value for $V_{s\ell}$. 2-The saturation reaches zero while $V_{s\ell} > 0$. The solution is restarted with a lower initial value for $V_{s\ell}$. 3-The solution reaches the bottom boundary with $V_{s\ell} > 0$. The solution is restarted with a lower initial value of $V_{s\ell}$. These iteration are done automatically using a Newton-Raphson type scheme. The solution is complete either when $V_{s\ell}$ and S satisfy the convergence criteria at any x or when $V_{s\ell}$ satisfies the convergence criterion at the bottom boundary. The convergence was defined as a factor of 400 reduction in $V_{s\ell}$ and $S \le 0.005$. These conditions were generally satisfied in less than five iterations.

RESULTS AND DISCUSSION

A number of test cases presented in References [1] and [2] were simulated with the present analytical procedure. The data in Reference [1] was obtained with a granular bed consisting of nickel particles. The porosity was given as 0.65 and the mean particle size as 28.3 μm. The corresponding permeability was obtained using Equation (5).

Figure 3 compares the Baum and Greaney data, [2], to predictions of the analytical model presented here for sludges from two commercial plants. The lower data points and the lower line correspond to the Plant A sludge. The fit for Plant A sludge is excellent using the porosity and permeability given in Reference [2]. The upper data points and line correspond to the Plant B sludge. Baum and Greaney reported fluidization of the Plant B sludge bed throughout the tests. The fluidization must have caused the longer measured wetted lengths, which may be the reason for lack of a fit here.

Figure 4 shows four data points for wetted length from Eckert, et al. [1]. The solid line represents predictions. It follows the trend quite well but falls about 2.0 to 3.0 cm short at all points. This is most probably due to the chimneys in the granular bed which were produced by large volumes of low pressure vapor.

The effect of vertical cracks in the granular material bed or of fluidization of the bed material is to make it easier for the liquid to penetrate and the vapor to escape. A similar effect can be produced by increasing the permeability of the porous bed in computer simulations. This was done for the Eckert data [1] and for the Plant B sludge of Baum and Greaney [2].

The effective permeability of the Plant B sludge was increased from the reported value of 2.47×10^{-13} to 9.3×10^{-12} m^2 throughout. The result of this exercise is shown in Figure 5. The Eckert tests were modeled by a highly permeable top section corresponding to the observed chimney region and a less permeable capillary region below it. The top section was assumed to extend 2.5 cm into the sludge and had a permeability of 4.6×10^{-11} m^2. The bottom section had a permeability of 1.2×10^{-11} m^2 which was calculated from the reported porosity and particle size in Reference [1]. The transition was assumed to be linear from 2.5 to 3.5 cm into the sludge. The net result of this was to raise the prediction line by about 3.0 cm, as shown in Figure 6. The top section showed little resistance to liquid penetration. The physical significance of this becomes clearer when we look at the liquid saturation profiles.

Figure 7 shows the saturation profile for a bed of uniform porosity, 0.65, and permeability, 1.2×10^{-11} m^2. The flatness of the profile in the last several centimeters may make accurate measurement of the wet length difficult especially at low heat flux. Figure 8 shows the change when a region of high permeability is assumed on top. The liquid saturation increases significantly at the start of the less permeable region. This would indicate a pool of liquid between the two layers. It is possible that the horizontal cracks observed by Eckert, et al., [1] in the granular bed were caused by liquid which collected on top of a region of low permeability. The saturation profile in Figure 9 was calculated using a higher heat flux. The discontinuity in the liquid saturation profile is less pronounced. At still higher heat flux values the discontinuity practically disappears. A typical plot of a normalized chemical concentration is shown in Figure 10.

It must be pointed out that the model presented here assumes among other things counter flow of liquid and vapor, continuous throughout the domain of solution. This condition may apply to crevices between tubes and tube support plates and to tubesheet sludge in steam generators where the tube pitch to diameter ratio's are not much greater than 1.0. As the distance from the heated surface increases the counterflow will be replaced by parallel flow where the vapor flows up near the heated surface and the liquid flows down away from it and the capillary pressure and the relative permeability functions are no longer valid.

CONCLUSIONS

An analytical procedure and a corresponding computer program have been developed to predict the thermal hydraulic behavior and chemical concentration in porous media, with emphasis on computing the liquid penetration into vertical steam generator sludge. The simulation of a granular bed with a highly permeable region on top of a less permeable one at the bottom predicts a highly liquid saturated region between the two layers and extends the wetted length deeper into the sludge. The chemical concentration profile does not show a similar discontinuity but the peak concentration occurs deeper in the sludge pile.

NOTATION

A_c	Flow area, m^2
C	Concentration ratio
D_o	Tube outside diameter, m
D_i	Tube inside diameter, m
d_p	Particle diameter, m
f	Normalized capillary pressure
g	Gravitational acceleration, m/s^2
h	Height, m
h_o	Secondary side film coefficient, W/m^2-K
h_i	Primary side film coefficient, W/m^2-K
k	Laminar permeability, m^2
K_{rg}	Vapor relative permeability in laminar flow, dimensionless
$K_{r\ell}$	Liquid relative permeability in laminar flow, dimensionless
k_t	Tube thermal conductivity, W/m-K
$k_{p\ell}$	Thermal conductivity of primary liquid, W/m-K
n	Turbulent permeability, m^2
n_{rg}	Vapor relative permeability in turbulent flow, dimensionless
$n_{r\ell}$	Liquid relative permeability in turbulent flow, dimensionless
P	Pressure, kPa unless stated otherwise
P_c	Capillary pressure, kPa
q"	Heat flux, W/m^2
P_r	Prandtl number
q	Heat transfer rate, W
R_e	Reynolds number
S	Normalized liquid saturation, dimensionless
S_e	Normalized liquid saturation, dimensionless
S_ℓ	Liquid saturation, dimensionless
$S_{\ell p}$	Immobile liquid saturation, dimensionless
S_m	Upper limit of mobile liquid saturation, dimensionless
T_p	Primary liquid temperature, K
T_s	Secondary temperature, K
T_w	Tube outside wall temperature, K
U_o	Overall thermal conductance, W/m^2-K
U_p	Thermal conductance, W/m^2-K
$V_{s\ell}$	Superficial liquid velocity, m/s
V_{sg}	Superficial vapor velocity, m/s
X	Distance into the sludge, m
ε	Porosity, dimensionless
$\lambda_{\ell g}$	Volumetric vaporization rate, m^3/s
ρ	Density, kg/m^3
σ	Surface tension, kg/m
θ	Angle
μ	Viscosity, Pa.s

Subscripts

g	Vapor
ℓ	Liquid

LITERATURE CITED

1. Eckert, E.R. G., et al., "A Study of the Boiling Processes in the Sludge Deposit of Steam Generators", EPRI NP-3018, (1983).

2. Baum, A. J. and P. K. Greaney, "An Experimental and Analytical Investigation of Boiling Heat Transfer in Porous Bodies", ASME/AIChE N.H.T.C., Milwaukee, (1981) ASME Paper No. 81-HT-44.

3. Chu, W., et al., AIChE Symposium Series 79, 225, 224 (1983).

4. Ergun, S., Chem. Eng. Prog., 1952, 48, 2, 89 (1952).

5. Verma, A. K., et al., "A Study of Two Phase Concurrent Flow of Steam and Water in an Unconsolidated Porous Medium", LBL-19084, DE-85014534, (1985).

6. Leverett, M. C., AIME Trans. 142, 152 (1941).

7. Thom, J.R.S., et al., Proc. Inst. Mech. Engrs., 180, 3C, 226 (1965).

8. Cleary, J. G., et al., "Diffusion and Hideout in Crevices", EPRI NP-2979, (1983).

9. Lipinski, R. J., Nuclear Technology, 65, 53 (1984).

10. Chung, M. O. and Ivan Cotton, AIChE Symposium Series, 84, 263, 177 (1988).

11. Udell, K. S., Int. J. Heat Mass Transfer, 28, 2, 485 (1985).

12. Dybbs, A. and R. W. Edwards, NATO ASI Series, Series E: Applied Sciences, 82, 199 (1984).

13. Kunz, K. S., Numerical Analysis, McGraw-Hill, New York (1957).

14. Rohsenow, W. M. and J. P. Hartnett, Handbook of Heat Transfer, McGraw-Hill, New York (1973).

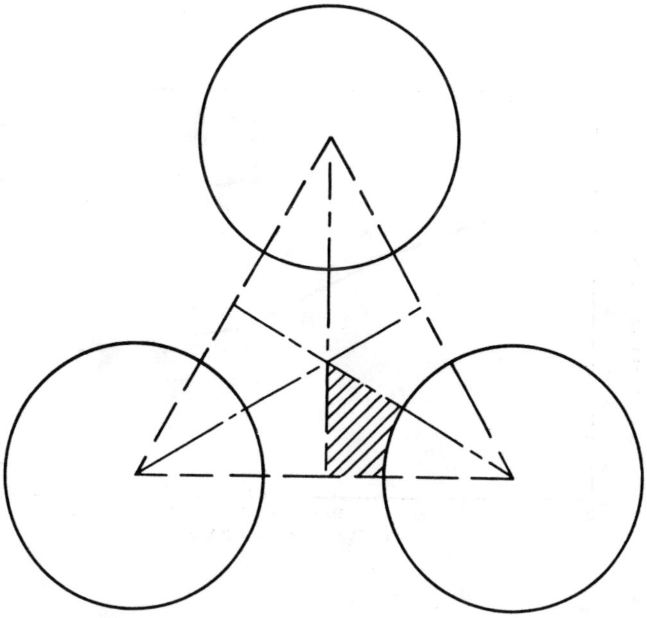

Figure 1. Tube/sludge geometry.

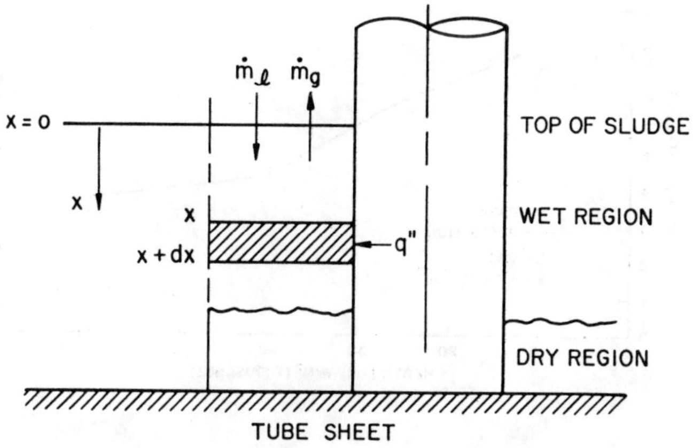

Figure 2. Schematic of flow field.

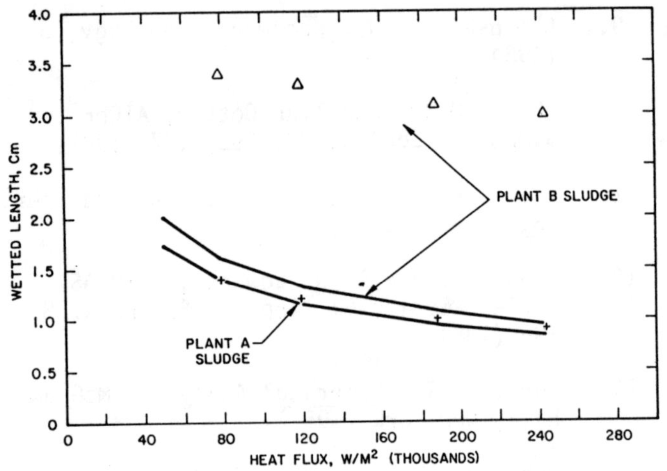

Figure 3. Comparison of model predictions to Baum and Greaney data (reference 2).

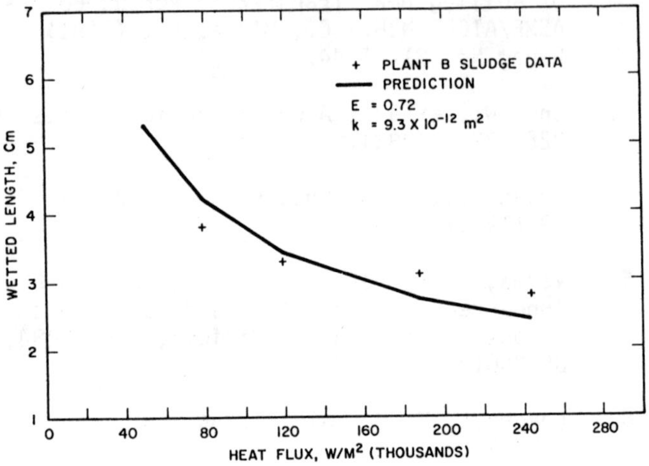

Figure 5. Comparison of model predictions with increased permeability to Baum and Greaney data.

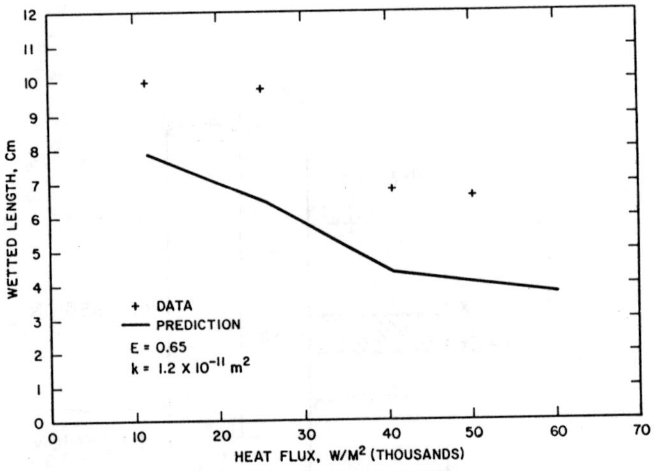

Figure 4. Comparison of model predictions to Eckert data (reference 1).

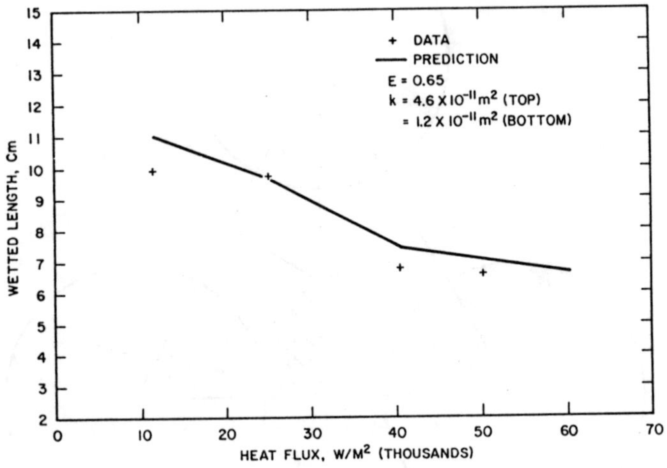

Figure 6. Comparison of two region simulation results to Eckert data.

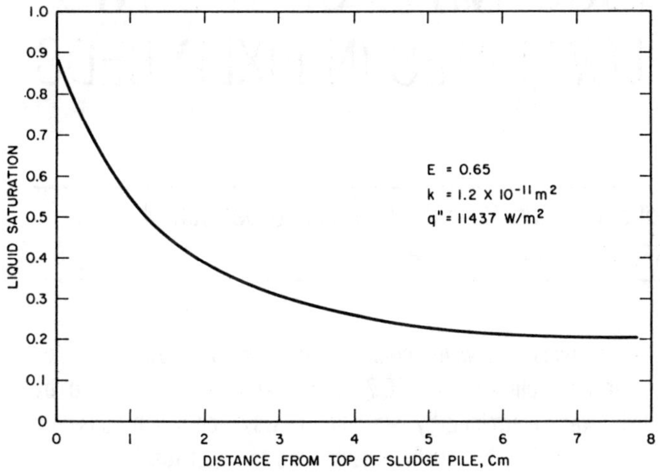

Figure 7. Liquid saturation profile in a porous bed of uniform properties.

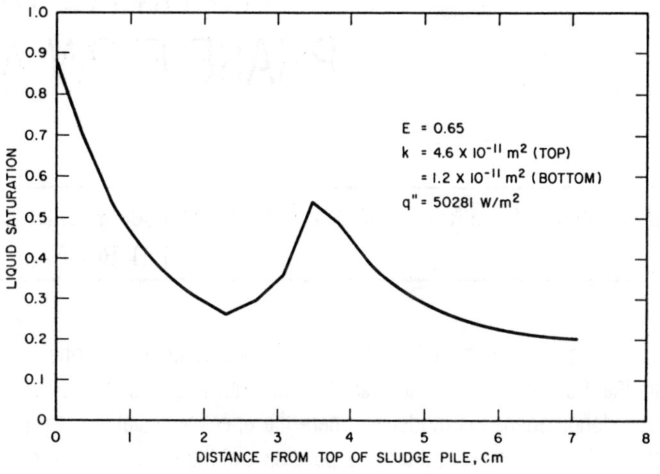

Figure 9. Liquid saturation profile in a two region bed.

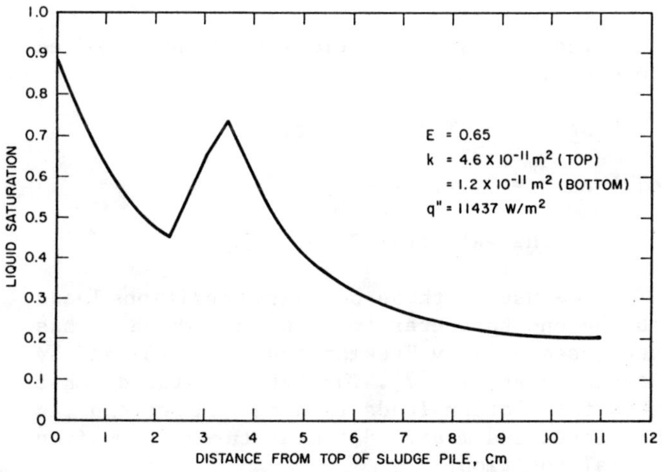

Figure 8. Liquid saturation profile in a porous bed of high permeability region on top.

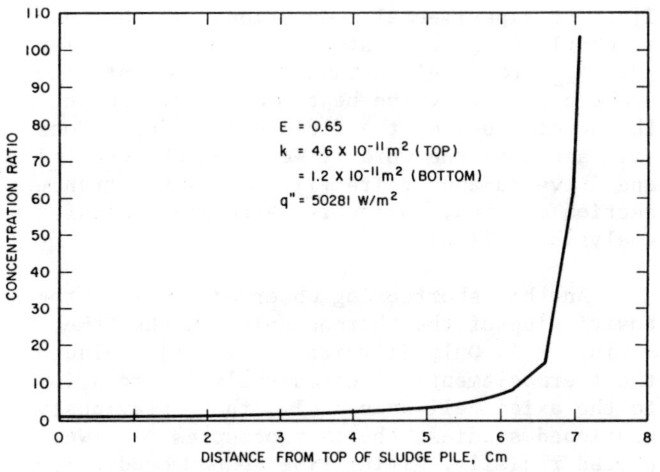

Figure 10. Chemical concentration ratio.

DETERMINATION OF THE EFFECTIVE RADIAL THERMAL CONDUCTIVITY FOR THE TWO-PHASE FLOW AT LOW RATES IN FIXED BEDS

J.T. Freire, A.M. Silveira, and J.C. Thoméo ■ Federal University of Sao Carlos, Chemical Engineering Department, P.O. Box 676, 13560-São Carlos-SP - Brazil

A study of the heat transfer in a fixed bed of glass spheres with a two-phase air-water cocurrent up-flow with low fluid rates in the bubble-flow regime was carried out. The bed to particle diameter ratio utilized was 26.7. The heat data were treated by pseudohomogeneous modeling where the effective radial thermal conductivity was determined by two different ways, on which considers this to be constant with the radial position and the other which assumes its variation. The first method having led to better results.

J.C. Thomeo is a student supported by FAPESP, research grant number 90/0059-1.

Despite its large application, Zhukova et al., [1], few reports about gas-liquid heat transfer in fixed beds have been published so far.

These reports normally present very distinct experimental conceptions, which lead to conflicting heat data. In some studies [2,3,4,5] the equipment consisted of an entrance section preceding the heat section to enhance the development of the fluid flow. This does not happen in the work of Weekman and Myers [6] and Silveira and Freire [7], where no entrance section existed, making isolated heat transfer analysis difficult.

Another shortcoming observed was the positioning of the thermocouples in the measuring cell. Only Silveira et al. [5] placed the thermoelements longitudinally in relation to the axial cell axis. In the previously mentioned studies, the thermocouples were placed radially, giving rise to heat conduction problems through these [8].

Regarding the calculus of the heat parameters concerning this system, in all the reports for two-phase flow the pseudohomogeneous model without the axial dispersion term was used. This may be given by the following energy equation:

$$(G_g Cp_g + G_l Cp_l) \frac{\partial T}{\partial z} = K_r \left(\frac{1}{r}\frac{\partial T}{\partial r} + \frac{\partial^2 T}{\partial r^2}\right) \qquad (1)$$

where G_l and G_g are the superficial mass rates of the liquid and the gas, respectively, assumed to have flat radial profiles; Cp_l and Cp_g are the specific heats of the liquid and the gas and K_r is the effective radial thermal conductivity, which is considered to be constant with this coordinate.

The boundary conditions frequently applied to equation (1) are:

c.1) at $z = 0$, $T = T_o$ (flat);

c.2) at $r = 0$, $\partial T/\partial r = 0$;

c.3) at the wall ($r = R$), $T = T_w$.

The use of these boundary conditions leads to the one parameter formulation, which has been used only by Weekman and Myers [6] and by Hashimoto et al. [2]. The latter stated that this formulation leads to a bad fit between the predicted and measured temperatures for a given radial position.

This deficiency was attributed to the existance of an additional heat resistance in the proximities of the bed wall, represented in the literature by the heat transfer coefficient between the wall and the bed (h_w), which was determined from Newton's "cooling law" as follows [9]:

c.4) at $r = R$, $-K_r\, \partial T/\partial r = h_w\,(T_w - T_R)$

In a recent work about one-phase flow, Tsotsas and Schlunder [10] stated that h_w only has physical meaning at high fluid rates and that the one parameter formulation represents much better situations of low flow rates. Howe ver, these authors note that this formulation should be modified in order to include the structural nonuniformities of the bed and the resulting radial variation in the flow rate which reaches a maximum close to the pipe wall [11], giving rise to a radial dependency of K_r.

The present work was carried out to contribute to the study of gas-liquid concurrent ascending flow at low fluid rates, a system frequently used in biochemical reactors [12]. Therefore, a packed bed with glass spheres percolated by a air-water mixture was used. The heat coefficients were calculated by the one parameter formulation and by means of the proposal of Kwong and Smith [13] which allows the radial variation of K_r.

EXPERIMENTAL METHOD AND CALCULATIONS

a) Experimental

The measuring cell, as shown in Figure 1, was made of glass to allow interior visualization so as to identify the flow regimen and the possible homogenization of the air-water mixture.

The inlet section has a length of 0.1 m and a diameter of 0.048 m. At the front of the inlet section a perforated sintered glass plate was installed to allow better flow distribution.

The heat section consists of an internal pipe with a 0.048 m diameter and a refrigerating jacket with a 0.060 m diameter, through which cooling water at a temperature of 17 °C flows.

In order to determine the temperature at the entrance of the heat section (z = 0 plane) five L-shaped thermocouples, which did not traverse the refrigerating jacket, were installed.

The temperature measurements at the desired heights of the bed were carried out by placing the thermocouples longitudinally. These elements were soldered to copper rings placed in five different radial positions. This assembly was fixed at a bed height of 0.014 m, the rings being covered by porous medium in order to avoid interference resulting from the exit of the bed. The efficiency of this assembly has been confirmed theoretically by Giudici [14].

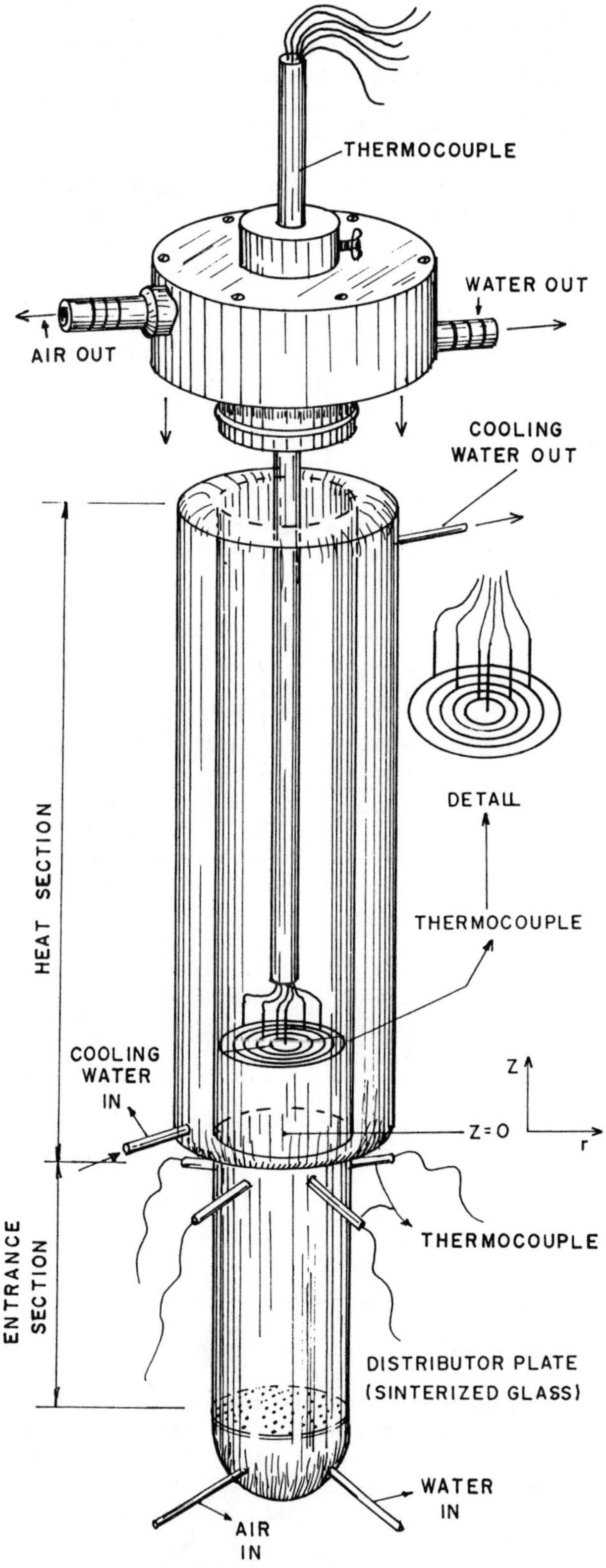

Figure 1. Measuring cell.

The particles used were glass spheres with diameters of 1.8×10^{-3} m, establishing in this way a bed to particle diameter ratio, D/d_p, of 26,7, sufficient to minimize the wall effects on the flow of the fluids [15].

Air and water flow rates varied from 0 to 0.156 $kg/m^2.s$ and from 0.287 to 0.677 $kg/m^2.s$, respectively. The modified Reynold's numbers used (defined as $N_{Re_g} = G_g.d_p/\mu_g$ for air and $N_{Re_l} = G_l.d_p/\mu_l$ for water, where μ_g and μ_l are the molecular viscosities of air and water, respectively) are listed in Table 1. The inlet temperatures of the fluids were fixed at 50.0 ^oC. A bubble flow regimen was visually observed at these fluid flow rates.

TABLE 1. K_r values (formulation A) as a function of N_{Re_l} and N_{Re_g} and the standard deviation of the coefficient

N_{Re_l}	N_{Re_g}	K_r (W/m.°C)	standard deviation (%)
	0	0.94	3.6
	5.4	1.77	4.5
	7.7	1.97	2.7
0.9	9.8	2.03	2.2
	11.6	1.69	4.3
	13.2	1.30	3.5
	14.3	1.24	3.7
	0	0.96	3.0
	5.4	1.96	2.5
	7.7	2.06	2.7
1.6	9.8	2.15	2.8
	11.6	2.09	2.8
	13.2	1.98	3.4
	14.3	1.81	2.8
	0	1.05	2.5
	5.4	2.10	2.7
	7.7	2.19	2.9
2.2	9.8	2.31	3.1
	11.6	2.23	3.3
	13.2	2.13	3.4
	14.3	2.01	3.1

b) Calculations

In order to calculate the effective thermal conductivity the previously mentioned one parameter formulation was used (formulation A), of which the analytic solution is:

$$\frac{T - T_w}{T_o - T_w} = 2 \sum_{n=1}^{\infty} \frac{J_o(a_n r/R)}{a_n J_1(a_n)} \exp\left[-\frac{K_r a_n^2 z}{G_l Cp_l + G_g Cp_g}\right] \quad (2)$$

where J_o and J_1 are zero and first order Bessel functions of the first kind, respectively and a_n is an eigenvalue defined by:

$$J_o(a_n) = 0.$$

The determination of K_r was done by using the minimum squares method, expressed as follows:

$$F = \sum_{1}^{n} (T_{exp} - T_{calc})^2 \quad (3)$$

where the F function was minimized using Marquardt's algorithm [16]; N is the number of radial positions where the experimental temperatures, T_{exp}, had been determined, and T_{calc} is the temperature predicted by the model. T_{calc} was determined using the first ten terms of equation 2.

Another method of calculating K_r (formulation B) was proposed by Kwong and Smith [13] which implies the radial variation of this coefficient. Hence, at the right-hand side of equation (1) the term $(\partial K_r(r)/\partial r).(\partial T/\partial r)$ should be added. In this way, $K_r(r)$ was calculated by the following equation:

$$K_r(r) = \frac{-b_1 \int_o^r T_L(r)\, G(r)\, C_p\, rdr}{r\, (\partial T/\partial r)\big|_{z=L}} \quad (4)$$

where $T_L(r)$ represents the radial temperature distribution at the exit of the bed, L, which was adjusted by a fifth order polynomial and b_1 is an eigenvalue defined as:

$$b_1 = \frac{1}{L} \frac{\int_o^R T_o(r) T_L(r) G(r) C_p\, rdr}{\int_o^R T_L^2(r) G(r) C_p\, rdr} \quad (5)$$

The equation used to determine G(r) was proposed by Fahien and Stankovic [17] for a one-phase flow. In order to suit this work a perfect mixture between the fluids was assumed.

In both formulations T_o was obtained from the relation:

$$T_o = \frac{\int_o^R T_o(r)\, rdr}{\int_o^R rdr} \quad (6)$$

where the integration was carried out numerically in the z = 0 plane

RESULTS AND DISCUSSION

The results obtained experimentally revealed the presence of a radial temperature profile at the entrance of the heat section (Figure 2). The T_o value normally used in the literature is based on the temperature at which

A good fit was achieved between the measured temperatures and those predicted by formulation A over the whole experimental range of this work, differences not being superior than 1.0 °C. This fit is contrary to the expectations observed in the literature, where Nakamura et al. [4] obtained very bad results using this formulation.

Since in this work the fluid rates were low, it is believed that a good mixture took place (evidenced visually) and an almost flat profile of this mixture exist. Therefore, the possibility for a preferential flow of the mixture to occur in the proximities of the pipe wall becomes smaller. This opposes the formation of an additional heat resistance in this region, which is attributed to deficiences of the A formulation.

The behaviour of K_r calculated from formulation A for the one-phase flow of water, Figure 5, showed to be increasing linearly with as increase of N_{Re1}. This result agrees with that of Yagi et al. [18], notwithstanding the fact these authors have used much superior liquid flow rates and a very simplified method of calculation.

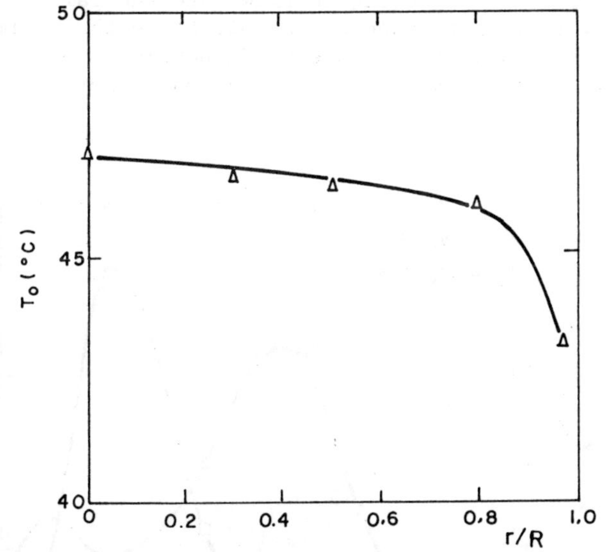

Figure 2. Temperature at the inlet (z = 0) as a function of radial position (N_{Re1} = 1.6).

As for the determination of $K_r(r)$ by the method of Kwong and Smith [13], the results were not good, as can be seen in Table 2 and Figure 6. The fit between the measured and predicted temperatures was very unsatisfactory over all the working range, differences being observed of up to 10.0 °C. These values contradict the expectations created by the work of Silveira [19] who obtained excellent K_r (r) values, Figure 7, for a system with a basically similar operation as the one used here and that was working with moderate fluid rates (N_{Re1} varying between 55 and 140 and N_{Reg} between 45 and 188).

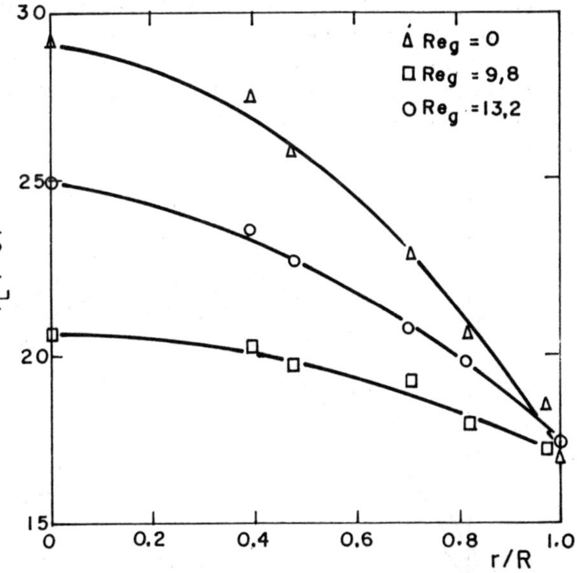

Figure 3. Temperature at the outlet of the bed (z = L) as a function of radial position for N_{Re1} = 0.9 and various N_{Reg}.

TABLE 2. $K_r(r)$ (formulation B) as a function of r/R for Re_1 = 0.9

r/R	K_r (W/m.°C)					
N_{Reg}	0,1	0,3	0,5	0,7	0,9	1,0
0	0.99	0.99	1.13	1.80	1.61	0.96
5.4	3.09	3.09	3.09	8.32	3.90	1.36
7.7	2.97	5.10	7.15	4.68	4.43	6.86
9.8	2.50	4.81	8.28	4.31	4.77	3.18
11.6	2.60	2.60	2.93	8.95	3.62	1.28
13.2	1.66	1.66	1.66	3.70	2.60	0.95
14.3	1.64	1.64	1.66	3.39	2.39	0.94

The major cause of error related to formulation B was believed to be the correlation used for the radial velocity distribuiton of the mixture. The equation proposed by Fahien and Stankovic [17] was obtained from one-phase air flow data at velocities superior to 0.128 m/s, corresponding to N_{Reg} bigger than 27.4. Therefore, the correct estimation of this formulation will depend on finding an equation in the literature that is applicable to two-phase flow at low flow rates.

the fluids enter the measuring cell [3,4]. In this way, the parameters used reflect this unreality compromising the estimated values especially near the inlet of the bed. On assuming the integrated mean value over the area for T_0 this deficiency is being reduced, since this mean value represents a heat boundary of the system with greater reliability. However, for the problem to be conveniently formulated it is necessary to include a function that represents this profile during modeling procedures, which will be carried out in future works.

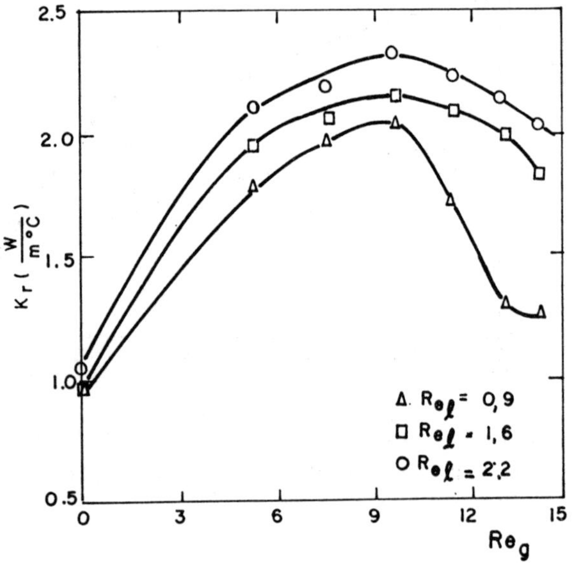

Figure 4. K_r (formulation A) as a function of N_{Re_g} for various N_{Re_1}.

The temperature at the exit of the bed showed a strong dependence on the gas saturation (Figure 3). With the introduction of air in the system there will be an increase of the lateral dispersion of the water and a better thermal homogenization, since water is a fluid with the capacity to transport a greater portion of energy than air. However, an increase of the gas flow resulted in an increase of the nonuniformity, as the air starts to occupy a significant part of volume, diminishing the capacity of heat exchange.

As a consequence of this fact, K_r calculated by formulation A (Table 1 and Figure 4) starts with low values for the one-phase flow of water, reaches a maximum with the introduction of air and falls again with an increase of the gas flow rate. This result agrees with that of Nakamura et al. [3], although these authors have worked with much higher flow rates.

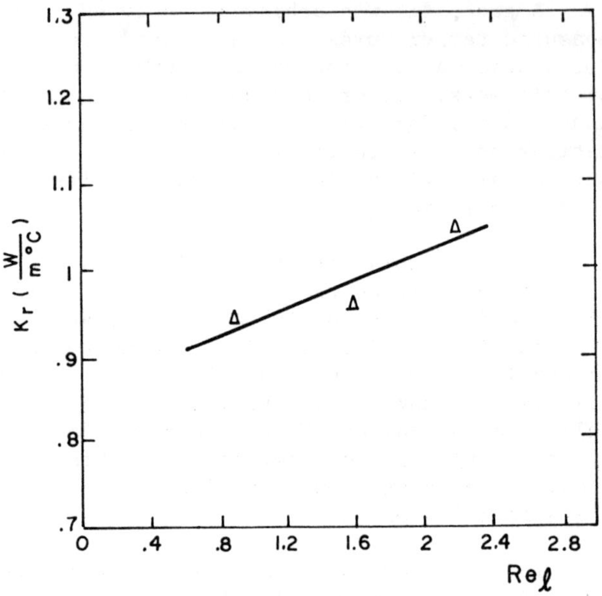

Figure 5. K_r (formulation A) as a function of N_{Re_1}.

In Figure 4 one observes that with the increase of the liquid flow rate the fall of K_r is less pronunced, consolidating the impression that the lateral dispersion of the water is responsible in great part for heat transport.

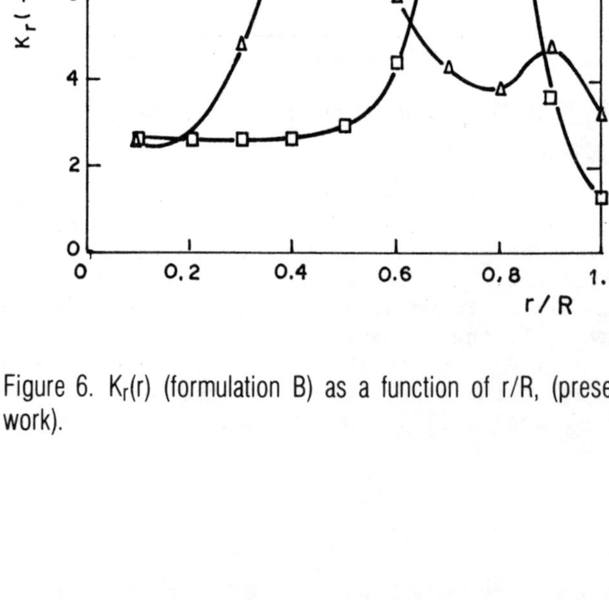

Figure 6. $K_r(r)$ (formulation B) as a function of r/R, (present work).

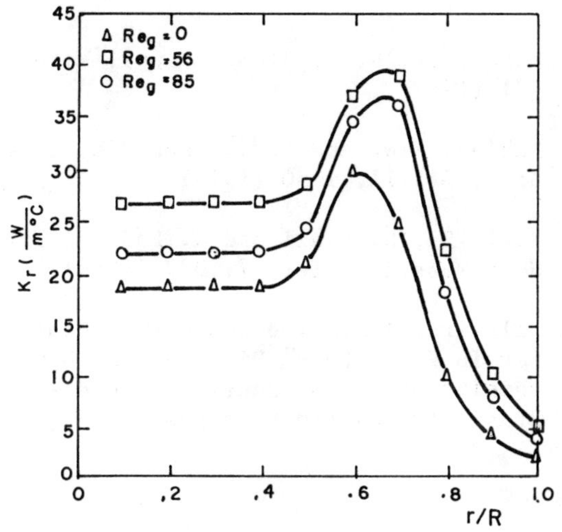

Figure 7. $K_r(r)$ (formulation B) as a function of r/R, air-water-glass spheres system, D/d_p = 19.0, N_{Re_1} = 55 [19].

In this manner, in spite of its following of the most recent tendencies of the literature, including information about structural and dynamic nonuniformities, results from formulation B were significantly inferior to those from formulation A, which imposes very restrictive hypotheses to the problem.

Thus, one should be cautions when using pseudohomogeneous modeling, since it is evident that at this moment the set of information necessary for a formulation to be applicable at any range of operation is not available.

CONCLUSIONS

In this work an apparatus capable of investigating the heat limits of a fixed bed percolated by an air-water mixture at low flow rates of the fluids in ascending flow in the bubble-flow regime was presented.

The results indicated the presence of a radial temperature profile at the inlet of the bed. The radial temperature distribution at the outlet of the bed also demonstrated to be strongly dependent on the lateral dispersion of the liquid, which tends to thermally homogenize the system. The lateral water dispersion was raised by introducing air to the system, improving energy transport up to the point where the air starts to occupy an expressive volume portion, prejudicing heat transfer.

The one parameter formulation used to determine the effective radial thermal conductivity, K_r, which considers velocity of the mixture and K_r to be constant in the radial direction, led to good fits between the measured and predicted temperatures.

The formulation proposed by Kwong and Smith [13] which admits the occurrence of a radial velocity and K_r distribution, did not give good fits, for no expression was encountered in the literature for the velocity of the mixture that represents the characteristics of the system.

More works should be brought into being in order to evaluate the potential of Kwong and Smith's formulation and others which follow the same tendency to include the nonuniformities of the system during modeling procedure. Therefore, information about structural, dynamic and thermal behaviour will be necessary over a wide operation range, in order to arrive at a general formulation.

NOTATION

a_n : eigenvalue defined by $J_o(a_n) = 0$

b_1 : eigenvalue defined in the equation 5

Cp : specific heat of the fluid

D : bed diameter

d_p : particle diameter

F : function defined in the equation 3

G : superficial mass flow rate of the fluid

h_w : wall do bed heat transfer coefficient

J_o, J_1: zero and first order Bessel function, respectively

K_r : effective radial thermal conductivity

L : bed height

r : radial coordinate

R : bed radius

T : temperature

T_L : outlet bed temperature

T_R : temperature at the position r = R

T_w : wall temperature

T_o : inlet bed temperature

z : axial coordinate

Greek Symbol

μ : fluid molecular viscosity

Subindex

g : gas

l : liquid

LITERATURE CITED

1. Zhukova, T.B., V.N. Pisarenko and V.V. Kafarov, Int. Chem. Eng., **30**, 1, 57 (1990).

2. Hashimoto et al., Int. Chem. Eng., **16**, 4, 720 (1976).

3. Matsuura et al., Heat Transfer Jpn. Res., **8**, 1, 44 (1979).

4. Nakamura, M. et al., Heat Transfer Jpn. Res., **10**, 92 (1981).

5. Silveira, A.M., G. Massarani and J.T. Freire, 3rd ENCIT, Itapema, SC, Brasil (1990) (in Portuguese).

6. Weekman, V.W. and J.E. Myers, AIChE J., **11**, 1, 13 (1965).

7. Silveira, A.M. and J.T. Freire, 12th ENEMP, Maringá, PR, Brasil (1985) (in Portuguese).

8. Dixon, A.G., Chem. Eng. J., **31**, 163, Lausanne (1985).

9. Coberly, C.A. and W.R. Marshall Jr., Chem. Eng. Prog., **47**, 141 (1951).

10. Tsotsas, E. and E.U. Schlunder, Chem. Eng. Sci., **45**, 4, 819 (1990).

11. Cohen, Y. and A.B. Metzner, AIChE J., **27**, 5, 705 (1981).

12. Nagashima, M. et al., Biotechnol. Bioeng., **26**, 992 (1984).

13. Kwong, S.S. and J.M. Smith, Ind. Eng. Chem. **49**, 5, 894 (1957).

14. Giudici, R., "Modeling of Ethilene Oxide Reactor: Heat Parameters Analysis and Catalyst Dilution Strategies", DSc. Thesis, Univ of São Paulo, Brasil (1990) (in portuguese).

15. Schwartz, C.E., and J.M. Smith, Ind. Eng. Chem., **45**, 1209 (1953).

16. Marquardt, D.W., SIAM J. Appl. Math., **11**, 431 (1963).

17. Fahien, R.W. and I.M. Stankovic, Chem.Eng. Sci., **34**, 11, 1350 (1979).

18. Yagi, S., D. Kunii and K. Endo, Int. J. Heat Mass Transfer, **7**, 333 (1964).

19. Silveira, A.M., "Two-phase Flow Heat Transfer in Fixed Beds", DSc. Thesis, Federal University of Rio de Janeiro, Rio de Janeiro, Brasil (1991) (in Portuguese).

ANALYSIS OF CONDENSATION ENHANCEMENT BY A POROUS/FLUID COMPOSITE SYSTEM

K.J. Renken and M. Aboye ■ University of Wisconsin-Milwaukee, Department of Mechanical Engineering, P.O. Box 784, Milwaukee, WI 53201

A preliminary numerical study on condensation enhancement by a porous/fluid composite system has been performed. The numerical simulations focus on laminar film condensation from an inclined porous coated plate which is contained in a pure water vapor environment. A mathematical model is formulated and the numerical results are compared with the classical study of Nusselt and uncoated experimental data. The dependence of the heat flux on the surface subcooling, the inclination, and the thickness of the porous/fluid composite system is documented. Preliminary experimental results of the average heat transfer coefficient (represented by the Nusselt number) for the limiting case of a vertical porous substrate system are also presented.

INTRODUCTION

Recent interest in the enhancement of film condensation has been generated by industrial demands for more efficient and more compact-size thermal equipment. Examples of thermal engineering applications in which condensation occurs and which stand to benefit from the enhancement of heat transfer are exemplified by condensers, dehumidifiers, heat pipes, heat exchangers, cooling electronic devices, nuclear reactors, and nuclear waste disposal systems. A general review of published works on condensation is summarized by Refs. (1-3).

The majority of studies using a porous coated surface to augment thermal communication have been devoted to film, nucleate or pool boiling. Relevant theoretical and experimental studies include those of Avedisian and Koplik (4), Avedisian et al. (5), Bergles and Chyu (6), Cheng and Verma (7), Ito et al. (8), Kajikawa et al. (9), Marto and Lepere (10), Nakayama et al. (11-13), Kovalev et al. (14), Styrikovich et al. (15), and Webb (16,17).

Two common conclusions drawn by these authors are that the employment of a porous surface structure (i) produces high heat transfer coefficients with small temperature differences when compared to the classical uncoated surface model and (ii) makes it a very attractive and viable heat transfer augmentation alternative.

Studies related to this work are very few in number. A remarkable lack of published results exist for the problem of enhanced condensation in a porous/fluid composite system. A few of the more interesting investigations related to the present include the work of Woodruff and Westwater (18) who discovered that gold surfaces promoted dropwise condensation (DWC) of steam at atmospheric pressure; Rifert et al. (19) who reported that wire-finned tubes enhanced the heat transfer coefficients in condensation by approximately 60 to 100 percent as compared with the case of smooth tubes; and others who have used organic coatings (20) or scratched rough surfaces (21) to promote dropwise and filmwise condensation. More relevant to the present study is the work of Renken et al. (22) who presented results showing that a conductive porous coating may yield a considerable heat transfer enhancement during condensation.

The present investigation summarizes the subsequent results of a numerical study of condensation enhancement by a porous/fluid composite system. The composite system is composed of a relatively thin, highly conductive and permeable porous coating which is attached to an inclined impermeable surface. The analysis is based on the usual approximations for thin film condensation and the Brinkman-Darcy model which is used to describe the flow and heat transfer processes in the porous region. Matching conditions are utilized at the interface of the porous/fluid layer in the solution of the condensation velocity and temperature distributions. A comparison of the results of the present study to Nusselt's correlation and to uncoated experimental data demonstrates the possibilities that exists for condensation enhancement with the employment of a porous/fluid composite system..

ANALYSIS

Mathematical Formulation

Figure 1 shows a schematic of the problem of interest, namely film condensation in a porous/fluid composite system. The coordinate system is also defined in Figure 1, where x and y are Cartesian coordinates in the horizontal and vertical directions with the positive y axis pointing toward the porous medium and the positive x axis pointing along the length of the plate. The origin of the coordinate system is at the top end of the plate. In more detail, we have a condensing surface of length L with a constant temperature of T_w which is inclined with the vertical axis by an angle of ϕ. The impermeable surface is coated with a very thermally conductive and permeable porous substrate of thickness H and permeability K. The mass flow rate of the condensing liquid which flows through the porous/fluid composite by gravitational forces is expressed as w and the film condensation thickness is represented by $\delta(x)$. The surrounding environment contains a pure vapor at saturation temperature T_s.

It is assumed that the flow is steady, laminar, incompressible, and two dimensional. In addition, we assume that: (i) a distinct boundary exists separating the liquid zone from the vapor zone, thus eliminating the difficulties associated with the use of relative permeability, (ii) the boundary layer is so thin that all the usual approximations common in thin film analysis can be adopted, (iii) the thermophysical properties of the condensate and the porous coating are assumed to be constant, (iv) the condensate-saturated porous substrate is considered homogeneous and isotropic and in local thermodynamic equilibrium with the fluid, and (v) the porous coating is thinner than the condensation film and, therefore covered by it.

In the liquid region, traditional laminar film condensation conservation equations of mass, momentum, and energy are used. They are not repeated here for brevity. The conservation equations for the porous region are based on the Brinkman-Darcy model which accounts for friction caused by macroscopic shear and satisfies the no-slip condition on a solid surface. The governing equations for the porous layer are

$$\nabla \cdot \underline{V} = 0 \tag{1}$$

$$\underline{V} \cdot \nabla \underline{V} = \frac{-1}{\rho_f} \nabla P + \nu_e \nabla^2 \underline{V} - \underline{V} \frac{\nu_f}{K} \tag{2}$$

$$\underline{V} \cdot \nabla T = \alpha_e \nabla^2 T \tag{3}$$

where $\alpha_e = k_e/\rho_f C_p$ and ρ_f and C_p refer to the density and the heat capacity of the fluid. The boundary conditions used for the present study are

$$x = 0: \quad u = 0, \quad v = 0, \tag{4}$$

$$y = 0: \quad u = 0, \quad v = 0, \quad T = T_w \tag{5}$$

$$y = \delta(x): \quad \frac{du}{dy} = 0, \quad v = 0, \quad T = T_s \tag{6}$$

The second and third boundary conditions refer to the no-slip condition at the isothermal wall and assumption of no interfacial shear at the condensate/vapor interface which is kept at a constant temperature of T_s, respectively. The following matching conditions at the interface of the porous/fluid layer are

$$y = H: \ u(y=H-) = u(y=H+) \tag{7a}$$

$$\mu_e \left(\frac{du}{dy}\right)\bigg|_{y=H-} = \mu_f \left(\frac{du}{dy}\right)\bigg|_{y=H+} \tag{7b}$$

$$T(y=H-) = T(y=H+) \tag{7c}$$

$$k_e \left(\frac{dT}{dy}\right)\bigg|_{y=H-} = k_f \left(\frac{dT}{dy}\right)\bigg|_{y=H+} \tag{7d}$$

These conditions include the continuity of velocity, shear, temperature, and heat flux at the condensate-saturated porous coating/liquid interface. In these equations, we have set the effective viscosity of the condensate-saturated porous substrate equal to the viscosity of the fluid and a volumetric averaging technique is used to calculate an effective thermal conductivity for the porous coating.

Numerical Solutions

The numerical scheme uses an iterative procedure whereby the solutions of the separate governing equations and the interface conditions are matched. To render compact results, the following dimensionless variables are introduced

$$x^* = \frac{x}{K^{1/2}} \qquad y^* = \frac{y}{K^{1/2}} \tag{8a}$$

$$H^* = \frac{H}{K^{1/2}} \qquad \delta^*(x^*) = \frac{\delta(x)}{K^{1/2}} \tag{8b}$$

$$\mu^* = \frac{\mu_e}{\mu_l} \qquad k^* = \frac{k_l}{k_e} \tag{8c}$$

$$u^* = u \frac{\mu_e}{[K(\rho_l - \rho_v)g \cos \phi]} \tag{8d}$$

$$\theta = \frac{(T - T_w)}{(T_s - T_w)} \tag{8e}$$

The dimensionless boundary and matching conditions are

$$x^* = 0: \qquad u^* = 0, \quad v^* = 0 \tag{9a}$$

$$y^* = 0: \qquad u^* = 0, \quad v^* = 0, \ \theta = 0 \tag{9b}$$

$$y^* = \delta^*(x^*): \qquad \frac{du^*}{dy^*} = 0, \quad v^* = 0, \ \theta = 1 \tag{9c}$$

$$y^* = H^*: \quad u(y^* = H^* -) = u^*(y^* = H^* +)$$

$$\mu^* \left(\frac{du^*}{dy^*}\right)\Bigg|_{y^*=H^*-} = \left(\frac{du^*}{dy^*}\right)\Bigg|_{y^*=H^*+}$$

$$\theta(y^* = H^* -) = \theta(y^* = H^* +) \tag{10}$$

$$k^* \left(\frac{d\theta}{dy^*}\right)\Bigg|_{y^*=H^*-} = \left(\frac{d\theta}{dy^*}\right)\Bigg|_{y^*=H^*+}$$

Upon inserting the above dimensionless variables into the governing equations of the porous and fluid regions and solving the system of differential equations with the above boundary and matching conditions in the same fashion as (22), the following velocity and temperature profiles result:

$0 < y^* < H^*$:

$$u^*(y^*) = \sinh y^*(\delta^*(x^*) - H^* + \sinh H^*)/\cosh H^* + 1 - \cosh y^* \tag{11}$$

$$\theta(y^*) = k^* y^* / (\delta^*(x^*) + H^*(k^* - 1) \tag{12}$$

$H^* < y^* < \delta^*(x^*)$:

$$u^*(y^*) = 0.5\ {}^*(H^{*2} - y^{*2}) + \mu^*\delta^*(x^*)(y^* - H^*) + \tanh H^*(\sinh H^* - H^* + \delta^*(x^*)) + 1 - \cosh H^* \tag{13}$$

$$\theta(y^*) = (y^* + H^*(k^* - 1))/(\delta^*(x^*) + H^*(k^* - 1) \tag{14}$$

The mass flow of condensate through any vertical position of the film is obtained by integrating the product of the density and velocity over each flow region. The resulting total mass flow rate through a cross-section of the film is a third-order polynomial in terms of the dimensionless thickness, $\delta^*(x^*)$.

The heat removed by the porous/fluid composite system over an incremental distance dx can be approximated by:

$$q'' = \lambda_{fg} b \frac{dw}{dx} \tag{15}$$

where λ_{fg} represents a modified heat of vaporization (1). From Fourier's Law and the assumption that we have a linear temperature distribution within the porous/condensate and pure liquid regions, the heat flux can be expressed as

$$q'' = \frac{-k_e}{H}(T_w - T_H) = \frac{-k_l}{(\delta - H)}(T_H - T_s) \tag{16}$$

where T_H is the temperature at the porous/fluid interface. Combining Equations (15) and (16) and integrating produces a fourth-order equation for $\delta^*(x^*)$ which is solved numerically using a Newton-Raphson method (22). For the special case of $H^* = 0$ and $\theta = 0$, the solution reduces to Nusselt's solution of an uncoated vertical surface.

The heat transfer results are reported by the local heat transfer coefficient and the local Nusselt number which are defined as

$$q'' = h_x(T_s - T_w) \tag{17}$$

$$Nu_x = \frac{h_x x}{k_l} \tag{18}$$

Using an energy balance, the local heat transfer coefficient is expressed in terms of the problem parameters as

$$h_x = \frac{k_l}{[K^{1/2}(\delta^*(x^*) + H^*(k^* - 1)]} \tag{19}$$

Integrating over the length of the surface produces the average convection coefficient which is used to define the average Nusselt number

$$Nu_L = \frac{h_L L}{k_l} \tag{20}$$

It should be noted that the Nusselt number is based on the thermal conductivity of the fluid to make for a more substantial comparison with the noncomposite system. It was also assumed that the effective dynamic viscosity of the condensate-saturated porous substrate (μ_e) was equivalent to the dynamic viscosity of the pure liquid (μ_l), or $\mu^* = 1$.

DISCUSSION OF RESULTS

The preliminary results of this investigation of film condensation enhancement by a porous/fluid composite system are illustrated in Figures 2-4. Figure 2 exhibits the numerical predictions of average Nusselt number as a function of surface subcooling temperature difference ($T_s - T_w$) and inclination angle. For these cases, the dimensionless porous substrate thickness (H^*) and the dimensionless thermal conductivity (k^*) are 0.25 and 3.26×10^{-3}, respectively. We note that the average Nusselt number decreases with increasing surface subcooling for all three test cases. It is observed that the value of average Nusselt number decreases with increased inclination angle for a particular value of surface subcooling. A comparison between a vertical surface and surfaces at 30 and 60 degree inclines reveal approximate increases in Nusselt number of 10 and 20 percent, respectively. This result is expected since the porous/fluid composite system is experiencing the effect of reduced gravity forces on film condensation heat transfer (3). Figure 2 also contains a comparison of the present simple model predictions with recent experimental data for the limiting case of film condensation of saturated steam at atmospheric pressure on a vertical isothermal porous coated plate (23). The porous substrate used in this experimental run was a thermospray self-bonding aluminum bronze copper coating of approximately 50 percent porosity that was adhered to a 7.62 x 12.7 x 1.27 cm (width x length x thickness) oxygen free 101 copper plate. The thickness of this porous coating was approximately 2.54×10^{-2} mm (0.001 in.). At the low end of the surface subcooling, we have as much as a 200 percent increase in the average Nusselt number as compared to the model prediction. At the higher end we have approximate agreement between the model and the experimental data.

In Figures 3 and 4, the numerical predictions and the available experimental data have been displayed in a compact form to compare the effect of reduced gravity forces on film condensation heat transfer rates for a specific value of surface subcooling, porous substrate thickness, and dimensionless thermal conductivity. More specifically, one comparison in Figures 3 and 4 is between the present model and Nusselt's correlation. The present model shows approximately a 25 percent heat transfer enhancement for all angles of inclination as compared to Nusselt's theory. There is also an increase in surface heat flux for the higher value of surface subcooling ($T_s - T_w = 33.3$ K) for both cases.

Nusselt's predictions were obtained under the assumption of laminar film condensation with no surface waves for an inclined uncoated surface (2). For the range of parameters used in our calculations and measurements, the laminar flow assumption is valid (Re ~ 10.5). We have utilized values of 0.25 for the dimensionless porous substrate thickness and 3.26×10^{-3} for the dimensionless thermal conductivity in these figures and have varied the effective body forces from 0.1g to 1g. The choice of surface subcooling values were chosen to match experimental data.

Another order of magnitude comparison that is made in Figures 3 and 4 is the data of Suryanarayana (3), who presented experimental results of film condensation on an inclined isothermal uncoated plane surface. His experiments condensed saturated steam at atmospheric pressure on one side of an inclined copper block. It is shown that his experimental results have heat transfer rates that are approximately 50 percent and 100 percent greater than the present model and Nusselt's theory, respectively. His results also show increases in surface heat flux for higher values of surface subcooling. This quantitative comparison shows magnitude verification of the present model predictions and the possible heat transfer augmentation that may occur with real permeable substrates as compared to theoretical models.

The final analysis is with the recent experimental data obtained by Renken and Mueller (23). Figure 3 shows the results of the limiting case of a vertical coated plate with a porous substrate thickness of 2.54 x 10^{-2} mm (0.001 in.) and surface subcooling of 16.6 K. This thermospray coating produced a noticeable increase (approximately 40 percent) in heat transfer as compared to the experimental data of Suryanarayana (3). Figure 4 depicts the experimental data of (23) for two different vertical porous substrates. Case A refers to the thermospray 2.54 x 10^{-2} mm data while case B is for a porous substrate thickness of 1.52 mm (0.060 in.). This second type of porous coating was a foam copper which was adhered to an identical oxygen free 101 copper plate. Both experimental cases were run at surface subcooling of 33.3 K. Again, the general trend is a heat transfer enhancement as compared to both the present model and Nusselt's theory.

A comparison of the uncoated experimental data (3) with (22) shows an equivalent enhancement for case A data and a slight decrease for case B data. The reason for this equivalent enhancement with Case A is hypothesized that the Suryanarayana (3) data was based on a Reynolds number of 135, which is at the probable onset of turbulence. Thus, higher heat transfer rates were produced than those predicted by the Nusselt correlation and by those of Case A. It should also be noted that Reference (3) observed surface waves which are not accounted for in the Nusselt correlation and not observed in Case A. The effect of such waves is again to increase heat transfer rates. The lower value of enhancement in Case B can be attributed to the concept of an optimal porous coating thickness which was suggested by (22). The physical significance of this result is that the porous substrate begins to act as an insulation. A final comparison of Figures 3 and 4 shows the thermospray experimental data has a 20 percent increase in heat transfer when the value of surface subcooling is doubled.

It is speculated by the authors that if several effects are taken into account in future theoretical models, the agreement between experiment and theory will improve over the range of governing parameters. These modifications include the effects of condensate Reynolds number and surface waves, use of a general flow model (Darcy-Brinkman-Forchheimer model), introduction of a more sensitive effective thermal conductivity (the effect of the conductivity ratio), introduction of vapor shear at the condensate/vapor interface, and the effect of the Darcy number.

CONCLUSIONS

Film condensation in an inclined porous/fluid composite system has been analyzed in the present work. Two sets of governing equations for the porous and fluid regions were employed and matching interface conditions were used to solve the problem. The characteristics of the velocity and temperature fields have been presented and exemplary heat transfer results were obtained. Comparisons of theoretical predictions of the porous/fluid composite system and the case of no porous substrate revealed a significant increase in surface heat flux. Recent experimental findings exhibit an order of magnitude verification of the present model. The effects of inclination, porous substrate thickness, and surface subcooling were also investigated. In conclusion, the results of this study demonstrate that the employment of a porous coating to a condensing surface needs to be considered as a viable alternative for heat transfer enhancement in condensation problems.

ACKNOWLEDGMENT

Financial support for this research provided by the National Science Foundation through Grant No. CTS-8909410 is greatly appreciated.

NOTATION

Roman Letters

b	width of surface [m]
C_p	specific heat of fluid [J/kg K]
g	gravitational acceleration [m/s^2]
H	thickness of porous coating [m]
h_L	average heat transfer coefficient [W/m^2 K]
h_x	local heat transfer coefficient [W/m^2 K]
k	thermal conductivity [W/m K]
k_e	effective thermal conductivity of porous coating [W/m K]
K	permeability of porous coating [m^2]
L	length of surface [m]
Nu_L	average Nusselt number
Nu_x	local Nusselt number
p	pressure [Pa]
q"	heat flux [W/m^2]
Re	condensate Reynolds number
T	temperature [K]
u	x-component velocity [m/s]
v	y-component velocity [m/s]
$\underline{V}$	velocity vector
w	mass flow rate [kg/s]
x	vertical Cartesian coordinate [m]
y	horizontal Cartesian coordinate [m]

Greek Letters

α	thermal diffusivity [m^2/s]
α_e	effective thermal diffusivity [m^2/s]
θ	dimensionless temperature
λ_{fg}	modified latent heat of vaporization [J/kg]
μ	dynamic viscosity [kg/m s]
ν	kinematic viscosity [m^2/s]
ρ	density [kg/m^3]
ϕ	plate inclination angle

Subscripts

e	pertaining to effective porous/fluid value
f	pertaining to fluid
H	pertaining to porous/fluid layer interface
l	pertaining to pure liquid layer
s	pertaining to saturation temperature
v	pertaining to pure vapor
w	pertaining to wall temperature

Superscripts

*	pertaining to dimensionless quantity

LITERATURE CITED

1. Burmeister, L.C., Convective Heat Transfer, Wiley & Sons, New York (1983).

2. Nusselt, W. "Die Oberflachenkondensation des Wasser dampfes," Zeitschrift des Vereins Deutshes Ingenieure, 60, 541-575 (1916).

3. Suryanarayana, N.V., "Condensation Heat Transfer Under Low and High Gravity Conditions," Report to NSF, Contract No. GK-27818, Michigan Technological University, Houghton, Michigan (1973).

4. Avedisian, C.T. and J. Koplik, "Liedenfrost Boiling of Methanol Droplets on Hot Porous/Ceramic Surfaces," Int. J. Heat Mass Transfer, 30, 379-393 (1982).

5. Avedisian, C.T., C. Ioffredo and M.J. O'Connors, "Film Boiling of Discrete Droplets of Mixtures of Coal and Water on a Horizontal Brass Surface," Chemi-

cal Engineering Science, 39, 319-327 (1984).

6. Bergles, A.E. and M.C. Chyu, "Characteristics of Nucleate Pool Boiling from Porous Metallic Coatings," J. Heat Transfer, 104, 279-285 (1982).

7. Cheng, P. and A. Verma, "The Effect of Subcooled Liquid on Film Boiling about a Vertical Heated Surface in a Porous Medium," Int. J. Heat Mass Transfer, 24, 1151-1160 (1981).

8. Ito, T., K. Nishikawa and K. Tanaka, "Enhanced Heat Transfer by Nucleate Boiling at Sintered Metal Layer - Discussion on Sintered Layer and Experiments by Piled Layer and Form Layer," Refrigeration, 57, 77-81 (1982).

9. Kajikawa, T., H. Takazawa and M. Mizuki, "Heat Transfer Performance of Metal Fiber Sintered Surfaces," Heat Transfer Engineering, 4, 57-66 (1983).

10. Marto, P.J. and V.J. Lepere, "Pool Boiling Heat Transfer from Enhanced Surfaces to Dielectric Fluids," J. Heat Transfer, 104, 292-299 (1982).

11. Nakayama, W., T. Daikoku and T. Nakajima, "Effects of Pore Diameters and System Pressure on Saturated Pool Nucleate Boiling Heat Transfer from Porous Surfaces," J. Heat Transfer, 104, 286-291 (1982).

12. Nakayama, W., T. Daikoku, H. Kuwahara and T. Nakajima, "Dynamic Model of Enhanced Boiling Heat Transfer on Porous Surfaces Part 1: Experimental Investigation," J. Heat Transfer, 102, 445-450 (1980).

13. Nakayama, W., T. Daikoku, H. Kuwahara and T. Nakajima, "Dynamic Model of Enhanced Boiling Heat Transfer on Porous Surfaces Part 2: Analytical Modeling," J. Heat Transfer, 102, 451-456 (1980).

14. Kovalev, S.A., S.L. Solov'yev and O.A. Ovodkov, "Liquid Boiling On Porous Surfaces," Heat Transfer - Soviet Research, 19, 109-120 (1987).

15. Styrikovich, M.A., S.P. Malyshenko, A.B. Andrianov and I.V. Talaev, "Investigation of Boiling on Porous Surfaces," Heat Transfer - Soviet Research, 19, 23-29 (1987).

16. Webb, R.L., "The Evolution of Enhanced Surface Geometries for Nucleate Boiling," Heat Transfer Engineering, 2, 46-69 (1981).

17. Webb, R.L., "Nucleate Boiling on Porous Coated Surfaces," Heat Transfer Engineering, 4, 71-82 (1983).

18. Woodruff, D.W. and J.W. Westwater, "Steam Condensation on Various Gold Surfaces," J. Heat Transfer, 103, 685-692 (1981).

19. Rifert, V.G., Y.Y. Trokoz and V.Y. Zadiraka, "Enhancement of Heat Transfer in Condensation of Ammonia Vapor on a Bundle of Wire-Finned Tubes," Heat Transfer - Soviet Research, 16, 36-41 (1984).

20. Holden, K.M., A.S. Wanniarachchi, P.J. Marto, D.H. Boone and J.W. Rose, "The Use of organic Coatings to Promote Dropwise Condensation of Steam," J. Heat Transfer, 109, 768-774 (1987).

21. Izumi, M., N. Yamakawa, T. Shinmura, Y. Isobe, S. Ohtani and J.W. Westwater, "Drop and Filmwise Condensation on Horizontally Scratched Rough Surfaces," Heat Transfer - Japanese Research, 1-14 (1989).

22. Renken, K.J., D.J. Soltykiewicz and D. Poulikakos, "A Study of Laminar Film Condensation on a Vertical Surface with a Porous Coating," Int. Comm. Heat Mass Transfer, 16, 181-192 (1989).

23. Renken, K.J. and C. D. Mueller, "Experimental Measurements of Enhanced Film Condensation on a Porous Coated Surface," The Fourth International Symposium on Transport Phenomena (ISTP-4) - Heat and Mass Transfer, Sydney, Australia, July 14-18 (1991).

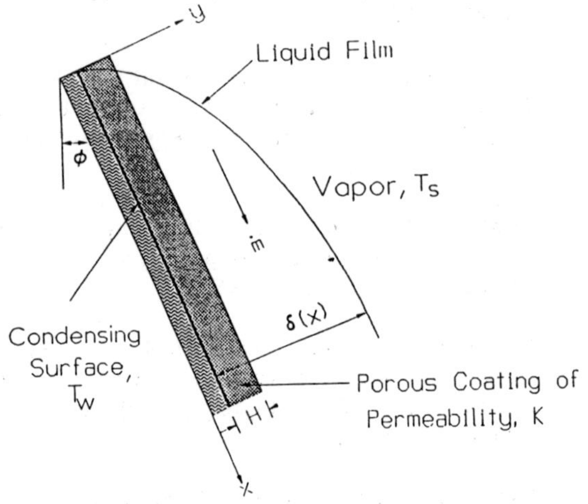

Figure 1 Schematic of problem: film condensation along an inclined surface coated with a porous material of permeability K and thickness H.

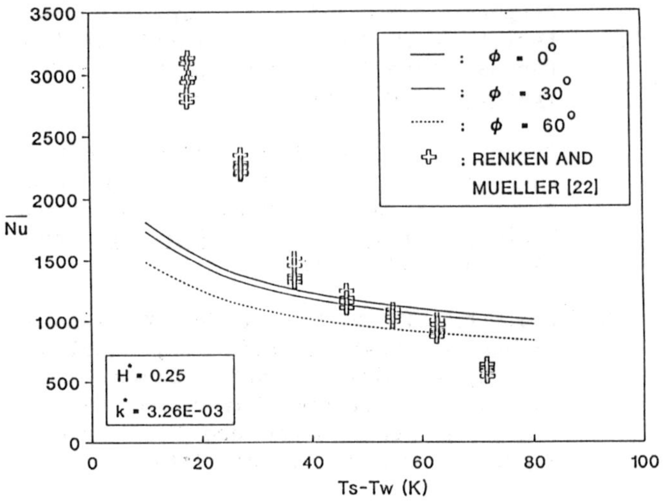

Figure 2 Comparison of Nusselt number variation with inclination and surface subcooling.

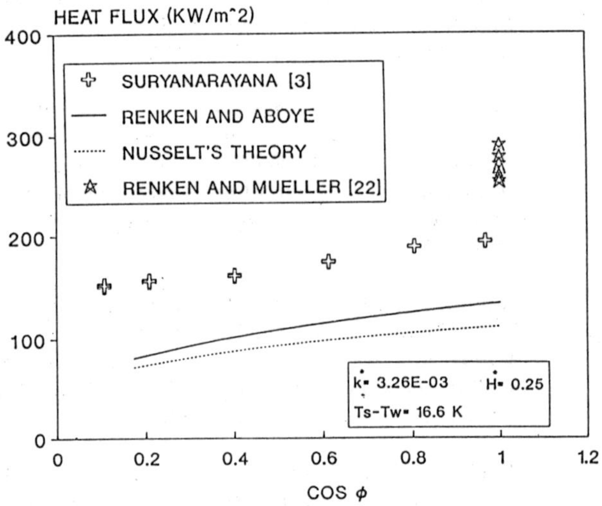

Figure 3 Comparison of heat transfer rate versus inclination for a surface subcooling (T_S-T_W) of 16.6 K.

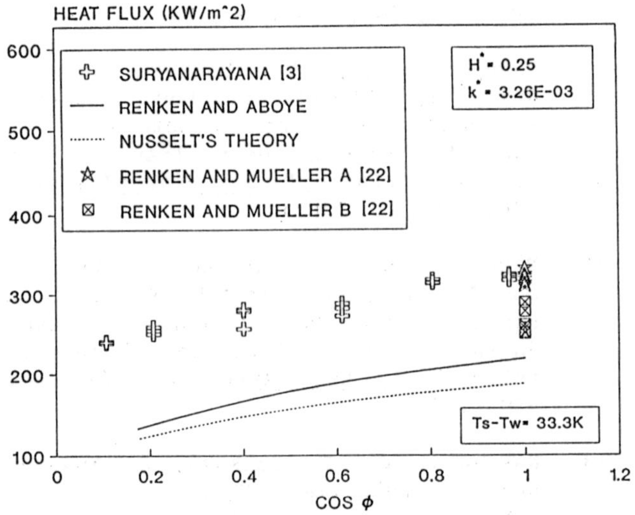

Figure 4 Comparison of heat transfer rate versus inclination for a surface subcooling (T_S-T_W) of 33.3 K.

HEAT TRANSFER AND PRESSURE DROP OF STUD FINNED TUBES

George Breber ■ Heat Transfer Research, Inc., Alhambra, CA

This paper presents the results of a review and evaluation of heat transfer and pressure drop predictive methods for bundles with stud-finned tubes in a staggered arrangement. Comparisons are made with published wind-tunnel data and open-literature correlations. The correlations that achieved the best agreement with data are identified.

Increased emphasis on the efficient use of fuels in recent years has resulted in the expanded application of heavy-duty finned tubes for boiler economizers, process heater convection sections, reformer furnaces, heat recovery boilers, etc. Due to the corrosive nature of the finside gas, thick carbon-steel fins or studs are usually welded to the base tube or pipe. Both the staggered and inline tube arrangements are used in heat-recovery heat exchangers. The staggered arrangement is more common. With the increased usage of these tubes has come the need for more accurate heat transfer and pressure drop predictive methods.

DATA SOURCES

There is a lack of data and suitable correlations in the area of stud finned tubes. Two data sources from open literature were used for comparison with available or suggested correlations. The authors of the data are Worley and Ross [1], and Ackerman and Brunsvold [2]. Twenty tube and bundle geometrical combinations were investigated. The tube outside diameters ranged from 38.1 mm (1.5 inches) to 63.5 mm (2.5 inches). The tube bundles were eight rows deep. The ratios of transverse pitch to outside tube diameter ranged from 1.87 to 3.2. Worley and Ross tested a broad variety of stud-finned tubes and bundle arrangements. Ackerman and Brunsvold tested only one kind of carbon steel stud-finned tubes, however, in five various staggered bundle layouts. The stud-finned tube geometry and bundle layout details are listed in Table 1.

With the Worley and Ross experiments, the gas temperature at the inlet ranged from 60 to 200°C and the water inlet temperature inside the tubes ranged from 10 to 55°C. Ackerman and Brunsvold operated a hot wind tunnel with the gas inlet temperature ranging from 90 to 217°C and with the water temperature ranging from 16 to 71°C.

HEAT TRANSFER

For comparison purposes, it is advantageous to present heat transfer correlations in the Colburn j-factor format defined as

$$j = \frac{h_c \, Pr^{2/3}}{C_p \, G_{max}} \tag{1}$$

The j-factor can also be presented as

$$j = \frac{Nu}{Re\ Pr^{1/3}} \qquad (2)$$

where the finside heat transfer coefficient, h_c, is based on 100 percent fin efficiency in order to normalize and correlate data taken with different types of fins. The correction was based on Gardner's paper [3]. G_{max} is maximum mass velocity, based on the minimum flow area in the tube bank.

The Colburn factor is plotted as a function of the Reynolds number defined as

$$Re = \frac{d_r\ G_{max}}{\mu} \qquad (3)$$

where d_r is the outside tube or root diameter of the finned tube.

Correlations

The open literature heat transfer correlations considered are those developed by Weierman [4], Escoa Co. [5], Worley and Ross [6], Mieth [6] and Biraghi Co. [7]. However, only the Worley and Ross and Biraghi empirical correlations are based on stud-finned tube data. The correlations of the other authors were developed for segmented-finned tube bundles which are geometrically close to the stud-finned tubes.

The Weierman [4] Correlation developed in 1976 and revised in 1977 was presented as follows

$$j = 0.23\ c_3\ c_5\ Re^{-0.35} \left(\frac{d_f}{d_r}\right)^{0.5} \left(\frac{T_b}{T_w}\right)^{0.25} \qquad (4)$$

$$c_3 = 0.55 + 0.45\ \exp\left(-\ 0.35\ \frac{\ell}{s}\right) \qquad (5)$$

where ℓ is the fin height and s is the fin spacing.

$$c_5 = 0.7 + \left[0.7 - 0.8\ \exp\left(-\ 0.15\ N_r^2\right)\right]$$

$$\left[\exp\left(-\ \frac{P_\ell}{P_t}\right)\right] \qquad (6)$$

The Escoa Engineering Manual [5] presents a correlation which is only slightly different from the Weierman [4] correlation defined by equations (4) through (6). The coefficient 0.23 in the basic equation (4) is substituted by the coefficient 0.25 in the Escoa method. All other coefficients and exponents are the same.

Worley and Ross [1] measured the heat transfer characteristics of a wide range of stud-finned tubes. The recommended correlation of this extensive research effort is

$$j = \frac{0.125}{Re^{0.3}} \qquad (7)$$

The investigators concluded that fin geometry correction terms in this correlation, such as the fin height, fin spacing, etc., were not necessary.

The Mieth [6] Correlation and its correction factors were published in the form of graph tables. Rabas et al. [8] converted the Mieth correlation to the following form

$$j = \frac{0.412}{Re^{0.41}} \left(\frac{s}{\ell}\right)^{0.3} \left(\frac{s}{t_f}\right)^{0.1} \qquad (8)$$

Biraghi Correlation. Biraghi Entrepose U.S., Inc. [7] presented a correlation for tubes with cylindrical studs in the graphical form. The correlation can also be presented as follows

$$j = 0.414\ Re^{-0.412} \qquad (9)$$

Comparisons

All experimental data sets, obtained from the tube bundles characterized in Table 1, were compared with the predicted values calculated from the correlations described in the previous paragraph and evaluated, statistically. Only typical comparisons of the selected data groups are presented for illustration of the magnitude of the deviations and the correlation trends.

Figure 1 compares the predicted Colburn factor curves to the Worley and Ross data obtained from Tube Bank No. 1. The Escoa predictions were in relatively good agreement with the data. The average error was -17.4 percent. The Weierman, Worley and Ross, and Biraghi correlations were slightly more conservative. The Mieth correlation predictions were about 50 percent of the data.

All mentioned correlations predicted the Ackerman and Brunsvold data in a similar way as illustrated in Figure 2, where the comparison with data from Tube Bank No. 17 are shown. The Escoa and Biraghi correlations predicted the data sets in the range of -1.5 to -19.5 percent. The other prediction methods were more conservative.

Generally, the Worley and Ross [1] data were predicted well by all the methods with an exception of the Mieth method, which was too conservative. The Escoa method predicted the data with an average error of only 0.6 percent; the range of the errors was from -17.4 to 28.5 percent. On the other hand, the Worley and Ross predictions were slightly conservative. The average error was -11 percent, however, the errors ranged only from -20.8 to -1.1 percent. The Ackerman and Brunsvold [2] data were predicted very well by the Escoa and Biraghi methods, with average errors of -8.5 and -10.6 percent respectively. The errors ranged from -11.9 to -3.7 percent for the Escoa method and from -19.5 to -1.5 percent for the Biraghi method.

PRESSURE DROP

The static pressure drop across a finned tube bank is composed of a friction loss and a momentum change for the gas as it flows through the bank:

$$\Delta P = \frac{4\,(G_{max})^2\,N_r}{2\,\rho}\,(f + a) \qquad (10)$$

where ΔP is the static pressure difference across the bundle, G_{max} is the maximum mass velocity, N_r is the number of tuberows, f is the friction factor, and a is the flow acceleration factor.

The flow acceleration factor, a, is small compared to f for air-cooler applications and can be neglected because the air density changes are relatively small. For heat recovery applications, this term is negative, indicating a pressure recovery, and may not be negligibly small.

Correlations

The Weierman [4] correlation for heat-recovery type segmented-finned tubes, revised in 1977, was presented as follows:

$$f = c_2\,c_4\,c_6\left(\frac{d_f}{d_r}\right)^{0.5} \qquad (11)$$

where

$$c_2 = 0.07 + 8.0\,Re^{-0.45} \qquad (12)$$

$$c_4 = 0.11\left(0.05\,\frac{P_t}{d_r}\right)^{\left[-0.7\left(\frac{\ell}{s}\right)^{0.23}\right]} \qquad (13)$$

$$c_6 = 1.1 + \left[1.8 - 2.1\exp\left(-0.15\,N_r^2\right)\right]\left[\exp\left(-2.0\,\frac{P_\ell}{P_t}\right)\right] - \left[0.7 - 0.8\exp\left(-0.15\,N_r^2\right)\right]\left[\exp\left(-0.6\,\frac{P_\ell}{P_t}\right)\right] \qquad (14)$$

Biraghi Correlation. The catalog of Biraghi Entrepose, U.S. Inc. [7] presents a correlation for cylindrical stud-finned tubes in the graphical form. The correlation can be also presented in the following form

$$f = 0.723\,Re^{-0.137} \qquad (15)$$

The data obtained by Worley and Ross from Tube Banks No. 8 and No. 9, and all the Ackerman and Brunsvold data were very poorly predicted by the Weierman and Biraghi correlations. Therefore, a new correlation was developed using the data trend analogy from bare tube bundles (Bell [9], ESDU [10], Zukauskas and Ulinskas [11]) and helical finned tube bundles (Stasiulevicius and Skrinska [12]), some Weierman correction factors, and with the help of the data discussed in this paper.

The new correlation can be presented as follows:

$$f = A_1 c_4 c_6 c_8 \left(\frac{d_f}{d_r}\right) \qquad (16)$$

where A_1 is the stud shape correction factor, defined in Figure 3. The c_4 and c_6 are correction factors from the Weierman method, defined by equations (13) and (14).

For $Re > 10^5$

$$c_8 = 0.07 + 0.12\, Re^{-0.05} \qquad (17)$$

For $Re < 10^4$

$$c_8 = 1.07\, Re^{-0.2} \qquad (18)$$

In the transition region, the two curves, expressed by the above equations, will cross at a certain Re_{lim}. The equation (18) should be applied for $Re < Re_{lim}$ and equation (17) for $Re > Re_{lim}$.

Comparisons

Pressure drop data from three sources and from 20 various bundles, characterized in Table 1, were compared with the predicted values and statistically evaluated. Four typical comparisons are presented in graphical form for illustration of the magnitude of deviations, data and correlation trends.

Figure 4 presents the comparisons between two sets of the Worley and Ross data received from Tube Banks No. 2 and No. 3 (two different numbers of the stud-rings per unit of the tube length and, therefore, two different stud spacings, s), and the predictions. The Weierman predictions were too conservative for both of the data sets and the correlation slope was too steep. The Biraghi method provided only one curve because the Biraghi method does not include the stud spacing as a parameter. This method predicted only data from Tube Bank No. 2 well. Equation (16) predicted both data sets very well with the average error less than -4 percent.

The Worley and Ross [1] data from Tube Bank No.1 are plotted in Figure 5. Again, the Weierman correlation was conservative for most of the data and the slope of the correlation was too steep. The Biraghi correlation predicted the data well, however, predictions calculated from Equation (16) were in best agreement with the data.

Figure 6 presents the comparisons between the Ackerman and Brunsvold [2] data from Tube Bank No. 17 and the predictions. Equation (16) best predicted the trend of the data. The Biraghi and Weierman correlations predicted the data too conservatively with the average errors of 40 and 93 percent.

Generally, the Worley and Ross [1] data were best predicted by Equation (16) with the average error of -6.6 percent and the error range from -31 to 13 percent. The Biraghi and Weierman methods predicted these data less accurately and in a very broad range of errors. The Ackerman and Brunsvold [2] data were predicted well by Equation (16) with the exception of data from Tube Bank No. 20 where the data were overpredicted by more than 50 percent. This data set was the only one that was predicted so conservatively by Equation (16). This data set from Tube Bank No. 20 was very poorly predicted by all methods. The Biraghi and Weierman methods predicted the Ackerman and Brunsvold [2] data very conservatively with errors from 40 to 263 percent.

CONCLUSIONS

Heat Transfer. The Biraghi [7], Escoa [5], Worley and Ross [1], and Weierman [4] correlations predicted the heat transfer data from all 20 bundles relatively well. A recommendation resulting from this study is to use the Biraghi method which provides slightly conservative, but very consistent predictions. The overall average error of prediction was -11 percent and the errors were very regularly distributed in the range of -26.9 to 4.3 percent.

Pressure Drop. The pressure drop data

were not as well predicted as the heat transfer data. Equation (16) provided the best predictions. The overall average error of the predictions was -1.2 percent and the errors were spread in the range of -30.5 to 54.5 percent. However, 81 percent of the data points were predicted by Equation (16) in the range of ±20 percent. The other methods did not predict the pressure data drop consistently.

Future Research. Further testing is needed in order to completely determine the general performance of a bank of heat-recovery and fired heater convection section stud-finned tubes. First, the effect of physical properties of gases at higher temperatures, above 500 C, requires further study, both experimentally and analytically. Second, there is a need for more pressure drop data from bundles with cylindrical stud-finned tubes.

NOMENCLATURE

A_1	stud shape correction factor, -
a	flow acceleration factor, -
C_p	specific heat of gas, J/kg K
c_n	coefficients, n = 2 to 8, -
d_f	fin outside diameter, m
d_r	tube root diameter, m
f	friction factor, -
G_{max}	maximum mass velocity, kg/s m
h_c	finside heat transfer coefficient, W/m^2 K
j	Colburn j-factor, -
ℓ	stud height, $(d_f - d_r)/2$, m
N_f	number of stud-rings, m^{-1}
N_r	number of rows, -
Nu	Nusselt number, -
ΔP	static pressure drop across tube bank, Pa
P_ℓ	longitudinal tube pitch, m
P_t	transverse tube pitch, m
Pr	Prandtl number, -
Re	Reynolds number, -
Re_{lim}	Reynolds number determining the limit of application of equations (17) and (18), -
s	stud-fin spacing, $(1/N_f - t_f)$, m
T_b	average bulk temperature of gas, K
T_w	average tube wall temperature, K
t_f	stud thickness, m
w_f	stud width, m
μ	average dynamic viscosity of gas, N s/m^3
ρ	average density of gas, kg/m^3

LITERATURE CITED

1. Worley, N. G. and Ross, W., "Heat Transfer and Pressure Drop Loss Characteristics of Crossflow Tubular Arrangements with Studded Surfaces," *Inst. Mech. Engrs.*, 5, (November 1960).
2. Ackerman, J. W. and Brunsvold, A. R., "Heat Transfer and Draft Loss Performance of Extended Surface Tube Banks," *Trans. ASME, J. Heat Transfer*, **92**, 215-220 (1970)
3. Gardner, K. A., "Efficiency of Extended Surface," *Trans. ASME*, **67**, 621-631 (1945).
4. Weierman, C., "Pressure Drop Data for Heavy-Duty Finned Tubes," *Chem. Eng. Progress*, **73**, No. 2, 69 (1977)
5. Escoa Engineering Manual, Heavy Duty Walded Fintube, Escoa, Pryor, Oklahoma (1977).
6. Mieth, H. C., "Method for Heat Transfer Calculation of Helical-Wound Fin Tubes," ASME Paper 75-Pet-4 (1970).
7. Biraghi Entrepose U.S., Inc., Heat Transfer Data for B. E. Studded Tubes, Pryor, Oklahoma (1978).
8. Rabas, T. J., Myers, G. A. and Eckels, P. W., "Comparison of the Thermal Performance of Serrated High-Finned Tubes Used in Heat-Recovery Systems," ASME Winter Annual Meeting, Vol. HT - 59, 33-40, Anaheim (1986).
9. Bell, K. J., "Final Report of the Cooperative Research Program on Shell and Tube Heat Exchangers," Bull. No. 5, U. of Delaware, (January 1963).
10. Engineering Sciences Data Unit, "Crossflow Pressure Loss over Banks of Plain Tubes in Square and Triangular

Arrays Including Effects of Flow Direction," Item No. 79034, Specialist Printing Serv. Ltd, London (1979).

11. Zukauskas, A. and Ulinskas, R., "Banks of Plain and Finned Tubes," in *Heat Exchanger Design Handbook*, Section 2.2.4. Hemisphere Publ. Corp., Washington (1982).

12. Stasinlevicius, J. and Skrinska, A., *Heat Transfer of Finned Tube Bundles in Crossflow*, Hemisphere Publ. Corp., Washington (1988).

Table 1. Stud-finned tube and tube bank geometry.

Tube Bank No.	d_r mm	d_f mm	ℓ mm	s mm	t_f mm	w_f mm	P_t mm	P_ℓ mm	Stud Shape*
				Worley and Ross Data (1960)					
1	50.8	101.6	25.4	4.0	2.4	25.4	127.0	88.9	R,Ch,E
2	50.8	101.6	25.4	10.3	2.4	25.4	104.8	88.9	Ch
3	50.8	101.6	25.4	4.0	2.4	25.4	104.0	88.9	Ch
4	50.5	88.9	19.1	4.0	2.4	25.4	104.8	82.6	Ch
5	50.8	88.9	19.1	2.7	2.4	25.4	104.8	82.6	Ch
6	38.1	76.2	19.1	3.2	3.2	19.1	101.6	63.5	Ch
7	38.1	76.2	19.1	4.0	2.4	19.1	101.6	63.5	Ch
8	50.8	101.6	25.4	4.0	3.2	12.7	104.8	88.9	E
9	38.1	76.2	19.1	4.0	3.2	12.7	79.4	69.9	E
10	63.5	101.6	19.1	9.5	2.4	31.8	133.4	88.9	Ch
11	38.1	76.2	19.1	9.5	2.4	19.1	79.4	69.9	Ch
12	50.8	114.3	31.8	4.0	2.4	25.4	177.8	88.9	Ch
13	50.8	88.9	19.1	4.0	2.4	25.4	127.0	76.2	Ch
14	50.8	82.6	15.9	4.0	2.4	25.4	114.3	76.2	Ch
15	50.8	88.9	19.1	4.0	2.4	25.4	95.3	69.9	Ch
				Ackerman and Brunsvold Data (1970)					
16	47.6	98.4	25.4	3.2	3.2	12.7	127.0	88.9	E
17	47.6	98.4	25.4	3.2	3.2	12.7	152.4	82.6	E
18	47.6	98.4	25.4	3.2	3.2	12.7	127.0	114.3	E
19	47.6	98.4	25.4	3.2	3.2	12.7	127.0	82.6	E
20	47.6	98.4	25.4	3.2	3.2	12.7	101.6	88.9	E

*R = Rectangular, Ch = Chamfered, E = Elliptical

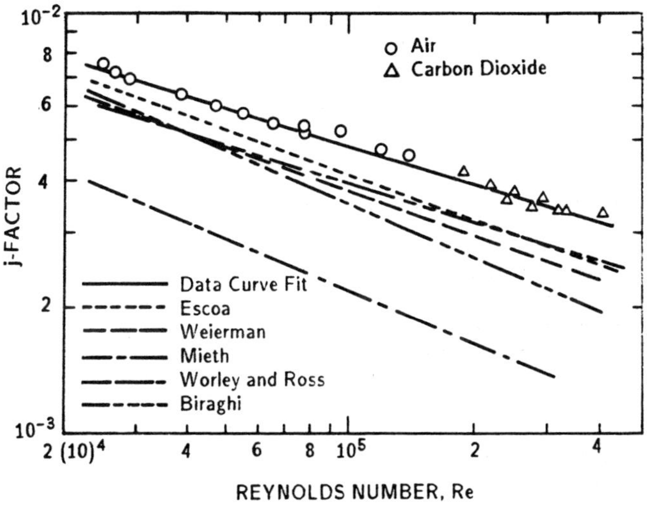

Figure 1. Comparison of the Worley and Ross [1] data with the correlations. The data were obtained from Tube Bank No. 1.

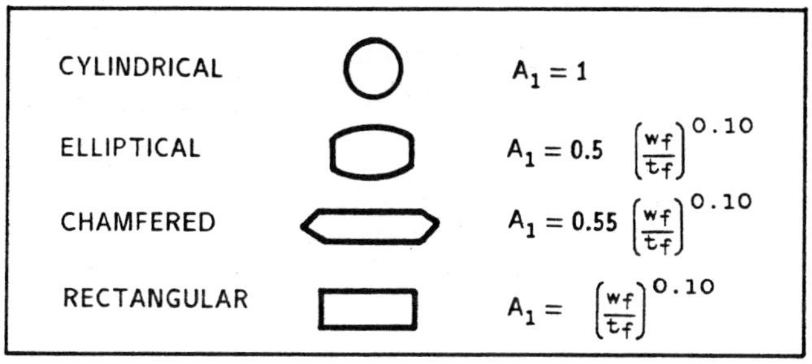

Figure 3. Various stud fin cross-sections and the stud shape correction factors.

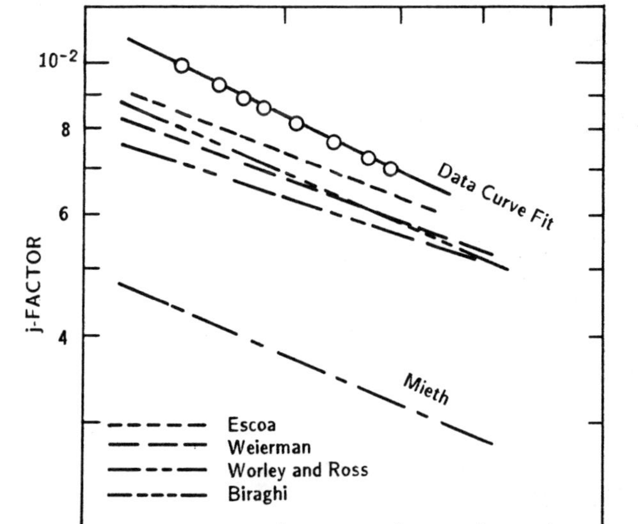

Figure 2. Comparison of the Ackerman and Brunsvold [2] data from Tube Bank No. 17 with the considered correlations.

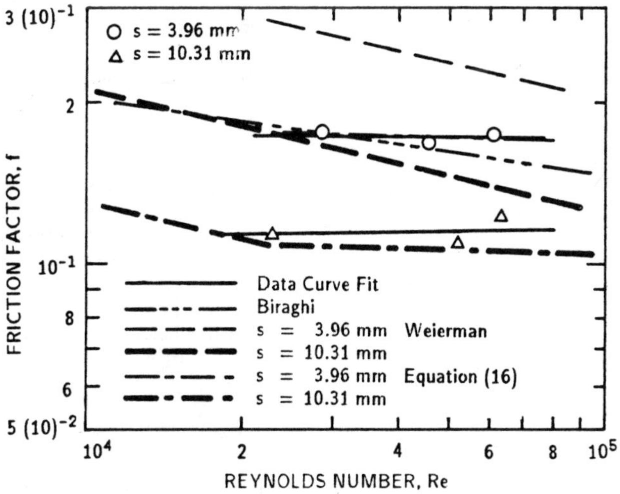

Figure 4. Comparisons of the Worley and Ross [1] data from Tube Banks No. 2 and No. 3 with the considered correlations.

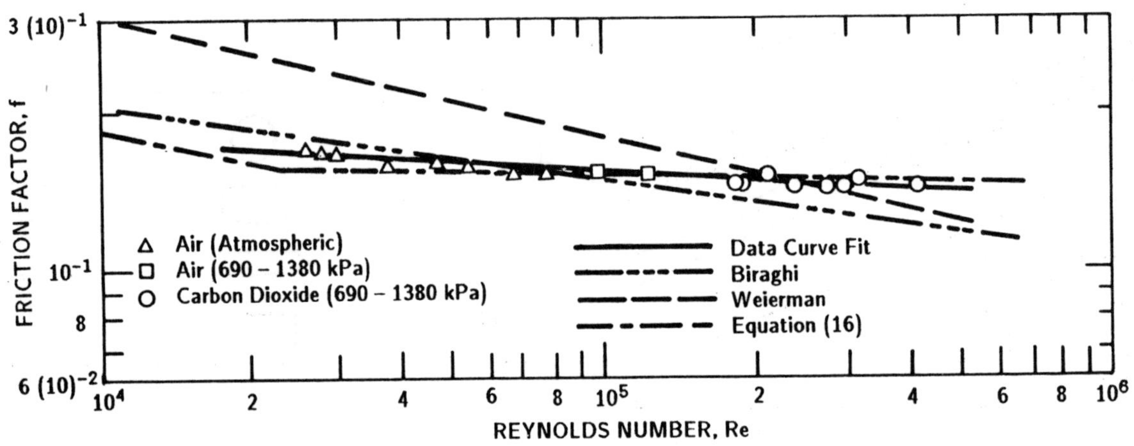

Figure 5. Comparison of the Worley and Ross [1] pressure drop data from Tube Bank No. 1 with the considered correlations.

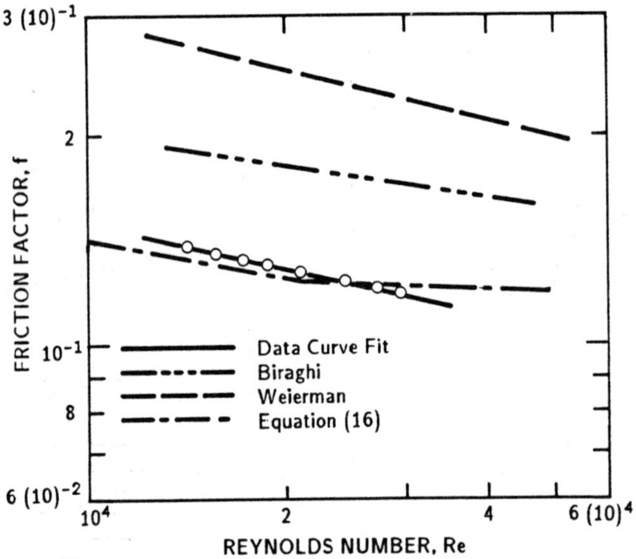

Figure 6. Comparison of the Ackerman and Brunsvold [2] data from Tube Bank No. 17 with the considered correlations.

A FORM OF CORRELATION FOR FORCED CONVECTIVE BOILING IN TUBES

D. Chisholm ■ Heat Transfer Research, Inc., P.O. Box 10009, College Station, TX 77842

Forms of correlation for convective boiling of single-component fluids in tubes are reviewed.
The present paper then defines a dimensionless group

$$\gamma = \frac{h_C}{h_L} \bullet \frac{1}{F_T}$$

where h_C is the convective boiling heat transfer coefficient; h_L is the convective heat transfer coefficient if the liquid component flows alone; and F_T is the Chen multiplier, corresponding to the two-phase Lockhart-Martinelli parameter, at the thermodynamic critical point.

It is shown that, for given physical properties, γ is, to the accuracy of other correlations, a constant. Values of γ are generally of the order of 2. Methods of predicting γ are compared.

This report concerns forced convective boiling at conditions where annular flow occurs and nucleate boiling is suppressed. These conditions normally occur at higher flow rates and dryness fraction.

The report is in three phases.

* Comparison of recent correlations
* Development of an alternative form of correlation
* A preliminary examination of a generalized equation for the single coefficient in the alternative form of correlation.

THE CHEN CORRELATION

Define a two-phase multiplier with respect to the heat transfer coefficient if the liquid component flows alone

$$\Psi_L^2 = \frac{h_c}{h_L} \quad (1)$$

Bennet and Chen (1) developed the following equation

$$\Psi_L^2 = Pr^{0.296} F \quad (2)$$

where F was given by Chen (2) in graphical form as a function of the Lockhart-Martinelli parameter. Kenning and Cooper (3) fitted the curve as follows

$$F = 1 + \frac{1.8}{X^{0.79}} \quad (3)$$

where the Lockhart-Martinelli parameter is

$$X = \left(\frac{Dp_L}{Dp_G} \right)^{0.5} \quad (4)$$

For turbulent flow of both components if flowing alone, this can be expressed

$$x = \left(\frac{1-x}{x} \right)^{(2-n)/2} \frac{1}{\Gamma} \quad (5)$$

where

$$\Gamma = \left(\frac{Dp_{Go}}{Dp_{Lo}} \right)^{0.5} \quad (6)$$

THE MOMENTUM-HEAT TRANSFER ANALOGY

Using the momentum-heat transfer analogy, Chen obtained

$$F = \left(\phi_L^{2/2-n}\right)^{0.8} \tag{7}$$

where

$$\phi_L^2 = \frac{Dp}{Dp_L} \tag{8}$$

Using the equation of Chisholm and Laird (4) in Equation 7, with n = 0.2, one obtains

$$F = \left(1 + \frac{21}{x} + \frac{1}{x^2}\right)^{0.444} \tag{9}$$

For refrigerants Jung et al. (5) obtained

$$\phi_L = \frac{3.58}{X^{0.735}} \tag{10}$$

Hence, using Equations 7 and 10, with n = 0.2

$$F = \frac{3.1}{X^{0.653}} \tag{11}$$

The conditions of the experimental data discussed in this paper are given in Table 1.

Table 2 compares Equations 3 ,9, and 11. The equations give similar order of magnitudes for the multiplier F.

For water, Kenning and Cooper (3), following an extensive series of tests, have recently obtained the equation

$$F = 1 + \frac{1.8}{X^{0.87}} \tag{12}$$

The Prandtl number correction in Equation 2 was taken as unity in obtaining this equation. Values of F obtained using this equation are also shown in Table 2. It can be seen that with decreasing value of the Lockhart-Martinelli parameter, corresponding to increasing dryness fraction, the Chen F curve, Equation 3, falls increasingly below values given by Equation 12. Recent data for refrigerant (6,7), which are discussed below, also show this trend.

Kenning and Cooper report that Equation 12 underestimates the heat transfer coefficient for churn-annular flow, and develop an equation for that regime.

Table 1. Test conditions.

Fluid	water	R22	R114	ethanol
Pressure/bar	1.7 - 5.9	4.0	2.6	1.5
Pr	1.4	2.66	5.38	7.5
Γ	13.8 - 24.8	6.47	6.17	13.5
G/[kg/(m^{2}s)]	123 - 630	250 - 720	250 - 720	145 - 290
γ	1.7	2.0	2.0	2.65
Diameter/mm	9.6	9.0	9.0	10
Material	s.s	s.s.	s.s	copper
Orientation	vertical	horizontal	horizontal	vertical

Table 2. Comparison with Chen multiplier.

1/X	1	5	10	50	100
Chen F, Eq. 3	2.8	7.42	12.10	40.70	69.47
Chisholm & Laird F, Eq. 9	4.02	8.71	12.78	37.70	64.90
Jung et al. F, Eq. 11	3.11	8.90	14.00	40.06	63.01
Kenning & Cooper F, Eq. 12	2.8	8.30	14.34	55.12	99.92
Equn 12/Equn 3	1.0	1.12	1.19	1.35	1.44

ANALYTIC SOLUTION OF KUNZ AND YERAZUNIS

Kunz and Yerazunis (8), from a detailed numerical study, obtained an equation for annular flow which can be expressed

$$h_c = 0.147 \frac{c_{pL} \left(\tau \rho_L\right)^{0.5}}{Re_L^{0.1} Pr_L^{0.6}} \quad (13)$$

The single-phase heat transfer coefficient where the liquid component flows alone is

$$h_L = 0.023 Re_L^{0.8} Pr_L^{0.4} \frac{k_L}{d} \quad (14)$$

Kunz and Yerazunis used a single-phase friction factor

$$f = \frac{0.046}{Re_L^{0.2}} \quad (15)$$

This gives the wall shear stress as

$$\tau = \frac{fG_L^2 \phi_L^2}{2\rho_L} = \frac{0.023}{\rho_L} \left(\frac{\mu_L}{d}\right)^2 Re_L^{1.8} \phi_L^2 \quad (16)$$

Combining Equations 13 to 16 gives

$$\Psi_L^2 = 0.968 \phi_L \quad (17)$$

There is little loss of accuracy in assuming

$$\Psi_L^2 = \phi_L \quad (18)$$

In that case from Equation 10

$$\Psi_L^2 = \frac{3.58}{X^{0.735}} \quad (19)$$

The measured heat transfer coefficients corresponding to these pressure drop tests were correlated (7) using

$$\Psi_L^2 = 2.37 \left(0.29 + \frac{1}{X}\right)^{0.85} \quad (20)$$

Table 3 shows that Equations 19 and 20 for 1/X equal and greater than unity agree to within +22%, -11%. This gives general confimation of the Kunz-Yerazunis analysis. A similar comparison cannot be made with the Kenning-Cooper equation as pressure drop measurements were not reported.

The analytic approaches of Chen and Kunz, and Yerazunis give, therefore, the general order of magnitude of the forced convective heat transfer coefficient; we discuss below the Bennet-Chen correlation. However, as the basis of a "correlation," these methods suffer from the disadvantage that the friction multiplier ϕ_L cannot be precisely predicted. An alternative basis for the correlation of data is now examined.

AN ALTERNATIVE FORM OF CORRELATION

The convective heat transfer coefficient if the mixture flows as liquid, and the coefficient if the liquid component flows alone, are related

$$h_{Lo} = \frac{h_L}{(1-x)^{0.8}} \quad (21)$$

Table 3. Comparison with Kunz-Yerazunis multiplier; refrigerants.

1/X	1	5	10	50	100
From pressure drop Ψ_L^2, Eq. 19	3.58	11.69	19.44	62.47	105.7
Heat transfer measurement Ψ_L^2, Eq. 20	2.94	9.76	17.19	66.22	119.1
Equn 19/Equn 20	1.22	1.20	1.13	0.94	0.89

Hence, at the thermodynamic critical point

$$\Psi_{Lt}^2 = F_t = \frac{1}{(1-x)^{0.8}} \quad (22)$$

As there is no Prandtl number correction, from Equation 2, Ψ_{lt}^2 is equal to F_t; the subscript t signifies conditions at the thermodynamic critical point or where the properties of each phase are identical. Equation 22 can also be obtained using Equation 7 noting that, where both phases have the same properties, $\phi_L^2 = 1/(1-x)^{2-n}$.

Combining Equations 5 and 22, noting that Γ is unity at the thermodynamic critical point, gives

$$\Psi_{Lt}^2 = \left[1 + \frac{1}{X^{2/2-n}} \right]^{0.8} \quad (23)$$

Define the term γ by

$$\gamma = \frac{\Psi_L^2}{\Psi_{Lt}^2} \quad (24)$$

then from Equations 23 and 24

$$\Psi_L^2 = \gamma \left[1 + \frac{1}{X^{2/2-n}} \right]^{0.8} \quad (25)$$

This can be transformed, where the multiplier is relative to the heat transfer coefficient if the mixture flows as liquid, to

$$\Psi_{Lo}^2 = \gamma \left[1 + x\ (\Gamma^{2/2-n} - 1)\right]^{0.8} \quad (26)$$

Table 4 shows values of γ estimated for water using Equations 12 and 25, and for refrigerants using Equations 20 and 25.

Where 1/X is equal to or greater than unity, for water, to an accuracy of +3%, -5%, γ = 1.7, and for refrigerants, to an accuracy of +4%, -16%, γ = 2; for 1/X equal or greater than 5, for refrigerants the accuracy is +4%, -2%.

Figure 1 compares the Kenning and Cooper data for water with Equation 24 using γ = 1.7, and Figure 2 compares the Jung data for R22 and R114 (8) with equation 24 using γ = 2.0. Figure 3 compares the data of Robertson and Wadekar (9) for ethanol using γ = 2.65.

An advantage of correlating data in terms of γ, rather than in terms of Ψ_L^2, is that the former term varies from 1 to 3, whereas that latter varies from 1 to greater than 100.

THE KLIMENKO CORRELATION

Using an extensive data base, Klimenko (10) obtained the following equation, which correlated the data base to within ±35%,

$$\Psi_{Lo}^2 = \gamma_k \left(\frac{\rho_\mu}{\rho_L}\right)^{0.2} \left\{1 + x\left(\frac{\rho_L}{\rho_\mu} - 1\right)\right\}^{0.6} \quad (27)$$

where

$$\gamma_k = \frac{3.79\ b^{0.4}}{Re_{Lo}^{0.2}\ \ Pr_L^{0.233}} \left[\frac{k_w}{k_L}\right]^{0.09} \quad (28)$$

and, a dimensionless diameter,

$$b = \left\{ \frac{g\ d^2\ (\rho_L - \rho_G)}{\sigma} \right\}^{0.5} \quad (29)$$

Table 4. Estimation of γ-coefficient

1/X	1	5	10	50	100
Water					
γ, Eq. 12	1.61	1.75	1.75	1.69	1.66
Refrigerants					
γ, Eq. 20	1.68	2.06	2.08	2.02	1.97

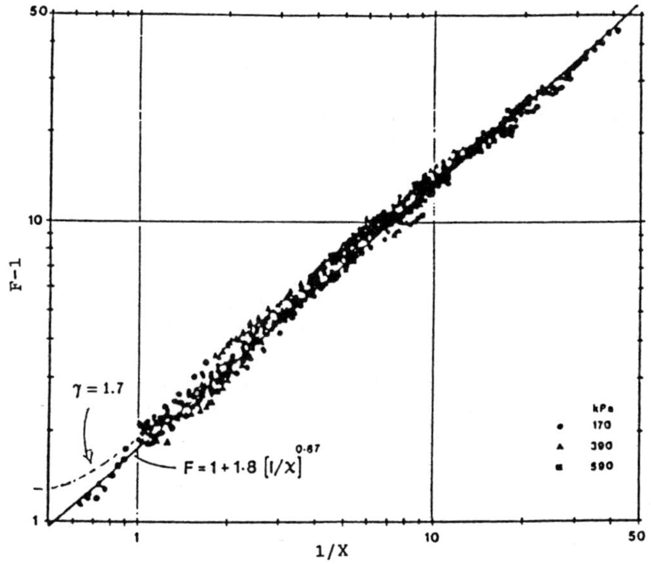

Figure 1. The group F-1 to a base of the reciprocal of the Lockhart-Martinelli parameter: water.

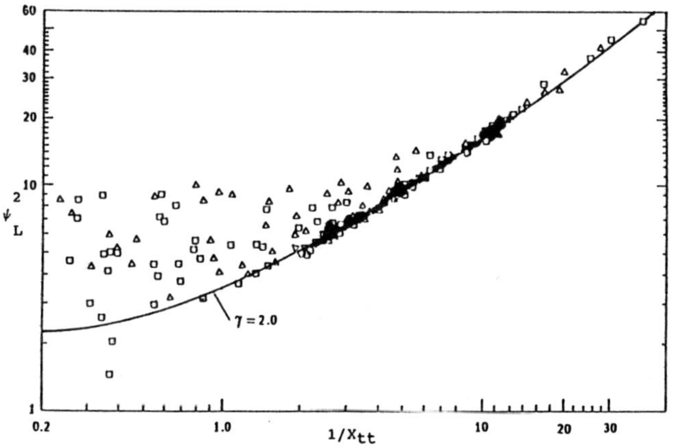

Figure 2. The heat transfer two-phase multiplier to the base of the reciprocal of the Lockhart-Martinelli parameter: refrigerants.

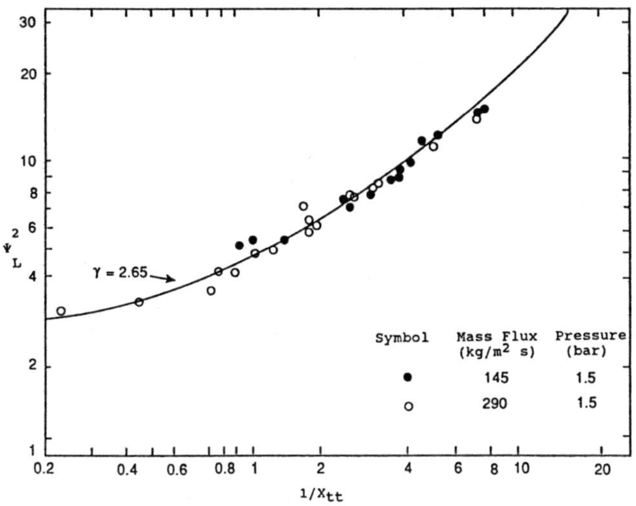

Figure 3. The heat transfer two-phase multiplier to the base of the reciprocal of the Lockhart-Martinelli parameter: ethanol.

Table 5 shows, for a range of density and viscosity ratios, the group $1.1\gamma_K/\gamma$, evaluated by combining Equations 26 and 27. Hence, approximately

$$\gamma = 1.1\ \gamma_K \qquad (30)$$

Thus, as demonstrated earlier (Table 4), the coefficient γ is again independent of the Lockhart-Martinelli parameter X.

THE COEFFICIENT γ

The data of Kenning and Cooper correspond to conditions with an average value of Prandtl number of 1.4. On the basis of Equation 2, then following Bennet and Chen,

$$\gamma = 1.54\ Pr^{0.296} \qquad (31)$$

Table 6 compares the various estimates of γ. Equation 31 gives better agreement with data than Equation 30. It should be remembered, however, that the Kunz and Yerazunis analysis implies that γ should be independent of Pr, and that Jung (7) did not observe an effect of Prandtl number; Prandtl numbers in the range 2.5 to 5.38 were tested by Jung.

CONCLUSIONS

The Chen correlation is shown to underestimate the heat transfer coefficient multiplier at higher dryness fractions.

An equation developed by Kunz and Yerazunis is transformed to a form indicating the multiplier is independent of Prandtl number, unlike the Bennet-Chen correlation.

A new form of correlation, Equation 25, is developed containing a single coefficient γ.

Using equations of Kenning and Cooper, and Jung et al., γ is shown to be essentially independent of the Lockhart-Martinelli parameter X, for values of 1/X greater than unity. This is further confirmed by comparison with ethanol data obtained b Robertson and Wadekar, and by a transformation of a correlation by Klimenko.

Further study is required to develop a generalized equation for the coefficient γ.

Table 5. The group $1.1\ \gamma_K/\gamma$, Eq. 30

1/X	1	5	10	50	100
ρ_L/ρ_G = 1000					
$\mu_L\mu_G$ = 20	0.96	0.89	0.94	1.08	1.11
$\mu_L\mu_G$ = 5	1.06	0.98	1.04	1.21	1.25
ρ_L/ρ_G = 100					
$\mu_L\mu_G$ = 20	1.07	0.96	0.98	1.02	1.03
$\mu_L\mu_G$ = 5	1.17	1.07	1.09	1.15	1.16
ρ_L/ρ_G = 10					
$\mu_L\mu_G$ = 10	1.14	1.10	1.00	0.99	0.99
$\mu_L\mu_G$ = 5	1.19	1.07	1.06	1.06	1.06

Table 6. The coefficient γ; comparison of methods.

Refrigerants		
2.0	1.5 - 2.4	2.0 - 2.6
Ethanol		
2.65	1.9 - 2.2	2.8

NOTATION

b	dimensionless diameter, Equation 29
c_{pL}	specific heat of liquid, Equation 13
d	internal diameter, Equation 16
F	Chen multiplier, Equation 2
h	heat transfer coefficient
h_c	convective vaporization, Equation 1
h_L	liquid component flows alone, Equation 1
h_{Lo}	mixture flows as liquid, Equation 21
n	Blasius exponent, Equation 7
Pr_L	Prandtl number of liquid, Equation 2
k	thermal conductivity, Equation 14
p	pressure
Dp	pressure gradient due to friction, Equation 8
Dp_G	vapor flows alone, Equation 4
DpG_o	mixture flows as vapor, Equation 6
Dp_L	liquid flows alone, Equation 4
Dp_{Lo}	mixture flows as liquid, Equation 6
Re_L	Reynolds number if mixture flows as liquid, Equation 13
X	Lockhart-Martinelli parameter, Equation 4
x	mass dryness fraction, Equation 5
Γ	group defined by Equation 6
γ	multiplier, Equation 24
γ_K	coefficient, Equation 28
σ	surface tension, Equation 29
τ	shear stress at tube wall, Equation 13
ϕ_L	friction two-phase multiplier, Equation 16
Ψ_L^2	heat transfer multiplier, Equation 1
Ψ_{Lo}^2	h_c/h_{Lo}, Equation 25

Subscripts

G	vapor
Go	mixture flows as vapor
L	liquid
Lo	mixture flows as liquid
t	thermodynamic critical point

LITERATURE CITED

1. Bennet, D.L. and Chen, J.C., "Forced Convective Boiling for Saturated Pure Components and Binary Mixtures," AIChE Journal, **26**, pp. 454-461, (1980).

2. Chen, J.C.,"Correlation for Boiling Heat Transfer to Saturated Fluids in Convective Flow," I&EC Process Design and Development, **5**, (3),pp. 322-329, (1966).

3. Kenning, D.B.R. and Cooper, M.G., "Saturated Flow Boiling of Water in Vertical Tubes." International Journal of Heat and Mass Transfer, **32**, (3), pp. 445-457, (1989).

4. Chisholm, D. and Laird, A.D.K., "Two-phase Flow in Rough Tubes," Transactions of the American Society of Mechanical Engineers, **80**, (2), pp. 276-286, (1958).

5. Jung, D.S. and Radermacher, R., "Prediction of Pressure Drop During Horizontal Flow Boiling of Pure and Mixed Refrigerants," International Journal of Heat and Mass Transfer, **32**, (12), pp. 2435-2445, (1989).

6. Jung, D.S., McLinden, M., Radermacher, R., and Didion D., "A Study of Flow Boiling Heat Transfer with Refrigerant Mixtures," International Journal of Heat and Mass Transfer, **32** (9), pp. 1751-1764, (1989).

7. Jung, D.S., McLinden, M., Radermacher, R., and Didion., D., "Horizontal Flow Boiling Heat Transfer Experiments with a Mixture of R22/R114," International Journal of Heat and Mass Transfer, **32**, (1),pp. 131-144, (1989).

8. Kunz, H.R. and Yarazunis, S., "An Analysis of Film Condensation, Film Evaporation, and Single-phase Heat Transfer for Liquid Prandtl Numbers From 10^{-3} to 10^{4}," Transactions of the American Society of Mechanical Engineers, Journal of Heat Transfer, pp. 413-420, (August 1969).

9. Robertson, J.M. and Wadekar, V.V., "Vertical Upflow of Boiling of Ethanol in a 10 mm Diameter Tube," Second UK National Heat Transfer Conference, Paper No. C198/88, pp.67-77, (1988).

10. Klimenko, V.M., "A Generalized Correlation for Two-phase Forced Flow Heat Transfer," International Journal of Heat and Mass Transfer, **31**, (3), pp.541-552, (1988).

TUBE VIBRATIONS IN KETTLE REBOILERS

A.E. Kassem ■ Heat Transfer Research Inc., College Station, TX 77843
F.S. Chavez ■ Technologia Industrial Gemminis, S.A. de C.V., Monterrey, Neuvo Leon, Mexico

Tube vibrations in kettle reboiler heat exchangers are the exception, not the rule, since, in general, they can be prevented by reducing the pitch of the support plates to a value below the TEMA recommended maximum span for the particular type of tubes used in the heat exchanger [1]. If, however, tube vibrations are detected in some exchangers, even though the support plate's pitch is less than the recommended span, attention must be directed towards either the inlet or exit regions of the tube bundle. TEMA recommends that the product of the density times the square of the velocity of the shell side fluid at the inlet and exit of the tube bundle must be less than 2250 kg/m s^2 [2]. Although exchangers may be designed either to satisfy, or in some cases to avoid, the implementation of TEMA entry and exit requirements at the process design conditions, under some partial operating conditions the TEMA shell side entry and exit requirements may be violated. The following example documents such conditions and shows how specific modifications could be incorporated in future design of kettle reboiler heat exchangers to eliminate the possibility of tube vibrations when the exchanger is operated far removed from the design process conditions.

In the process industry there are myths that are mistakenly held as truth in the operation of process heat exchangers. One of the myths is the belief that tube vibrations in kettle reboilers are eliminated simply by reducing the support plate pitch. Another is the belief that if the exchanger operates satisfactorily at the process design conditions and full capacity, it will continue to operate satisfactorily at partial capacity. This paper documents a case in which the exchanger operated satisfactorily at full capacity and suffered damage from tube vibration when operated at partial capacity. The objective of this exercise is to convince both, the designer, and the operator of the heat exchanger that the safe operation of the heat exchanger is a balance of competing forces which has been maintained at the process design conditions and full capacity. Operating the heat exchanger outside the process design conditions and full capacity, i.e. at partial capacity, may disturb the balance in favor of the damaging forces. From the operating engineer's point of view, it may be necessary to understand fully the implications of the design features incorporated in the heat exchanger in order to avoid damaging the exchanger when operated outside of the full capacity or design process conditions. From the design engineer's point of view, it seems that, as a good engineering practice, design features that restore the balance in favor of safe operation under partial capacity as well must be identified and incorporated in the heat exchanger in the early stages of design.

A.E. Kassem is presently with
The Ralph M. Parsons Co., Pasadena, CA.

PROBLEM STATEMENT

A kettle reboiler, AKU, is used to boil

a mixture of a 50 percent Ucarsol (a trade mark of Union Carbide Corporation), and 50 percent water on the shell side of the exchanger using 3.72 kPa steam at 142.22 C as the heating medium. Ucarsol is a natural gas treating agent used to absorb carbon dioxide and sulfur dioxide. The tube bundle is made of 19.05 mm low integral finned tubes 5.915 m straight length, with 748 fins/m. The tube bundle contains 798 U-tubes at 30 degree layout angle and 23.8 mm pitch. The bundle diameter is 1.041 m. The total surface area of the tube bundle is about 1393.5 m^2. The exchanger is equipped with a passive liquid level controller. A schematic diagram of the exchanger is shown in Figure 1. Based on the above process conditions and bundle geometry the exchanger is capable of generating up to 0.907 $(10)^5$ kg/hr of vapor [3]. These operating conditions are referred to henceforth as the full capacity operating conditions.

Due to refinements in the process, the operating requirements of the exchanger were modified. The exchanger is now operated at a partial capacity of 0.4712 $(10)^5$ kg/hr. At the partial capacity operating conditions, the bundle suffered from tube vibration which ultimately led to tube failure. Tube failure was confirmed when traces of the shell side stream were found in the tube side stream. Plugging the damaged tube continued for a while, after which a decision to re-tube the bundle was made.

Upon opening the exchanger, a detailed picture of the damaged tube bundle could be seen. Figures 2 through 7 show the shell, tube bundle, and the locations of the damaged tubes. Figure 2 shows the shell of the exchanger as well as three of the tubes that broke off the bundle during dismantling the heat exchanger. Figure 3 shows the tube bundle from the U-tube end. Notice that the damage is only to the outermost tube rows in the bundle. The damage also extends over more than one support plate in the bundle. Figure 4 shows a view of the tubesheet. The plugs used to prevent inter-leakage between the shell side and the tube side stream, are good indications of the tube failure location. It seems that the damaged tubes are concentrated on an arc extending from the one up to the five o'clock position of the tube bundle cross section. Figure 5 shows the damage extends to the support plates as well as the tubes. There are two types of damage to the support plates. The first type is the enlarged tube holes due to tube vibration. The second is due to the shearing off of parts of the support plate. Figure 6 shows the progressive damage to the support plate before the portion shown in Figures 5 and 7, was sheared off by tube vibration. Figure 7 shows the stages which the damaged tube goes through before it breaks completely from the bundle. Notice that some of the tubes lost their fins and look like plain tubes under the baffle. From these figures, it is clear that the damage to the bundle is caused partially by tube vibration and is probably augmented by erosion, corrosion, and metal fatigue.

ANALYSIS

Knowing how boiling occurs inside this type of heat exchanger from movies recorded at Heriot-Watt college [4], it can be assumed that, if the damage was related to local velocities during the boiling process, the damage to the tubes due to tube vibration would be localized in the top central portion of the tube bundle around the twelve o'clock position, not on the arc described earlier. The top of the bundle location is chosen, since the local velocity of the two-phase mixture caused by the density differential between the generated vapor and the liquid is maximum only there. Excluding the boiling process from being the cause of the damage, one is left only with the entry region of the cold stream into the bundle as being the location of the damaged tubes. Notice that most of the damage is located on the same side where the nozzle feeding the liquid to the bundle is located. However, a check on the velocity of the liquid stream in the nozzle indicates that it was around 1.22 m/s for the partial capacity operating conditions when the damage to the tubes occurred. This velocity translates to around 1500 kg/m s^2 for the product of the density times the square of the velocity of the fluid entering the bundle. This quantity is within the limit specified by TEMA for the entry criterion.

If the bundle was operated at full capacity, which is assumed to be the process design conditions, the velocity of the liquid entering the bundle would be around 2.03 m/s. The entry parameter would be in the range of 4350 kg/m s^2 which is larger than the TEMA recommended value. This value is high enough to cause damage similar to the damage seen in the bundle if the entering fluid directly impacts the tube bundle. It is safe to assume that the decision to locate the nozzle in its current position, above the tube bundle, may have been arrived at for economical reasons. It is assumed that at the process design conditions the bundle is fully covered by liquid up to the liquid level controller. The fluid entering the bundle then will not impact the tubes since it is discharged in the disengagement space, and, hence, no special design consideration was given to the entry region of the bundle at full capacity operating conditions.

Before dismissing the entry point of the cold stream into the bundle as being the main cause of the damage in the bundle, it is important to recall that the bundle was operated under partial capacity operating conditions and that the liquid level in the bundle was below the entry nozzle. Notice the level of the liquid inside the exchanger shell shown in Figure 2. Notice also that the tube bundle, Figure 7, shows the same demarcation line indicating the level of the liquid in the tube bundle during operations at partial capacity. It is also common practice for kettle reboilers operating under partial capacity to let the liquid level drop below the top of the bundle to reduce the effective surface area used in boiling the shell side liquid. An alternative way to operate at partial capacity is to keep the liquid level above the top of the bundle, i.e., make sure that liquid is being discharged from the liquid level controller and instead reduce the effective mean temperature difference between the heating medium and the boiling fluid. One way to accomplish this is by reducing the pressure of the steam used in the heat exchanger and, hence, reduce the saturation temperature of the heating medium. Commonly, this is done by throttling the heating medium through a throttle valve.

From Figure 1, it can be deduced that the liquid level is about 0.61 m below the top of the nozzle. Using a free-falling body under constant acceleration to estimate the impact velocity when it reaches the tube bundle, it can be deduced that the vertical component of velocity is approximately 3.35 m/s. The maximum resultant velocity obtained from the addition of the horizontal and vertical components of velocity is approximately 3.66 m/s. Similarly, the minimum resultant velocity is 2.13 m/s. The product of density times the square of the maximum and minimum velocities for a falling body of liquid impacting the tube bundle are approximately 12,750 and 4050 kg/m s^2 respectively. Either values is larger than TEMA criterion for entry into the bundle. Without the installation of an impingement device, these conditions are sufficient to excite tube vibration which would lead to the damage shown in the tube bundle. Notice, however, that if the exchanger was operated with the liquid level covering the top of the bundle coupled with lower heating stream temperature, the damage seen to the tube bundle could have been avoided with no changes in the design of the exchanger.

Another way of judging whether the terminal velocity calculated above is large enough to produce fluid-elastic instability in the tubes impacted by the incoming shellside liquid is to calculate Conners critical velocity [5]. Knowing that the distance between support plates was 1.47 m wide, the cross flow critical velocity is approximately, 0.914, and 9.14 m/s for liquid and gas respectively. This may seem to be a small value. However, if the the liquid level is above the inlet nozzle, the cross flow velocity of the liquid reaching the bundle will be much less than this velocity. However, if the liquid level is below the inlet nozzle, the impact velocity is large enough to cause tube vibrations. In addition, although the baffle spacing is less than the TEMA recommended span for plain tubes, it is larger than both the TEMA recommended span for finned tubes or the maximum recommended span for tube vibrations [6].

From the design engineer's point of view, it is preferable to incorporate design features which reduces the possibility of damage to the heat exchanger at partial operating conditions irrespective of how these conditions are achieved and maintained.

PROPOSED REMEDIES

Once the cause of the exciting force necessary to cause tube vibrations has been determined, remedial steps can be prescribed. Most of the following modifications can be implemented in the design stage of the exchanger to accommodate partial as well as full capacity operating conditions and to make safe and flexible operating parameters available to the operating engineer at minimum cost.

* When the kettle reboiler is operating at full capacity and the liquid is at the liquid

level controller, i.e., covering the bundle, having the nozzle location around the two o'clock position apparently did not cause damage to the tube bundle. However, at partial capacity when the liquid level in the bundle was such that the falling liquid acquired enough energy to excite the impacted tubes, damage to the tube bundle did occur. The rationale behind the location of the nozzle is sound only for the full capacity operating conditions. The rationale may also be sound if certain modifications to the process design were instituted to permit safe operations at partial capacity with the tube bundle fully covered with liquid. If, however, the location of the nozzle is changed from the two o'clock to the four or five o'clock position, the entry parameter of the cold stream is within the TEMA limits at the partial capacity operating conditions.

* At full capacity operating conditions however, the entry parameter to the bundle will be above the acceptable limit, around 4200 kg/m s^2. It is then necessary to modify the design in order to reduce the velocity of the liquid entering the shell side of the exchanger. Possible modifications could be: using a larger nozzle; splitting the flow among multiple entry nozzles while keeping the existing nozzle size; a distribution device; or an impingement device. Two nozzles of the existing size are recommended in this case.

* A further enhancement for the tube bundle performance against tube vibration in the entry region of the exchanger is achieved by replacing two or three rows of tubes directly in front of the nozzle with solid rods of identical size. These rods are less likely to vibrate due to the increased stiffness of the rod cross section over the tube cross section. These rods can also act as an impingement device protecting the tubes in the interior of the bundle. The location of the rods is on an arc extending from the two o'clock position to the six o'clock position.

* To comply with the recommendations presented in reference [7], the distance between the support plates is reduced to 1.184 m. Five support plates are recommended.

* Since the over fin diameter of the finned tubes is usually less than tube plain end, the clearance between the tube and the baffle hole is larger than that of plain tubes. Hence, the support experienced by a finned is much less than that of a plain tube. Whenever possible finned tubes with land under the support plate are used instead of regular finned tubes.

* To improve the performance of the tubes in the entry region of the heat exchangers in general, HTRI introduced FIVERS [8]. FIVERS, in their original form, were introduced for segmentally baffled heat exchangers. For kettle reboilers, however, the concept can be modified to fit this type of exchanger. A partial support plate in the form of a crescent two to three rows of tubes deep extending from the six o'clock position to the two o'clock position is used to provide support for the tubes facing the entry

nozzles. One partial support plate facing each nozzle may be used in this case. This will improve the frequency response for the tubes in the entry region by a factor of four. The shape of the FIVERS, as well as the location of the rods described earlier, are shown in Figure 8. The rods are shown as solid circles.

* In case the temperature differential between the boiling fluid and the heating medium is large (this is not the case for this process) a perforated impingement device could also be added. The perforated impingement device is usually preferable to the solid device since it allows for better distribution and mixing of the incoming cold stream and the boiling fluid inside the shell of the exchanger.

Figure 9 shows a schematic diagram of the exchanger incorporating all of the above modifications. These modifications can be incorporated in the design and manufacturing stages of similar exchangers. It will result in a better exchanger design with safe operations under partial as well as full capacity operating conditions. The economical impact on the exchanger cost is outside the range of this paper; however, the impact will be negligible and will no doubt be covered during the service life of the exchanger.

CONCLUSION

Damage to the tube bundle due to tube vibration in kettle reboilers, although rare, does occur every now and then, as shown in the case documented herein. Using good design practices, similar to the ones mentioned in this example, will prevent damage to the tube bundle over a wide range of operating conditions. Although these modifications will probably increase the initial cost of the exchanger, it is the authors' belief that the increased cost will be covered through safe, flexible, and longer trouble free life for the heat exchanger.

LITERATURE CITED

1. Standards of Tubular Exchangers Manufacturers Association, 6th ed., TEMA, New York, 1978
2. Ibid.
3. Hewitt, G. F. (editor), Hemisphere Handbook of Heat Exchanger Design, pp 3.6.5-2,3, Hemisphere Publishing Corporation, New York, 1990
4. A movie made at Heriot-Watt College, entitled "boiling in Tube Bundles", available through the HTRI Administrative Office, College Station, Texas
5. Conners, H. J., Jr., "Fluidelastic Vibration of Tube Arrays Excited by Cross Flow" Presented at the ASME Winter Annual Meeting, New York, December 1970
6. Chenoweth, J. M. and Kistler, R. S., "Tube Vibrations in Shell-and-Tube Heat Exchangers", Presented at the AICHE-ASME 15th National Heat Transfer Conference, San Francisco, California, August 1975
7. Ibid.
8. Chenoweth, J. M., ASME Symp. on Flow-Induced Vibrations, 3, 33 1984

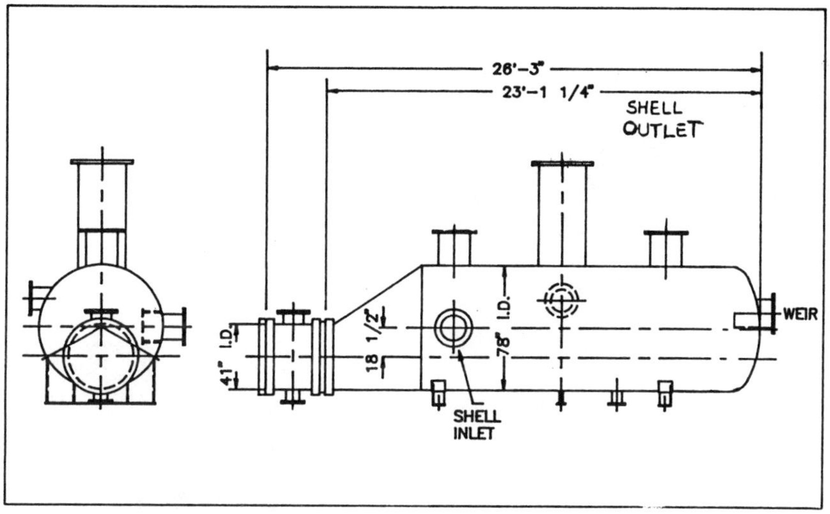

Figure 1. A schematic diagram of the kettle reboiler.

Figure 2. The shell of the kettle reboiler.

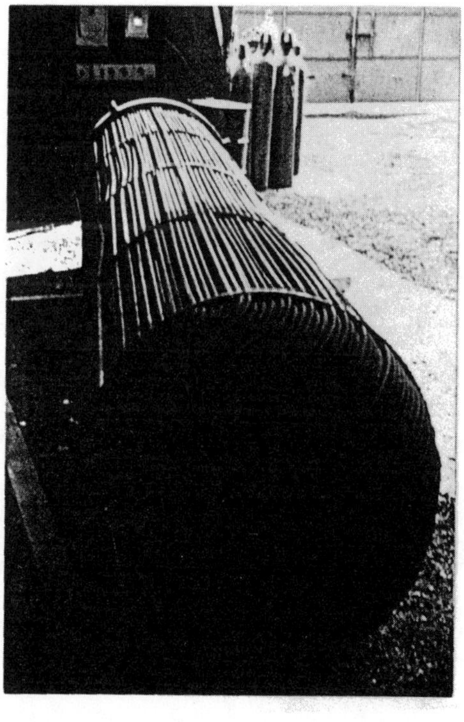

Figure 3. The tube bundle of the kettle reboiler shown from the *U*-bend end.

Figure 4. The tube bundle of the kettle reboiler shown from the tubesheet end.

Figure 6. A damaged support plate.

Figure 5. A close up of the damaged tubes.

Figure 7. A close up of a damaged support plate.

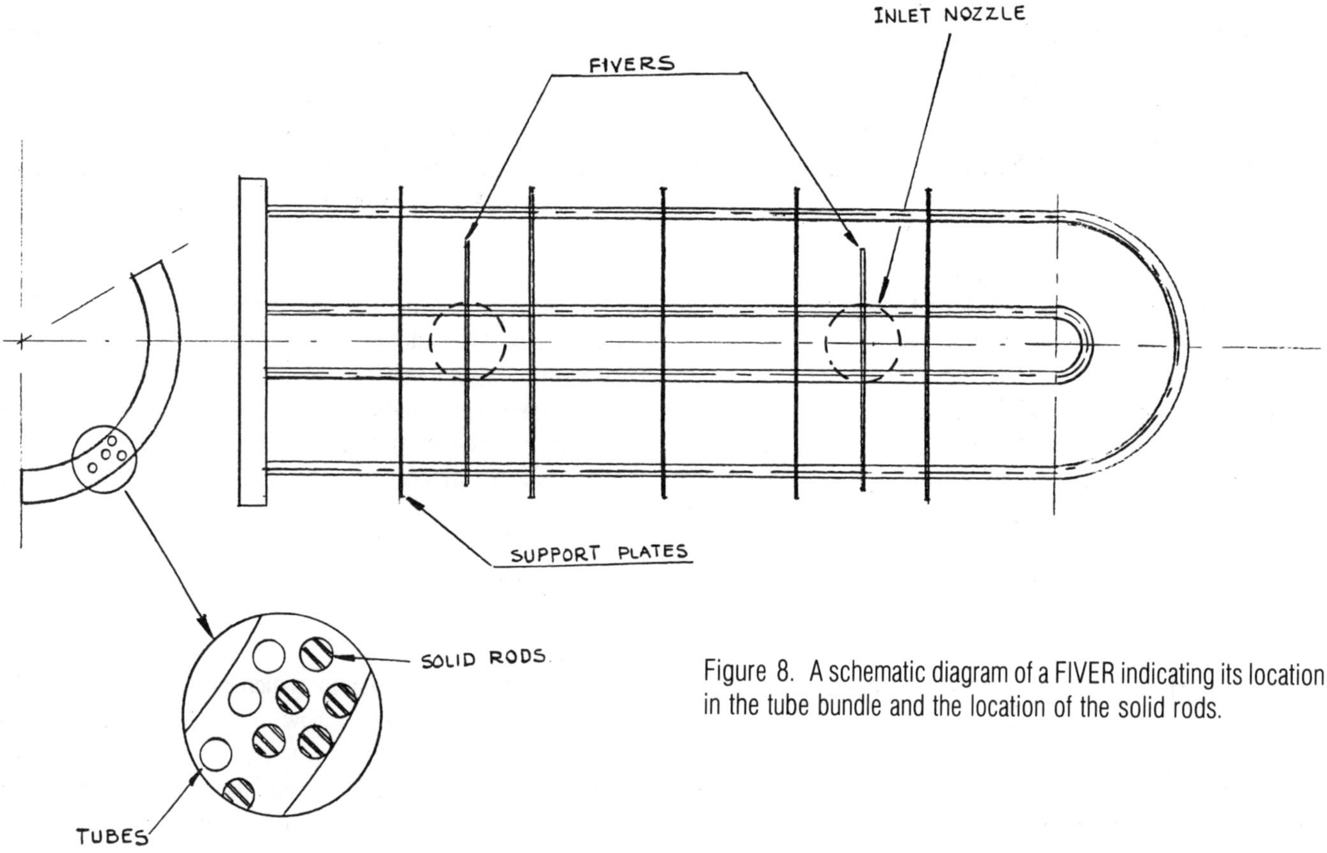

Figure 8. A schematic diagram of a FIVER indicating its location in the tube bundle and the location of the solid rods.

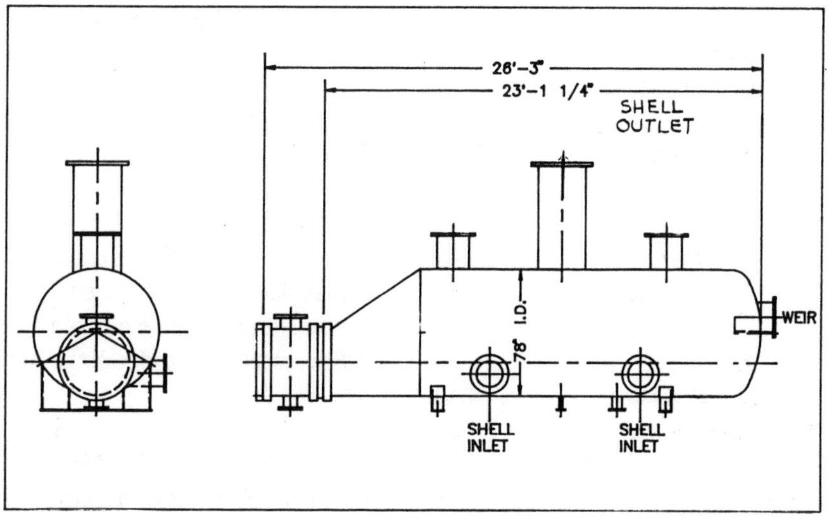

Figure 9. A schematic diagram of the modified kettle reboiler.

INDEX

T

V

W

Z

SYMPOSIUM SERIES

ADSORPTION

96 Developments in Physical Adsorption
117 Adsorption Technology
219 Recent Advances in Adsorption and Ion Exchange
230 Adsorption and Ion Exchange—'83
233 Adsorption and Ion Exchange—Progress and Future Prospects
242 Adsorption and Ion Exchange: Recent Developments
264 Adsorption and Ion Exchange: Fundaments and Applications

AEROSPACE

33 Rocket and Missile Technology
52 Chemical Engineering Techniques in Aerospace

BIOENGINEERING

69 Bioengineering and Food Processing
84 The Artificial Kidney
86 Bioengineering ... Food
93 Engineering of Unconventional Protein Production
99 Mass Transfer in Biological Systems
108 Food and Bioengineering—Fundamental and Industrial Aspects
114 Advances in Bioengineering
163 Water Removal Processes: Drying and Concentration of Foods and Other Materials
172 Food, pharmaceutical and bioengineering—1976/77
181 Biochemical Engineering Renewable Sources of Energy and Chemical Foodstocks

CRYOGENICS

224 Cryogenic Processes and Equipment 1982
251 Cryogenic Properties, Processes and Applications 1986

CRYSTALLIZATION

110 Factors Influencing Size Distribution
193 Design Control and Analysis of Crystallization Processes
215 Nucleation, Growth and Impurity Effects in Crystallization Process Engineering
240 Advances in Crystallization From Solutions
253 Fundametnal Aspects of Crystallization and Precipitation Processes

DRAG REDUCTION

11 Drag Reduction
130 Drag Reduction in Polymer Solutions

ENERGY

Conversion and Transfer

5 Heat Transfer, Atlantic City
57 Heat Transfer, Boston
59 Heat Transfer, Cleveland
75 Energy Conversion Systems
79 Heat Transfer with Phase Change
87 Advances in Cryogenic Heat Transfer
113 Convective and Interfacial Heat Transfer
118 Heat Transfer—Tulsa
119 Commercial Power Generation
138 Heat Transfer—Research and Design
162 Energy and Resource Recover from Industrial and Municipal Solid Wastes
174 Heat Transfer, Research and Application
189 Heat Transfer—San Diego 1979
202 Transport with Chemical Reactions
208 Heat Transfer—Milwaukee 1981
216 Processing of Energy and Metallic Minerals
225 Heat Transfer—Seattle 1983
236 Heat Transfer—Niagra Falls 1984
245 Heat Transfer—Denver 1985
257 Heat Transfer—Pittsburgh 1987
263 Heat Transfer—Houston 1988
269 Heat Transfer—Philadelphia 1989
283 Heat Transfer—Minneapolis 1991

Nuclear Engineering

53 Part XIII
56 Part XIV
95 Part XX
104 Part XXI
106 Part XXII
119 Commercial Power Generation
168 Heat Transfer in Thermonuclear Power Systems
169 Developments in Uranium Enrichment
191 Nuclear Engineering Questions Power Reprocessing, Waste, Decontamination Fusion
221 Recent Developments in Uranium Enrichment

ENVIRONMENT

78 Water Reuse
97 Water—1969
115 Important Chemical Reactions in Air Pollution Control
122 Chemical Engineering Applications of Solid Waste Treatment
124 Water—1971
126 Air Pollution and its Control
133 Forest Products and the Environment
137 Recent Advances in Air Pollution Control
139 Advances In Processing and Utilization of Forest Products
144 Water—1974: I. Industrial Wastewater Treatment
145 Water—1974: II. Municipal Wastewater Treatment
146 Forest Product Residuals
147 Air: I. Pollution Control and Clean Energy
148 Air: II. Control of NO_{xx} and SO_x Emissions
149 Trace Contaminants in the Environment
151 Water—1975
156 Air Pollution Control and Clean Energy
157 New Horizons for the Chemical Engineer in Pulp and Paper Technology
165 Dispersion and Control of Atmospheric Emissions, New-Energy-Source Pollution Potential
170 Intermaterials Competition in the Management of Shrinking Resources
171 What the Filterman Needs to Know About Filtration
175 Control and Dispersion of Air Pollutants: Emphasis on NO_x and Particulate Emissions
177 Energy and Environmental Concerns in the Forest Products Industry
184 Advances in the Utilization and Processing of Forest Products
188 Control of Emissions from Stationary Combustion Sources Pollutant Detection and Behavior in the Atmosphere
195 The Role of Chemical Engineering in Utilizing the Nation's Forest Resources
196 Implications of the Clean Air Amendments of 1977 and of Energy Considerations for Air Pollution Control
198 Fundamentals and Applications of Solar Energy
200 New Process Alternatives in the Forest Products Industries
201 Emission Control from Stationary Power Sources: Technical, Economic and Environmental Assessments
207 The Use and Processing of Renewable Resources—Chemical Engineering Challenge of the Future
209 Water—1980
210 Fundamentals and Applications of Solar Energy II
211 Research Trends in Air Pollution Control: Scrubbing, Hot Gas Clean-up, Sampling and Analysis
213 Three Mile Island Cleanup
223 Advances in Production of Forest Products
232 Applications of Chemical Engineering in the Forest Products Industry
239 The Impact of Energy and Environmental Concerns on Chemical Engineering in the Forest Products Industry
243 Separation of Heavy Metals and Other Trace Contaminants
246 Advances in Process Analysis and Development in the Forest Products Industries.
265 Resource Recovery of Municipal Solid Wastes

FLUIDIZATION

101 Fundamental Processes in Fluidized Beds
105 Fluidization Fundamentals and Application
116 Fluidization: Fundamental Studies Solid-Fluid Reactions, and Applications
176 Fluidization Application to Coal Conversion Processes
205 Recent Advances in Fluidization and Fluid-Particle Systems
234 Fluidization and Fluid Particle Systems: Theories and Applications
241 Fluidization and Fluid Particle Systems: Recent Advances
255 New Developments in Fluidization and Fluid-Particle Systems
262 Fluidization Engineering: Fundamentals and Applications
270 Fluidization and Fluid Particle Systems: Fundamentals and Application
276 Advances in Fluidization Engineering
281 Advances in Fluidized System

HISTORY OF CHEMICAL ENGINEERING

100 The History of Penicillin Production
235 Diamond Jubilee Historical/Review Volume

ION EXCHANGE

79 Adsorption and Ion Exchange Separations
219 Recent Advances in Adsorption and Ion Exchange
230 Adsorption and Ion Exchange—'83
233 Adsorption and Ion Exchange—Progress and Future Prospects
259 Recent Progress in Adsorption and Ion Exchange

KINETICS

25 Reaction Kinetics and Unit Operations
73 Kinetics and Catalysis

MINERALS

15 Mineral Engineering Techniques
85 Fossil Hydrocarbon and Mineral Processing
173 Fundamental Aspects of Hydrometallurgical Processes
180 Spinning Wire from Molten Metals
216 Processing of Energy and Metallic Minerals

PETROCHEMICALS

49 Polymer Processing
127 Declining Domestic Reserves—Effect on Petroleum and Petrochemical Industry
135 The Petroleum/Petrochemical Industry and the Ecological Challenge
142 Optimum Use of World Petroleum
212 Interfacial Phenomena in Enhanced Oil Recovery
280 Enhanced Oil Recovery

PETROLEUM PROCESSING

103 C_4 Hydrocarbon Production and Distribution
127 Declining Domestic Reserves—Effect on Petroleum and Petrochemical Industry
135 The Petroleum/Petrochemical Industry and the Ecological Challenge
155 Oil Shale and Tar Sands
226 Underground Coal Gasification: The State of the Art
282 Tar Sand and Oil Upgrading Technology

PHASE EQUILIBRIA

2 Pittsburgh and Houston
6 Collected Research Papers
88 Phase Equilibria and Gas Mixtures Properties

PROCESS DYNAMICS

36 Process Dynamics and Control
46 Process Systems Engineering
55 Process Control and Applied Mathematics
214 Selected Topics on Computer-Aided Process Design and Analysis
267 Process Sensing and Diagnostics

SEPARATION

120 Recent Advances in Separation Techniques
192 Recent Advances in Separation Techniques—II
250 Recent Advances in Separation Techniques—III

SONICS

109 Sonochemical Engineering

MISCELLANEOUS

48 Chemical Engineering Reviews
70 Small-Scale Equipment for Chemical Engineering Laboratories
112 Engineering, Chemistry, and Use of Plasma Reactors
125 Vacuum Technology at Low Temperatures
143 Standardization of Catalyst Test Methods
182 Biorheology
183 The Modern Undergraduate Laboratory Innovative Techniques
185 Electro Organic Synthesis Technology
186 Plasma Chemical Processing
187 Chronic Replacement of Kidney Function
194 Hazardous Chemical—Spills and Waterborne Transportation
203 A Review of AIChE's Design Institute for Physical Property Data (DIPPR) and Worldwide Affiliated Activities
204 Tutorial Lectures in Electrochemical Engineering and Technology
206 Controlled Release Systems
217 New Composite Materials and Technology
220 Uncertainty Analysis for Engineers
228 Problem Solving
229 Tutorial Lectures in Electrochemical Engineering and Technology—II
231 Data Base Implementation and Application
237 Awareness of Information Sources
238 New Developments in Liquid-Liquid Extractors: Selected Papers From ISEC '83
244 Experimental Results from the Design Institute for Physical Property Data. I: Phase Equilibria
247 Chemical Engineering Data Sources
248 Industrial Membrane Processes
249 Measurement of High Temperatures in Furnaces and Processes
252 Thin Liquid Film Phenomena
254 Electrochemical Engineering Applications
256 Experimental Results From the Design Institute for Physical Property Data: Phase Equilibria and Pure Component Properties
258 Fiber Optics: Processing and Applications
261 New Membrane Materials and Processes for Separation
266 Diffusion and Convection in Porous Catalysts
268 Membrane Reactor Technology
271 Experimental Results From the Design Institute for Physical Property Data: Phase Equilibria and Pure Component Properties—Part II.
272 Membrane Separations in Chemical Industries
273 Fundamentals of Resid Upgrading
274 Competitiveness of the U.S. Chemical Industry in International Markets
275 Design Institute for Physical Property Data: Ten Years of Accomplishment
277 Food Emulsion and Foams: Theory and Practice
278 Perspectives on Corrosion
279 Results from the Design Institute for Physical Property Data: Experimental Results and Data Compilation Procedures

MONOGRAPH SERIES

3 The Manufacture of Nitric Acid by the Oxidation of Ammonia—The DuPont Pressure Process by Thomas H. Chilton
4 Experiences and Experiments with Process Dynamics by Joel O. Hougen
5 Present, Past, and Future Property Estimation Techniques by Robert C. Reid
6 Catalysts and Reactors by James Wei
7 The 'Calculated' Loss-of-Coolant Accident by L.J. Ybarrondo, C.W. Solbrig, H.S. Isbin
8 Understanding and Conceiving Chemical Process by C. Judson King
9 Ecosystem Technology: Theory and Practice by Aaron J. Teller
10 Fundamentals of Fire and Explosion by Daniel R. Stull
11 Lumps, Models and Kinetics in Practice by Vern W. Weekman, Jr.
12 Lectures in Atmospheric Chemistry by John H. Seinfeld
13 Advanced Process Engineering by James R. Fair
14 Synfuels from Coal by Bernard S. Lee
15 Computer Modeling of Chemical Processes by J.D. Seader
16 "High-Tech" Materials by Sheldon Isakoff
17 Separations: New Directions for an Old Field
128 Biotechnology: Status and Perspectives